Microgrid Dynamics and Control

Microgrid Dynamics and Control

Hassan Bevrani
University of Kurdistan, Kurdistan, Iran

Bruno Francois
Centrale Lille, France

Toshifumi Ise
Osaka University, Osaka, Japan

WILEY

This edition first published 2017

Registered Office
John Wiley & Sons, Inc., 111 River Street, Hoboken, NJ 07030, USA

Editorial Office
111 River Street, Hoboken, NJ 07030, USA
For details of our global editorial offices, customer services, and more information about Wiley products visit us at www.wiley.com.

Wiley also publishes its books in a variety of electronic formats and by print-on-demand. Some content that appears in standard print versions of this book may not be available in other formats.

Library of Congress Cataloging-in-Publication Data

Names: Bevrani, Hassan, author. | Francois, Bruno, author. | Ise, Toshifumi, 1957- author.
Title: Microgrid dynamics and control / by Hassan Bevrani, Bruno Francois, Toshifumi Ise.
Description: First edition. | Hoboken, NJ : John Wiley & Sons, 2017. | Includes bibliographical references and index. |
Identifiers: LCCN 2017016748 (print) | LCCN 2017027057 (ebook) | ISBN 9781119263692 (pdf) | ISBN 9781119359357 (epub) | ISBN 9781119263678 (cloth)
Subjects: LCSH: Microgrids (Smart power grids)
Classification: LCC TK3105 (ebook) | LCC TK3105 .B48 2017 (print) | DDC 621.31–dc23
LC record available at https://lccn.loc.gov/2017016748

Cover image: Courtesy of Hassan Bevrani
Cover design by Wiley

Set in 10/12pt WarnockPro by SPi Global, Chennai, India

Printed in the United States of America

10 9 8 7 6 5 4 3 2 1

Dedicated to our families and students.

Contents

Foreword

The electric power industry is in the midst of a critical period in its evolution. Today's high-voltage transmission network is reliable and controllable but suffers from cascading failures and volatility. Its efficiency and use of resources are also poor. Generation converts only one-third of fuel energy into electrical power, wasting the generated heat. Furthermore, another 6% is lost in the transmission and distribution lines. Approximately 20% of the generation capacity exists to meet peak demand 5% of the time. These issues become compounded with high penetration of renewable sources due to their intermittent behavior. Revolutionary changes are not expected in the transmission network, but improvement through continued evolution can greatly reduce events such as the 2003 blackout in the United States. On the other hand, the distribution system provides major opportunities for huge improvements. We need to rethink about our distribution system including the integration of high levels of distributed energy resources (DER) to provide a smarter and more flexible distribution system. Basic objectives are as follows:

- Improved reliability
- High penetration of renewable sources
- Self-healing
- Increased system efficiencies
- Lower cost.

Using DER in the distribution system reduces the physical and electrical distance between generation and loads. Bringing sources closer to loads contributes to the enhancement of the voltage profile, to the reduction of distribution and transmission bottlenecks, to improved reliability, to lower losses, and to the enhancement of the use of waste heat. A basic issue for high penetration of DER is the technical difficulties related to control of a significant number of distributed energy sources. This issue is complex, but the call for extensive development in fast sensors and complex control has the potential for disaster. The fundamental problem with a complex control system in addition to cost is that failure of a control component or a software error could bring the system down. One way to realize the emerging potential of DER is to take a system approach that views clusters of DER and associated loads as a grid resource or a "microgrid." The sources and loads can operate in parallel to the grid or as an island. A microgrid can provide for the customer's critical needs while providing services to the distribution system.

Smarter and more flexible distribution systems can be realized in practical applications when we understand how to use the dynamic resources of microgrids. This implies that there are many microgrids working together to achieve smart grid objectives. Control concepts and coordination requirements to create such a distribution system has yet to be elucidated. The problems related to achieving a smarter distribution system are new, and performance objectives are still being explored. We are moving from a passive 100-year-old distribution system to the one that can provide many services that have yet to be fully understood.

The research presented in this book is an excellent start toward understanding control approaches, but we must keep the basic issues foremost. Individual and clustered microgrids need to be simple to design and construct, robust, stable, and scalable with the ability to gracefully degrade during major events.

Bob Lasseter
University of Wisconsin-Madison
Madison, WI, USA

Acknowledgments

Most of contributions, outcomes, and insights presented in this book were achieved through long-term teaching and research conducted by the authors and their research groups on the microgrid operation and control issues over the years. The experimental tests given in this book have been performed at the *Smart/Micro Grids Research Center* (*SMGRC*) at the University of Kurdistan, *Laboratory of Electrical Engineering and Power electronics* (*L2EP*) at the Centrale Lille and the *Power Electronics and Electrical Energy* (*PE&EE*) *Laboratory* at the Osaka University. It is a pleasure to acknowledge the received supports and awards from all mentioned sources and their sponsors.

The authors also would like to thank their colleagues Dr M. Fathi, Dr Q. Shafiee, Dr S. Bahramara, Dr H. Golpira, Dr Y. Batmani, Dr Y. Miura, Dr H. Kakigano, Dr J. Liu; and postgraduate students M. R. Feizi, Y. Khayat, S. Shokoohi, O. Sarchami, M. Naderi, S. Ahmadi, S. Attaee, N. Hajimohammadi, M. Azimi, H. Ahmadi, G. Delille, Y. Wang, D. Lu, P. Li, T. Zhou, H. Kanchev, X. Yan, T. Shintai, J. Alipoor, Y. Hirase, and K. Sakimoto for their active role and continuous supports.

Special thanks go to Prof. B. Lasseter, the inventor of microgrid concept, for reviewing the manuscript and including his observation as a part of final product. Last but not least, the authors offer their deepest personal gratitude to their family for all their patience and help during the preparation of this book.

Preface

The *microgrid* (MG) concept provides a quite appealing solution for overcoming the challenges of integrating renewable energy sources (RESs) and distributed generators (DGs) into power grids. The MGs are small electrical distribution systems that interconnect multiple customers, DGs, and storage systems. The MGs are typically characterized by multipurpose electrical power services to communities that are connected via low-voltage networks. However, in order to allow seamless deployment of MGs, the relevant dynamic stability and control challenges should be solved. Compared to conventional power systems with synchronous generators (SGs), MGs with DGs/RESs have either small or no inertia, which is the main source of stability. With growing penetration level of DGs/RESs, the negative impact of low inertia and damping effect on the MG dynamic performance and stability increases.

MGs are known as the main building blocks of the future smart grids. The increasing penetration of renewable energy power into the power system among MGs makes serious challenges in control design and operation and highlights the importance of these systems. Understanding the dynamics and using appropriate analytic methodologies are significant issues for MG stability analysis and control synthesis. Despite the small scale of an MG, it has many of the complexities of a large-scale conventional power system. In conventional power grids, stability analysis is well established with standard models of SGs, governors, and excitation systems of varying orders that are known to capture the important modes for particular classes of problems. This does not yet exist for the MGs and may be difficult to achieve because of the wide range of power technologies that might be deployed.

As an example, today, the massive use of power electronic converters for grid connection of generators, DC loads, and high-voltage direct-current (HVDC) power transmission leads to imagine the operation of electrical networks with no synchronous machines, that is, with 100% power electronics penetration, in the near future. As proposed in this book, MGs propose new high-level control strategies and management rules when operating, for example, solutions to ensure system reliability (adequacy and security) when there is no longer a physical link between load/generation imbalance and the frequency in the electrical network. Since ancillary services are expected to be different in such a grid, high-level management rules must also be adjusted.

Current efforts are mostly being made into the design of more effective control strategies and special protection schemes in different control levels that ensure stable, reliable, secure, and economical operation of MGs in either grid-connected or islanded operation mode.

Control is one of the key enabling technologies for the deployment of future power grids. Various control loops must be used to improve the MG stability and performance. Current, voltage/amplitude, frequency/angle, and active and reactive power are the main feedback variables used in the existing MG control loops in both grid-connected and islanded operation modes. Similar to conventional power grids, the MG has a hierarchical control structure with different operation layers. The hierarchical control structures of MGs are responsible for providing proper load sharing and DG coordination, voltage/frequency regulation in both operating modes, MG resynchronization with the main grid, operating cost optimization, and power flow control between the MG, neighborhood grids, and the main grid.

Advances in the MG control improve the MG potential to be integrated into the conventional electrical systems in a higher capacity. This improvement not only covers their internal control performance and connection/disconnection procedures in order to limit the negative dynamic impacts in the host network and other connected MGs but also includes the grid support functionalities to enhance the global operation of distribution networks.

In the context of smart grids, tested innovative solutions on MGs can be easily adapted to offer new possibilities to increase flexibilities in distribution networks in order to operate them with higher dynamics and efficiency. MG control techniques and management system make possible a new global planning strategy of power systems leading to a local economic optimization of electric energy with satisfied generation/load balancing and technical operation constraints (ancillary services) while participating in energy and service exchanges with the surrounding stakeholders.

This book provides a comprehensive coverage of MG dynamics understanding, stability analysis, control synthesis, and realization. The book covers the authors' long-term research, teaching, and practical experiences on the MG control and operation. The materials given in this book are mainly the research outcomes and original results of three laboratories: *Smart/Micro Grids Research Center* (*SMGRC*) at the University of Kurdistan (Sanandaj, Iran), *Laboratory of Electrical Engineering and Power Electronics* (*L2EP*) at the Centrale Lille (France), and the *Power Electronics and Electrical Energy* (*PE&EE*) *Laboratory* at the Osaka University (Japan).

This book could be useful for engineers and operators in power grids and MGs planning, control, and operation, as well as for students and academic researchers. The book describes MG dynamics modeling and control issues from introductory to the advanced steps. The book could also be useful as a textbook for university students in electrical engineering at both undergraduate and postgraduate levels in standard courses of MGs, smart grids, and modern power system control. This book is organized into 12 chapters.

Chapter 1 describes renewable power generation briefly and presents the principles for the integration of wind and solar power to the grid. The modeling procedure and control structures of the grid-connected wind and photovoltaic (PV) energy systems are emphasized. The PV and wind energy sources in hybrid configurations are also discussed. To investigate the proposed modeling and control schemes for the grid-connected wind and PV sources, some simulations and experimental results are provided.

Chapter 2 investigates the contribution of wind power generators in inertial response and primary frequency control. A statistical approach to assess the impact of wind power variations on the size of primary reserve provided by the wind power generators is presented. Then, in the same way, the capability of solar power generators in the grid control support is also discussed. Finally, the MG concept as a solution to integrate more RESs into the power system is introduced.

Chapter 3 addresses the main structure, fundamentals, and concepts of MGs. Then, the MG operating modes and the overall control mechanisms for the MG's distributed generators are explained. The possible role of MGs to support the main grid ancillary services, particularly frequency regulation, is emphasized, and finally, the technologies of the existing MG laboratories are introduced.

Chapter 4 presents a dynamic model for the MG, main grid, and connecting components, after giving an overview on the recent developments in MG dynamic representation and modeling. The main grid is represented as an aggregated group of diesel generators. The model is then extended by including the MG system to form an overall representation structure for the grid-connected MG. Modeling of some components (distributed generators, energy storage systems, power converters, and loads) in the MG is emphasized, and a simplified frequency response model using low-order transfer functions is introduced. Then, a state-space model using dynamic details is presented, and the MG modeling and dynamic analysis as a multivariable system is discussed.

Chapter 5 emphasizes the most important issues and challenges in MG control. A general overview of the main control loops is presented. This chapter classifies MG control strategies into four control levels: local, secondary, central and emergency, and global, where the first three levels are associated with the operation of the MG itself, and the fourth level (global control) demonstrates the coordinated operation of the MG and neighbor grids as well as the host grid. The mentioned control levels are discussed in detail, and a synthesis example is given at the end.

Chapter 6 addresses the control and operation in DC MGs. To show the DC MG characteristics and for a smooth introduction of numerous DGs to satisfy high-efficiency and high-quality power supply, a low-voltage bipolar-type DC MG is introduced. To confirm the fundamental characteristics, a laboratory-scale experimental system is constructed. The results of voltage sag in the utility grid and disconnection and reconnection procedures are presented. Then, a DC distribution voltage control system for the DC/DC converters with an energy storage unit is designed. The proposed control combines a gain-scheduling technique with droop control. The experimental results indicate that DC voltage regulation and stored energy balancing control are realized, simultaneously.

Chapter 7 examines a basic virtual synchronous generator (VSG) structure with the main components. The VSG control performance is compared with the well-known droop control performance. Small-signal and state-space models are performed to study the frequency transient response and active power oscillation, respectively. An analytical approach for active power performance assessment in an MG with multiple VSGs is presented. Then, the stabilizing property of the VSG system is addressed, and a VSG-based oscillation damping approach is explained. Finally, a different VSG topology emulating the dynamic characteristics of a synchronous machine via modeling the SG properties is introduced.

Chapter 8 introduces several VSG control schemes to support the MG stability and regulation performance. First, an enhanced VSG control is proposed to achieve oscillation damping and proper transient active power sharing by the virtual stator reactance adjustment and common AC bus voltage estimation. Then, for a parallel operation of VSG and SG, to prevent unbalanced SG current and inverter overcurrent, a new control mechanism is developed. An alternating inertia concept is then introduced to enhance the VSG control performance. The VSG support for the serious voltage sags is also assessed in this chapter. Finally, the impacts of a VSG topology on the MG stability are examined. All given control approaches and synthesis methodologies are verified analytically and examined through computer simulations and laboratory experiments.

Chapter 9 addresses the application of robust control in MGs. The robust frequency control in islanded MGs is emphasized. The H_2, H_∞, and μ-synthesis robust control techniques are used to develop the MG frequency control loop. In the proposed control scheme, some microsources (diesel engine generator, microturbine, and fuel cell) are assumed to be responsible for balancing the load and power in the MG system. The synthesized H_∞ and μ controllers are examined on an MG test platform, and the controllers' robustness and performance are evaluated in the presence of various disturbances and parametric uncertainties. The results are compared with two well-tuned proportional–integral control methodologies. The robust control theory is also used for the sake of optimal tuning of VSGs.

Chapter 10 focuses on some application examples of intelligent systems in MG operation and control. After an introduction, the most important intelligent control technologies for application in MGs are addressed. Artificial-neural-network-based power and load forecasting examples in MGs are briefly described. Then, several examples for intelligence-based frequency and voltage control synthesis using fuzzy logic, particle swarm optimization, and multiagent systems are given. The explained examples are supplemented by simulations and real-time experimental tests.

Chapter 11 addresses the load shedding as the most important and well-known emergency control strategy, following a general introduction on emergency control in MGs. Two load shedding algorithms are explained, and undervoltage–frequency load shedding (UVFLS) is emphasized. The potential benefits of the mentioned load shedding strategy compared to conventional underfrequency load shedding (UFLS) and undervoltage load shedding (UVLS) approaches are examined, and finally, some important factors for the voltage and frequency performance during the load shedding process are discussed.

Chapter 12 addresses some significant aspects in MG planning and management and relevant topics. After an introduction, an MG planning example for real MG case study is briefly explained. The PV power and demand forecasting are given in detail. Energy management concerning the technical, environmental, and economic issues, supplemented by a real-time experimental studies, is described. The day-ahead optimal operation and power reserve dispatching are emphasized, and a robust energy consumption scheduling methodology for interconnected MGs in the presence of demand uncertainty is presented.

Hassan Bevrani, University of Kurdistan, Kurdistan, Iran
Bruno Francois, Centrale Lille, Lille, France
Toshifumi Ise, Osaka University, Osaka, Japan
February 2017

1

Grid-connected Renewable Energy Sources

In the face of energy challenge, the demand for new primary energy sources worldwide is evolving. But reserves of oil and fossil fuels in our planet are going to be exhausted in the near future. On the other hand, global warming becomes a more serious issue due to the greenhouse effect. The production of electrical energy from the oil, natural gas, coal, and nuclear sources is one of the main sources of greenhouse gases. For reducing the greenhouse gas emission and assuring the energy security, renewable energy sources (RESs) perform effective alternative solutions for clean and sustainable electrical power production.

In this chapter, the renewable power generation is briefly described. The principles for integration of wind power and solar power to the grid are presented. Then, the modeling procedure and control structures of the grid-connected wind and photovoltaic (PV) energy systems are emphasized. The problem of the primary energy intermittency is tackled and the PV and wind energy sources in hybrid configurations are also discussed. To investigate the proposed modeling and control schemes for the grid-connected wind and PV sources, some simulations and experimental results are provided, and finally the chapter is summarized.

1.1 Introduction

Today, global warming becomes more serious due to the greenhouse effect. Some emissions of greenhouse gases come from the human activities. The production and processing of electrical energy is one of the main sources of greenhouse gases. Directly or indirectly, environmental concerns dominate the thrust for an expanded deployment of renewable energy technologies. Climate change concerns, which arose during the late 1980s, have created a new input for clean, low-carbon energy technologies, such as renewable energy technologies. In December 1997, the Kyoto protocol was established

Microgrid Dynamics and Control, First Edition. Hassan Bevrani, Bruno Francois, and Toshifumi Ise.

in order to reduce global emissions of greenhouse gases. In the area of power generation, this protocol promotes RESs [1]. In the 11th session of the Paris to the 1997 Kyoto protocol, the 2015 United Nations Climate Change Conference in Paris drafted an agreement to set a goal of limiting global warming to less than 2 °C [2]. In the adopted version of the Paris agreement, the party will also pursue efforts to limit the temperature increase to 1.5 °C, which may require zero emissions sometime between 2030 and 2050.

Renewable energy is the energy generated from natural renewable resources such as sunlight, wind, ocean, hydropower, biomass, geothermal resources, biofuels, and hydrogen. The contribution potential of renewable energies in all countries is growing as the technologies mature. The RESs contribute to the diversity of energy supply portfolio and reduce the risks of continued/expanded use of fossil fuels and nuclear power. The RESs provide interesting options to consumers and are also the most environmentally benign energy supply options available in current and near-term markets. Moreover, the RESs contribute to a healthy economy, both in their contribution to the efficiency of the energy system and in the employment/investment opportunities that arise from continued rapid market growth.

Nowadays, governments and public organizations are concerned about the production of energy with technologies as clean as possible [3, 4]. As a consequence, the guidelines for future energy production are established according to the Kyoto protocol in 1997 and reemphasized by the Paris agreement in 2015. The energy production technologies based on hydro, wind, PV, and geothermal energies can be considered to be clean and renewable alternatives to the nonclean conventional technologies based on fossil fuels and nuclear fission. Among the clean technologies, PV and wind turbines have experienced a tremendous growth in the last years.

Wind power generators (WPGs) and solar power generators (SPGs) are able to provide the electrical power to supply a grid load. The energy conversion systems provided by wind turbines and PVs enable to extract the maximum wind/solar power by adjusting the wind turbine's rotational speed and solar plate effective surface. The obtained electrical power must be adapted before being sent to the grid. Therefore, an effective grid-following power balancing strategies must be used, while the WPGs/SPGs are working in the maximum power point tracking (MPPT) operating mode.

In order to reduce RES power variations, the energy storage systems (ESSs) can be used to build a hybrid power system (HPS). It is known that the HPS as a microgrid (MG) can be a good solution for the integration of distributed RESs in a power network. In this chapter, the WPG- and SPG-based hybrid systems modeling and control are presented. Some proper power control strategies such as "grid-following" and "power dispatching" strategies are also briefly addressed. In the performed power control strategies, the DC-bus voltage is regulated with the powers from the WPG/SPG and the ESSs.

1.2 Renewable Power Generation

The RESs convert natural energy sources (sunlight, wind, ocean, hydropower, biomass, geothermal resources, biofuels, and hydrogen) into consumable energy forms (electricity and heat), which are suitable to transport and to use. Solar, wind, water, biofuel, and geothermal power are known as the main sources of the renewable electric power.

Solar power: Solar power comes from the radiant light and heat from the sun. Sunlight can be converted directly into electricity by using PV panels or indirectly with concentrating solar power. The concentrating solar power normally focuses on the solar energy to boil water, which is then used in a steam turbine to provide electrical power. The PV technologies use semiconductor materials to convert sunlight into electricity. They have gradually dropped in price in the last two decades. The PV is now widely considered as a cost competitive option for many grid-connected, building-integrated, and off-grid applications as in telecommunications and local power supply.

Wind power: Wind power is considered as one of the most promising technologies for electricity production and the costs, with good wind regimes, that are comparable to fossil alternatives, particularly when the environmental benefit is considered. Airflow can run wind turbines for generating electricity. The rated power of present wind turbines ranges from 0.5 to 10 MW. The power output of a wind turbine depends on the wind speed and so, as the wind speed increases, the power output increases. The location of wind turbine installation is usually chosen in the areas with strong and constant winds, such as offshore and high-altitude sites. The offshore wind power experience shows that the mean wind speed is about 90% greater than the onshore one, so it may contribute more significantly to the supply of future power. Globally, the long-term technical potential of wind energy is believed to be several times the total current global energy production.

Water power: Water power can be exploited in a form of kinetic energy. Since water is about 800 times denser than the air, a slow flowing stream of water can yield considerable amount of energy. Water power exists in many forms. Hydroelectric energy is a term which is usually reserved for large-scale hydroelectric dams. Microhydro systems are hydroelectric power installations that typically produce up to 100 kW power, which are often used in water-rich areas as a remote area power supply. Ocean energy describes all the technologies to harness energy from the ocean and the sea including marine current power, ocean thermal energy conversion, and wave power.

Biofuel power: Plants use photosynthesis to grow and produce biomass. The biomass can be used directly as fuel or to produce biofuels. Agriculturally produced biomass fuels such as biodiesel, ethanol, and bagasse can be burned in internal combustion engines or boilers. The biofuel is typically burned to release its stored chemical energy.

Geothermal power: Geothermal power comes from the heat of the earth itself, from hundreds of meters deep into the earth crust in some places of the globe or from tens of meters in geothermal heat pump in all the places of the planet. Geothermal technology is mostly used for thermal power production; the space heating is becoming increasingly important. Geothermal electricity production is a base load technology, which can be a low-cost option if the hot water or steam resource is at a high temperature and near the earth surface.

The *energy security*, *environment protection*, and *economic growth* are the main primary benefits of renewable energy options. Dependency of power energy on the limited stock of fossil fuels over an extended period is unsustainable and unsecure. Renewable energy may relieve the increasing need of the fossil fuels and reduces the mentioned dependency. On the other hand, the distributed capability of renewable energies brings the power production closer to the end-use, and minimizes energy transportation concerns and costs. Moreover, a greater use of RESs in the power energy portfolio may decrease overall generation costs relative to the probable risks [2]. Energy policies should focus on developing efficient generating portfolios that do not solely rely on stand-alone costs but also on expected portfolio risk, including year-to-year cost fluctuations.

Renewable energy has several important economic benefits. In developed countries, some economic benefits are employment creation and increased trade of technologies and services. An additional advantage is the insensitivity to fuel prices since they are free natural resources, which decreases the operational cost of renewable energy systems and reduces economic operation risks.

The initial investment in renewable-energy-based power generating systems can be considered as a drawback. It is often more expensive to build renewable energy systems than conventional energy systems, since the environmental deterioration has not yet been taken into account for the cost calculation. However, this investment cost will be reduced with the fast developing technologies (such as the development of computer industry and communication systems in the last three decades). Specific requirements of the site and the unpredictability of the generated power can be considered as other disadvantages of the RESs. The intermittent availability of the RESs means to pay a higher cost for regulation issues and maintaining reserve capacity. These quality and security problems have already been encountered in some countries with a high penetration of wind turbines.

1.2.1 Renewable Energy Development

Different kinds of renewable energy technologies have been established in the world markets. Some renewable energy technologies are becoming quickly competitive in growing markets, and some are widely recognized as the lowest cost options for the stand-alone and off-grid applications. The capital costs of certain renewable energy technologies have been obviously reduced over the last two decades and it is possible to be drastically reduced again over the next decade. During the last few years, the installed PV and wind powers dramatically increased.

The development of different renewable sources is limited by different constraints depending on their intrinsic characteristics. The hydropower and geothermal powers are naturally limited because of the lack of geographic sites. Biomass requires large storage places for the natural resources. That is why, a large development in PVs and wind turbines can be seen. But it is noteworthy that these RESs are intermittent power sources (Fig. 1.1a). The production of electricity from solar sources depends on the amount of light energy in a given location. Solar output varies throughout the days and seasons and is affected by cloud cover. Wind-generated power is also a variable resource, and the amount of produced electricity depends on wind speed, air density, and turbine characteristics. If the wind speed is too low (less than about 2.5 m/s), then wind turbines are not able to generate electricity. If it is too high (more than about 25 m/s), the turbines have to be shut down to avoid damage. Since the primary sources of renewable energy are uncertain, the RESs may produce a large amount of power when loads in the grid are very low, and they are not always available, such as solar power in the nighttime and wind power when the wind is not blowing.

The addition of intermittency increases the complexity of the use of RESs. The renewable-energy-based generations with intermittency decrease the reliability of a power system. The nature of intermittency is different for the respective renewable energy technologies, and this difference could be a relevant factor as far as mitigating the impacts of intermittency is concerned. As the percentage of intermittent generation capacity in a grid increases and becomes more significant, additional uncertainty to be created in the management of real-time demand and generation balance. This may require the increasing amount of conventional power reserve (spinning reserve) to manage the grid securely [5].

Most of RESs such as wind and PV generators may not individually participate to the grid management effectively, because they are dependent on the availability of the primary renewable sources. Most of the time, they may not work in their nominal capacity (Fig. 1.1a), and their maximum power variations

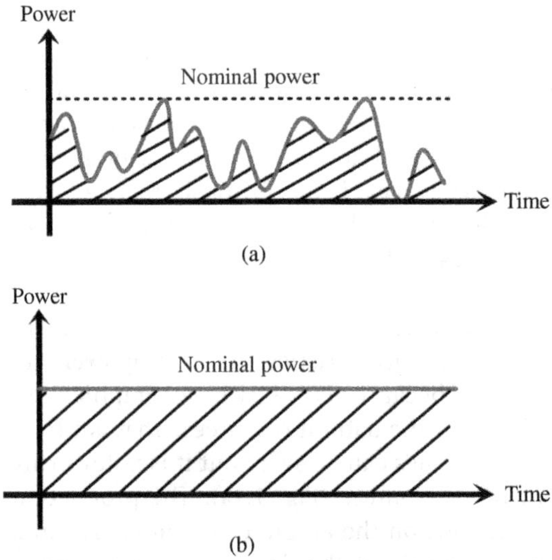

Figure 1.1 Power characteristics of renewable and conventional generators: (a) renewable power and (b) conventional power.

degrade the system reliability and efficiency. So without ESSs and more power generation systems, a high reliability and efficiency for a power system cannot be ensured.

In contrary, conventional power generators are easily controllable and can supply necessary power to satisfy the grid requirements. They can usually provide some ancillary services to the grid. They are mostly fossil and nuclear fueled and rely on the abundant fuel supply such as coal, oil, natural gas, or nuclear fuels. Most of the time, they can work at any power level below the nominal power (Fig. 1.1b) by controlling the fuel supply.

Energy storage devices can serve as backup power plants. They can be used to store or release electrical power such as an energy buffer, supporting the operation of sources, transmission, distribution, and loads. Therefore, as will be discussed later, they are useful to solve the problems of power intermittency and system low inertia in an RES-based power grid.

The combination of energy storage devices and RESs constitutes a hybrid power generator system, which can provide power not only for the local load but also can supply ancillary services to the main grid as the conventional generators [6]. Among all the RESs, the PV and wind turbines are known as the most popular renewable power sources.

1.3 Grid-connected Wind Power

In general, the WPGs can be classified into three categories [7].

1.3.1 Wind Power Generator Without Power Electronic Converters

Most of these topologies are based on a squirrel-cage induction machine (SCIM), which is directly connected to the grid. A soft starter is usually used to reduce the inrush current during start-up. Moreover, a capacitor bank is necessary for the reactive power compensation (Fig. 1.2a). By adding power electronic converters into the wind generator, a variable-speed wind generator can be achieved. Although the system complexity and the solution cost are increased, a better control of the converted primary power and of the grid connection can be obtained. For example, maximum power can be extracted from a large variation of wind speed. The use of power electronic converters in the wind generator can be further divided into two categories: systems using partial- and full-scale power electronic converters.

1.3.2 Wind Power Generator Using Partial-Scale Power Electronic Converters

A particular structure is based on a wounded rotor induction machine. An extra resistor controlled by power electronic converters is added in the rotor and gives a variable speed range of 2–4%. The power converter for the rotor resistor control is for low voltage but high currents. This solution also needs a soft starter and a reactive power compensator. Another solution is to use a back-to-back power electronic converter with the wounded rotor induction machine (WRIM), as shown in Fig. 1.2b.

In this case, a power converter connected to the rotor through slip rings controls the rotor currents. If the generator is running supersynchronously, the electrical power is delivered through both the rotor and stator. If the generator is running subsynchronously, the electrical power is only delivered into the rotor from the grid. A speed variation of 60% around synchronous speed may be obtained by the use of a power converter of 30% of the nominal power.

1.3.3 Wind Power Generator Using Full-Scale Power Electronic Converters

By implementing a full-scale power converter between the electrical machine and the utility grid, additional technical performances of the wind generator can be achieved. Normally, as shown in Fig. 1.2c, an SCIM or a synchronous machine (SM) is used in this configuration. By using a multipole wound rotor (or permanent magnet) synchronous machine, a high frequency of electrical quantities is generated and the gearbox can be eliminated (Fig. 1.2d).

In this section, a variable-speed WPG, as shown in Fig. 1.2c, is considered for extracting the maximum available wind power. Such kind of WPG supplies continuously varying powers, which depend on the intermittent and fluctuant wind velocity. When a large scale of WPGs is connected to the grid, stability problems occur due to the dependence of the power production on the wind condition [8–10].

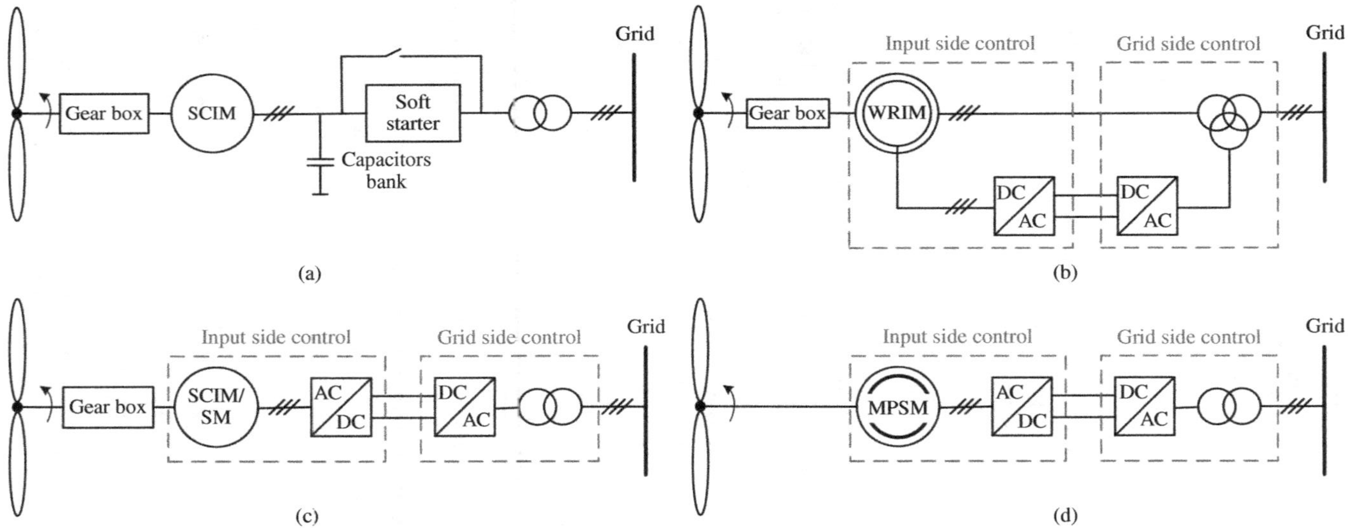

Figure 1.2 Wind generator with power electronics: (a) minimum electronics unit, (b) partial power converter, (c) full-scale power converter with gearbox, and (d) full-scale power converter without gearbox.

1.3.3.1 Wind Energy Conversion System: Modeling, Control, and Analysis

A classical wind energy conversion system consists of a three-blade turbine, a gearbox, an electrical machine, a three-phase rectifier, a DC-bus capacitor, a three-phase inverter, and line filters that are connected to the grid through a grid transformer (Fig. 1.3).

When the wind energy conversion system works in the MPPT strategy, a fluctuated power is delivered to the grid. A typical wind-speed-power profile recorded in a wind farm in the north of France is shown in Fig. 1.4 [11].

1.3.3.1.1 Electrical Conversion Chain Model Using equivalent average modeling of power electronic converters, the average model of the electrical power conversion chain can be obtained for the wind energy conversion system (Fig. 1.5). The grid with transformer is considered as three-phase voltage sources and the electrical machine is considered as three-phase current sources. Two back-to-back voltage source converters introduce control inputs for the power control. As the DC bus has a relatively slow dynamic, three different

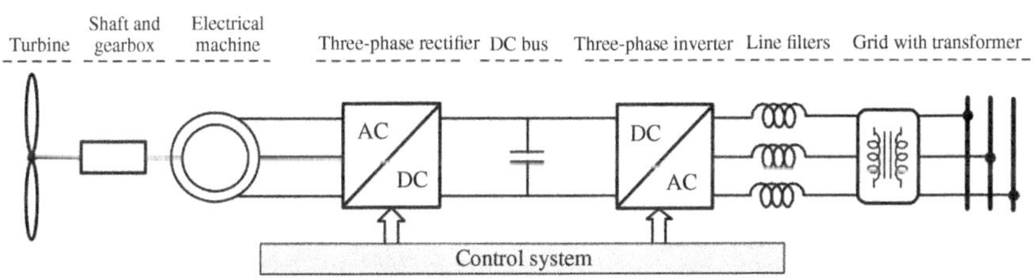

Figure 1.3 A conventional variable-speed WPG.

Figure 1.4 Real recorded wind speed-power pattern: (a) wind speed and (b) wind power.

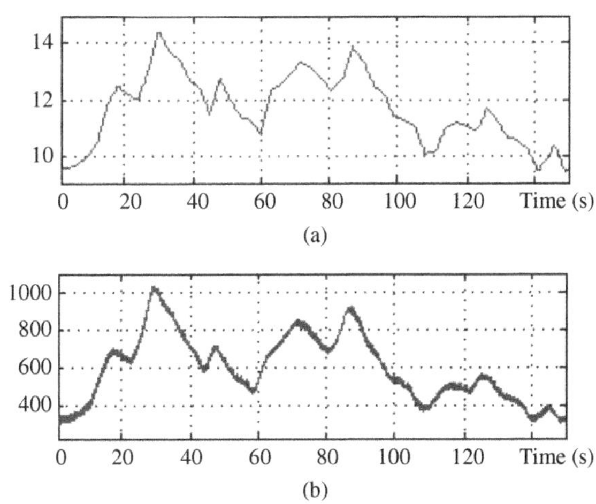

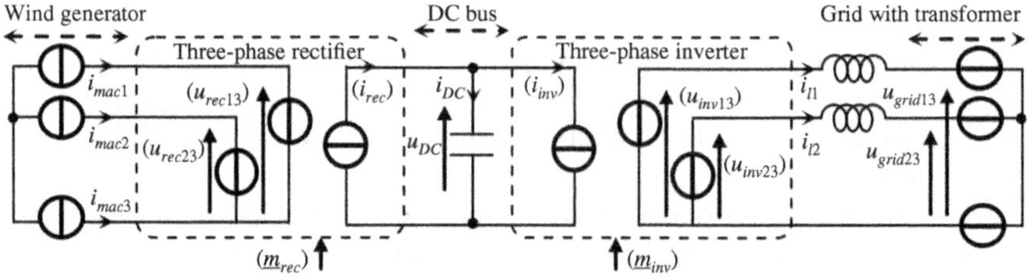

Figure 1.5 Equivalent average model of the power electronic converters.

subsystems with their inner dynamic and control tasks (wind generator, grid connection system, and the DC bus) can be considered.

Three-phase rectifier: The equivalent average model is used with average modulation functions. It yields the average values ($\langle \underline{u}_{rec}\rangle = [(u_{rec13}, u_{rec23}]^T$) of the modulated voltages from the DC-bus capacitor voltage (u_{DC}) and the average value ($\langle i_{rec}\rangle$) of the modulated current from the machine currents ($\underline{i}_{mac} = [i_{mac1}, i_{mac2}]^T$):

$$\begin{cases} \langle \underline{u}_{rec}\rangle = \langle \underline{m}_{rec}\rangle u_{DC} \\ \langle i_{rec}\rangle = \langle \underline{m}^{\mathrm{T}}_{rec}\rangle \underline{i}_{mac} \end{cases} \tag{1.1}$$

where $\underline{m}_{rec}$ is the vector of modulation functions of the grid inverter.

Three-phase inverter: The three-phase inverter is modeled in the same way. The average value of the modulated voltages ($\langle \underline{u}_{inv}\rangle = [(u_{inv13}, u_{inv23}]^T$) is calculated from the DC-bus voltage (u_{DC}) and the average value ($\langle i_{inv}\rangle$) of the modulated current from the line currents ($\underline{i}_l = [i_{l1}, i_{l2}]^T$):

$$\begin{cases} \langle \underline{u}_{inv}\rangle = \langle \underline{m}_{inv}\rangle u_{DC} \\ \langle i_{inv}\rangle = \langle \underline{m}^{\mathrm{T}}_{inv}\rangle \underline{i}_l \end{cases} \tag{1.2}$$

where $\underline{m}_{inv}$ is the vector of modulation functions of the grid inverter.

1.3.3.1.2 Wind Generator Model

The modeling of the wind energy generation system is presented in detail in the block diagram shown in Fig. 1.6.

Wind: The wind is modeled by a mechanical source, which sets the wind velocity (v_{wind}) to the blades.

Turbine: The turbine is modeled as a mechanical converter. The torque (T_{tur}), which is produced by the turbine, depends on the wind velocity (v_{wind}) and the blade pitch angle (β):

$$T_{tur}(t) = \frac{1}{2}C_T(\lambda, \beta)\rho S_b R_b v^2_{wind}(t) \tag{1.3}$$

where S_b is the area that is swept by the blades, R_b is the blade length, ρ is the air density, C_T is the torque coefficient that is a nonlinear function of the

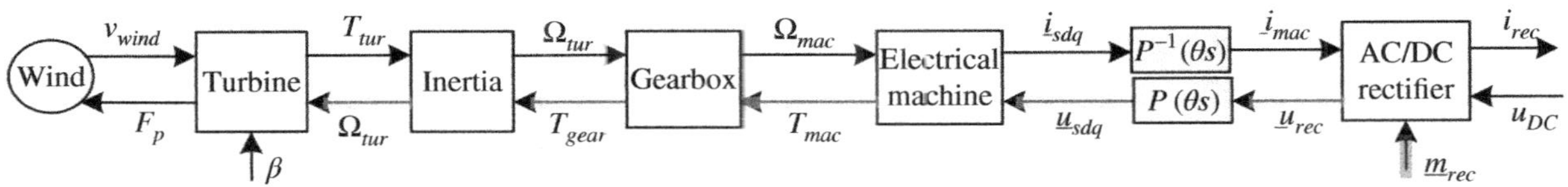

Figure 1.6 Block diagram of the considered wind energy generation system.

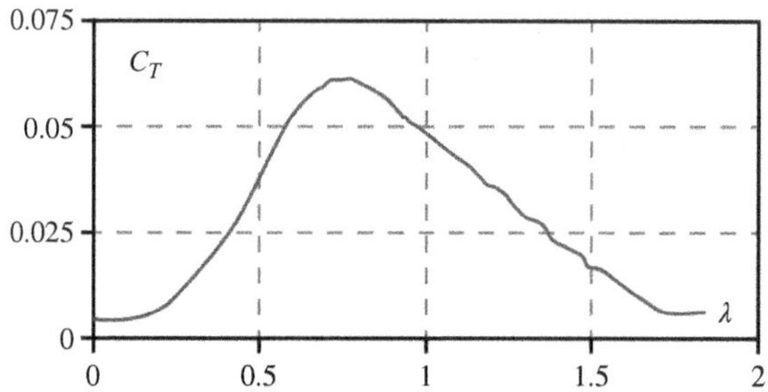

Figure 1.7 Blade characteristic: C_T versus λ for a fixed blade angle.

tip-slip ratio (λ) (Fig. 1.7), and λ is the tip-slip ratio depending on the wind velocity and the rotational speed (Ω_{tur}).

$$\lambda(t) = \frac{R_b \Omega_{tur}(t)}{v_{wind}(t)} \tag{1.4}$$

For the present study, a normal turbine operating with a constant pitch angle is considered.

Inertia: The shaft is an element with energy accumulation, which imposes the rotational speed (Ω_{tur}) with the torque difference between the blade torque (T_{tur}) and the gear torque (T_{gear}),

$$J_{shaft} \frac{d\Omega_{tur}(t)}{dt} = T_{tur}(t) - T_{gear}(t) - f_{shaft}\Omega_{tur}(t) \tag{1.5}$$

where the J_{shaft} is the equivalent inertia moment of the shaft and the f_{shaft} is the friction coefficient of the equivalent shaft.

Gearbox: The gearbox is a mechanical converter and adapts the low speed of the turbine with the high speed of the electrical machine. It yields the rotational speed (Ω_{mac}) and the torque (T_{gear}) through the gear ratio (m_{gear}):

$$\begin{cases} \Omega_{mac}(t) = m_{gear}\Omega_{tur}(t) \\ T_{gear}(t) = m_{gear} T_{mac}(t) \end{cases} \tag{1.6}$$

Electrical machine: The electrical machine can be generally modeled as an electromechanical converter with the rectifier voltages ($u_{rec} = [u_{rec13}, u_{rec23}]^T$) and the gear speed ($\Omega_{mac}$) as the inputs. The stator currents ($i_{mac} = [i_{mac1}, i_{mac2}]^T$) and the machine torque (T_{mac}) are the outputs.

For understanding the electrical machine dynamic, first one needs to use a mathematical transformation from phase-to-phase voltages ($\underline{u}_{rec}$) to machine line voltages. Then, the Park transformation expresses stator voltages and currents in a d–q rotational frame ($\underline{u}_{sdq}$):

$$\begin{cases} \underline{u}_{sdq} = P(\theta_s)\underline{u}_{rec} \\ \underline{i}_{mac} = P^{-1}(\theta_s)\underline{i}_{sdq} \end{cases} \tag{1.7}$$

where θ_s (or θ_{sdq}) is the angle of the rotating d–q frame with respect to the stator stationary frame. In the d–q frame, the equivalent stator windings set the stator currents ($\underline{i}_{sdq} = [i_{sd}, i_{sq}]^T$) as state variables, which are calculated with the stator voltages ($\underline{u}_{sdq} = [u_{sd}, u_{sq}]^T$) and the electromotive force (emf) ($\underline{e}_{sdq} = [e_{sd}, e_{sq}]^T$),

$$\begin{cases} L_s \dfrac{di_{sd}}{dt} = u_{sd} + e_{sd} - R_s i_{sd} \\ L_s \dfrac{di_{sq}}{dt} = u_{sq} - e_{sq} - R_s i_{sq} \end{cases} \tag{1.8}$$

R_s is the resistor of the stator winding and L_s is the cyclic inductor of the stator winding.

Finally, such as an electromechanical converter, it leads to the machine torque (T_{mac}) and the emf ($\underline{e}_{sdq}$) from the stator currents and the rotor's electrical angular speed (Ω_{mac}).

$$T_{mac} = \frac{3}{2}\rho(\phi_{rd} i_{sq} - i_{sd}\phi_{rq})$$

The rotor flux orientation is achieved by aligning the d-axis of the synchronous reference frame with the rotor flux vector ϕ_r. The resultant d- and q-axis rotor flux components are $\phi_{rq} = 0$ and $\phi_{rd} = \phi_r$.

$$T_{mac} = \frac{3}{2}\rho\phi_{rd} i_{sq} \tag{1.9}$$

$$\begin{cases} e_{sd} = L_s \Omega_{mac} i_{sq} \\ e_{sq} = L_s \Omega_{mac} i_{sd} \end{cases} \tag{1.10}$$

where ρ is the number of pole pairs and ϕ_{rd} is constant rotor flux of the permanent magnet.

1.3.3.1.3 Modeling of the Grid Connection

The modeling of the grid connection system is presented in Fig. 1.8.

Grid filter: The line currents ($\underline{i}_l$) are calculated from dynamic equations of the filter with the inverter phase-to-phase voltages ($\underline{u}_{inv}$) and the phase-to-phase grid voltages ($\underline{u}_{grid}$):

$$L_{line}\frac{d\underline{i}_l}{dt} = \frac{1}{3}\begin{bmatrix} 2 & -1 \\ -1 & 2 \end{bmatrix}(\underline{u}_{inv} - \underline{u}_{grid}) - r_{line}\underline{i}_l \tag{1.11}$$

where L_{line} is the equivalent inductor of the grid filter and r_{line} is the equivalent resistor in series of the grid filter. The electrical network with the grid transformer is considered as ideal sinusoidal phase-to-phase voltage sources $\underline{u}_{grid} = [u_{grid13}, u_{grid23}]^T$.

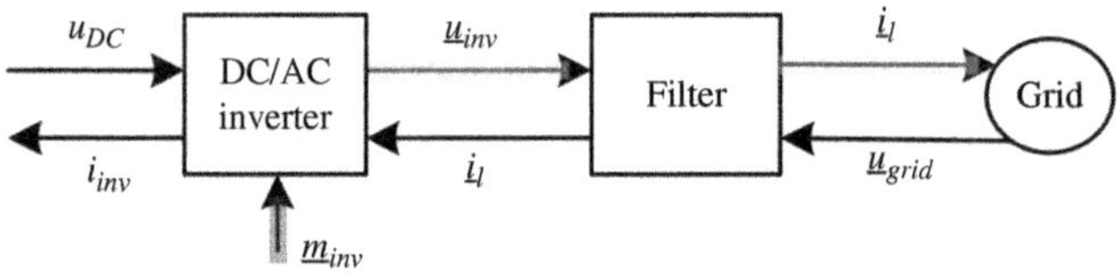

Figure 1.8 Block diagram of the grid connection system.

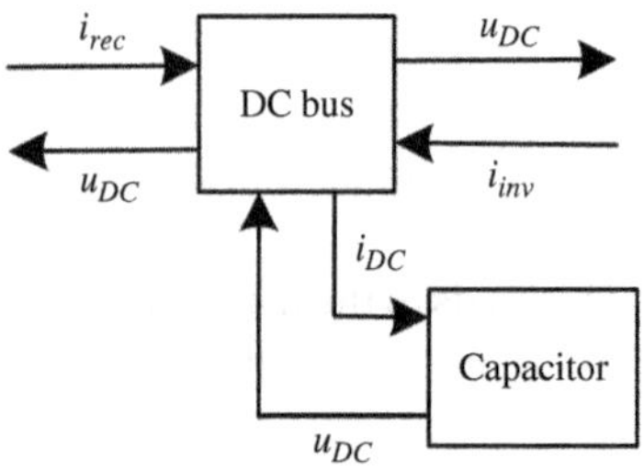

Figure 1.9 Block diagram of the DC bus.

1.3.3.1.4 Modeling of the DC Bus The model of the DC bus is presented in Fig. 1.9. In order to control the power exchanges around the DC bus, the DC coupling should be modeled in detail:

$$i_{DC} = i_{rec} - i_{inv} \tag{1.12}$$

where i_{rec} is the rectifier's modulated current and i_{inv} is the inverter's modulated current.

The DC-bus capacitor is an element with energy accumulation and its voltage is calculated from the dynamic equation:

$$C_{DC}\frac{du_{DC}}{dt} = i_{DC} \tag{1.13}$$

where C_{DC} is the DC-bus capacitor and u_{DC} is the DC-bus voltage.

1.3.3.1.5 Modeling of the Entire Wind Energy Conversion System The model of the entire wind energy conversion system can be obtained by combining all block diagrams (Figs 1.6, 1.8, and 1.9), as shown in Fig. 1.10.

1.3.3.2 Hierarchical Control Structure

The wind energy conversion system transfers power from the WPG to the electrical grid. Two power converters are used to regulate the power exchange. A hierarchical control structure, such as shown in Fig. 1.11, can be considered to implement the control system. Two switching control units (SCUs) and two automatic control units (ACUs) are used separately in the control system for the mentioned power converters. A common power control unit (PCU) and a common mode control unit (MCU) are used for the instantaneous power

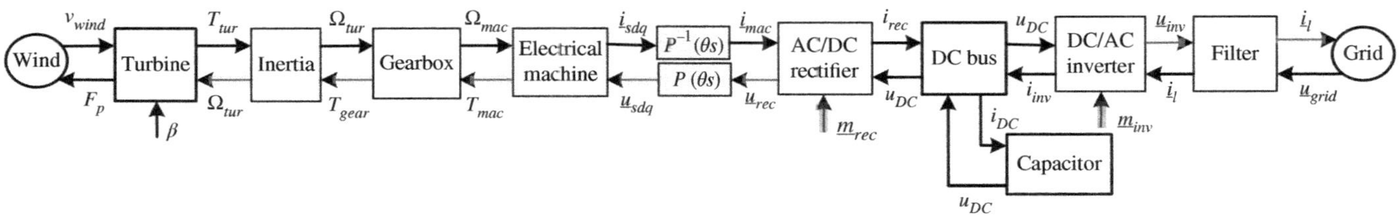

Figure 1.10 Block diagram of the entire wind energy conversion system.

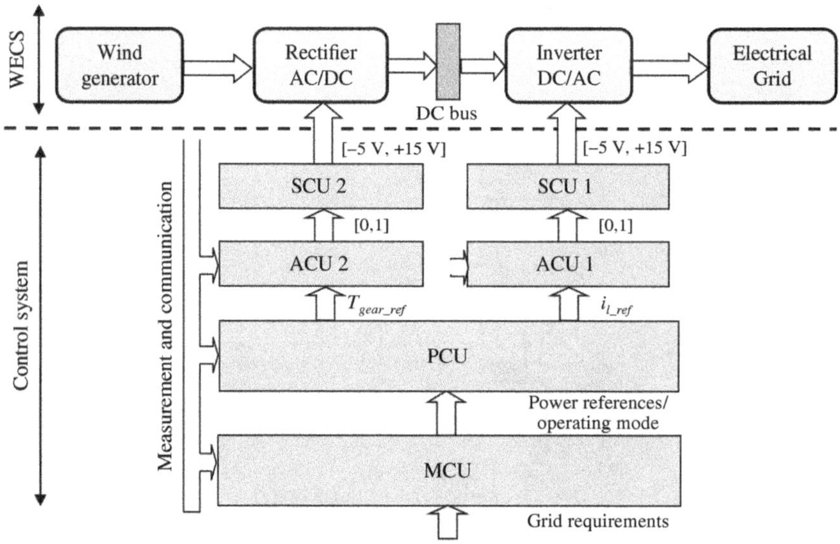

Figure 1.11 Hierarchical control structure of the wind energy conversion system.

balancing and the long-term energy management of the entire power system, respectively.

In the SCU of each converter, the insulated-gate bipolar transistors (IGBT) drivers and pulse width modulation (PWM) techniques are used to control the semiconductors in power electronic converters. The control algorithms in the ACU should be presented in order to highlight the physical quantities, which can be used for the power flow control among the different energy sources.

1.3.3.2.1 ***Wind Generator Control*** The electrical power versus speed curves of a typical wind turbine is given in Fig. 1.12. For example, if the wind velocity is v_1, the output power can be raised to the maximum value at point A by setting the mechanical speed to Ω_1. If the wind speed changes to v_2, the power output jumps to point B. For this wind velocity, the maximum power can be extracted by setting the speed at point C (Ω_2). This shows that, as the wind speed changes, the generator speed should track these changes in order to extract the maximum power. This is called MPPT strategy.

The block diagram of the wind energy conversion system modeling (Fig. 10) shows that the speed (Ω_{tur}) can be controlled by acting on two inputs: aerodynamic torque (T_{tur}) and torque of the generator (T_{gear}). As here we consider a normal operation with a constant pitch angle, the aerodynamic torque must be considered as a perturbation input (linked to the wind speed) for the system. So the turbine speed can be controlled by acting on the gearbox torque (T_{gear}) via the control input ($\underline{m}_{rec}$) of the power electronic converter.

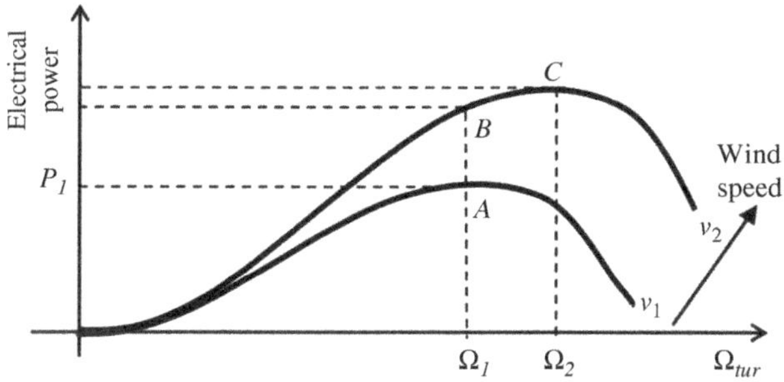

Figure 1.12 Turbine power-speed characteristic.

From the block diagram of the wind energy conversion system, an action chain appears from the control inputs ($\underline{m}_{rec}$) of the rectifier to the gear's mechanical torque (T_{gear}) (Fig. 1.10). The control scheme of the wind energy generation system is obtained by inverting this action chain (Fig. 1.13). It involves calculating the reference of the rectifier's duty ratios ($\underline{m}_{rec}$_ref) according to a torque reference (T_{gear_ref}). It is composed of a torque control, a field-oriented control, and a rectifier control.

Torque control: The calculated mechanical torque reference (T_{gear_ref}) from an MPPT strategy is converted into the machine torque reference (T_{mac_ref}) by inverting (1.6):

$$T_{mac\_ref} = \frac{1}{m_{gear}} T_{gear\_ref} \tag{1.14}$$

Field-oriented control: A standard field-oriented control is used to control the electrical machine (Fig. 1.14) [12]. The inversion of (1.9) leads to the current references ($\underline{i}_{sdq\_ref} = [i_{sd\_ref}\ i_{sq\_ref}]^T$), i_{sq_ref} is obtained from the torque reference (T_{mac_ref}) with the constant rotor flux (ϕ_{rd}) of the permanent magnet and i_{sd_ref} is set to zero with a properly chosen frame orientation:

$$\begin{cases} i_{sq\_ref} = \frac{(2/3)}{p\phi_{rd}} T_{mac\_ref} \\ i_{sd\_ref} = 0 \end{cases} \tag{1.15}$$

As the stator windings are accumulation elements, closed-loop controllers are needed to invert (1.8):

$$\underline{u}_{sdq\_ref} = PI(\underline{i}_{sdq\_ref} - \widehat{\underline{i}}_{sdq}) + \tilde{e}_{sdq} \tag{1.16}$$

where $PI(x_{ref}-x)$ is the controller of the variable x. In practice, the emf ($\underline{e}_{sdq}$) cannot be measured, but it can be estimated through the sensed rotor's rotational speed (Ω_{mac}) by using (1.10):

$$\begin{cases} \tilde{e}_{sd} = L_s \Omega_{mac} \widehat{i}_{sq} \\ \tilde{e}_{sq} = L_s \Omega_{mac} \widehat{i}_{sd} \end{cases} \tag{1.17}$$

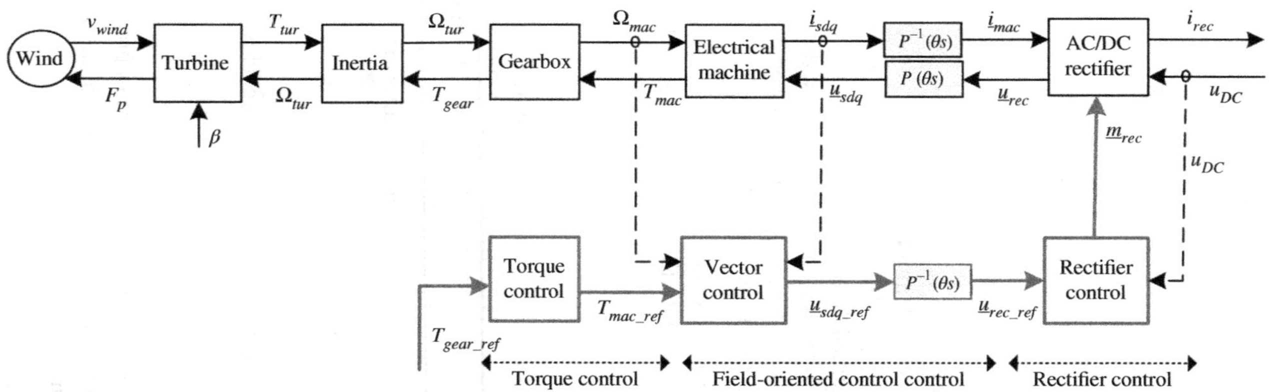

Figure 1.13 Control scheme of the wind energy generation system.

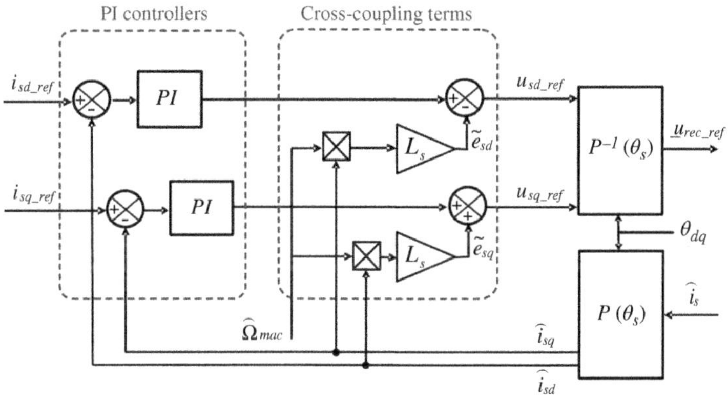

Figure 1.14 Block diagram of the oriented field control of the electrical machine.

Finally by inverting (1.7), the inverse Park transformation ($P^{-1}(\theta_s)$) leads to the references of the rectifier voltages ($\underline{u}_{rec\_ref}$):

$$\underline{u}_{rec\_ref} = P^{-1}(\theta_s)\underline{u}_{sdq\_ref} \tag{1.18}$$

Rectifier control: The reference modulation functions $\underline{m}_{rec\_ref}$ are obtained by inversion of (1.1) through the measurement of the DC-bus voltage u_{DC}:

$$\underline{m}_{rec\_ref} = \frac{1}{\hat{u}_{DC}}\underline{u}_{rec\_ref} \tag{1.19}$$

1.3.3.2.2 Grid Connection Control Considering the block diagram of the grid connection system (Fig. 1.8), a path from the control inputs ($\underline{m}_{inv}$) of the inverter to the line currents ($\underline{i}_l$) can be seen. The control scheme of the grid connection system is obtained by inverting this path (Fig. 1.15). It involves calculating the reference of the inverter's duty ratios ($\underline{m}_{inv\_ref}$) according to the line currents' references ($\underline{i}_{l\_ref}$).

Line current control: The grid is a voltage source ($\underline{u}_{grid}$), so the line current should be controlled in order to regulate the exchanged power with the grid. In order to control this current, a current controller is needed with a Park transformation (P) and an inverse Park transformation (P^{-1}) as shown in Fig. 1.16:

$$\underline{u}_{inv\_ref} = P^{-1}(\theta_{dq})\,\underline{u}_{inv\_dq\_ref}$$

$$\begin{cases} u_{inv\_d\_ref} = PI(i_{l\_d\_ref} - i_{l\_d}) + \hat{u}_{grid\_d} - \tilde{e}_{lq} \\ u_{inv\_q\_ref} = PI(i_{l\_q\_ref} - \underline{i}_{l\_q}) + \hat{u}_{grid\_q} + \tilde{e}_{ld} \end{cases} \tag{1.20}$$

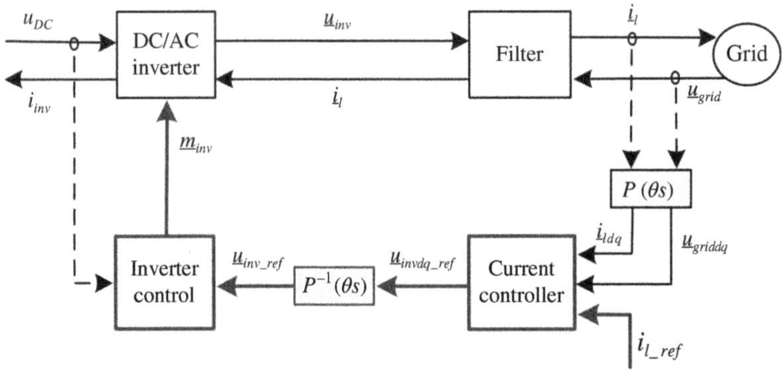

Figure 1.15 Control scheme of the grid connection system.

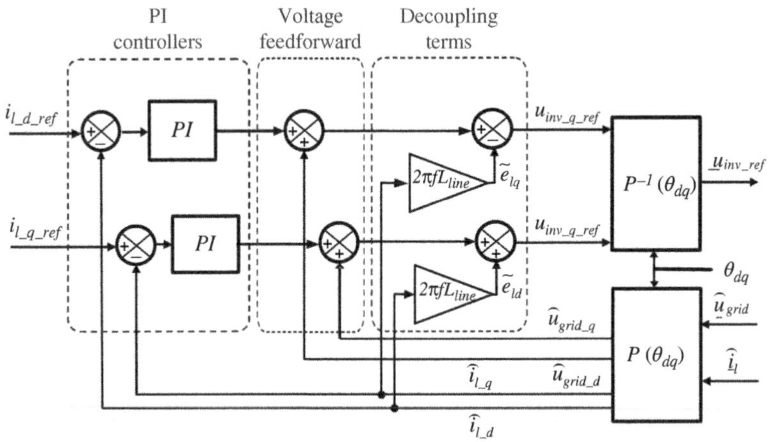

Figure 1.16 Block diagram of the line current control in the grid connection system.

where $\underline{i}_{l\_dq\_ref} = P(\theta_{dq})\underline{i}_{l\_ref}$, $\hat{\underline{i}}_{l\_dq} = P(\theta_{dq})\hat{\underline{i}}_{l}$, $\hat{\underline{u}}_{grid\_dq} = P(\theta_{dq})\hat{\underline{u}}_{grid}$, $\tilde{e}_{lq} = 2\pi f L_{line}\hat{i}_{l\_q}$, $\tilde{e}_{ld} = 2\pi f L_{line}\hat{i}_{l\_d}$, and f is the frequency of the grid voltage.

Inverter control: The modulation functions of the inverter are obtained by inverting (1.2):

$$\underline{m}_{inv\_ref} = \frac{1}{\hat{u}_{DC}}\underline{u}_{inv\_ref} \tag{1.21}$$

1.3.3.2.3 ***DC Bus Control*** The wind energy conversion system can be decomposed into three independent subsystems if the DC-bus voltage is constant. The DC-bus voltage should be well regulated for the stability of the grid connection

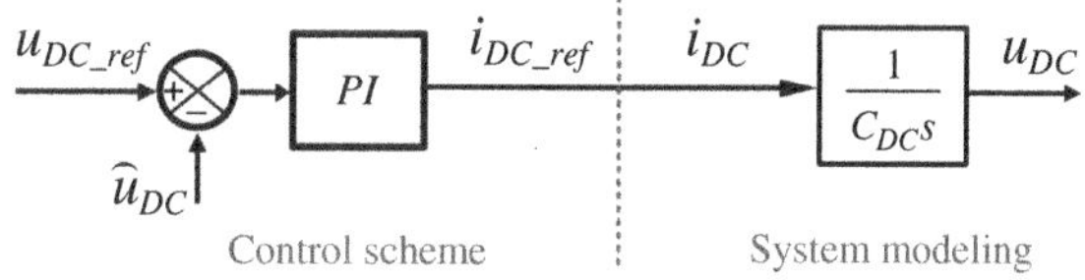

Figure 1.17 Control scheme of the DC bus.

because it is used in a division operator in the converter control algorithms ((1.19) and (1.21)), so a voltage controller is needed (Fig. 1.17):

$$i_{DC\_ref} = PI(u_{DC\_ref} - \widehat{u}_{DC}) \tag{1.22}$$

1.3.3.2.4 Control of the Entire Wind Energy Conversion System The control scheme of the entire wind energy conversion system is obtained by combining all the control schemes presented earlier. The result is drawn in Fig. 1.18. The corresponding block diagram of the ACUs is shown in Fig. 1.19.

1.3.3.2.5 Power Control Unit In the studied wind energy conversion system, all power exchanges are performed via the DC-bus (Fig. 1.20) and have an impact on the DC-bus voltage:

$$\frac{dE_{DC}}{dt} = C_{DC}u_{DC}\frac{du_{DC}}{dt} = u_{DC}i_{DC} = p_{DC} = p_{wg} - p_{gc} \tag{1.23}$$

where E_{DC} is the stored energy in the DC-bus capacitor, p_{DC} is the exchanged power with the DC-bus capacitor, p_{wg} is the injected power into the DC bus from the wind generator, and p_{gc} is the extracted power from the DC bus into the grid.

- **General layout**
  The PCU can be divided into two levels: the power control level (PCL) and the power sharing level (PSL). The PCL involves calculating the reference of the related quantities (T_{gear_ref}, $\underline{i}_{DC\_ref}$, $\underline{i}_{l\_ref}$) from the power references (p_{wg_ref}, p_{DC_ref}, p_{g_ref}, q_{g_ref}). The PSL coordinates the power flow exchanges among the different energy sources.
- **Power control level**
  Each controlled quantities implies a power, which is calculated in Table 1.1. For the wind generator, an MPPT strategy is used to extract the maximum power. The power reference (p_{wg_ref}) can be set by calculating the corresponding torque reference (T_{gear_ref}) with the sensed value of the rotational speed (Ω_{tur}) according to the inverse equation (iii_c). The powers, which are exchanged with the grid, can be calculated through the "two-wattmeter" method with the equation (ii) and the line current references are calculated by the inverse equations (ii_c). The output of the DC-bus voltage control loop is a reference for the required DC current (i_{DC_ref}), and its product with

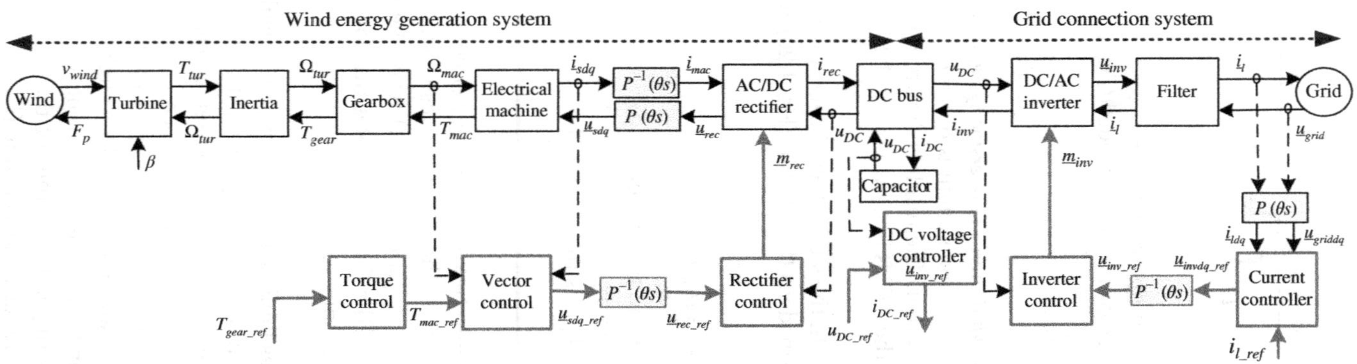

Figure 1.18 Control scheme of the entire wind energy conversion system.

the measured DC-bus voltage (u_{DC}) gives the necessary power reference (p_{DC_ref}) for the DC-bus voltage regulation according to (i_c).

- **Power sharing level**
  The choke filters are sized to obtain a small voltage drop across them and their losses and reactive powers are small. Moreover, these powers will be considered as disturbances and can be attenuated by the various closed-loop controls used. So we will not discuss in detail the power estimation of losses and compensation algorithms in order to focus on the power balancing algorithms.

We can assume that the wind power (p_{wg}) is divided into two parts (Fig. 1.21). One part (p_{DC}) is sent to the DC-bus capacitor. The other part is sent to the grid (p_g). The power exchange can be expressed as

$$p_g = p_{wg} - p_{DC} \tag{1.24}$$

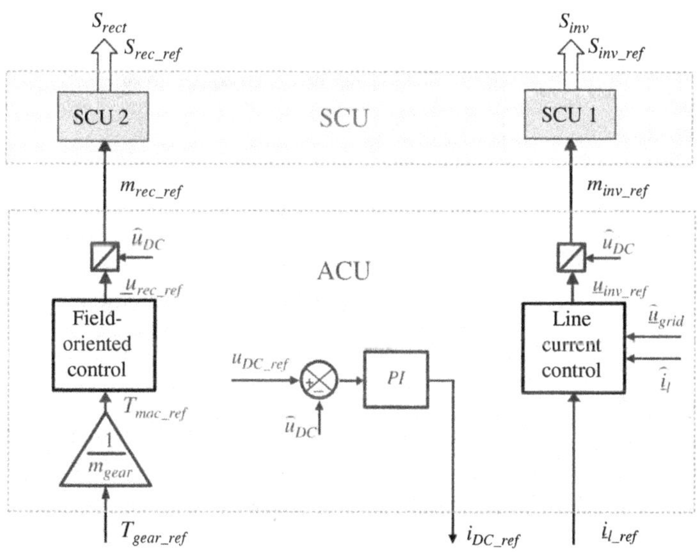

Figure 1.19 Block diagram of the automatic control units for the wind energy conversion system.

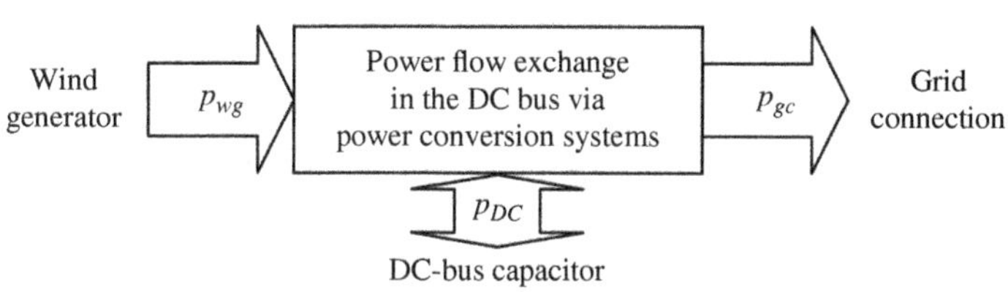

Figure 1.20 Power flow exchanges around the DC bus.

Table 1.1 Power calculation and control algorithms for the wind energy conversion system.

Energy source	Power calculation	Power control
DC-bus capacitor	$p_{DC} = u_{DC} i_{DC}$ (i)	$p_{DC\_ref} = \widehat{u}_{DC} i_{DC\_ref}$ (i_c)
Grid connection	$\begin{cases} p_g = u_{13} i_1 + u_{23} i_2 \\ q_g = \sqrt{3}(u_{13} i_1 - u_{23} i_2) \end{cases}$ (ii)	$\begin{cases} i_{l1\_ref} = \dfrac{(2\widehat{u}_{13} - \widehat{u}_{23}) p_{g\_ref} + \sqrt{3}\widehat{u}_{23} q_{g\_ref}}{2\widehat{u}_{13}^2 - 2\widehat{u}_{13}\widehat{u}_{23} + 2\widehat{u}_{23}^2} \\ i_{l2\_ref} = \dfrac{(2\widehat{u}_{23} - \widehat{u}_{13}) p_{g\_ref} - \sqrt{3}\widehat{u}_{13} q_{g\_ref}}{2\widehat{u}_{13}^2 - 2\widehat{u}_{13}\widehat{u}_{23} + 2\widehat{u}_{23}^2} \end{cases}$ (ii_c)
Wind generator	$p_{wg} = \Omega_{tur} T_{gear}$ (iii)	$T_{gear\_ref} = \dfrac{1}{\widehat{\Omega}_{tur}} p_{wg\_ref}$ (iii_c)

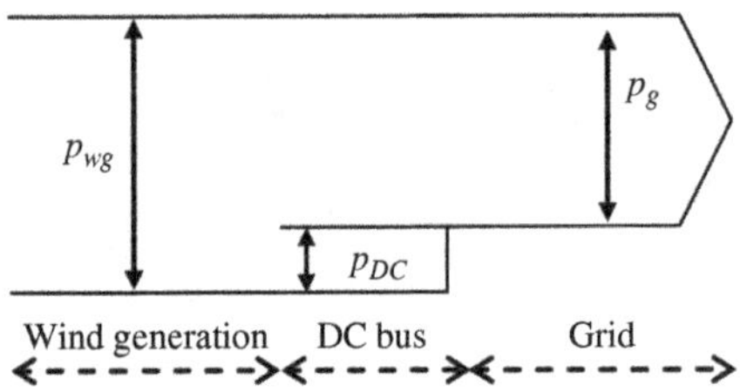

Figure 1.21 Power flow exchange inside the wind energy conversion system.

The wind generator is connected to a three-phase rectifier and various control strategies can be used. For example, when the wind velocity is not too high, the wind generator can work with an MPPT strategy to improve the global energy efficiency. When the wind velocity becomes too high, the wind generator can work with the rated power strategy by reducing the power efficiency for security reasons. In our study, we assume that the wind velocity is medium and we use an MPPT strategy in the form of a searching algorithm for the maximum power as shown in Fig. 1.12. Therefore, with the fluctuant wind, the wind power is very fluctuant. This fluctuant power is rectified and sent to the DC bus. Hence, a "grid-following" power balancing strategy should be used (Fig. 1.22) because the availability of the wind power is not ensured for the DC bus control, and the DC-bus voltage is regulated by the line current controller through the three-phase inverter. The wind power (p_{wg}) must be seen as a fluctuant disturbance. In order to regulate the DC-bus voltage, the only way is to use the grid power (p_g), as shown in the closed loop ($u_{DC\_ref} \rightarrow p_{DC\_ref} \rightarrow p_{g\_ref} \rightarrow m_{inv} \rightarrow p_g \rightarrow p_{DC} \rightarrow u_{DC}$) in Fig. 1.22.

So the grid power reference (p_{g_ref}) is obtained by taking into account the DC-bus power reference (p_{DC_ref}) for the voltage regulation and the estimated wind power $\tilde{p}_{wg}$. In practice, we can set $\tilde{p}_{wg} = p_{wg\_ref}$. The hierarchical control

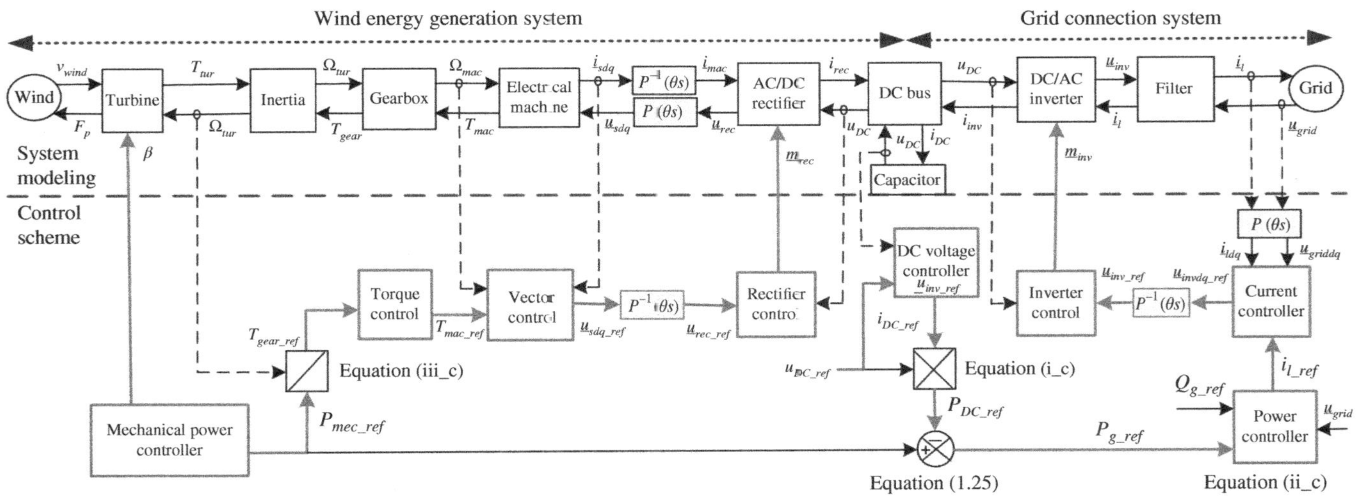

Figure 1.22 Multilevel representation of the wind energy conversion system.

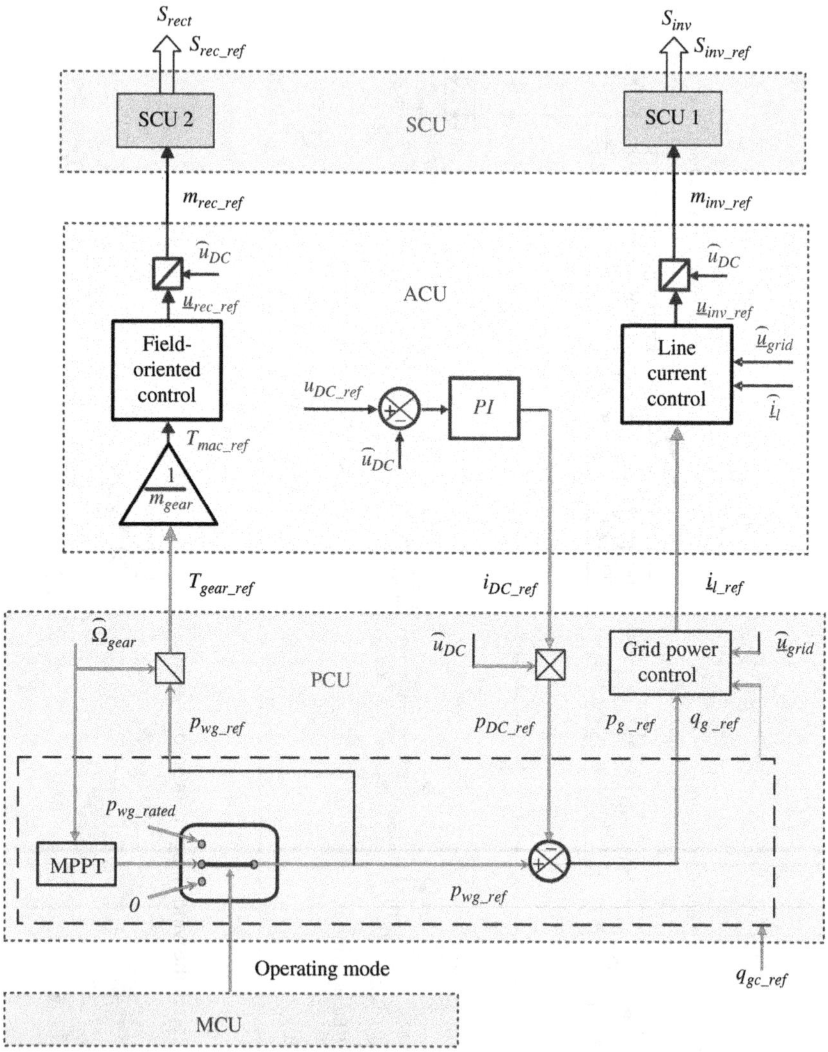

Figure 1.23 Block diagram of the hierarchical control for the wind energy conversion system.

system of the wind energy conversion system is then extended as shown in the block diagram of Fig. 1.23.

$$p_{g\_ref} = \tilde{p}_{wg} - p_{DC\_ref} \tag{1.25}$$

1.3.3.2.6 Mode Control Unit The operating mode of the wind generator depends on the wind speed condition and the grid requirements (Fig. 1.23).

The wind generator can work normally while the grid capacity is large enough to receive the fluctuant wind power without much impact. If the wind speed is low or medium, the wind generator works in an MPPT strategy. Otherwise, the wind generator should be limited with the rated power value (p_{wg_rated}) with high wind speed, or even be shut down with extremely high wind speed for security reasons.

$$p_{wg\_ref} = \begin{cases} \mathrm{MPPT}(\Omega_{tur}) & \text{with weak or medium wind} \\ p_{wg\_rated} & \text{with strong wind} \\ 0 & \text{with extremely strong wind} \end{cases} \quad (1.26)$$

These strategies can be switched in the PCU by a signal (*operating mode*) coming from the MCU (Fig. 1.23).

1.3.3.3 Simulation and Experimental Examination

1.3.3.3.1 Hardware and Software Implementation To show the validity of the proposed model, wind generator emulators have been used in the laboratory. In this section, a simplified version is presented to obtain the same power variation as from the real wind generator (Fig. 1.4). In order to have a flexible and "easy-to-use" wind energy conversion system for testing control algorithms, an emulator with a reduced rated power (1.2 kW) has been developed.

The wind power emulator is a controllable power source, which must provide the same power profile as the wind energy generation system. So, the wind generator in Fig. 1.5 is replaced experimentally by the wind power emulator as shown in the equivalent average modeling (Fig. 1.24).

In this case, the average value of the modulated current from the chopper ($\langle i_{m\_wg} \rangle$) in Fig. 1.24 is proportional to the average value of the modulated current from the three-phase rectifier ($\langle i_{rec} \rangle$) in Fig. 1.5. The power electronic stage of the wind power emulator is implemented with a step-up power conversion circuit (Fig. 1.25), including a constant DC voltage source (120 V), a choke filter (20 mH), and a DC chopper. By controlling the current (i_{wg}) of the filter

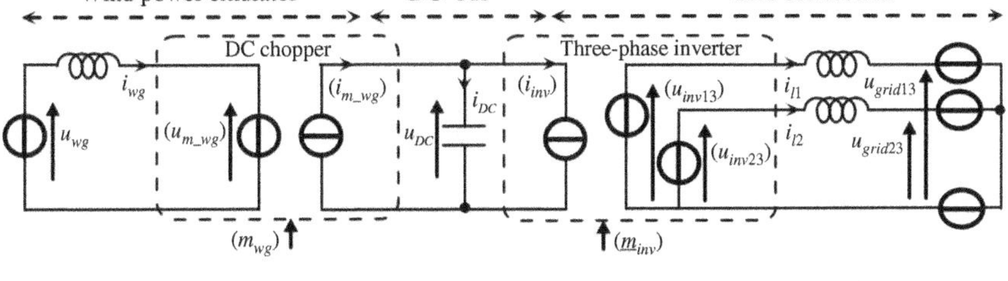

Figure 1.24 Equivalent average modeling of the power conversion chain with a wind power emulator.

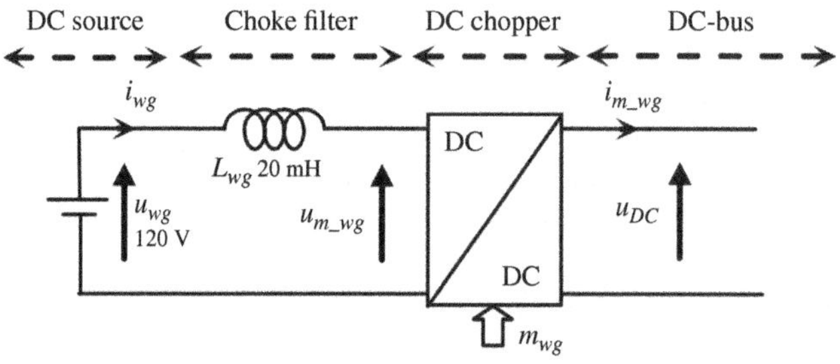

Figure 1.25 Power electronic stage of the wind power emulator.

inductor, the power of the emulator can be well controlled through the duty ratio (m_{wg}) as control input.

The dynamic representation and relevant equations can be easily obtained as described in the previous sections. For modeling the DC chopper, the choke filter is an element with accumulation and the state variable is the current (i_{wg}):

$$L_{wg}\frac{di_{wg}}{dt} = \widehat{u}_{wg} - u_{m\_wg} \tag{1.27}$$

where L_{wg} is the inductor and u_{m_wg} is the modulated voltage of the chopper. The chopper is a conversion element:

$$\begin{cases} u_{m\_wg} = m_{wg}\widehat{u}_{DC} \\ i_{m\_wg} = m_{wg}\widehat{i}_{wg} \end{cases} \tag{1.28}$$

where m_{wg} is the duty ratio of the chopper and u_{DC} is the DC-bus voltage.

The power electronic stage of the emulator performs a path between the control input (m_{wg}) and the choke current (i_{wg}). The objective is to control this current. The control scheme of the wind power emulator is obtained by inverting this path. So a converter controller and a current controller are required.

A current controller is needed to make equal the inductor current (i_{wg}) to a reference value (i_{wg_ref}):

$$u_{m\_wg\_ref} = \widehat{u}_{wg} - PI(i_{wg\_ref} - \widehat{i}_{wg}) \tag{1.29}$$

A converter controller is obtained by inverting (1.28):

$$m_{wg\_ref} = \frac{1}{\widehat{u}_{DC}}u_{m\_wg\_ref} \tag{1.30}$$

The generated power can be described as follows:

$$p_{wg} = u_{wg}i_{wg} \tag{1.31}$$

Then, this power reference (p_{wg_ref}) for the emulator leads to a current reference (i_{wg_ref}) for the control system.

$$i_{wg\_ref} = \frac{1}{\widehat{u}_{wg}} p_{wg\_ref} \tag{1.32}$$

In normal operation, the DC-bus voltage is regulated to a prescribed constant value and then from (1.24), we obtain

$$p_g = p_{wg} \tag{1.33}$$

So the objective is to make the produced power from the emulator equal to a previously recorded wind power profile (Fig. 1.4),

$$p_{wg\_ref} = p_{g\_record} \tag{1.34}$$

The "wind power profile," which is implemented in the digital signal processor (DSP), sets this power reference (p_{wg_ref}) according to the recorded wind speed profile (Fig. 1.4). The experimental test bench is built with a 1.2 kW rated power; thus, the coefficient (k_{wg}) is adapted to have a wind power profile ($p_{wg\_ref} = k_{wg} p_{wg\_ref}$) below 1.2 kW. The grid connection system is experimentally implemented with a wind power emulator through a DC-bus capacitor (2300 μF), a three-phase inverter, three line filters (10 mH), and a three-phase grid transformer. The implementation of the wind energy conversion experimental test bench is shown in Fig. 1.26.

The modeling and control is obtained by replacing the wind energy generation system by the wind power emulator using the proposed modeling and control scheme (Fig. 1.22). The representation model of the entire experimental test bench can be drawn as shown in Fig. 1.27. The proposed hierarchical control system (Fig. 1.23) is experimentally applied and tested with the real DC bus and grid connection system by taking into account the fluctuant wind power.

1.3.3.3.2 Simulation and Experimental Results In order to validate our mathematical modeling and control design, it is first simulated for the given case study in the MATLAB/Simulink™ software environment. It can be seen that the obtained simulated active power (p_g) in Fig. 1.28a is very close to the recorded grid active power (Fig. 1.4). Hence, this mathematical model gives us the same power dynamics as from a real wind generator. Moreover, we have now some knowledge about the internal physical quantities and also about the different control functions.

The grid connection test of the wind energy conversion system is performed with the same wind power profile during 150 s. The experimental results are compared with the previous simulation results. We can see that the similar power profile (Fig. 1.28b) can be generated as the recorded wind power profile (Fig. 1.4). The DC-bus voltage is well regulated (around 400 V) by the line current control loop in a "grid-following" power balancing strategy. The emulated

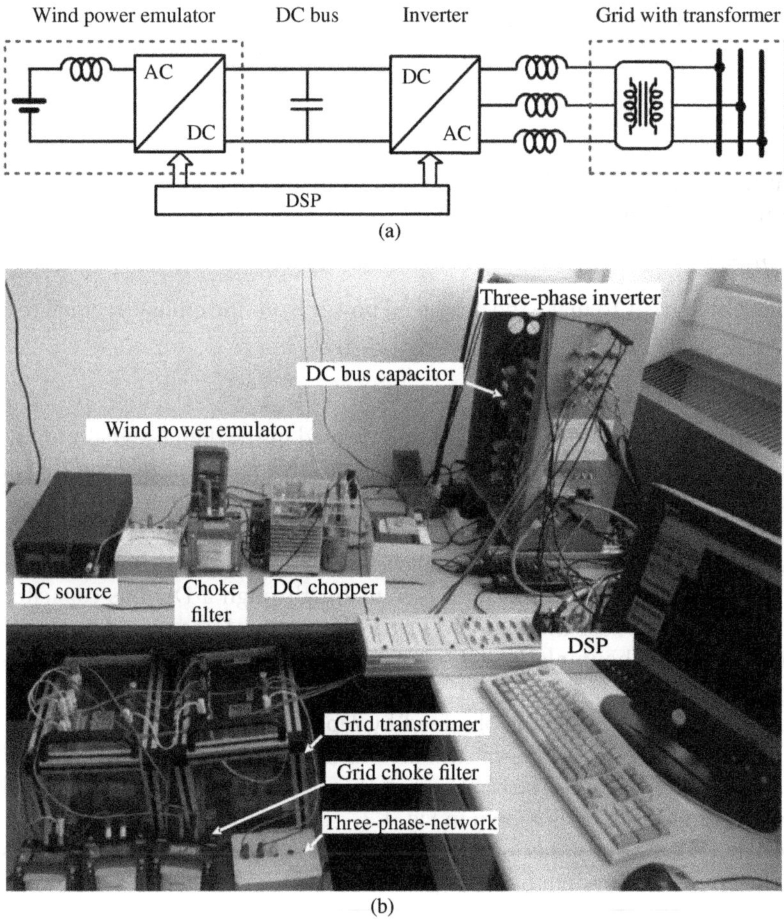

Figure 1.26 Implementation of the wind energy conversion experimental test bench; (a) block diagram and (b) laboratory experiment.

fluctuant wind power is totally delivered to the grid through the three-phase inverter.

This experimental test bench enables us to have similar power dynamics and characteristics as a real wind generator. We will use it to validate our proposed improvement in order to design a wind-based active generator. The fluctuant power from the wind generator depends entirely on the wind condition, but not on the grid's requirements. Therefore, the wind energy conversion system working in MPPT strategy behaves like a passive generator. It cannot supply smooth powers to the grid and cannot supply any ancillary services for the power system. It can only generate continuously varying powers depending on

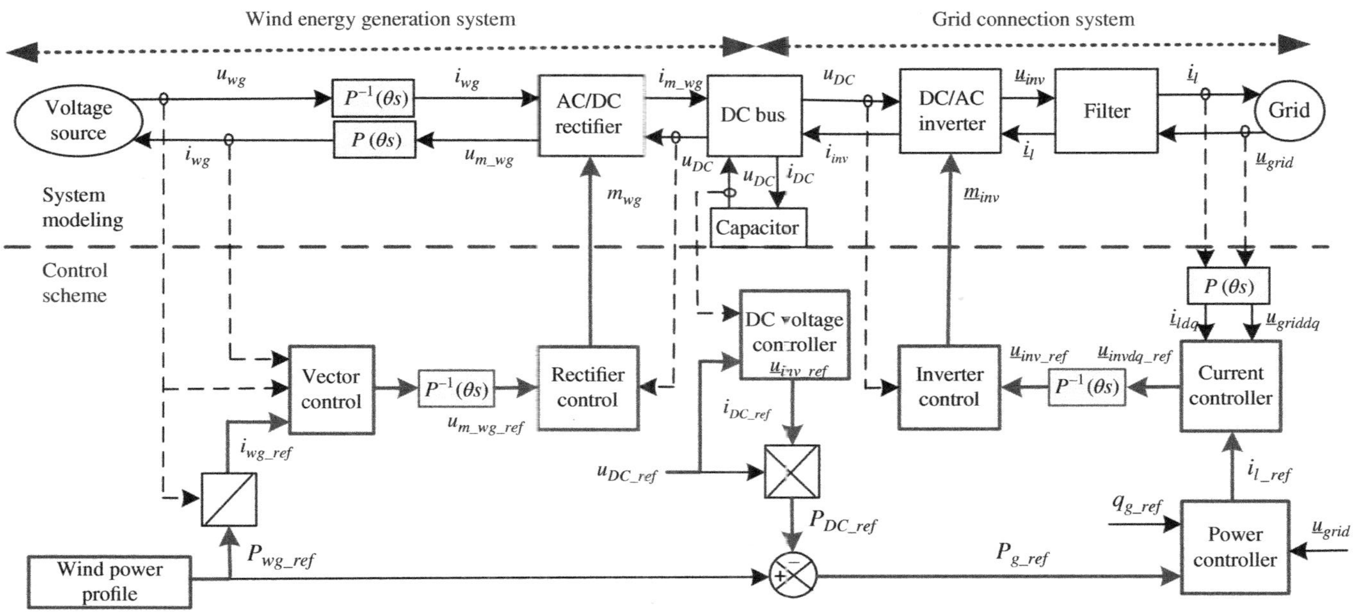

Figure 1.27 Model representation of the wind energy conversion experimental test bench.

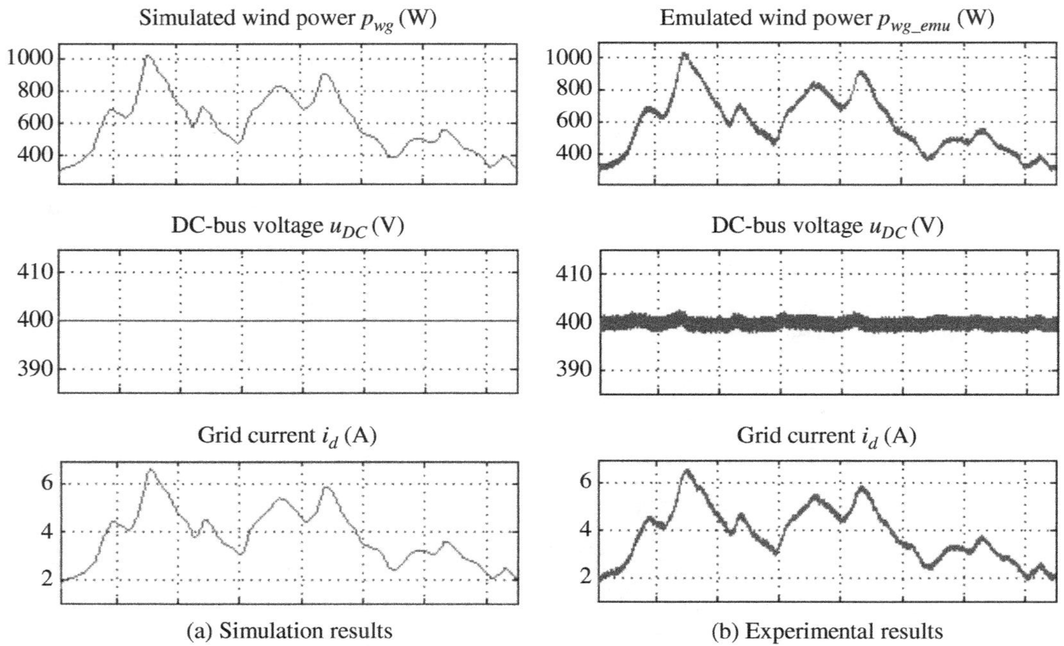

Figure 1.28 Test results of the wind energy conversion experimental test bench.

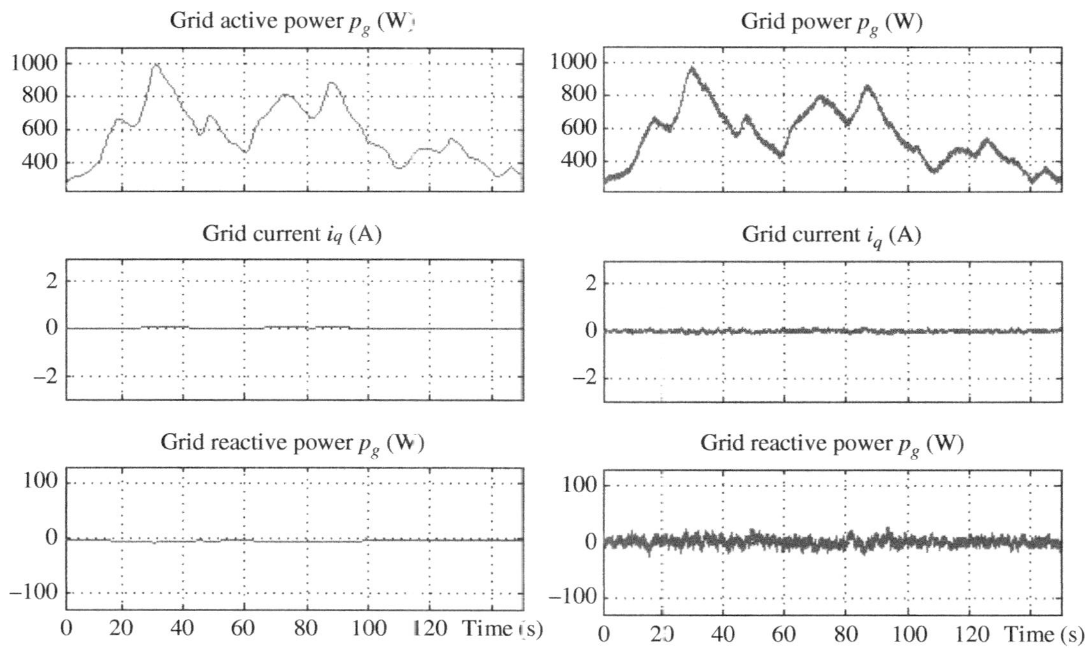

(a) Simulation results

(b) Experimental results

Figure 1.28 (*Continued*)

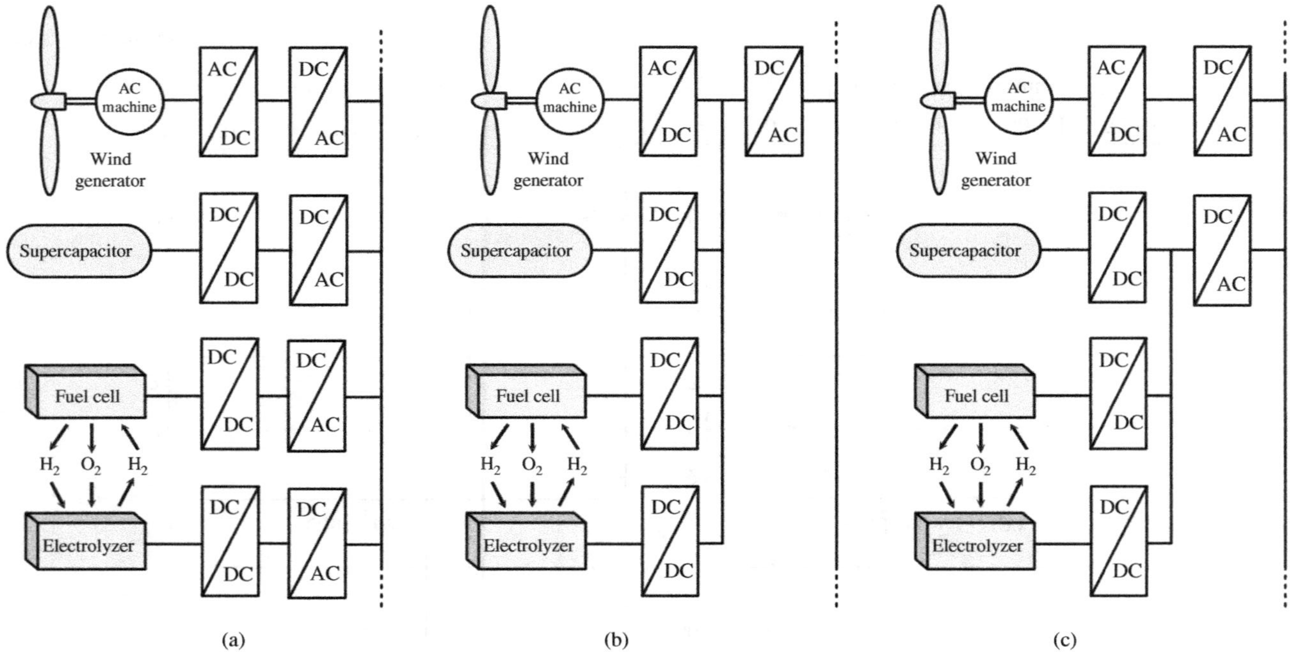

Figure 1.29 Structures of hybrid power systems for distributed generation.

meteorological conditions. Moreover, it becomes a considerable disturbance input for the grid power quality if many wind generators are used. ESSs can help to solve the fluctuation problem of the wind power and can ensure a good energy availability. However, additional control functions should be added to coordinate the different sources.

1.3.3.4 Wind Power Generators with Embedded Energy Storage Units in Hybrid Power Systems

The WPGs can work as distributed generators (DGs) together with other RESs and ESSs to perform an HPS. The DGs are usually smaller than 50 MW and are connected to the distribution network (directly or through an HPS), which has a low or medium operating voltage level.

The structures of HPSs can be classified into two categories: AC- and DC-coupled. In an AC-coupled HPS, all sources are connected to a main AC-bus before being connected to the grid (Fig. 1.29a) [13]. In AC-coupled structure, different sources can be located anywhere in the MG with a long distance from each other. However, the voltage and the frequency of the main AC bus should be well controlled in order to ensure the stability of the DG and the compatibility with the utility network.

In a *DC-coupled HPS*, all sources are connected to a main DC-bus before being connected to the grid through a main inverter (Fig. 1.29b). In a DC-coupled structure, the voltage and the frequency of the grid are independent from those of each source. The DC-coupled structure is flexible and expandable since the number and the type of the energy sources may be freely chosen. Even more, the grid frequency is independent from the sources through the use of the DC bus. The grid voltage is also independent from the DC-bus voltage and each source's voltage through the use of different power converters. So even if both control structure and power management are developed properly for a specified HPS, the number and type of the power sources do not alter the global control structure of the HPS and the main idea of the power management. However, not all HPSs can be classified into AC- or DC-coupled system, since it is possible to have both coupling methods (Fig. 1.29c) [14], then a mixed HPS is obtained. In this case, some advantages can be taken from both structures.

1.4 Grid-Connected PV Power

Large PV plants are being built all over the world, commonly having ratings up to 1 MW and reaching the level of 250 MW. These plants could be an alternative energy source for conventional ones and solving the energy dilemma of human community. However, the associated electronic systems, mainly inverters used for interfacing with the grid, have to solve some problems related to the operation in the electric distribution system. These systems should be reliable, robust, and manageable. This section presents and discusses the most existing critical points in actual inverters, summarizing, explaining, and proposing approaches to solve or mitigate them.

Similar to the WPGs, the PV systems as SPGs are becoming economically viable, even without government subsidies. The PV system capacity is increasing continuously over the world reaching values of hundreds of megawatts; thus, making these plants would be a crucial part of future electric energy systems, MGs, and smart grids. These large/medium SPGs operate as distributed energy resources (DERs) mostly in rural areas where require enough surface for installing. They can be connected to radial distribution grids as shown in the example of Fig. 1.30.

1.4.1 Solar Power Generators with Embedded Energy Storage Systems

Electrical systems must ensure a balance between production and consumption at all times, while maintaining a satisfactory voltage. Grid reliability can be mainly ensured by having excessive capacity in the system with unidirectional flow from centrally dispatched large power plants to dispersed consumers. Dispatched production refers to sources of electricity that can be dispatched at the request of system operators. They are able to change their power production upon demand.

The large-scale development of intermittent PV sources causes large amount of variable power [15]. The RESs with intermittency decrease the reliability of a power system. As the percentage of intermittent generation capacity increases and becomes more significant, an additional uncertainty is appearing in the real-time management of the electrical system balance between demand and generation. This requires increasing amounts of conventional power reserve capacity (spinning reserve) that can be available immediately and of plants capable of providing ancillary services (e.g., frequency response and voltage control), which are required to manage the electrical power system securely [16]. However, an electrical generation system depending entirely on the RESs is not reliable because the availability of the RESs cannot be constantly ensured.

Because of the intermittency of PV power generation, the SPGs may not be used as a stable, reliable, and controllable power source and may not provide ancillary services such as conventional generators. One solution is to upgrade SPG with an embedded ESS and a local energy management system (LEMS) for the coordination of inner power flows. Storage technologies are varying and first it is essential to characterize the required need and complementary performances they must offer. Typically, an energy reserve must be provided to the electrical system and can be implemented by long-term energy storages. Moreover, the supply of power with fast dynamics is also mandatory to smooth the generated PV power, compensate the power gap, and absorb instantaneous high power peaks. At the present state of the art, a high energy storage unit

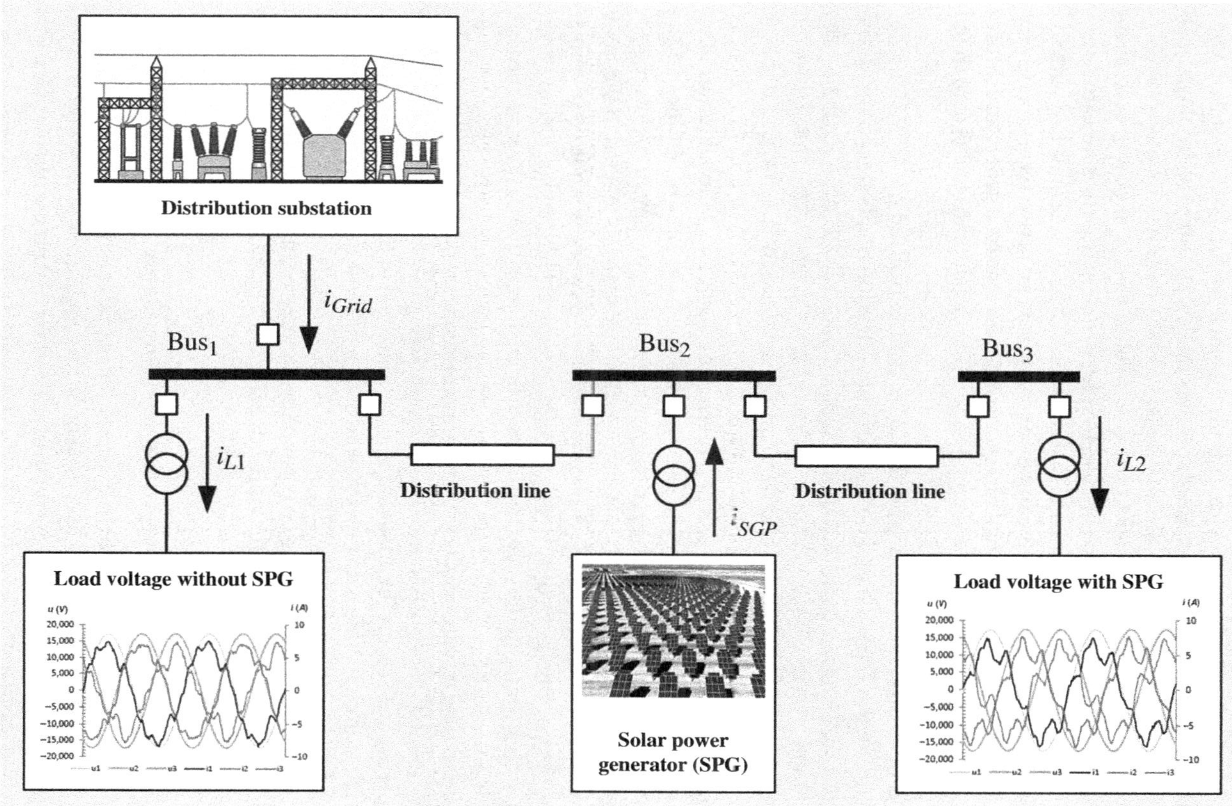

Figure 1.30 A distribution grid example with connected SPGs.

delivering high power with fast dynamics in a reduced volume does not exist and a mix of storage technologies has to be considered. Lead acid batteries for long-term energy storage device and ultracapacitors for fast dynamic power regulation are today realistic and economic choices, but other technologies can also be considered, for example, fuel cells, REDOX batteries, and flywheels.

Storage devices can be used to store or release electrical power such as an energy buffer. Therefore, they can help to solve problems due to renewable energies' intermittent availabilities and fast transients. The SPGs combined with the ESSs perform an HPS, which can be considered as an active PV generator for the grid since it can supply ancillary services as conventional generators. Moreover, it can be dispatched and then provide a power reference that is demanded by the grid operator [17].

A general structure that has been widely used in isolated power systems for integrating the ESSs is based on the direct connection of a battery bank to the DC-bus of the grid-connected inverter. A PV controller is used to extract the maximum power from PV panels and send it to the battery bank. However, the stochastic nature of PV and demand powers lead to fast charge/discharge actions of batteries and a fast battery aging.

To enable a more efficient use of batteries, AC- and DC-coupled power electronic converters can be considered in order to have control abilities of the exchanged powers with the batteries, thanks to the development of power electronics. In an AC-coupled HPS structure, all sources are connected to the main AC network (Fig. 1.31a) [17]. A communication network is required to implement the coordination of this set. In a DC-coupled HPS, all sources are connected to a common inner DC-bus before being connected to the grid through a main inverter (Fig. 1.31b) [18]. One advantage is that the battery bank is connected to the DC-bus via a DC/DC converter, which can be used to implement an optimized charge/discharge operation mode. The second advantage is that a supercapacitor bank is added and is also connected to the DC bus via another DC/DC converter. Hence, a fast power compensation can be performed. The PV array is connected to the DC bus via a PV converter. Similar to the WPGs (Fig. 1.29), apart from these two coupling structures, a mixed structure can be also used (Fig. 1.31c) with some advantages of both DC- and AC-coupled structures.

1.4.2 Solar Energy Conversion System: Modeling, Control, and Analysis

The interest for a PV-based hybrid active generator and the need for grid operator to get more dispatched DGs have been already justified. As mentioned, in order to have a local energy reserve and to filter fast PV power fluctuations, ESS is used to build a PV/ESS HPS in a DC-coupled structure.

The power flows between the different sources must be controlled in order to supply the real and reactive power required by the grid operator

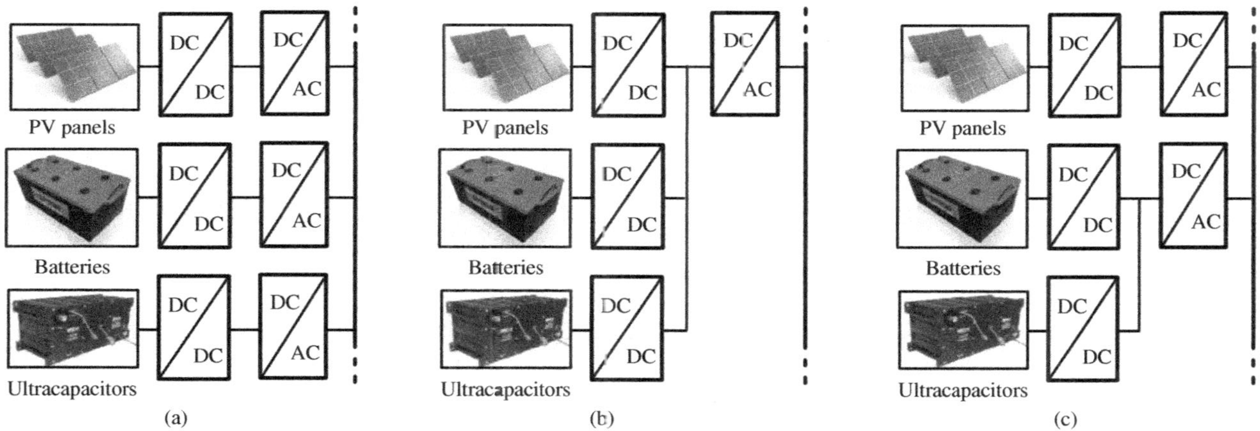

Figure 1.31 Various structures of HPSs: (a) AC-coupled, (b) DC-coupled, and (c) mixec structure.

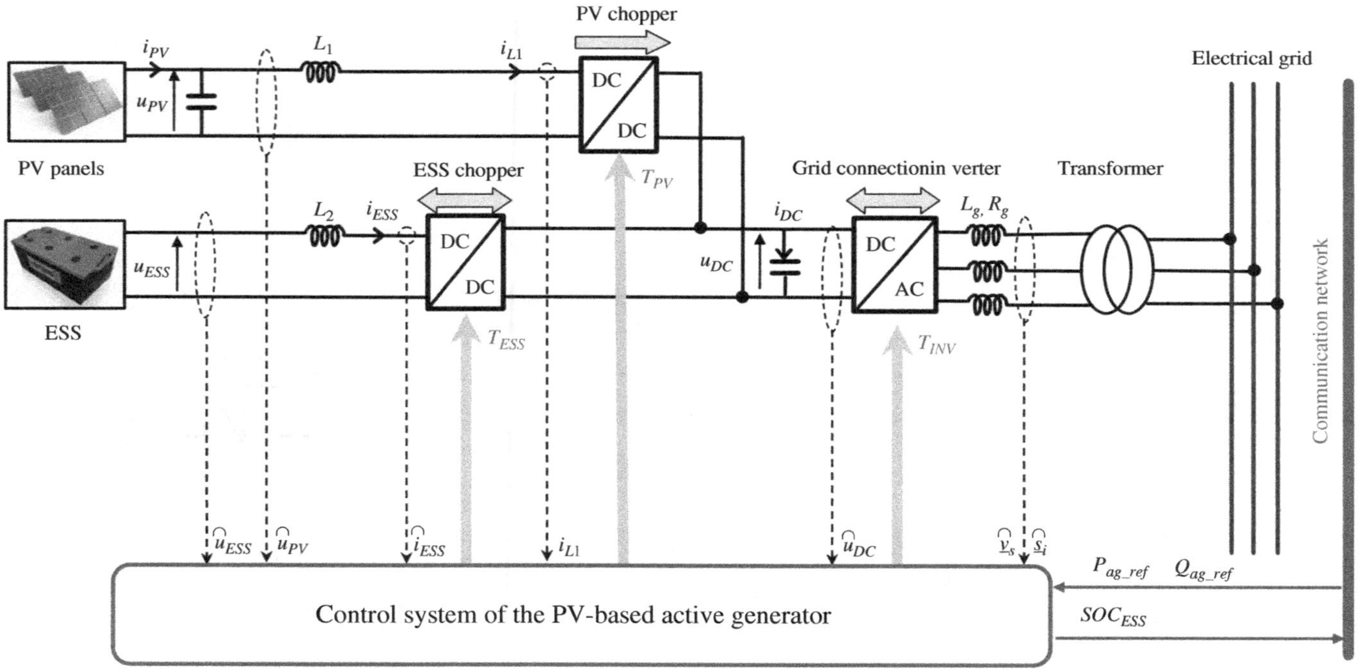

Figure 1.32 Grid-connected PV-based active generator with the control system.

(P_{ag_ref}, Q_{ag_ref}). This is performed by the control of different power electronic converters, which must be then coordinated (Fig. 1.32). Different control strategies have been presented to design the energy management of HPSs and for various applications.

In this section, first the entire PV-based active generator is modeled and then a hierarchical control structure is used and the design of the control system, including the power balancing and energy management strategies are detailed. Similar to the WPGs modeling method, in order to design the control system of the PV active generator, the model of the entire system has to be analyzed. For the modeling of the PV-based active generator, all the modulated values can be replaced by their average values during the modulation period. So each power electronic converter can be replaced by an equivalent double modulated generator (Fig. 1.33).

The equivalent electrical diagram with the equivalent average modeling of power electronic converters makes appear three parts corresponding to the PV conversion system: the ESS, the grid connection, and the DC bus. The three power converters are used to introduce control inputs for each power conversion system, in order to control the power generated by the PV panels, to maintain a constant DC-bus voltage, to supply the required power exchange with the grid, and to ensure the power buffering of each energy storage unit.

1.4.2.1 PV Power Conversion System

In the PV power conversion model, the PV panels are usually considered as a current source (i_{PV}) and it must be supplied by a voltage (u_{PV}) [19]. The voltage comes from a filter (C_1, L_1), which is fed by the modulated voltage (Fig. 1.34).

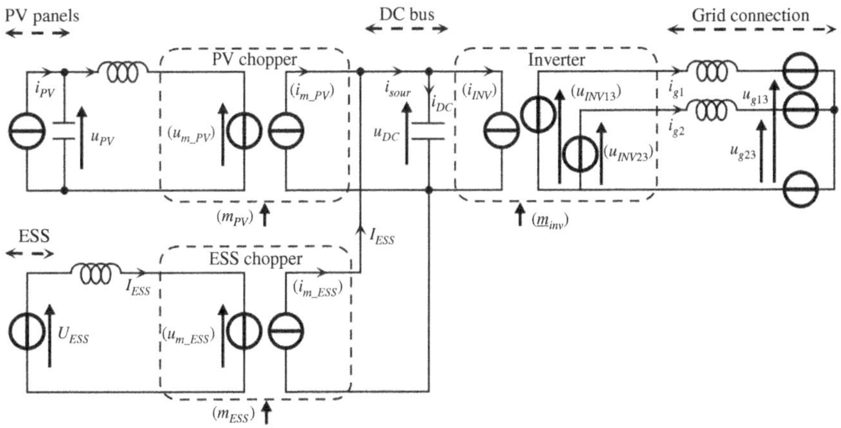

Figure 1.33 Equivalent electrical diagram of the PV-based active generator.

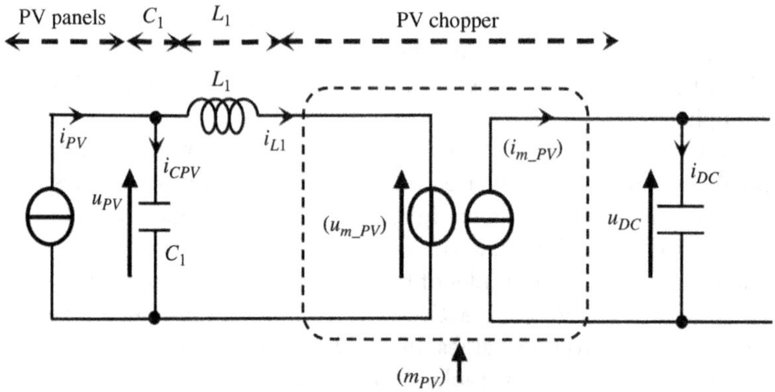

Figure 1.34 Equivalent electrical diagram of PV power conversion system.

The choke is also modeled as a current source (i_{L1}). This current depends on the PV voltage and the modulated voltage of the chopper output (u_{m_PV}):

$$\frac{di_{L1}}{dt} = \frac{1}{L_1}(u_{m\_PV}(t) - u_{PV}(t)) \tag{1.35}$$

Losses in the filter and the capacitor are neglected. The capacitor (C_1) can stabilize the voltage (u_{PV}) across the terminals of the PV panels. This capacitor can be modeled by using the PV current (i_{PV}) and the filtered current (i_{L1}):

$$\begin{cases} \dfrac{du_{PV}}{dt} = \dfrac{1}{C_1} i_{CPV}(t) \\ i_{CPV}(t) = i_{PV}(t) - i_{L1}(t) \end{cases} \tag{1.36}$$

i_{CPV} is the injected current in capacitor. The mean value of the terminal voltage of the chopper (u_{m_PV}) is obtained from the DC voltage (u_{DC}) and the duty cycle ratio ($m_{\_PV}$):

$$\begin{cases} \langle u_{m\_PV} \rangle = m_{PV} \cdot u_{DC}(t) \\ \langle i_{m\_PV} \rangle = m_{PV} \cdot i_{L1}(t) \end{cases} \tag{1.37}$$

It is noteworthy that the notation $\langle x \rangle$ means "the average values of the instantaneous electrical quantities" $x(t)$, but since here all quantities are in mean values, the notation $\langle x \rangle$ will not be used later.

1.4.2.2 Energy Storage System

In the models of ESSs, an ESS is usually considered as a voltage source (u_{ESS}), which is connected to a choke filter (L_2) (Fig. 1.35).

By neglecting losses in the filter, the dynamic equation of the filtered current (i_{ESS}) is expressed with the ESS voltage (u_{ESS}) and the modulated

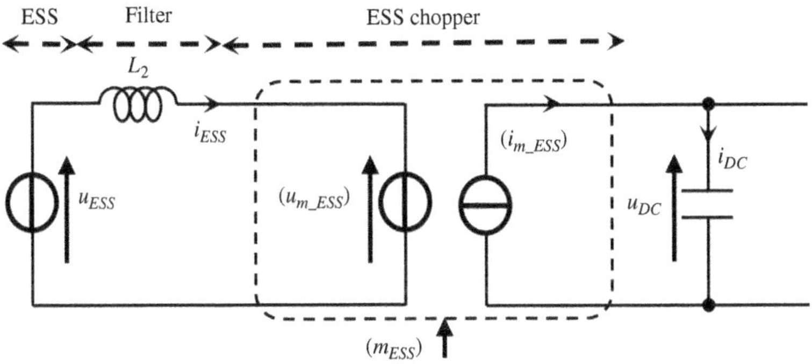

Figure 1.35 Equivalent electrical diagram of the batteries energy storage system.

voltage (u_{m_ESS}):

$$\frac{di_{ESS}}{dt} = \frac{1}{L_2} \cdot (u_{ESS}(t) - u_{m\_ESS}(t)) \tag{1.38}$$

where L_2 is the inductor of the filter.

The mean value of the terminal voltage of the ESS chopper (u_{m_ESS}) is obtained from the DC voltage and the duty cycle ratio (m_{ESS}):

$$\begin{cases} u_{m\_ESS} = m_{ESS} \cdot u_{DC}(t) \\ i_{m\_ESS} - m_{ESS} \cdot i_{ESS}(t) \end{cases} \tag{1.39}$$

In the same way, the filtered current (i_{ESS}) is modulated by the chopper, and then this modulated current (i_{m_ESS}) is injected into the common DC bus. The average value of the output current (i_{m_ESS}) of the chopper is equal to the ESS current (i_{ESS}) multiplied by the duty cycle ratio (m_{ESS}).

1.4.2.3 Grid Connection

A three-phase inverter within a choke as filter is used for the grid connection. Hence, an equivalent mean modeling of this three-phase inverter is sufficient for representing fundamental components of voltage/current (Fig. 1.36) as dependent phase-to-phase voltage sources (u_{INV_13} and u_{INV_23}) with the DC-bus voltage (u_{DC}) through modulation indexes (m_{INV_13} and m_{INV_23}) and a dependent current source (i_{INV}) with AC currents through the same modulation indexes.

Then mean values of modulated phase-to-phase voltages and of the average currents are expressed as

$$\begin{cases} u_{INV\_13} = m_{INV\_13} \cdot u_{DC} \\ u_{INV\_23} = m_{INV\_23} \cdot u_{DC} \end{cases} \tag{1.40}$$

$$\langle i_{INV} \rangle = m_{INV\_13} \cdot i_{g1} + m_{INV\_23} \cdot i_{g2} \tag{1.41}$$

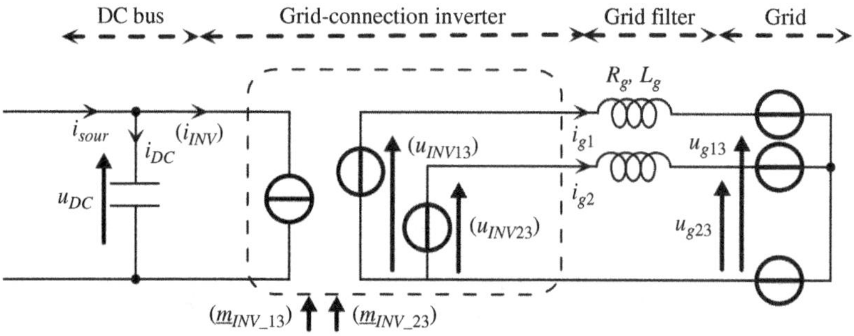

Figure 1.36 Electrical diagram of the grid connection.

By assuming that grid voltages are balanced, line voltages are obtained through

$$\begin{cases} v_{INV1} = \frac{2}{3} \cdot u_{INV\_13} - \frac{1}{3} \cdot u_{INV\_23} \\ v_{INV2} = -\frac{1}{3} \cdot u_{INV\_13} + \frac{2}{3} \cdot u_{INV\_23} \end{cases} \tag{1.42}$$

The filter currents are deduced from the following differential equations:

$$\begin{cases} \frac{di_{g1}}{dt} = \frac{1}{L_g} (v_{INV1} - R_g \cdot i_{g1} - v_{g1}) \\ \frac{di_{g2}}{dt} = \frac{1}{L_g} (v_{INV2} - R_g \cdot i_{g2} - v_{g2}) \end{cases} \tag{1.43}$$

Three-phase inverter voltages, grid voltages, currents, and duty cycles can be expressed as vectors, respectively, by

$$\underline{v_{INV}} = \begin{bmatrix} v_{INV1} \\ v_{INV2} \\ v_{INV3} \end{bmatrix}, \quad \underline{v_g} = \begin{bmatrix} v_{g1} \\ v_{g2} \\ v_{g3} \end{bmatrix}, \quad \underline{i_g} = \begin{bmatrix} i_{g1} \\ i_{g2} \\ i_{g3} \end{bmatrix}, \quad \underline{m_{INV}} = \begin{bmatrix} m_{INV\_13} \\ m_{INV\_23} \end{bmatrix} \tag{1.44}$$

1.4.2.4 DC Bus

In this hybrid generating system, three energy sources (PV panels, ESS, and the electrical grid) are all connected to the common DC bus via different power electronic converters (Fig. 1.33). So according to this DC-coupling, the capacitor current of the DC bus (i_{DC}) is expressed as

$$i_{sour}(t) = i_{ESS}(t) + i_{m\_PV}(t) \tag{1.45}$$

$$i_{DC}(t) = i_{sour}(t) - i_{INV}(t) \tag{1.46}$$

where i_{m_PV} is the modulated current from the PV chopper, i_{m_ESS} is the modulated current from the ESS chopper, and i_{INV} is the modulated current from the grid inverter.

DC-bus voltage is expressed as

$$\frac{du_{DC}}{dt} = \frac{1}{C_{DC}} i_{DC}(t) \tag{1.47}$$

The C_{DC} is the capacitor of the DC bus.

1.4.2.5 Modeling of the Entire PV Energy Conversion System

By combining the all mentioned sources' power conversion system models, which is discussed in the previous sections, the block diagram of the entire active PV generator can be obtained as shown in Fig. 1.37. Three action paths appear from the control inputs of the three power converters to the different electrical quantities (u_{PV}, i_{ESS}, $\underline{i}_g$): from the PV chopper (m_{PV}) to the PV panels terminal voltage (u_{PV}) in yellow, from the ESS chopper (m_{ESS}) to the batteries current (i_{BAT}) in green, and finally from the grid connection three-phase inverter ($\underline{m}_{INV}$) to the line currents ($\underline{i}_g$) in blue.

1.4.2.6 Hierarchical Control Structure

Like WPG hierarchical system (Fig. 1.11), a hierarchical structure of the control system is proposed for this active PV generator. The structure of this hierarchical control system includes four levels (Fig. 1.38). Each one has precise control tasks depending on its hierarchical position: SCU, ACU, PCU, and MCU.

The MCU decides the operating mode for the whole hybrid generator according to the availability of the PV production, the states of each storage unit and the actual power demand from the grid. The PCU calculates the power reference for each source according to the sensed values and the selected operating mode from the MCU. The ACU applies the control algorithm to meet the current or voltage references. The SCU level implements the modulation technique to each converter and generates the semiconductor signals ({−5, +15}) to apply wished ideal states ({0, 1}).

In the studied active PV generator, three sources are considered: the PV panels (PV), the ESS, and the grid connection. Three power electronic converters are used to regulate the power exchanges among them. So in the control system, three SCUs and three ACUs are used for the control of the three sources, a common PCU and a common MCU are used for the power dispatching among the different sources, the real-time power balancing and the long-term energy management of the entire active generator (Fig. 1.38).

In the SCU of each power electronic converter, the IGBT and PWM techniques are used to control the switches. The control algorithms in the ACU level should be presented in order to highlight the physical quantities, which can be used for the power flow control among the different energy sources.

1.4.2.6.1 Automatic Control Unit The task of the ACU is to calculate duty cycles for each SCU in order to set dynamical quantities equal to their references (current and voltage), which are coming from the PCU.

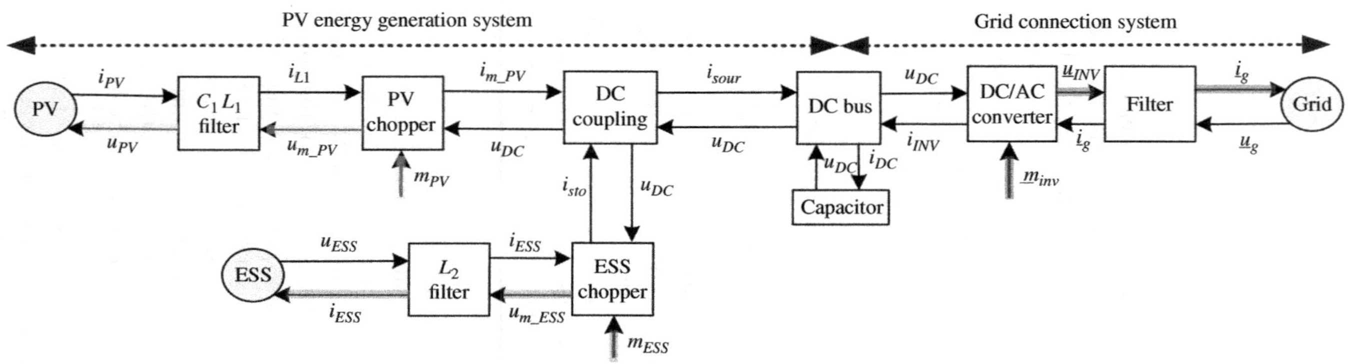

Figure 1.37 The block diagram of the PV active generator.

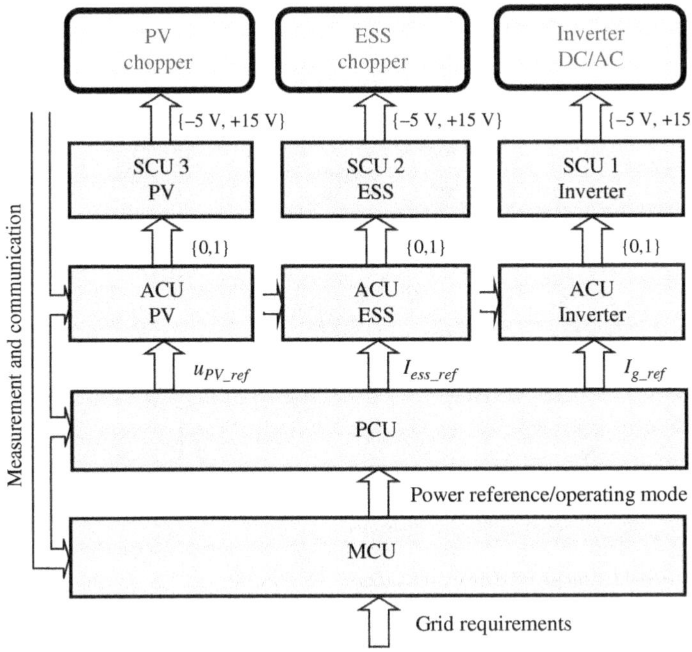

Figure 1.38 Hierarchical control structure for the active PV generator.

PV Controller The ACU of the PV power conversion system must regulate the terminal voltage across the PV panels in order to implement an MPPT strategy or a power limitation (PL) strategy. The variation of this voltage modifies the generated PV power (Fig. 1.39).

The PV power conversion system provides an action chain from the duty cycle ratio (m_{PV}) and pointing to the PV terminal voltage (u_{PV}). According to the inversion rules of the CSU, the automatic control system is achieved by

- a closed-loop control of the filtered current (i_{L1}) with a simple controller (K_1),

$$u_{m\_PV\_ref} = K_1(i_{L1\_ref} - i_{L1\_mes}) + u_{PV\_mes} \tag{1.48}$$

- a closed-loop control of the terminal voltage (u_{PV}) with a simple controller (K_2),

$$i_{L1\_ref} = K_2(u_{PV\_ref} - u_{PV\_mes}) + i_{PV\_mes} \tag{1.49}$$

- a converter control for the calculation of the duty cycle, which will be sent to the SCU:

$$m_{PV\_reg} = \frac{u_{m\_PV\_ref}}{u_{DC\_mes}} \tag{1.50}$$

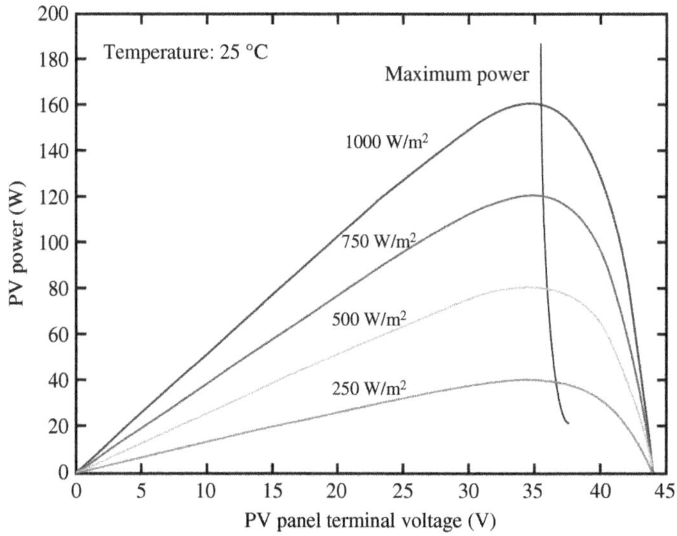

Figure 1.39 Power characteristics of one PV module.

EES Controller From the ESS, an action chain from the duty cycle ratio (m_{ESS}) and pointing to the ESS current (i_{ESS}) can be considered. A control chain is obtained by inversing this action chain. The inversion rules of the CSU yields two control functions:

- A closed-loop control of the current with a controller (K_3)

$$u_{m\_ESS\_ref} = K_3(i_{ESS\_ref} - i_{ESS\_mes}) \tag{1.51}$$

- A converter controller to calculate the duty cycle ratio, which will be sent to the SCU:

$$m_{ESS\_reg} = \frac{u_{m\_ESS\_ref}}{u_{DC\_mes}} \tag{1.52}$$

Grid Connection Control The block diagram of the grid connection system modeling makes appear a path in blue from the control inputs ($\underline{m}_{inv}$) of the inverter to the line currents ($\underline{i}_g$). The control scheme of the grid connection system is obtained by inverting this path. It involves calculating the reference of the inverter's duty ratios ($\underline{m}_{inv\_ref}$) according to the line currents' references ($\underline{i}_{g\_ref}$). The objective of the grid connection control is to regulate the delivered active power and reactive power to the grid. The grid is a voltage source ($\underline{u}_{grid}$), so the line current should be controlled in order to regulate the exchanged power with the grid. Hence, the delivered power control is based on the grid current control. According to the inversion rules of the CSU, the automatic control system is achieved by an inverter controller and a closed-loop control of currents.

For the design of controllers, a Park transform is used. The voltages and currents in the three-phase frame (a,b,c) can be transformed to two voltages and two currents in the two-phase rotating frame (d,q) with synchronization with the first line voltage:

$$\begin{cases} \underline{v_{inv\_dq}} = P(\theta)\underline{v_{INV}} \\ \underline{v_{g\_dq}} = P(\theta)\underline{v_g} \\ \underline{i_{g\_dq}} = P(\theta)\underline{i_g} \end{cases} \tag{1.53}$$

$P(\theta)$ is the matrix of the Park transformation.

$$P(\theta) = \frac{2}{3}\begin{bmatrix} \cos\theta & \cos\left(\theta - \frac{2\pi}{3}\right) & \cos\left(\theta + \frac{2\pi}{3}\right) \\ \sin\theta & \sin\left(\theta - \frac{2\pi}{3}\right) & \sin\left(\theta + \frac{2\pi}{3}\right) \\ \frac{1}{2} & \frac{1}{2} & \frac{1}{2} \end{bmatrix} \tag{1.54}$$

Hence in this frame, filter current equations are expressed as

$$\begin{cases} \dfrac{di_{g\_d}}{dt} = \dfrac{1}{L}(v_{inv\_d} - v_{g\_d} - R_g i_{g\_d} - L_g \cdot \omega_s \cdot i_{g\_q}) \\ \dfrac{di_{g\_q}}{dt} = \dfrac{1}{L}(v_{inv\_q} - v_{g\_q} - R_g i_{g\_q} + L_g \cdot \omega_s \cdot i_{g\_d}) \end{cases} \tag{1.55}$$

where $\omega_s = 2\pi f$ and f is the grid frequency.

Therefore, similar to the block diagram shown in Fig. 1.16, the control of these currents is organized in three parts: proportional–integral (PI) control, grid voltage compensation, and current decoupling. The modulated voltage references of the inverter outputs can be expressed as

$$\begin{cases} v_{inv\_d\_ref} = PI_1(i_{g\_d\_ref} - i_{g\_d\_mes}) + v_{g\_d\_mes} - L_g \cdot \omega_s \cdot i_{g\_q\_mes} \\ v_{inv\_q\_ref} = PI_2(i_{g\_q\_ref} - i_{g\_q\_mes}) + v_{g\_q\_mes} + L_g \cdot \omega_s \cdot i_{g\_d\_mes} \end{cases} \tag{1.56}$$

with two current PI correctors (PI_1, PI_2).

The obtained voltage references in the *dq* frame can be transformed to the three voltage references in the *abc* frame with an inverse Park transformation:

$$\underline{v_{inv\_ref}} = P^{-1}(\theta)\underline{v_{inv\_dq\_ref}} \tag{1.57}$$

Phase-to-phase voltages are obtained with the following equations:

$$\begin{cases} u_{inv13\_ref} = v_{INV1\_ref} - v_{inv3\_ref} \\ u_{inv23\_ref} = v_{INV2\_ref} - v_{inv3\_ref} \end{cases} \tag{1.58}$$

Mean values of the modulation functions are calculated by using the inverse equation (1.40):

$$m_{inv13\_reg} = \frac{u_{inv13\_ref}}{u_{DC\_mes}}, \quad m_{INV23\_reg} = \frac{u_{inv23\_ref}}{u_{DC\_mes}} \tag{1.59}$$

DC Bus Control The decomposition into the different subsystems relies on the assumption that the DC-bus voltage (u_{DC}) is constant. The control scheme of the DC bus is obtained by inverting the block diagram of the system modeling (similar to Fig. 1.17) and requires a closed-loop control of the DC-bus voltage (u_{DC}) with a PI corrector (PI):

$$i_{DC\_ref} = PI(u_{DC\_ref} - u_{DC\_mes}) \tag{1.60}$$

The current flowing through the capacitor of the DC bus (i_{DC}) can be expressed as

$$i_{DC} = i_{m\_PV} + i_{m\_ESS} - i_{inv} \tag{1.61}$$

Thus, the regulation of the current in the capacitor can be achieved by controlling the currents of each source (PV panels and ESS) corresponding to the power control of each source.

Control of the Entire PV Energy Conversion System The global control scheme of the entire active PV generator can be obtained by combining all control schemes, which have been explained in the previous parts, as shown in Fig. 1.40. A block diagram of all control functions inside the ACU is also presented in Fig. 1.41. Four references (i_{DC_ref}, i_{g_ref}, u_{PV_ref}, and i_{ESS_ref}) must be set to interface the automatic control level with the PCL.

1.4.2.6.2 Power Control Unit The PCU is divided into two levels: the PCL and the PSL. The PCL involves calculating the reference quantities (i_{g_ref}, u_{PV_ref}, i_{ESS_ref}) from power references. The PSL coordinates the power flow exchanges among the different energy sources with different power balancing strategies.

To implement the power control, the inner power flow of the active generator has to be calculated. In order to focus on the power exchanges with the different sources around the DC bus, the instantaneously exchanged power with the choke, the losses in the filters and the losses in the power converters are neglected. Only the sources' powers and the exchanged power with the DC-bus capacitor are taken into account here (Fig. 1.42). Powers in the DC part are easily calculated by multiplying DC currents and DC voltages (Table 1.2). All the variables with hat symbol are the sensed variables. The expressions of the powers are inverted to obtain the control equations (Fig. 1.42).

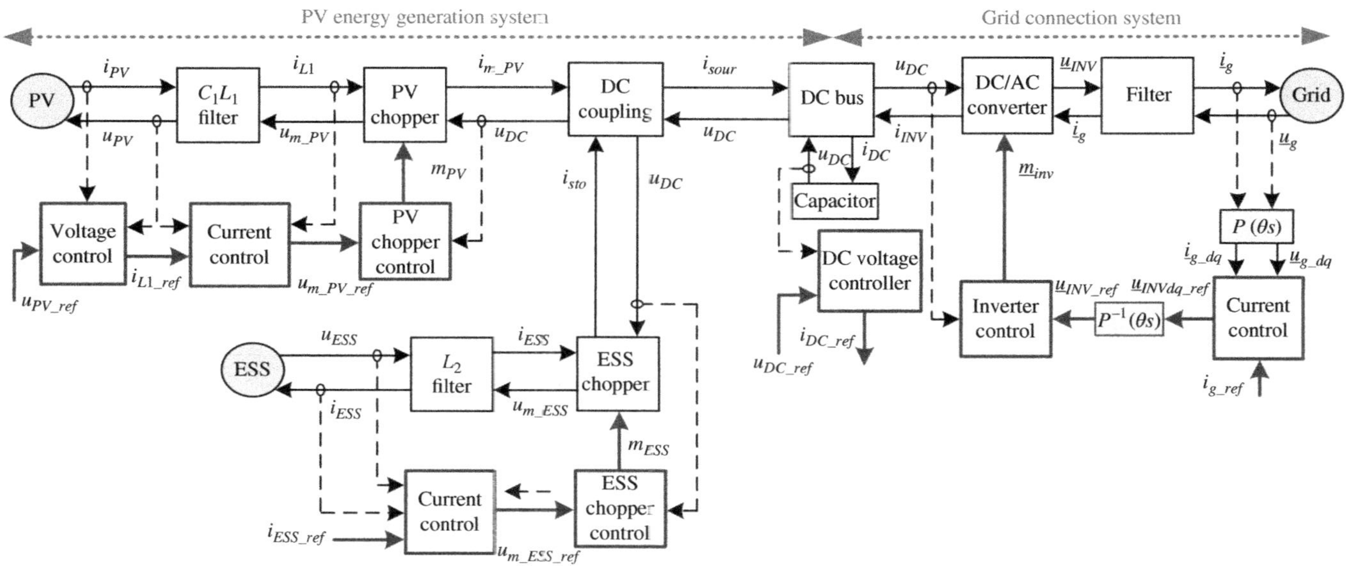

Figure 1.40 Block diagram of a grid-connected PV system with control loops.

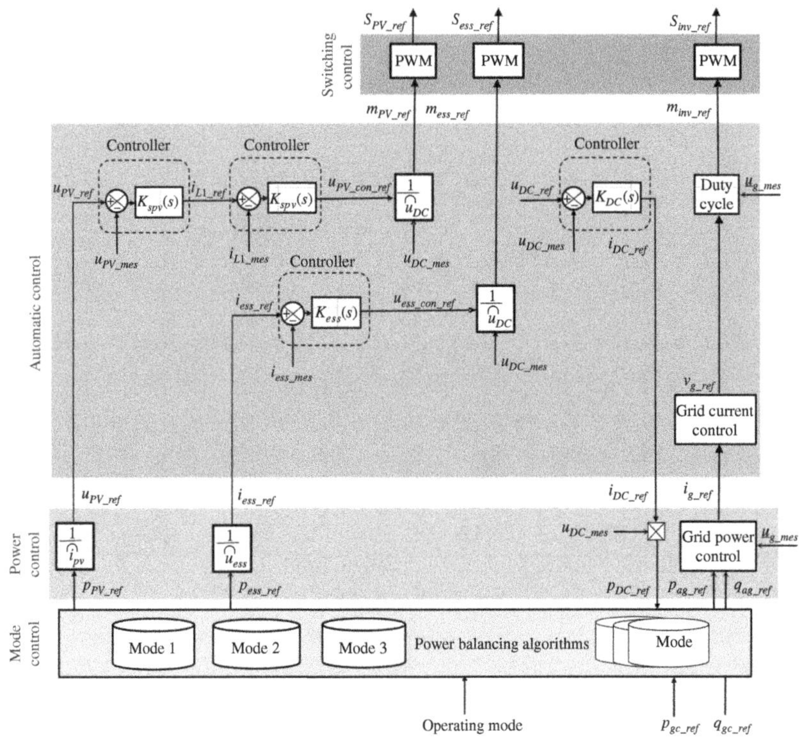

Figure 1.41 Block diagram of the automatic control units for the active PV generator.

Table 1.2 Power calculation and power control algorithms for the active PV generator.

Energy source	Power calculation	Power control
DC-bus capacitor	$p_{DC} = u_{DC} i_{DC}$ (iv)	$p_{DC\_ref} = \widehat{u}_{DC} i_{DC\_ref}$ (iv_c)
Grid connection	$\begin{cases} p_{ag} = u_{g13} \cdot i_{g1} + u_{g23} \cdot i_{g2} \\ q_{ag} = \sqrt{3}(u_{g13} \cdot i_{g1} - u_{g23} \cdot i_{g2}) \end{cases}$ (v)	$\begin{cases} i_{g1\_ref} = \dfrac{(2\widehat{u}_{13} - \widehat{u}_{23}) p_{ag\_ref} + \sqrt{3}\widehat{u}_{23} q_{ag\_ref}}{2\widehat{u}_{13}^2 - 2\widehat{u}_{13}\widehat{u}_{23} + 2\widehat{u}_{23}^2} \\ i_{g2\_ref} = \dfrac{(2\widehat{u}_{23} - \widehat{u}_{13}) p_{ag\_ref} - \sqrt{3}\widehat{u}_{13} q_{ag\_ref}}{2\widehat{u}_{13}^2 - 2\widehat{u}_{13}\widehat{u}_{23} + 2\widehat{u}_{23}^2} \end{cases}$ (v_c)
PV panels	$p_{PV} = u_{PV} i_{PV}$ (vi)	$u_{PV\_ref} = \dfrac{1}{\widehat{i}_{PV}} p_{PV\_ref}$ (vi_c)
ESS	$p_{ESS} = i_{ESS} u_{ESS}$ (vii)	$i_{ESS\_ref} = \dfrac{1}{\widehat{u}_{ESS}} p_{ESS\_ref}$ (vii_c)

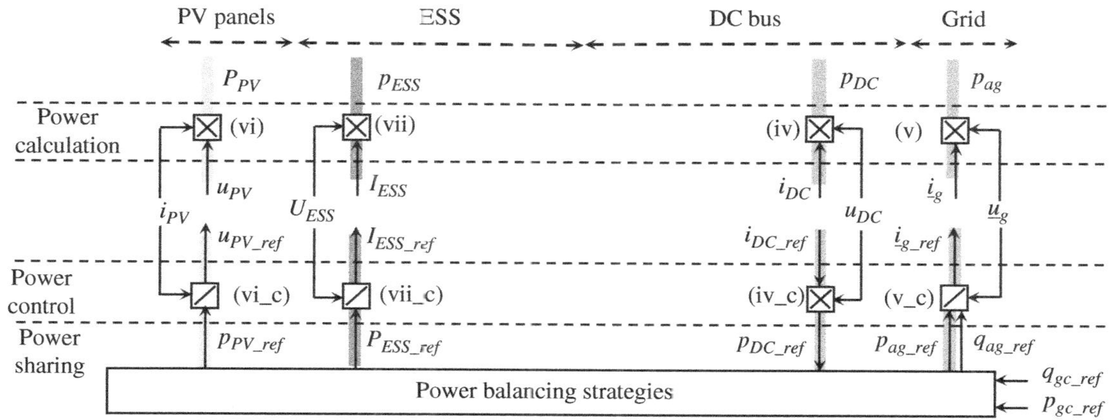

Figure 1.42 Power modeling and control.

Power control of PV panels: A reference value for the PV power can be set by calculating the PV voltage reference (u_{PV_ref}) with the sensed PV current:

$$u_{PV\_ref} = \frac{p_{PV\_ref}}{i_{PV\_mes}} \tag{1.62}$$

Power control of storage units: For the storage systems, the powers are calculated by multiplying currents with the voltages ((vii) in Table 1.2). The current references are given by dividing corresponding power references with the sensed voltages.

Power control of the grid connection: The powers, which are exchanged with the grid, can be calculated through the "two-wattmeter" method ((v) in Table 1.2). According to the grid power reference and the measured phase-to-phase voltages of the grid, the current references for the inverter can be deduced by inverting this equation ((v_c) in Table 1.2).

Power control of the DC bus: The output of the DC-bus voltage control loop is the current reference (i_{DC_ref}) of the DC-bus capacitor, and its product with the measured DC-bus voltage gives the power reference (p_{DC_ref}) for the DC-bus voltage regulation (iv_c). This regulation power for the DC-bus voltage control can be satisfied by the difference between the sources power and the grid power.

The DC bus directly leads to the stability of the generator. All power exchanges are performed via the DC-bus and have an impact on the DC-bus voltage (u_{DC}):

$$C_{DC} u_{DC} \frac{du_{DC}}{dt} = \frac{dE_{DC}}{dt} = p_{DC} = p_{sour} - p_{ag} = p_{ESS} - p_{ag} \tag{1.63}$$

where E_{DC} is the energy stored in the DC-bus capacitor; p_{DC} is the resulted power into the DC-bus capacitor; p_{PV} is the power injected into the DC bus from the PV generator; p_{ESS} is the power exchanged between the EES and the DC bus; p_{ag} is the power extracted from the DC bus into the grid; and p_{sour} is the total power from the sources. So to get a constant DC-bus voltage, the instantaneous powers must be balanced. For different studied modes, the power flow must be then modeled in order to be controlled in a second step.

Power sharing: The PSL is used to implement the power balancing strategies in order to coordinate the various sources in the HPS. The power references for each source are set according to different power control strategies, which are specific to different operating modes. Operating modes are defined according to the various conditions (the power demand from the grid, the availability of the PV power, the state of the energy storage units, etc.). In order to set the power reference of each source, the power flow between each source should be determined and balanced. So a modeling of the power flow in the active generator for each mode is necessary.

1.4.3 Experimental Results

Figure 1.43a outlines the used DC-coupled active PV generator as an HPS prototype test system. The grid connection is performed by a three-phase inverter. Chokes and capacitors are used to filter the modulated electrical waveforms. Here, the EES includes two sets of lead-acid batteries (48 V, 106 Ah) and ultracapacitors (48 V, 112 kW min). A DC current source is used for simulating the PV panels. The control system is implemented into a DSP card. Figure 1.43b shows the experimental laboratory systems.

In order to ensure an optimal operation and coordination, an LEMS of the active PV generator must calculate and send control signals to each power electronic converter in order to enable production of power demand for the grid operator, management of renewable energy intermittency, management of storage state of charges (SOC), power system protection, and provision of grid ancillary services.

A hierarchical control structure for the given HPS system (Fig. 1.38) similar to the one in Fig. 1.41 can be considered. The SCU implements the modulation technique to each converter and generates the switching signals. The ACU performs control algorithms in order to meet the current or voltage references. The PCU calculates these references according to the power references from the MCU and measured values. The MCU decides the operating mode for the whole hybrid generator according to the availability of the PV production, the state of each storage unit, and the actual power demand from the grid.

The power reference for the PV generator, $p_{PV\_ref}(t)$, is classically calculated from a maximum PV power tracking algorithm. The inner power balancing shows that powers from the PV, $p_{PV}(t)$, the battery, $p_{BAT}(t)$, the ultracapacitor, $p_{UC}(t)$ must be decreased by the required power to regulate the DC bus, $p_{DC}(t)$, and constitutes the total generated power [21]:

$$p_{ag}(t) = p_{BAT}(t) + p_{UC}(t) + p_{PV}(t) - p_{DC}(t) \tag{1.64}$$

There are two operation strategies for an active PV generation system: grid-following strategy (GFS) and source supplying strategy (SSS). The GFS and SSS are conceptually shown in Fig. 1.44a,b, respectively. Using the GFS, the reference for the power (generated by the inverter) is deduced using the measured power and contributes to the regulation of the DC bus:

$$p_{ag\_ref}(t) = \widehat{p}_{BAT}(t) + \widehat{p}_{UC}(t) + \widehat{p}_{PV}(t) - p_{DC\_ref}(t) \tag{1.65}$$

Then the required power reference from the grid operator is provided by the storage units taking into account the available PV power.

$$p_{sto\_ref}(t) = p_{gc\_ref}(t) - \widehat{p}_{PV}(t) \tag{1.66}$$

A simple method to dispatch the power between batteries and ultracapacitors can be realized using a slope-limited low-pass filter [22]. Ultracapacitors are controlled to supply the required transient power from storage units.

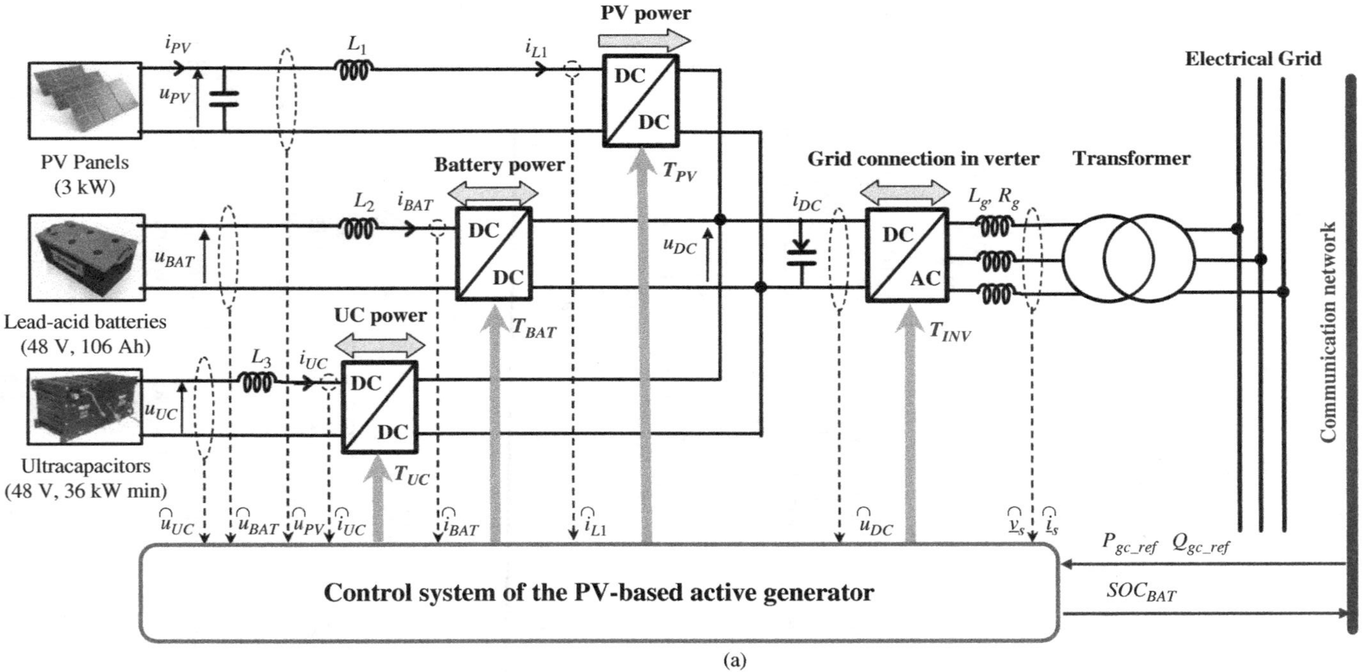

Figure 1.43 The grid-connected active PV generator (HPG) test system: (a) system configuration and (b) experimental units.

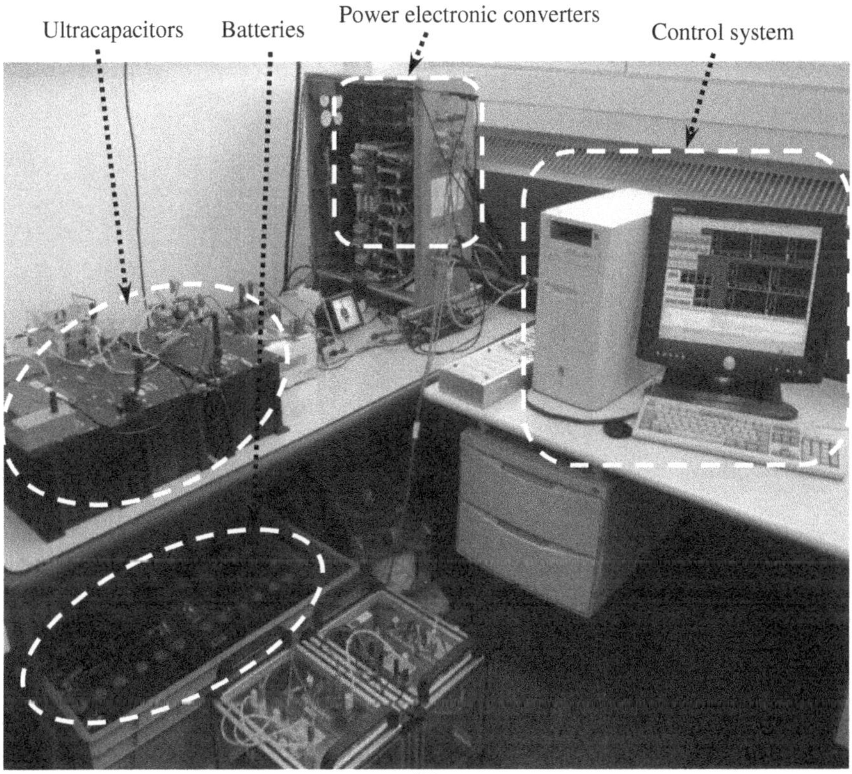

(b)

Figure 1.43 (*Continued*)

Using the SSS, the power from the storage units is adjusted according to the sensed PV power to regulate the DC bus. The required power reference is fixed by the grid operator, which can be directly provided by the inverter.

With both strategies, the PV generator can be dispatched by the grid operator since it is now able to deliver the prescribed active and reactive powers ($p_{gc\_ref}(t)$, $q_{gc\_ref}(t)$), but only the SSS is able to work without a connection to the grid because the DC bus is regulated by the inner power. This strategy is preferred since autonomous/isolated operation is possible and may be used to store PV energy without grid connection. Another advantage is the possibility to stay connected in case of undervoltage grid because the DC-bus voltage is autonomously regulated by inner sources (PV panels and storage units). Hence, in this situation, the real and reactive power generation is possible to help the electric network [23].

Using the mentioned strategy, a test scenario is performed in the daytime having a step change in the injected reference power into the grid (p_{gc_ref}). The

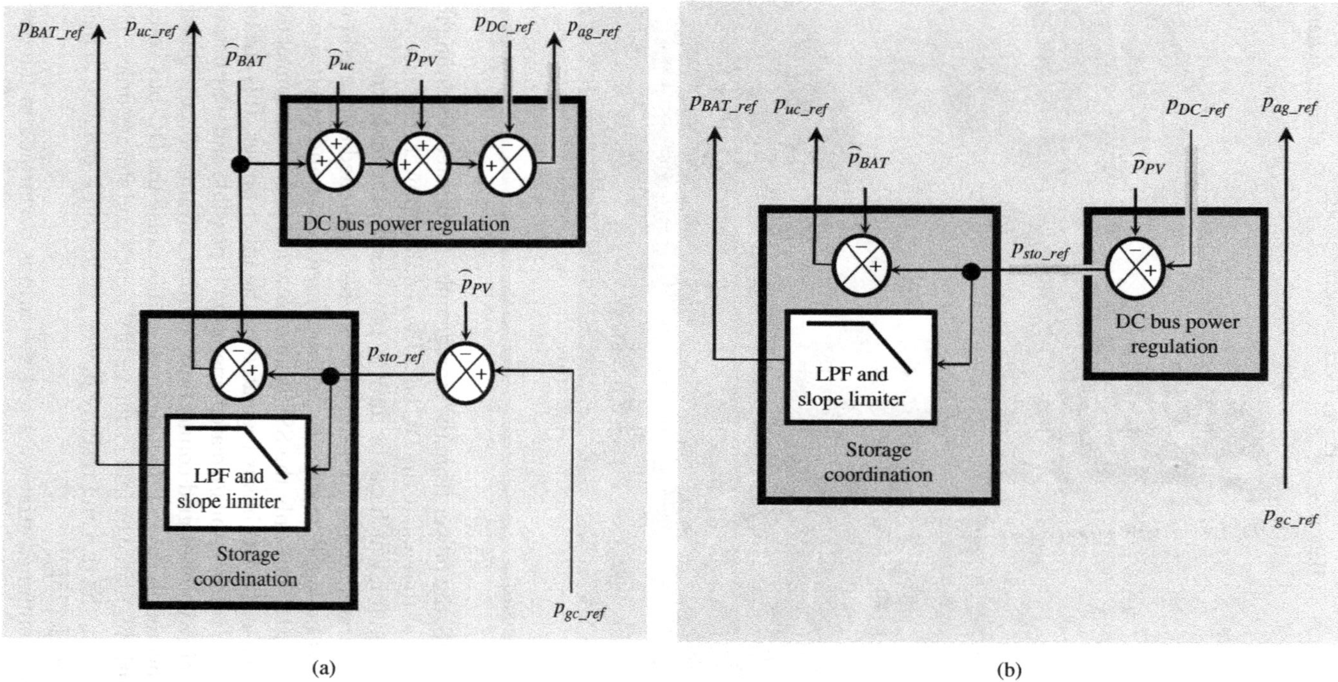

Figure 1.44 Operation strategies: (a) grid-following strategy and (b) source supplying strategy.

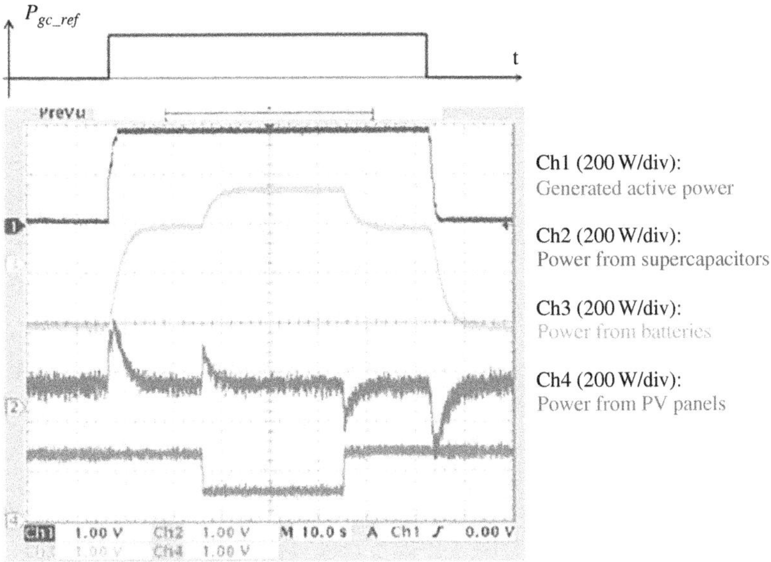

Figure 1.45 Experimental results for an active PV generator.

records are shown in Fig. 1.45. The PV power production changes between fourth and seventh second from 250 to 100 W. During the test, the batteries cannot immediately supply all the surplus production when the PV power production changes or when the PV production decreases fast, so the ultracapacitors help to perform the power balancing.

Next scenario is done at night, when there is no electrical production from PV power. Figure 1.46 shows the variation of the grid power and the dynamic currents from both storage units. When a step change from 0 to 200 W occurs in the grid load demand (p_g) at the 19 s, the batteries are discharged with a slow power increase and, at the same time, the ultracapacitors are instantaneously discharged with a high current to meet the grid load demand (Fig. 1.46).

1.4.4 Control of Grid-Connected Solar Power Inverters: A Review

Over the years, the PWM converters have drastically increased their importance on the market of energy conversion for PV applications. Two technology breakthroughs enabled these remarkable developments: (i) innovations in the field of power electronics, which brings improvements in efficiency (reduction of switching losses), power density, power quality, common mode voltage, electromagnetic interference, and so on; (ii) innovations in the field of control schemes, which currently contain many sophisticated control functions (Fig. 1.47), for example, inner current/power and outer DC-link controls, MPPT, monitoring and grid synchronization, special control functions for

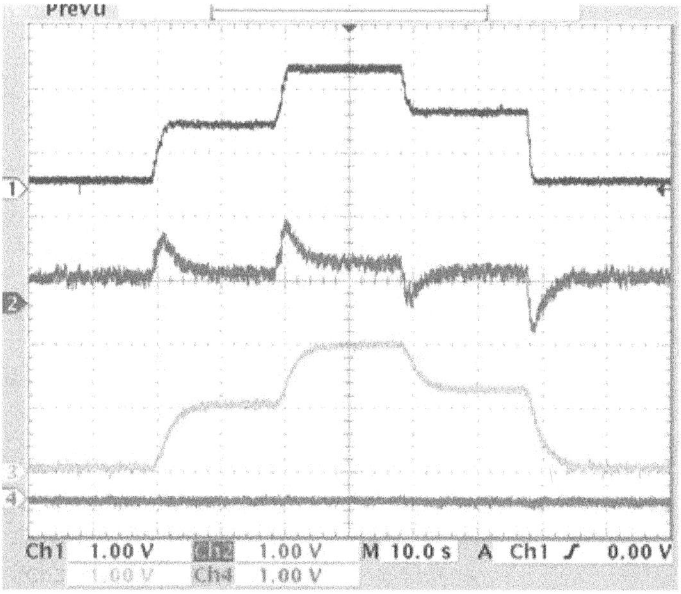

Ch1 (200 W/div): p_g; Ch2 (200 W/div): p_{uc}; Ch3 (200 W/div): p_{bat}; Ch4 (200 W/div): p_{PV}

Figure 1.46 Experimental results in the nighttime.

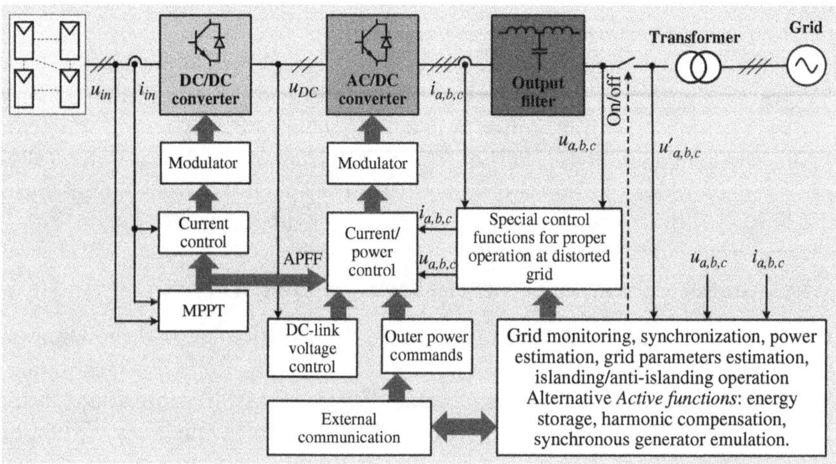

Figure 1.47 General control structure of a grid-connected SPG. *Source*: Cadaval *et al.*, 2015 [20]. Reproduced with permission of IEEE.

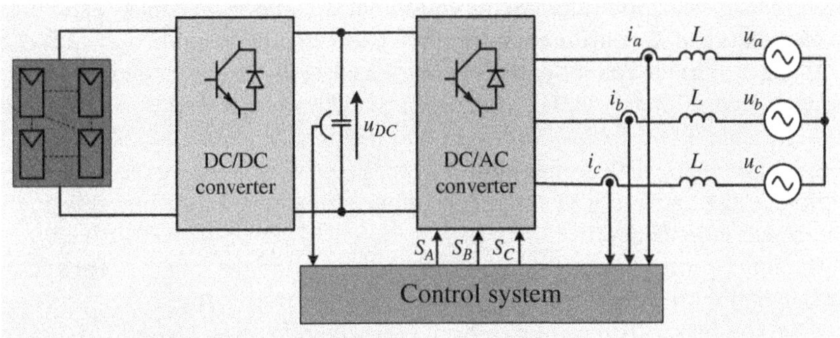

Figure 1.48 Simplified DC/AC PWM converter structure.

proper operation at significantly distorted grid, islanding/anti-islanding operation, active power feedforward, outer control loop for active/reactive power control to support power system, energy storage, harmonic compensation, and synchronous generator emulation [24–28].

Recently, various inner control algorithms have been proposed in recent works for the type of DC/AC PWM converters in MW PV applications [24, 29], but many of them are still under way and they have not yet placed in industry. Therefore, using the simplified system structure shown in Fig. 1.48, this section only describes the most promising and attractive methods for industry such as voltage-oriented control (VOC), direct power control space vector modulated (DPC-SVM), and predictive control (PC) approaches.

The conventional VOC uses the closed-loop current control in the rotating reference frame. A characteristic feature for this current controller is the processing of signals in two coordinate systems (Fig. 1.49), after converting

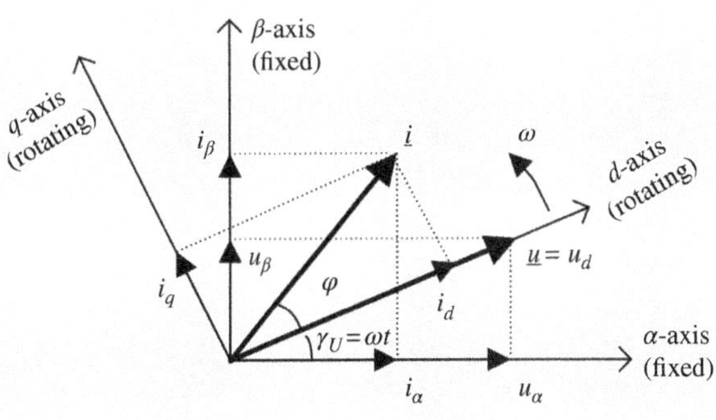

Figure 1.49 Coordinate transformation of line current and voltage from the stationary α–β to the rotating d–q coordinates.

three-phase measured values to the equivalent two-phase system $\alpha-\beta$ and then to the rotating $d-q$ coordinate system via a $\alpha-\beta/d-q$ block [29].

As the $d-q$ frame rotates with the same speed as the electrical quantities, the projected coordinates are DC signals. In the voltage-oriented $d-q$ frame, the AC grid current vector $\underline{i}$ is split into two rectangular components $\underline{i}=[i_d, i_q]$. The component i_q determines indirectly the reactive power, whereas i_d decides indirectly the active power flow. Thus, the reactive and active powers can be controlled independently.

The unity power flow (UPF) condition is met when the grid current vector $\underline{i}$ is aligned with the grid voltage vector $\underline{u}$, which means that the q-axis current should be set to zero in all cases, while the reference current i_d is set by the DC-link voltage controller and controls the active power flow. As shown in Fig. 1.50a, the output signals from PI controllers after a $dq/\alpha\beta$ transformation are used to generate switching signals via a space vector modulator (SVM).

An important drawback of the VOC is the algorithm complexity (many coordinate transformations are needed) as well as the sensitivity to the grid voltage distortion. The last drawback can be eliminated, but it significantly complicates the algorithm [26]. The VOC can be also realized in the stationary $\alpha-\beta$ coordinate system, but PI controllers cannot be used for current regulation because they result in a steady-state error. However, this can be solved by replacing PI current regulators with proportional/resonant (PR) controllers, which consist of a proportional gain and a resonant integrator. The transfer function of a PR controller contains a pair of conjugate poles tuned at the fundamental grid frequency ω, which allows perfect tracking of sinusoidal signals without any error.

A basic block diagram of the VOC in the stationary coordinate system is shown in Fig. 1.50b and it is based on an inner current control loop with PR and an outer voltage control loop with PI. The commanded DC-link voltage u_{DC_ref} is compared with the measured u_{DC} voltage. The error is delivered to the PI controller, which generates the amplitude of the commanded current i_{DC_ref}. Next, this current is multiplied by the angle of the line voltage u_L as $\cos(\omega t)$ and $\sin(\omega t)$ to get i_{α_ref} and i_{β_ref}. These reference current signals in the stationary coordinate system are compared with the measured grid currents i_α and i_β and the errors are delivered to the PR controller. The outputs of the PR controller are directly (without transformation) used to generate switching signals by an SVM. An interesting feature of this control is the proper operation under distorted grid voltages by connecting multiple PR compensators, which are tuned at specific high-order harmonics, for example, fifth, seventh, in parallel [30].

Another less known method is direct power control-space vector modulated (DPC-SVM), which uses inner loops of active and reactive power control (Fig. 1.50c) [31]. The commanded reactive power q_{ref} (set to zero for UPF operation) and active power p_{ref} values (delivered from the outer PI-DC voltage controller) are compared with the estimated q and p values, respectively. The errors are delivered to PI controllers to eliminate steady-state errors because

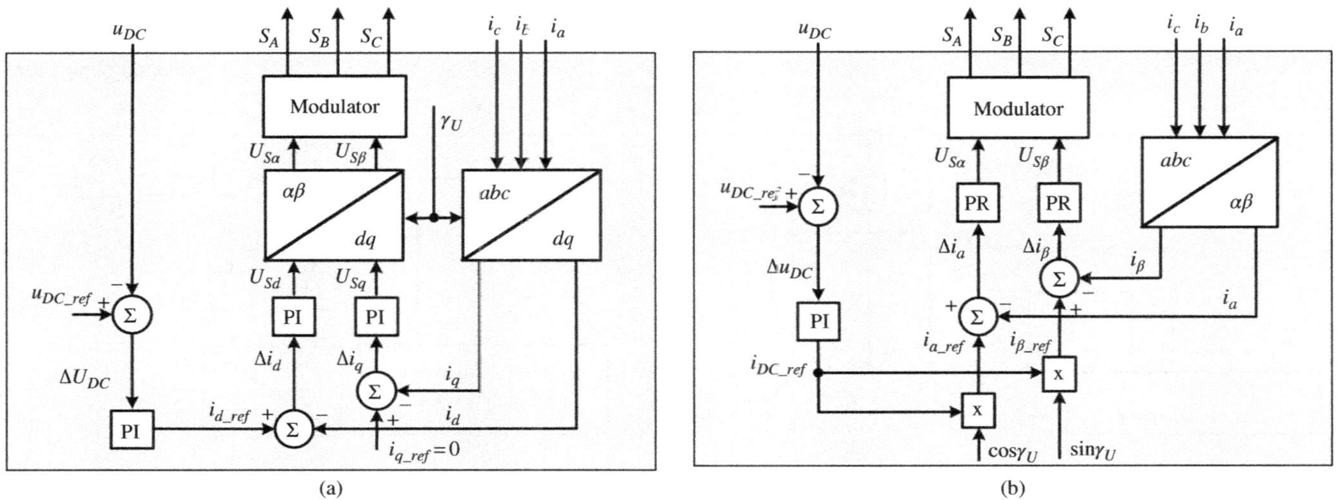

Figure 1.50 Control system schemes: (a) VOC in the synchronous rotating coordinate system, (b) VOC in the stationary coordinate system, (c) direct power control with the space vector modulator (DPC-SVM), and (d) model predictive control (MPC). *Source*: Cadaval *et al.*, 2015 [20]. Reproduced with permission of IEEE.

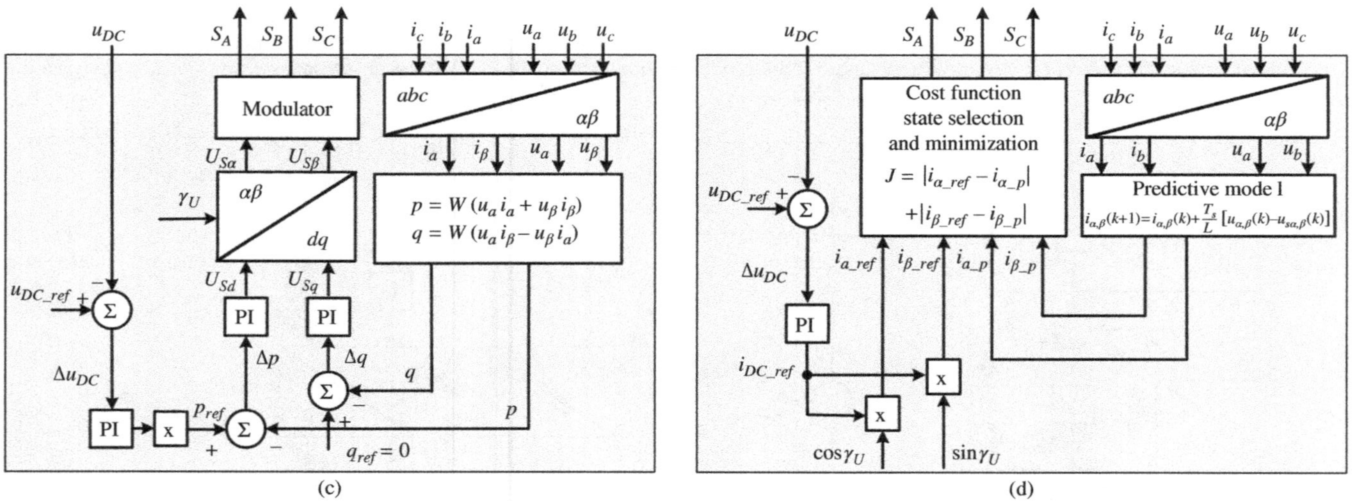

Figure 1.50 (*Continued*)

they are DC quantities. The output signals from the PI controllers after transformation from the stationary to synchronous rotating coordinates are used to generate switching signals by an SVM.

Recently, a control scheme called model predictive control (MPC) (Fig. 1.50d) has been applied to control inverters. It is based on the mathematical model of the controlled system and the calculation (prediction) of future values of the state variables. It is assumed that the system can represent a finite number of states in every time period [32]. Thus, first of all, the model must be as accurate as possible because the control performance is highly dependent on the parameters. Using measured values of grid currents, their values in the forthcoming sampling can be calculated, with respect to available control states. It must be done for all switching states in each iteration of the algorithm.

Some features of PC are very interesting, for example, fast dynamics and the capability of dealing with multiple constraints, which is attractive especially for high power converters operating at very low switching frequency (500–1000 Hz). Similar to other methods, the MPC has also some drawbacks, for example, variable switching frequency (causing difficulties for the design of the LC input EMI filter), high sampling frequency needed for digital implementation (demand of fast microprocessor), and sensitivity to parameter variations of the grid filter, which cause difficulties in implementing MPC in industry. The advantages and features of the control schemes described above are summarized in Table 1.3 [20].

Table 1.3 Advantages and features of control schemes for DC/AC converter in PV applications.

<table>
<tr><th colspan="2"></th><th>VOC (Fig. 50a)</th><th>VOC (Fig. 1.50b)</th><th>DPC-SVM (Fig. 1.50c)</th><th>MPC (Fig.1.50d)</th></tr>
<tr><td colspan="2">Operation in stationary coordinate system</td><td>No</td><td>Yes</td><td>Yes/no</td><td>Yes</td></tr>
<tr><td colspan="2">Power control – indirect</td><td>Yes</td><td>Yes</td><td>No</td><td>Yes</td></tr>
<tr><td colspan="2">Power control – direct</td><td>No</td><td>No</td><td>Yes</td><td>No</td></tr>
<tr><td colspan="2">Constant switching frequency</td><td>Yes</td><td>Yes</td><td>Yes</td><td>No</td></tr>
<tr><td colspan="2">Low algorithm complexity</td><td>No</td><td>Yes/no</td><td>Yes/no</td><td>Yes</td></tr>
<tr><td colspan="2">Low computation intensity</td><td>Yes</td><td>Yes</td><td>Yes</td><td>No</td></tr>
<tr><td colspan="2">Low sensitivity to line inductance variation</td><td>Yes</td><td>Yes</td><td>Yes</td><td>No</td></tr>
<tr><td rowspan="2">Low sensitivity to line voltage distortion</td><td>THD of line current</td><td>No</td><td>Yes</td><td>Yes</td><td>No</td></tr>
<tr><td>Power factor</td><td>Yes</td><td>No</td><td>No</td><td>Yes</td></tr>
</table>

1.5 Summary

This chapter addressed the integration of RESs and described the fundamentals of PV and wind energy conversion systems. The relevant characteristics and control structures were discussed. The average models for the wind and PV-based active generators (WPG and SPG) were obtained. The modeling of these complex generation systems were led by considering each source with the dedicated power electronic converters. For the design of relevant control schemes in each energy conversion system, a hierarchical control structure was used.

The results were evaluated using simulations and laboratory experiments. The performed simulations showed the performances of the modeling and control structures. Experimental test results verified the effectiveness of the outcomes and showed desirable dynamic performances.

References

1 United Nations Environment Program (1997) Kyoto Protocol, http://www.kyotoprotocol.com.

2 International Energy Agency (IEA) (2015) Energy and Climate Change, World Energy Outlook Special Report, http://www.iea.org/publications/freepublications/publication/weo-2015-special-report-energy-climate-change.html.

3 Zhong, Q.-C. and Hornik, T. (2013) *Control of Power Inverters in Renewable Energy and Smart Grid Integration*, Wiley, Hoboken, NJ.

4 Bevrani, H., Watanabe, M., and Mitani, Y. (2014) *Power System Monitoring and Control*, IEEE-Wiley Press, New York.

5 Bevrani, H., Ghosh, A., and Ledwich, G. (2010) Renewable energy sources and frequency regulation: survey and new perspectives. *IET Renewable Power Generation*, **4** (5), 438–457.

6 Robyans, B., Francois, B., Delille, G., and Saudemont, C. (2015) *Energy Storage in Electric Power Grids*, Wiley.

7 Blaabjerg, F., Teodorescu, R., Liserre, M., and Timbus, A.V. (2006) Overview of control and grid synchronization for distributed power generation systems. *IEEE Transactions on Industrial Electronics*, **53** (5), 398–1409.

8 Ataee, S., Khezri, R., Feizi, M. R., and Bevrani, H. (2015) *Impacts of Wind and Conventional Power Coordination on the Short-Term Frequency Performance*. 23rd Iranian Conference on Electrical Engineering ICEE, Tehran, Iran.

9 Khezri, R. and Bevrani, H. (2015) Voltage performance enhancement of DFIG-based wind farms integrated in large-scale power systems: coordinated AVR and PSS. *International Journal of Electrical Power and Energy Systems*, **73**, 400–410.
10 Saleh, M. and Bevrani, H. (2011) Dynamic analysis and stability improvement concerning the integration of wind farms: Kurdistan electric network case study, in *Innovation in Power, Control and Optimization: Emerging Energy Technologies* (eds P. Vasant, N. Barsoum, and J. Webb), IGI Global, pp. 198–219. Chapter 6.
11 Bouscayrol, A. and Delarue, P. (2002) Simplifications of the maximum control structure of a wind energy conversion system with an induction generator. *International Journal of Renewable Energy Engineering*, **4** (2), 479–485.
12 Li, P. (2010) Design and control of a PV active generator with integrated energy storages: application to the aggregation of producers and consumers in an urban micro smart grid. PhD thesis, Ecole Centrale de Lille, France.
13 Onar, O.C., Uzunoglu, M., and Alam, M.S. (2006) Dynamic modeling design and simulation of a wind/fuel cell/ultra-capacitor-based hybrid power generation system. *International Journal of Power Sources*, **161** (1), 707–722.
14 Zhou, T. (2009) Control and energy management of a hybrid active wind generator including energy storage system with super-capacitors and hydrogen technologies for microgrid application. PhD thesis, Ecole Centrale de Lille, France.
15 Robyns, B., Davigny, A., François, B. *et al.* (2012) *Electricity Production from Renewables Energies*, Wiley.
16 Bevrani, H. (2014) *Robust Power System Frequency Control*, 2nd edn, Springer.
17 Li, P., Degobert, P., Robyns, B., and Francois, B. (2008) Implementation of interactivity across a resilient microgrid for power supply and exchange with an active distribution network. CIRED Seminar 2008: SmartGrids for Distribution, Frankfurt.
18 Fakham, H., Lu, D., and Francois, B. (2011) Power control design of a battery charger in a hybrid active PV generator for load-following applications. *IEEE Transactions on Industrial Electronics*, **58** (1), 95–104.
19 Lu, D. (2010) Design and control of a PV active generator with integrated energy storages: application to the aggregation of producers and consumers in an urban micro smart grid. PhD thesis, Ecole Centrale de Lille, France.
20 Cadaval, E.R., Francois, B., Malinowski, M., and Zhong, Q.C. (2015) Grid-connected photovoltaic plants: an alternative energy source, replacing conventional sources. *IEEE Industrial Electronics Magazine*, **9** (1), 18–32.

21 Kanchev, H., Lu, D., Colas, F. *et al.* (2011) Energy management and power planning of a microgrid with a PV-based active generator for smart grid applications. *IEEE Transactions on Industrial Electronics,* **58** (10), 4583–4592.

22 Lu, D., Fakham, H., Zhou, T., and Francois, B. (2010) Application of Petri nets for the energy management of a photovoltaic based power station including storage units. *Renewable Energy,* **35** (6), 1117–1124.

23 Xingyu, Y., Abbes, D., Bevrani, H., and Francois, B. (2016) *Day-Ahead Optimal and Reserve Power Dispatching in PV Based Urban Microgrid.* 18th European Conference on Power Electronics and Applications-EPE'16 ECCE, Karlsruhe, Germany, 5–9 September 2016.

24 Wilamowski, B. and Irwin, D. (2011) *Industrial Electronics Handbook,* Taylor & Francis Group.

25 Bevrani, H. (2012) Microgrid controls, in *Standard Handbook for Electrical Engineers,* 16th edn (ed. H. Wayne Beaty), McGraw-Hill, pp. 160–176 Section 16.9.

26 Kazmierkowski, M.P., Jasinski, M., and Wrona, G. (2011) DSP-based control of grid-connected power converters operating under grid distortions. *IEEE Transactions on Industrial Informatics,* **7** (2), 204–211.

27 Teodorescu, R., Liserre, M., and Rodríguez, P. (2011) *Grid Converters for Photovoltaic and Wind Power Systems,* Wiley-IEEE Press.

28 Guerrero, J.M., Chandorkar, M., Lee, T.-L., and Loh, P.C. (2013) Advanced control architectures for intelligent microgrids – Part I: decentralized and hierarchical control. *IEEE Transactions on Industrial Electronics,* **60** (3), 1254–1262.

29 Malinowski, M., Kazmierkowski, M.P., and Trzynadlowski, A. (2003) A comparative study of control techniques for PWM rectifiers in AC adjustable speed drives. *IEEE Transactions on Power Electronics,* **18** (6), 1390–1396.

30 Abu-Rub, H., Malinowski, M., and Al-Hadad, K. (2014) Power electronics for renewable energy systems, transportation and industrial applications, in *Photovoltaic Energy Conversion Systems* (eds S. Kouro, B. Wu, H. Abu-Rub, and F. Blaabjerg), Wiley, Hoboken, NJ, Chapter 7.

31 Malinowski, M., Jasinski, M., and Kazmierkowski, M.P. (2004) Simple direct power control of three-phase PWM rectifier using space-vector modulation (DPC-SVM). *IEEE Transactions on Industrial Electronics,* **51** (2), 447–454.

32 Rodriguez, J. and Cortes, P. (2012) *Predictive control of power converters and electrical drives,* Wiley, Hoboken, NJ.

2

Renewable Power for Control Support

Due to deregulation of the electricity supply industry, the power grid requires to achieve regulation and control support through some services that the system operator has to obtain from network stakeholders, known as ancillary services. Ancillary services are needed to support the reliable delivery of electricity and operation of transmission systems. These services may contribute to frequency regulation and operating reserves, in different timescales. They are also applied to power balance (e.g., scheduling and dispatch of balancing energy), voltage control (e.g., tap changer control and reactive power control), restoration of supply (e.g., black start capability and island operation), or system management (e.g., power quality assurance, operation, and asset management).

To ensure power system security with high renewable energy sources (RESs) penetration, the participation of RESs, particularly wind generators and photovoltaic (PV) systems, in regulation and ancillary services is essential. Previous studies have shown the technical capability of RESs to participate in frequency regulation at a large-scale level.

In this chapter, after an introduction on the relevant literature, contribution of wind power generators (WPGs) in inertial response and primary frequency control is investigated. A statistical approach to assess the impact of wind power variation on the size of primary reserve provided by the WPGs is presented. Then, in the same way, the capability of solar power generators (SPGs) in the grid control support is also discussed. Finally, the microgrid (MG) concept as a solution to integrate more RESs into the power system is introduced.

2.1 Introduction

Renewable power generation capacity has been increasing over the last years due to political incentives for development of low carbon generation. The high penetrations of RESs, because of their variability, uncertainty, and nonsynchronous grid connection, pose several technical challenges. The variability and uncertainty of wind power and solar power generation increase

Microgrid Dynamics and Control, First Edition. Hassan Bevrani, Bruno Francois, and Toshifumi Ise.

the need for operational flexibility and reserve requirements. Moreover, during periods of high RESs, less conventional synchronous generators are required to balance generation and demand. Replacement of conventional synchronous generators by the RESs reduces both the total system inertia and the number of groups available to provide regulation services. This situation is more critical in island systems that have a limited number of conventional power plants and a relatively small kinetic energy base.

In areas where the renewable energy penetration is relatively significant, system operators encounter serious power, frequency, and voltage control concerns. The situation is worse in isolated island systems that already have small kinetic energy from their base-load generators [1]. Especially when renewable power production is relatively large during off-peak hours, conventional synchronous generators may be switched off to balance generation and demand. This results in less system inertia and regulation power reserve in the system. Therefore, some utilities have already updated their grid codes to ensure power system security and reliability so that RESs are now expected to provide ancillary services and regulation/control supports [2, 3]. Thus, recently many research works have been oriented to the design of new control loops for regulation support of the RESs and energy storage systems (ESSs). For example, for the WPGs, two solutions are particularly interesting for frequency regulation with variable speed wind turbines (VSWTs): the inertial control and the primary speed-droop control [4].

A considerable number of research activities are focused on the integration of large amounts of RESs in the electrical system. The attention is now oriented toward the use of distributed generators (DGs) for improving grid operation by contributing to ancillary services, increasing the energy reserve and reducing CO_2 emissions. In practice, new facilities are expected to reduce congestion to minimize the production costs and to regulate the frequency and voltage of the grid.

In this chapter, the implementation of inertial response and primary frequency control in a wind turbine (WT) control unit are investigated. Main factors affecting the performances of frequency regulation are identified and characterized. The influence of control parameters and the turbine operating point (OP) on the inertial response are analyzed through the obtained performances in an islanded power system. The combined control scheme using both inertial and primary controllers is also developed and the potential of the obtained grid services at partial load is discussed. Then, the possibility of using WTs in secondary frequency control is addressed.

The design of the inertial control is discussed in [5] and then further studied in [1, 6–8]. A "virtual" wind inertia is created to respond to the frequency drops by using the kinetic energy stored in rotating masses of the WT. Reference [9] discusses factors affecting the provision of inertial response, such as the converter maximum currents and controller parameters. Furthermore,

[10] shows that the WT inertial response capability can be higher than that of a synchronous machine for the same inertia value, since greater speed variations are acceptable for the VSWT and so more kinetic energy can be converted into electrical energy.

The capability of providing long-term primary frequency regulation for doubly fed induction generators (DFIGs) and full-converter WTs have been, respectively, studied in [11] and [12]. As for conventional power plants, the proposed control strategy requires WTs to preserve a power margin. The primary droop controller adjusts wind active power according to the frequency variation (Δf) through the pitch regulation. In [13], both frequency control schemes, described before, are exploited for further grid frequency improvements. The contribution in secondary frequency control is also investigated in [1, 14, 15].

The implementation of a frequency control response into the VSWT has been investigated and proven in many works. However, some important points still require further studies. First, only the rated OP of wind farms (WF) has been simulated and analyzed, where the full-load operation period represents only a small part of total operating time for most WFs. Secondly, the limitation on the rate of change of WT electrical power (dP/dt) has hardly been considered in previous studies. References [16, 17] showed that (dP/dt) should be limited by a maximum value to avoid fast variations of the turbine power output and to reduce the mechanical stresses. On the other hand, to enhance the WT frequency response, a minimum response rate of each online WT is required by the system operators. Therefore, the impacts of the (dP/dt) value on the WT frequency response as well as on the turbine behavior should be characterized as explained in [18] and this chapter.

In order to integrate RESs and ensure system security and reliability, newly connected WPGs and SPGs are required to contribute to primary reserve. This capability is currently required by the grid code of several systems. Moreover, if WPG/SPG should be curtailed because of system constraints such as minimum inertia, reserve provision, and network congestion, the curtailed energy can be used to provide upward reserves [19].

In recent years, significant research efforts have been carried out in the design of primary frequency controllers for wind plants [20, 21]. As a result, the technical capability of WTs for providing primary frequency regulation has been proven by different studies and demonstration projects [11, 12]. Different control methods, such as speed-droop control [6] and fuzzy logic control, are used as primary frequency controllers that enable WTs to maintain a power margin and to release the stored reserve to the grid during low frequency conditions by adjusting their active power according to the frequency variation [22]. Furthermore, the obtained fast response from WTs increases the system robustness by reducing frequency excursions following the grid disturbances [18]. Previous research works about the contribution of the WT in primary frequency control are based on dynamic simulations performed over a time horizon of several minutes and have effectively confirmed the technical capability of WPG's participation in frequency control.

The problem of quantifying the aggregated volume of primary reserve that can be provided by the wind power fleet has received less attention. In fact, in order to quantify the contribution of WPG primary frequency control to the system security, one needs to estimate the amount of conventional generation primary reserve that can be displaced by WPG reserve.

The variable nature of WPGs' output impacts the amount of reserve that can be provided by WTs at different periods. Moreover, the uncertainty in the forecast of wind power at different timescales adds further uncertainty to the reserve provided by WTs. The impact of wind forecast errors depends on the lead time between the decisions of reserve placement and can be reduced by the possibilities of redispatch closer to real time and the increase in the wind forecasting accuracy [23].

Recent works have investigated the wind reserve allocation at a single farm level. For example, [24] proposed to allocate primary reserve to each turbine according to the available wind speed. However, the ancillary services, especially the frequency control services, should be managed from a system point of view by taking into account the wind smoothing effect. In addition to the frequency response dynamics, other requirements such as the minimum time interval that the primary reserve service requires to be held for the needs to be taken into account. For example, in the European Network of Transmission System Operators for Electricity (ENTSO-E) synchronous system, the deployed primary reserve must be held about at least 15 min for security concerns [25]. This time interval is likely to increase to 30 min in the future. The variability of wind power generation during the reserve deployment period will impact the volume of primary reserve that can be expected from variable generation sources, such as WPGs.

The allocation of primary reserve by the WPG cannot be done in the same way as by conventional generation since wind output experiences significant variability. For example, the reserve allocation obtained from a security-constrained unit commitment (SCUC), which is performed a day ahead for hourly time steps [26], needs to be converted into a "firm" wind reserve level, given wind intrahourly variability. Failing to do so can lead to a risk of overestimating the volume of primary reserve that can actually be provided from WPGs.

In this chapter, as an RES example, the impact of wind power variability between two SCUC time steps is evaluated, using an approach based on statistical analysis of wind power variability. In order to analyze the effect of wind variability, a perfect forecast of wind output is assumed. This allows us the assessment of the maximum amount of "firm" wind primary reserve that meets the expected technical requirements.

A practical example of the estimation of "firm" wind primary reserve from wind power is quantified by using the proposed approach and real wind data measurement for the French island of Guadeloupe. The efficiency of using different reserve allocation strategies to handle wind power intrahourly variability and maximize the amount of wind primary reserve is also analyzed. In this chapter, the capability of SPGs to support the regulation service is also discussed and finally the MG as a useful solution for penetration of more and more RESs is introduced. Some parts of this chapter are detailed in [1, 14, 26–28].

2.2 Wind-Energy-based Control Support

2.2.1 Wind Turbines Inertial Response

As explained in Chapter 1, the control strategy of a wind generator is a combination of the pitch control and the static power electronic converter control. The inertial controller consists in adjusting the torque reference as a function of the derivative of the grid frequency. This mechanism is shown in Fig. 2.1. When the grid frequency falls, the electromagnetic power set point increases, leading to a deceleration of the rotor speed and thus an extraction of the kinetic energy stored in rotating masses. The required power can therefore be released to the grid for dynamic frequency control support. The frequency can be estimated by using a phase-locked loop (PLL) synchronization. In order to minimize noise impacts, a low-pass filter is introduced in the control loop [29], with a T_f time constant (about 100 ms according to the performance of the equipment). The controller gain K_{ic} determines the quantity of additional injected wind power in the case of a grid fault. To reduce mechanical stresses, WTs must not generate abrupt variations in power output. Consequently, the rate of change of power (ROCOP) injection is limited by the parameter $(dP/dt)_{max}$ in order to reduce mechanical stresses on the drive train. According to manufacturer's data, the value of $(dP/dt)_{max}$ should not exceed 0.45 pu/s [16].

The dynamic behavior of a WT providing inertial response is sensitive to initial operating conditions. If the wind speed is above rated, the turbine is initially operating at full load and the pitch actuator is activated to maintain the rotor speed at its maximum value.

Wind turbine inertial response at different OPs is simulated in Fig. 2.2. In the case of an underfrequency event (at $t = 10$ s), additional power is demanded (Fig. 2.2a), which decreases the turbine speed (Fig. 2.2c). Then, the pitch angle is reduced to raise the rotor speed to its reference value (Fig. 2.2b). After the power surge, the inertial control adjusts progressively the total wind power to the predisturbance level (Fig. 2.2a). As the electrical power exceeds for a short period the steady-state rating of the turbine, the magnitude of the incremental power will depend on the power electronic sizing.

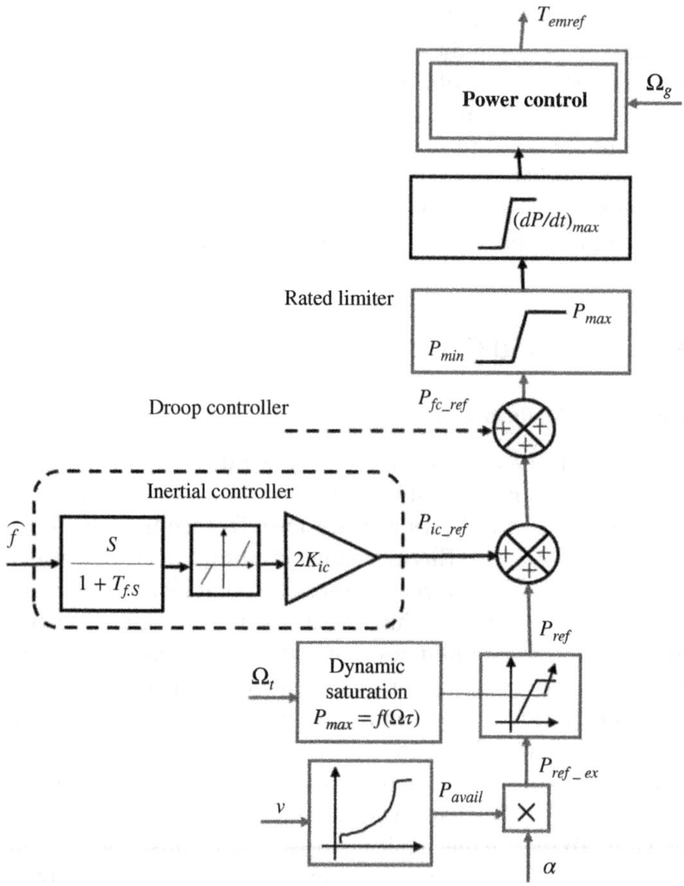

Figure 2.1 Integration of the inertial controller in the control system. *Source:* Wang *et al.*, 2013 [18]. Reproduced with permission of IEEE.

For moderate or low wind speeds, the pitch angle is maintained at zero to maximize the wind production (Fig. 2.2e). When the inertial controller is activated, additional power is first injected to the grid at full load (Fig. 2.2d), leading to a decrease in the rotor speed. Then the turbine can no longer operate at the optimum speed issued from the maximum power point tracking (MPPT) control (Fig. 2.2f). Therefore, compared to the initial production, less power will be delivered to the grid after the power surge until the turbine speed returns to its initial optimum value (Fig. 2.2d). The duration of the speed recovery period depends on the turbine mechanical dynamics and lasts generally several tens of seconds.

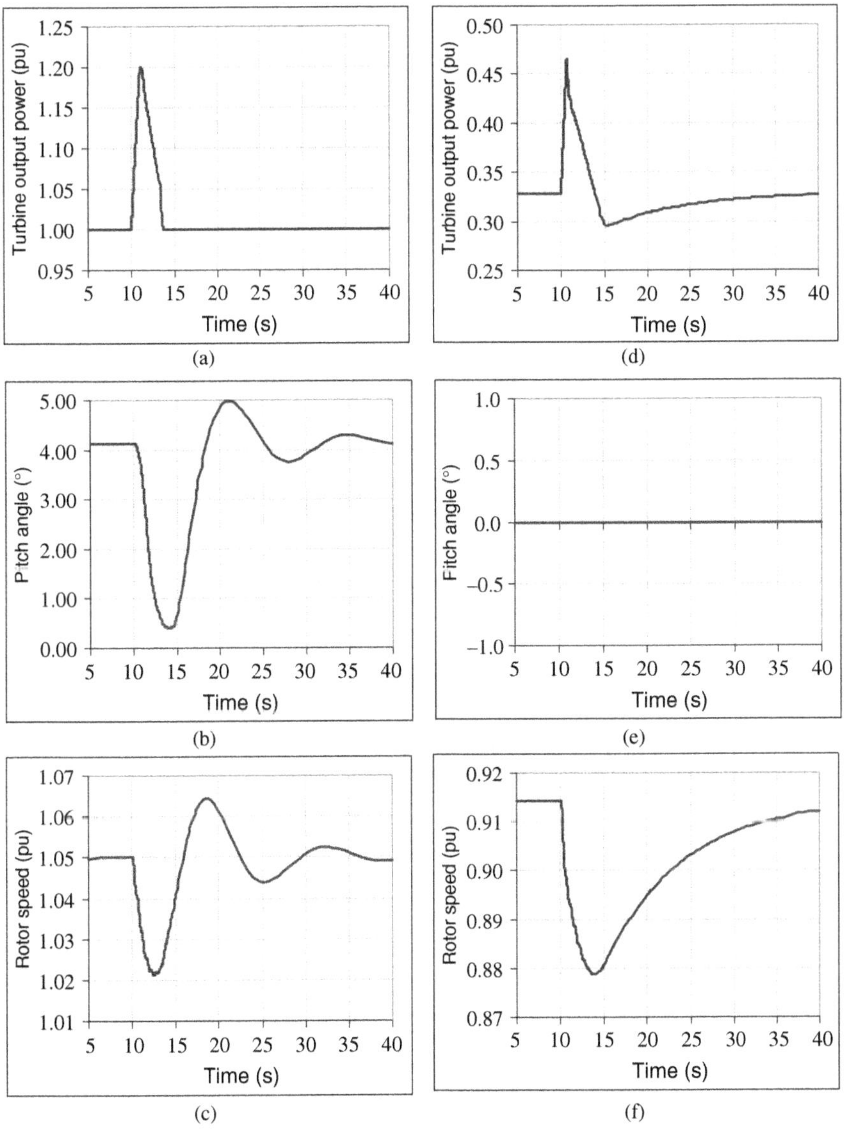

Figure 2.2 Wind turbine inertial response at different operating points ((a)–(c) at full load and (d)–(f) at partial load).

Similar to the inertial behavior of synchronous machines, the WT inertial response at partial load is essentially energy neutral, meaning that the period of increased power is followed by a period of decreased power, during which the released kinetic energy is recovered by "withdrawing" power from the grid. However, the inertial response at full load is fundamentally different. From an energy point of view, the incremental electric power is supplied by the power available from the wind and the turbine speed is recovered via pitch control. Therefore, there is not any decrease in the wind output power after the disturbance.

In the first seconds following a fault, the electrical power from a grid-connected synchronous generator varies naturally in order to counteract frequency deviations. This natural inertial reaction of a synchronous machine can be characterized by the following equation:

$$P_e(s) = -2H \cdot \widehat{f}(s) \cdot s \tag{2.1}$$

where H is the synchronous machine inertia constant, s is the Laplace variable, and $\widehat{f}$ is the measured per-unit frequency. Typical value of H lies between 2 and 6 MWs/MVA according to the type of generating unit [1]. The power response is therefore instantaneous and does not depend on the dynamics of speed control loops.

The WT inertial response is dominated by the additional control function presented in Fig. 2.1 and so does not have the same recovery phase as conventional machines. The power set-point variation in the first seconds after a fault is therefore characterized by the following formula:

$$P_{ic\_ref}(s) = -2K_{ic}\frac{s}{1 + T_f \cdot s} \cdot \widehat{f}(s) \quad \text{if} \quad \frac{dP_{ic\_ref}}{dt} \leq \left({}^{dP}\!/_{dt}\right)_{max} \tag{2.2}$$

Because of the measurement and filter delays, the frequency responses are different from those obtained using synchronous generators. Besides, protections may saturate the WT inertial response. In this case, the derivative of the power is forced to be a constant as shown in (2.3), and then the inertial contribution is limited.

$$P_{ic\_ref}(t) = \left({}^{dP}\!/_{dt}\right)_{max} \cdot t \quad \text{if} \quad \frac{dP_{ic\_ref}}{dt} > \left({}^{dP}\!/_{dt}\right)_{max} \tag{2.3}$$

A dead band is used to restrict the WT inertial control to large events (Fig. 2.1). The continuous small perturbations in frequency that characterize normal grid operation are not passed through to this controller. Theoretical analysis states that the performances of the inertial response are mainly affected by the value of the K_{ic} controller gain and the $(dP/dt)_{max}$ ROCOP limitation parameter, but not by the turbine inertia constant.

2.2.2 Study on a Real Isolated Power System

Impact on the power reserve has been studied and validated for a real isolated power system in France, located in the eastern Caribbean Sea, the Guadeloupe archipelago, which is composed of five main (interconnected) islands. It covers 1600 km^2 and its population is about 404,000 inhabitants. The peak demand reached 260 MW in 2010. Figure 2.3 presents the structure of the high-voltage (HV) grid as well as the main generation sites. The transmission system is operated at 63 kV and includes 13 substations, each consisting of two step-down 63/21 kV transformers (10, 20, or 36 MVA) equipped with on-load tap changers. The model of the Guadeloupe power system including HV lines, transformers, and power plants has been previously simulated with the Eurostag software package [30] and validated, and the obtained results are much close to sensed system responses [31].

Currently, according to the French feed-in tariff and purchase obligation scheme, the RESs have the highest priority level in the dispatch order and do not participate in frequency control. The primary reserve is mainly provided by diesel engines, bagasse/coal-fired units, and combustion turbines during peak demand times.

2.2.2.1 Simulated Scenarios

As shown in Table 2.1, the 24.7-MW coal-fired plant at "Le Moule" does not provide governor frequency response but contributes inherently to the system

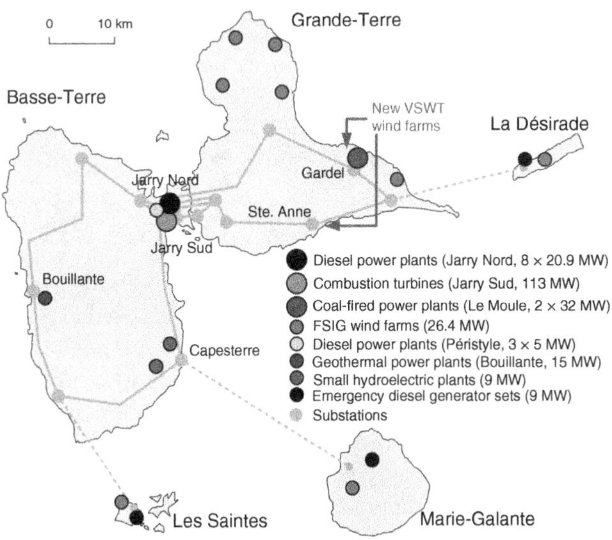

Figure 2.3 The HV grid, main generation plants, and DG locations in Guadeloupe (France).

Table 2.1 Set points for the reference scenario with 11% wind power.

Power station	Units online	Power set-point (MW)	Reserve
Jarry Nord	4	55.8	27.8 MW
Le Moule	1	21.3	Disconnected
	1	24.7	0 MW
Péristyle	2	9.4	0.8 MW
VSWT	0	0	0 MW
Other renewables	Various	27.3	0 MW
Total	–	138.5	28.6 MW

inertia. The power response of this plant after a grid disturbance will only be determined by its inertial nature and thus can be compared with the WT inertial response. The considered fault was based on the tripping of another "Le Moule" power plant producing 21.3 MW before the disturbance. This scenario with only 11% wind penetration is considered as the reference scenario.

In the second scenario with VSWT, the "Le Moule" power plant does not participate in the primary frequency control, which is replaced by the VSWT generation with the same power output, leading to a 29.5% wind penetration. Power set points of other generators remain unchanged. Hence, more than a quarter of the total rotating masses of the grid are lost, since the steam turbines provide a significant part of the kinetic energy.

Three different scenarios are studied and compared: the reference case, the wind case with and without inertial control. Since the focus of the simulations is the study of short-term dynamic performances, no long-term models (e.g., automatic generation control [32]) have been included. All automatic load-shedding (ALS) relays were also removed to provide a clear comparison between various cases.

2.2.2.2 Impacts of the Controller Gain

Each manufacturer has its own maximum ROCOP parameter $(dP/dt)_{max}$. For the presented simulations, the 0.2 pu/s $(dP/dt)_{max}$ value has been used in order to get a general model adequate with possible larger $(dP/dt)_{max}$ values. Moreover, further exposed results will prove that larger values for the $(dP/dt)_{max}$ parameter do not improve the inertial response performance. As the dynamic behavior of a WT providing inertial response changes according to initial wind conditions, both full-load and partial-load OPs have been studied.

At high wind speeds, all VSWTs are supposed to operate at full load because of the small size of Guadeloupe island. In Fig. 2.4a, the dynamic behaviors of the frequency for the reference case and the wind cases are shown. In the absence of inertial control, the frequency nadir of the wind case is deeper than that

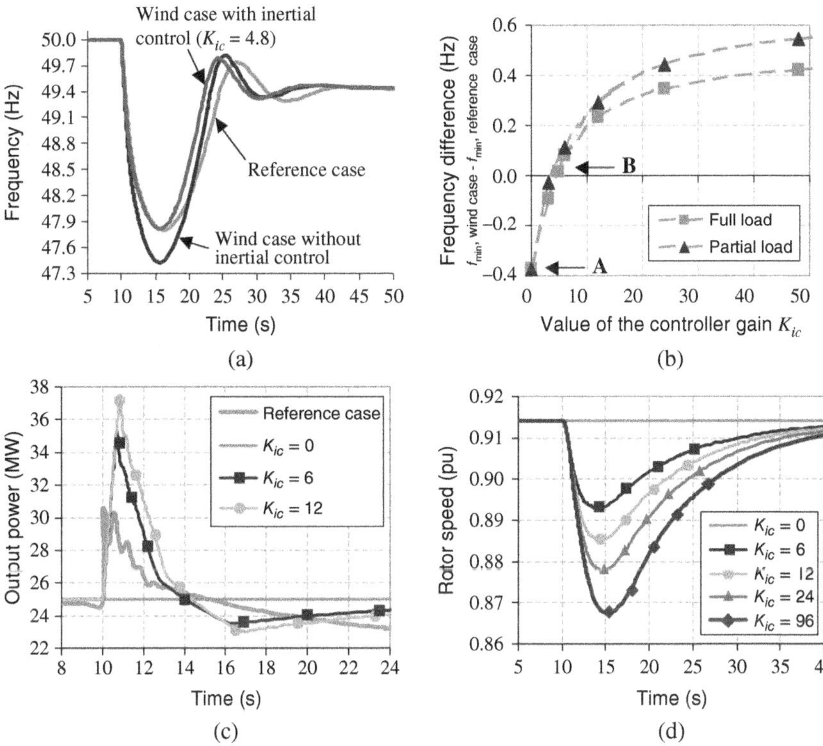

Figure 2.4 (a) Frequency response, (b) frequency nadir difference, (c) inertial response at partial load, and (d) rotor speed. *Source:* Wang *et al.*, 2013 [18]. Reproduced with permission of IEEE.

of the reference case due to the reduction in system inertia. However, when the inertial control is enabled, the minimum frequency can be maintained at the same level as for the reference case with a pertinent controller gain.

Different values of the controller gain were implemented and tested. The minimum grid frequency is used as the key performance metric in this analysis, since it can be correlated to trip thresholds in the specified ALS schemes. The curve with squares in Fig. 2.4b describes the difference of frequency nadirs between wind cases and the reference case as a function of the controller gain for the full-load operating range. With wind inertial response, the grid minimum frequency is reduced while the controller gain is increased.

Two interesting OPs can be identified. Point A corresponds to the case without inertial control ($K_{ic} = 0$). An important 400 mHz deviation from the reference case was observed; hence, the reduction in system inertia can change significantly the system frequency response. Therefore, the possibility of providing an inertial response seems to be essential for wind integration, as it would help to avoid load-shedding in some high-wind cases. At point B, WTs have an "equivalent" contribution to system inertia as the replaced conventional plant. With higher controller gain, the WT inertial response could even have better performances.

The same studies were also performed for the partial-load operating range (with 0.33 pu initial production of WFs) in the same condition. Hence, more WTs are used to produce the same total wind power. According to Fig. 2.4b, with a given controller gain, the minimum frequency in the partial-load case is close to that in the full-load case, especially in the case of low gain values. Since three times more turbines are connected to the grid at partial load, it seems that the inertial response of a single WT is less efficient in this case. In fact, at full load, additional power can be extracted from the wind via pitch regulation to provide inertial response (Fig. 2.4b); therefore, with the same controller gain, more power is injected by a single turbine to the grid in the first seconds for frequency support compared with the partial-load case (Fig. 2.2a,d). Moreover, a WT operating at partial load requires recovering its optimum speed after discharging kinetic energy. Hence, the inertial response is always followed by a negative power spike (Fig. 2.2d), which offsets the contribution of the power surge.

Figure 2.4c shows the inherent inertial response of the coal-fired generator (reference case) and the controlled inertial responses of WTs for different controller gains at partial loads. As demonstrated by theoretical analysis, the synchronous generator reacts almost instantaneously after the disturbance, while the WT inertial response is delayed and limited by the admissible rate of change of electrical power. More power is injected to the grid by WFs after the grid fault, as the wind nominal power in the partial-load case is much higher than that of the replaced conventional power plant (75 MW vs 32 MW).

Large values of K_{ic} lead to a faster drop in the turbine speed. Generally, the rotor speed of the VSWT should not fall below 0.7–0.8 pu in order not to disrupt its normal operation. According to our tests, even with very high controller gain values, the minimum turbine speeds remain far above 0.8 pu (Fig. 2.4d). Thus, the WTs still have the potential to extract more kinetic energy for inertial response.

However, as shown in Fig. 2.4a, the benefit of increasing the controller gain is not proportional to its value as the WT inertial response is saturated for high values of K_{ic}. Thus, it is not effective to set high controller gains, which could induce more wear and tear on WTs without significant increase in performances.

2.2.2.3 Influence of the Rate of Change of Power Limit

In order to reduce mechanical stresses of drive trains and so increase the WT lifetime, it might be desirable to set small values of $(dP/dt)_{max}$ inferior to 0.2 pu/s. The "wind case" scenario has been simulated. The controller gain was kept constant ($K_{ic} = 12$), whereas the value of $(dP/dt)_{max}$ had been progressively decreased, from 0.2 to 0 pu/s (which means that all power variations are eliminated, i.e., the inertial control is not enabled).

Both full-load and partial-load cases were tested. As shown in Fig. 2.5a,d, the frequency nadirs become deeper while the $(dP/dt)_{max}$ value is reduced. This can be explained by the fact that the inertial response of wind farms is limited with low values of $(dP/dt)_{max}$ (Figs 2.5b,e). If $(dP/dt)_{max}$ is decreased, less incremental power will be injected to the grid via inertial control, that is, less kinetic energy will be extracted from rotating masses. This results in a slower drop in the rotor speed (Fig. 2.5c,f). In consequence, the fatigue of mechanical parts will be moderated.

Figure 2.6 shows the maximum frequency deviation from the normal value after the fault as a function of the ROCOP limit. Even with a very low slope of the power variation (0.02 pu/s), the inertial response still has remarkable effects compared to the case without any inertial control ($(dP/dt)_{max} = 0$ pu/s), since a reduction of about 400 mHz in frequency nadir can be obtained. However, the benefit is not linear. It was observed that above a certain limit (0.12 pu/s), the increase in $(dP/dt)_{max}$ has very slight contributions to improve the performance of the inertial response. It is therefore not necessary to set very high $(dP/dt)_{max}$ values.

2.2.3 Primary Frequency and Inertial Controls

2.2.3.1 Principles of the Primary Frequency Control

The WT inertial control can only be used as a short-term dynamic frequency support. For effective long-term frequency regulations, an additional frequency droop control, which takes into account quasi-steady-state frequency deviations, must be integrated into the active power control loop of the VSWTs [11, 33]. The implemented droop control is similar to the one usually in use in synchronous generators and generates a power reference that is added into the WT power control system (Fig. 2.1). When frequency deviations exceed a specified dead band, the active power increment is proportional to the grid frequency variation and is defined as

$$P_{fc\_ref} = \frac{1}{R_w}(f_{ref} - \widehat{f}) \tag{2.4}$$

The droop parameter R_w is expressed in percentage of the WT rating. In order to increase the wind plant active power output in response to an underfrequency condition, some active power production must be kept in reserve, that is, the plant must be operated below the available power. At deloaded OPs, the WT rotor speed is accelerated until its maximum value is reached. For further power curtailment, the pitch control is activated to reduce mechanical power extracted from wind.

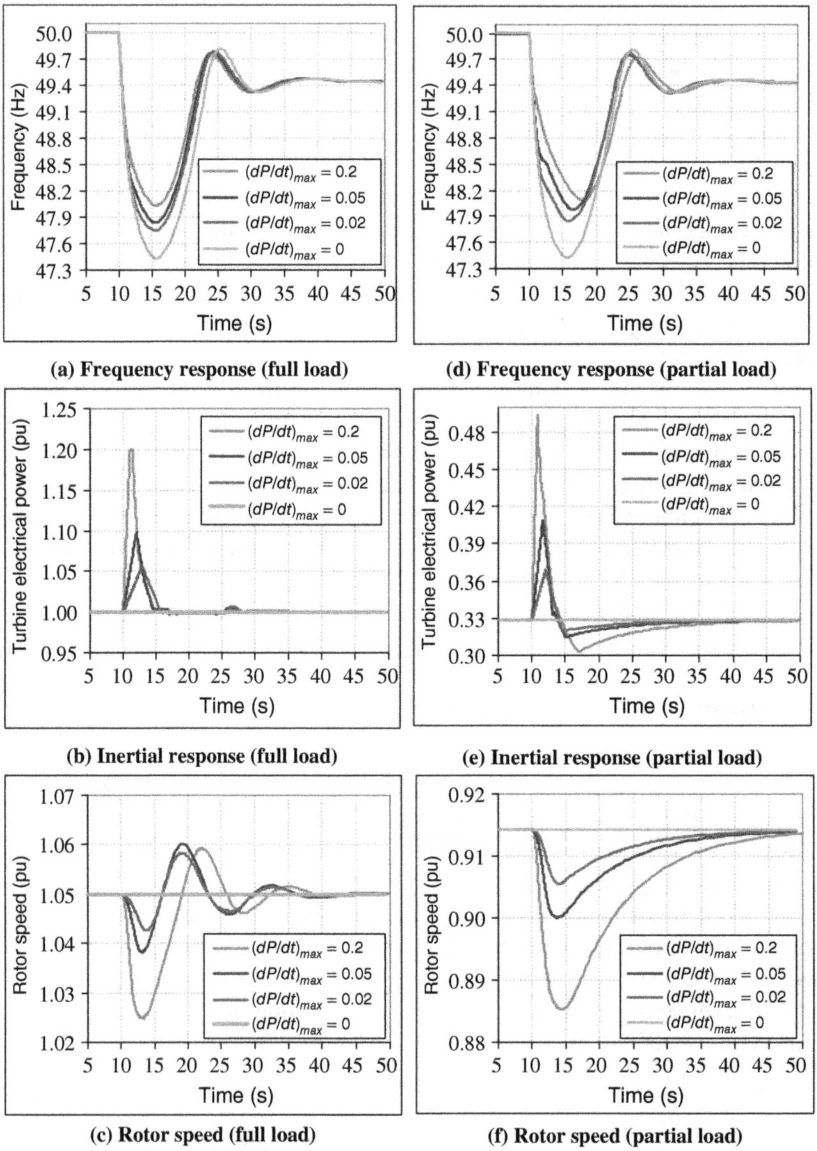

Figure 2.5 Influences of the limit of the ROCOP. *Source:* Wang *et al.*, 2013 [18]. Reproduced with permission of IEEE.

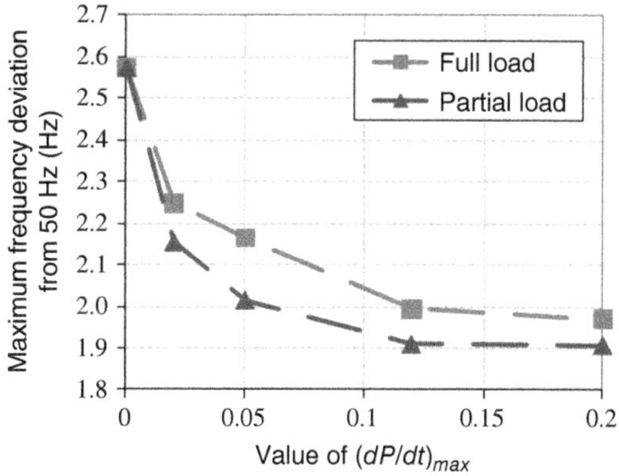

Figure 2.6 Influence of the limit of the ROCOP on frequency response. *Source:* Wang *et al.*, 2013 [18]. Reproduced with permission of IEEE.

The primary frequency response for the given real case study (the Guadeloupe power system) has been studied. The wind primary frequency control has been applied in the scenarios corresponding, respectively, to 20.2% and 29.2% wind penetration rates, as load-shedding was observed after the grid fault (loss of the largest online generator) in the absence of wind participation in frequency regulation. The 29.2% wind penetration case with a high wind speed is taken as an example in this paragraph to illustrate dynamic behaviors of the frequency and WT responses. The power set points of diesel and wind power plants as well as the power reserve distribution are shown in Table 2.2. Without a primary frequency control, all VSWTs produce the maximum available power (25 MW) and the online diesel plants provide 25.2 MW primary reserve. In the second case, 10% of the available wind power is kept in reserve and the VSWT production is deloaded (22.5 MW), which induces an increase in the conventional production, thus a decrease in conventional power reserve (22.7 MW).

As the fixed-speed induction generator (FSIG) wind farms are not able to participate in primary frequency control, their production remains unchanged

Table 2.2 Power reserve distribution for the 29.2% wind case.

	Without primary frequency response	**With primary frequency response**
Power set point of diesel power plants	37.5 MW	40 MW
Reserve provided by diesel power plants	*25.2 MW*	*22.7 MW*
Power set point of wind farms	25 MW (VSWT) + 16 MW (FSIG)	22.5 MW (VSWT) + 16 MW (FSIG)
Reserve provided by wind farms	*0 MW (VSWT) + 0 MW (FSIG)*	*2.5 MW (VSWT) + 0 MW (FSIG)*

(16 MW). This transfer of power reserve from diesel plants to the VSWT causes a slight reduction in wind penetration (from 29.2% to 27.5%). A 4% droop setting is used in our studies. This value should be carefully chosen, as a very low droop setting may induce unstable operating conditions of WTs and an unsuccessful frequency support [14].

Although the total power reserve remains constant in both cases, the minimum frequency is reduced by 250 mHz with a WT primary frequency response (Fig. 2.7a). Then load-shedding previously observed can therefore be avoided (the first stage of load shedding scheme in the Guadeloupe power system is triggered when the frequency decreases below 48.5 Hz). As it can be seen in Fig. 2.7b, the total wind power reserve (0.1 pu = 10%. 25 MW wind power) can be released into the grid in 0.5 s thanks to the fast response of power electronic converters and following the 0.2 pu/s ROCOP setting.

Thus, the active power from WTs can be controlled almost instantaneously compared with diesel plants. As the dynamics of the WT mechanical part is much slower than that of converters (Fig. 2.7d), the mechanical torque stays lower than the electromagnetic torque during the first seconds after the grid fault. In consequence, a transient drop appears in the rotor speed (Fig. 2.7c). In steady state, the turbine speed stabilizes to its rated value while the pitch is kept lower than the initial angle (Fig. 2.7d).

2.2.3.2 Using Primary and Inertial Controllers

The WT primary frequency control can be applied, together with the inertial control, to further improve grid frequency responses [13]. Figure 2.8 shows the maximum transient frequency deviation from the rated value for different wind penetration rates. The simulated scenario is based on the outage of the largest generation infeed during off-peak hours and with a high wind speed. In Guadeloupe, load-shedding starts when the system frequency drops to 48.5 Hz, which is the underfrequency protection relay setting.

Figure 2.8 shows that the inertial control and the primary droop control have "equivalent" contributions for the system, as grid frequency nadirs are almost identical with only one controller enabled. One level of load-shedding in the defense plan could be avoided with either inertial or primary frequency response. So, fewer consumers would experience the outage when the WTs participate in frequency control. If both controllers are simultaneously activated, transient frequency excursions will be further reduced (e.g., two stages of ALS are avoided at 20.2% wind penetration).

The same studies have also been performed for medium wind speeds, which are more realistic in practice. In order not to change the grid OP in this case, the wind power is held constant, and thus, more turbines are connected to produce the same power. As shown in Fig. 2.9, when the inertial control and the primary frequency control are both enabled, the improvement in frequency response is much better compared with the full-load case. At 29.2% wind penetration rate,

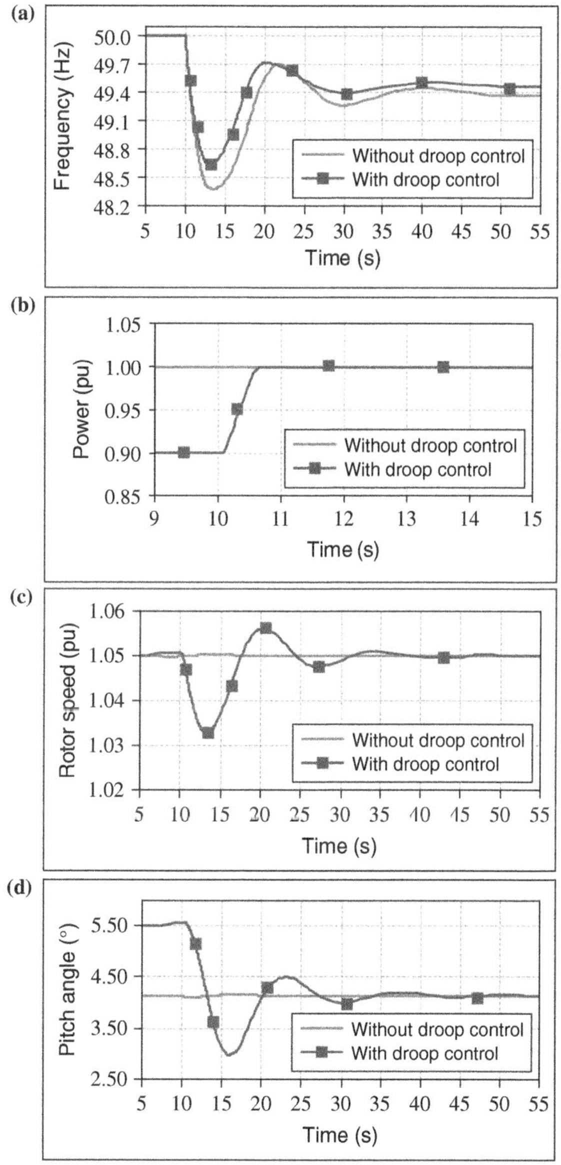

Figure 2.7 Grid frequency and wind turbine dynamic behavior. *Source:* Wang *et al.*, 2013 [18]. Reproduced with permission of IEEE.

the wind combined-control service can help to bring the minimum frequency (after the disturbance) up to about the level of the no-wind case (0% wind penetration, i.e., maximum system inertia).

At partial load and with an inertial response, the turbine speed would need to come back to its optimum value after the extraction of kinetic

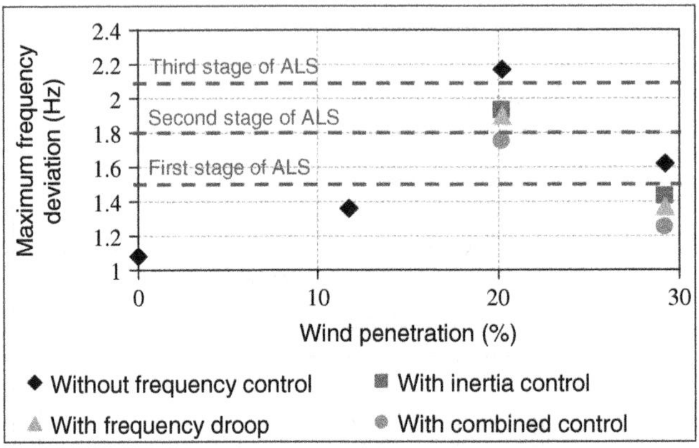

Figure 2.8 Contribution of wind turbine frequency control (full-load case). *Source:* Wang *et al.*, 2013 [18]. Reproduced with permission of IEEE.

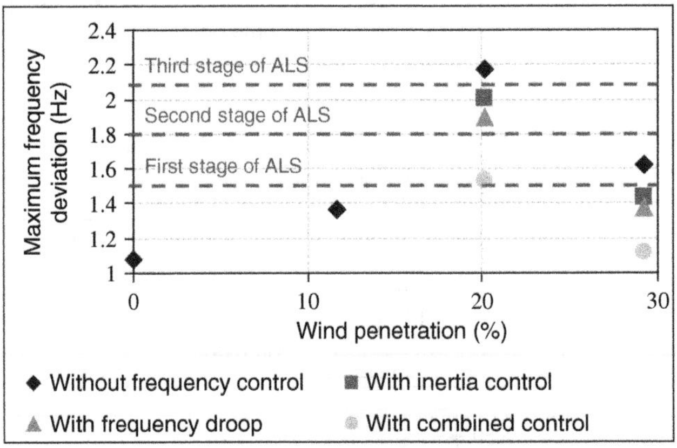

Figure 2.9 Contribution of wind turbine frequency control (partial-load case). *Source:* Wang *et al.*, 2013 [18]. Reproduced with permission of IEEE.

energy (Fig. 2.2d–f). This reduces the output power of the WT as the rotor accelerates. The performance of the inertial response could be modified by this speed recovery phase since other generators must compensate for this temporary – but necessary – reduction in wind active power output. However, the combination with the primary frequency control could solve this problem.

The combined-control response of the WT with both frequency control functions is presented in Fig. 2.10. In fact, when wind power is deloaded, the turbine is forced to operate away from the maximum power extraction curve. Unlike the case in which only the inertial controller is enabled, the initial turbine speed under combined control is higher than the optimum speed (Fig. 2.10b).

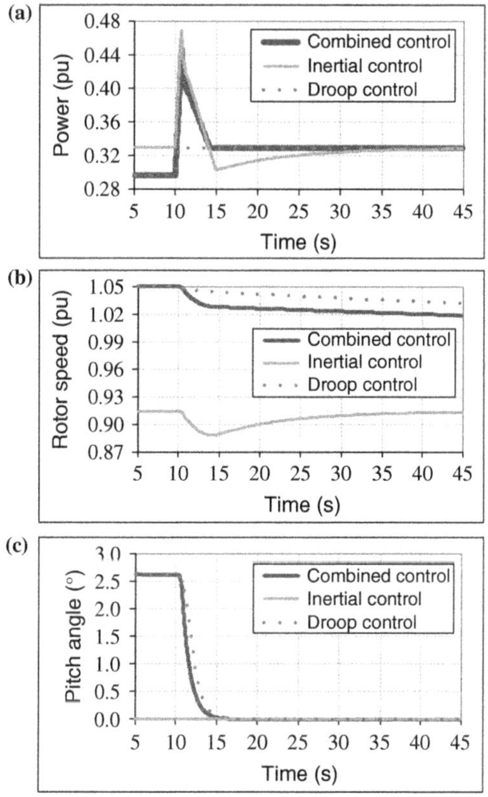

Figure 2.10 Wind turbine dynamic behavior with various control schemes. *Source:* Wang *et al.*, 2013 [18]. Reproduced with permission of IEEE.

During the grid fault, the rotor speed decreases, as a part of kinetic energy is released for frequency support. Then the WT OP will move toward the optimal point with the decrease in the rotor speed (Fig. 2.10b). Consequently, there is no need to recover the rotor speed and the wind power will not be reduced after the inertial response (Fig. 2.10a). From an energy point of view, at partial-load OPs, as the stored kinetic energy is increased when the WT is deloaded, the inertial response is much improved, which leads to a better performance of the combined control scheme compared with the full-load case.

Although natural delays appear for the deployment of the synthetic inertial response of the VSWTs in comparison with the inherent response of classical direct-connected synchronous generators, studies presented in this section have shown satisfying performances and even better contribution to frequency stabilization after the loss of a major generation infeed than with synchronous machines. The performance of the WT inertial response can be improved by either increasing the auxiliary controller gain or relaxing the limit on the rate of change of wind electrical power. However, the benefit is not linear and the WT will suffer more wear and tear in these cases; the economic consequences are still to be assessed. Therefore, a good compromise should be made for appropriate controller parameter settings, which allow WTs to satisfy the grid code but do not impact seriously their own stability and lifetime. Moreover, the capability of WTs to stand repetitive mechanical stresses resulted from inertial contribution should be carefully examined by analysis over a long period before industrial applications.

The initial OP of the WT also influences the inertial response performance. Future works should consider more OPs to better investigate the limitation of inertial response at partial load and quantify performances of inertial response.

The WT primary droop control could also help to increase system robustness by reducing frequency excursions following grid disturbances. The total wind power reserve can be released into the system in less than 1 s thanks to the fast response of power electronic converters. The combined control with both inertial and primary frequency controllers improves the frequency behavior and mainly when the WT is operating at partial load. In this case, the deloading of the turbine leads to an increase in rotor speed. Thus, more kinetic energy will

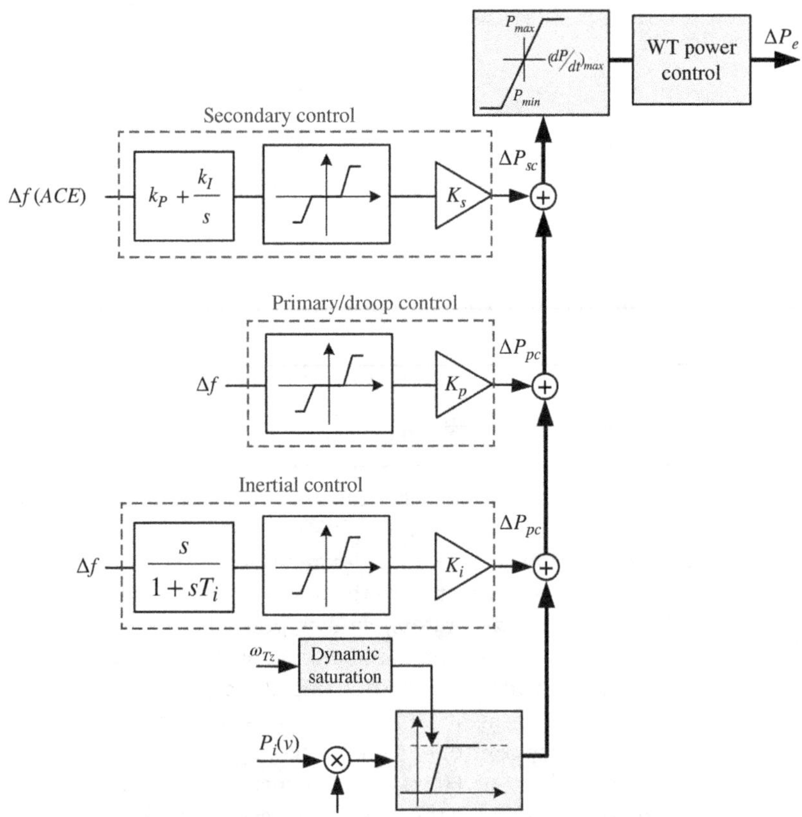

Figure 2.11 Inertial, primary, and secondary frequency control supports by WT.

be stored in rotating masses and the contribution of the inertial response can be much improved.

2.2.4 Using Secondary Control

In addition to inertial and primary/droop control, the large WTs may also support the secondary frequency control [1]. Concerning this issue, the WT control structure shown in Fig. 2.1 can be updated as depicted in Fig. 2.11. Figure 2.11 shows a conceptual framework for integration of possible frequency regulation loops in the control system of a variable-speed WT (e.g., rotor-side converter of a DFIG). This framework includes inertial control loop, primary frequency control loop (responding to frequency changes using the droop control concept), and a secondary or external supervisory control loop [1]. Parameter α determines the amount of reserve power for frequency control support and $0 \leq \alpha \leq 1$. The secondary (external supervisory) control is required to respond to the request of the system operator. The *ACE*, v, P_i, ω_T, ΔP_{ic}, ΔP_{pc}, and ΔP_{sc} are area control error, wind speed, available power, turbine blade speed, inertial control, and secondary control, respectively. The mentioned control loops can be implemented as shown in the dashed boxes shown in Fig. 2.11. Interested reader can follow the present issue in [1].

2.3 Renewable Primary Power Reserve

In order to analyze the contribution of wind/solar power to primary frequency regulation at the system level, one needs to quantify the amount of primary reserve from conventional sources that can be displaced. This amount of reserve depends on the aggregated variability of wind/solar generation during each reserve provision time interval. This section presents a statistical approach to assess the impact of intrahour wind power variability on the volume of primary reserve that can be provided from wind generation. Furthermore, the effectiveness of different reserve allocation strategies is compared. The proposed approach is applied to the French Guadeloupe island power system case study based on real wind data measurements. Results show that for a small isolated system neglecting wind generation intrahourly variability leads to an overestimation of its contribution to primary reserve. Same study can be easily done for the solar primary power reserve.

2.3.1 Instantaneous Wind Power Reserve

Wind upward reserve strategies use the part of the maximum output that is curtailed, as shown in Fig. 2.12, to displace some of primary reserve from other sources. The first reserve allocation strategy proposed is the "proportional curtailment strategy" (PCS). This strategy leads to a curtailed power ($P_{curtailed}$) proportional to the maximum available power (P_{avail}) (Fig. 2.12a). The

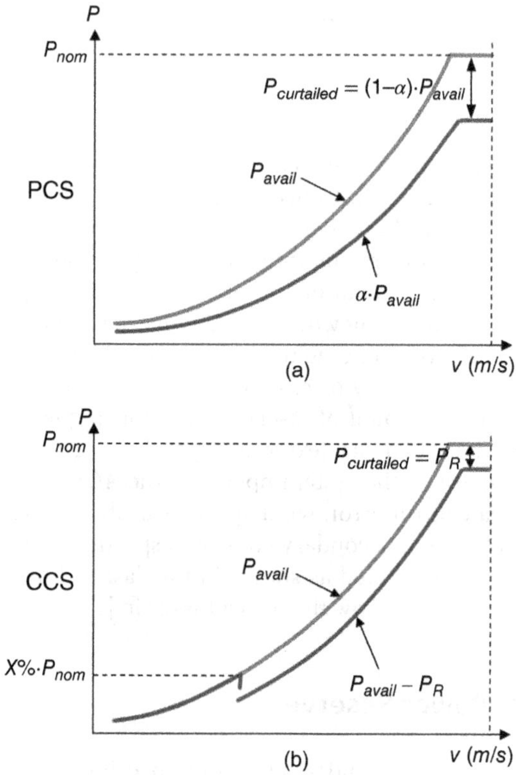

Figure 2.12 Operating principle of both reserve allocation strategies.

parameter α may vary theoretically from 1 (no reserve) to 0 (total curtailment).

$$P_{curtailed}(t) = (1 - \alpha) \cdot P_{avail}(t) \tag{2.5}$$

Dynamic simulations have shown that with an adequate controller; the WTs are able to keep a percentage of their available power as primary reserve throughout all the operating zones. The value of α can be defined so that a certain amount of wind reserve is provided. This amount of reserve can be obtained from system stability studies or the SCUC. The PCS is easy to implement and provides a continuous variation of the reserve power. The curtailed power changes according to wind variability.

The second strategy is called "constant curtailment strategy" (CCS). The wind power is curtailed in order to get a constant reserve power as for conventional generators (Fig. 2.12b):

$$P_{curtailed} = P_R \tag{2.6}$$

This power curtailment is possible over a certain output level. For example, if the wind generation is less than $X\%$ of the nominal power (P_{nom}), no reserve will be provided in this low-power operating zone. The technical feasibility of the CCS should not raise problems, as it is part of the regulation capacities required by the TSO for the specified offshore wind farm [34]. The CCS enables a finer control of the curtailed WG; however, its application leads to a discontinuity of the wind farm power generation when enabled. Therefore, further comparison of these two reserve allocation strategies should be performed by considering the impact of their application on the instantaneous reserve availability.

Traditionally, primary reserve is provided from conventional generators. In some power grids [35], this reserve is scheduled a day ahead for every half-hour period. The day-ahead scheduled power set points of power plants and primary reserve allocation can be modified in intraday if necessary.

Regarding the technical requirements of primary reserve, some grid codes [35] require that primary power reserve is fully available in less than 30 s and its provision must be held for 15 min, after a power imbalance. The 15-min duration of the primary frequency control is considered necessary to enable the restoration of the primary power reserve by the tertiary frequency control [32, 36].

The application of current rules needs to be analyzed when WTs contribute to the primary reserve, as their curtailed power is variable during each dispatch step, especially when the PCS is applied. In order to ensure the availability of wind reserve during at least 15 min, it is essential to consider the minimum value of the curtailed wind power during each dispatch time step, as illustrated in Fig. 2.13. Thus, the notion of the instantaneous available reserve (R_{avail_H}) for the corresponding scheduling period (H) can be defined as

$$R_{avail\_H} = \text{MIN}(P_{curtailed}[t_0, t_0 + H]) \tag{2.7}$$

This means that for a given time interval (15 min, 30 min, 1 h, etc.), a constant amount of wind reserve should be scheduled by taking into account the lowest wind power output expected during each time interval.

Considering the example of Fig. 2.13, it is possible to see that instantaneous curtailed wind power cannot be directly translated as the "firm" available

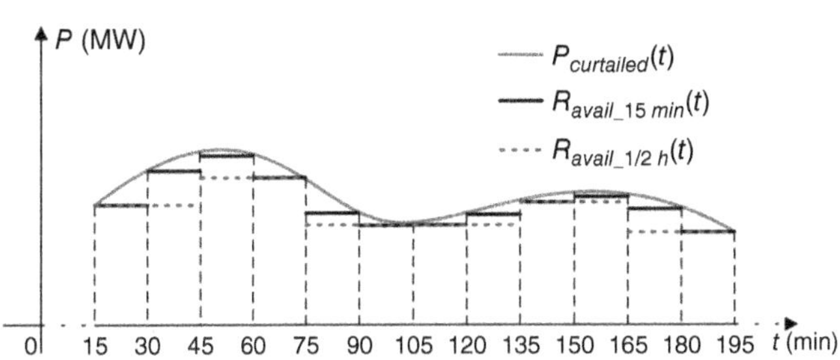

Figure 2.13 Wind variability while providing reserve from wind farms with the PCS (for 15- and 30-min time steps).

reserve that meets the technical requirements. The difference between the curtailed power (gray line) and the "firm" available reserve (Fig. 2.13) (solid and dotted step lines) represents the amount of curtailed generation that does not contribute to the "firm" available reserve ($MWh_{curtailed,unused}$). This results in a loss to the wind farm operators and to the system. In order to measure the quality of the WT reserve allocation strategies, an efficiency indicator (η) can be defined as follows:

$$\eta = \left(1 - \frac{MWh_{curtailed,unused}}{MWh_{curtailed,total}}\right) \tag{2.8}$$

where $MWh_{curtailed,total}$ is the total curtailed WG during the scheduling interval.

For a given reserve allocation strategy, the higher corresponding efficiency indicator and the lower "unused" curtailed energy are desirable. The efficiency is used in the following section as a metric to compare the performance of the mentioned two proposed strategies.

2.3.2 An Evaluation on the Real Case Study

The approaches described in the previous section are applied to the French Guadeloupe island case study (Fig. 2.3). The installed wind capacity was about 27 MW in 2010 and the load demand is 260 MW. In order to benefit from the high potential for wind energy in this island, previous studies showed that some primary reserve has to be supplied by wind farms. This is true if we want to operate the system above a certain "critical" wind penetration rate. This is needed since for these penetration rates, the remaining conventional generators cannot provide enough reserve to cover the most critical incident and the system will encounter dynamic frequency problems due to the reduction in inertia [18].

The statistical studies described hereafter are based on the analysis of historical instantaneous wind power data from three wind farms. Data were recorded with 15-min steps from January 1 to March 31, 2011. It should be noted that the performed study is based on a rather limited data set and the results presented have the purpose of validating the methodologies by using real measurements. The data set is enough for characterizing the potential of the instantaneous available reserve that can be provided from a single wind farm and from a group of farms (referred to as "aggregated wind farm"). These three farms with 2.1, 3, and 3.3 MW rated powers present a good geographical dispersion in the archipelago. This diversity leads to an important smoothing effect at the aggregated farm level. The installed capacity (IC) of the "aggregated wind farm" (8.4 MW) represents about one-third of the wind power capacity of Guadeloupe (27 MW).

The cumulative distribution function (CDF) of the Guadeloupe's 15-min wind variability is shown in Fig. 2.14. The results reveal that 99.5% of the 15-min power fluctuations of an individual wind farm can reach more than

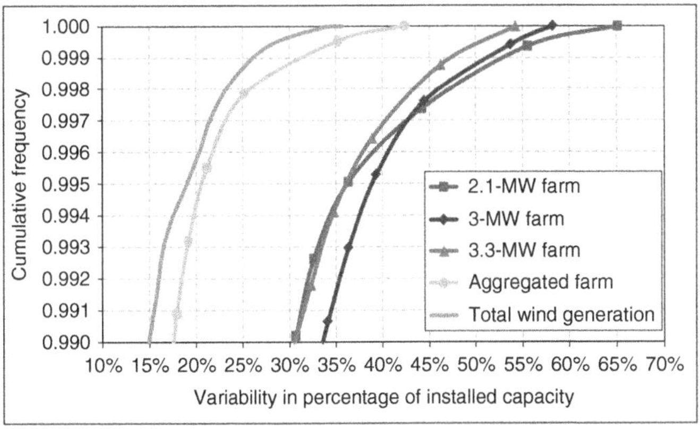

Figure 2.14 The CDF of wind power fluctuations at 15-min intervals (zoom on the last percentile).

35% of its IC. The 15-min fluctuations are reduced to 20.5% and 19% of the IC for the aggregated farm and for the whole island, respectively. The variability of Guadeloupe's wind power is significant due to the small size of the island.

The variability of the "aggregated farm" is close to that of the total grid-connected wind power and much lower than that of a single WF. Hence, in order to take into consideration the geographical smoothing effect, it is important to perform system-wide studies, even for small isolated power grids.

The performance of the WT primary frequency control is characterized by assessing the maximum instantaneous "firm" available reserve, obtained by using different strategies. This "firm" reserve takes into account the possible "waste" of curtailed energy due to wind variability inside a scheduling interval. The objective is to find out which strategy leads to a higher instantaneous "firm" reserve, while incurring the same energy curtailed to the "aggregated" wind farm.

The PCS was first applied by considering a curtailment of 10% of the maximum instantaneous available WG of the "aggregated farm." As wind power varies, the obtained curtailed power also varies during each 30-min dispatch time step. In order to ensure the "firmness" of wind reserve, the available reserve was calculated by considering the minimum value that is observed for each 30-min period (Fig. 2.13 and (2.7)). The amount of reserve is therefore smaller than that of the curtailed power, as shown in Fig. 2.15.

The analysis of the "firm" reserve contribution of the aggregated wind farm (Fig. 2.15) with the PCS with a curtailment of 10% of the IC shows that the amount of reserve (that can be allocated to wind) is higher than 2.5% of the IC

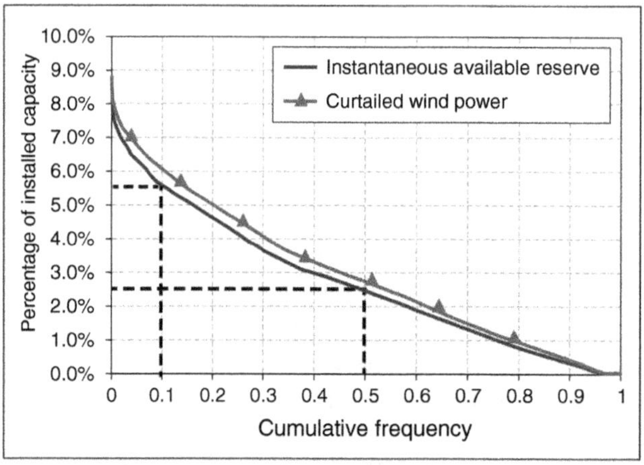

Figure 2.15 Duration curve of the instantaneous available reserve and the curtailed power of the aggregated wind farm (application of the PCS).

of the aggregated farm during 50% of the time and higher than 5.5% of the IC during 10% of the time.

In order to be able to compare the CCS strategy with the PCS, the strategy parameters are defined so that the same amount of curtailed energy is obtained. One possible combination of parameters is to curtail a constant power equal to 5% of the rated capacity of each wind farm whenever its power output is higher than or equal to 20% of its rated power, that is, 0.2 pu. Thus, the total curtailment from the three wind farms, during the considered period (from January to March), is almost identical using both reserve allocation strategies (542 MWh with the PCS, 539 MWh with the CCS).

With the CCS, the amount of reserve provided by a single wind farm is binary. It is equal to either 5% of IC (i.e., the value of the curtailed power) or zero. Since the three sites are subjected to different wind speeds, the reserve availability is switched at different frequencies. Therefore, the duration of the "firm" available reserve of the aggregated farm includes 2^3 "steps," which correspond to the number of discrete states of individual farm's reserve. However, in practice, wind reserve should be allocated to each turbine. For a given number of turbines (n), the "real" reserve characteristic would contain 2^n "steps" and would thus be much smoother. From the instantaneous available reserve curve (shown in Fig. 2.16), we can conclude that the implementation of the CCS leads to a reserve amount of the aggregated farm superior to 3% of its rated power during 50% of the time and a constant reserve amount (equal to 5% of IC) during 37% of the time.

Since the wind power variability of the "aggregated farm" is quite close to that of the total wind generation in the system (Fig. 2.14), it is assumed that

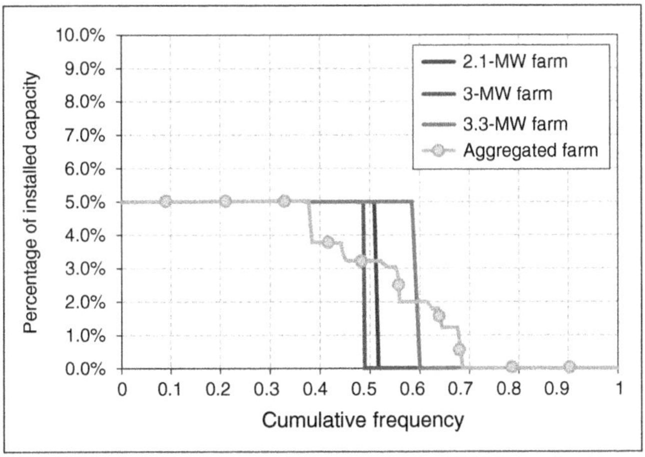

Figure 2.16 Duration of the instantaneous available reserve of the three wind farms and of the aggregated farm (application of the CCS).

Table 2.3 Estimated amount of instantaneous available wind reserve on the Guadeloupe island.

Frequency and reserve	PCS		CCS	
Cumulative frequency (%)	50	10	50	37
Instantaneous available reserve (pu)	>2.5	>5.5	>3.0	5.0
Reserve amount in (MW = with 70 MW of IC)	>1.75	>3.85	>2.1	3.5

the results could be generalized to all WFs in the Guadeloupe island (under the same average wind speeds). For example, as shown in Table 2.3, for the future scenario of 70 MW wind capacity, the potential of the instantaneous available reserve that could be provided by wind power depends on the applied strategy. Using the PCS, the reserve amount is greater than 3.85 MW for 10% of the time. This contribution represents more than 15% of the minimum primary reserve currently required (about 25 MW). With the CCS, the reserve amount is equal to 3.5 MW for 37% of the operation time, that is, 14% of the primary reserve requirement.

This study shows that wind power can provide a significant potential of instantaneous primary reserve in spite of its short-term variability. With an expected considerable IC, the contribution of wind power to power reserve could therefore cover, at least in part, the growth of the grid minimum reserve requirements.

2.3.3 Comparison of the Reserve Allocation Strategies

2.3.3.1 Reserve Quantification Over an Operating Period

As a first step, it is assumed that power reserve is distributed to WTs all the time throughout the whole studied period. As explained before, in order to curtail the same amount of energy in the "aggregated farm" with both strategies, a curtailment of 10% of the instantaneous available power with PCS and 5% of each individual farm IC are required for periods when its output exceeds 0.2 pu with CCS.

The efficiency indicators for each strategy (the aggregated and individual WFs) were first calculated according to (2.8) and are shown in Table 2.4. With the PCS, the reserve allocation efficiency, at the aggregated farm level, is higher compared to that at a single farm level. With the CCS, the efficiency is almost identical for the aggregated and the WF levels. This shows that the PCS benefits more from the smoothing effect; therefore, reserve allocation at a large power system level using this strategy would be more efficient.

The duration characteristic of the instantaneous "firm" available reserve provided with both reserve allocation strategies, depicted in Figs 2.15 and 2.16, cannot be directly compared, as they represent only a statistical distribution of the amount of reserve for a given frequency of occurrence. In order to perform this comparison, an average reserve ($\overline{R}$) is defined as follows:

$$\overline{R} = \frac{\sum R_{avail\_H} \times H}{\text{total hours}} \tag{2.9}$$

where H is the dispatch time step.

For an equivalent amount of wind energy curtailment, both strategies lead to a similar amount of primary reserve, during the studied period (Table 2.5) and present almost identical efficiency at the scale of several farms. This comparison is based upon a reserve allocation throughout the whole operating period and shows that both strategies are equally efficient.

Nevertheless, the participation of WTs in primary frequency control can be discussed when their output is low, as other conventional generators are connected to the grid and would provide sufficient primary reserve. However, the system may require an enhancement of primary frequency control due

Table 2.4 Efficiency indicators of the reserve allocation strategies.

Efficiency indicator	2.1-MW farm (%)	3-MW farm (%)	3.3-MW farm (%)	Aggregated farm (%)
η_{PCS}	86	85	88	91
η_{CCS}	90	88	91	91

Table 2.5 Comparison of the reserve allocation strategies within the aggregated wind farm during the overall studied period.

Strategy	Curtailed production (MWh)	Average reserve (of installed capacity) (%)	Efficiency indicator (%)
PCS	542	2.7	91
CCS	539	2.7	91

to dynamic constraints and then sources that have a faster response, such as wind farms, are required.

2.3.3.2 Reserve Potential with Wind Reserve Participation

In this section, the performance of both strategies when wind reserve is required only during "critical" periods with an instantaneous wind penetration above a critical value is assessed. The core of method is to identify the required period of wind power reserve allocation, from the overall operating time of wind farms.

In a power system, for a given demand ($d(t)$ in MW), the system operator can identify the corresponding critical instantaneous wind penetration rate τ_c. Above this rate, the contribution of WTs to primary reserve is essential for system security. In fact, when the penetration level of wind power is increased, some of the conventional power plants that would have been online without wind power could be disconnected. Beyond an accepted "maximum" penetration rate, the required minimum reserve may no longer be guaranteed with a small number of conventional generators and should be partially allocated to WTs. In order to generalize results, the method to identify the periods for which wind contribution to primary reserve is critical and detailed in [37] is used.

From an operational point of view, if the power produced by the wind farms (P_w) exceeds the critical wind generation ($d(t)\cdot\tau_c$), a certain amount of primary reserve should be provided by the WTs. The overall *OP* of the total wind farms can be defined as the ratio of the produced wind power (P_w) to the total IC (P_{inst}):

$$OP(t) = \frac{P_w(t)}{P_{inst}} \tag{2.10}$$

Hence, it can be deduced that WTs should provide primary reserve when the overall $OP(t)$ exceeds a critical value ($OP_c(t)$) that can be defined as follows:

$$OP_c(t) = \frac{d(t) \cdot \tau_c}{P_{inst}} \tag{2.11}$$

Otherwise, the contribution of wind farms to the primary frequency control could not be mandatory. This method enables the identification of periods corresponding to high WG during which the primary reserve provided by WTs has a greater value for the system operator.

Ideally, from an operational point of view, the critical wind penetration rate should be calculated for each scheduling period by using a multiperiod SCUC solved with a fine time resolution (<15 min). For small systems, this should be coupled with stability studies. The simplified approach used in this work is, however, deemed sufficient to perform the analysis on reserve allocation strategies that minimize the impact of intrahourly variability.

The considered scenario corresponds to a recorded off-peak consumption. As shown in [37], the wind penetration rate in Guadeloupe should be limited to 29% ($\tau_c = 29\%$) in order to guarantee the minimum reserve required for system security. This critical rate is obtained by considering a simplified optimized unit commitment of the Guadeloupe system with linearized frequency limits. This "worst" case corresponds to the minimum load with $d(t) = 140$ MW. According to (2.11), the critical OP of the Guadeloupe wind farms can be obtained from

$$OP_c = \frac{140 \times 29\%}{P_{inst}} \text{ (Guadeloupe)} \tag{2.12}$$

The obtained results for different installed wind capacities are shown in Table 2.6. As almost the same 15-min wind variability was observed for the "aggregated farm" and for the total WG (Fig. 2.14), the critical OP of the "aggregated farm" is assumed to be equal to that of the total wind farms in Guadeloupe for the sake of simplicity. For a given wind IC, primary reserve will be allocated to three WFs, whenever the OP of the aggregated farm is higher than the above-calculated OP_c. The increase in the wind IC leads to a higher possibility of having a strong wind power output. Accordingly, the OP_c is smaller for large wind power capacities. This means that the higher the IC the higher the number of scheduling periods when WTs should contribute to primary reserve.

To compare PCS and CCS reserve allocation strategies, the same previous parameters are used: 10% of the maximum instantaneous available wind power is curtailed when PCS is applied, while the amount of constant reserve kept with the CCS is adapted for each installed wind power capacity. This leads to a total

Table 2.6 Critical operating points of the Guadeloupe system as a function of the total installed wind capacity.

Installed wind capacity (MW)	70	80	90	100	110	120
OP_c (%)	58.0	50.8	45.1	40.6	36.9	33.8

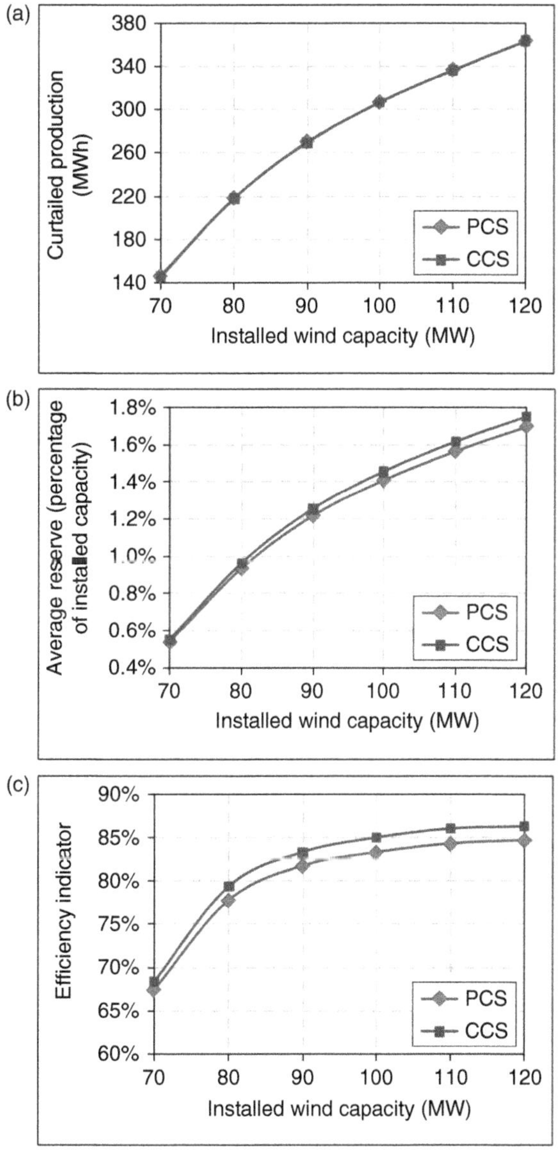

Figure 2.17 Comparison of the reserve allocation strategies during the required period: (a) total curtailed WG, (b) average reserve in percentage of the installed capacity of the aggregated farm, and (c) efficiency indicator.

curtailed wind power energy, during the required reserve allocation period, equivalent for both strategies (see Fig. 2.17a). Figure 2.17b,c presents the average upward reserve provided by wind power and the efficiency indicators for each strategy. These values increase with the total wind power capacity. The results shown in Fig. 2.17 indicate that both strategies have similar efficiencies.

The obtained results show that both strategies seem equally "efficient" for real-time reserve allocation and allow providing almost the same amount of average reserve by curtailing identical values of wind power. The performance of the strategies has been confirmed for both cases with reserve provided for all scheduling periods or only during periods of high winds. Each strategy seems to have its own interesting features. It raises the question whether a combination of both strategies can benefit from the advantages of each strategy and thus leads to a more efficient reserve allocation.

In this section, a "combined" approach is proposed, in which the CCS is applied at high wind speeds and the PCS when the power output of WTs is relatively low. This strategy presents the following advantages:

1) It allows a good control of the wind reserve quantity at each dispatch time step during high wind situations, that is, during the period when the participation of WTs in the primary frequency control is required.
2) Reserve could be provided from WTs at any time without incurring a discontinuous wind power output, which would also maximize the wind farm frequency response characteristics (MW/Hz) [1] with the participation of all turbines.
3) The application of the PCS at lower levels of production can take advantage of the smoothing effect and leads to a better global efficiency of the reserve allocation.

This combined strategy is applied to the Guadeloupe case study in order to compare its performance with single strategies (PCS or CCS alone). The parameters of the combined strategy are 5.5% of the rated capacity (P_{nom}) of each farm curtailed when its power output exceeds 0.5 pu and 11% of the maximum instantaneous power curtailed when its production is below 0.5 pu (Fig. 2.18). In this way, the total wind energy curtailed in all the three WFs remains the same as in previous cases (about 540 MWh in 3 months).

Figure 2.19 illustrates the duration characteristic of the "firm" reserve, which can be provided by the "aggregated wind farm" when the combined strategy is implemented. From the figure, an amount of reserve higher than 2% and 4% of the IC can be provided during, respectively, 60% and 30% of the time.

The efficiency indicator of the combined reserve allocation strategy is calculated by using (2.8) (Table 2.7) and is compared with the PCS and the CCS (Table 2.8). The results show that with the proposed combined strategy, it is possible to obtain a higher "firm" wind reserve from the "aggregated farm" while curtailing less WPG.

Although only a specific combination of the PCS and CCS has been studied in this section, improvement in terms of efficiency can be observed. This combined approach seems interesting for enlarging wind reserve potential and

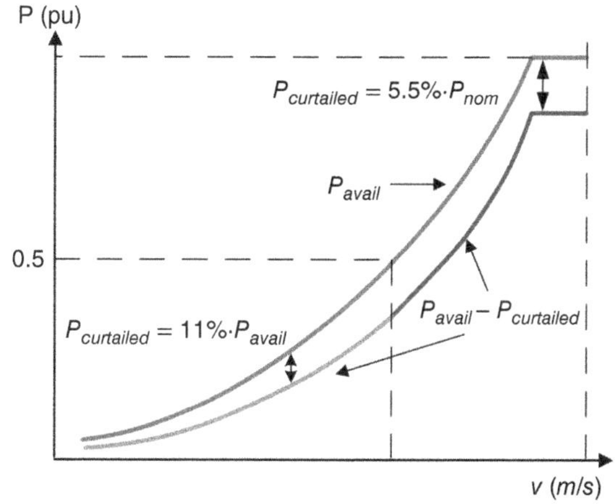

Figure 2.18 Proposed combined strategy.

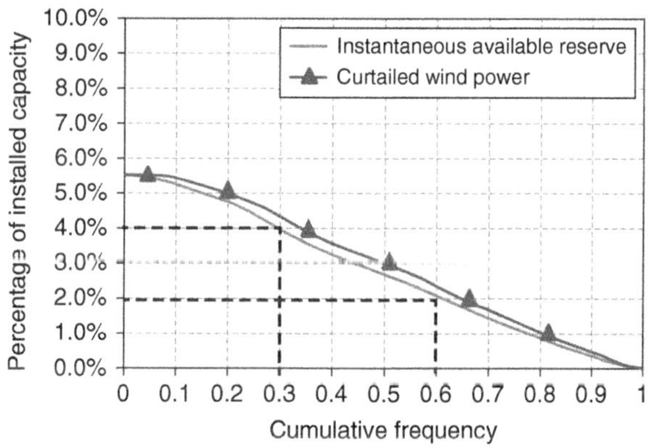

Figure 2.19 Duration curve of the instantaneous available reserve and of the curtailed power of the aggregated wind farm with a 30-min time step (application of the combined reserve allocation strategy).

Table 2.7 Summary of the main characteristics of the proposed combined strategy.

	Curtailed production	**Average reserve (of installed capacity) (%)**	**Efficiency indicator (%)**
Combined strategy	535 MWh	2.75	93

worth being taken into account while choosing the reserve allocation strategy to be applied to wind farms. However, it should be noted that the actually implemented WT controllers are generally not sophisticated enough and in most cases a very precise parameter setting of different reserve allocation strategies could not be implemented.

Table 2.8 General comparison of the three reserve allocation strategies.

Strategy	Efficiency	Implementation simplicity	Suitability for wide range sites	Control of curtailed WG
PCS	+	++	++	−
CCS	+	+	+	++
Combined strategy	++	−	−	+

2.4 PV-Energy-Based Control Support

In order to maintain power system reliability, the natural intermittent characteristics of primary renewable energy must be taken into account, requiring a precise estimation of their ancillary service provision in advance. If necessary, other utilities, such as ESSs or fast dynamic generators, must be planned. Hence, the integration of SPGs and WPGs into an electrical system implies the extension of the actual hierarchical power management system that takes into account different timescales and economic dispatch. It has to pave the way to develop new interactions between real-time pricing, physical constraints, and demand response (DR)-based loads.

At present, and even more in future smart grid (SG) scenarios [27], the SPGs can receive operation references, mainly active and reactive power references to generate/consume energy for delivering ancillary services, from the corresponding electric system agent. The SPG central control systems will have to translate these global references into individual operation references for each inverter within the plant and taking into account the actual state of all the inverters and weather forecast. Therefore, inverters will be the main devices that must be able to interact with the grid and with the ESSs for providing these services in a proper way. It is shown that the SPGs are able to provide the ancillary services with the right topology, additional systems, power or current ratios, proper control algorithms and communication systems [38].

As discussed in Chapter 1, most SPGs operate as current sources and inject the energy produced by PV cells tracking the maximum power point (MPP) into the grid [38]. The electronic converters generate a reference current that is sinusoidal, in phase with the grid voltage and with the required root mean squared (RMS) value to achieve the operation in the MPP. To be used as an alternative to conventional energy sources, SPGs also have to provide ancillary services including regulation of active power and frequency, reactive power and voltage, as well as harmonic compensation. In order to achieve this, most of the distribution system operators (DSOs) are establishing various codes to regulate how these plants should operate.

If it is assumed that the distribution grid voltage is sinusoidal, then the following current components can be defined, according to power definitions given by IEEE-Standard 519 [39]:

- An active or direct current component, i_P, which is related to the average active power (P)
- A reactive or in-quadrature component, i_Q, related to the fundamental reactive power (Q)
- A harmonic component, i_H, related to the harmonic power (H).

Usually, only the active current component is controlled in the conventional SPGs and others are set to zero.

In a modern power system context, where using active distribution grids is a strategic priority, the SPGs offer a unique opportunity to increase the operation possibilities of energy generators as they are distributed sources that have free capacity (because most of the time they operate under its nominal power, mainly at night) to be used to help grid management. If we consider an ideal PV generation day where the maximum power reaches the 90% of the nominal power ratio of the inverter, we can discount the active component from the nominal current obtaining the available current (nonactive current), I_{NA}, given by

$$I_{NA}^2 = I_N^2 - I_P^2 = I_Q^2 + I_H^2 \tag{2.13}$$

In the proposed case of Fig. 2.20a, I_{NA} (Fig. 2.20b) could reach values that, depending on the time of the day, vary from 40% to 100% of the nominal current. Chapter 1 discussed how the active component can be controlled by using ESS. In this section, the attention is focused on how this nonactive current component availability can be used to generate other current components, such as reactive or harmonic ones.

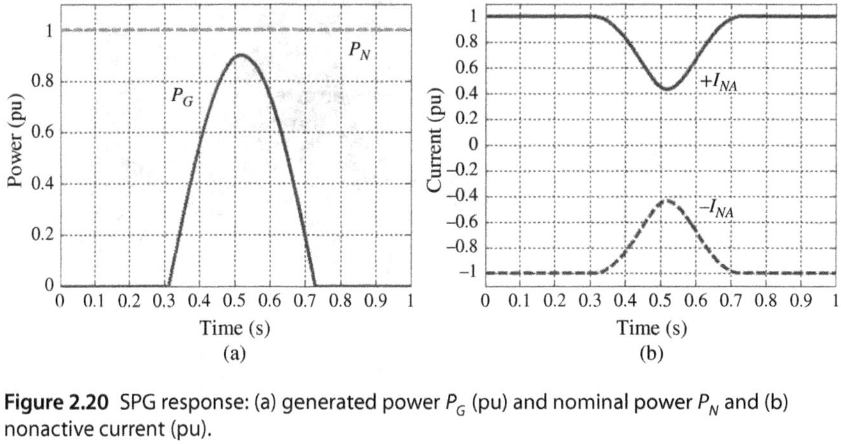

Figure 2.20 SPG response: (a) generated power P_G (pu) and nominal power P_N and (b) nonactive current (pu).

By controlling the reactive component, it is possible to stabilize the grid via controlling the grid frequency [40] or the voltage [41] at the point of common coupling (PCC) where the SPG is connected. By controlling the voltage at this point, it is also possible to prevent overvoltages and reduce grid losses [28]. Grid losses can be reduced further if an optimization strategy is applied in multibus systems [42].

Therefore, the reactive component can be used during abnormal grid or transient states, mainly during faults or after-faults, in order to compensate voltage sags. In three-phase balanced voltage sags, the SPG should inject reactive power in order to raise the voltage in all phases. In one- or two-phase faults, the main concern of the SPG is to equalize voltages by reducing the symmetric negative and, if possible, the homopolar (or zero) sequences [41].

The SPGs can generate harmonic components to prevent the effects of the current harmonics produced by nonlinear loads connected to the grid, operating as integrated active power filters [43, 44]. The SPGs could even compensate current unbalance, control the leakage and earth current [44, 45] avoiding the propagation into the grid. For example, in the case presented in Fig. 2.21, the current produced by the SPG is not sinusoidal because it contains the harmonic components demanded by the load, achieving in this way a sinusoidal current flowing into the grid (and so a harmonic compensation). In some cases, dual or collaborative topologies are used, associating the different current components under control to different converters [44]. When the active, reactive, and harmonic components generated by the SPG are greater than the available non-active components, some kind of priority has to be established. This operation

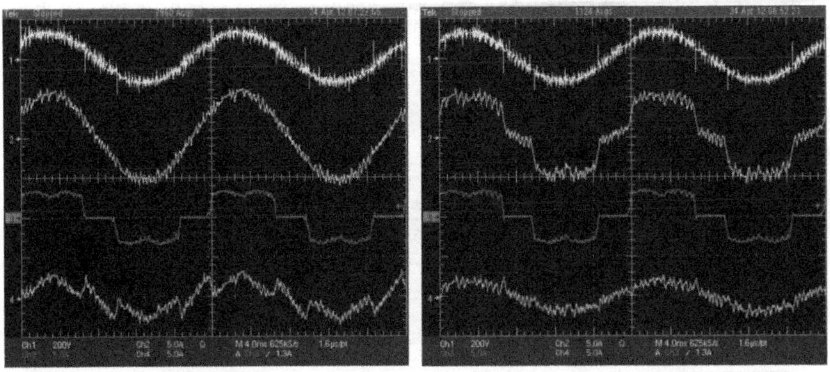

(a) (b)

Figure 2.21 An example for injecting the harmonic component: (a) harmonic injection turned OFF and (b) harmonic injection turned ON. Curves from top to bottom: phase-to-neutral grid voltage, inverter current, local load current, and current flowing to the substation.

priority could be dependent on the status of the grid, being different in normal and abnormal states.

The possibilities of SPG as a manageable distributed generation are much more interesting than the conventional centralized energy resources. For example, in the given system of Fig. 1.30, the SPG could generate the reactive power for controlling the voltage at the PCC or to compensate the reactive power consumed by connected loads. The SPG can also generate the harmonic component needed to compensate the harmonic components of current i_{L1} and i_{L2}, which prevent these harmonic currents flowing from the substation. By properly controlling the SPG, we can assure that the current demanded from the grid, i_{Grid}, will be sinusoidal and in phase with the substation voltage, achieving a unity power factor, reducing distribution losses and preventing any undesirable effect on the substation transformers.

As previously discussed, most of the SPGs operate as current sources to inject the energy produced into the grid. When the penetration level of SPG reaches a certain level, the SPGs should also provide ancillary services. As demonstrated, this can be achieved by controlling the current injected into the grid. Another way is to control the grid-connected inverters to behave like conventional synchronous generators so that SPGs can interact with the grid in an effective way [46–48].

2.5 Integration of Renewable Energy Systems Through Microgrids

2.5.1 A Solution for Renewable Power Penetration

Today, central electricity production still dominates electricity production because of many issues such as economy of scale, efficiency, fuel capability, and lifetime. However, the advantage of the central production's economy of scale is decreasing, because fossil fuels, which are economically suitable for central production, are not abundant enough for the next century (without steady supply or stable price). Recently, the world has shown its great interest and ambition of integrating RESs and DGs into the transmission and distribution network (electrical grid) because of many advantages. These advantages include additional energy-related benefits (improved security of supply, avoidance of overcapacity, peak load reduction, reduction of grid losses) and network-related benefits (transmission network infrastructure cost, power quality support, reliability improvement). Moreover, RESs/DGs can increase the overall fuel efficiency of the plant considerably as the heat can be used locally.

Except for large-scale hydro, offshore wind farms and cocombustion of biomass in conventional (fossil fueled) power plants, most RESs are small DGs that are connected to a grid through an MG. In the last decades, electric power systems undertook several modifications toward a more decentralized energy system paradigm, allowing an increase in RESs integration.

The MGs comprise dispersed energy resources (such as WTs, PV panels, and microgas turbines (MGTs)), storage devices (such as flywheels, supercapacitors, and batteries), and controllable loads in order to offer considerable control capabilities to the local network operation. These systems are interconnected to the low-voltage distribution network, but they can also be operated in the islanding mode in case of faults in the upstream network. From the customer point of view, the MGs provide both thermal and electricity supplies, as well as enhance local reliability, reduce emissions, improve power quality by supporting voltage and reducing voltage dips, and potentially reduce costs of energy supply.

With the fast development of distributed RESs, many efforts should be done in domains of (i) modernization of *transmission and distribution networks* (such as SGs) for integration of distributed RESs and MGs, (ii) innovation of the local system's *management and control concepts* (such as MGs) for distributed RESs' local demand and supply optimization, and (iii) implementation of *controllable and reliable generation* (such as active generators) to overcome the intermittent and fluctuant availability of renewable energies and to supply ancillary services to the electrical network.

Research activities are more and more headed toward solutions for meeting the ever-growing energy demand. In order to ensure a continuous development in a sustainable way, a considerable portion of the electrical energy has to be generated by the RESs. One of their main drawbacks is the nonconstant nature of the primary energy sources (e.g., solar irradiation and wind changes). Hence, the increasing penetration of renewable generators into the energy mix could cause difficulties for system operators in matching the power production and demand thus degrading the quality of power supplied to the customers and further causing disruptions in power supply. Moreover, further investments have to be made in conventional generators to create additional power reserve for compensating the cyclic and stochastic nature of renewable energy.

Alternative to grid reinforcement could be a restructuring of the power system architecture and an increase in the share of RESs that generate electricity at a local scale. Hence, a combination of DGs such as MGTs-based combined heat production (CHP), diesel generators, wind and PV generators and ESSs plays an essential role in domestic small-scale electricity generation. This combination performs a hybrid power system, which is known as MG system.

The MG can operate in both connected mode with a distribution network and islanding mode. The microgrid central control (MGCC) controls the MG operation through the local generator controllers (Fig. 2.22). In the connected operation mode, a timing power planning is established between the DSO and the MGCC. The MGCC adapts the power reference of some DGs (e.g., gas microturbine and diesel generator) in order to fulfill this contract. In the islanded

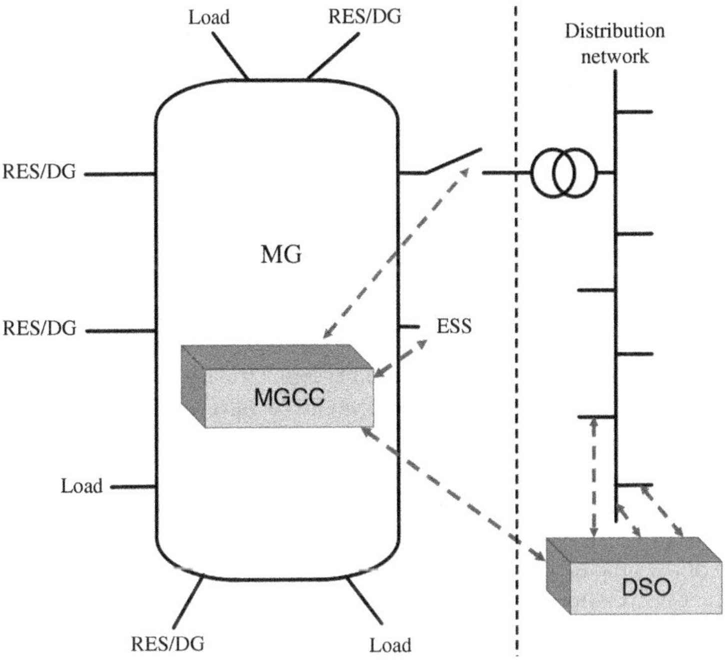

Figure 2.22 A microgrid example.

operation mode, the MG should ensure the local electricity supply and energy security.

To perform the power balance between the production and consumption, the MGCC should predict the local production capacity and the local consumption needs for the next period. If the production is more than the consumption, the power production of the undercontrolled DGs must be reduced, and during this transient state, the excess power is dissipated in dummy resistor loads in case of overvoltage. If the production is less than the consumption, two solutions exist. The specified DGs will be prepared to supply the deficit power or a part of loads will be disconnected from the MG (if necessary).

Some wind generators cannot be dispatched by the MGCC, because their output active and reactive powers are not controllable. In an isolated MG if a wind generator is used, the electrical production may exceed the consumed load power. This event may occur at night when the loads are very low. This is a typical problem, which is a barrier for the integration of wind generator in the grids.

For example, one desirable control action is to be able to reduce the wind power production when it is not useful for the MG operation. In order to increase the value of renewable energy, it is interesting to store locally the nonused electrical power. Another advantage of having load storage capability is to make it able to supply more power (as the WT can) if the MG operator asks it.

Moreover, the generated wind power is very fluctuant. So some DGs have to perform a real-time compensation of power transients and it implies abnormal stresses. Important collateral effects may decrease expected economic and environmental benefits. So a great improvement is to eliminate the wind-caused power variations by using a local storage unit with fast dynamic abilities. It should smooth the produced power. A local control system must manage the storage to supply or absorb the power difference between the generated wind power and the power reference set by the MGCC.

In the connected mode, it is very important to have a constant power exchange with the grid during each period (15 or 30 min) between the MG and the distribution grid. If renewable energy-based generators are connected, the MGCC must estimate the production capacity with the climate forecasts as well as the local consumption demands for the next period τ_n. To have a great benefit of "clean" electricity production, we must imagine a possible scenario. The MGCC should send to the DSO the average power, which will be exported during the period τ_n. Hence, the DSO has enough time in advance to gather and to send set points to coordinate the power production for his interest. At the same time, the MGCC assigns the power references to each local generator for the period τ_n according to the climate condition. Then each active generator should respect its power reference during the period τ_n, against the possible large fluctuations of weather conditions. These problems and possible solutions show the necessity to get a wind generator, which is capable of exchanging information with the MGCC and of supplying the power requirements of the MG.

2.5.2 Microgrids in Future Smart Grids

Today's electrical grids have to take the challenges to match the modern digital economy and information age, which requires higher load demands, uninterruptible power supplies, and other high-quality, high-value services. Especially, the integration of more and more intermittent and fluctuant RESs will lead to reliability problems of ancillary services, power quality disturbances, brownouts, and blackouts. The original power grid technology has its control systems embedded in the generating plants. The utilities attempt to meet the demand and succeed or fail with varying degrees (brownout, rolling blackout, uncontrolled blackout). The total amount of power loaded by the users can have a very wide probability distribution, which requires a lot of spare generating plants in standby mode to respond to the rapidly changing power usage. Thus, the clusters generating capacity should usually be oversized, so it is very expensive for the power producers, and the resulting brownouts and outages can be also very costly for consumers.

New electricity transmission and distribution network is required for integrating the newly emerging distributed RESs. SG is a modernized "grid" that uses robust two-way communications, advanced sensors, and distributed computers to improve the efficiency, reliability, and safety of power delivery and use. With the application of communication and information technologies to the electric grid, many smart digital meters can be integrated in the modernized grid to replace analog mechanical meters. The SG system operator (SGSO) is able to (i) control the electrical powers down to the residential level, small-scale DGs, and storage devices through the MGs; (ii) communicate information on operating status and needs and to collect information on prices and grid conditions; and (iii) transform the grid under central control into a collaborative network.

The current electrical transmission and distribution networks are rather passive and centralized from the supervision point of view. This situation makes difficult the coordination of distributed RESs in the grid. In order to coordinate power generation in an optimal way and to improve efficiency, reliability, and security, the SG organization has to incorporate distributed intelligence and interactive communication at all levels of the electric network [49, 50].

A step toward the SG is to integrate locally RESs, conventional generators, and loads in the MGs. These MGs may be operated in the islanded or connected mode (with the distribution grid) and can also provide ancillary services to the grid. They must be locally aggregated and controlled by the MGCC. This aggregation can be considered by the DSO as a stakeholder, which is able to locally control a cluster of generators and flexible loads. A problem to overcome is to match locally the power demand and the production in an optimal way while minimizing the use of non-RESs for electricity generation and decreasing economic costs during operation [51].

The DR is another important issue for the SG and MG development [52]. This can be very simple like timers to switch off electric water heaters during peak demand periods, but such systems are unable to respond to contingencies. The full SG allows generators and loads to interact in real time, by using modern communication and information technology. So the demand managements can eliminate the cost of generators, can cut the wear and extend the life of equipment, and allow users to get more value from the system by putting their most important needs first.

Continuously micro and SG energy systems are under development to integrate massively RESs, DGs, ESSs, critical, and noncritical loads in the control system of grids. In future SGs, the interests for MGs are exposed specifically for integrating renewable and nonconventional energy resources.

As discussed in Chapter 1, the RESs/DGs use a power electronic converter for the grid connection and so dedicated advanced and robust control functions are required for a better integration into the MG management system [53]. Different control schemes are reviewed and classified in order to highlight the required interface with the grid control system to make it operate as dispatchable DGs. Then fundamental principles of grid management systems and classical practices are recalled in order to better highlight constraints, required control functions for the RESs/DGs as well as for the grid control system and practical implementation.

Various scenarios for future architecture of electricity systems recognize the fundamental fact that with increased levels of RES penetration, the distribution network can no longer be considered as a passive appendage to the transmission network. The entire distributed system has to be designed and operated as an integrated unit. In addition, this increased complex operation must be undertaken by a system under a multilevel management electrical system.

In a modern power grid, the MGs provide their own supply with limited exchange of energy with the rest of the distribution network [54]. The distribution network operator operates more like a coordinating agent between separate MGs rather than a controller of the system. This kind of electrical network is smarter and can provide reserve and more different services based on intelligent appliances [55].

With a large number of widely dispersed DGs, the real-time communication and control are difficult to perform for the whole power system, especially with extremely large amounts of information and long transmission distance. Therefore, the optimized real-time control and management of DGs and loads should be implemented within local power grids. The necessary information should be first gathered and then be exchanged with the grid system operator for the whole power grid control and optimization. As a new control and management concept of organizing the distributed RESs and local loads, the MGs have attracted great attention all over the world. An example of organizing all DGs and loads by means of SGs and MGs in a hierarchical structure is shown as in Fig. 2.23.

The SG interfaces many MGs, and each MG combines locally the dispersed generators and loads, which are totally or partially controllable. The advanced communication technologies are required for connecting the SGSO, the MGCC, and the controllers of DGs and loads.

Many technical challenges are associated with the operation and the control of MGs. Energy management is very important for the achievement of good energy efficiencies by optimizing production and consumption [56]. With various conflicting requirements and limited communication techniques among a large number of distributed energy sources, the management of instantaneous active and reactive power balancing is a key challenge of the MGs [57]. Another key challenge of the MGs is to ensure stable operation during faults and various network disturbances. Transitions from interconnected operation

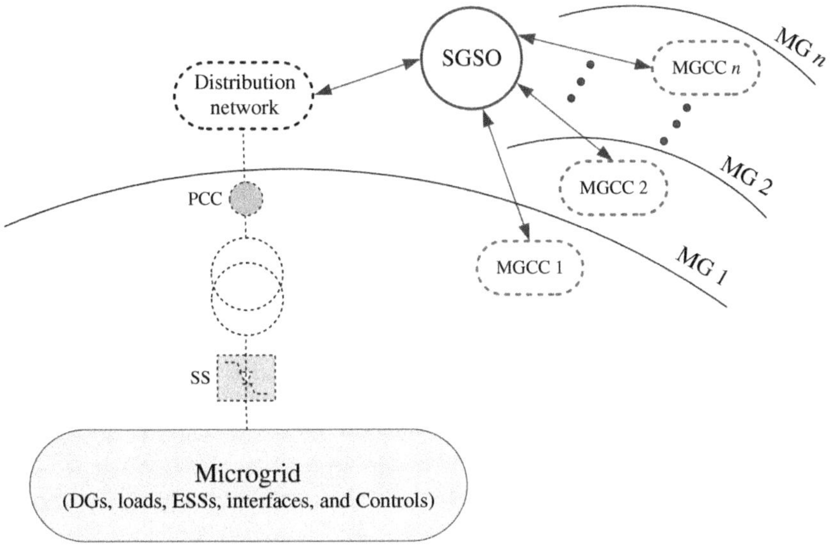

Figure 2.23 Example for organizing interconnected MGs in a smart grid.

to islanding operation are likely to cause large mismatches between generations and loads and to cause severe frequency and voltage control problems. Maintaining stability and power quality in the islanding mode of operation requires the development of sophisticated control strategies.

The MGs may also provide ancillary services for the main grid, if needed, such as the frequency regulation. So the study of the MGs in a grid-connected mode includes research studies not only on the MGs but also on the distribution network. Some important requirements and research needs in MG operation and control area can be summarized as follows:

- Define and develop control strategies to ensure efficient, reliable, and economic operation and management of MGs.
- Define appropriate protection and grounding policies to assure safety, fault detection, separation, and islanded operation.
- Develop RESs/DGs models and steady-state and dynamic analysis tools enabling simulation of LV asymmetrical, inverter-dominated MG performance.
- Develop hierarchical and distributed (agent-based) control algorithms for both islanded and interconnected operating principles.
- Definitions of RESs/DGs interface response and intelligence requirements grounding and protection schemes.
- Investigation of new RESs/DGs controllers to provide effective and efficient operation of MGs.

- Development of alternative control strategies using next-generation information and communications technology.
- Creation of alternative network designs, including application of modern protection methods, modern solid-state interfaces, and operation at variable frequencies.
- Technical and commercial integration of multiple MGs, including interface of several MGs with upstream distribution management systems, plus operation of decentralized markets for energy and ancillary services.
- Executing extensive field trials of alternative control strategies in actual installations, with experimental validation of various MG architectures in interconnected and islanded modes, and during transition testing of power electronics components and interfaces and of alternative control strategies, communication protocols, and so on.

2.6 Summary

Although natural delays appear for the deployment of the control support of the RESs in comparison with the regulation services of the classical direct-connected synchronous generators, studies presented in this chapter have shown satisfying performances of the RESs to provide control/regulation supports. Among the RESs, this chapter is focused on wind and SPGs using recorded real data.

The performance of the WT to contribute in inertial and primary controls is presented. The combined control with both inertial and primary loops could help to increase system robustness by reducing frequency excursions following grid disturbances. However, as wind power generation is of variable nature and hard to forecast, the power reserve provided by WTs would logically not be as reliable as that procured from conventional generators. The impact of intrahour wind variability (inside each scheduling interval) on the amount of "firm" wind reserve is characterized, and the comparison of different wind reserve strategies is discussed. The overall reserve allocation to each generator for each scheduling period can be obtained from an SCUC. The notion of instantaneous available reserve only characterizes the maximum potential of real-time primary reserve that can be provided from WTs. Nowadays, forecasting errors of WG cannot be neglected, and so a way to ensure a satisfactory level of wind reserve availability during the day takes into account a power margin the day ahead for covering forecast errors. The result is, of course, a change in the amount of wind reserve that can actually be dispatched in real time.

The MGs as small hybrid power systems provide a suitable solution for integration of RESs into the distribution systems. In this chapter, the MG philosophy, general structure, and operating modes were briefly addressed, and the role of MGs in future power grids was emphasized.

References

1 Bevrani, H. (2014) Wind power and frequency control, in *Robust Power System Frequency Control*, 2nd edn, Springer, Chapter 10.

2 Nordic Grid Code 2007 (Nordic Collection of Rules) (2007) Nordel. Available at: https://www.entsoe.eu/.

3 EirGrid Grid Code v3.5, Eirgrid (2011). Available at: http://www.eirgrid.com.

4 Ataee, S., Khezri, R., Feizi, M.R., and Bevrani, H. (2015) *Impacts of Wind and Conventional Power Coordination on the Short-Term Frequency Performance.* 23rd Iranian Conference on Electrical Engineering (ICEE), Tehran, Iran.

5 Ekanayake, J. and Jenkins, N. (2004) Comparison of the response of doubly fed and fixed-speed induction generator wind turbines to changes in network frequency. *IEEE Transactions on Energy Conversion*, **19**, 800–802.

6 Morren, J., de Haan, S.W.H., Kling, W.L., and Ferreira, J.A. (2006) Wind turbines emulating inertia and supporting primary frequency control. *IEEE Transactions on Power Systems*, **21** (1), 433–434.

7 Ramtharan, G., Ekanayake, J.B., and Jenkins, N. (2007) Frequency support from doubly fed induction generator wind turbines. *IET Renewable Power Generation*, **1** (1), 3–9.

8 Gautam, D., Goel, L., Ayyanar, R. *et al.* (2011) Control strategy to mitigate the impact of reduced inertia due to doubly fed induction generators on large power systems. *IEEE Transactions on Power Systems*, **26** (1), 214–224.

9 Kayikçi, M. and Milanovic, J. (2009) Dynamic contribution of DFIG-based wind plants to system frequency disturbances. *IEEE Transactions on Power Systems*, **24** (2), 859–867.

10 Keung, P.K., Li, P., Banakar, H., and Ooi, B.T. (2009) Kinetic energy of wind-turbine generators for system frequency support. *IEEE Transactions on Power Systems*, **24** (1), 279–287.

11 de Almeida, R.G. and Peças Lopes, J.A. (2007) Participation of doubly fed induction wind generators in system frequency regulation. *IEEE Transactions on Power Systems*, **22** (3), 944–950.

12 Conroy, J.F. and Watson, R. (2008) Frequency response capability of full converter wind turbine generators in comparison to conventional generation. *IEEE Transactions on Power Systems*, **23** (2), 649–656.

13 Ma, H.T. and Chowdhury, B.H. (2010) Working towards frequency regulation with wind plants: combined control approaches. *IET Renewable Power Generation*, **4** (4), 308–316.

14 Badmasti, B., Bevrani, H., and Naghshbandi, A. (2012) Impacts of high wind power penetration on the frequency response considering wind power reserve. *International Journal of Energy Optimization and Engineering*, **1** (3), 32–47.

15 Ataee, S., Khezri, R., Feizi, M.R., and Bevrani, H. (2014) *Investigating the Impacts of Wind Power Contribution on the Short-Term Frequency Performance*. Smart Grid Conference, Tehran, Iran.

16 Clark, K., Miller, N.W., and Sanchez-Gasca, J.J. (2010) *Modeling of GE Wind Turbine-Generators for Grid Studies, Version 4.5*. General Electric International. Inc. Available at: http://www.gepower.com/.

17 Kanellos, F.D. and Hatziargyriou, N.D. (2009) Control of variable speed wind turbines equipped with synchronous or doubly fed induction generators supplying islanded power systems. *IET Renewable Power Generation*, **3** (1), 96–108.

18 Wang, Y., Delille, G., Bayem, H. *et al.* (2013) High wind power penetration in isolated power systems – assessment of wind inertial and primary frequency responses. *IEEE Transactions on Power Systems*, **28** (3), 2412–2420.

19 Bevrani, H., Ghosh, A., and Ledwich, G. (2010) Renewable energy sources and frequency regulation: survey and new perspectives. *IET Renewable Power Generation*, **4** (5), 438–457.

20 Vidyanandan, K.V. and Senroy, N. (2013) Primary frequency regulation by deloaded wind turbines using variable droop. *IEEE Transactions on Power Systems*, **28** (2), 837–846.

21 Zertek, A., Verbic, G., and Pantos, M. (2012) A novel strategy for variable-speed wind turbines' participation in primary frequency control. *IEEE Transactions on Sustainable Energy*, **3** (4), 791–799.

22 Ruttledge, L., Miller, N.W., O'Sullivan, J., and Flynn, D. (2012) Frequency response of power systems with variable speed wind turbines. *IEEE Transactions on Sustainable Energy*, **3** (4), 683–691.

23 Chen, N., Qian, Z., Nabney, I.T., and Meng, X. (2013) Wind power forecasts using gaussian processes and numerical weather prediction. *IEEE Transactions on Power Systems*, **29**, 656–665.

24 Chang-Chien, L.R., Hung, C.-M., and Yin, Y.-C. (2008) Dynamic reserve allocation for system contingency by DFIG wind farms. *IEEE Transactions on Power Systems*, **23** (2), 729–736.

25 Rebours, Y., Kirschen, D., Trotignon, M., and Rossignol, S. (2007) A survey of frequency and voltage control ancillary services – part I: technical features. *IEEE Transactions on Power Systems*, **22** (1), 350–357.

26 Bertsimas, D., Litvinov, E., Sun, X.A. *et al.* (2013) Adaptive robust optimization for the security constrained unit commitment problem. *IEEE Transactions on Power Systems*, **28** (1), 52–63.

27 Calderaro, V., Galdi, V., Lamberti, F., and Piccolo, A. (2015) A smart strategy for voltage control ancillary service in distribution networks. *IEEE Transactions on Power Systems*, **30** (1), 494–502.

28 Ueda, Y., Kurokawa, K., Tanabe, T. *et al.* (2008) Analysis results of output power loss due to the grid voltage rise in grid-connected photovoltaic

power generation systems. *IEEE Transactions on Industrial Electronics*, **55** (5), 2744–2751.

29 Morren, J., Pierik, J., and de Haan, S.W.H. (2006) Inertial response of variable speed wind turbines. *Electric Power Systems Research*, **76** (11), 980–987.

30 Eurostag software. Available at: www.eurostag.be. Accessed in March 2017.

31 Delille, G., Francois, B., and Malarange, G. (2012) Dynamic frequency control support by energy storage to reduce the impact of wind and solar generation on isolated power system's inertia. *IEEE Transactions on Sustainable Energy*, **3** (4), 931–939.

32 Bevrani, H. and Hiyama, T. (2011) *Intelligent Automatic Generation Control*, CRC Press, New York.

33 El Mokadem, M., Courtecuisse, V., Saudemont, C. *et al.* (2009) Fuzzy logic supervisor-based primary frequency control experiments of a variable-speed wind generator. *IEEE Transactions on Power Systems*, **24** (1), 407–417.

34 Energinet (2004) Wind turbine connected to grids with voltage above 100 kV. Technical regulation TF 3.2.5.

35 Wang, Y., Bayem, H., Girald-Devant, M. *et al.* (2015) Methods for Assessing available wind primary power reserve. *IEEE Transactions on Sustainable Energy*, **6** (1), 272–280.

36 ENSTO-E (2012) Operational reserve AD HOC team report. Final version. Available at: www.entsoe.eu.

37 Wang, Y. (2012) Evaluation de la performance des réglages de fréquence des éoliennes à l'échelle du système électrique: application à un cas insulaire. PhD report, L2EP, Ecole Centrale de Lille.

38 Cadaval, E.R., Francois, B., Malinowski, M., and Zhong, Q.C. (2015) Grid-connected photovoltaic plants: an alternative energy source, replacing conventional sources. *IEEE Industrial Electronics Magazine*, **9** (1), 18–32.

39 IEEE (2010) IEEE standard definitions for the measurement of electric power quantities under sinusoidal, nonsinusoidal, balanced, or unbalanced conditions. IEEE Std 1459-2010 (Revision of IEEE Std 1459-2000), 2010, pp. 1–50.

40 Bialasiewicz, J.T. (2008) Renewable energy systems with photovoltaic power generators: operation and modeling. *IEEE Transactions on Industrial Electronics*, **55** (7), 2752–2758.

41 Camacho, A., Castilla, M., Miret, J. *et al.* (2013) Flexible voltage support control for three-phase distributed generation inverters under grid fault. *IEEE Transactions on Industrial Electronics*, **60** (4), 1429–1441.

42 Cagnano, A., Tuglie, E., Liserre, M., and Mastromauro, R.A. (2011) Online optimal reactive power control strategy of PV inverters. *IEEE Transactions on Industrial Electronics*, **58** (10), 4549–4558.

43 Lo, Y.-K., Lee, T.-P., and Wu, K.-H. (2008) Grid-connected photovoltaic system with power factor correction. *IEEE Transactions on Industrial Electronics*, **55** (5), 2224–2227.

44 Wang, X., Zhuo, F., Li, J. *et al.* (2013) Modeling and control of dual-stage high-power multifunctional PV system in *d–q–o* coordinate. *IEEE Transactions on Industrial Electronics*, **60** (4), 1556–1570.

45 Barater, D., Buticchi, G., Lorenzani, E., and Concari, C. (2014) Active common-mode filter for ground leakage current reduction in grid-connected PV converters operating with arbitrary power factor. *IEEE Transactions on Industrial Electronics*, **61** (8), 3940–3950.

46 Zhong, Q.-C. and Weiss, G. (2011) Synchronverters: inverters that mimic synchronous generators. *IEEE Transactions on Industrial Electronics*, **58** (4), 1259–1267.

47 Liu, J., Miura, Y., Bevrani, H., and Ise, T. Enhanced virtual synchronous generator control for parallel inverters in microgrids. *IEEE Transactions on Smart Grids*, In Press, DOI:10.1109/TSG.2016.2521405.

48 Bevrani, H., Ise, T., and Miura, Y. (2015) Virtual synchronous generators: a survey and new perspectives. *International Journal of Electrical Power and Energy Systems (IJEPES)*, **54**, 244–254.

49 Kanchev, H., Colas, F., Lazarov, V., and Francois, B. (2014) Emission reduction and economical optimization of an urban microgrid operation including dispatched PV-based active generators. *IEEE Transactions on Sustainable Energy*, **5** (4), 1397–1405.

50 Li, F., Qiao, W., Sun, H. *et al.* (2010) Smart transmission grid: vision and framework. *IEEE Transactions on Smart Grid*, **1** (2), 168–177.

51 Zhang, Y., Gatsis, N., and Giannakis, G. (2013) Robust energy management for microgrids with high-penetration renewables. *IEEE Transactions on Sustainable Energy*, **4** (4), 944–953.

52 P. Babahajiani, H. Bevrani, Q. Shafiee, Intelligent demand response contribution in frequency control of multi-area power systems. *IEEE Transactions on Smart Grids*, 2016, In Press, DOI:10.1109/TSG.2016.2582804.

53 Morel, J., Bevrani, H., Ishii, T., and Hiyama, T. (2010) A robust control approach for primary frequency regulation through variable speed wind turbines. *IEEJ Transactions on Power & Energy*, **130** (11), 1002–1009.

54 Fathi, M. and Bevrani, H. (2013) Statistical cooperative power dispatching in interconnected microgrids. *IEEE Transactions on Sustainable Energy*, **4** (3), 586–593.

55 Xingyu, Y., Abbes, D., Bevrani, H., and Francois, B. (2016) *Day-Ahead Optimal and Reserve Power Dispatching in PV Based Urban Microgrid*. 18th European Conference on Power Electronics and Applications-EPE'16 ECCE, Karlsruhe, Germany, 5–9.

56 Fathi, M. and Bevrani, H. (2013) Adaptive energy consumption scheduling for connected microgrids under demand uncertainty. *IEEE Transactions on Power Delivery*, **28** (3), 1576–1583.
57 Fathi, M. and Bevrani, H. (2015) *Wireless Networking of Smart Meters in Next Generation Power Systems.* 2nd International Scientific Conf-University of Human Development (UHD Comp15), Sulaimaniyah, Iraq.

3

Microgrids: Concept, Structure, and Operation Modes

Small and smart grid energy systems are under continuous development to integrate massively renewable resources, microgenerators, and small energy storage systems (ESSs), as well as critical and noncritical loads in the control system of grids. In the previous chapters, these grid energy systems are called hybrid energy systems or microgrids (MGs). The MGs are small electrical distribution systems that connect multiple customers to multiple distributed sources of generation and storage. The MGs are typically characterized by multipurpose electrical power services to communities that are connected via low-voltage networks. It is of great interest is that these hybrid power systems have the potential to provide reliable power supply to remote communities where connection to transmission supply is uneconomic.

The concept of MG provides an appropriate solution to integrate more and more RESs in the existing distribution networks. At the same time, the MGs can supply the local and sensitive loads, and then this feature together with the capability of operation in both islanded and connected modes increases the reliability of the power grid from the viewpoint of loads/customers.

In this chapter, following an introduction on the recent developments on the MGs, the main structure, fundamentals, and concepts are described. Then, the MG operating modes and the overall control mechanisms for the distributed generators of MGs are explained. The possible role of MGs to support the main grid ancillary services, particularly frequency regulation, is emphasized; and finally, the technologies of the existing MG laboratories are introduced.

3.1 Introduction

Different scenarios for future architectures of electricity systems recognize a fundamental fact that with increased levels of distributed generation (DG) penetration the distribution network can no longer be considered as a passive appendage to the transmission network. The entire distributed system has to be designed/operated as an integrated unit. In Chapter 1, the connection of

Microgrid Dynamics and Control, First Edition. Hassan Bevrani, Bruno Francois, and Toshifumi Ise.

renewable energy sources (RESs) such as wind turbines and photovoltaic (PV) units to the power grids is discussed. It is shown that the power fluctuations of RESs can be compensated by integrating ESSs. The local controllers have been enhanced in order to perform the independent and simultaneous control of the generated active and reactive powers. Some new control capabilities have been developed in order to create new possibilities to manage a power system in the presence of renewable-energy-based generators.

In addition, this increased complex operation must be undertaken by a system under a multilevel management electrical system. The growing share of DGs/RESs in the electricity system may evolve in three distinct stages [1]: accommodation step, decentralization, and dispersal. First, in the accommodation step, the DGs are accommodated in the current market. Distributed units are running free while the network is operated through the conventional production units. In this scheme, the RESs must not be dispatched and so classical wind generators and PV generators are connected. If the share of DGs increases massively, the decentralization step is achieved. Then virtual utilities optimize the services of decentralized providers through the use of common information and communication technology (ICT) systems [2]. A virtual utility as a flexible collaboration of independent and market-driven entities provides efficient energy service demanded by consumers without necessarily owning the corresponding assets. Here, central monitoring and control is needed. In this scheme, the RESs/DGs must be dispatched with a virtual utility and received power references. In the last dispersal stage, local low-voltage network segments provide their own supply with limited exchange of energy with the rest of the distribution network. These electrical networks are called MGs. The distribution network operator (DNO) operates more like a coordinating agent between separate power systems (different MGs) rather than a controller of the system. This kind of electrical network is smarter; it can provide more different services based on numerous intelligent appliances.

In this chapter, the context of future electricity system development and the interests for MGs are exposed specifically for integrating renewable and non-conventional energy resources. The DGs/RESs use power electronic converters for the grid connection and so dedicated control functions are required for better integration into the MG management system.

In general, the MG concept assumes a cluster of loads and DGs/RESs operating as a controllable system that provides both power and heat to its local area. The benefits of MGs, such as enhancement of the local reliability, reduction of feeder losses, and control of the local voltage, increase the efficiency by using the waste heat from combined heat and power (CHP) generators, the voltage sag correction, or the provision of uninterruptible power supply functions.

Generally, an MG is centrally controlled and managed by a microgrid central controller (MGCC) installed at the medium-/low-voltage (MV/LV) substation [3]. The MGCC includes several key functions, such as economic managing functions and control functionalities, and is the head of the MG hierarchical control systems. The MG is intended to operate in the following two different operating conditions: the normal interconnected mode with a distribution network and the emergency mode in islanding operation via a central switch.

Most DGs/distributed energy resources (DERs) that can be installed in an MG are not suitable for the direct connection to the electrical network due to the characteristics of the produced energy. Therefore, power electronic interfaces (DC/AC or AC/DC/AC) are required. Inverter control is thus an important concern in MG operation [4].

When the total number of MGs reaches a significant high share in the LV substations, similar technical benefits can be expected in upstream grids as a consequence of multi-MG operation [5]. The basic MG network is assumed to be radial with several feeders and a collection of loads as shown in Fig. 3.1. The radial system is connected to the distribution system through a separation device, usually a static switch (SS) at the point of common coupling (PCC). Each feeder has a circuit breaker (CB) and a power flow controller. The MG architecture consists of an LV network, loads (some of them are interruptible), both controllable and noncontrollable power generators, energy storage units, and a hierarchical-type management. The control scheme is supported by a communication infrastructure used to monitor and control power generators and loads. The center of the hierarchical control system is the MGCC. At the low hierarchical control level, load controllers (LCs) and microsource controllers (MCs) exchange information with the MGCC for managing the MG operation by providing set points to the LCs and MCs. The exchanged information includes mainly messages about power reference setting for each

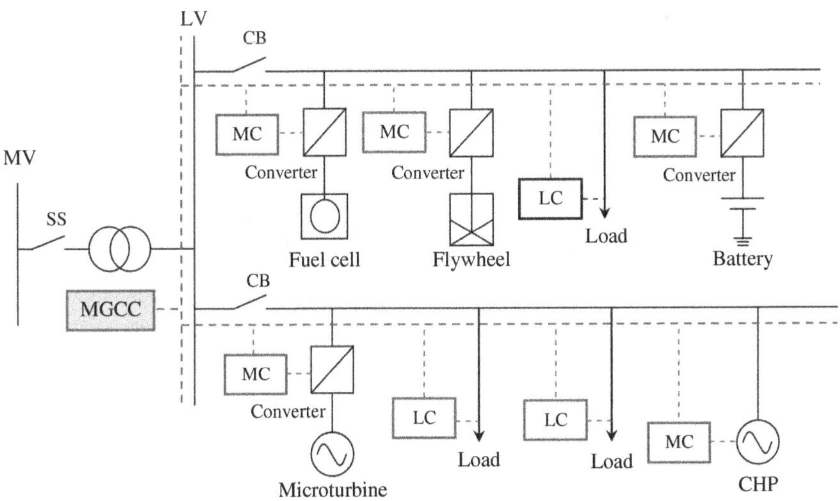

Figure 3.1 Basic MG architecture with an MGCC.

source; MG switches orders that are sent by the MGCC to the local controllers (MCs and LCs) and the sensed voltage/current information to the MGCC from each local capture. The MG structure and relevant fundamentals are explained in Section 3.2.

Intensive research activities on various aspects of MGs have been launched worldwide. In the European Union (EU), during 1998–2006, two major research efforts entitled "Microgrids" and "More Microgrids" have been devoted exclusively to MGs and up to now these activities have been continued among numerous projects in the EU countries. In the EU's fifth framework program, the project "Microgrids: large scale integration of micro-generators for low-voltage networks" was organized by more than 10 partners from 7 EU countries [6]. The main objectives of this project were as follows: studying the MGs operation to increase penetration of RESs/DGs while reducing carbon emissions; studying the MGs operation in parallel with the grid and in islanded mode; defining and developing control strategies to ensure efficient, reliable, and economic operation and management of MGs; defining appropriate protection and grounding policies to assure safety, fault detection, separation, and islanded operation; identifying and developing the required telecommunication infrastructures and protocols; determining the economic benefits of MG operation and proposing systematic methods to quantify them; and finally, simulating and demonstrating the MG operation in laboratory scales.

Several levels of centralized and decentralized control have been developed by participating EU laboratories. This project has proposed some solutions and innovative techniques such as DGs/RESs models and steady-state and dynamic analysis tools enabling simulation of LV asymmetrical, inverter-dominated MG performance; islanded and interconnected operating principles for hierarchical and distributed (agent-based) control algorithms; local black-start strategies; definitions of DGs/RESs interface response and intelligence requirements; grounding and protection schemes; methods for quantification of reliability benefits, and laboratory MGs of various complexities and functionalities.

A follow-up project titled "More microgrids: advanced architectures and control concepts for more microgrids" within the EU's sixth framework program was aiming to increase microgeneration systems in electric power systems by broadening the original concept of MG. The project objectives include

- Investigation of new DG controllers to provide effective and efficient operation of MGs;
- Development of alternative control strategies using next-generation information and communication technologies;
- Creation of alternative network designs, including application of modern protection methods, modern solid-state interfaces, and operation at variable frequencies;

- Technical and commercial integration of multiple MGs, including interface of several MGs with upstream distribution management systems, plus operation of decentralized markets for energy and ancillary services;
- Standardization of technical and commercial protocols and hardware to allow easy installation of DGs with plug-and-play capabilities;
- Study of impacts of MGs on power system operation, including benefits quantification of MGs at regional, national, and EU levels of reliability improvements, reduction of network losses, and environmental benefits;
- Exploration of the impact on the development of electricity network infrastructures, including quantification of the MGs benefits to the overall network and to the reinforcement and replacement strategy of the aging EU electricity infrastructure; and
- Execution of extensive field trials of alternative control strategies in actual installations, with experimental validation of various MG architectures in connected and islanded modes, and during transition testing of power electronic components and interfaces, and of alternative control strategies, and communication protocols.

Japan is the current world leader in MG demonstration projects. The Japanese government has set ambitious targets for increasing the contribution of the RESs, such as wind turbines and PVs, but the fluctuating power of RESs might degrade the country's outstanding power quality requirement (PQR). An MG can contribute to the load following capability for a utility grid by balancing its own energy requirement using controllable prime movers to balance fluctuating load and renewable output. For example, an MG with electrical storage and/or generator set can potentially fully compensate its intermittent renewable supply and so is presented itself to the grid as a constant load. This principle has motivated much of the MGs research and development in Japan and has led to focus on controls and electrical storage.

The Aomori MG in Hachinohe was put into operation in October 2005 and was being evaluated for PQR, cost-effectiveness, and greenhouse gases emission reductions over a planned demonstration period lasting until March 2008 [7]. A central feature of the system is based only on the RESs. The MG serves numerous buildings in a wide area, and it is connected to the commercial grid at a single PCC. The energy management system developed through this project optimally meets building demands for electricity and heat by controlling the output of the generator sets and boilers, together with charging and discharging of the battery bank. The control objective is to minimize operating costs and CO_2 emissions while maintaining a constant power flow at the PCC.

The first New Energy and Industrial Technology Development Organization (NEDO) demonstration (Aichi) project started its operation at the site for the world exposition in March 2005 [7]. The system was moved to the central Japan airport city near Nagoya in 2006, where it began operation in early 2007.

It now supplies a Tokoname city office building and a sewage plant via a private distribution line. Its main feature is a combination of the fuel cells as the main sources. Experiment results of intentional islanding mode have also been obtained. The efficiency of this MG with storage has been improved thanks to batteries in reliability, economy, and environment assessments.

The United States has a modest but slowly expanding MGs research program for many years, supported both by the US Department Of Energy under the Office of Electricity Delivery and Energy Reliability and by the California Energy Commission through its Public Interest Energy Research Program. Heightened demand for high PQR in the United States, primarily to match the high end of heterogeneous end-use requirements, has naturally led to increased focus on enhancing PQR locally using MGs.

The most well-known United State's MG effort has been pursued under the Consortium for Electric Reliability Technology Solutions (CERTS) [8], which was established in 1999 to explore implications for power system reliability of emerging technological, economic, regulatory–institutional and environmental influences. The specific concept of the CERTS MG (CM) was fully developed in 2002. Subsequently, building physical examples was undertaken.

The CM, as for other MG paradigms, is intended to, as seamlessly as possible, separate from normal utility service during a disruption and continues to serve its critical internal loads until acceptable utility service is restored. The CM provides this function for relatively small sites (~<2 MW peak power) without the need for expensive fast controls or expensive site-specific engineering.

As another relevant effort, General Electric (GE) developed an MG energy management (MEM) framework for a broad range of MG applications that provides a unified control, protection, and energy management platform [6]. At the asset level, the MEM is intended to provide advanced controls for both generation and load assets that are robust with respect to low-inertia environments. At the supervisory level, the coordinated operation of interconnected assets in the MG is optimized by the MEM to meet customer objectives such as maximizing operational efficiency, minimizing cost of operation and emissions impact, and is also intended to enable integration of the RESs and MG dispatchability.

The program was executed in two phases. The first phase of the program was focused on the fundamental controls and energy management technology development guided through the use of case studies considered to have market potential. These technologies were validated in simulation on a detailed model of an MG. In the second phase, a multibuilding campus was selected to demonstrate the technologies in a real-world application. Upon installation of equipment, validation and verification experiments to prove the advanced MG functionality was executed in mid-2008 [6]. Recently, many projects in the area of MG planning, operation, and control have been also conducted in other parts of the world, especially in China, India, South Korea, Iran, and South America.

The rest of this chapter is organized as follows: following an introduction on the recent MG developments, the MG structure and concepts are described in Section 3.2. Then, the MG operation modes are briefly discussed in Section 3.3. The overall control mechanisms for the given DGs inside an MG are explained in Section 3.4. Section 3.5 emphasizes the possible role of MGs to support the main grid ancillary services, particularly frequency regulation. The technologies of the existing MG laboratories across the world are introduced in Section 3.6; and finally, the chapter is summarized in Section 3.7.

3.2 Microgrid Concept and Structure

The MGs comprise dispersed energy resources, such as wind turbines, PV panels, fuel cells, and microgas turbines, storage devices such as flywheels, supercapacitors, and batteries, and controllable loads in order to offer considerable control capabilities to the local network operation. These systems can be connected to the LV distribution network, but they can also be operated in islanded mode in case of faults in the main network. From the customer's point of view, the MGs provide both thermal and electricity supplies, and, in addition, enhance local reliability, reduce emissions, improve power quality by supporting voltage and reducing voltage dips, and potentially reduce costs of energy supply.

Islanded operation may be expected to happen sometimes and it is obvious that the main concern is to "keep the lights on" in such periods. In the connected operation mode, the MG system operator or the MGCC should ensure the maximization of renewable energy generation and the optimization of the MG's operation. Controller functions have to be considered in order to achieve optimal operation of the MG in the connected mode. The MG system operator might use load forecasts (electric and possibly heat) and production capacity forecasts (from local generators). It uses the market prices of electricity, gas costs, local production capability, local load demands, grid security concerns, and distribution network's requests to determine the amount of power that the MG should draw from its owned DGs and the amount of power that should be exchanged with the main grid network. The defined optimized operating profile can be achieved by controlling the local generators and loads in the MG by sending control signals to their controllers. In this framework, noncritical controllable loads can be cut off, when necessary. Furthermore, it is required to monitor the actual active and reactive power balancing. These techniques can be considered equivalent to the secondary control of the conventional power grid [9, 10].

Because of the intermittent and fluctuant availability of the RESs, an MG as a small hybrid power system provides a high level of energy security through the mix of various generation systems and often incorporates ESSs to ensure maximum reliability of power supply. Several kinds of hybridization of power sources are presented in an MG. Due to the intermittent availability of RESs, backup power units are usually integrated for a high level of local energy security. For example, diesel generator, microgas turbine, and fuel cells are generally used for uninterrupted power supplies. Some renewable primary sources can be associated for complementary advantages. For example, in some areas, the hybrid PV-wind system may be proposed because the PV panels generate power only in the day time and wind generators produce usually more power with stronger wind at night. The association of energy storage devices with RESs can ensure reliability and security of the distributed power generation system while maximizing the benefit from renewable energies. For these systems, excess and deficit energy production can be optimally adjusted by the energy storage units to increase the energy efficiency [11].

A general structure of an MG that is able to operate in both connected and islanded operation modes is shown in Fig. 3.2. The grid-connected mode is a

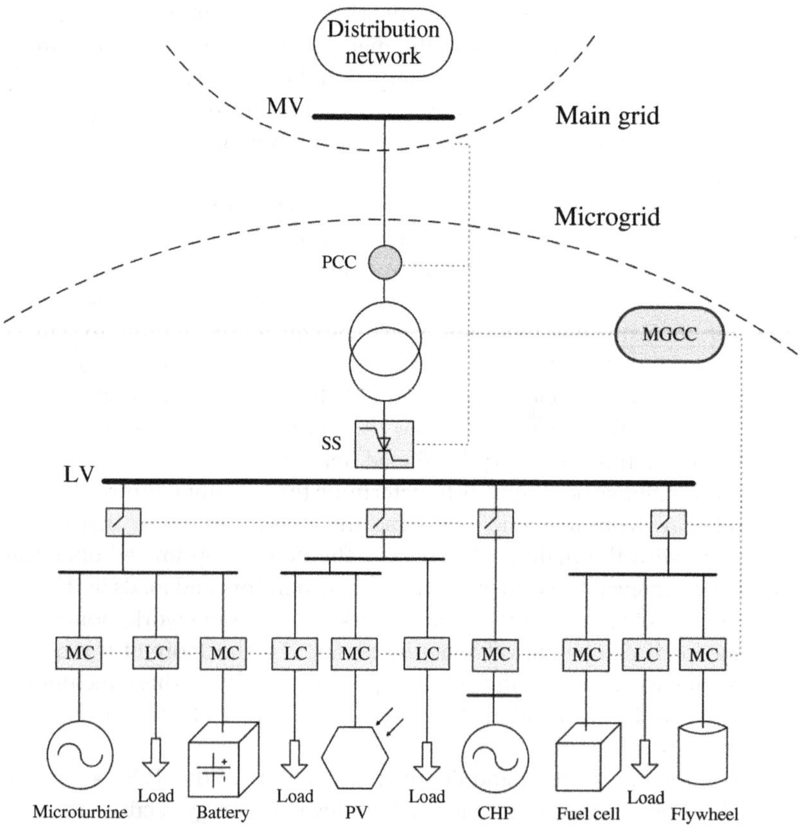

Figure 3.2 A typical MG structure.

particular complex operating mode since local loads inside the MG have to be supplied as well as power has to be sent to the main grid. In order to manage the DG units for the grid-connected mode, an MGCC is necessary. The communication bus between DGs and the MGCC is supposed to be ideal; that is, the signal transmission among the communication interface is propagated without time delay. The external loads can influence the distribution network as well as the MG power exchange by creating voltage droops or frequency fluctuations at the grid-connected bus.

The basic MG network is assumed to be radial with several feeders and a collection of loads, as shown in Fig. 3.2. In this radial system, each feeder has a CB and a power flow controller [12]. The CB is used to disconnect the corresponding feeder (and the associated unit) to avoid the impacts of severe disturbances through the MG. The MG is connected to the distribution system (main grid) at the PCC via a SS. The SS is capable to island the MG for maintenance purposes or when a fault or contingency occurs.

Generally, an MG is centrally controlled and managed by an MGCC installed at the MV/LV substation. The MGCC includes several key functions, such as economic managing and control functionalities as the head of MG hierarchical control systems. As mentioned, the MG is intended to operate in the normal connected and islanded operation modes. Before reconnection, the MGCC must implement the synchronization between the MG and main grid. Power electronic interfaces are often required to connect the existing DGs to the MG electric network. When the total number of MGs reaches a significant high share in LV substations, similar technical benefits can be expected in the main grids as a consequence of multi-MG operation. An interconnected MG network with a distribution power grid is shown in Fig. 3.3.

The MG operational architecture consists of an LV network, loads (some of them are interruptible), both controllable and noncontrollable power generators, energy storage units, and a hierarchical-type management. The control scheme is supported by a communication infrastructure to monitor/control power generators and loads. The center of the hierarchical control system is the MGCC. At the low hierarchical control level, local controllers of loads and microsources (LCs and MCs) exchange information with the MGCC for managing the MG operation by providing set points to the local controllers. The exchanged information includes mainly messages about power reference setting for each source; the MGCC command signals are finally sensed by the MG switches and power electronic devices.

In emergency conditions, for example, following a problem for the main grid, the MG can be separated from the main grid via an SS in about a cycle, as smoothly as possible. The MG can be also islanded intentionally for specific reasons even though there is no disturbance or serious fault in the main grid side. The balance between generation and demand of power is one of the most important requirements of MG management in both connected and islanded operation modes [13].

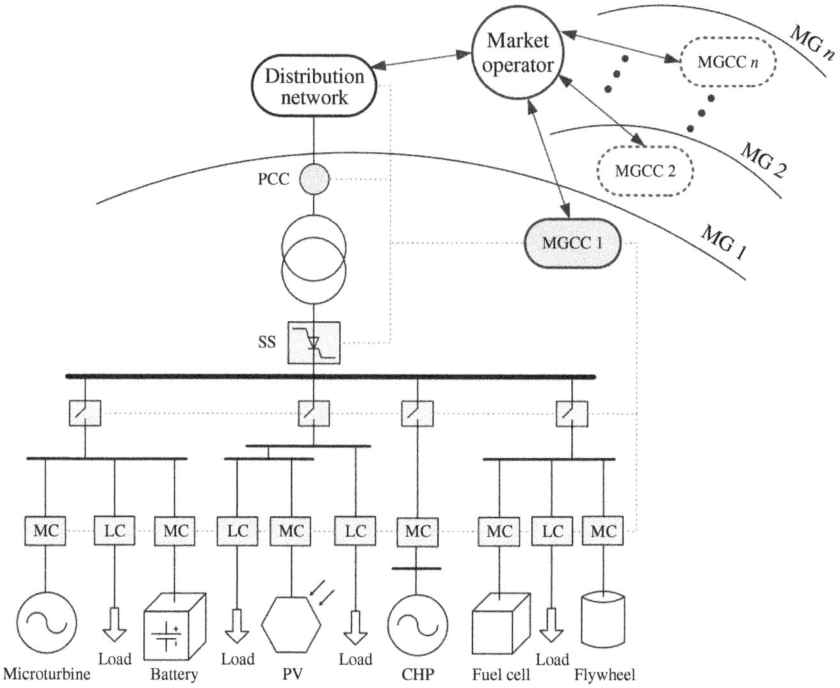

Figure 3.3 MG in an interconnected distribution power grid.

The MG has a hierarchical control structure with different layers [10]. The MGs require effective use of advanced control techniques at all control layers. The secure operation of MGs in connected and islanded operation modes and successful disconnection or reconnection processes depend upon MG controls. The controllers must guarantee that the processes occur seamlessly and the system is working in the specified operating points. In an interconnected distribution network (Fig. 3.3), each MG is locally controlled by the MGCC via its MCs and LCs. The LCs are installed at the controllable loads to provide load control capabilities. For each MG, the MGCC interfaces between the DNO or market operator and the MG. The DNO has the responsibility of managing the operation of medium- and low-voltage areas where more than one MG may exist.

Similar to large-scale power systems [14], the MGs can operate by using various control loops. As described in Chapter 5, these control loops can be mainly classified into four control groups: local, secondary, central/emergency, and global controls. In contrast to the local control, operating without communication, secondary, global, and emergency controls may need communication channels. While the local controls are known as *decentralized* controllers, the global, and to some extent, secondary and emergency controllers are operating as *centralized* controllers.

3.3 Operation Modes

MG flexibility can be achieved by operating under two different conditions. In the normal operation, the MG is connected to a main MV distribution grid being either partially supplied from it or injecting some amount of power into it. Depending on the demand request, in the grid-connected mode, the main grid and local DGs may send power to the loads. As the MV grid sets the root mean square (RMS) voltage, all DGs inside the MG can only generate currents but can be dispatched by the MGCC in order to provide power references. The control system in the local controller of the DGs (MCs) is known as "PQ inverter control" and the DG is said to be in "PQ mode."

If an event in the main grid appears, an islanded operation mode can be implemented because the electrical system is organized in the form of an MG with an MGCC. The MG islanding process may result from an intentional disconnection from the MV grid (due to maintenance needs) or from a forced disconnection (due to a fault in the MV network such as voltage dips). As mentioned earlier, the disconnection is performed by a static bypass switch opening the MG as a controllable small grid.

The emergency operation of an MG consists of disconnection and then the operation in the islanded operation mode. Hence, the MG frequency and voltage must be maintained by the MG DGs. In a DG, the corresponding local controller must implement a voltage source inverter (VSI) control with an inner RMS voltage reference and frequency set point thereby compensating the unbalancing between the MG load and power. In this case, the mentioned DG is said to be in "VF mode" and is defined as the "master source." Other DGs are known as "slave sources," whose local controllers must operate in the PQ mode. The MGCC shares the total amount of real power to produce among the "slave sources" and send them PQ references. It is important to avoid the overload of inverters and to ensure that load changes are controlled in a proper form. So in islanded mode, one DG may be operated in the "VF mode" and the others in the "PQ mode."

When the MG is operating in islanded mode, any fault current must be covered by generators inside the MG. These fault currents may have low values. But, compared with the current limit of the power electronic converters for the generators, these fault currents are relatively high, and may be sufficient to trip the conventional overcurrent protection relays and therefore other protection techniques have to be considered. One approach to develop a real-time fault location technique is proposed in [15] that identifies the exact location of the fault much more accurately than the classical relaying. Low-cost approaches such as current transformer (CT)-based zero sequence detection and differential current and/or voltage methods are also promising.

An active islanding detection technique for a DG at the distribution voltage level is presented in [16], which uses a three-phase, voltage source converter

(VSC) as an interface unit. The proposed method is based on injecting a disturbance signal into the system through either the direct-axis (d-axis) or the quadrate-axis (q-axis) current controllers of the interfaced VSC. Signal injection through the d-axis controller modulates the amplitude of the voltage at the PCC, whereas signal injection through the q-axis controller causes a frequency deviation at the PCC, under islanded conditions.

A fault in an MG may cause a ground potential rise, even if the energy sources operate at the LV level. Thus, grounding the distributed energy sources and the transformer connecting the MG to the utility network must be carefully analyzed and appropriate rules need to be developed. The LV earthing systems are defined according to the earthing techniques of the secondary side of the MV/LV transformer and the frame of the load equipment. Besides, a grounding system for a typical MG is designed in [17] and its adequacy is analyzed during fault conditions from an electrical safety point of view.

3.4 Control Mechanism of the Connected Distributed Generators in a Microgrid

Today, large centralized facilities provide most of the electrical power around the world with fossil fuel, nuclear, and hydropower plants. Normally, these plants have good economies of scale, but they usually have to transmit electricity to customers far away, hence may affect the environment negatively.

Using the RESs/DGs that are installed near the customer end reduces relatively electrical power losses on transmission lines because electricity is generated very close to the point of end use. This also decreases the size and number of power lines that must be constructed. The RESs/DGs appear as faster and less expensive option to the construction of large/central power plants with high-voltage transmission lines. They offer consumers the potential for lower cost, higher service reliability, high power quality, increased energy efficiency, and energy independency. The use of RES technologies and green power can also provide a significant environmental benefit due to less emissions.

As discussed in Chapter 1, DGs can be structurally divided into two types: conventional DGs and inverter-based DGs. In this section, the general control mechanisms for controlling these DGs, when they are used in an MG, are explained [5].

3.4.1 Speed Control of Classical Distributed Generators

Industrial MGs are relatively small power systems that can be powered by only one AC single generator, which is driven by a gas turbine or a diesel engine generator. The frequency of a synchronous AC generator is directly proportional to the speed of its rotating electrical field [18]. Hence, the power management

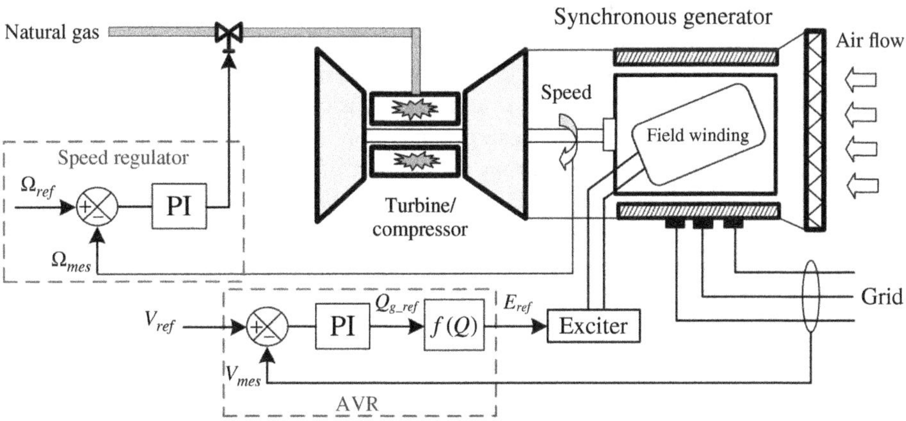

Figure 3.4 Scheme of an isochronous speed control system of gas turbine.

relies on an isochronous speed control, which keeps the turbine at a constant speed. The energy being admitted in the prime mover is regulated in response to the load changes, which would tend to cause changes in the speed. This control mechanism for a gas turbine is shown in Fig. 3.4 and the heat recuperating unit and the associated heat water network are not shown. Any increase in load decreases the speed (Ω). The kinetic energy in the large inertia contributes to compensate the speed in few instants. Moreover, energy is quickly admitted to the prime mover to maintain the speed at the set point (Ω_{ref}) by the speed regulation by using a simple proportional–integral (PI) controller. Any decrease in load increases the speed, but energy is quickly reduced to the prime mover to maintain the speed at the set point. An automatic voltage regulator (AVR) is used to control the output RMS voltage of the machine (V_{mes}) by regulating the voltage across the field exciter (V_{ref}) [19].

In case of a load change in the network, the power system's inertia can perform energy balance with detection of the variation in system frequency. It is pointed out that a system with clusters of small power DGs (with small inertia) designed to operate in islanded mode must provide some form of storage to ensure initial energy balance [20, 21]. For this purpose, some fast dynamic storage units may provide the amount of power required to balance the system following disturbances and/or significant load changes. These storage units must act as controllable AC sources to face sudden system changes.

3.4.2 Control of Inverter-based Distributed Generators

Because of internal characteristics, as mentioned in Chapter 1, most DGs are not suitable for direct connection to an electrical network. Therefore, power electronic interfaces are required. The most conventional power electronic topology is the back-to-back structure with two VSCs (one for the generator-side converter and one for the grid-side converter). In PV generator and wind turbine applications, as explained in Chapter 1, the generator-side converters are usually controlled to implement a maximum power point tracking (MPPT), and the grid-side converter is usually a three-phase inverter.

Power electronic converters can provide more flexible operation with higher dynamics compared to the direct connection of synchronous and induction generators to the grid [22, 23]. Functions that must be implemented by the control system of inverter must be distinguished according to the operating mode of the MG. For a grid-connected mode, the AC voltage and frequency are supplied by the grid. So all dispatchable DGs' inverters must be controlled in a "P/Q mode" to inject powers. In islanded operation mode, the inverter must control the RMS value and the frequency of the AC voltage in a "VSI mode."

3.4.2.1 Control Structure in Grid-connected Mode

3.4.2.1.1 Grid-following Strategy for Passive Generators With a *grid-following strategy*, the grid-side converter can control the voltage across the DC-link capacitor (Chapter 1) and control the exchanged reactive power with the grid. This is a useful strategy for nondispatchable DGs/RESs, such as wind and PV generators, which due to uncertainty in the primary energy resource cannot enable a correct DC bus voltage control. The mechanism of grid-following strategy for a PV with a variable DC bus voltage is shown in Fig. 3.5. Different DC voltage values (u_{PV}) can be provided in order to implement an MPPT of PV or wind turbines [24].

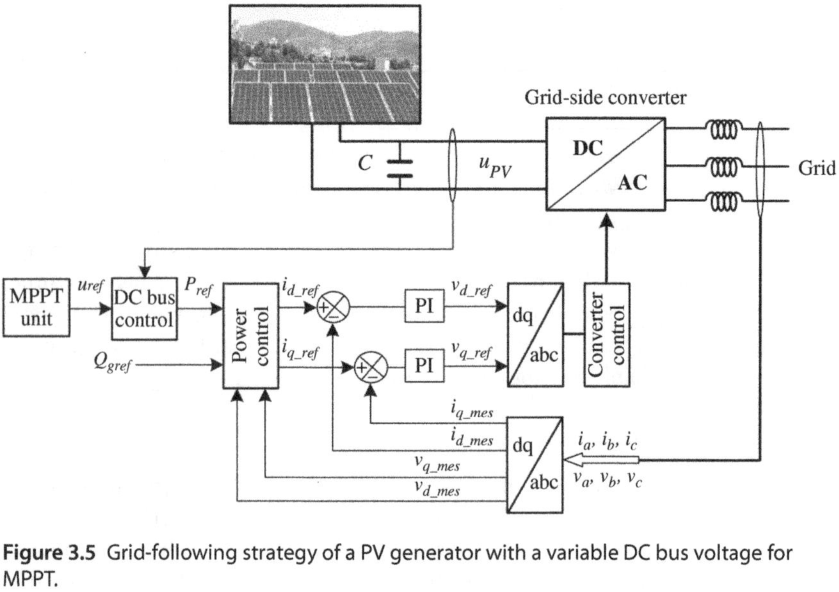

Figure 3.5 Grid-following strategy of a PV generator with a variable DC bus voltage for MPPT.

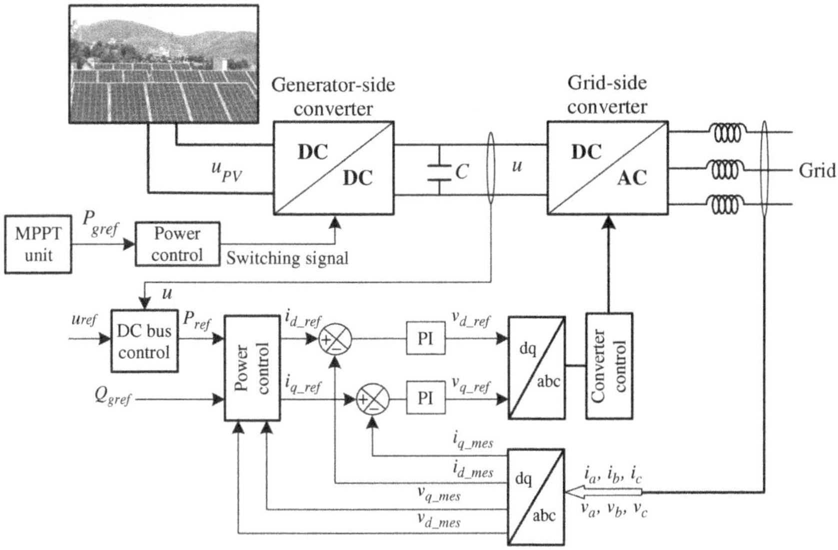

Figure 3.6 Grid-following strategy of a PV generator with a generator-side converter for MPPT.

This function can also be done by an additional generator-side converter as depicted in Fig. 3.6. With a grid-following strategy, using of a choke filter, the grid-side inverter is controlled as a current-controlled source. Direct and quadrature components of currents are calculated by a Park transformation for power calculation. Generated real power variations of the DG cause a DC-bus voltage error, which is corrected via the DC bus control by adjusting the reference of the real power (P_{ref}) injected to the grid. The reactive power output is also controlled using the PI regulator by adjusting the magnitude of the inverter reactive current output. This inverter can operate with a unit power factor or can receive a set point (from the MGCC) for the output reactive power.

3.4.2.1.2 Grid-following Strategy for the "PQ Mode" With a grid-following strategy, the inverter operates by injecting the available power at the primary source into the grid because the DC bus voltage is constant. For the RESs such as PV and wind generators, the output power is fluctuant. In the DGs with the possibility of increasing/decreasing the amount of primary power, an additional active power reference can be used by the local controller (P_{gref}), or the MGCC to make the generated active power variable. These kinds of DGs are dispatchable and their control system is known as "PQ control," as described in Fig. 3.7.

3.4.2.1.3 Power Dispatching Strategy for the "PQ Mode" With a power dispatching strategy, the voltage of the DC-link capacitor is maintained by controlling the generated power from the primary source (Fig. 3.8). The primary source is

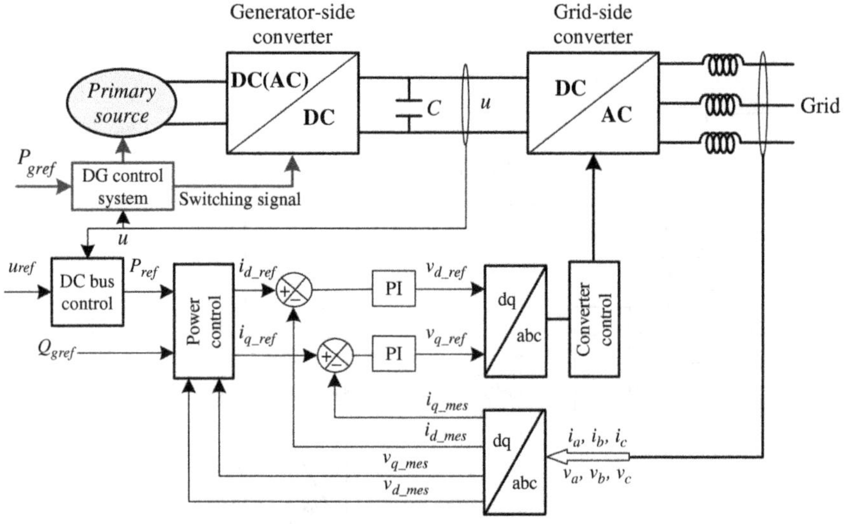

Figure 3.7 Grid-following strategy of a dispatchable generator in the PQ mode.

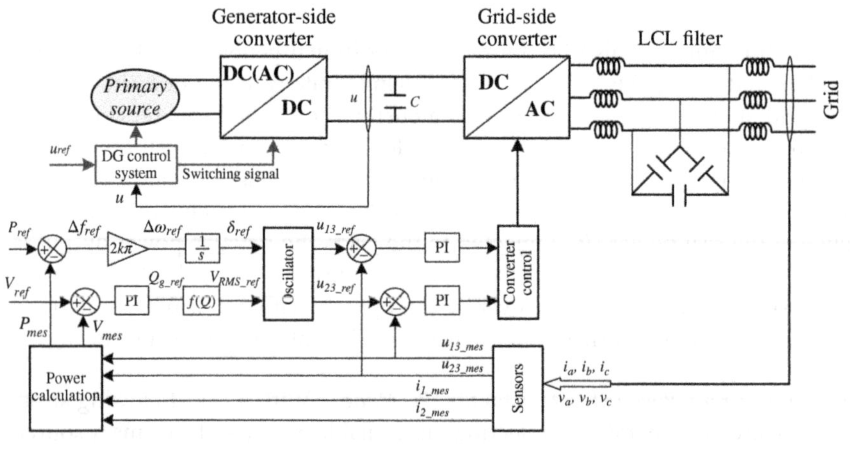

Figure 3.8 Power dispatching strategy of a gas microturbine for a VSI control.

responsible for responding to this power demand in order to maintain a constant DC voltage bus. This is suitable for applications with controllable power generation such as hydroturbines, gas microturbines, and fuel cells as the availability of the primary energy is guaranteed. Then the voltage across the DC bus can be considered as constant and the grid-side converter can control the active and reactive power output; this kind of DGs is dispatchable. The "PQ inverter control" consists of an outer power control loop that calculates current references for the current control inner loop.

This control scheme is similar to the isochronous speed control of the classical DGs and the role of DC bus is equivalent to the inertia in providing a constant DC voltage. In practice, a small DC bus time response is resulted and the performance is not similar to the classical DG.

3.4.2.2 Control Structure in Islanded Mode

In islanded operation mode, a DG must feed the MG with predefined values for the system voltage and frequency variables. In order to generate AC voltages, AC capacitors are required and so an LCL filter is used for the grid connection as illustrated in Fig. 3.8. With this power structure, the control strategy of the grid-side converter can control the voltage and the frequency at the PCC and thus can emulate the behavior of a synchronous machine [22, 24, 25]. This control scheme is known as the "VSI control." The phase-to-phase voltages are controlled by a closed-loop control according to the following references:

$$\begin{cases} u_{13\_ref}(t) = V_{RMS\_ref}\sqrt{3}\sqrt{2}\sin\left(2\pi ft - \frac{\pi}{6} + \delta_{ref}\right) \\ u_{23\_ref}(t) = V_{RMS\_ref}\sqrt{3}\sqrt{2}\sin\left(2\pi ft - \frac{\pi}{2} + \delta_{ref}\right) \end{cases} \tag{3.1}$$

Voltage references are generated by an oscillator, which is fed by the shift between the grid voltage and the modulated voltage (δ_{ref}) and the desired RMS value of line voltages (V_{RMS_ref}). In practice, the inductance of the grid-connected choke is low to minimize the voltage drop and so voltages across AC capacitors are nearly equal to the grid voltages.

By neglecting the filter losses (and so the equivalent resistor), the single-phase equivalent circuit of the grid connection side is obtained as given in Fig. 3.9.

In order to generate a current (i), the modulated voltage (V_m) must be higher than the AC capacitor voltage (V). The vectors V_m and V are corresponding to the vector of (three) modulated voltages and (three) capacitor voltages, respectively. The powers at the capacitor connection side are expressed as follows [26]:

$$P_g = 3V\frac{V_m \sin\delta}{L\omega} \tag{3.2}$$

$$Q_g = 3V\frac{(V_m \cos\delta - V)}{L\omega} \tag{3.3}$$

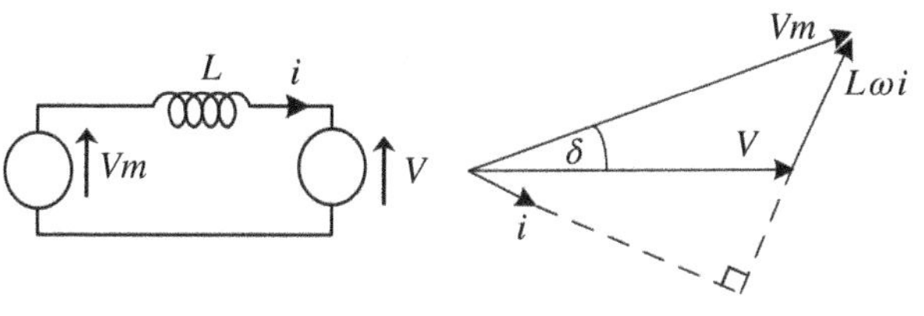

Figure 3.9 Equivalent single-phase circuit of the grid connection side and vector diagram.

where V is the line grid voltage RMS value, V_m is the fundamental modulated voltage RMS value, and $L\omega$ is the reactance of the coupling reactor.

In practice, $3\frac{V}{L\omega}V_m$ is large and so $\sin\delta$ is small. In this condition, one can assume

$$\sin\delta = \delta, \quad \cos\delta = 1 \tag{3.4}$$

Thus, the real power can be expressed as follows:

$$P_g = 3\frac{V}{L\omega}V_m\delta \tag{3.5}$$

The real power is classically controlled by a droop controller, which calculates the required frequency; the speed variation is calculated and the shift is obtained by an integration operator. Figure 3.10 shows that for a constant V_m magnitude the increase of the shift induces an increase in the generated power. Moreover, the capacitor voltage (V) must remain constant, then the grid current (i) and the reactive power must change.

The closed-loop control of the grid voltage is used to calculate the required reactive power to be produced. The reactive power can be expressed as

$$Q_g = 3\frac{V}{L\omega}(V_m - V) \tag{3.6}$$

The RMS value of the modulated voltage is obtained by inverting the following equation:

$$V_{RMS\_ref} = \frac{1}{3}\frac{L\omega}{V}Q_{g\_ref} + V \tag{3.7}$$

For dispatchable DGs, the power reference is sent to the local controller of the generator (MC). Inverter control is thus a very important main concern in MG operation since it may or may not give new control flexibilities to the grid operator. As described in Chapter 1, the PV and wind power generators are controlled with the grid-following strategy since their primary energy is volatile. For instance, in normal operation (connected mode), the grid-following control strategy has been applied for the given PV-based active generator in Chapter 1. In a disconnected mode, a "power dispatching strategy" must be used to control the DC bus voltage.

As the PV active generator has embedded ESSs, the availability of enough energy can be guaranteed as long as the storage units are not empty and can

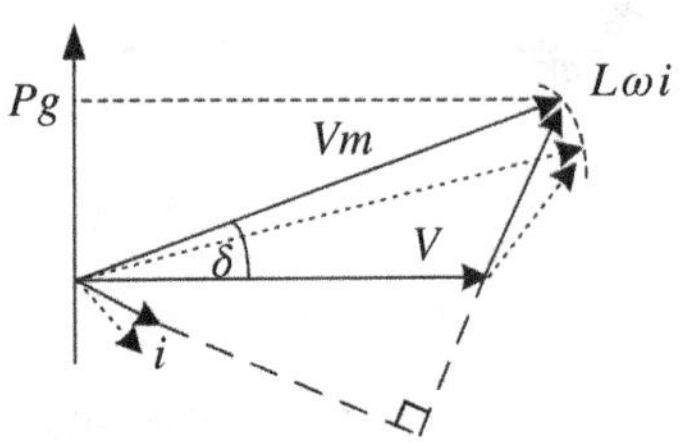

Figure 3.10 Increase of the shift.

be used. In this condition, a power dispatching strategy can be implemented for the DGs and the availability of the embedded energy power must be known and managed by the MGCC.

3.5 Contribution in the Upstream Grid Ancillary Services: Frequency Control Support Example

The grid-connected mode is a particular complex operating mode since local loads inside the MG have to be supplied as well as the extra power may have to be sent to the (main) upstream grid. Therefore, the MG may also provide ancillary services for the main grid, if needed, such as the frequency regulation [11]. In order to manage the DG units for the connected operation mode, the MGCC is necessary. The modeling and control structure of the DGs (solar and wind power generators) have been presented in Chapter 1 and explained in detail in Chapter 4. In this book, it is assumed that the required communication lines between the DGs and the MGCC are available; it means that the signal transmission among the communication interfaces is propagated without time delay.

In order to describe the capability of an MG to contribute to the provision of ancillary services (here, frequency regulation support) of the connected main grid, a case study is illustrated and is shown in Fig. 3.11. This grid includes a diesel group (20 kV, 200 kW) as the main grid, a HV transmission line, a transformer between the HV bus and an MG bus [27]. The total HV loads outside the MG are also modeled as an equivalent passive load (Load 2). The external loads can influence the distribution network as well as the MG power exchange (P_{mg}, Q_{mg}) by creating voltage droops or frequency fluctuations at the HV bus.

Based on this modeling, the central supervision functions are designed and presented. The details of each function level are given. Finally, simulation results using MATLAB/Simulink software are presented to validate the proposed MGCC structure. In this part, MG case study is connected to a distribution grid as shown in Fig. 3.12.

For the DNO, an interconnected MG can be considered as a potential power reserve contributor. An advanced interface control system inside the MGCC must be developed in order to provide the available regulation power in compliance with the distribution network requirement ($P_{mg\_dno\_ref}$). The block of ancillary services implements the grid integration and coordination of the MG with the main grid (distribution network system). It includes frequency and voltage control algorithms at the PCC within a grid power flow assessment, protections, and additional measurements. In the following section, first the fundamental principle of the primary and secondary frequency control of the grid is recalled and then the necessary functions inside the MGCC to adapt the operating of the MG with the distribution network are summarized.

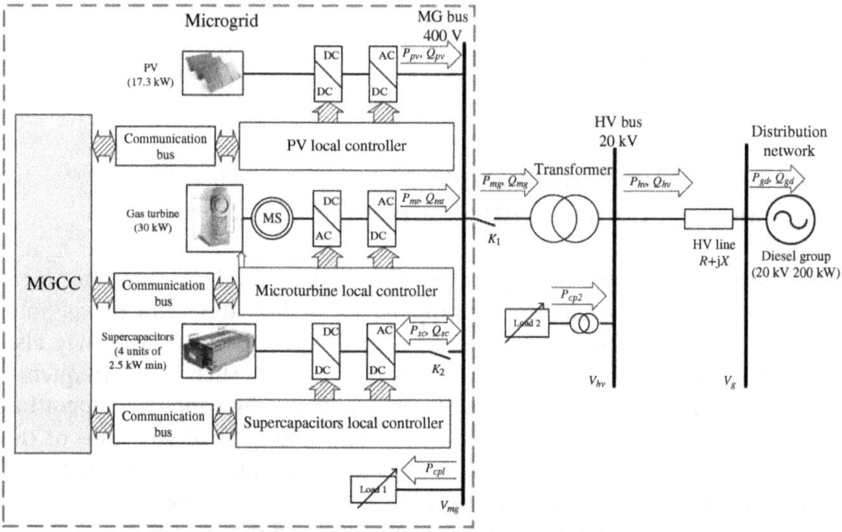

Figure 3.11 The MG case study for supporting the main grid frequency regulation.

3.5.1 Participation in the Frequency Regulation

Nowadays, the synchronous operation of conventional power plants and the power balance is maintained by the grid frequency control. In this section, the mentioned control principle is briefly described [28, 29]. After a power variation, conventional power plants will immediately release or absorb the kinetic energy from their rotating mass according to the following equation:

$$E = \frac{1}{2} J \Omega^2 \tag{3.8}$$

where J is the inertia of the machine and Ω is the rotational speed of the machine. As a result, the frequency changes and the response are determined by the movement equation that is called inertial response:

$$\frac{d\left(\frac{1}{2} J \Omega^2\right)}{dt} = \Delta P_m - \Delta P_l \tag{3.9}$$

where ΔP_m is the generated power and ΔP_l is the equivalent load power. For the diesel group (main grid), the load power (ΔP_l) is the sum of the MG power (ΔP_{mg}) and all load powers $\sum \Delta P_{li}$ and generation powers $\sum \Delta P_{gi}$ outside the diesel group:

$$\Delta P_l = \Delta P_{mg} + \sum \Delta P_{gi} + \sum \Delta P_{li} \tag{3.10}$$

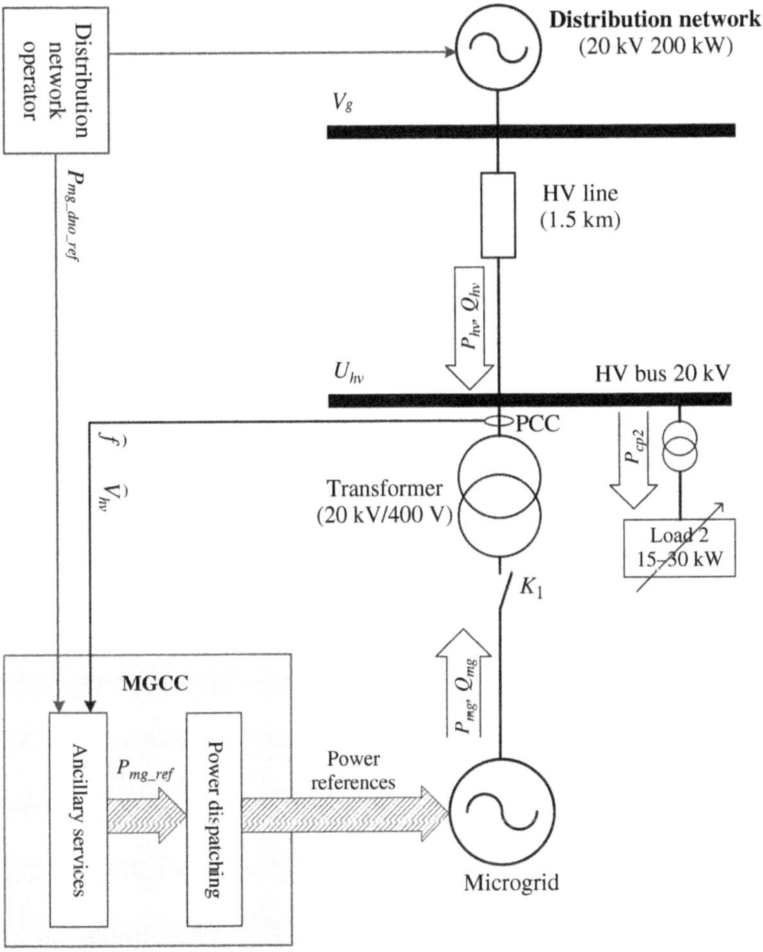

Figure 3.12 General organization of the system.

ΔP_{dm} is assumed to be the small change in the difference between the generated power from the diesel group (P_m) and the load power (P_l):

$$\Delta P_{dm} = \Delta P_m - \Delta P_l \tag{3.11}$$

According to (3.9), the frequency variation can be expressed by neglecting the change from the voltage effect as

$$\Delta\omega = \frac{1}{D + 2Hs}\Delta P_{dm} \tag{3.12}$$

where D, H, and s are damping constant, inertia, and Laplace variable, respectively. The small change in the frequency (Δf) can be expressed as

$$\Delta f = \frac{\Delta\omega}{2\pi} = \frac{1}{2\pi(D + 2Hs)}\Delta P_{dm} \tag{3.13}$$

The frequency of the grid is then expressed with the constant grid frequency ($f_{ref} = 50$ Hz):

$$f = f_{ref} + \Delta f \tag{3.14}$$

When the frequency deviation exceeds a predefined threshold value, the primary frequency controllers will be activated to increase/decrease the power from the prime movers to restore the power balance. The primary frequency control contribution of the generators is based on a droop constant, which gives the additional power that is supplied as a function of the frequency deviation [30]. This traditional control scheme can be realized concerning the relation (3.13). A proportional controller is often used here.

$$\Delta P_{dm\_ref} = k_{dg}(\Delta f_{ref} - \Delta\widehat{f}) = -k_{dg}\Delta\widehat{f} \tag{3.15}$$

where k_{dg} is the ratio of the compensator. The small change of the frequency $\Delta\widehat{f}$ is calculated with the sensed frequency $\widehat{f}$. Therefore, the estimation of the relation (3.14) is used:

$$\Delta\tilde{f} = f_{ref} - \widehat{f} \tag{3.16}$$

Since other powers such as $\sum \Delta P_{li}$ and $\sum \Delta P_{gi}$ are difficult to sense a unique predesigned dispatching ratio (k_{dm}), the ΔP_{mg_ref} and ΔP_{l_ref} must be calculated from ΔP_{dm_ref} through the inversion of relations (3.10) and (3.11).

$$\Delta P_{mg\_ref} = k_{dm}\Delta P_{dm\_ref} \tag{3.17}$$

Therefore, by combining (3.15)–(3.17), a traditional frequency droop control expression can be found as follows:

$$\Delta P_{mg\_ref} = k_{dm}k_{dg}(f_{ref} - \widehat{f}) = k(f_{ref} - \widehat{f}) \tag{3.18}$$

where $k = k_{dm}k_{dg}$ is the total ratio of the participation to the primary frequency control. For example, if the frequency changes from f_{ref} to f_1, the reference of the generated power will move in normal conditions from P_{mg_ref0} to another value $P_{mg\_ref\_1}$ as shown in Fig. 3.13.

After restoration of the power balance by the primary control, the system is stable (point 2 in Fig. 3.13) but at another frequency (f_1). The secondary frequency control brings the frequency back to its normal value (f_{ref}) and the power operating point is changed (point 3 in Fig. 3.13).

For the given application, the DNO sends a wished dispatching system control power reference signal $P_{mg\_dno\_ref}$, which is considered as the exchanged

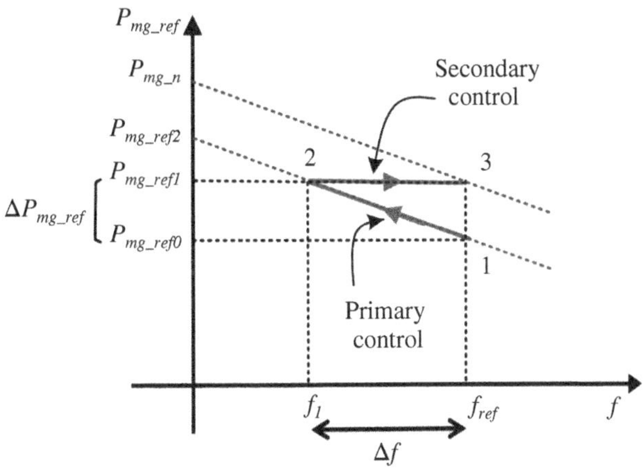

Figure 3.13 Idealized power frequency control characteristic.

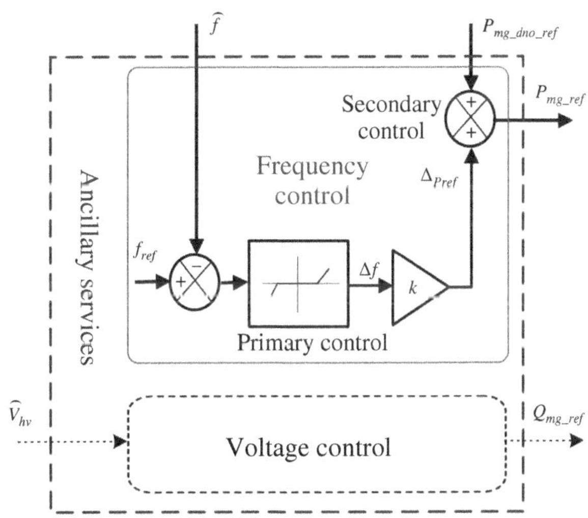

Figure 3.14 Block diagram representation for the ancillary services provided by the MG.

power from the MG to the distribution network in a long term. The conventional power/frequency control principle can be also used inside the MGCC in order to create the MG contribution to the main grid primary frequency control (Fig. 3.14).

The power/frequency parameter (k) is calculated as follows:

$$k = \frac{1}{S}\frac{P_{mg\_max}}{f_{ref}} \tag{3.19}$$

where S is the droop slope and P_{mg_max} the maximum available power, which can be exported to the distribution network.

3.5.2 Power Dispatching

For the example at hand, the purpose of the power dispatching is to drive the three DGs (PV, gas microturbine, and supercapacitors shown in Fig. 3.11) to supply local loads in an optimal way for the electrical distribution and production. Here, the MG operation in the context of a connected mode is studied, and the necessary control functions for this purpose are explained. For gas microturbines, the economic interest can be mathematically expressed as a power generation over a minimal power value. Because of the physical capacity, supercapacitors can be used to smooth fast power variations resulting from loads and PV.

The power dispatching is based on a power management scheme whose purpose is to implement the real-time power balancing (Fig. 3.15). Before sending the dispatching power set points to each unit, some protection and

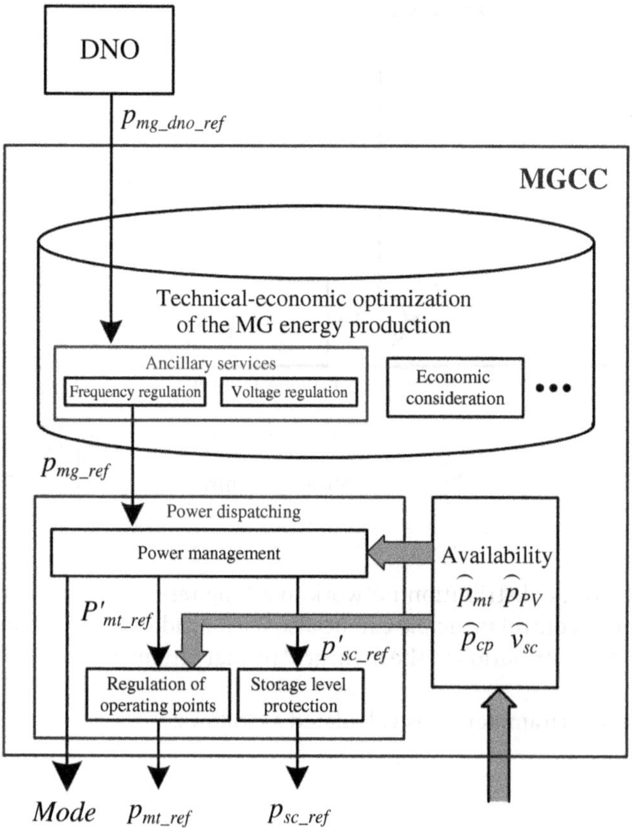

Figure 3.15 The proposed power management scheme for the given case study.

regulation strategies must be added. Then, the calculated power references may be modified by the regulation of the operating points of DGs in static state and the protection of supercapacitors storage level.

3.5.2.1 Power Management

For the given case study, the balancing condition implies that the exchanged power with the distribution network has to be produced from all sources in the MG:

$$p_{mg}(t) = p_{mt}(t) + p_{PV}(t) + p_{sc}(t) + p_{cp}(t) \tag{3.20}$$

where p_{mg} is the total power of the MG, p_{mt} is the power from the microturbine, p_{PV}, is the power from the PV generation unit, p_{sc} is the power from supercapacitors in generation mode, and p_{cp} is the total consumed power by the local loads.

Since the microturbine has a slow dynamic response time, the power management strategy is to use it to provide the power for a long time range and to ensure the long-term energy management. The average power during a time range T is expressed as

$$P = \{p(t)\}_T = \frac{1}{T}\int_0^T p(t)dt \tag{3.21}$$

During such a long time range, the average value of fast power variations, which are exchanged with supercapacitors, can be neglected. The power balancing condition (3.20) can be rewritten as follows:

$$P_{mg} = \{p_{mg}\}_T = \{p_{mt}(t) + p_{PV}(t) + p_{sc}(t) + p_{cp}(t)\}_T = P_{mt} + P_{PV} + P_{cp} \tag{3.22}$$

A great advantage of MGs is the facility of using a communication bus. By assuming that the total load power ($\widehat{p}_{cp}$) and the PV-generated power ($\widehat{p}_{PV}$) are sensed and received by the MGCC, the power reference for the microturbine P'_{mt_ref} can be calculated by the inversion of the average equation (3.22):

$$P'_{mt\_ref} = P_{mg\_ref} - \widehat{P}_{PV} - \widehat{P}_{cp} \tag{3.23}$$

Supercapacitors have the dynamic ability to master in real-time fast variations of the power flow, and here they are used to ensure the short-term power balancing. The power reference for supercapacitors p'_{sc_ref} is calculated by the inversion of the real-time equation (3.20) as follows:

$$p'_{sc\_ref}(t) = p_{mg\_ref}(t) - \widehat{p}_{mt}(t) - \widehat{p}_{PV}(t) - \widehat{p}_{cp}(t) \tag{3.24}$$

This power reference is sent to the local controller of the supercapacitors bank. This one injects or absorbs in real-time active power whenever the frequency deviation differs from zero. In order to simplify the MGCC for the

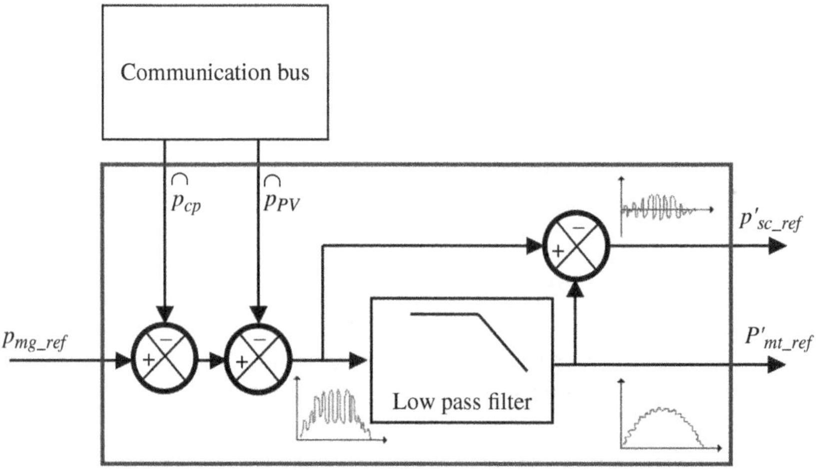

Figure 3.16 Power management.

islanded mode, the sensed power from the microturbine can be considered equal to its reference power. Therefore,

$$p'_{sc\_ref}(t) = -P'_{mt\_ref} - \hat{p}_{PV}(t) - \hat{p}_{cp}(t) \tag{3.25}$$

The power management can be also described with the block diagrams shown in Fig. 3.16, where the average power calculation is based on a low-pass filter.

3.5.2.2 Storage Level Protection

Supercapacitors have a finite storage capacity. The terminal voltage of a supercapacitor represents its energy storage level. For security reasons, it should be between the maximum allowed value (which represents the maximum storage energy E_{sc_max}) and 50% of this value (which represents the minimum storage energy E_{sc_min}) for efficiency reasons. In order to limit the terminal voltage, an additional control function has to be used. For example, if this voltage is under V_{sc_pmin} (105 V for four supercapacitor modules in series, which represents 30% of the maximum storage energy in the given case study), the available energy for generation decreased in a linear manner as shown in Fig. 3.17a. If this voltage is under V_{sc_min} (96 V, which represents 25% of the maximum storage energy), supercapacitors cannot operate in a generation mode (Fig. 3.17a). The limitation mode for the accumulation mode is designed in a similar way (Fig. 3.17b).

The final power reference that should be sent to the local controller of supercapacitors can be obtained as follows:

$$p_{sc\_ref} = \beta p'_{sc\_ref} \tag{3.26}$$

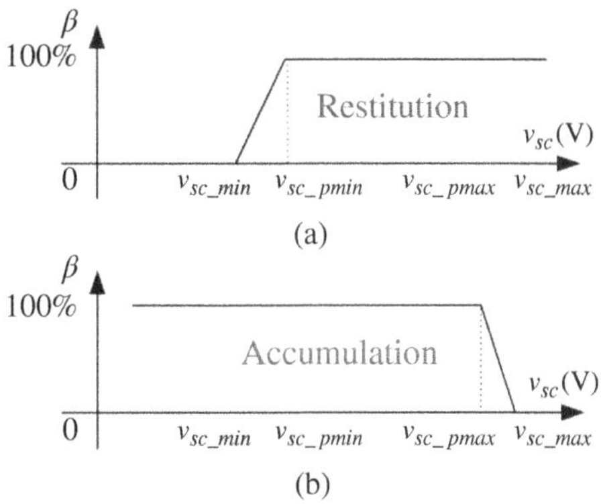

Figure 3.17 Energy limitation for the supercapacitor storage level: (a) restitution mode and (b) accumulation mode.

where

$$\beta = \begin{cases} 0 \quad \text{for} \quad \widehat{v}_{sc} < v_{sc\_min} \\ \dfrac{\widehat{v}_{sc} - v_{sc\_min}}{v_{sc\_pmin} - v_{sc\_min}} \quad \text{for} \quad v_{sc\_min} < \widehat{v}_{sc} < v_{sc\_pmin} \\ 1 \quad \text{for} \quad v_{sc\_pmin} < \widehat{v}_{sc} < v_{sc\_pmax} \\ \dfrac{v_{sc\_max} - \widehat{v}_{sc}}{v_{sc\_max} - v_{sc\_pmax}} \quad \text{for} \quad v_{sc\_pmax} < \widehat{v}_{sc} < v_{sc\_max} \\ 0 \quad \text{for} \quad v_{sc\_max} < \widehat{v}_{sc} \end{cases} \tag{3.27}$$

3.5.2.3 Regulation of Operating Points

Another task of the power management is to supervise the storage level of supercapacitors. If its level reaches a high level, a reduction in the generated power by the microturbine can be used. Hence, more power will be extracted from supercapacitors and will reduce the e_{sc_ref} stored energy. So the stored energy can be estimated as

$$\tilde{e}_{sc}(t) = \int_{\Delta t} \Delta p_{sc}(t)dt = \frac{1}{2}C\widehat{v}_{sc}^2 \tag{3.28}$$

Then, the microturbine power adjustment is calculated to minimize the difference between this value and an achieved energy level (e_{sc_ref}).

$$\Delta P_{mt\_ref}(t) = k_{pe}(e_{sc\_ref} - \tilde{e}_{sc}(t)) \tag{3.29}$$

Therefore (3.23) can be modified to regulate the storage level v_{sc}.

$$P_{mt\_ref}(t) = P'_{mt\_ref}(t) + \Delta P_{mt\_ref}(t) \tag{3.30}$$

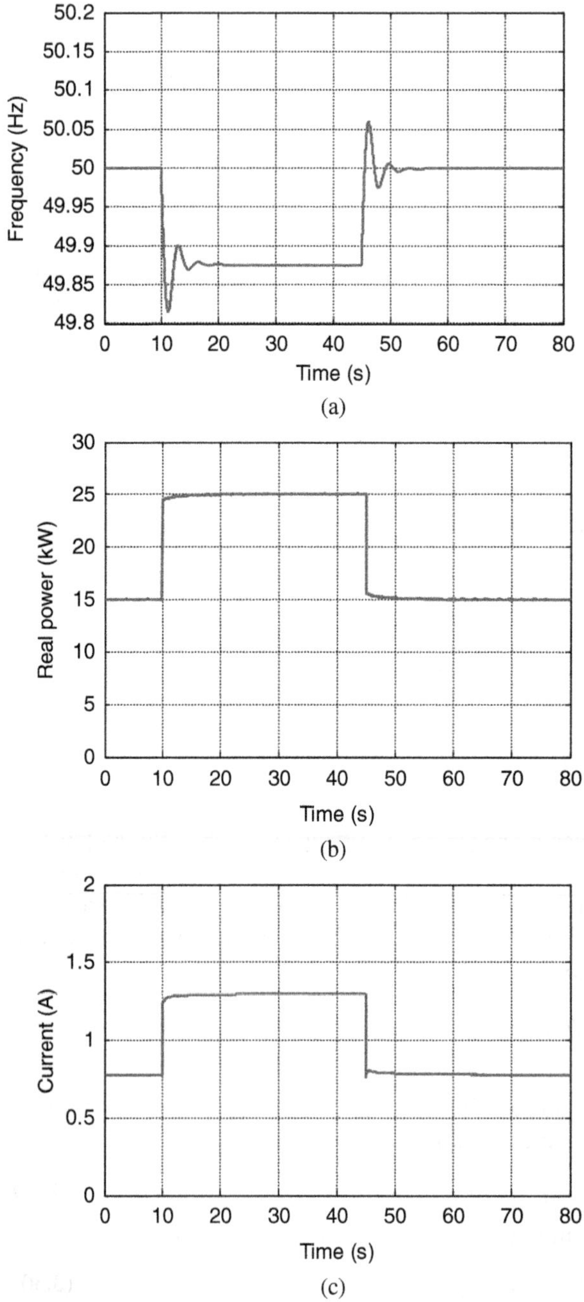

Figure 3.18 System response without MG support: (a) frequency, (b) P_{hv}, and (c) HV line current.

This power reference is sent to the local controller of the microturbine. The presented control strategy can be integrated into the MGCC.

3.5.3 Simulation Results

To highlight possible contributions of the MG, a load step change from 15 to 25 kW between 10 and 45 s is applied in the distribution network [31]. As the first scenario, the MG is disconnected from the distribution network (K_1 is open in Fig. 3.11) and the reactive power (Q_{hv}) is set to zero. Verification has been done with MATLAB–Simulink™ simulation tool. The obtained simulation results in Fig. 3.18a shows the frequency variations, which are caused by the step load change. Figure 3.18b presents the power production and distribution inside the network at the PCC. Without the MG utility, this power is equal to the required power by load 2. Currents in HV lines (Fig. 3.18c) induce losses.

For the second test scenario, the MG is considered in the connected mode, and the DNO asks a change of the spinning power reserve 20 s after the load transient (at 30 and 65 s in the Fig. 3.19d). This power reserve is used to participate in the secondary frequency regulation. By comparison, it can be seen that the frequency deviation (Fig. 3.19a) and the power inside the distribution network (Fig. 3.19b) are less. Especially in Fig. 3.19a, the two intervals between 10–30 and 45–65 s show the contribution of the primary frequency control. The time intervals between 30–45 and 65–80 s highlight the interests of the secondary frequency regulation. In consequence, the HV line current (Fig. 3.19c) has been decreased.

The real-time variations in the output power of the gas microturbine (p_{mt}), supercapacitors (p_{sc}), and PV generating system (p_{PV}) are shown in Fig. 3.19e. The PV system produces an intermittent power; the gas microturbine matches the power requirements in a long time range; and the supercapacitor system performs the transient power management. Figure 3.19f shows a good contribution of MG in both primary and secondary frequency controls. Figure 3.19g represents the variation of the supercapacitor terminal voltage (v_{sc}). This voltage is well controlled between its limits v_{sc_max} and v_{sc_min}. In consequence, the supercapacitor storage energy varies between E_{sc_max} (nominal energy, 330 kJ) and E_{sc_min} (25% of the nominal energy) as shown in Fig. 3.19h.

3.6 Microgrids Laboratory Technologies

In the last two decades, numerous laboratories (Labs) have been established to conduct valuable research on the MG planning, operation, and control all over the world. Laboratories have been mostly funded by national and private institutes, universities, and industry. In structural point of view, the existing active MG laboratories can be divided into four classes [32]: real MG

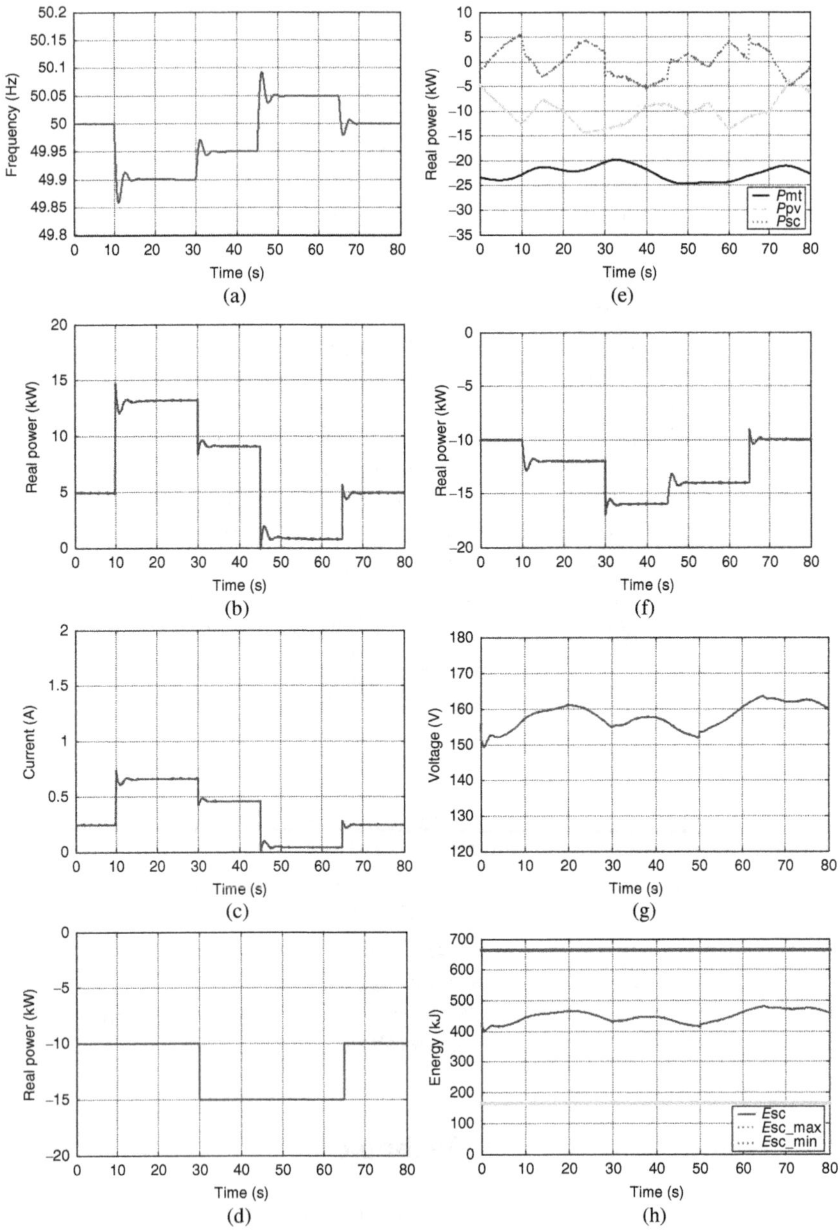

Figure 3.19 System response with MG support: (a) frequency, (b) P_{hv}, (c) HV line current, (d) $P_{mg\_dno\_ref}$, (e) powers from sources inside the MG, (f) injected power at the PCC, (g) supercapacitor terminal voltage, and (h) supercapacitor storage energy.

Figure 3.20 A real MG under High Tech Green Campus project at the Kyushu Institute of Technology (Kitakyushu, Japan, September 2012).

laboratory, simulation software-based MG laboratory, hardware-in-the-loop (HIL)-based MG laboratory, and hybrid MG laboratory.

In a real MG laboratory, a real-world MG exists. In addition to supply the local loads in a building, campus, institute, or complex, this type of MGs can be also used for the research activities/projects (Fig. 3.20). In a real MG laboratory, it is clear that the duration for research activities, the range of applied test scenarios, and change in configurations/topologies are strict and limited. Usually, the recorded data from daily operation of the MG, its DGs, and modules are usually useful for analysis research projects.

The simulation software-based MG laboratory (Fig. 3.21) comprises one or more computers with appropriate simulation software. In this laboratory, the MG is constructed in the simulation environment and there are no any real DGs, power electronic units, and other external digital or analog modules. Unlike the real MG laboratory, this laboratory is cheap and there is no limit on the type and number of scenarios implementation and application ideas.

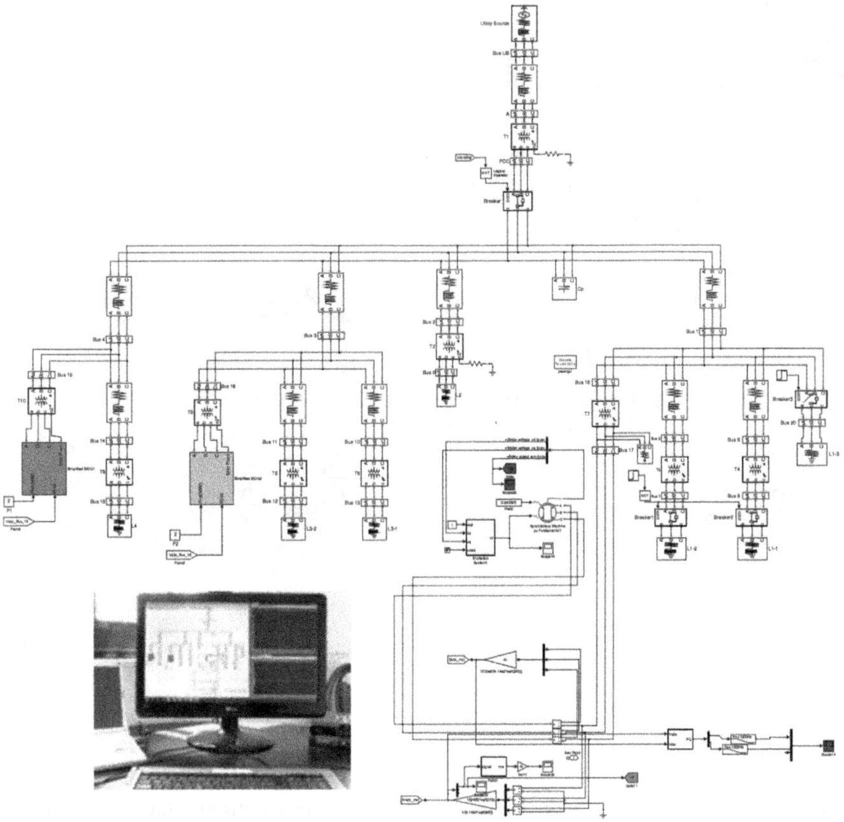

Figure 3.21 A simulation software-based MG laboratory (an MG example in MATLAB–Simulink™ environment).

However, since there is usually a big gap between reality (physical environment) and simplicity (of modeling/simulation environment), sometimes the validity of research results may be questionable.

The HIL-based MG laboratory enables the test of real devices such as ESSs or DGs/RESs, by connecting them to a grid model and control system that are simulated in a real-time manner [33, 34]. This way, the studied equipment behaves as if it is connected to the real network; it can thus be tested over a wide range of grid configurations, including fault conditions, without any risk. In recent years, the HIL-based MG laboratory (Fig. 3.22) gives promising results within the framework of various MG research projects across the world. It has made it possible to move from a classical design approach to a real-time process [30]. Necessarily, it is not needed to use real RESs/DGs or ESSs in the HIL-based MG laboratory, and usually appropriate emulators and analog/digital simulators can be applied instead of the real units. However, it is noteworthy that by simulating

Figure 3.22 Analog power system simulator as an HIL-based power system laboratory in Kyushu Power Electric Co. (Fukuoka, Japan, June 2010).

Figure 3.23 A hybrid laboratory (Power System Control laboratory, Kumamoto University, Kumamoto, Japan, August 2010).

a complete network/unit in a single processor of the real-time simulator (RTS), the required computation time may be far beyond the maximum time step to achieve real-time simulation.

The hybrid MG laboratory is a combination of real MG and simulation software-based MG laboratories. Both real and simulator units and simulation software are available (Fig. 3.23). In the MG laboratories, for the sake of MG planning, control synthesis/analysis, and operation, several software and simulation environments such as MATLAB, Simulink, SimPowerSystems,

DIgSILENT [35], PSS-SINCAL [36], PSCAD [37], Hommer [38], and Eurostag [39] may be used. Eurostag is a time-domain simulation program useful for dynamic studies of isolated power grids. It uses phasor representation for power system dynamic simulation (transient, mid-, and long-term stability).

The SimPowerSystems library blocks are used to simulate basic elements of the grid, such as standard synchronous machine, wind turbine, PV, loads, and transmission lines. The Simulink blocks are useful for modeling the control feedback systems such as PI controllers, speed governor systems, and automatic voltage regulators. The DIgSILENT, PSS-SINCAL, and Hommer are useful professional simulation environments for the power grids planning, control synthesis, and operation analysis. The PSCAD software is mostly suitable for research on power electronics devices and systems.

3.6.1 Hardware-in-the-loop-based Microgrid Laboratory

Here, due to simplicity and wide applicability, the HIL-based MG laboratory is explained in more detail. But first of all, it is better to discuss the main core of this type of laboratory, that is, "HIL simulation."

3.6.1.1 Hardware-in-the-loop Simulation

The HIL simulation has been intensively used for controller assessment over the years [40]. The aerospace industry uses this technique since flight control system is a safety-critical aspect. This methodology yields exhaustive testing of a control system to prevent costly and damageable failures. Since 1990, many groups in automotive industry have employed the HIL simulation for testing embedded electronic control units. This methodology avoids intense and complex integration tests on the actual vehicle. The development time is reduced and a high-quality assurance is obtained. The HIL simulation is increasingly used to develop new components and actuators in many fields, such as vehicle component evaluation, assessment of drive controls, power electronics and MGs, servo control and robotics, railway traction systems for trains and subways, and education applications [41].

In electric industry applications, software simulation is always an essential preliminary step to test the required performances of the drive and its control by using simple models of the power system. The HIL simulations are sometimes used for validation tests before implementation on actual processes. In contrast to software simulation, the HIL simulation uses one or more actual devices and the other parts of the process are simulated in a controller board with digital signal processor (DSP) or in parallel computers. The HIL simulation enables to check availability and reliability of drives (machines, power electronics, and control) before their insertion on a whole system. Moreover, many implementation constraints are taken into account such as sensor accuracy, sampling period, modulation frequency, active limitations, and so on. More specifically, electrical generators of wind energy conversion systems can be tested by using HIL simulation, and some small-rate power systems can be used to validate control algorithms and maximum power point tracking strategies before implementation on a full-rate power system. Power system stabilizers and performance controllers can be also tested by the HIL simulation using analog power system simulators, and the proposed control loops can be tested before integration to the real power grids [42–45].

Here, as an example for better understanding of the HIL simulation, let us discuss on the power propulsion system testing for electric vehicles and hybrid electric vehicle. An electrical drive can be decomposed into several subsystems (Fig. 3.24a): the process control, the power electronics set, the electrical machine, and the mechanical load to move (the mechanical power train of a vehicle for example). Power devices are connected according to the action and reaction principle. A controller board contains the process control and yields the switching orders of the power electronics converter.

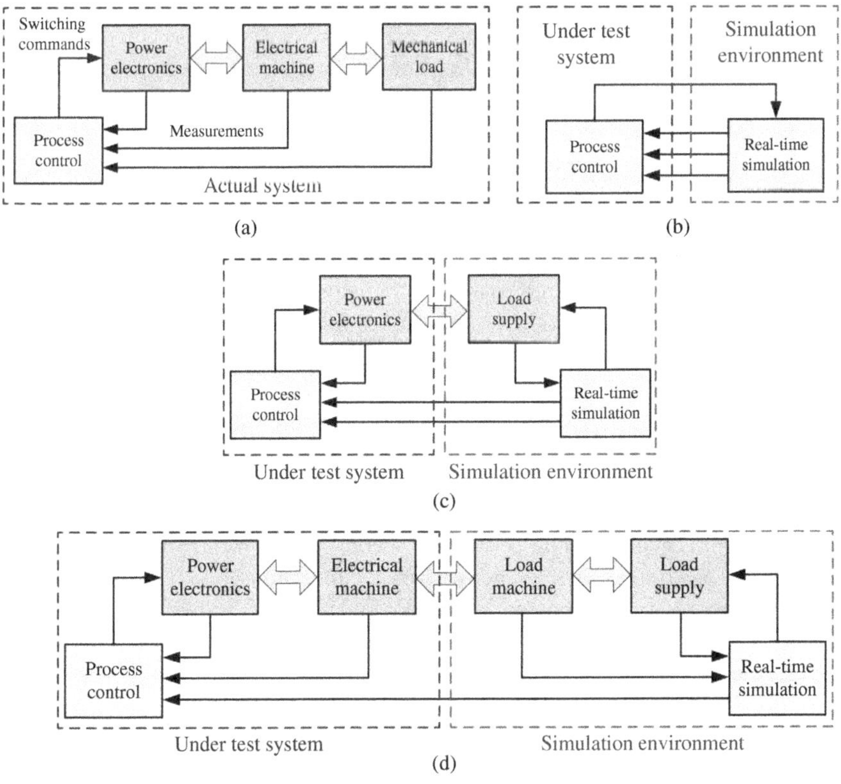

Figure 3.24 HIL simulation structure: (a) subsystems of an electric drive, (b) signal level HIL simulation, (c) power level HIL simulation, and (d) mechanical level HIL simulation.

Measurements of all power parts are inputs for this controller board. In some cases, several controller boards are used. In other cases, analog devices as field-programmable gate array (FPGA) boards are used to control the faster dynamics and to achieve high-frequency modulations of power electronic circuits.

The first step of the study is the simulation of the whole system including its control. All parts are simulated in the same simulation environment (software). In order to reduce the computation time, simple models and other simplifications are considered. For instance, the sampling period of the control is often neglected. For these reasons, simulation is not always accurate enough to enable a direct real-time implementation of the control. Before implementation on the actual system, different validations have to be made. The HIL simulation could be a very useful intermediary step. One of the simulated electrical parts can be replaced by its hardware device, which is then powered by an amplifier, that is derived through a digital to analog (D/A) converter by the simulation environment. By this way, the real constraints of this hardware subsystem are taken into account in the simulation loop. Three levels of HIL simulation can be considered: signal level, power level, and mechanical level.

In the first level, only the controller board (which contains the process control) is tested (Fig. 3.24b). The other parts (power electronics, machine, and mechanical load) are simulated in real time. The simulation system must manage inputs and outputs of the controller board under test. A second controller board is thus used to simulate in real time the power parts of the system. A specific signal conditioning is required to impose the same inputs and outputs as imposed by the power parts. This method can be called "signal level HIL simulation" because only signals are used at the interface between the system under test and the simulation environment. This kind of HIL has been very often employed in aerospace and automotive applications for assessment of controller boards.

In the second level, the actual controller board and the power electronics converter are evaluated. The other parts (electrical machine and mechanical load) are simulated. The simulation system must impose inputs and outputs for the power electronics and the controller board under test. The simulation environment is generally composed of a second power electronics set (electric load) and a second controller board (real-time simulation) (Fig. 3.24c). This method can be called "power level HIL simulation." Indeed, the interface between the system under test and the simulation environment require signal and power variables.

In the last level, the whole drive (control, power electronics, and electric machine) is tested and the mechanical part is simulated. The simulation system must impose mechanical inputs and outputs to the electrical machine under test. Moreover, measurements on the mechanical part have to be sent to the controller board under test. Another electrical machine (load machine) is often used as controlled mechanical load. It is supplied by a second power electronics set (load supply). A second controller board (real-time simulation) is required to control the load machine and to send fictitious mechanical "measurements" to the controller board under test (Fig. 3.24d). This method can be called "mechanical level HIL simulation." Indeed, the interface between the system under test and the simulation environment corresponds to mechanical variables.

3.6.1.2 The Hardware-in-the-loop-based Microgrid Laboratory

In the HIL-based MG laboratory, to validate simulation results and the robustness of the control system designs, some parts of MG will be simulated on the real-time simulation software environment. This allows the user to convert MATLAB/Simulink and SimPowerSystems models and then to run real-time simulations of those models on multiple target computers equipped with multi-core DSPs. This is used particularly for HIL and rapid control prototyping applications. Then, a solver must be utilized to optimize the simulation in discrete mode. It is a fundamental constraint of real-time simulation that the model must use fixed-step integration solvers.

The complexity of a simulated system may require much computational resources to be simulated in real time (meshed network with up to a dozen of DGs and devices). When simulating a complex system in a single processor of the RTS, the corresponding computation time for the original model greatly exceeded the maximum admissible computing time step. It is therefore not guaranteed that the studied system would be simulated in a real-time manner: possible overruns would have caused communication problems as real equipment is interfaced to the simulator. To meet real-time requirements, some solutions are required to be tested either to distribute calculations in several processors or to simplify the system structure (and the associated set of equations) by reducing the number of DGs and devices.

The distributed configuration enables the distribution of complex models over a cluster of multicore PCs running in parallel, thus reducing computation time to achieve real-time performances. The target nodes in the cluster communicate between each other with low latency protocols and the real-time cluster is linked to the simulator command station through a communication network protocol. Parallel distributed solving implies a separation of the studied system and the allocation of the subsystems to different cores of the target PC clusters, each one being solved by a unique core. A DSP card (e.g., dSPACE card) interfaces the control system in the simulation environment and the external hardware and analog/digital simulators. For example, it allows controlling the injected current from the controller (e.g., in Simulink environment) through the real inverter in the laboratory.

As the control system was being phased in the HIL-based MG platform, many adjustments were necessary to make it operational: notably soft-start sequence, emergency stop command, and improvements of the control algorithm to account for nonideal behavior of real equipment and sensor noise. This laboratory is particularly helpful to improve the realism of the developed control system far beyond what could have been done using only dynamic simulations.

In the HIL-based MG platform, real-time analysis consists in using real parts of the electrical component under operating conditions very close to the reality [41]. However, in general, it is difficult to make the testing of real system because of the risks involved (disruption of service, etc.), high costs involved (staff, equipment, etc.), and lack of flexibility (limited number and type of tests). Figure 3.25 shows an HIL-based MG platform. The setup is divided into three blocks: power block (real DGs, equipment, and loads), control block (MCs, LCs, and MGCC), and RTS block (to simulate the rest of the power grid). The HIL is used for real time simulation of the operating conditions. Consequently, this type of analysis provides more realistic results as the off-line simulation regarding the operation of the under-study devices.

The power block may include real PV, ESSs, and other DGs. Virtual devices are simulated in the RTS block. It may include unavailable DGs, lines, and loads. A power amplifier (Amp.), which is used as an interface to create the point of coupling, interfaces the RTS and control blocks to the power block via analog-to-digital and digital-to-analog conversion cards. The required controllers to produce appropriate control action signals for using in the RTS and power blocks are implemented in the control block, which may include several PCs. A communication bus is required to exchange the signals between the MGCC, LCs, and other units. An Ethernet with a communication protocol is used to transfer data and to enable the connection of various devices (for example, a system that measures currents and voltages and the communication of data to a computer).

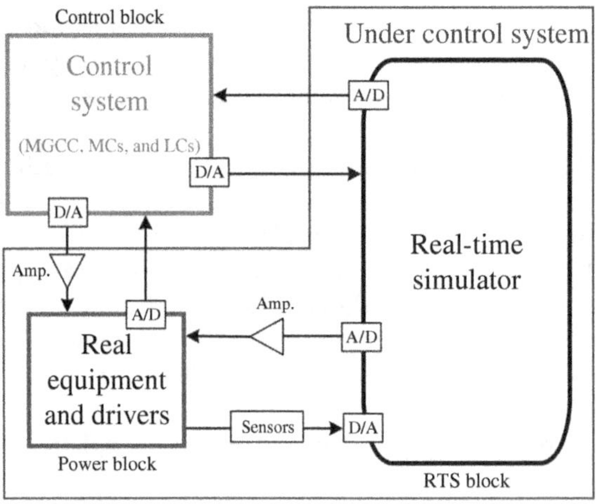

Figure 3.25 The HIL test setup.

3.6.2 Participant Laboratories to Provide the Present Book

The materials given in the present book are mainly the research outcomes and original results of three laboratories: *Smart/Micro Grids Research Center (SMGRC)* at the University of Kurdistan (Sanandaj, Iran), *Laboratory of Electrical Engineering and Power electronics (L2EP)* at Lille (France), and the *Power Electronics and Electrical Energy (PE&EE) Laboratory* at the Osaka University (Osaka, Japan). The SMGRC and the L2EP can be classified in the hybrid laboratories group, while the PE&EE is a HIL-based MG laboratory.

3.6.2.1 Smart/Micro Grids Research Center

The SMGRC provides a rich source of training, testing, and experimental laboratory facilities for various smart/micro grids projects, especially in the area of advanced robust and intelligent control synthesis and analysis methodologies. In addition to the modeling and simulation environment for training and educational purposes, a wide range of microsources (e.g., PV panel, wind turbine, diesel generator), ESSs, grid components, modules, and subsystems are available for different classes of testing and validation capabilities. The laboratory flexibilities allow testing of multiple configurations in the view point of MG types (DC, AC, and hybrid), operating modes (grid-connected and islanded), and multiple MGs clusters. A view of some facilities is shown in Fig. 3.26. The University of Kurdistan and West Regional Electric Company (WREC) are the main domestic supporters of the SMGRC. In the near future, the laboratory is

Figure 3.26 A view of SMGRC laboratory (University of Kurdistan, Sanandaj, Iran, May 2016).

going to be configured to accommodate a variety of research activities on MG subsystems such as smart monitoring, control, and management.

3.6.2.2 Laboratory of Electrical Engineering and Power Electronics

At the Laboratory of Electrical Engineering and Power Electronics (L2EP) of Lille, an experimental platform for MG and distributed energy sources has been developed to study electrical networks and systems in the presence of small dispersed generators. A view of L2EP facilities is shown in Fig. 3.27. It is composed of several generation systems, storage units, and loads, such as a monitored PV power plant, super capacitors and static/dynamic hardware in the loop emulators. Furthermore, the central core of the platform is an RTS that can be connected to hardware equipment through power amplifiers.

Figure 3.27 A view of the L2EP laboratory (Lille, France, January 2016).

Using this experimental platform, various control techniques for the MG control synthesis and analysis as well as for the coordination of different generation systems can be studied. For this purpose, the L2EP includes a hybrid RTS, PV generators, CHP microturbines, fuel cells and other ESSs, controllable loads, wind turbine emulator, and associated control systems.

3.6.2.3 Power Electronics and Electrical Energy Laboratory

The Power Electronics and Electrical Energy (PE&EE) Laboratory is focused on developing basic technologies based on power electronics and applied superconductivity for future power systems and MGs considering environmental and energy resource problems. From MG point of view, this laboratory is a HIL-based MG laboratory with a wide range of hardware and software and an RTS platform. Computer simulators are used to investigate the proposed systems and control methods by modeling power grids, circuits, and control

Figure 3.28 A view of the PE&EE laboratory (Osaka University, Osaka, Japan, August 2015).

systems as well as by using various software such as PSCAD/EMTDC, PSIM, MATLAB–Simulink, Saber, and PLECS.

Solar power units, wind power simulators, superconducting magnet energy storage (SMES), transmission lines, and load units are also available. Inverters with DSP controllers demonstrate various experiments, such as laboratory-scale DC MGs, matrix converters for gas engine cogeneration, soft switching inverters for FACTS, and bidirectional DC/DC converters for DC power system. Some of the laboratory's facilities are shown in Fig. 3.28.

3.7 Summary

MGs are small electrical distribution systems that connect multiple customers to multiple distributed sources of generation and storage. MGs are typically characterized by multipurpose electrical power services to communities with overall energy demands ranging up to several thousand kilowatt hours per day and are connected via low-voltage networks. A great interest is that these hybrid power systems have the potential to provide reliable power supply to remote communities where connection to transmission supply is uneconomic.

In this chapter, following an introduction on the recent MG developments, the MG structure and concepts were described. Then, the MG operation modes were briefly discussed and the overall control mechanisms for the given DGs inside an MG were explained. The possibility of MGs' contribution in the upstream grid's ancillary services, particularly frequency regulation, was emphasized. Finally, the technologies of the existing MG laboratories across the world as well as the participating laboratories in this book materials were introduced.

References

1 Kok, J.K., Warner, C.J., and Kamphuis, I.G. (2005) *Power Matcher: Multi-agent Control in the Electricity Infrastructure*. Proceedings of Autonomous Agents and Multi-Agent Systems, 25–29 July, Utrecht, Netherlands.

2 Bevrani, H., Watanabe, M., and Mitani, Y. (2014) An introduction on power system monitoring, in *Power System Monitoring and Control* Chapter 1, IEEE-Wiley Press, New York.

3 Fathi, M. and Bevrani, H. (2013) Statistical cooperative power dispatching in interconnected microgrids. *IEEE Transactions on Sustainable Energy*, **4** (3), 586–593.

4 Rocabert, J., Luna, A., Blaabjerg, F., and Rodriguez, P. (2012) Control of power converters in AC microgrids. *IEEE Transactions on Power Electronics*, **27** (11), 4734–4748.

5 Lu, D. (2010) Design and control of a PV active generator with integrated energy storages: application to the aggregation of producers and consumers in an urban micro smart grid. PhD thesis. Ecole Centrale de Lille, France.

6 Hatziargyriou, N., Asano, H., Iravani, R., and Marnay, C. (2007) Microgrids – an overview of ongoing research, development and demonstration projects. *IEEE Power & Energy Magazine*, **5** (4) 78–94.

7 Marnay, C. and Firestone, R. (2007) *Microgrids: An Emerging Paradigm for Meeting Building Electricity and Heat Requirements Efficiently and with Appropriate Energy Quality*. The European Council for an Energy Efficient Economy 2007 Summer Study, 4–9 June, La Colle sur Loup, France.

8 Web site of CERTS http://certs.lbl.gov (accessed May 15, 2016).

9 Tsikalakis, A. and Hatziargyriou, N. (2005) Economic scheduling functions of a microgrid using a central controller and applying different market and demand side options. *CIGRE Symposium*, 13–16 April, Athens.

10 Bevrani, H., Watanabe, M., and Mitani, Y. (2012) Microgrid controls, in *Standard Handbook for Electrical Engineers*, 16th edn (ed. H. Wayne Beaty) Section 16.9, McGraw-Hill, pp. 160–176.

11 Bevrani, H., Ghosh, A., and Ledwich, G. (2010) Renewable energy sources and frequency regulation: survey and new perspectives. *IET Renewable Power Generation*, **4** (5), 438–457.

12 Huang, J.Y., Jiang, C.W., and Xu, R. (2008) A review on distributed energy resources and microgrid. *Renewable and Sustainable Energy Reviews*, **12** (9), 2472–2483.

13 Bevrani, H. (2014) Frequency control in microgrids, in *Robust Power System Frequency Control*, 2ndChapter 11 edn, Springer, Switzerland.

14 Bevrani, H., Watanabe, M., and Mitani, Y. (2014) *Power System Monitoring and Control*, IEEE-Wiley Press, New York.

15 Lasseter, R. and Paigi, P. (2004) MicroGrid: a conceptual solution. *IEEE Annual Power Electronic Specialists Conference*, **6** (1), 4285–4290.

16 Guillermo, H.G. and Reza, I. (2006) Current injection for active islanding detection of electronically-interfaced distributed resources. *IEEE Transactions on Power Delivery*, **21** (3), 1698–1705.

17 Jayawarna, N., Jenkins, N., Barnes, M. *et al* (2005) Safety analysis of a Microgrid. *International Conference on Future Power Systems*, 1–7.

18 Bevrani, H. (2014) Frequency control and real power compensation, in *Robust Power System Frequency Control*, 2ndChapter 2 edn, Springer, Switzerland.

19 Golpira, H., Bevrani, H., and Naghshbandi, A.H. (2012) An approach for coordinated AVR-PSS design in large scale interconnected power systems considering wind power penetration. *IET Generation, Transmission, and Distribution*, **6** (1), 39–49.

20 Lasseter, R., Akhil, A., Marnay, C. *et al.* (2002) Integration of distributed energy resources – the microgrid concept. *CERTS Microgrid Review*, White paper, pp. 1–27.

21 Bevrani, H., Feizi, M.R., and Ataee, S. (2016) Robust frequency control in an islanded microgrid: Hinf and Mu synthesis approaches. *IEEE Transaction on Smart Grids*, **7** (2), 706–717.

22 Lasseter, R. and Abbas, A. (2003) *Integration of Distributed Energy Resources: The CERTS Microgrid Concept*. Consortium for Electric Reliability Technology Solutions, California Energy Commission, P50003-089F.

23 Zhou, T. and Francois, B. (2011) Electrical energy systems energy management and power control of an hybrid active wind generator for distributed power generation and grid integration. *IEEE Transactions on Industrial Electronics*, **58** (1), 95–104.

24 Baroudi, J.A., Dinavahi, V., and Knight, A.M. (2007) A review of power converter topologies for wind generators. *Renewable Energy*, **32** (14), 2369–2385.

25 Awad, B., Wu, J., and Jenkins, N. (2008) Control of distributed generation. *Elektrotechnik & Informationstechnik*, **125** (12), 409–414.

26 Bouhali, O., Francois, B., Berkouk, M., and Saudemont, C. (2009) Power sizing and control of a three-level NPC converter for grid connection of wind generators. *Electromotion Journal*, **16**, 38–48.

27 Li, P. (2010) Formalisme pour la Supervision des Systèmes Hybrides Multi-Sources de Générateurs d'Energie Répartie: Application à la Gestion d'un Micro Réseau. PhD thesis. Ecole Centrale de Lille, France.

28 Bevrani, H. (2014) *Robust Power System Frequency Control*, 2nd edn, Springer, Switzerland.

29 Courtecuisse, V., Robyns, B., Francois, B. *et al.* (2008) Variable speed wind generators participation in primary frequency control. *Wind Engineering*, **32** (3), 299–318.

30 Bevrani, H. and Hiyama, T. (2011) *Intelligent Automatic Generation Control*, CRC Press, NY, USA.

31 Li, P., Degobert, P., Robyns, B., and Francois, B. (2009) Participation in the frequency regulation control of a resilient microgrid for a distribution network. *International Journal of Integrated Energy Systems*, **1** (1), 1–5.

32 Bevrani, H. (2016). *Study on Design and Implementation of Smart Micro-grids in West Electric Industry of Iran: Challenges and Practical Solutions (in Persian)*. Technical report, Final version, University of Kurdistan, January.

33 Wang, Y., Delille, G., Guillaud, X., Colas, F., and Francois, B. (2010) Real-time Simulation: the Missing Link in the Design Process of Advanced Grid Equipment. IEEE PES General Meeting, Minneapolis, MN, July.

34 Ren, W., Steurer, M., and Baldwin, T.L. (2009) An effective method for evaluating the accuracy of power hardware-in-the-loop emulations. *IEEE Transactions on Industry Applications*, **45** (4), 1484–1490.

35 DIgSILENT software official website: http://www.digsilent.de (accessed May 15, 2016).

36 PSS®SINCAL Database Interface and Automation, SIEMENS, April 2016. http://sincal.s3.amazonaws.com/doc/Misc/SINCAL_DatabaseInterface.pdf (accessed May 15, 2016).

37 PSCAD software official website: https://hvdc.ca/pscad/ (accessed May 15, 2016).

38 Homer software official website: http://www.homerenergy.com/ (accessed May 15, 2016).

39 Eurostag software official website: http://www.eurostag.be (accessed May 15, 2016).

40 Bouscayrol, A., Guillaud, X., and Delarue, P. (2005) *Hardware-in-the-Loop Simulation of a Wind Energy Conversion System Using Energetic Macroscopic Representation*. IEEE-IECON'05, Raleigh (USA), November.

41 Bevrani, H. (2012) *Engineering Education System in Japan: Observations in Study, Teaching and Research*. Proceeding of 1st Engineering Education Conference, University of Duhok, Duhok, Iraq, 17–19 April 2012.

42 Bevrani, H., Daneshfar, F., and Hiyama, T. (2012) A new intelligent agent-based AGC design with real-time application. *IEEE Transaction on Systems, Man, and Cybernetics, Part C*, **42** (6), 994–1002.

43 Kakigano, H., Miura, Y., and Ise, T. (2013) Distribution voltage control for DC microgrids using fuzzy control and gain-scheduling technique. *IEEE Transactions on Power Electronics*, **28** (5), 2246–2258 May.

44 Bevrani, H. and Hiyama, T. (2009) On load-frequency regulation with time delays: design and real time implementation. *IEEE Transactions on Energy Conversion*, **24** (1), 292–300.

45 Bevrani, H., Hiyama, T., and Mitani, Y. (2008) Power system dynamic stability and voltage regulation enhancement using an optimal gain vector. *Control Engineering Practice*, **16** (9), 1109–1119.

4

Microgrid Dynamics and Modeling

Understanding the dynamics and using appropriate modeling methodologies are significant issues for microgrid (MG) control synthesis and stability analysis. Although the MG is of small scale, it has many of the complexities of a large-scale conventional power system. Dynamic analysis is therefore required to ensure that the MG operates in a stable manner with controlled voltage and frequency fluctuations. Particularly, transition from the grid-connected to the islanded operation mode may pose severe challenges. Depending on the configuration, type, and components, the MG dynamics may change; and for different applications, different modeling methodologies may be required. Several approaches for the MG dynamics representation and modeling can be found in the literature.

In this chapter, following an overview on the recent developments in MG dynamic representation and modeling, a dynamic model is given for the main grid and connecting components. The main grid is represented as a group of diesel generators. The model is then extended by including the MG system to perform an overall representation structure for the grid-connected MG. Modeling of some components (distributed generators, energy storage systems (ESSs), power converters, and loads) in the MG is emphasized, and a simplified frequency response model using low-order transfer functions is introduced. Then, a state-space model using more dynamic details is presented, and the modeling and dynamic analysis for an MG as a multivariable system are discussed. Finally, the chapter is summarized.

4.1 Introduction

MGs as the vital building blocks of future smart grids incorporate various distributed generators (DGs)/renewable energy sources (RESs) into the power grid and solve many problems of conventional power systems. Despite their many benefits, due to (i) close electrical proximity, fast dynamics, and short response time of DGs/RESs; (ii) inherent unbalanced nature of the MG; (iii) low energy

Microgrid Dynamics and Control, First Edition. Hassan Bevrani, Bruno Francois, and Toshifumi Ise.

storage capacity and lack of inertia; (iv) high number and diversity of used power microsources, power electronic converters, and other circuits/devices; and (v) high degree of parametric and topological uncertainties, the modeling and dynamic stability analysis of MGs may be more difficult than conventional power systems and are highly challenging. In response to this challenge, several approaches and timely research works can be found in the literature.

One of the important reasons for necessity of dynamic modeling is the small-signal stability analysis in MGs. In conventional power grids, stability analysis is well established with standard models of synchronous machines, governors, and excitation systems of varying orders that are known to capture the important modes for particular classes of problems. This does not yet exist for MGs and may be hard to achieve because of the wide range of power technologies that might be deployed [1]. To analyze the dynamic behavior and small-signal stability of an MG, DGs can be considered as the voltage source converters/voltage source inverters (VSCs/VSIs). VSCs are the main components in an MG power control and voltage regulation. In dynamic modeling of an MG, too slow and too fast dynamics are usually neglected. For instance, the conventional generator/source dynamics is much slower than the VSC dynamics, so the generator dynamics can be neglected. The generator bus voltage can be considered as a constant parameter and the associated dynamics is not also modeled. On the other hand, due to high frequency of switching elements in converters/inverters, the relevant dynamics are also negligible.

In [2–6], the dynamic order of each MG's DG is reduced to a first-order linear model with a time constant and a gain factor while the network dynamics are neglected. In some works, an MG is represented by a DC source with a VSC, which is connected to the main grid by means of an RL filter, step-up transformer, and a circuit breaker [7, 8]. Then, a low-order dynamic model for the mentioned equivalent system is obtained and can be easily used as an aggregated model for the connected MG for stability analysis and control synthesis purposes.

Some modeling approaches for MGs with inverter-based DGs are given in [9]. The modeling approaches considered the full dynamic model of the complete network rather than an inverter. In [9], the MG system is divided into three subsystems, namely, inverter, network, and loads. The inverter model embeds dynamics of controller, output filter, and coupling inductor. The state equations of network and load are represented on one of the inverter's reference frame, which is assumed to be the common reference. Then, by using the transformation technique, all the other inverters are transformed to this common frame. Each subsystem is modeled in the state-space form and combined together on the mentioned common reference frame.

A small-signal dynamic model of an MG including synchronous and power electronically interfaced DGs, as well as the relevant network is presented in

[10]. The same MG structure is taken into account in [11], and the stability analysis for various transient conditions such as energizing load, transition from grid-connected mode to islanded mode and vice versa is performed. An algorithm for simulating the dynamic behavior of low-voltage (LV) MGs is presented in [12]. The algorithm follows the stability approach and focuses on low-frequency dynamics. Inverter droops are used to interface DGs and loads are modeled as constant powers. This model has an advantage of modeling small unbalanced networks but lacks the analytic details required for stability analysis. A method to form the system matrices of large MGs in islanded mode is discussed in [13], where the considered DGs are inverter-based, and hence the dynamics are similar to that shown in [9].

A linear time-invariant (LTI) state-space model is provided in [14], and then an eigenvalue analysis is performed to investigate the MG dynamic behavior, while the electrical parameters and control gains are changing. To carry out eigenvalue analysis, a linear state-space model of the given MG is first developed. Then, the state-space model of the open-loop MG model is augmented with existing controllers to obtain the state-space model of the closed-loop system.

To investigate MG dynamic stability, a small-signal model of a typical MG containing asynchronous and synchronous DGs, power electronic-based energy storage, and power network is proposed in [1]. The small-signal model of each subsystem is established, respectively, and then the global model is set up in a global reference axis frame. Eigenvalues distributions of the MG system under certain steady operating status are identified to indicate the damping of the oscillatory terms and its effect on the system stability margin.

The small-signal models containing synchronous DGs, inverter-based DGs, and power network are established in [10, 15], and the dynamic characteristics in autonomous operation mode are analyzed based on the individual models. The global small-signal model of the inverter-based MG (without considering the asynchronous wind turbine) is set up in [16] to design the controller of inverter-based DGs.

Some stability aspects of MGs depending on the modes of operation, control topology, types of microsources, and network parameters are investigated in [17]. Similar to a conventional system, the stability issues in an MG can be divided as small-signal (steady-state), transient, and voltage stability (and frequency stability in AC/hybrid MGs). Small-signal stability in an MG is related to some issues such as feedback systems/controllers, continuous load switching, and power limit of the DGs. The transient analysis of an MG can ensure system operability after large disturbances, and voltage stability problems may happen due to reactive power limits, load dynamics, and tap changers.

The dynamic aspects and modeling methodologies presented in this chapter facilitates an organized way to plan the microsource operation, MG stability analysis and controller design, islanding procedure, frequency/voltage control in normal and emergency conditions, and the load shedding criteria. The rest of the chapter is organized as follows: a dynamic model representation for the main grid and connecting components is given in Section 4.2. Section 4.3 extends the obtained model by including the MG system to perform an overall model for the grid-connected MG. Section 4.4 introduces the modeling of some components in the MG, and a simplified frequency response model using low-order transfer functions is given in Section 4.5. Section 4.6 presents a state-space model using more dynamic details, and the modeling and dynamic analysis for the MG as a multivariable system are discussed in Section 4.7. Finally, the chapter is summarized in Section 4.8.

4.2 Distribution Network (Main Grid) and Connection Modeling

The grid-connected mode is a particular complex operation mode since local loads inside the MG have to be supplied as well as powers have to be sent to the main grid [18]. Moreover, the MG can also provide ancillary services for the main grid, if needed, such as the frequency regulation. So the study of the MG in a grid-connected operation mode not only includes the MG but also covers the distribution network. Figure 4.1 shows an MG in a grid-connected mode. The subscripts of "*g* or *gd*," "*mv*," "*mg*," and "*cp*" represent grid, medium voltage, MG, and consumption power, respectively.

Before focusing on the MG components such as DGs and local loads, modeling of the distribution network (main grid) and the existing devices between the main grid and the MG are necessary since the static and dynamic characteristics are very important for the validation (such as its reaction in response to the MG behaviors). Therefore, first modeling of the main grid is introduced. It includes the distribution network, MV line, transformer between the MV bus and the MG bus. The total MV loads outside the MG are also modeled as an equivalent passive load (as shown in Fig. 4.1). The external loads may influence the distribution network as well as the MG power exchange (P_{mg}, Q_{mg}) by creating voltage droops or frequency fluctuations at the MV bus. The coupling method at each bus is also presented at the end.

4.2.1 Distribution Network Modeling

In many research studies, the MG is assumed to be connected to an infinite network, which is implemented with an ideal voltage source with a constant frequency value and a constant voltage root mean square (RMS) value. This cannot be appropriate in the context of researches on the grid-connected MGs, especially for highlighting the participation of ancillary services, such as the primary frequency regulation. Therefore, a simplified model of a bulk synchronous

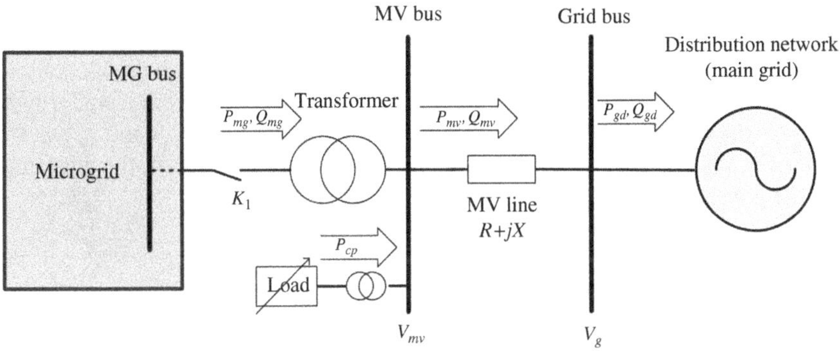

Figure 4.1 Global scheme of an MG in a grid-connected mode.

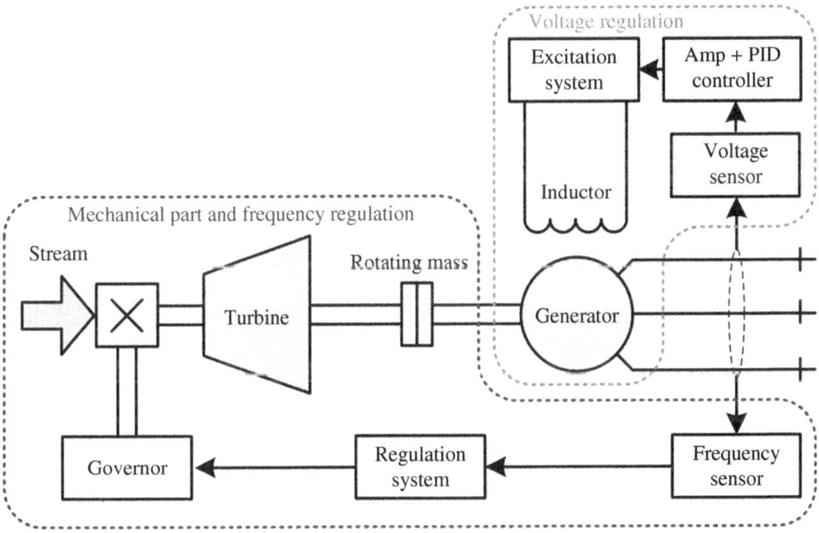

Figure 4.2 Diesel group structure.

generator or an aggregated model of a group of diesel generators is used with control systems of the frequency and the voltage RMS value [19, 20].

The purpose of the model is to mathematically describe the transient stability of the electrical quantities from the diesel group (such as the frequency variation or the voltage variation). So the mathematical equations are established with only variation quantities (Δ) of physical variables. The structure of a diesel group that is presented in Fig. 4.2 is composed of four parts [19]: (i) the mechanical part and frequency regulation loop, (ii) the voltage regulation, (iii) the coupling between these two parts, and (iv) additional equations between the per units and the international system (SI) units.

4.2.1.1 Mechanical Part and Frequency Regulation Loop

The mechanical part is composed of a governor, a turbine, a rotating mass including equivalent loads, a regulation system, and the power angle calculation [21]. The difference between the reference and the real power is transformed through the hydraulic amplifier to the steam valve position command (ΔP_v). By assuming a linear relationship and considering a time constant τ_g, the equation for the governor can be expressed as

$$\Delta P_v = \frac{1}{1+\tau_g s}(\Delta P_{ref} - \Delta P_{reg}) \tag{4.1}$$

where ΔP_{ref} is the change of reference real power in per unit (pu) and ΔP_{reg} is the change of the real power in pu.

A simple prime mover model of a nonreheat steam turbine can be approximated with a single time constant τ_t:

$$\Delta P_m = \frac{1}{1+\tau_t s}\Delta P_v \tag{4.2}$$

where ΔP_m is the change of mechanical power output in pu.

Therefore, the speed–load characteristic (in pu) is approximated by considering the rotating mass and electrical load real power, which is approximately equivalent to the electromechanical power consumed by the generator (details can be found in [4]):

$$\Delta\omega = \frac{1}{D+2Hs}(\Delta P_m - \Delta P_c - \Delta P_e) \tag{4.3}$$

where ΔP_c is the change of active load power in pu, ΔP_e is the change due to the effect of voltage upon real power in pu, H is the inertial constant, and D is expressed as a percent change of the load divided by a percent change in the frequency.

The small change of the power angle ($\Delta\delta$) is obtained by integration of the small change in the speed:

$$\Delta\delta = \frac{1}{s}\Delta\omega \tag{4.4}$$

The regulation system is composed of a governor speed regulation with (or without) a load frequency control (LFC) [4], which is based on an integral controller:

$$\Delta P_{reg} = -\left(\frac{1}{R} + \frac{K_I}{s}\right)(\Delta\omega_{ref} - \Delta\omega) \tag{4.5}$$

where R is the speed regulation ratio, K_I is the integral controller gain for the LFC, and $\Delta\omega_{ref}$ is the variation reference of grid frequency. Without the LFC loop, the regulation system is expressed as (4.6). Organization of the modeling equations is represented in Fig. 4.3.

$$\Delta P_{reg} = \frac{1}{R}\Delta\omega \tag{4.6}$$

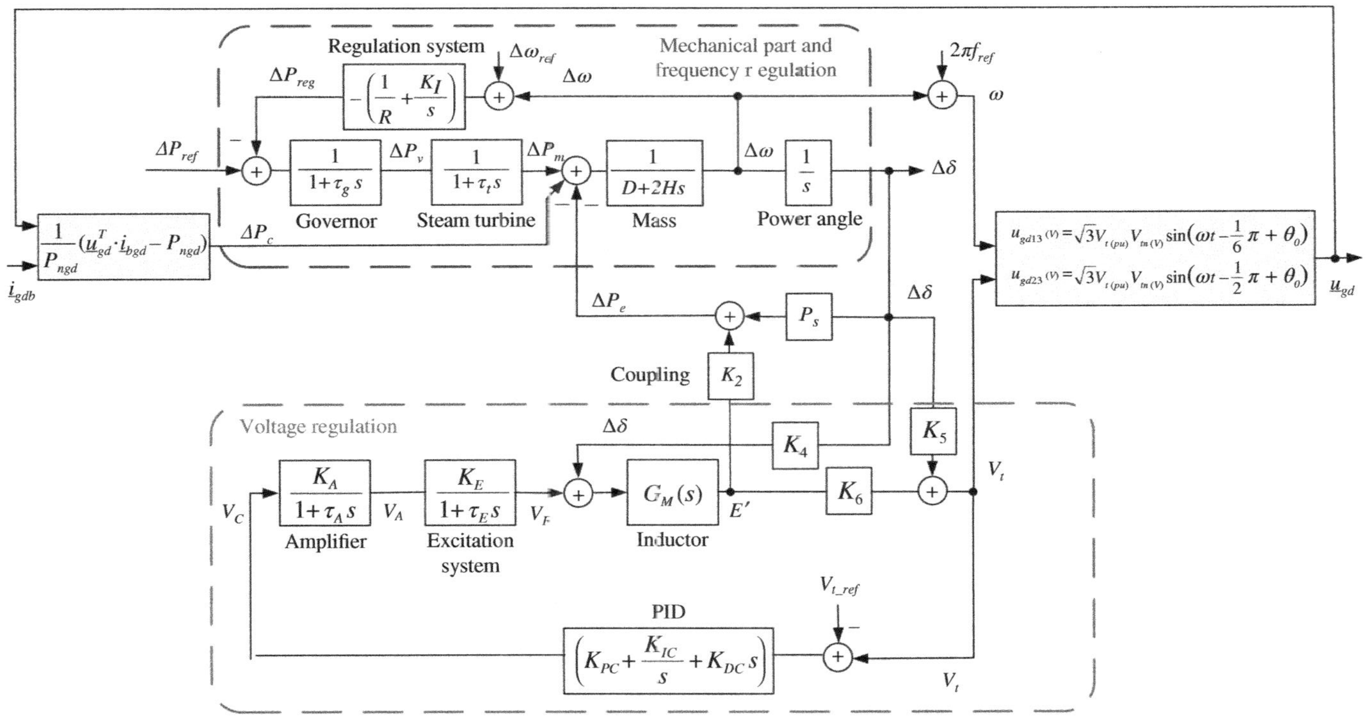

Figure 4.3 Block diagram of the diesel group standard model.

4.2.1.2 Voltage Regulation

The voltage regulation of the diesel group is composed of an amplifier, an exciter, a fourth-order generator model, the calculation of the terminal voltage and a proportional–integral derivative (PID) controller. The amplifier is represented by a transfer function with a gain K_A and a time constant τ_A:

$$V_A = \frac{K_A}{1 + \tau_A s} V_C \tag{4.7}$$

where V_A is the terminal voltage of the amplifier in pu and V_C is the terminal voltage reference given by the PID controller in pu. The transfer function of a modern exciter can be simplified as

$$V_F = \frac{K_E}{1 + \tau_E s} V_A \tag{4.8}$$

where V_F is the terminal voltage of the exciter in pu, K_E is the exciter gain, and τ_E is a time constant.

A fourth-order generator model [22] is used to obtain a better accuracy of the generator field calculation than with a simple first-order generator model:

$$\begin{aligned} E' &= \frac{(1 + T_{z1}s)(1 + T_{z2}s)(1 + T_{z3}s)(1 + T_{z4}s)}{(1 + T_{p1}s)(1 + T_{p2}s)(1 + T_{p3}s)(1 + T_{p4}s)}(V_F - K_4\Delta\delta) \\ &= G_M(s)(V_F - K_4\Delta\delta) \end{aligned} \tag{4.9}$$

where T_{p1} and T_{z1} are the pole-zero time constants of the first-order model, T_{p2} and T_{z2} are the pole-zero time constants of the second-order model, T_{p3} and T_{z3} are the pole-zero time constants of the third-order model, T_{p4} and T_{z4} are the pole-zero time constants of the fourth-order model, and K_4 is a gain.

Including the small effect of the rotor angle upon the generator terminal voltage, the generator terminal voltage can be written as follows [20]:

$$V_t = K_5\Delta\delta + K_6E' \tag{4.10}$$

where K_5 is the gain of the change in the terminal voltage for a small change in rotor angle with a constant stator electromotive force (emf) and K_6 is the gain of the change in the terminal voltage for a small change in the stator emf with a constant rotor angle.

A PID controller is used to improve the dynamic response as well as to reduce or eliminate the steady-state error:

$$V_C = \left(K_{PC} + \frac{K_{IC}}{s} + K_{DC}s\right) V_t \tag{4.11}$$

where K_{PC} is proportional gain, K_{IC} is integral gain, and K_{DC} is derivative gain of the PID controller.

4.2.1.3 Coupling Between the Mentioned Two Parts

Equations (4.4) and (4.12) represent the interaction between the frequency regulation and the voltage regulation. Small changes in real power due to the effect of voltage can be expressed as

$$\Delta P_e = P_s \Delta\delta + K_2 E' \tag{4.12}$$

where P_s is the synchronizing power coefficient and K_2 is the change in electrical power for a small change in the stator emf.

4.2.1.4 Adaptation Between the Per Units and SI Units

The adaptation between per units and SI units is derived by Equations (4.13)–(4.16). The change for the grid frequency is calculated as

$$\omega_{(\mathrm{rad/s})} = 2\pi f_{ref\,(\mathrm{Hz})} + \Delta\omega_{(\mathrm{rad/s})} \tag{4.13}$$

$$\Delta\omega_{(\mathrm{rad/s})} = 2\pi f_{ref\,(\mathrm{Hz})} \Delta\omega_{(\mathrm{pu})} \tag{4.14}$$

where f_{ref} is the grid frequency (e.g., 50 Hz). The change from per unit to SI unit (V) for the voltages is given as

$$\begin{cases} u_{gd13\,(\mathrm{V})} = \sqrt{3} V_{t\,(\mathrm{pu})} V_{tn\,(\mathrm{V})} \sin\left(\omega t - \frac{1}{6}\pi + \theta_0\right) \\ u_{gd23\,(\mathrm{V})} = \sqrt{3} V_{t\,(\mathrm{pu})} V_{tn\,(\mathrm{V})} \sin\left(\omega t - \frac{1}{2}\pi + \theta_0\right) \end{cases} \tag{4.15}$$

Here $\underline{u}_{gd} = [u_{gd13}, u_{gd23}]^T$ is the vector of phase-to-phase terminal voltages, V_{tn} is the nominal voltage value of the diesel group, and θ_0 is the initial angle. The change from SI units (W) to pu of load power fluctuations is expressed as follows:

$$\Delta P_c = \frac{1}{P_{ngd}} (\underline{u}_{gd}^T \cdot \underline{i}_{gdb} - P_{ngd}) \tag{4.16}$$

where P_{ngd} is the nominal generated real power by the diesel group and $\underline{i}_{gdb} = [i_{gdb1}, i_{gdb2}]^T$ is the vector of diesel group currents. As a result, the description of the diesel group as the main grid in a single block can be given as shown in Fig. 4.4.

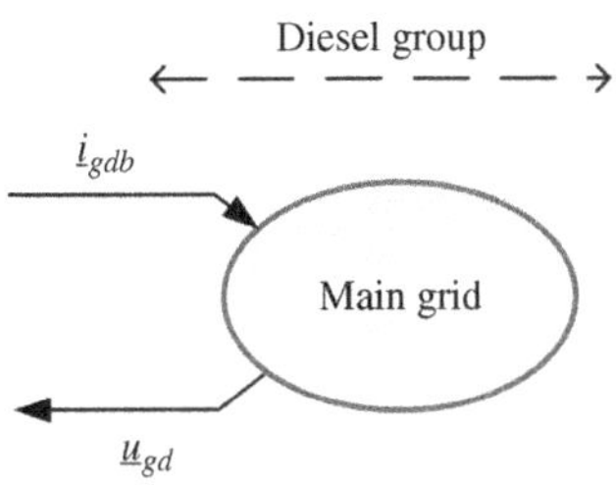

Figure 4.4 The main grid (diesel group) representation as a single block.

4.2.2 Modeling of Connection Between the Main Grid and the Microgrid

4.2.2.1 Modeling of the Medium Voltage Transmission Lines

Classical modeling of the MV transmission line uses a π model, which is composed of a resistor, a reactance, and two capacitors for each line (Figs. 4.5 and 4.6). The resistors are small and model the power losses in transmission line. Capacitors are neglected in this research work due to their small value and poor influence on the MG [18].

Therefore, similar to the modeling of a three-phase filter, the modeling of the MV transmission lines can be represented as follows:

$$\frac{d\underline{i}_{lmv}}{dt} = \frac{1}{L_{mv}}\underline{v}_{lmv\_l} \tag{4.17}$$

$$\underline{v}_{lmv\_l} = \underline{v}_{lmv\_rl} - \underline{v}_{lmv\_r} \tag{4.18}$$

$$\underline{v}_{lmv\_rl} = C_{ucs}\cdot(\underline{u}_{gdb} - \underline{u}_{mvb}) \tag{4.19}$$

$$\underline{v}_{lmv\_r} = R_{mv}\underline{i}_{lmv} \tag{4.20}$$

where $\underline{i}_{lmv} = [i_{lmv1}, i_{lmv2}]^T$ is the vector of the MV transmission line currents, L_{mv} and R_{mv} are the inductor and resistor of the MV transmission line model, $\underline{v}_{lmv\_l} = [v_{lmv\_l1}, v_{lmv\_l2}]^T$ is the vector of modeled inductor terminal voltages, $\underline{v}_{lmv\_r} = [v_{lmv\_r1}, v_{lmv\_r2}]^T$ is the vector of modeled resistor terminal voltages, $\underline{v}_{lmv\_rl} = [v_{lmv\_rl1}, v_{lmv\_rl2}]^T$ is the vector of modeled element terminal

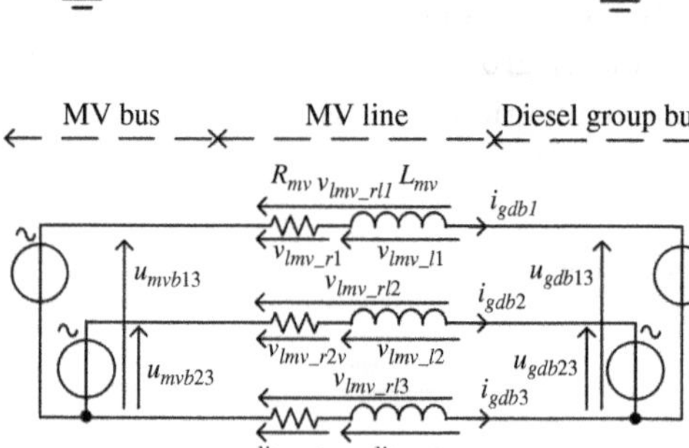

Figure 4.5 Model of one phase of an MV transmission line.

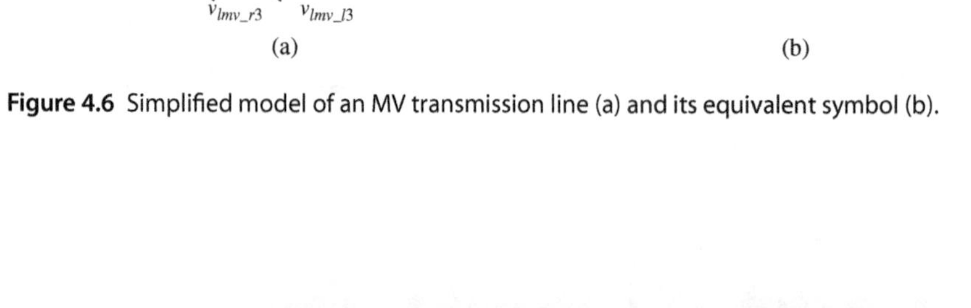

Figure 4.6 Simplified model of an MV transmission line (a) and its equivalent symbol (b).

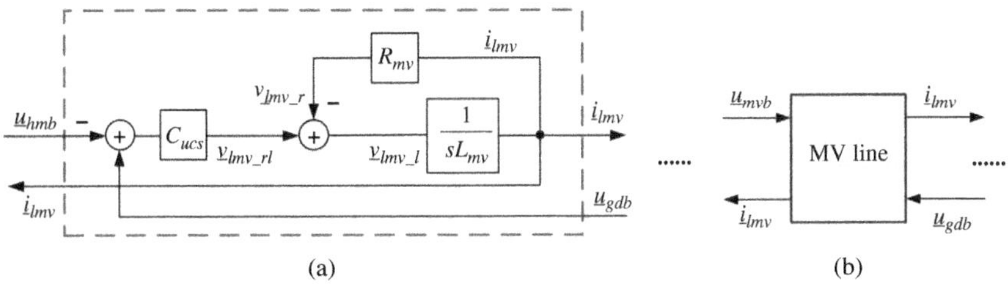

Figure 4.7 MV line model: (a) block diagram and (b) single block representation.

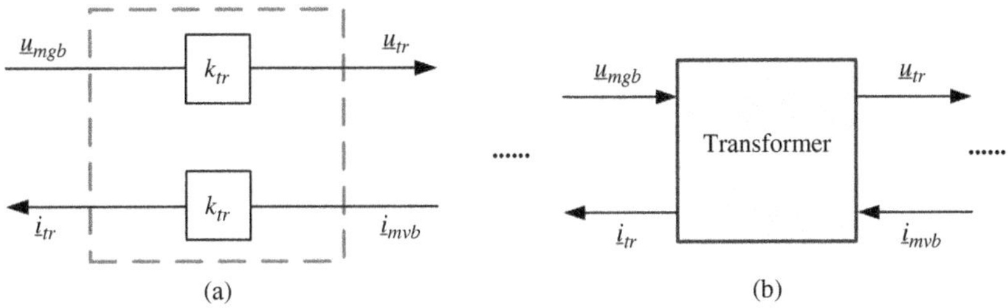

Figure 4.8 Transformer model: (a) block diagram and (b) single block representation.

voltages, $\underline{u}_{gdb} = [u_{gdb13}, u_{gdb23}]^T$ is the vector of phase-to-phase voltages of the diesel group bus, $\underline{u}_{mvb} = [u_{mvb13}, u_{mvb23}]^T$ is the vector of phase-to-phase voltage of the MV bus, and C_{ucs} is the matrix for the passage from phase-to-phase voltages to the single-phase voltages.

The block diagram of an MV transmission line is given in Fig. 4.7a. The modeling of MV transmission lines in a single block is shown in Fig. 4.7b.

4.2.2.2 Modeling of the Three-Phase Transformer

Here an ideal three-phase transformer is used with a transmission ratio k_{tr}.

$$\underline{u}_{tr} = k_{tr}\underline{u}_{mgb} \tag{4.21}$$

$$\underline{i}_{tr} = k_{tr}\underline{i}_{mvb} \tag{4.22}$$

where $\underline{u}_{tr} = [u_{tr13}, u_{tr23}]^T$ is the vector of phase-to-phase voltages, $\underline{u}_{mgb} = [u_{mgb13}, u_{mgb23}]^T$ is the vector of input phase-to-phase voltage, $\underline{i}_{tr} = [i_{tr1}, i_{tr2}]^T$ is the vector of transformer currents, and $\underline{i}_{mvb} = [i_{mvb1}, i_{mvb2}]^T$ is the vector of input currents. The full block diagram and single block representation of the three-phase transformer is given in Fig. 4.8a,b, respectively.

4.2.2.3 Modeling of Passive Loads

For modeling of passive loads, two types of model can be considered: voltage and current receptor types. A "voltage receptor-type" load model is considered

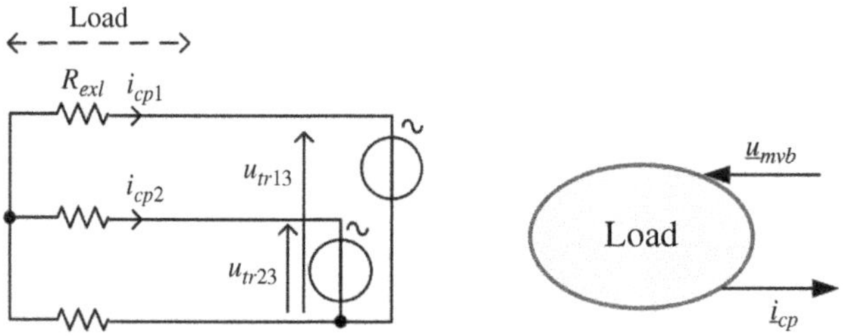

Figure 4.9 Modeling of a load as a current-type source.

as a voltage receptor and outputs a current to the grid (Fig. 4.9). So it can be also viewed as a current-type source:

$$\underline{i}_{cp} = \frac{1}{R_{exl}} C_{ucs} \cdot \underline{u}_{mvb} \tag{4.23}$$

where $\underline{u}_{mvb} = [u_{tr13}, u_{tr23}]^T$ is the vector of phase-to-phase voltages, $\underline{i}_{cp} = [i_{cp1}, i_{cp2}]^T$ is the vector of load currents, and R_{exl} is the resistor corresponding to the real load power (P_{exl}).

$$R_{exl} = \frac{U^2_{nmvb}}{P_{exl}} \tag{4.24}$$

where U_{nmvb} is the nominal value of the load voltages.

A "current receptor-type" load model is considered as a current receptor and outputs a voltage to the grid (Fig. 4.10). So it can be viewed also as a (phase-to-phase) "voltage-type" source:

$$\underline{u}_{mvb} = C_{usc} \cdot (R_{exl} \underline{i}_{cp}) \tag{4.25}$$

where R_{exl} is the resistor corresponding to the real load power (P_{exl}) and C_{usc} is the calculated matrix from single-phase voltages to phase-to-phase voltages.

$$R_{exl} = \frac{P_{exl}}{I^2_{nmvb}} \tag{4.26}$$

I_{nmvb} is the nominal value of the load currents. The choice of the model type depends on the network architecture modeling.

4.2.2.4 Modeling of Relevant Buses

The method for modeling the grid architecture is based on the characterization of coupling bus. Assuming "*the voltage at a bus must be set by a unique voltage-type source unit*" simplifies the modeling of connected buses between an MG and the main grid. As a result, among the various units (connected to one bus) only one unit must be a voltage-type source and others must be

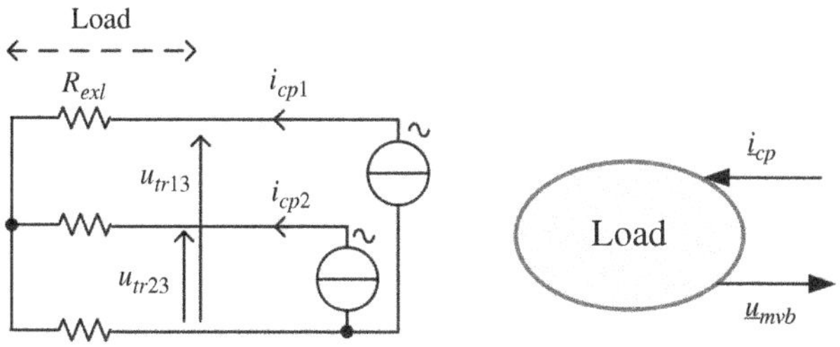

Figure 4.10 Modeling of a load as a voltage-type source.

considered as "current-type" sources. As shown in Fig. 4.1, there are usually two buses between the main grid and the MG (MG bus): the grid bus and the MV bus.

4.2.2.4.1 ***The Grid Bus*** Only two units are connected to the grid (diesel group) bus: the main grid and the MV lines (Fig. 4.1). The diesel group is a voltage-type source since it controls its terminal voltages. Hence, the MV lines are modeled as a current-type source:

$$\underline{i}_{lmv} = \underline{i}_{gdb} \tag{4.27}$$

$$\underline{u}_{gdb} = \underline{u}_{gd} \tag{4.28}$$

The block diagram description of the grid bus is given in Fig. 4.11.

4.2.2.4.2 ***MV Bus*** Three units are connected to the MV bus: MV lines, equivalent MV loads, and the transformer (Fig. 4.1). The MV lines are current-type sources as shown in the previous analysis. Therefore, either the MV loads or the transformer must be modeled as voltage-type sources. It is interesting to choose the transformer as a voltage-type source since it is always connected to the network. Therefore, the transformer can be modeled as a voltage-type

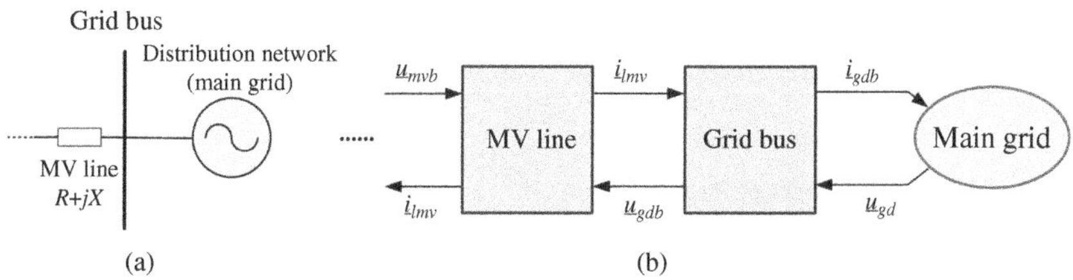

Figure 4.11 Coupling at the grid bus: (a) single-line diagram and (b) block diagram representations.

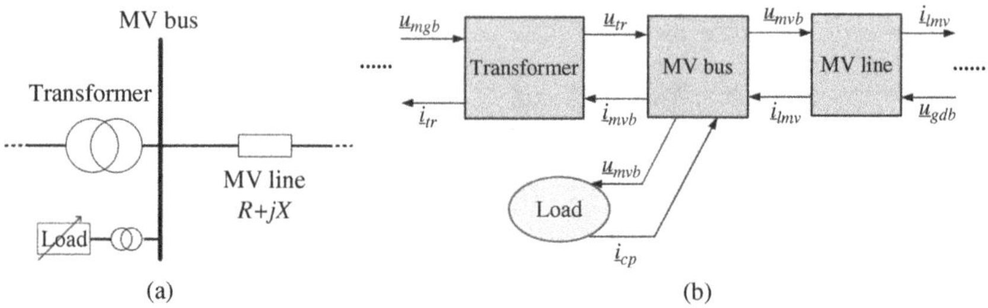

Figure 4.12 Coupling at the MV bus: (a) single-line diagram and (b) block diagram representations.

source and the MV loads as current-type sources (the model description of the MV bus is given in Fig. 4.12).

$$\underline{u}_{mvb} = \underline{u}_{tr} \tag{4.29}$$

$$\underline{i}_{mvb} = -\underline{i}_{lmv} - \underline{i}_{cp} \tag{4.30}$$

4.3 Overall Representation of the Grid-Connected Microgrid

4.3.1 Microgrid Bus

Many DGs and components such as the microturbines (MTs), PVs, storage systems, transformer, and the local loads are connected to the MG bus. As previously discussed, the transformer is modeled as a voltage-type source at the MV line side. Therefore, it is considered as a current-type source for the MG bus. All DG units are connected to the MG bus through three-phase filters. These filters behave always as current-type sources. In consequence, local loads must be modeled as a voltage-type source at the MG bus.

$$\underline{u}_{mgb} = \underline{u}_{cp} \tag{4.31}$$

$$\underline{i}_{cp} = -\underline{i}_{tr} - \underline{i}_{DG1} - \underline{i}_{DG2} - \cdots - \underline{i}_{DGn} \tag{4.32}$$

The model description of the MG bus and connected units is given in Fig. 4.13.

4.3.2 Global Architecture Representation

Finally, the global architecture modeling takes into consideration all presented block diagrams (Figs 4.11–4.13), and the result is depicted in Fig. 4.14.

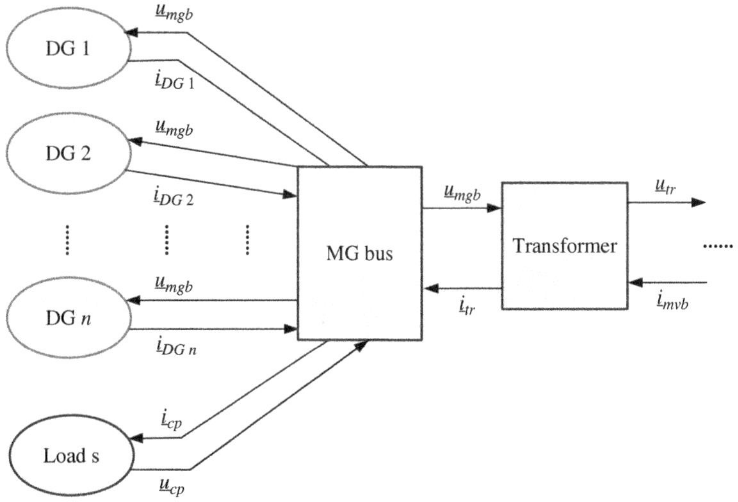

Figure 4.13 Coupling at the MG bus.

Using the proposed model, it is easy to study the dynamic behavior of the interconnected MG and main grid network. For instance, it could be useful to study the contribution of an MG in the main grid frequency regulation, which is described in Section 3.5. Recalling (3.10)–(3.17), the frequency response model for the frequency control loop of the interconnected MG-main grid system can be realized as shown in Fig. 4.15. If the frequency changes from f_{ref} to f_1, the reference of the generated power will move from the value in normal conditions $P_{mg\_ref\_0}$ to another value $P_{mg\_ref\_1}$ (see Fig. 3.13).

4.3.3 Microgrid Representation in the Islanded Operation Mode

When the MG operates in the islanded operation mode, the modeling of the MG must take into account only the production and storage units as well as loads within the MG. The architecture modeling of the MG is achieved by considering only the MG bus. Let us explain the representation of the MG in the islanded operation mode by using the given MG in Fig. 3.11, which is redrawn in Fig. 4.16a.

The principle of architecture modeling is the same as that of our previous analysis. All production and storage units are connected at the MG through three-phase filters. Therefore, they are considered as current-type sources for modeling of the MG bus. So the local loads must be modeled as voltage-type sources in the MG bus modeling. The overall representation is given in Fig. 4.16b.

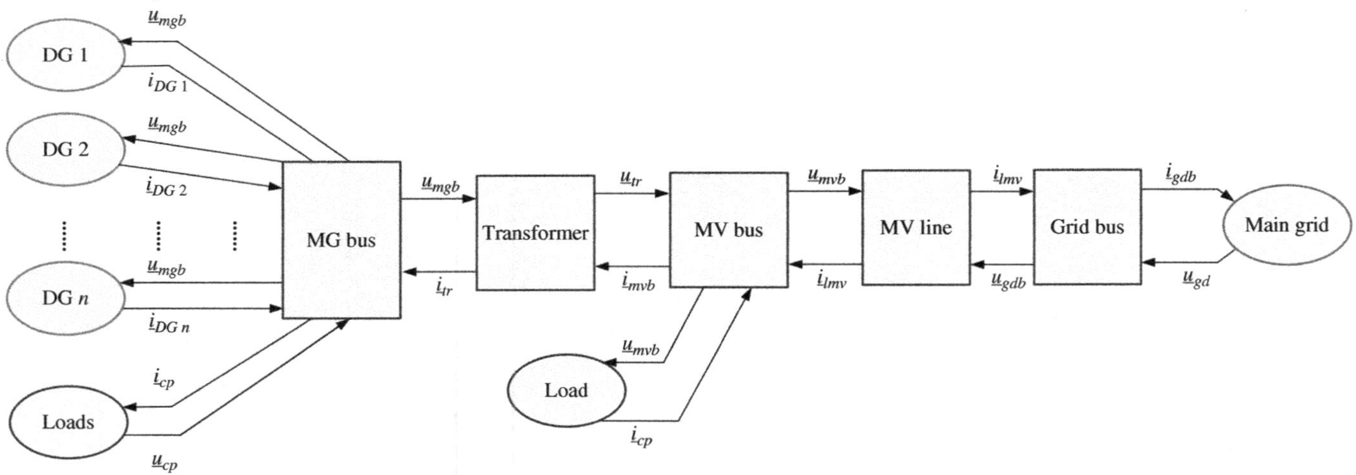

Figure 4.14 Global architecture representation.

The DG units are connected in parallel at the MG bus. So the load terminal voltage $\underline{u}_{cp}$ imposes the MG voltage $\underline{u}_{mgb}$ (4.33). The currents of all DG units are calculated by Kirchhoff's current law (4.34).

$$\underline{u}_{mgb-mt} = \underline{u}_{mgb-PV} = \underline{u}_{mgb-sc} = \underline{u}_{cp} \tag{4.33}$$

$$\underline{i}_{cp} = -\underline{i}_{mt} - \underline{i}_{PV} - \underline{i}_{sc} \tag{4.34}$$

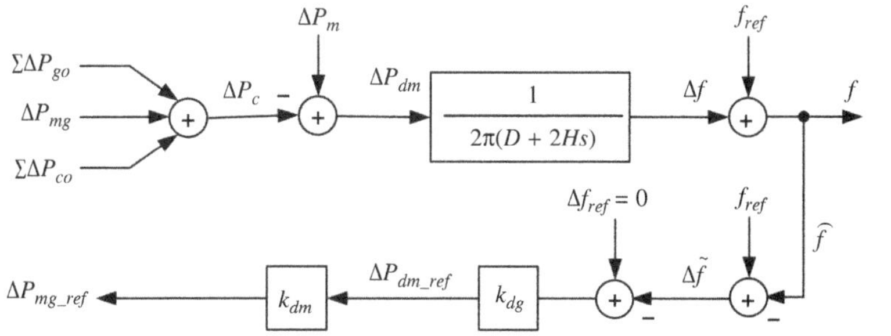

Figure 4.15 Global architecture modeling for contribution of MG in the power grid frequency regulation.

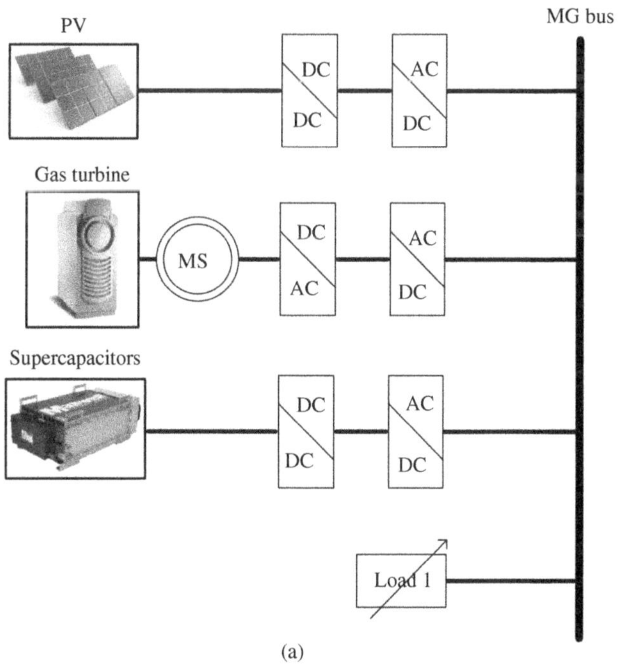

Figure 4.16 The islanded MG described in Fig. 3.11: (a) case study and (b) overall representation.

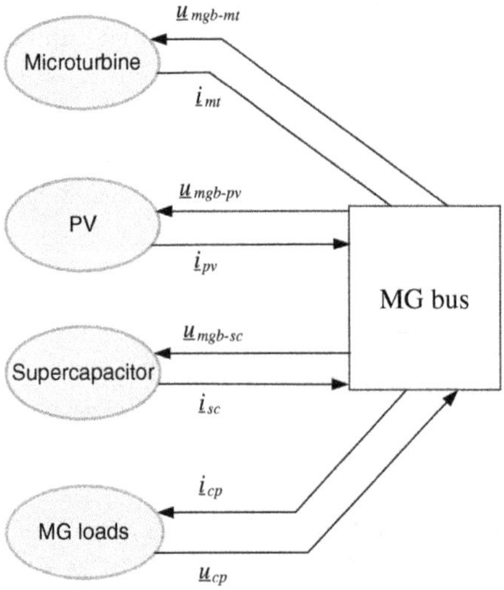

(b)

Figure 4.16 (*Continued*)

4.4 Microgrid Components Dynamics and Modeling

An MG contains several basic components: DGs, ESSs, power converters, and loads. The modeling of the components helps us to better understand the dynamic behavior of MGs. It also simplifies the control synthesis and closed-loop dynamic analysis. The modeling of most well-known renewable DGs, wind turbine and PV, in the grid-connected operation mode is discussed in Chapter 1. The general dynamic model of a synchronous DG such as diesel generator and gas MT is also explained in Section 4.2.

The dynamic modeling of MG's components is explained in many published works. In this section, from DGs, the modeling of PV and from ESSs, the modeling of lead-acid batteries and supercapacitors (SCs), and from other MG components, the modeling of converters are emphasized.

4.4.1 PV Model

4.4.1.1 PV Panels

In order to analyze the dynamic behavior of a PV, an electrical equivalent model is considered. An ideal PV cell may be considered as a current source in parallel with a diode. In practice, a PV cell is not ideal, so a shunt and a series resistance components are added to the model [23]. The resulting equivalent circuit is shown in Fig. 4.17.

In this equivalent circuit, the PV cell is considered as a current source with the photovoltaic effect. The output current is expressed as

$$I(t) = I_L(t) - I_D(t) - I_{SH}(t) \tag{4.35}$$

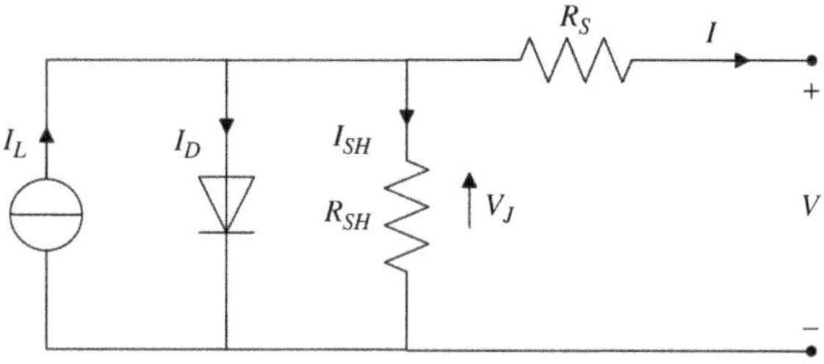

Figure 4.17 Equivalent circuit of a PV cell.

where I is output current, I_L is photo generated current, I_D is diode current, and I_{SH} is shunt current. The current through these elements is governed by the voltage across them:

$$V_j(t) = V(t) + I(t)R_s \tag{4.36}$$

where V_j is voltage across the diode and the shunt resistor R_{SH}, V is voltage across the output terminals, and R_S is series resistor.

With the Shockley diode equation, the current through the diode is expressed as

$$I_D(t) = I_0 \left\{ \exp\left[\frac{qV_j(t)}{nkT}\right] - 1 \right\} \tag{4.37}$$

where I_0 is reverse saturation current, n is diode ideality factor (1 for an ideal diode), q is elementary charge ($1.6022 \times 10\text{–}19$ C), k is Boltzmann's constant ($1.3806 \times 10\text{–}23$ J/K), and T is absolute temperature and $kT/q \approx 0.0259$ V at 25 °C. With Ohm's law, the current through the shunt resistor is expressed as

$$I_{SH}(t) = \frac{V_j(t)}{R_{SH}} \tag{4.38}$$

Substituting (4.36)–(4.38) into (4.35) gives the characteristic equation of a PV cell, which relates PV cell parameters to the output current and voltage:

$$I(t) = I_L(t) - I_0 \left\{ \exp\left[\frac{q(V(t) + I(t)R_s)}{nkT}\right] - 1 \right\} - \frac{V(t) + I(t)R_s}{R_{SH}} \tag{4.39}$$

For a given voltage $V(t)$, the equation may be solved to determine the output current $I(t)$. Because the equation involves the current on both sides in a transcendental function, the equation has no general analytical solution. However, it is easily solved by using numerical methods. Since the parameters I_0, n, R_S, and R_{SH} cannot be measured directly, a characteristic equation is generally used with a nonlinear regression to extract the values of these parameters on the basis of their combined effect on the PV cell behavior.

When the cell is operating in short circuit, $V = 0$ and the current $I(t)$ through the terminals is defined as the short-circuit current ($I_{SC}(t)$). For a high-quality PV cell (low value of R_S and I_0, and high value of R_{SH}) the short-circuit current I_{SC} is expressed as

$$I_{SC}(t) \approx I_L(t) \tag{4.40}$$

By taking into account the effect of the irradiance and the temperature, the current I_{SC} is expressed as

$$I_{SC}(t) = I_{SCS}\frac{G(t)}{G_s}[1 + \Delta I_{SC}(T(t) - T_s)] = I_L(t) \tag{4.41}$$

where G is irradiance (W/m^2), T is cell temperature (K), I_{SCS} is short circuit current measured in standard test conditions (STCs), G_s is standard illumination (1000 W/m^2), T_s is standard temperature (298.15 K), and ΔI_{SC} is temperature coefficient of short circuit current.

When the cell is operating in open circuit and $I = 0$, the voltage across the output terminals is defined as the open-circuit voltage (V_{OC}). Assuming the shunt resistor (R_{SH}) is high enough to neglect the final term of the characteristic equation (4.39), the open-circuit voltage is expressed as

$$V_{OC}(t) \approx \frac{kT}{q}\ln\left(\frac{I_L(t)}{I_0} + 1\right) \tag{4.42}$$

In an open-circuit condition, $V = V_{OC}$ and $I = 0$, Equation (4.35) can be rewritten as

$$I_L(t) = I_D(t) + I_{SH}(t) = I_{sat}\left[\exp\left(\frac{V_{oc}(t)}{V_t}\right) - 1\right] + \frac{V_{oc}(t)}{R_{SH}} \tag{4.43}$$

where $V_t = AkT/q$ and A is the ideal factor of the diode, the saturation current is defined as

$$I_{sat} = \frac{I_L(t) - \dfrac{V_{oc}(t)}{R_{SH}}}{\exp\left(\dfrac{V_{oc}(t)}{V_t}\right) - 1} \tag{4.44}$$

The open-circuit voltage is expressed as

$$V_{OC} = V_{OCS} + \Delta V_{OC}(T - T_s) \tag{4.45}$$

where V_{OCS} is open-circuit voltage in STC, and ΔV_{OC} is temperature coefficient of open circuit voltage. It is noteworthy that the resistors and the ideal factor are influenced by temperature [24]. In order to simplify the model, these values are set in STC.

4.4.1.1.1 Considering the MPPT A PV cell may operate over a wide range of voltages (V) and currents (I). By increasing the voltage of an irradiated cell continuously from zero (a short circuit) to a maximum value (an open circuit), the maximum-power point (MPP) can be determined. The point is achieved when the cell delivers the maximum electrical power for the received level of irradiation. For calculating the MPP, an algorithm may use the following equation:

$$P_{PV}(t) = V_{PV}(t) \cdot I_{PV}(t) \tag{4.46}$$

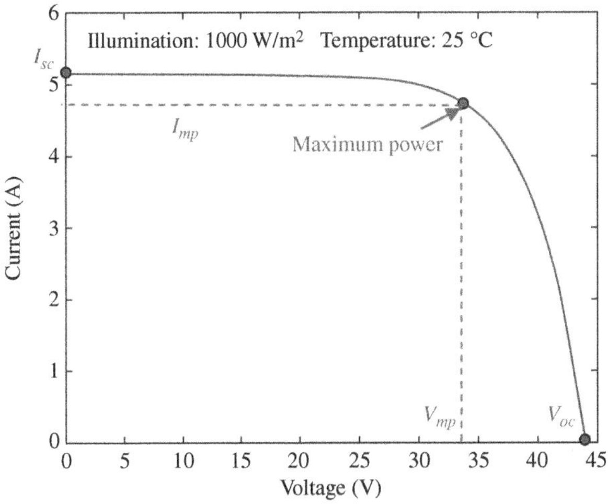

Figure 4.18 $I\,(V)$ characteristic of a PV panel.

By differentiating Equation (4.46) with respect to voltage V_{PV}:

$$\frac{dP_{PV}}{dV_{PV}} = I_{PV}(t) + V_{PV}(t) \cdot \frac{dI_{PV}}{dV_{PV}} \tag{4.47}$$

and at the MPP:

$$\frac{dP_{PV}}{dV_{PV}} = 0 \tag{4.48}$$

$$\frac{dI_{PV}}{dV_{PV}} = -\frac{I_{PV}}{V_{PV}} \tag{4.49}$$

Therefore, the algorithm seeks the point at which the conductance (I_{PV}/V_{PV}) is equal to the conductance increment (dI_{PV}/dV_{PV}). The $I\,(f,V)$ characteristic shown in Fig. 4.18 consists of four important values: short-circuit current (I_{SC}), open-circuit voltage (V_{oc}), current at the MPP (I_{mp}), and voltage at the MPP (V_{mp}).

The required equations to generate the I–V characteristic curve of a cell are developed as follows [25]:

$$I(t) = I_{sc}\left[1 - C_1\left(\exp\left(\frac{V(t)}{C_2 V_{oc}}\right) - 1\right)\right] \tag{4.50}$$

$$C_1 = \left(1 - \frac{I_{mp}}{I_{sc}}\right)\exp\left(-\frac{V_{mp}}{C_2 V_{oc}}\right) \tag{4.51}$$

$$C_2 = \frac{\dfrac{V_{mp}}{V_{oc}} - 1}{\ln\left(1 - \dfrac{I_{mp}}{I_{sc}}\right)} \tag{4.52}$$

The current corresponding to the maximum power is expressed as

$$I_{mp} = I_{mps}\frac{G}{G_s}[1 + \Delta I_{mp}(T - T_s)] \tag{4.53}$$

where I_{mps} is the current at MPP, which is measured in STCs, and ΔI_{mp} is the temperature coefficient of the current at the MPP. The voltage corresponding to the maximum power is calculated by using [26]

$$V_{mp} = V_{mps}[1 + \Delta V_{mp}(T - T_s)] + K_1 V_t \ln\left(\frac{G}{G_s}\right) + K_2\left[V_t \ln\left(\frac{G}{G_s}\right)\right]^2 \tag{4.54}$$

where V_{mps} is the voltage at the point of maximum power in STC, ΔV_{mp} is temperature coefficient of the voltage at the MPP, and K_1 and K_2 are constant parameters. In a normal operation of the PV system, the MPPT inverter searches always the MPP. This allows the simplification of this model with only the current and voltage at the MPP (I_{mp} and V_{mp}).

4.4.2 Energy Storage Systems Modeling

4.4.2.1 Lead-acid Battery

An equivalent electrical circuit of a typical battery model [26] is shown in Fig. 4.19.

This model consists of a voltage source E_b and an internal resistor R_i. This model includes the main variables of the system: the battery state of charge (SOC), current flowing across the battery I_{bat}, temperature T, and number of cells in series n_b. The circuit equation is expressed as

$$V_{bat}(t) = n_b E_b(t) + n_b I_{bat}(t) R_i \begin{cases} E_b = f(SOC) \\ R_i = f(I_{bat}, SOC, T) \end{cases} \tag{4.55}$$

The voltage source E_b represents the voltage in open circuit across the battery terminals. This voltage is due to the stored energy into the battery through the electrochemical reactions. Obviously, this term directly depends on the stored energy. R_i is a resistor and represents the losses. This resistor value includes the effects of the working point (I, SOC, T) and the health

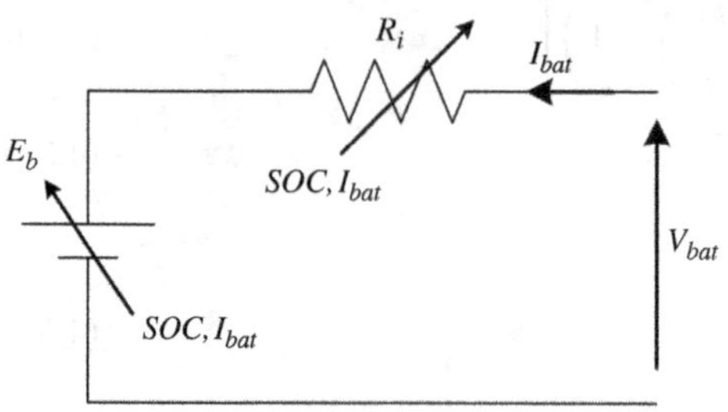

Figure 4.19 Equivalent electrical circuit of a typical battery model.

of the battery. A damaged battery has a high value of resistor irrespective of its working point. Also, the resistor is inversely proportional to the SOC. Furthermore, while the battery is discharging, the resistor value is increasing. Note that all physical effects can be expressed into electrical equations and can be modeled by equivalent electric circuit.

The SOC must be considered as an indicator of the electrical charge stored in the battery. The value range is $0 < SOC < 1$. The following equations describe the behavior of this indicator.

$$SOC(t_i) = \frac{1}{C(t_i)} \int_{t_0}^{t_i} \eta_c I_{bat}(\tau) d\tau + SOC(t_0) \tag{4.56}$$

$$C(t) = \frac{C_n C_t}{\left(1 + A_{cap}\left(\frac{|I_{bat}(t)|}{I_n}\right)^{B_{cap}}\right)(1 + \alpha_c \Delta T(t) + \beta_c \Delta T(t)^2)} \tag{4.57}$$

$$I_n = \frac{C_n}{n} \tag{4.58}$$

Here the involved terms are $C(t)$ the battery capacity, η_c the charging efficiency, and $I_{bat}(t)$ the current flowing through it. C_n is the rated battery capacity (at n hours); C_t, A_{cap}, and B_{cap} are model parameters; ΔT is the temperature variation from the reference value at 25 °C; I_n is the discharge current corresponding to the C_n; n is the time in hours; and α_c and β_c are the temperature parameters [27].

The SOC must be understood as the relation between the accepted energy and the available capacity at all times. The inner integral term models the accepted energy over the battery working life. Also, the outer integral term models the battery capacity due to the working point environment at any given time. Both terms are functions of time and are evolving continuously. When the SOC is unity, the battery cannot accept more energy from the system because the stored energy fills all the battery capacity, and when the SOC is null the battery has no energy.

For a more intuitive use of the SOC concept in PV applications, a new indicator named as level of energy (LOE) is introduced. This indicator shows the amount of energy available in the battery under normal working conditions. Equations (4.59) and (4.60) define the LOE, where T_1 and T_2 determine the known temperature range of battery operation. It is noteworthy that the LOE only depends on the constitutive parameters of the device and the accumulated charge over time, not on the working environment of the battery. The LOE calculus may be carried out by considering the maximum available battery capacity C_n, obtained by taking into account the range of possible current and temperature values. So, C_n is evaluated from (4.60), with battery current equal to zero and maximum temperature value (if a standard temperature range (−40, 40 °C) is considered for a first theoretical approach, then $T = 40$ °C). Thus, LOE is not limited to the higher limit of unity, but *LOE* values near or greater than unity are undesirable in order to avoid damaging the battery.

$$LOE(t_i) \equiv \frac{1}{C_n}\int_{t_0}^{t_i} \eta_c I_{bat}(\tau)d\tau + LOE(t_0) \tag{4.59}$$

$$C_n \equiv \max(C(t))\Bigg|\begin{matrix} I_{bat}=0 \\ T\in[-T_1,T_2]\end{matrix} \tag{4.60}$$

The *LOE* directly represents the stored energy in the battery (C_n is constant). Therefore, the *SOC* and *LOE* are complementary. Furthermore, a battery can have $SOC=1$ and $LOE\neq 1$; this indicates that the battery is saturated but only at a percentage of the nominal capacity.

Here a simplified model is introduced by considering a constant internal resistor R_i. The resistor is supposed to be constant during the charge and discharge cycles and does not vary with the magnitude of the current. When the preset model is used in the *SimPowerSystem* simulation software, a generic value is loaded, corresponding to the nominal voltage and the rated capacity of the battery. The battery voltage is expressed as

$$E_b(t) = E_0 - KQ\frac{1}{(Q-i_t(t))} + Ae^{-Bi_t(t)} \tag{4.61}$$

$$i_t = \int_{t_0}^{t} i_{bat}(\tau)d\tau + i_{bat}(i_0) \tag{4.62}$$

where E_0 and Q denote fully charged voltage (V) and battery capacity (Ah), respectively. A is exponential voltage (V), B is exponential capacity (Ah), and K is polarization voltage (V). All parameters of this equivalent circuit can be identified by considering the discharge characteristics with a nominal current. A typical discharge curve is composed of three sections (Fig. 4.20).

The first section represents the exponential voltage drop if the battery is initially fully charged. The width of this region depends on the battery type.

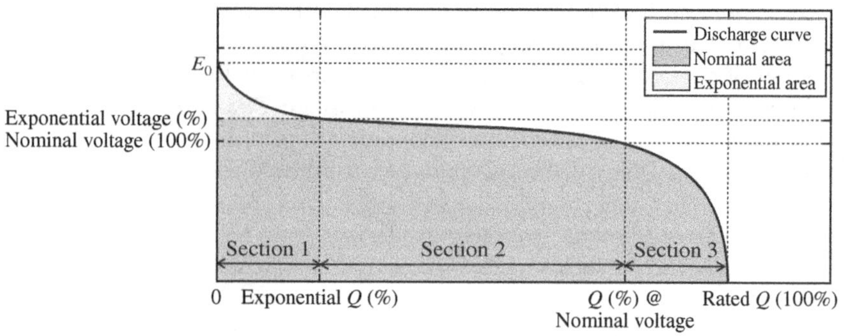

Figure 4.20 Discharge curve (*Q–V*) of a battery.

The second section represents the charge that can be extracted from the battery until the voltage drops below the battery nominal voltage. Finally, the third section represents the total discharge of the battery, when the voltage drops quickly.

4.4.2.2 Supercapacitor

SCs or ultracapacitors are electrochemical double-layer capacitors (EDLCs) that have a high energy density compared to common capacitors. There are several different models describing the characteristics of the SCs. The various equivalent electrical circuits of SC model can be ranged according to the complexity.

4.4.2.2.1 Supercapacitor Modeling The SC exhibits a nonideal behavior, mainly because of the porous material, which is used to form the electrodes. This causes the resistance and capacitance to be distributed in such a way that the electrical response of the SC can be described by the behavior of a transmission line [28].

Transmission Line Model The transmission line model is currently the most used model for the SC. A five-stage model has a good accuracy for most applications and accounts for frequencies up to 10 kHz (Fig. 4.21). The transmission line model physically mimics the SC distribution and has a good accuracy over a wide range of frequencies. The capacitance in the highly porous electrodes is modeled as a circuit of transmission lines. The model gives the double-layer capacitance and the electrolyte resistor that extends deep into the pores of the material. The resistor of the electrode material is taken to be much smaller than the electrolyte resistance and the capacitors behave like small impedance elements at high frequencies. The current mainly flows among R_1 and C_1 into the electrode material and almost no current flows deep into the pore. The resistor and the double-layer capacitance are reduced at high frequencies.

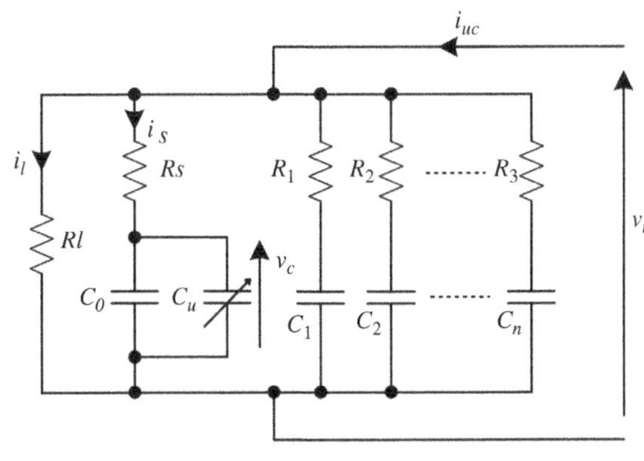

Figure 4.21 Transmission line model of a supercapacitor.

Branch Model The number of RC branches chosen for the model depends on the duration of the required time response. The three-branch model covers a time span of up to approximately 30 min (Fig. 4.22) [29]. Each branch has a different time constant. The first or immediate branch behaves in the order of a few seconds. The second (delayed branch with R_{del} and C_{del}) gives the behavior over the range of a few minutes, while the third (long-term branch with R_{long} and C_{long}) behaves of the SC for a time longer than 10 min. A leakage resistance R_l is also included in the equivalent circuit and represents the self-discharge of the SC. Finally, the voltage dependency of the SC is modeled by a voltage-dependent capacitor in the first branch of the model for simplicity.

Each branch has a different time constant; therefore, the transient process of each branch can be observed independently from the others by measuring the terminal voltage as a function of time. This allows the model parameters to be calculated by using a single fast current-controlled charge [29].

The classical equivalent circuit model in Fig. 4.23a is used to describe the capacitor performances in fast discharge applications (in the order of a few seconds). The variable capacitor (C_u) depends on the technology and the used equipment, and it can be neglected in some applications. It consists of an equivalent series resistance (R_s), which represents losses and models the internal heating in the capacitor. R_s is important during charging and discharging and

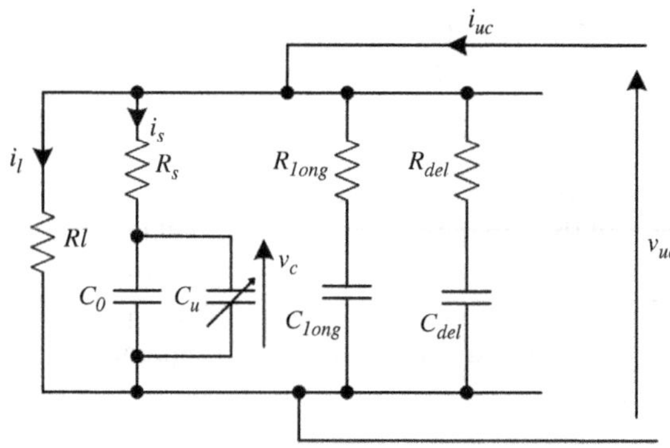

Figure 4.22 Three-branch model of a supercapacitor.

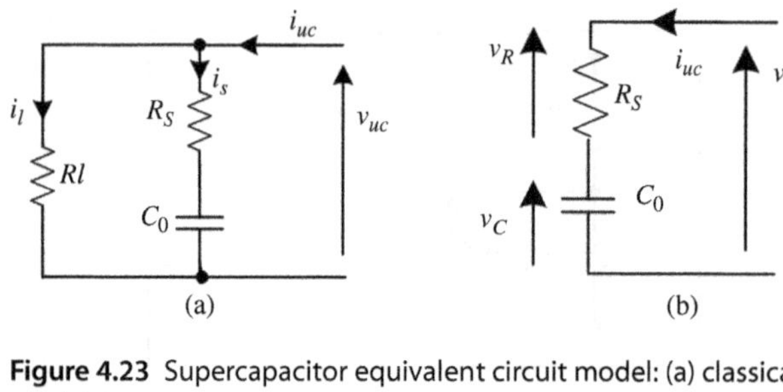

Figure 4.23 Supercapacitor equivalent circuit model: (a) classical model and (b) simplified model.

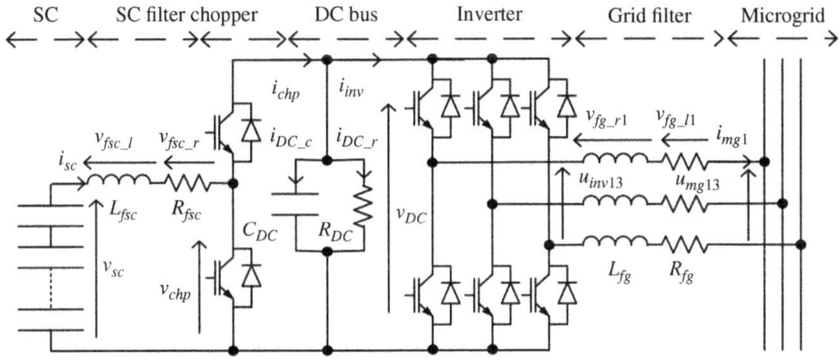

Figure 4.24 Supercapacitor power conversion system.

the equivalent parallel resistance (R_l) models the current leakage effect. C_0 is the series capacitance.

Simplified Model In many applications, the charging and discharging cycles are fast enough to ignore the phenomenon of relaxation. The leakage resistance R_l is also neglected because of its great value. The simplified model (Fig. 4.23b) includes the final capacitor C_0 in series with a resistance R_s [30]. All internal electrical relations are expressed in the following equation:

$$\frac{dv_c}{dt} = \frac{1}{C_0} i_{uc}(t) \tag{4.63}$$

$$v_R(t) = R_s i_{uc}(t) \tag{4.64}$$

$$v_{uc}(t) = v_c(t) + v_R(t) \tag{4.65}$$

with the SC current i_{uc} and the output voltage v_{uc}.

4.4.2.2.2 Connecting to the MG Here a supercapacitor bank (SCB) power conversion system (as shown in Fig. 4.24) is considered to explain the modeling description of a connected SC to the MG.

The SCB is composed of the SC, SC filter, chopper, DC bus, inverter, and three-phase grid filter (as shown in Fig. 4.24). The first step of the SCB dynamic modeling is to obtain a dynamical model of each element as a single block (*Bi*) and then gathering all blocks to obtain an overall block diagram.

- *Modeling of the SC (B1)*

The SC module is modeled as a voltage source. For the study of many power system applications, the simplified model shown in Fig. 4.23b can be applied (the model with a resistor R_s and an ideal capacitor C_0 in series).

$$\frac{dv_C}{dt} = \frac{1}{C_0} i_{sc}(t) \tag{4.66}$$

$$v_R(t) = R_s i_{sc}(t) \tag{4.67}$$

$$v_{sc}(t) = v_C(t) + v_R(t) \tag{4.68}$$

- *Modeling of the filter (B2)*

The SC filter is modeled as an inductance (L_{fsc}) and a resistance (R_{fsc}) in series.

$$\frac{di_{sc}}{dt} = \frac{1}{L_{fsc}} v_{fsc\_l}(t) \tag{4.69}$$

$$v_{fsc\_l}(t) = v_{sc}(t) - v_{chp}(t) - v_{fsc\_r}(t) \tag{4.70}$$

$$v_{fsc\_r}(t) = R_{fsc} i_{sc}(t) \tag{4.71}$$

- *Modeling of the chopper (B3)*

The chopper adapts the low voltage across the SC to the desired voltage for the DC bus. An equivalent continuous model of the chopper is used by a mean value modulation function $m_{chp} \in [0, 1]$:

$$\begin{cases} v_{chp}(t) = m_{chp} v_{dc}(t) \\ i_{chp}(t) = m_{chp} i_{sc}(t) \end{cases} \tag{4.72}$$

- *Modeling of the grid-side DC bus (B4)*

The DC bus is considered as a capacitor (C_{DC}) and a resistance (R_{DC}) in parallel.

$$\frac{dv_{DC}}{dt} = C_{DC} i_{DC\_c}(t) \tag{4.73}$$

$$i_{DC\_c}(t) = i_{chp}(t) - i_{inv}(t) - i_{DC\_r}(t) \tag{4.74}$$

$$i_{DC\_r}(t) = \frac{1}{R_{DC}} v_{DC}(t) \tag{4.75}$$

- *Modeling of the inverter (B5)*

An equivalent mean modeling of the power converters is mostly sufficient. It represents fundamental phase-to-phase voltage $u_{inv} = [u_{inv13}, u_{inv23}]^T$ and line currents $i_{mg} = [i_{mg1}, i_{mg2}]^T$ components as

$$\underline{u}_{inv} = \underline{m}_{inv} \cdot v_{DC} \tag{4.76}$$

$$i_{inv} = \underline{m}_{inv}^T \cdot \underline{i}_{mg} \tag{4.77}$$

where $m_{inv} = [m_{inv1}, m_{inv2}]^T$ is the modulation index vector. In some figures in this section, the notation of v_{DC} is used instead of u_{DC}. Line voltages

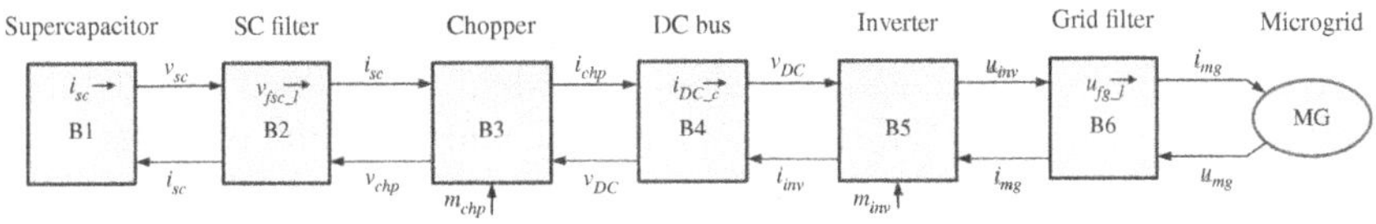

Figure 4.25 Block diagram representation of the connected SBC to the MG.

$v_{inv} = [v_{inv1n}, v_{inv2n}]^T$ are obtained by

$$\underline{v}_{inv} = \frac{1}{3}\begin{bmatrix} 2 & -1 \\ -1 & 2 \end{bmatrix} \underline{u}_{inv} \tag{4.78}$$

- *Modeling of the three-phase filter (B6)*

The line currents i_{mg} are deduced from the inverter voltages u_{inv} and the grid voltages $u_{mg} = [u_{mg13}, u_{mg23}]^T$.

$$\frac{d}{dt}\underline{i}_{mg} = \frac{1}{L_{fg}}\underline{v}_{fg\_l} \tag{4.79}$$

$$\underline{v}_{fg\_l} = \frac{1}{3}\begin{bmatrix} 2 & -1 \\ -1 & 2 \end{bmatrix} (\underline{u}_{inv} - \underline{u}_{mg}) - \underline{v}_{fg\_r} \tag{4.80}$$

$$\underline{v}_{fg\_r} = R_{fg}\underline{i}_{mg} \tag{4.81}$$

where L_{fg} and R_{fg} are the inductance and resistance of the filter, the $v_{fg\_l} = [v_{fg\_l1}, v_{fg\_l2}]^T$ and $v_{fg\_r} = [v_{fg\_r1}, v_{fg\_r2}]^T$ are the voltages, respectively, across L_{fg} and R_{fg}.

- *Modeling of the MG*

The grid voltages u_{mg} can be modeled by

$$\underline{u}_{mg} = \begin{bmatrix} u_{mg13} \\ u_{mg23} \end{bmatrix} = \sqrt{2}E\begin{bmatrix} \sin(2\pi ft - \pi/6 + \theta_0) \\ \sin(2\pi ft - \pi/2 + \theta_0) \end{bmatrix} \tag{4.82}$$

where E is the *rms* value of the grid phase-to-phase voltage, f is the grid frequency, and θ_0 is the initial angle of the grid voltage. The line currents i_{mg} are considered as disturbances for the MG. The final model representation of the connected SBC is shown in Fig. 4.25

4.4.3 Power Electronic Converters

Equivalent continuous models of the power electronic converters are sufficient for the most MG control synthesis and analysis studies. Three types of power converters are generally in use: DC chopper, three-phase inverter, and three-phase rectifier. The equivalent continuous modeling of power converters are carried out under the assumptions of (i) switches are ideal, (ii) switching is instantaneous, and (iii) switches are considered as short circuits in ON state and as open circuits in OFF state.

A switching function (s_{ij}) is defined for each power switch. It represents the ideal commutation order and takes the values 1 when the switch is closed (ON) and 0 when it is opened (OFF). In s_{ij}, i is the number of the commutation circuit and j is the number of switches in the commutation circuit.

As ideal power switches are considered, the switches in the same commutation circuit are in complementary states:

$$s_{i1} + s_{i2} = 1 \quad \forall i \in \{1, 2, 3\} \tag{4.83}$$

DC Chopper Modeling In the given power electronic structure, a DC chopper is located between a current source and a voltage source. For the case of SC storage system, the current source is the choke L_{sc} and generates the current i_{sc}. The DC-bus capacitor is the voltage source and generates the voltage (u_{DC}) (Fig. 4.26). The i_{sc} and u_{DC} can be obtained by solving (4.69) and (4.73), respectively.

The modulation functions of the DC chopper can be expressed from the switching function (s_{sc11}) of the first switch in the commutation circuit: $m_{sc} = s_{sc11}$. The switches K and K' in Fig. 4.26 represent s_{sc11} and s_{sc12}, respectively. Then, the modulated voltage u_{m_sc} and the modulated current i_{m_sc} of the DC chopper are obtained as follows:

$$\begin{cases} u_{m\_sc} = m_{sc} u_{DC} \\ i_{m\_sc} = m_{sc} i_{sc} \end{cases} \tag{4.84}$$

In practice, a connection controller is used to create two complementary switching functions (s_{sc11} and s_{sc12}) with necessary dead times from the modulation function m_{sc}. Then the switching functions are converted into ON/OFF signal for each semiconductor switch through some drivers and optocouplers. The modulation function m_{sc} is obtained by comparing an average modulation function $\langle m_{sc} \rangle$ with a triangular signal ξ (Fig. 4.27).

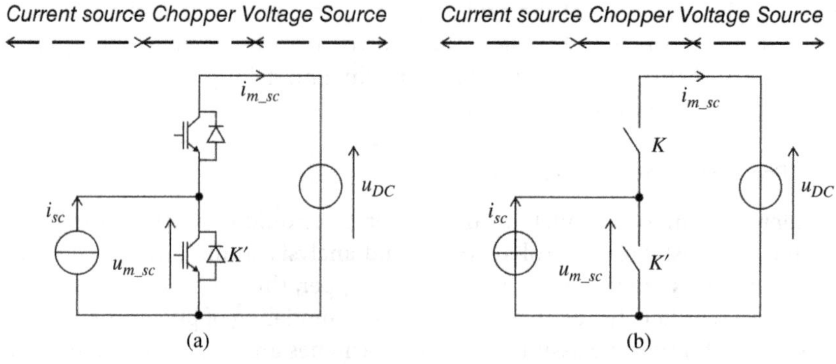

Figure 4.26 Diagram of the DC chopper in the supercapacitor storage system: (a) electric diagram of the DC chopper and (b) DC chopper with ideal switches.

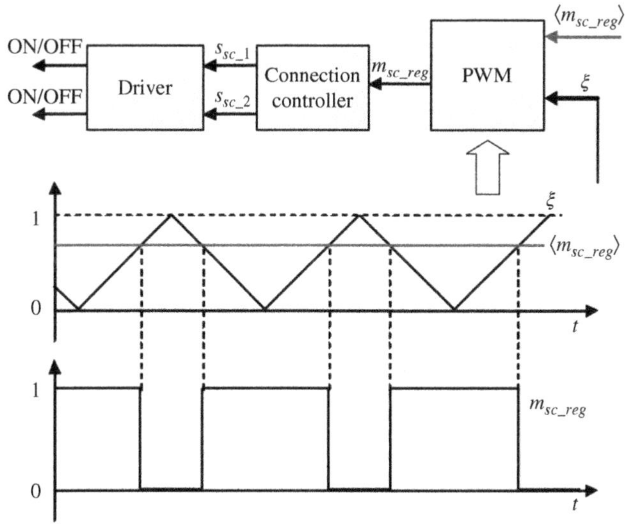

Figure 4.27 Classical PWM method.

In theory, when the modulation frequency of the carrier signal ξ is much higher than the frequency domain of the control signal (which is a continuous value for the DC chopper), the average modulation function can be obtained as follows:

$$\langle m_{sc} \rangle = \frac{1}{\Delta t} \int_{t_0}^{t_0+\Delta t} m_{sc}(t)dt + m_{sc}(t_0) \tag{4.85}$$

By considering that the voltage u_{DC} and the current i_{sc} are constant during the time interval Δt, the average values of the modulated voltage $\langle u_{m\_sc} \rangle$ and the modulated current $\langle i_{m\_sc} \rangle$ are expressed as follows:

$$\begin{cases} \langle u_{m\_sc} \rangle = \langle m_{sc} \rangle u_{DC} \\ \langle i_{m\_sc} \rangle = \langle m_{sc} \rangle i_{sc} \end{cases} \tag{4.86}$$

As a result, the equivalent average model of the DC chopper is obtained and the electrical diagram is shown in Fig. 4.28. Finally, the variable $\langle m_{sc} \rangle$ can be used to model the DC chopper.

Three-phase Inverter Modeling A three-phase VSI is usually used to connect the DC bus to the AC grid [31]. The task is to invert the DC voltage into AC modulated voltages in real time. The three-phase VSI is located between a three-phase current source and a voltage source. For the grid power conversion system, the current sources come from the choke filters and is set to generate the AC line current ($\underline{i}_{line} = [i_{line\_1}, i_{line\_2}]^T$) and the voltage source comes from

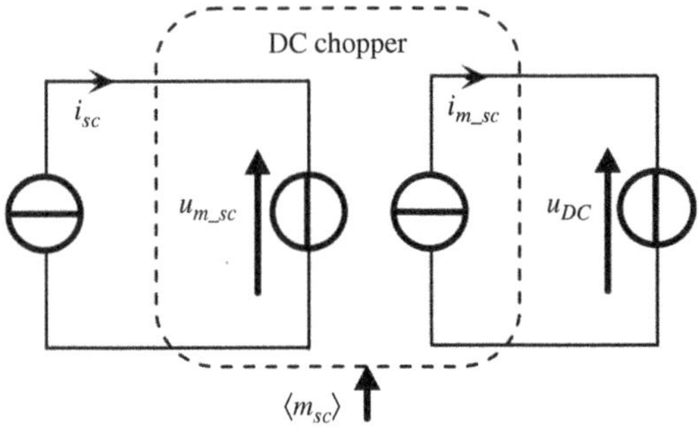

Figure 4.28 Equivalent average electrical diagram of the DC chopper.

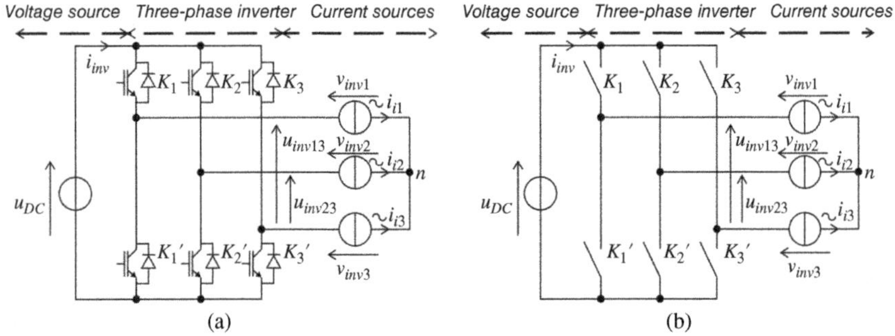

Figure 4.29 Diagram of the three-phase inverter in the grid power conversion system: (a) electrical diagram of the three-phase inverter and (b) three-phase inverter with ideal switches.

the DC bus and is set to generate the DC-bus voltage (u_{DC}) (Fig. 4.29). The switches K_i and K_i' in Fig. 4.29 represent s_{inv1i} and $\bar{s}_{inv1i}$, respectively.

The modulation functions ($\underline{m}_{inv} = [m_{inv13}, m_{inv23}]$) of the three-phase inverter can be expressed from the switching functions ($\underline{s}_{inv} = [s_{inv11}, s_{inv21}, s_{inv31}]$) of first switches of the three commutation circuits:

$$\underline{m}_{inv}(t) = \begin{bmatrix} m_{inv13}(t) \\ m_{inv23}(t) \end{bmatrix} = \begin{bmatrix} 1 & 0 & -1 \\ 0 & 1 & -1 \end{bmatrix} \begin{bmatrix} s_{inv11}(t) \\ s_{inv21}(t) \\ s_{inv31}(t) \end{bmatrix} \tag{4.87}$$

Then, the modulated voltage ($\underline{u}_{inv} = [u_{inv13}, u_{inv23}]^T$) and the modulated current i_{inv} of the three-phase inverter are obtained from the DC-bus voltage and the line currents ($\underline{i}_l = [i_{l1}, i_{l2}]$), which are considered constant during the time interval Δt:

$$\begin{cases} \underline{u}_{inv}(t) = \underline{m}_{inv}(t) u_{dc} \\ i_{inv}(t) = \underline{m}_{wg}^T(t) \underline{i}_l \end{cases} \tag{4.88}$$

In practice, a connection controller is used to create the six switching functions $\{s_{inv11}, s_{inv12}, s_{inv21}, s_{inv22}, s_{inv31}, s_{inv32}\}$ with necessary dead times from

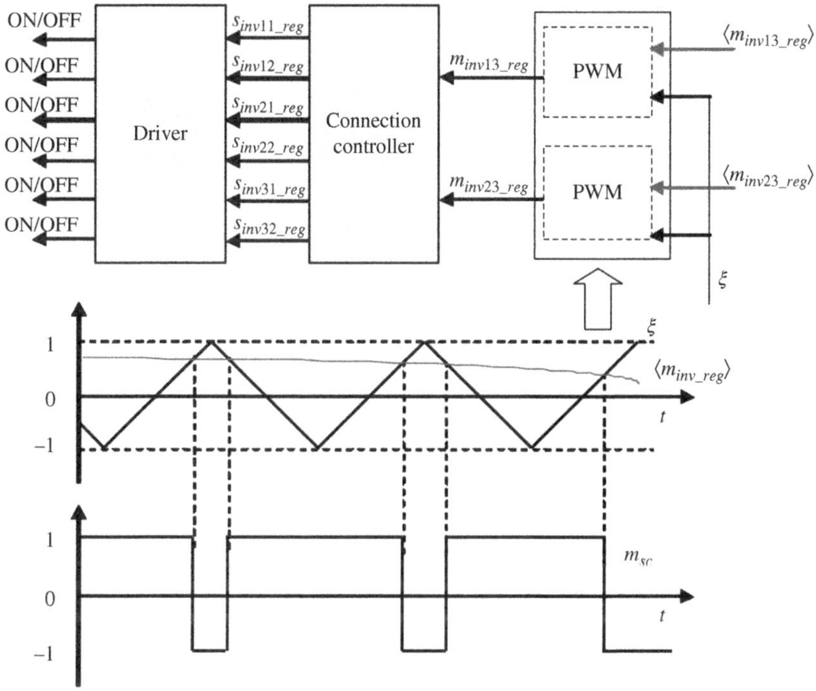

Figure 4.30 Classical sinusoidal PWM method.

the modulation function m_{inv}. Then the switching functions are converted into ON/OFF state of each switch through some drivers and optocouplers. The modulation function m_{inv} can be obtained by comparing an average modulation function $\langle m_{inv} \rangle$ with a triangular signal ξ (Fig. 4.30).

In theory, when the modulation frequency of the carrier signal ξ is much higher than the frequency domain of the control signals (which are sinusoidal values for the inverter) the average modulation function can be obtained as follows:

$$\langle \underline{m}_{inv} \rangle = \frac{1}{\Delta t} \int_{t_0}^{t_0+\Delta t} \underline{m}_{inv}(t)dt + \underline{m}_{inv}(t_0) \tag{4.89}$$

Then the average values $\langle \underline{u}_{inv} \rangle$ and $\langle i_{inv} \rangle$ of the modulated voltage and the modulated current are expressed as follows:

$$\begin{cases} \langle \underline{u}_{inv} \rangle = \langle \underline{m}_{inv} \rangle u_{DC} \\ \langle i_{inv} \rangle = \langle \underline{m}_{inv}^{T} \rangle \underline{i}_{l} \end{cases} \tag{4.90}$$

The voltage u_{DC} and the currents $\underline{i}_l$ are assumed to be constant during the switching period. As a result, the equivalent continuous model of the

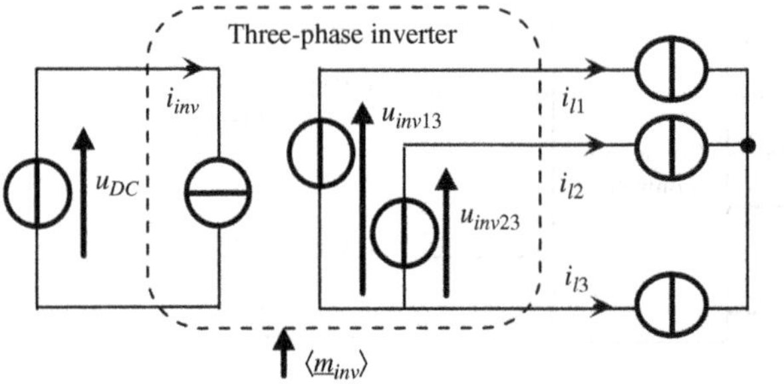

Figure 4.31 Equivalent electrical diagram of the three-phase inverter.

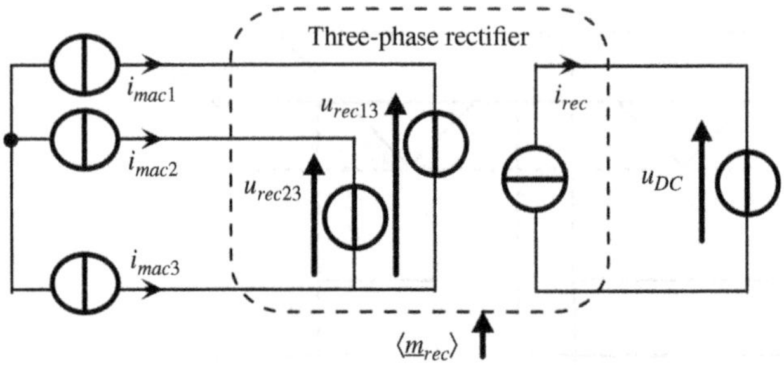

Figure 4.32 Equivalent electrical average diagram of the three-phase rectifier.

three-phase inverter is obtained and the electrical diagram is shown in Fig. 4.31. Finally, a vector $\langle \underline{m}_{inv} \rangle$ can be used to model the three-phase inverter.

Three-phase Rectifier Modeling A three-phase rectifier is used to rectify three-phase sinusoidal currents from the machine of the wind generator. For the three-phase rectifier, similar relations are obtained for the average values of the modulated voltage ($\langle \underline{u}_{rec} \rangle = [u_{rec13}, u_{rec23}]^T$) and the modulated current ($\langle i_{rec} \rangle$) from the DC-bus voltage u_{DC} and the currents ($\underline{i}_{mac} = [i_{mac1}, i_{mac2}]^T$) of the electrical machine, with the averaged vector ($\langle \underline{m}_{rec} \rangle = [m_{rec13}, m_{rec23}]^T$) of the rectifier modulation functions:

$$\begin{cases} \langle \underline{u}_{rec} \rangle = \langle \underline{m}_{rec} \rangle u_{DC} \\ \langle i_{rec} \rangle = \langle \underline{m}_{rec}^T \rangle \underline{i}_{mac} \end{cases} \tag{4.91}$$

As a result, the equivalent continuous model of the three-phase rectifier is obtained and the electrical diagram is shown in Fig. 4.32. Finally, a vector $\langle \underline{m}_{rec} \rangle$ can be used to model the three-phase rectifier.

4.5 Simplified Microgrid Frequency Response Model

To precisely simulate the dynamic behaviors of practical DGs and other MG's components, one should use high-order mathematical models with nonlinearities and physical constraints. For many studies such as steady-state analysis and MG power-frequency control synthesis/analysis issues, however, simplified models or transfer functions are generally employed. Moreover, the transfer functions of MG components such as DGs and ESSs can be represented by a low-order system and even a first-order lag as

$$G_i(s) = \frac{K_i}{1 + sT_i} \tag{4.92}$$

where K_i and T_i are the gain and time constants of the component. Dynamic behavior of each component is determined by its gain and time constants. For instance, in EESs that play an important role to effectively supply insufficient energy of power generation subsystems of the MG within a very short time, the time constant of a battery is larger than that of a flywheel. Because the battery takes time to charge energy to the battery cells, its time constant is limited to several seconds [32–34]; while the flywheel stores the energy in the form of kinetic energy, it has the ability to store surplus energy during off-peak periods and quickly release energy during the peak loads [34, 35].

Similar to conventional power grids, the relationship between load-generation deviation power and system frequency in an MG can be represented by a first-order transfer function. To maintain a stable operation of an MG, the total power generation must be effectively controlled to meet the total power demand by minimizing the amount of power imbalance (ΔP_e) as follows:

$$\Delta P_e = \Delta P_L - \Delta P_G \tag{4.93}$$

where ΔP_L and ΔP_G are the deviations of total power demand and power generation, respectively. Since system frequency is changed with net power variation, the system frequency variation Δf can be calculated by

$$\Delta f = \frac{\Delta P_e}{K_{MG}} \tag{4.94}$$

where K_{MG} is the MG frequency characteristic constant. As an inherent time delay exists between the MG power and frequency deviations, the transfer function for the MG frequency variation per unit power deviation can be expressed as follows:

$$G_{MG}(s) = \frac{\Delta f}{K_{MG}} = \frac{1}{K_{MG}(1 + sT_{MG})} = \frac{1}{D + sM} \tag{4.95}$$

where M and D are the equivalent inertia constant and damping constant of the MG, respectively [4].

4.5.1 Example 1

As the first example, an islanded AC MG including AC loads, wind turbine generator (WTG), PV, diesel engine generator (DEG), MT, fuel cell (FC), and

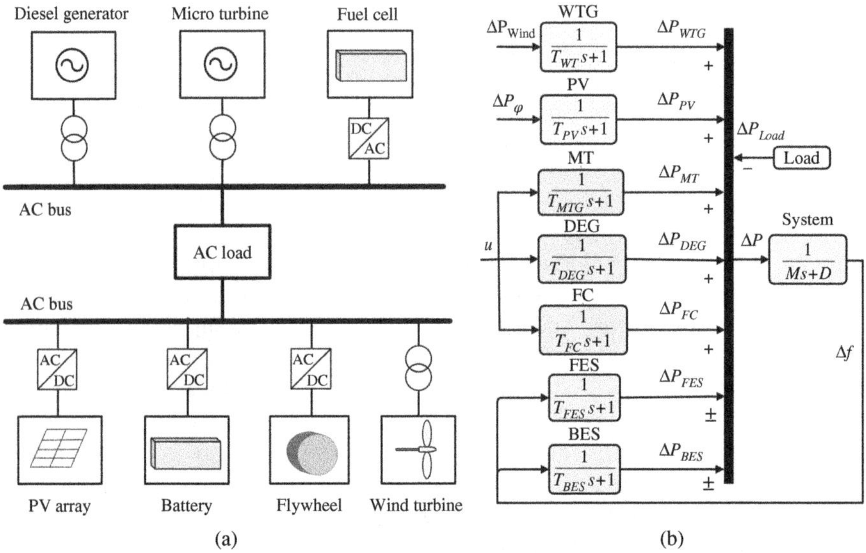

Figure 4.33 Simplified schematic of an islanded MG: (a) single-line diagram and (b) dynamic frequency response model.

energy storage devices such as flywheel energy storage system (FESS) and battery energy storage system (BESS) are considered as shown in Fig. 4.33a [6]. The distributed resources are connected to the AC bus by power electronic devices used for synchronization of the AC sources such as DEG and WTG, and they are used to invert DC voltages into AC in sources such as PV arrays and FCs. A converter is also considered for the BESS to convert AC into DC in charging mode and DC into AC in discharging mode.

The total power generation of distributed resources for supplying demand side comprises the output power of DEG, MT, WTG, PV, FC, and exchange power of flywheel energy storage (FES) and battery energy storage (BES).

$$P_{Load} = P_{DEG} + P_{MT} + P_{WTG} + P_{PV} + P_{FC} \pm P_{BES} \pm P_{FES} \tag{4.96}$$

Since the produced power by the RESs such as PVs and WTGs depends on the environmental condition, they are not commonly used for the regulation issues; so for control purposes, MT, DEG, and FC can be considered. An expression for changes in the MG power resources can be presented as follows:

$$\Delta P_{Load} + \Delta P_{DEG} + \Delta P_{MT} + \Delta P_{FC} + \Delta P_{PV} + \Delta P_{WTG} + \Delta P_{BES} + \Delta P_{FES} = 0 \tag{4.97}$$

Using the simplified dynamical model representation for DGs/storage units (4.92) and MG's load-inertia system (4.95), a dynamic frequency response model can be obtained as shown in Fig. 4.33b. The linearized state-space model is the most useful representation for the application of linear, modern, and robust control theorems. Using appropriate definitions and state variables, the linearized state-space realization of the MG system (Fig. 4.33b) can be easily obtained in the form of (4.98) [4].

$$\begin{aligned} \dot{x}(t) &= Ax(t) + B_1 w(t) + B_2 u(t) \\ y(t) &= Cx(t) \end{aligned} \tag{4.98}$$

where

$$\begin{aligned} x^T &= [\Delta P_{WTG}\ \Delta P_{PV}\ \Delta P_{DEG}\ \Delta P_{FC}\ \Delta P_{MT}\ \Delta P_{BES}\ \Delta P_{FES}\ \Delta f] \\ w^T &= [\Delta P_{Wind}\ \Delta P_{\varphi}\ \Delta P_{Load}] \\ y &= \Delta f \end{aligned}$$

and

$$\dot{x} = \begin{bmatrix} -1/T_{WTG} & 0 & 0 & 0 & 0 & 0 & 0 & 0 \\ 0 & -1/T_{PV} & 0 & 0 & 0 & 0 & 0 & 0 \\ 0 & 0 & -1/T_{DEG} & 0 & 0 & 0 & 0 & 0 \\ 0 & 0 & 0 & -1/T_{FC} & 0 & 0 & 0 & 0 \\ 0 & 0 & 0 & 0 & -1/T_{MT} & 0 & 0 & 0 \\ 0 & 0 & 0 & 0 & 0 & -1/T_{BES} & 0 & 1/T_{BES} \\ 0 & 0 & 0 & 0 & 0 & 0 & -1/T_{FES} & 1/T_{FES} \\ 0 & 0 & 0 & 0 & 0 & 0 & 0 & -2D/M \end{bmatrix} \begin{bmatrix} \Delta P_{WTG} \\ \Delta P_{PV} \\ \Delta P_{DEG} \\ \Delta P_{FC} \\ \Delta P_{MT} \\ \Delta P_{BES} \\ \Delta P_{FES} \\ \Delta f \end{bmatrix}$$

$$+ \begin{bmatrix} 1/T_{WTG} & 0 & 0 \\ 0 & 1/T_{PV} & 0 \\ 0 & 0 & 0 \\ 0 & 0 & 0 \\ 0 & 0 & 0 \\ 0 & 0 & 0 \\ 0 & 0 & 0 \\ 0 & 0 & 2/M \end{bmatrix} \begin{bmatrix} \Delta P_{Wind} \\ \Delta P_{\varphi} \\ \Delta P_{Load} \end{bmatrix} + \begin{bmatrix} 0 \\ 0 \\ 1/T_{DEG} \\ 1/T_{FC} \\ 1/T_{MT} \\ 0 \\ 0 \end{bmatrix} u, \quad y = \begin{bmatrix} 0 & 0 & 0 & 0 & 0 & 0 & 0 & 0 & 1 \end{bmatrix} x$$

Here u is the control input signal, x is the state vector, and w is the input disturbance vector.

In the mentioned MG case study (Fig. 4.33b), ΔP_{Wind}, ΔP_{φ}, and ΔP_{Load} are considered as MG disturbance signals.

4.5.2 Example 2

As second example, an isolated AC MG system shown in Fig. 4.34a is considered [2]. The MG system contains conventional DEG, PV panel, WTG, FC system, BESS, and FESS. As shown in Fig. 4.34b, the DGs are connected to the MG by power electronic interfaces, which are used for synchronization in AC sources such as DEG and WTG and to reverse voltage in DC sources such as PV panel, FC, and energy storage devices. The FC contains three fuel blocks, an

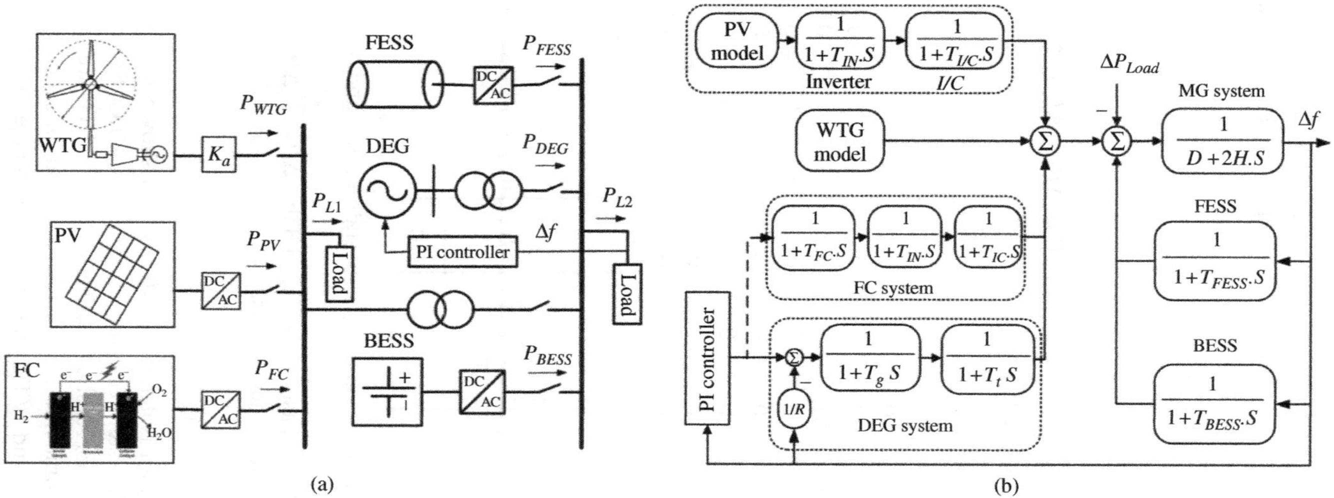

Figure 4.34 An MG case study: (a) single-line diagram and (b) frequency response model.

inverter for converting DC to AC voltage, and an interconnection device (IC). Here, unlike Example 1, third- and second-order models are considered for the FC and DEG units, respectively.

4.6 A Detailed State-Space Dynamic Model

In this section, a state-space model with more dynamic details is introduced for MGs [14]. Let us consider an MG with N dispatchable DGs that its schematic diagram is shown in Fig. 4.35a. Also assume that the MG includes N local loads and $N-1$ distribution line segments. Each DG unit is represented by a voltage source, a voltage source converter (VSC), and a series RL filter, and is interfaced to the DG through a step-up transformer (with a proper power rating). In the grid-connected operation mode, the main grid can be represented by an AC voltage source through series R and L elements. The simplified equivalent electric circuit for the given MG is shown in Fig. 4.35b.

4.6.1 Mathematical Modeling

Here, each DG is represented by a three-phase controlled voltage source and a series RL branch, and each load is represented by a parallel RLC network. Each distribution line is represented by lumped series RL elements. Now, the MG can be virtually partitioned into N subsystems. The mathematical model of subsystem 1 in the *abc* three-phase base can be written as [14]

$$\begin{cases} i_{1,abc} = i_{t1,abc} + C\dfrac{dv_{1,abc}}{dt} + i_{L1,abc} + \dfrac{v_{1,abc}}{R_1} \\ v_{t1,abc} = L_{f1}\dfrac{di_{1,abc}}{dt} + R_{f1}i_{1,abc} + v_{1,abc} \\ v_{1,abc} = L_1\dfrac{di_{L1,abc}}{dt} + R_{l1}i_{L1,abc} \\ v_{1,abc} = L_{t1}\dfrac{di_{t1,abc}}{dt} + R_{t1}i_{t1,abc} + v_{2,abc} \end{cases} \tag{4.99}$$

In order to find an LTI model, the most common method is converting the three-phase circuit equations into the *dq0* frame using the Park transformation [21, 36]. The Park transformation converts three-phase voltages and currents into their *dq0* components, respectively. This transformation can be represented as follows:

$$f_{dq} = \frac{2}{3}\begin{bmatrix} \cos\theta & \cos\left(\theta - \frac{2}{3}\pi\right) & \cos\left(\theta - \frac{4}{3}\pi\right) \\ -\sin\theta & -\sin\left(\theta - \frac{2}{3}\pi\right) & -\sin\left(\theta - \frac{4}{3}\pi\right) \\ \frac{1}{\sqrt{2}} & \frac{1}{\sqrt{2}} & \frac{1}{\sqrt{2}} \end{bmatrix} f_{abc} \tag{4.100}$$

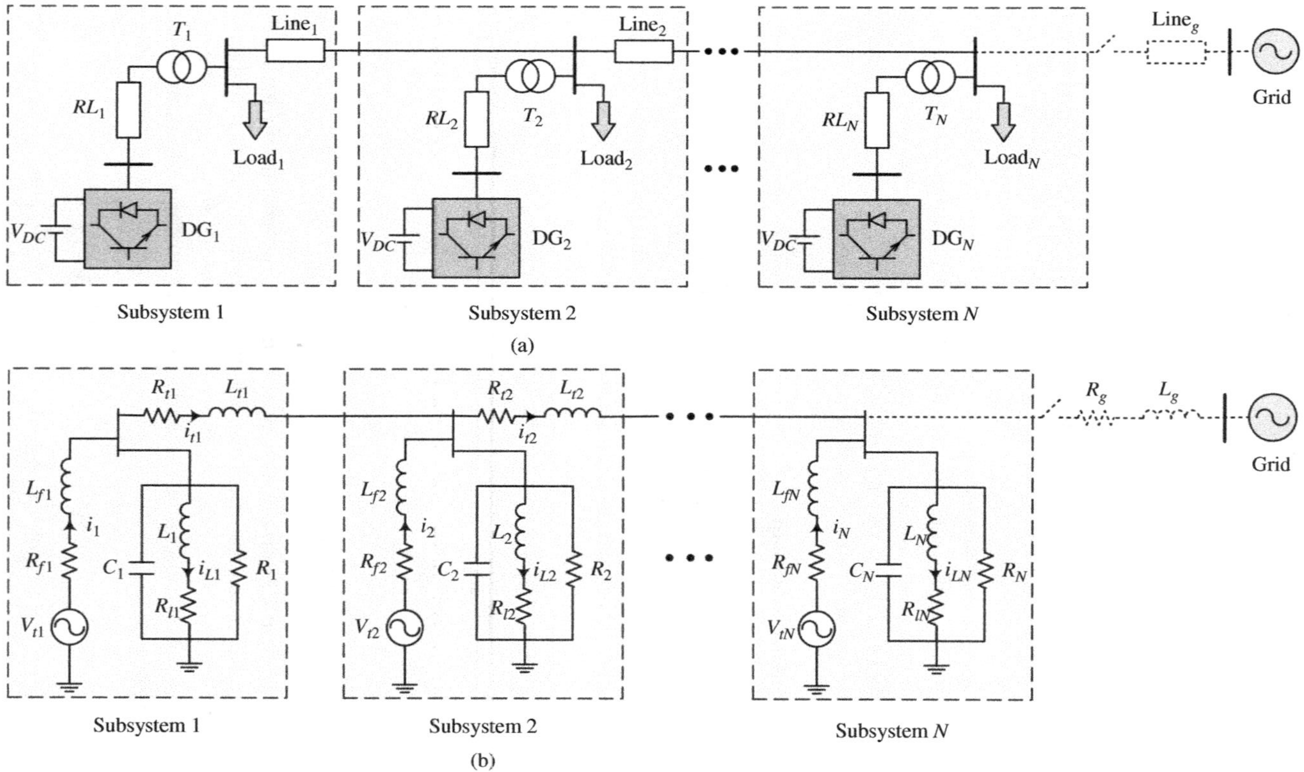

Figure 4.35 An MG with N DGs: (a) schematic diagram and (b) equivalent circuit.

where f_{dq} shows the direct and quadratic components and f_{abc} represents the balanced three-phase components. The θ is rotation angle and can be expressed as

$$\theta(t) = \int_0^t \omega(\tau)d\tau + \theta(t_0) \tag{4.101}$$

where ω is angular frequency. Using (4.99) and (4.100) and Fig. 4.35, the dynamical equations of subsystem 1 in dq-frame are obtained as follows:

$$\begin{cases} \dfrac{dV_{1,dq}}{dt} = \dfrac{1}{C_1}I_{1,dq} - \dfrac{1}{C_1}I_{t1,dq} - \dfrac{1}{C_1}I_{L1,dq} - \dfrac{1}{R_1C_1}V_{1,dq} - j\omega V_{1,qd} \\ \dfrac{dI_{1,dq}}{dt} = \dfrac{1}{L_{f1}}V_{t1,dq} - \dfrac{R_{f1}}{L_{f1}}I_{1,dq} - \dfrac{1}{L_{f1}}V_{1,dq} - j\omega I_{1,qd} \\ \dfrac{dI_{L1,dq}}{dt} = \dfrac{1}{L_1}V_{1,dq} - \dfrac{R_{L1}}{L_1}I_{L1,dq} - j\omega I_{L1,qd} \\ \dfrac{dI_{t1,dq}}{dt} = \dfrac{1}{L_{t1}}V_{1,dq} - \dfrac{R_{t1}}{L_{tt1}}I_{t1,dq} - \dfrac{1}{L_{t1}}V_{2,dq} - j\omega I_{t1,qd} \end{cases} \tag{4.102}$$

Similarly, the dq-frame-based models of other subsystems (2, 3, …, N) can be also performed to construct the state-space model of the overall system

$$\begin{aligned} \dot{x} &= Ax + Bu \\ y &= Cx \end{aligned} \tag{4.103}$$

where

$$\begin{aligned} x^T = (&V_{1,d}, V_{1,q}, I_{1,d}, I_{1,q}, I_{L1,d}, I_{L1,q}, I_{t1,d}, I_{t1,q}, V_{2,d}, V_{2,q}, I_{2,d}, I_{2,q}, I_{L2,d}, I_{L2,q}, \\ &I_{t2,d}, I_{t2,q}, V_{3,d}, V_{3,q}, I_{3,d}, I_{3,q}, I_{L3,d}, I_{L3,q}, I_{t3,d}, I_{t3,q}, \ldots, V_{N,d}, V_{N,q}, \\ &I_{N,d}, I_{N,q}, I_{LN,d}, I_{LN,q}) \end{aligned}$$

$$u^T = (V_{t1,d}, V_{t1,q}, V_{t2,d}, V_{t2,q}, V_{t3,d}, V_{t3,q}, \ldots, V_{tN,d}, V_{tN,q})$$

$$y^T = (V_{1,d}, V_{1,q}, V_{2,d}, V_{2,q}, V_{3,d}, V_{3,q}, \ldots, V_{N,d}, V_{N,q})$$

$$A \in R^{(8N-2)\times(8N-2)}, \quad B \in R^{(8N-2)\times 2N}, \quad C \in R^{2N\times(8N-2)}$$

For example, the state-space model of an MG with three DGs can be obtained as follows:

$$
A = \left[\begin{array}{cccccccccccccccccccccc}
\frac{-1}{R_1C_1} & -\omega & \frac{1}{C_1} & 0 & \frac{-1}{C_1} & 0 & \frac{-1}{C_1} & 0 & 0 & 0 & 0 & 0 & 0 & 0 & 0 & 0 & 0 & 0 & 0 & 0 & 0 & 0 \\
-\omega & \frac{-1}{R_1C_1} & 0 & \frac{1}{C_1} & 0 & \frac{-1}{C_1} & 0 & \frac{-1}{C_1} & 0 & 0 & 0 & 0 & 0 & 0 & 0 & 0 & 0 & 0 & 0 & 0 & 0 & 0 \\
\frac{-1}{L_{f1}} & 0 & \frac{-R_{f1}}{L_{f1}} & \omega & 0 & 0 & 0 & 0 & 0 & 0 & 0 & 0 & 0 & 0 & 0 & 0 & 0 & 0 & 0 & 0 & 0 & 0 \\
0 & \frac{-1}{L_{f1}} & -\omega & \frac{-R_{f1}}{L_{f1}} & 0 & 0 & 0 & 0 & 0 & 0 & 0 & 0 & 0 & 0 & 0 & 0 & 0 & 0 & 0 & 0 & 0 & 0 \\
\frac{1}{L_1} & 0 & 0 & 0 & \frac{-R_{l1}}{L_1} & \omega & 0 & 0 & 0 & 0 & 0 & 0 & 0 & 0 & 0 & 0 & 0 & 0 & 0 & 0 & 0 & 0 \\
0 & \frac{1}{L_1} & 0 & 0 & 0 & \frac{-R_{l1}}{L_1} & 0 & 0 & 0 & 0 & 0 & 0 & 0 & 0 & 0 & 0 & 0 & 0 & 0 & 0 & 0 & 0 \\
\frac{1}{L_{t1}} & 0 & 0 & 0 & 0 & 0 & \frac{-R_{t1}}{L_{t1}} & \omega & \frac{-1}{L_{t1}} & 0 & 0 & 0 & 0 & 0 & 0 & 0 & 0 & 0 & 0 & 0 & 0 & 0 \\
0 & \frac{1}{L_{t1}} & 0 & 0 & 0 & 0 & -\omega & \frac{-R_{t1}}{L_{t1}} & 0 & \frac{-1}{L_{t1}} & 0 & 0 & 0 & 0 & 0 & 0 & 0 & 0 & 0 & 0 & 0 & 0 \\
0 & 0 & 0 & 0 & 0 & 0 & \frac{1}{C_2} & 0 & \frac{-1}{R_2C_2} & \omega & \frac{1}{C_2} & 0 & \frac{-1}{C_2} & 0 & \frac{-1}{C_2} & 0 & 0 & 0 & 0 & 0 & 0 & 0 \\
0 & 0 & 0 & 0 & 0 & 0 & 0 & \frac{1}{C_2} & -\omega & \frac{-1}{R_2C_2} & 0 & \frac{1}{C_2} & 0 & \frac{-1}{C_2} & 0 & \frac{-1}{C_2} & 0 & 0 & 0 & 0 & 0 & 0 \\
0 & 0 & 0 & 0 & 0 & 0 & 0 & 0 & \frac{-1}{L_{f2}} & 0 & \frac{-R_{f2}}{L_{f2}} & \omega & 0 & 0 & 0 & 0 & 0 & 0 & 0 & 0 & 0 & 0 \\
0 & 0 & 0 & 0 & 0 & 0 & 0 & 0 & 0 & \frac{-1}{L_{f2}} & -\omega & \frac{-R_{f2}}{L_{f2}} & 0 & 0 & 0 & 0 & 0 & 0 & 0 & 0 & 0 & 0 \\
0 & 0 & 0 & 0 & 0 & 0 & 0 & 0 & \frac{1}{L_2} & 0 & 0 & 0 & \frac{-R_{l2}}{L_2} & \omega & 0 & 0 & 0 & 0 & 0 & 0 & 0 & 0 \\
0 & 0 & 0 & 0 & 0 & 0 & 0 & 0 & 0 & \frac{1}{L_2} & 0 & 0 & -\omega & \frac{-R_{l2}}{L_2} & \omega & 0 & 0 & 0 & 0 & 0 & 0 & 0 \\
0 & 0 & 0 & 0 & 0 & 0 & 0 & 0 & \frac{1}{L_{t2}} & 0 & 0 & 0 & 0 & 0 & \frac{-R_{t2}}{L_{t2}} & \omega & \frac{-1}{L_{t2}} & 0 & 0 & 0 & 0 & 0 \\
0 & 0 & 0 & 0 & 0 & 0 & 0 & 0 & 0 & \frac{1}{L_{t2}} & 0 & 0 & 0 & 0 & -\omega & \frac{-R_{t2}}{L_{t2}} & 0 & \frac{-1}{L_{t2}} & 0 & 0 & 0 & 0 \\
0 & 0 & 0 & 0 & 0 & 0 & 0 & 0 & 0 & 0 & 0 & 0 & 0 & 0 & \frac{1}{C_3} & 0 & \frac{-1}{R_2C_3} & \omega & \frac{1}{C_2} & 0 & \frac{-1}{C_2} & 0 \\
0 & 0 & 0 & 0 & 0 & 0 & 0 & 0 & 0 & 0 & 0 & 0 & 0 & 0 & 0 & \frac{1}{C_3} & -\omega & \frac{-1}{R_2C_3} & 0 & \frac{1}{C_2} & 0 & \frac{-1}{C_2} \\
0 & 0 & 0 & 0 & 0 & 0 & 0 & 0 & 0 & 0 & 0 & 0 & 0 & 0 & 0 & 0 & \frac{-1}{L_{f3}} & 0 & \frac{-R_{f3}}{L_{f3}} & \omega & 0 & 0 \\
0 & 0 & 0 & 0 & 0 & 0 & 0 & 0 & 0 & 0 & 0 & 0 & 0 & 0 & 0 & 0 & 0 & \frac{-1}{L_{f3}} & -\omega & \frac{-R_{f3}}{L_{f3}} & 0 & 0 \\
0 & 0 & 0 & 0 & 0 & 0 & 0 & 0 & 0 & 0 & 0 & 0 & 0 & 0 & 0 & 0 & \frac{1}{L_3} & 0 & 0 & 0 & \frac{-R_{l3}}{L_3} & \omega \\
0 & 0 & 0 & 0 & 0 & 0 & 0 & 0 & 0 & 0 & 0 & 0 & 0 & 0 & 0 & 0 & 0 & \frac{1}{L_3} & 0 & 0 & -\omega & \frac{-R_{l3}}{L_3}
\end{array}\right]
$$

$$B^T = \begin{bmatrix} 0&0&0&0&0&0&0&0&0&0&0&0&0&0&0&0&0&0&0&0&\frac{1}{L_{f3}}&0 \\ 0&0&0&0&0&0&0&0&0&0&0&0&0&0&0&0&0&0&\frac{1}{L_{f3}}&0&0&0 \\ 0&0&0&0&0&0&0&0&0&0&0&\frac{1}{L_{f2}}&0&0&0&0&0&0&0&0&0&0 \\ 0&0&0&0&0&0&0&0&0&0&\frac{1}{L_{f2}}&0&0&0&0&0&0&0&0&0&0&0 \\ 0&0&0&\frac{1}{L_{f1}}&0&0&0&0&0&0&0&0&0&0&0&0&0&0&0&0&0&0 \\ 0&0&\frac{1}{L_{f1}}&0&0&0&0&0&0&0&0&0&0&0&0&0&0&0&0&0&0&0 \end{bmatrix}$$

$$C = \begin{bmatrix} 1&0&0&0&0&0&0&0&0&0&0&0&0&0&0&0&0&0&0&0&0&0 \\ 0&1&0&0&0&0&0&0&0&0&0&0&0&0&0&0&0&0&0&0&0&0 \\ 0&0&0&0&0&0&0&0&1&0&0&0&0&0&0&0&0&0&0&0&0&0 \\ 0&0&0&0&0&0&0&0&0&1&0&0&0&0&0&0&0&0&0&0&0&0 \\ 0&0&0&0&0&0&0&0&0&0&0&0&0&0&0&0&1&0&0&0&0&0 \\ 0&0&0&0&0&0&0&0&0&0&0&0&0&0&0&0&0&1&0&0&0&0 \end{bmatrix}$$

4.6.2 Simulation Example

For the sake of simulation, a simple MG with three DGs, three loads, and two transmission lines as its single-line diagram is shown in Fig. 4.36. The simulation parameters are also given in Table 4.1. Figure 4.37 shows the complete eigenvalues of the system for the initial conditions in the presence of R, RL, and RLC loads. Figure 4.38 shows eigenvalue trajectories as VSC filter impedance changes from 0.1 to 0.8 Ω for different loads.

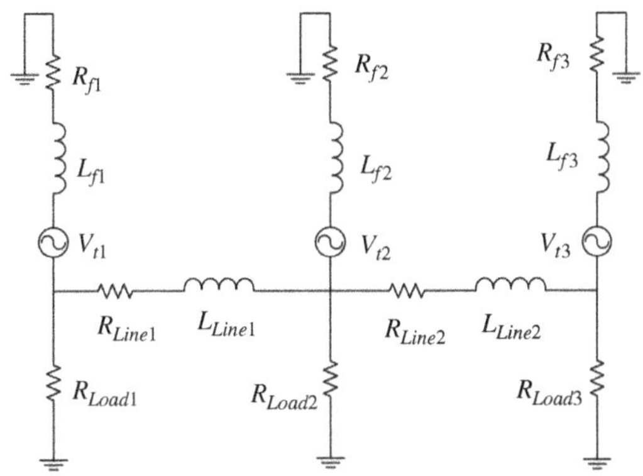

Figure 4.36 Single-line diagram of the simulated MG.

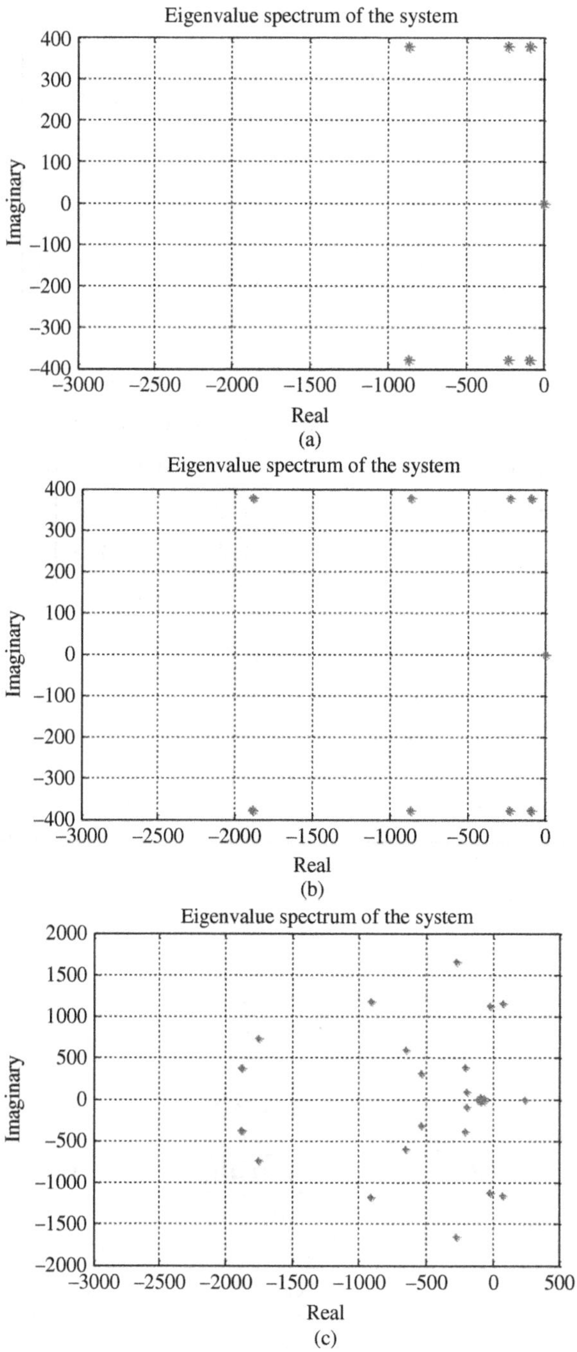

Figure 4.37 System eigenvalues for initial conditions with (a) R load, (b) RL load, and (c) RLC load.

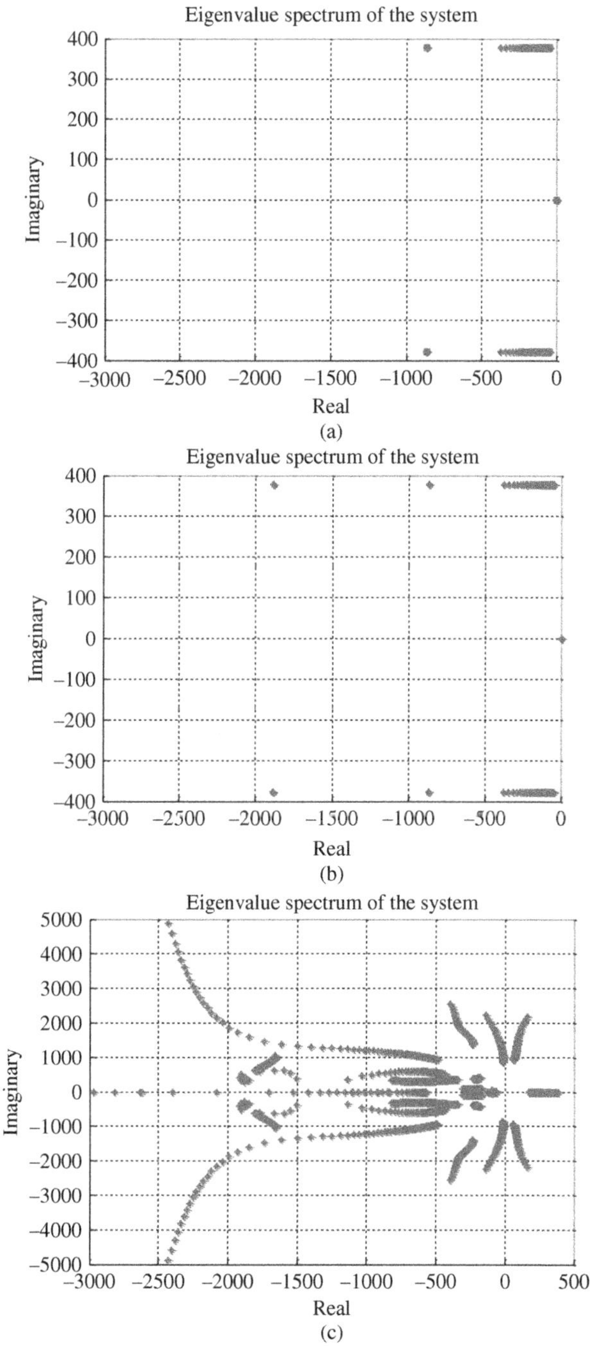

Figure 4.38 System eigenvalues for changes in VSC filter impedance with (a) R load, (b) RL load, and (c) RLC load.

Table 4.1 Parameters of the MG case study.

Parameter	Value	Parameter	Value
R_f	[0.1 0.1 0.1] Ω	$I_{line\,d}$	[−3.8 7.6] A
L_f	[1.1 1.1 1.1] mH	$I_{line\,q}$	[0.4 −1.3] A
R_{line}	[0.23 0.35] Ω	$V_{o\,d}$	[380.8 381.8 380.4] V
L_{line}	[0.26 1.5] mH	$V_{o\,q}$	[0 0 0] V
R_{load}	[25 20 20] Ω	$I_{o\,d}$	[11.4 11.4 11.4] A
L_{load}	[5 4 4] Ω	$I_{o\,q}$	[0.4 −1.45 1.25] A
C_{load}	[4 4 4] Ω	δ_0	[0 1.9e−3 −0.0113] rad
ω_c	31.4 rad/s	ω_s	314 rad/s

4.6.3 Closed-Loop State-Space Model

Considering the dynamics of existing controllers with the state-space model of other components of the MG provides closed-loop state-space model. This model can be useful for dynamic and stability analysis, rather than control synthesis. Recently, several approaches are given to determine the MGs' closed-loop dynamic models [1, 9]. Here, the given method in [9] for modeling an inverter-based MG is briefly described.

It is assumed that each DG inverter uses an outer power loop based on droop control to share the fundamental real and reactive powers with other DGs. Inverter internal controllers including voltage and current controllers are used to reject high-frequency disturbances and damping undesirable oscillations due to connection with the other circuits and the external network. The modeling approach presented in [9] divides the whole system into three major subsystems: inverter, network, and loads (Fig. 4.39). The inverter model includes the power sharing controller dynamics, output filter dynamics, coupling inductor dynamics, and voltage and current controller dynamics. The small-signal flow among the subsystems is shown in Fig. 4.39.

The state equations of the network and the loads are represented on the reference frame of one of the individual inverters. This reference frame is considered as the common reference frame. All other inverters are translated to this common reference frame using an appropriate transformation technique. By combining individual subsystems, the overall state-space model of MG can be obtained as follows:

$$\begin{bmatrix} \dot{x}_{inv} \\ \dot{x}_{net} \\ \dot{x}_{load} \end{bmatrix} = A_{cl} \begin{bmatrix} x_{inv} \\ x_{net} \\ x_{load} \end{bmatrix} \tag{4.104}$$

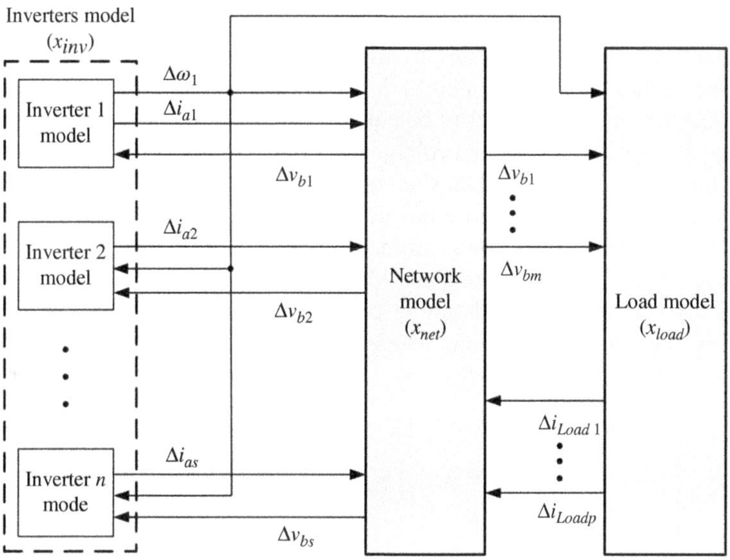

Figure 4.39 Closed-loop block diagram.

The subscripts of "*inv*," "*net*," and "*cl*" represent inverter, network, and closed-loop systems, respectively. Detailed derivation and information on the state-space matrices can be found in [9].

4.7 Microgrid Dynamic Modeling and Analysis as a Multivariable System

Since an MG is a multi-input and multi-output system with numerous variables, multivariable analysis tools are expected to be useful for employment. For instance, the generalized Nyquist diagram and the relative gain array (RGA) concept can be used, respectively, for the MG stability assessment and solving the pairing problem among the inputs and outputs. Using these, the droop control dependency on the X/R ratio of the MG's DGs can be also recognized easily.

This section presents the mathematical equations of simple MG with two DGs in dq rotating frame similar to the approach explained in the previous section. Then, it is considered as a multivariable system. and a systematic approach is presented for system recognition from the controllability and observability point of view. For a given dynamic model example, the generalized Nyquist stability criterion, and input–output sets selection are also outlined.

4.7.1 State-space Modeling

The schematic block diagram of a simple MG with two DGs, which is used as the case study, is depicted in Fig. 4.40 [37]. As mentioned, in order to find a linear model of the MG, each DG can be considered as a VSC (in the form of a controllable three-phase sinusoidal voltage source with a static DC link). The VSC output filter, transformer, and line between the VSC and point of common coupling (PCC) are modeled by an equivalent inductance and an equivalent resistance. The load is considered as a common resistive in the PCC. Figure 4.40 shows the single-line diagram of the given MG.

Using the Park transformation, the three-phase voltages and currents can be converted into their *dq0* components, respectively. For this purpose, the following transformation is used here:

$$\begin{bmatrix} f_d \\ f_q \\ f_0 \end{bmatrix} = \frac{2}{3}\begin{bmatrix} \cos(\theta) & \cos\left(\theta - \frac{2\pi}{3}\right) & \cos\left(\theta + \frac{2\pi}{3}\right) \\ -\sin(\theta) & -\sin\left(\theta - \frac{2\pi}{3}\right) & -\sin\left(\theta + \frac{2\pi}{3}\right) \\ \frac{1}{2} & \frac{1}{2} & \frac{1}{2} \end{bmatrix}\begin{bmatrix} f_a \\ f_b \\ f_c \end{bmatrix} \tag{4.105}$$

where f_d, f_q, and f_0 are the direct, quadratic, and zero components, respectively; θ is rotation angle, f_a, f_b, and f_c are the balanced three-phase components. Under the balanced conditions, the zero-sequence of the three-phase variable $\bar{f}_{abc} = \begin{bmatrix} f_a & f_b & f_c \end{bmatrix}^T$ in (4.105) is zero ($f_0 = 0$). In this case, the Park transformation is composed of two transformations as follows:

$$\bar{f}_{\alpha\beta} = \frac{2}{3}\begin{pmatrix} e^{j0} & e^{j\frac{2\pi}{3}} & e^{-j\frac{2\pi}{3}} \end{pmatrix}\bar{f}_{abc} \tag{4.106}$$

$$\bar{f}_{dq} = \bar{f}_{\alpha\beta} \cdot e^{-j\theta} \tag{4.107}$$

where $\bar{f}_{abc}$ is a 3×1 vector representing a balanced three-phase variable, $\bar{f}_{\alpha\beta} = f_\alpha + jf_\beta$ and $\bar{f}_{dq} = f_d + jf_q$. In (4.107), $\theta = \omega_0 t + \theta_0$ is the phase-angle of any three-phase voltage in the system, which is usually obtained using a phase-locked loop (PLL). Figure 4.41 shows *abc*, *αβ*, and *dq* frames in relation

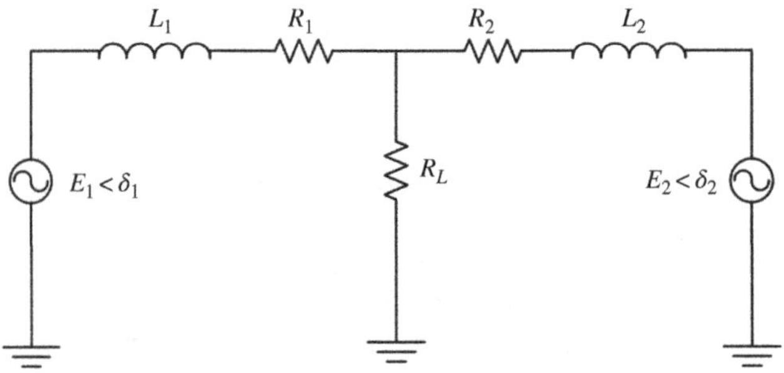

Figure 4.40 Single-line diagram of the MG example.

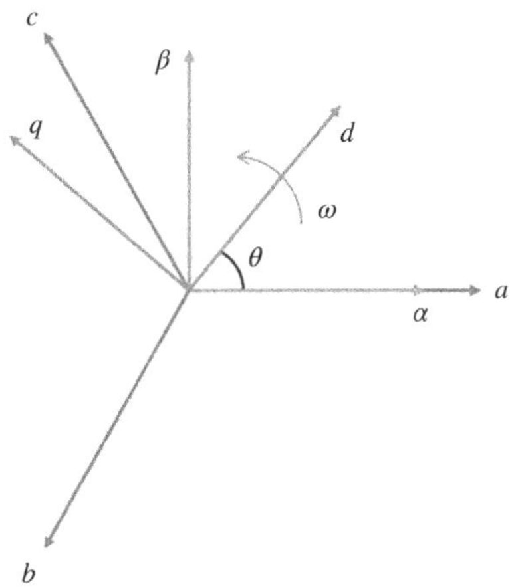

Figure 4.41 *abc*, *αβ*, and *dq* frames in relation to each other.

to each other. Using Kirchhoff's voltage and current laws (Kirchhoff's voltage law, KVL and Kirchhoff's current law, KCL) and from Fig. 4.40, the dynamical equations of the MG in *abc*-frame are obtained as follows:

$$e_{abc1} = R_1 i_{abc1} + L_1 \frac{di_{abc1}}{dt} + R_L(i_{abc1} + i_{abc2}),$$
$$e_{abc2} = R_2 i_{abc2} + L_2 \frac{di_{abc1}}{dt} + R_L(i_{abc1} + i_{abc2}) \tag{4.108}$$

where e_{abc1}, e_{abc2}, i_{abc1}, and i_{abc2} are 3×1 vectors containing the three-phase variables. The state-space equations of the system in the $\alpha\beta$-frame using (4.106) are obtained as follows:

$$L_1 \frac{di_{\alpha\beta1}}{dt} = -(R_1 + R_L) i_{\alpha\beta1} - R_L i_{\alpha\beta2} + e_{\alpha\beta1},$$
$$L_2 \frac{di_{\alpha\beta2}}{dt} = -R_L i_{\alpha\beta1} - (R_2 + R_L) i_{\alpha\beta2} + e_{\alpha\beta2} \tag{4.109}$$

Substituting (4.107) in (4.109) leads to the following state-space representation of the MG:

$$L_1 \frac{di_{dq1}}{dt} + j\omega_0 L_1 i_{dq1} = -(R_1 + R_L) i_{dq1} - R_L i_{dq2} + e_{dq1},$$
$$L_2 \frac{di_{dq2}}{dt} + j\omega_0 L_2 i_{dq2} = -R_L i_{dq1} - (R_2 + R_L) i_{dq2} + e_{dq2} \tag{4.110}$$

Finally, by decoupling the real and imaginary equations of (4.110), the real form of the state-space equations is as follows:

$$\frac{di_{d1}}{dt} = -\left(\frac{R_1 + R_L}{L_1}\right) i_{d1} + \omega_0 i_{q1} - \frac{R_L}{L_1} i_{d2} + \frac{1}{L_1} e_{d1},$$

$$\frac{di_{q1}}{dt} = -\omega_0 i_{d1} - \left(\frac{R_1 + R_L}{L_1}\right) i_{q1} - \frac{R_L}{L_1} i_{q2} + \frac{1}{L_1} e_{q1},$$
$$\frac{di_{d2}}{dt} = -\frac{R_L}{L_2} i_{d1} - \left(\frac{R_2 + R_L}{L_2}\right) i_{d2} + \omega_0 i_{q2} + \frac{1}{L_2} e_{d2},$$
$$\frac{di_{q2}}{dt} = -\frac{R_L}{L_2} i_{q1} - \omega_0 i_{d2} - \left(\frac{R_2 + R_L}{L_2}\right) i_{q2} + \frac{1}{L_2} e_{q2} \tag{4.111}$$

Using (4.105), the direct and quadratic components of the VSC output voltage are obtained as

$$e_d = E \sin(\delta), \quad e_q = -E \cos(\delta) \tag{4.112}$$

where E and δ are the VSC voltage amplitude and phase angle, respectively. Since these voltages are dependent on each other for any VSC, their components E and δ should be selected as control inputs. Hence, the standard state-space model of the MG as a multivariable system is obtained by substituting (4.112) in (4.111). Linearization of the resulting equation around the equilibrium points can be obtained as follows:

$$\frac{d}{dt}\begin{bmatrix} \Delta i_{d1}(t) \\ \Delta i_{q1}(t) \\ \Delta i_{d2}(t) \\ \Delta i_{q2}(t) \end{bmatrix} = A \begin{bmatrix} \Delta i_{d1}(t) \\ \Delta i_{q1}(t) \\ \Delta i_{d2}(t) \\ \Delta i_{q2}(t) \end{bmatrix} + B \begin{bmatrix} \Delta \delta_1(t) \\ \Delta E_1(t) \\ \Delta \delta_2(t) \\ \Delta E_2(t) \end{bmatrix}$$
$$\begin{bmatrix} \Delta i_{d1}(t) \\ \Delta i_{q1}(t) \\ \Delta i_{d2}(t) \\ \Delta i_{q2}(t) \end{bmatrix} = I_4 \begin{bmatrix} \Delta i_{d1}(t) \\ \Delta i_{q1}(t) \\ \Delta i_{d2}(t) \\ \Delta i_{q2}(t) \end{bmatrix} + 0_4 \begin{bmatrix} \Delta \delta_1(t) \\ \Delta E_1(t) \\ \Delta \delta_2(t) \\ \Delta E_2(t) \end{bmatrix} \tag{4.113}$$

where

$$A = \begin{bmatrix} \dfrac{-(R_1 + R_L)}{L_1} & \omega_0 & \dfrac{-R_L}{L_1} & 0 \\ -\omega_0 & \dfrac{-R_1}{L_1} & 0 & 0 \\ \dfrac{-R_L}{L_2} & 0 & \dfrac{-(R_2 + R_L)}{L_2} & \omega_0 \\ 0 & 0 & -\omega_0 & \dfrac{-R_2}{L_2} \end{bmatrix},$$

$$B = \begin{bmatrix} \dfrac{E_{10}\cos\delta_{10}}{L_1} & \dfrac{\sin\delta_{10}}{L_1} & 0 & 0 \\ \dfrac{E_{10}\sin\delta_{10}}{L_1} & -\dfrac{\cos\delta_{10}}{L_1} & 0 & 0 \\ 0 & 0 & \dfrac{E_{20}\cos\delta_{20}}{L_2} & \dfrac{\sin\delta_{20}}{L_2} \\ 0 & 0 & \dfrac{E_{20}\sin\delta_{20}}{L_2} & -\dfrac{\cos\delta_{20}}{L_2} \end{bmatrix}$$

In these equations, δ_{10} and δ_{20} are the steady-state values of $\delta_1(t)$ and $\delta_2(t)$, respectively.

4.7.2 Dynamic Analysis

The given MG system has four inputs, four outputs, and four state variables. Therefore, the transfer function matrix is a 4×4 frequency variable matrix where each entity is a fourth-order transfer function. For the MG parameters presented in Table 4.2, the state and input matrices and the related transfer function matrix are obtained as follows:

$$A = \begin{bmatrix} -128.4 & 314.16 & -0.093 & 0 \\ -314.16 & -128.3 & 0 & 0 \\ -0.125 & 0 & -107.05 & 314.16 \\ 0 & 0 & -314.16 & -106.92 \end{bmatrix},$$

$$B = \begin{bmatrix} 0.0853 & 0.0081 & 0 & 0 \\ 0.0086 & -0.0812 & 0 & 0 \\ 0 & 0 & 0.1118 & 0.0088 \\ 0 & 0 & 0.0089 & -0.1094 \end{bmatrix}$$

$$G(s) = [g_{ij}(s)], \quad i,j = 1:4$$

where the arrays of the transfer function are represented as follows:

$$g_{11} = \frac{0.085s^3+31.87s^2+12310s+1.5\times10^6}{s^4+470.7s^3+2.8\times10^5s^2+5.3\times10^7s+1.27\times10^{10}} \quad g_{12} = \frac{-22.72s^2-4337s-2.69\times10^6}{s^4+470.7s^3+2.8\times10^5s^2+5.3\times10^7s+1.27\times10^{10}}$$

$$g_{13} = \frac{-0.01s^2-2.7s-175.3}{s^4+470.7s^3+2.8\times10^5s^2+5.3\times10^7s+1.27\times10^{10}} \quad g_{14} = \frac{3s+397.5}{s^4+470.7s^3+2.8\times10^5s^2+5.3\times10^7s+1.27\times10^{10}}$$

$$g_{21} = \frac{-23.86s^2-4554s-2.83\times10^6}{s^4+470.7s^3+2.8\times10^5s^2+5.3\times10^7s+1.27\times10^{10}} \quad g_{22} = \frac{-0.08s^3-30.36s^2-11720s-1.43\times10^6}{s^4+470.7s^3+2.8\times10^5s^2+5.3\times10^7s+1.27\times10^{10}}$$

$$g_{23} = \frac{3.25\,s+429.2}{s^4+470.7s^3+2.8\times10^5s^2+5.3\times10^7s+1.27\times10^{10}} \quad g_{24} = \frac{0.2554s-973.4}{s^4+470.7s^3+2.8\times10^5s^2+5.3\times10^7s+1.27\times10^{10}}$$

$$g_{31} = \frac{-0.01s^2-2.83s-181.6}{s^4+470.7s^3+2.8\times10^5s^2+5.3\times10^7s+1.27\times10^{10}} \quad g_{32} = \frac{2.94s+326}{s^4+470.7s^3+2.8\times10^5s^2+5.3\times10^7s+1.27\times10^{10}}$$

$$g_{33} = \frac{0.11s^3+43.4s^2+16640s+1.7\times10^6}{s^4+470.7s^3+2.8\times10^5s^2+5.3\times10^7s+1.27\times10^{10}} \quad g_{34} = \frac{-31.2s^2-7573s-3.85\times10^6}{s^4+470.7s^3+2.8\times10^5s^2+5.3\times10^7s+1.27\times10^{10}}$$

$$g_{41} = \frac{3.34s+533.6}{s^4+470.7s^3+2.8\times10^5s^2+5.3\times10^7s+1.27\times10^{10}} \quad g_{42} = \frac{0.32\,s-957.9}{s^4+470.7s^3+2.8\times10^5s^2+5.3\times10^7s+1.27\times10^{10}}$$

$$g_{43} = \frac{-31.81s^2-7725s-3.93\times10^6}{s^4+470.7s^3+2.8\times10^5s^2+5.3\times10^7s+1.27\times10^{10}} \quad g_{44} = \frac{-0.12s^3-42.56s^2-16320s-1.67\times10^6}{s^4+470.7s^3+2.8\times10^5s^2+5.3\times10^7s+1.27\times10^{10}}$$

Using the obtained state-space representation of the MG, one can see that the system is asymptotically stable. Since there are no input, output, and input–output decoupling zeros, the utilized state-space representation is both controllable and observable. Nevertheless, there are four element zeros in the right open half plane, which cause some difficulties in the sequential controller design procedure.

Table 4.2 MG parameters and linearization data.

Parameters	Values	Parameters	Values
VSCs nominal power	7 MW	E_{10}	1.05 pu
VSCs nominal voltage	4140 V	E_{20}	1.02 pu
Nominal frequency	50 Hz	δ_{10}	0.1 pu
L_1	95.5 mH	δ_{20}	0.08
L_2	71 mH	X_1/R_1	1
R_L	2.78 Ω	X_2/R_2	1.2

Since the MG is a MIMO system, for designing decentralized controllers, the first step in the design procedure is to pair each input with a proper output. Indeed, it should be determined for any output which control input must be used. The *dq* currents exporting is one of the most common output set selections [38], which leads to a strictly proper transfer function matrix. Another choice is $\{V_{d,PCC}, V_{q,PCC}\}$ in the islanded mode where the MG needs a local voltage and frequency control [39]. In addition to input and output sets selection, the coupling between them is a key point in control design issue. The RGA is an index for input and output coupling, which can be used to solve the pairing problem. Let us present its definition in advance and then use it in our problem.

Definition: The RGA of a nonsingular square complex matrix G is a square complex matrix defined as

$$RGA(G) = \Lambda(G) = G \times (G^{-1})^T \tag{4.114}$$

where × denotes element-by-element multiplication [40].

A desirable input–output couple has a positive RGA's element close to one. The obtained results, represented in Table 4.3, show that the solution of the pairing problem depends on the X/R ratio. Indeed, in the first scenario, $\{\delta, i_q\}$ and $\{E, i_d\}$ are the best couples according to the RGA rule. However, for the

Table 4.3 Three MG output X/R with the corresponding RGAs.

Scenario	1	2	3
$(X/R)_1$,$(X/R)_2$	0.3, 0.4	1, 1.2	4, 5
RGA	$\begin{pmatrix} 0.39 & 0.61 & 0 & 0 \\ 0.61 & 0.39 & 0 & 0 \\ 0 & 0 & 0.28 & 0.72 \\ 0 & 0 & 0.72 & 0.28 \end{pmatrix}$	$\begin{pmatrix} 0.78 & 0.22 & 0 & 0 \\ 0.22 & 0.78 & 0 & 0 \\ 0 & 0 & 0.84 & 0.16 \\ 0 & 0 & 0.16 & 0.84 \end{pmatrix}$	$\begin{pmatrix} 0.96 & 0.04 & 0 & 0 \\ 0.04 & 0.96 & 0 & 0 \\ 0 & 0 & 0.97 & 0.03 \\ 0 & 0 & 0.03 & 0.97 \end{pmatrix}$

last scenario, the RGA proposes to use couples $\{\delta, i_d\}$ and $\{E, i_q\}$ in a control design procedure. This case study is used for designing robust controllers in Chapter 9.

4.8 Summary

The modeling and dynamic analysis issues in an MG have been investigated by many researchers in the recent years, mostly focusing on a particular aspect. Depending on the type of MG, network parameters, DGs diversity, and the purpose of use, the modeling may vary. With more and more penetration of renewable energy sources, new devices, and functionalities, the modeling of an MG highly depends on the control level and type, stability issue, and MG configuration.

After an introduction on the recent achievements in MG dynamic representation and modeling, a dynamic model is developed for the main grid and connecting components. The mentioned model is extended to an overall system by including the MG system for the grid-connected operation mode. Modeling of some components in the MG is introduced, and a simplified frequency response model using low-order transfer functions is presented. The state-space model representation using more dynamic details is emphasized, and the MG modeling and dynamics are analyzed as a multivariable system.

References

1 Tang, X., Deng, W., and Qi, Z. (2014) Investigation of the dynamic stability of microgrid. *IEEE Transactions on Power Systems*, **29** (2), 698–706.

2 Bevrani, H., Habibi, F., Babahajyani, P. *et al.* (2012) Intelligent frequency control in an AC microgrid: on-line PSO-based fuzzy tuning approach. *IEEE Transactions on Smart Grids*, **3** (4), 1935–1944.

3 Lee, D.J. and Wang, L. (2008) Small-signal stability analysis of an autonomous hybrid renewable energy power generation/energy storage system, part I: time-domain simulations. *IEEE Transactions on Energy Conversion*, **23** (1), 311–320.

4 Bevrani, H. (2014, Chapter 11) Frequency control in microgrids, in *Robust Power System Frequency Control*, 2nd edn, Springer, Switzerland.

5 Senjyu, T., Nakaji, T., Uezato, K., and Funabashi, T. (2005) A hybrid power system using alternative energy facilities in isolated island. *IEEE Transactions on Energy Conversion*, **20** (2), 406–414.

6 Bevrani, H., Feizi, M.R., and Ataee, S. (2016) Robust frequency control in an islanded microgrid: hinf and mu synthesis approaches. *IEEE Transactions on Smart Grids*, **7** (2), 706–717.

7 Karimi, H., Nikkhajoei, H., and Iravani, R. (2008) Control of an electronically-coupled distributed resource unit subsequent to an islanding event. *IEEE Transactions on Power Delivery*, **23** (1), 493–501.

8 Gao, F. and Iravani, R. (2008) A control strategy for a distributed generation unit in grid-connected and autonomous modes of operation. *IEEE Transactions on Power Delivery*, **23** (2), 850–859.

9 Pogaku, N., Prodanovic, M., and Green, T.C. (2007) Modeling, analysis and testing of autonomous operation of an inverter-based microgrid. *IEEE Transactions on Power Electronics*, **22** (2), 613–625.

10 Katiraei, F., Iravani, M.R., and Lehn, P.W. (May 2007) Small-signal dynamic model of a microgrid including conventional and electronically interfaced distributed resources. *IEE Generation, Transmission & Distribution*, **1** (3), 369–378.

11 Katiraei, F., Iravani, M.R., and Lehn, P.W. (2005) Micro-grid autonomous operation during and subsequent to islanding process. *IEEE Transactions on Power Delivery*, **20** (1), 248–257.

12 Soultanis, N.L., Papathanasiou, S.A., and Hatziargyriou, N.D. (2007) A stability algorithm for the dynamic analysis of inverter dominated unbalanced LV microgrids. *IEEE Transactions on Power Systems*, **22** (1), 294–304.

13 Guzmán, D. *et al.* (2009) Complex-valued state matrices for simple representation of large autonomous microgrids supplied by PQ and Vf generation. *IEEE Transactions on Power Systems*, **24** (4), 1720–1730.

14 Etemadi, A.H. and Iravani, R. (2012) Eigenvalue and robustness analysis of a decentralized voltage control scheme for an islanded multi-DER microgrid, in *IEEE Power and Energy Society General Meeting*, San Diego, CA.

15 Katiraei, F. (2005) Dynamic analysis and control of distributed energy resources in a microgrid. PhD dissertation. University of Toronto, Toronto, ON, Canada.

16 Lopes, J.A.P., Moreira, C.L., and Madureira, A.G. (2006) Defining control strategies for microgrid ESS islanded operation. *IEEE Transactions on Power Systems*, **21** (2), 916–924.

17 Majumder, R. (2013) Some aspects of stability in microgrids. *IEEE Transactions on Power Systems*, **28** (3), 3243–3252.

18 Li, P. (2010) Design and control of a PV active generator with integrated energy storages: application to the aggregation of producers and consumers in an urban micro smart grid. PhD thesis. Ecole Centrale de Lille, France.

19 Bevrani, H., Watanabe, M., and Mitani, Y. (2014) *Power System Monitoring and Control*, IEEE-Wiley Press, New York.

20 Saadat, H. (1999) *Power System Analysis*, McGraw-Hill International Editions, pp. 527–569.

21 Kundur, P. (1994) *Power System Stability and Control*, McGraw-Hill, pp. 389–417.

22 Bevrani, H., Hiyama, T., and Mitani, Y. (2008) Power system dynamic stability and voltage regulation enhancement using an optimal gain vector. *Control Engineering Practice*, **16**, 1109–1119.

23 Xiao, W., Dunford, W.G., and Capel, A. (2004) *A Novel Modeling Method for Photovoltaic Cells*. IEEE Power Electrics Specialists Conference, Aachen, Allemagne.

24 Shawn, F. (2004) *A Method for Predicting PV Module and Array Performance at Other than Standard Reporting Conditions*, North Carolina Solar Center.

25 Rauschenbach, H.S. (1980) *Solar Cell Array Design Handbook*, Van Nostrand-Reinhold, New York.

26 Gergaud, O., Robin, G., Multon, B., and Ben Ahmed, H. (2003) Energy modelling of a lead-acid battery within hybrid wind/photovoltaic systems. SATIE – Brittany Branch, EPE 2003, Toulouse.

27 Guasch, D. and Silvestre, S. (2003) Dynamic battery model for photovoltaic applications. *Revue: Progress in Photovoltaics: Research and Applications*, **11** (3), 193–206.

28 Zubieta, L. and Bonert, R. (2000) Characterization of double-layer capacitors for power electronics applications. *IEEE Transactions on Industry Applications*, **36** (1), 199–200.

29 Spyker, R. and Nelms, R.M. (2000) Analysis of double layer capacitors supplying constant power loads. *IEEE Transactions on Aerospace and Electronic Systems*, **36** (4), 1439–1443.

30 Li, P., Francois, B., Degobert, Ph, and Robyns, B. (2008) Multi-level representation for control design of a super capacitor storage system for a microgrid connected application. ICREPQ'08, Santander, March 12–14.

31 François, B. and Hautier, J.P. (1999), *Pulse Position and Pulse Width Modulation of Electrical Power Conversions: Application to a Three-Phase Voltage-Fed Inverter*. 3rd International Symposium on Advanced Electromechanical Motion Systems: Electromotion, vol. 2, pp. 653–658, Patras, Greece, July 8–9.

32 Agbossou, K., Kolhe, M., Hamelin, J., and Bose, T.K. (2004) Performance of a stand-alone renewable energy system based on energy storage as hydrogen. *IEEE Transactions on Energy Conversion*, **19** (3), 633–640.

33 Duryea, S., Islam, S., and Lawrance, W. (2001) A battery management system for stand-alone photovoltaic energy systems. *IEEE Transactions on Industry Applications*, **7**, 67–72.

34 Ribeiro, P.F., Johnson, B.K., Crow, M.L. *et al.* (2001) Energy storage systems for advanced power applications. *Proceedings of the IEEE*, **89** (12), 1744–1756.

35 Dettmer, R. (Apr. 1990) Revolutionary energy – a wind/diesel generator with flywheel storage. *IEE Review*, **36** (4), 149–151.

36 Radwan, A.A.A. and Mohamed, Y.A.-R.I. (2014) Integrating VSCs to weak grids by nonlinear power damping controller with self-synchronization capability. *IEEE Transactions on Power Systems*, **29** (2), 805–814.

37 Naderi, M., Khayat, Y., Batmani, Y., and Bevrani, H. (2016) *Multivariable Control Based Modeling, Analysis and Robust Control Synthesis for an Islanded Microgrid*. 3rd International Conference on Power and Energy System Engineering (CPESE 2016), Kitakyushu, Japan, September.

38 Mahdi, A. and Mohamed, Y.A.-R. (2015) Multivariable droop control of synchronous current converters in weak grids/microgrids with decoupled *dq*-axes currents. *IEEE Transactions on Smart Grid*, **6** (4), 1610–1620.

39 Babazadeh, M. and Karimi, H.R. (2013) A robust two-degree-of-freedom control strategy for an islanded microgrid. *IEEE Transactions on Power Delivery*, **28** (3), 1339–1347.

40 Albertos, P. and Sala, A. (2004) *Multivariable Control Systems: An Engineering Approach*, Springer-Verlag, London.

5

Hierarchical Microgrid Control

The increasing interest in penetration of renewable energy sources (RESs) into the power system via microgrids (MGs) presents some significant challenges in both hosts and MGs from stability and control point of view. This chapter addresses a general overview on the existing technologies and major challenges in MG control. In order to ensure stable, reliable, secure, and economical operation in either the grid-connected or islanded operation mode, many efforts are being put into the design of MGs control and protection systems, in the last two decades.

Advances in MG control have improved the potential of MGs to be integrated into the conventional electrical systems in a higher capacity. This improvement not only covers their internal control performance and connection/disconnection procedures in order to limit the negative dynamic impacts in the network and other connected MGs but also includes the grid support functionalities to enhance the global operation of distribution networks.

In this chapter, the most important issues and challenges in MG control are presented, and a general overview of the main control loops is also given. The chapter classifies MG control strategies into four control levels: local, secondary, central and emergency, and global, where the first three levels are associated with the operation of the MG itself, and the fourth level (global control) demonstrates the coordinated operation of the MG and neighbor grids as well as the main/host grid. The mentioned control levels are discussed in detail and a synthesis example is given at the end.

5.1 Introduction

The MG concept provides a quite appealing alternative for overcoming the challenges of integrating RESs and distributed generators (DGs) into power systems. However, in order to allow seamless deployment of MGs, the stability and control challenges should be solved. Current efforts are mostly being put into the design of more effective control strategies in different control levels and special protection schemes that ensure reliable, secure, and economical operation of MGs in either grid-connected or islanded operation mode. A general overview of the existing MG control technologies and remaining challenges is addressed in [1–3].

Low inertia, uncertainties, dynamic modeling and stability, and bidirectional power flow issues are known as the most relevant challenges in MGs control and protection. On the other hand, DGs' output voltage and current control, active/reactive power balancing and frequency/voltage regulation, demand-side management, economic dispatch, and transition between operation modes are mentioned as the most desirable features of the MG control system. While in the grid-connected operation mode, main concerns are on the interaction with the main grid, reliability and control issues are more significant in the islanded operation mode. In the islanded mode, the operation also is more challenging [1].

As described in Chapter 4 and emphasized in [4] as well, similar to a conventional power grid, the instability issues in an MG can be divided into small signal instability, transient instability, voltage instability, and frequency instability (in AC/hybrid MG types). There are many recurring reasons for small signal instability in an MG such as dynamic impacts of feedback controllers, continuous load switching, oscillation modes, and DGs' power limit. Unexpected islanding, DG outage, large and sudden load change, and cascaded faults in an MG are known as the most important reasons of the transient instability problems. Reactive power limits, load dynamics, and tap changers create most of the voltage stability problems; and finally serious load-generation imbalance, underfrequency load shedding, and active power limits can be considered as the main reasons for frequency instability in an MG.

Various control loops must be used to improve the MG's stability and performance. Current, voltage/amplitude, frequency/angle, and active and reactive powers are the main feedback variables used in the existing MG control loops in both grid-connected and islanded operation modes. The hierarchical control structure of MGs are responsible to provide proper load sharing and DGs coordination, voltage/frequency regulation in both operating modes, MG resynchronization with the main grid, operating cost optimization, and power flow control between the MG, neighborhood grids, and the main grid.

Hierarchical control strategy consists of four levels [5], namely the *local* (primary), *secondary, central/emergency,* and *global* controls, as shown in Fig. 5.1. The *local control* that includes fundamental control hardware, comprises DGs internal voltage and current control loops, maintains DGs stability by measuring and controlling the local signals. It is essential to

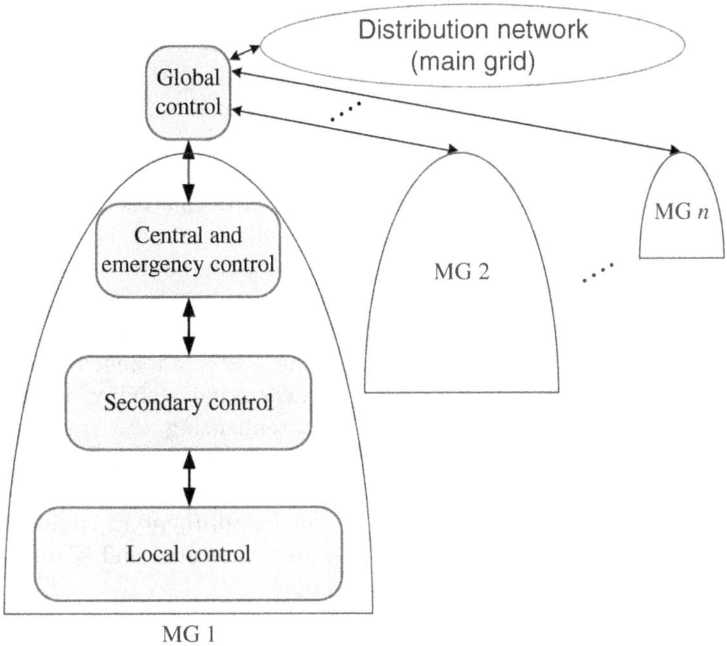

Figure 5.1 Hierarchical control levels in MGs.

provide independent active and reactive power sharing controls and avoiding undesired circulating currents for the DGs/RESs using the available current, voltage and frequency feedback signals. The *secondary control* provides power sharing as a communication-based method for parallel configuration of DGs and compensates the voltage and frequency deviations caused by the load variation and local control operation. The *central/emergency control* level facilitates MG supervision activities, and its role is particularly important in the islanded operation mode. It operates as an MG energy management system (EMS) and monitors the MG's local and secondary controllers. It is also responsible for islanding detection and connection/disconnection to/from the main grid, as well as emergency control and overall protection schemes. Finally, the *global control* manages the power flow between the given MG, other interconnecting MGs and the main grid, and facilitates an economically optimal operation. This level is working in the distribution network area, outside the MGs.

Most of MG components may attend in the existing control features. Dispatchable DGs/RESs can be fully controlled; however, nondispatchable units cannot, and are typically operated to extract the maximum possible power. Loads within the MGs can be also controlled using either a conventional circuit breaker (CB) or a more sophisticated power electronic (PE) interface to allow more flexible control functionality. The energy storage systems (ESSs) may also improve the dynamic stability, voltage and frequency regulation, and transient stability following a disturbance in an MG.

Furthermore, MGs can competently participate in regulating the distribution regulation services, by means of controlling the active and reactive power delivery through droop control algorithms. This regulation capability can be further improved if ESSs are also integrated in the MG, as the control capability becomes less dependent on the availability of the primary energy resource [4].

Over the years, extensive research has been carried out on the development of MG control strategies; however, the following areas can benefit from further research [1]: improving robust performance and stability against the structured and unstructured uncertainties; enhancing the transient response of the closed-loop systems; accounting for imbalance and harmonics; improving scalability of the control frameworks; incorporating the DC-side dynamics in the control synthesis; enhancing fault ride-through capabilities; and developing new control schemes for both grid-connected and islanded operation modes, and providing a smooth transition.

Finally, it is noteworthy that MGs can offer other coordinated services related to power quality improvement, in both grid-connected and island operation modes. Harmonics and unbalance compensation, flicker reduction, transient voltage support during grid-faults, or reactive power compensation could be also considered as common functionalities in modern interconnected MG systems [6].

Regarding the recent development of MGs and their future role in the new electrical networks, the route map of the new developments in MGs can be grouped into five main areas, which are closely linked to each others [3, 7]: (i) grid integration of ESSs in MGs and power generation facilities, (ii) uncertainty analysis and robust control synthesis in the presence of parameters perturbation and power intermittency, (iii) active management of the demand (demand response), (iv) improvement of the MG controllability and monitoring, and (v) enhancement of MGs' capability to support the distribution network in regulation point of view.

This chapter presents an overview of the control levels and existing control technologies and challenges in MGs. Section 5.2 describes the hierarchical control systems applied to MGs and discusses the classification of the control features in different control level. Droop characteristic/control in inverter-based DGs and virtual impedance concept are addressed in Section 5.3. Sections 5.4 and 5.5 explain the role of control levels in MGs' power management and control. The discussion is supplemented by a real-time simulation, and finally, the chapter is summarized in Section 5.6.

5.2 Microgrid Control Hierarchy

Control is one of the key enabling technologies for the deployment of MG systems. Similar to conventional power grids, MGs have a hierarchical control structure with different layers. MGs require effective use of advanced control techniques at all levels. The secure operation of MGs in connected and islanding operation modes and successful disconnection or reconnection processes depend upon MG controls. The controllers must guarantee that the processes occur seamlessly and the system is working in the specified operating points [8].

As already mentioned, MGs should be able to not only operate autonomously but also interact with the main grid. In the grid-connected operation mode, the MGs are integrated into a constantly varying electrical grid with changing tie-line flow, voltages, and frequency. To cope with those variations, to respond to grid disturbances, and to perform active power/frequency regulation, as well as reactive power/voltage regulation, MGs need to use proper control loops. Furthermore, suitable islanding detection feedbacks/algorithms are needed for ensuring a smooth transition from the grid-connected to islanded mode to avoid cascaded failures.

A general scheme for operating controls in an MG is shown in Fig. 5.2. Each MG is locally controlled by a microsource control (MC) system. The load controllers (LCs) are installed at the controllable loads to provide load control capabilities. For each MG, there is a central controller (MGCC) that interfaces between the distribution management system (DMS) or distribution network operator (DNO) and the MG. The DMS/DNO has responsibility to manage the operation of medium voltage (MV) and low-voltage (LV) areas in which more than one MG may exist.

Similar to conventional power grids [9], MGs are operating by using various control loops, which can be mainly classified in four levels in a hierarchical control scheme: *local, secondary, central/emergency,* and *global* controls. In fact, this hierarchical control scheme provides a compromise between fully centralized and decentralized control schemes, and the performed control levels are mainly different in their infrastructure requirements and dynamic timescale in which they are operating.

The *local control* deals with initial primary controls such as current and voltage control loops in the DGs/RESs. The *secondary control* ensures that the frequency and average voltage deviation of the MG are regulated toward zero after every change in load or supply. It is also responsible for local ancillary services. The *central/emergency control* covers all possible emergency control schemes and special protection plans to maintain the MG stability and availability in the face of contingencies. The emergency controls identify proper preventive and corrective measures that mitigate the effects of critical contingencies. The *global control* allows MG operation at an economic

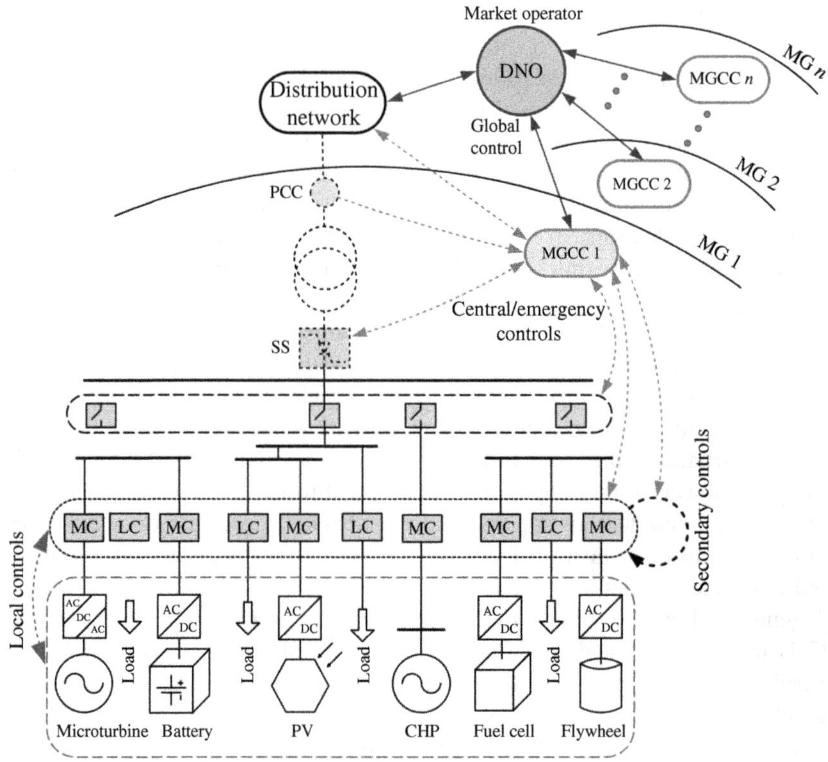

Figure 5.2 A general scheme for MG controls. *Source:* Bevrani *et al.*, 2014 [7]. Reproduced with permission of IEEE-Wiley Press.

optimum and organizes the relation between an MG and distribution network as well as other connected MGs.

The DNO allows the distribution grid and the connected MGs to operate at an economic optimum and it organizes the relation between the connected MGs and distribution network as well as other connected grids. As shown in Fig. 5.2, the global control center interfaces MGCCs of MGs as well as the distribution network (main grid) and also supervises the power flow control and market operating. This control unit controls power dispatching between the MGs to maintain close to the scheduled values.

In contrast to the local control, operating without communication, secondary, global, and emergency controls may need communication channels. While the local controls are known as *decentralized* controllers, the global, and, to some extent, secondary and emergency controllers are operating as *centralized* controllers.

The global control level typically operates in the order of several of minutes to hour, providing signals to the central level controls at MGs and other subsystems that form the interconnected distribution grid. Central controls, on the other hand, coordinate internal secondary and local controls within the MGs in the span of a few minutes. Secondary control reacts in the order of few seconds to minute. Finally, local controls are designed to operate independently and operate in predefined ways instantaneously to local events.

5.2.1 Local Control

Local (also called primary or internal) controls perform the first level in the control hierarchy, featuring the fastest response, and they appear in different forms depending on the type of DGs, which can be addressed based on their technologies such as induction generators, synchronous generators, and PE inverters/converters. In comparison of synchronous and induction generating units, the Power Electronic (PE) inverters/converters provide more flexible operation.

The local controllers deal with the inner control of the DG units that usually do not need the communication links and result in simple circuitry and low cost. Local controls are the basic category of the MG controls. The main usage of local controllers is to control DGs to operate in normal operation. Figure 5.3 shows the local controllers for a DG that operates as a voltage-controlled voltage source converter (VSC), where the voltage reference is provided by the conventional droop controllers. Here, the droop controllers, which are described in the following section, are also working as local controllers. The nested voltage and frequency control loops in the voltage control mode are also shown in Fig. 5.3. This controller feeds the current signal as a feedforward term via a transfer function (e.g., virtual impedance that is introduced later). The LPF blocks show the *low-pass filter* to provide active and reactive powers from the DG's output voltage and current signals.

The proportional–integral (PI) controllers are widely used to the design of the control loops in the local control level. The PI controllers are used in the

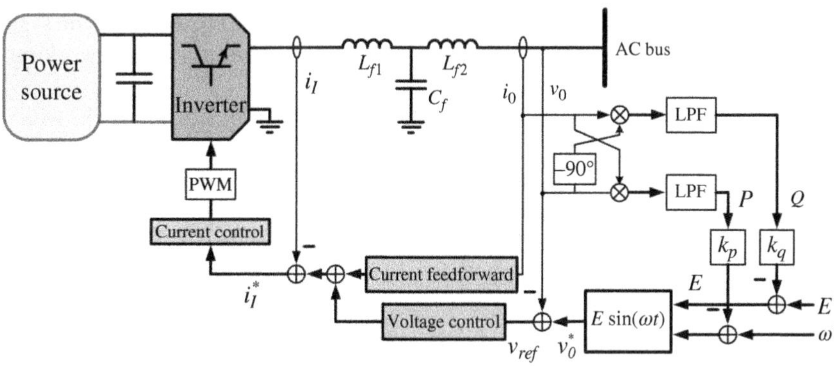

Figure 5.3 Local control loops in a typical voltage-controlled VSC-based DG.

control loop alone or with additional feedforward compensation to enhance the performance of closed-loop systems. As shown in Fig. 5.3, this type of controllers is aimed to control operating points of the DGs and their PE interfaces. Besides the primary voltage and frequency controls, DGs must control active and reactive powers. The droop-based active and reactive power controls are the most common methods to control these powers. The k_p and k_q are constant droop parameters.

Similar to the DG's output control, the power sharing control between DGs could be also included in this level. DG's output control usually includes an inner and outer control loops for current and voltage regulation (Fig. 5.3), respectively. Power sharing control in an MG is responsible for the adequate share of active and reactive power mismatches. Both control functions can be performed by using primary active power–frequency and reactive power–voltage droops based on local measurement and without need for communication.

The local controllers design usually should be based on a detailed dynamic model of the DGs, including the resistive, reactive, and capacitive local load and the distribution system. This model should be adapted to the practical operating conditions of MGs in order to guarantee that the controllers respond properly to the system's inherent dynamics and transients.

For the sake of modeling, stability analysis, and local control synthesis in inverter-based DGs, there are three reference frames: natural (abc), stationary ($\alpha\beta$), and synchronous (dq). The natural reference frame utilizes controllers realized with simple structure (e.g., PI) and is useful to time domain response analysis of the DGs and MGs. The stationary reference frame is mostly associated with sinusoidal variables, and, finally, the synchronous reference frame is associated with DC variables and controllers.

5.2.2 Secondary Control

Secondary controls as second layer control loops complement the task of inner control loops to improve the power quality inside MGs and to enhance the system performance by removing the steady-state errors. They are closely working with local and central control groups.

In the grid-connected operation mode, all the DGs and inverters in MGs use the grid electrical signals as references for voltage and frequency regulation. But, in the islanding operation mode, the DGs lose the reference signal provided by the main grid. In this case, they may coordinate to manage a simultaneous operation using single/multimaster operation methods. Secondary controls also cover some of controls needed to improve the parallel operation performance for DGs (or inverters). There are many control techniques in the literature to make a successful parallel operation of DGs/inverters via master/slave, current/power sharing, and generalized frequency/voltage droop control techniques [5].

Similar to the secondary control in conventional power systems, secondary controls in MGs are responsible to provide steady-state voltage and frequency deviations in the presence of load changes and action of the local controllers in the islanded operation mode. For example, the voltage/frequency reference signals (E^* and ω^*) in Fig. 5.3 are provided by secondary control loops. Secondary control operates on a slower timescale compared to the local control. Improving the power quality for a set of DGs connected to a common bus [10] can be also considered as a secondary control loop. In contrary to the local control, in secondary control, it may need to use low bandwidth communications. Several design examples on MG secondary control are given in [11–13].

5.2.3 Central/Emergency Control

Central/emergency control refers to the MG central EMS which is responsible for the reliable, secure, and economical operation of the MG in either grid-connected or islanded operation mode. The main objectives of this control level are to find the optimal unit commitment (UC), reactive power supply, and voltage and frequency control regulation (in coordination with secondary control), black-start restoration, and dispatch of the available DGs/RESs in normal conditions and to perform load shedding and special protection schemes in off-normal and emergency conditions. Central/emergency control is the highest level in the MG hierarchical control structure with a significant role in the islanded operation mode. The central control can be also used to synchronize the MG before connecting to the main grid, to facilitate the transition from islanded to grid-connected mode. This issue can be usually performed in coordination with MGCC as the MG supervisor.

Figure 5.4 shows a block diagram showing the role and cooperation of local, secondary, and central controls in the MG's frequency and voltage regulation. In the case of supporting the main grid ancillary services by the MG, the global control also comes to play. As shown in Fig. 5.4, in the grid-connected mode, the power flow between MG and main grid can be managed by adjusting the amplitude and frequency of DGs' voltages. First, the active and reactive output powers of the MG (P_{MG} and Q_{MG}) are measured. Then, these quantities are compared with the corresponding reference values ($P_{MG\text{-}ref}$ and$Q_{MG\text{-}ref}$) to obtain the frequency and voltage references (f_{ref} and v_{ref}). These reference values are then used as the reference values to the secondary control. The α_i is a *participation factor* of the *i*th DG in the MG frequency or voltage regulation. Following a load disturbance within the MG, the produced appropriate secondary control signal is distributed among DGs according to their participation rate, to compensate the generation-load imbalance. In a given MG, the sum of participation factors is equal to 1.

$$\sum_{i=1}^{n} \alpha_i = 1, \quad 0 \leq \alpha_i \leq 1$$

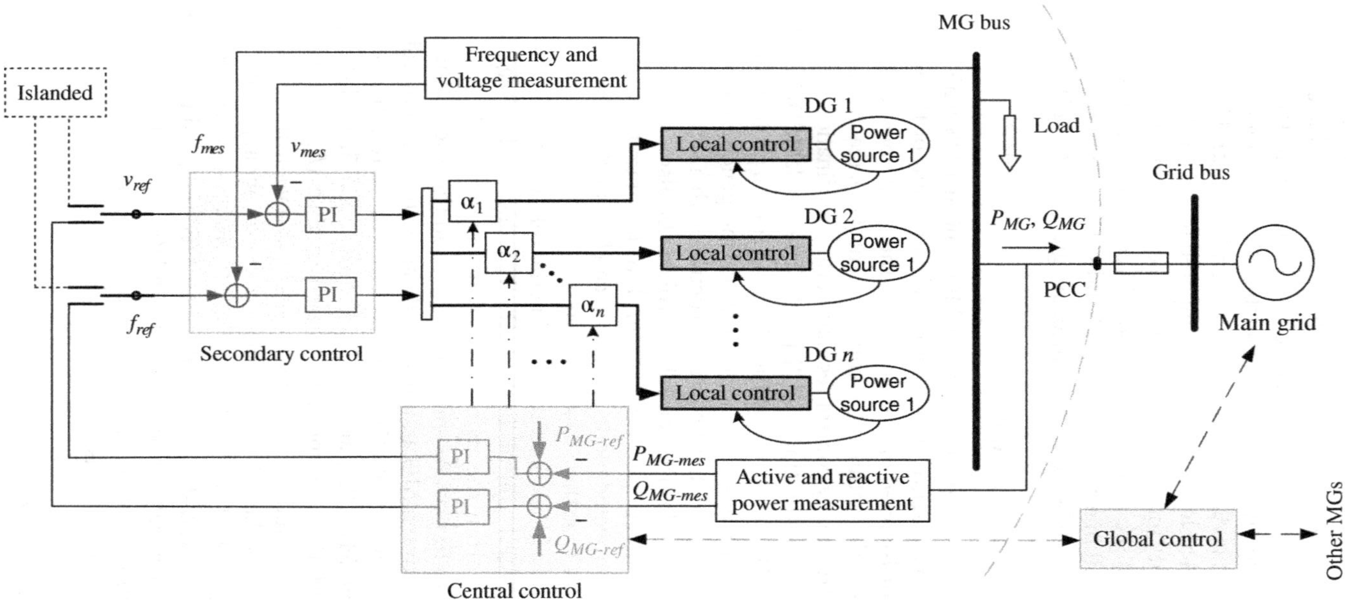

Figure 5.4 Local, secondary, central, and global controls.

In addition to the constraints for the individual DGs, the whole MG should also take advantage of operating in the islanded mode, during power outage, blackout, or emergency condition in the main grid, to increase the overall reliability of the power supply. In the emergency condition, an immediate change in the output power control of the MG is required, as it changes from a dispatched power mode to one controlling frequency and/or voltage of the islanded section of the network.

The islanding plan can be considered as the most important emergency control scheme in the MG systems. When an MG system is islanded, the voltage/frequency might go beyond the power quality limits. In order to ensure system survival following islanding, it is necessary to exploit controllable DGs, storage devices, local load as well as load shedding schemes and special protection plans in a cooperative way [14]. The load shedding can be considered as an effective emergency control scheme in MGs, which is to be started following a significant drop in frequency and/or voltage. The emergency control, as discussed in Chapter 11, is also organized by the MGCC.

In an MG, the consequences of an immediate tripping of DG units may become adverse when a sudden change in a power index is seen by other DG units. Using advanced communication/networking technologies as another important issue has a significant role in MGs' operation and control. Therefore, the design and implementation of new communication infrastructures and networking technologies for MGs are key factors to realize robust/intelligent control strategies, specifically in emergency and central control loops.

In summary, the central/emergency controller is required to ensure that the MG operation is as seamless as possible during major disturbances such as transition between grid-connected and islanded operation modes. It is responsible for economic optimization of the MG in normal operation, as well as maintaining reliable, secure, and safe operation in emergency conditions. The central control level is also in charge of restoring the regulation reserve, managing eventual congestions, and giving support to the secondary control among MGs, if necessary.

5.2.4 Global Control

The global control is the highest level of control for coordinating the operation of multiple interconnected MGs, and communicating requirements with the main grid. For instance, coordination features for active/reactive power management of a grid comprising the main grid and interconnecting MGs could be accomplished by the global control. On the other hand, in the global control point of view, the MG can be controlled to interact with the distribution grid as a dispatchable and constant impedance load.

Global control deals with some overall responsibilities for an MG, such as interchange power with the main grid and/or other connected MGs.

These controls, which are mainly done by a central controller, are acting in an economical-based energy management (EM) level between an MG and the neighbors similar to the existing supervisors for power exchanges and economic dispatch in a conventional multiarea power system. The MGCC interfaces the MG and the main grid and also supervises the entire MG units for operations, such as disconnection, reconnection, power flow control, fault level control, market operating, and load shedding. The MGCC may also generate the power output set points for the DGs using gathered local information. Moreover, the MGCC controls power flow at the PCC to maintain closed to the scheduled value.

In an MG, identifying the optimal generation schedule to minimize production costs and balancing the demand and supply, which comes from both DGs and the distribution feeder, as well as online assessment of MGs security and reliability are the responsibilities of global control. Global control supervises the MG's market activities such as buying and selling active and reactive powers to the grid and possible network congestions not only in the MG itself but also by transferring energy to nearby feeders of the distribution network and other MGs. The global control performs an EMS for the MG to ensure a subset of basic functions such as load and weather forecasting, economic scheduling, security assessment, and demand-side management.

In general, generation scheduling optimization, enhanced overall system control and dispatch services, energy imbalance compensation, and spinning reserve operation can be highlighted as the most significant objectives of the global control. MGs can be also controlled in a coordinated way with the MGCC to provide some ancillary services oriented to enhance the performance of the main grid [15]. Moreover, these services can be further extended if EESs are integrated in the MG. In such a case, functionalities such as the extension of the operational reserve capability, overall frequency regulation, peak shaving, backup of intentional electrical islands, and optimized management of daily renewable energy cycles [16, 17], might be implemented by the global control as well [3].

In addition to the grid voltage and frequency regulation, new grid-supporting power converters in an MG should be able to contribute to damp power oscillations, either in the grid-connected or islanded operation mode [18]. Power oscillations deteriorate power quality in an MG and can even give rise to serious problems regarding the stability of the system. The occurrence of active power oscillations in MGs are not only due to the inherent intermittent nature of the primary RESs used in DGs, but they are also produced as a consequence of the characteristic electromechanical response of conventional synchronous generators when a sudden load variation happens [19]. The coordinated control of the power converters forming an MG allows optimizing the attenuation of the power oscillations generated by DGs and setting the most effective power damping strategy according to the grid layout and operating conditions. It is noteworthy that the installation of EESs in MGs also benefits the damping of active power oscillations [3].

In summary, the global control level is responsible for optimizing the MG operation and setting its interaction with the distribution network and neighborhood MGs by controlling the active and reactive power references of DGs through the MGCCs. This optimization is usually based on economic criteria, which considers the demand-generation balance, together with economic aspects. The MGCCs and global control level coordinately manage the exchange powers and some ancillary services. These functionalities are discussed in Sections 5.4 and 5.5.

5.3 Droop Control

5.3.1 Droop Characteristic in Conventional Power Systems

For understanding droop characteristic concept in a conventional power grid, consider a single machine infinite bus system as shown in Fig. 5.5 [20]. The well-known swing equation of a synchronous generator, which describes the dynamics of load-generation power imbalance in terms of frequency deviations, is given by

$$M\frac{d}{dt}\omega + D(\omega - 1) = P_m - P_c \tag{5.1}$$

where ω is the angular velocity, M is the inertia constant, D is the damping coefficient, P_m is the mechanical input to the generator, and P_e is the electrical power output. The P_e can be calculated as follows:

$$P_e = \frac{V_s V_g \sin\delta}{x'_d + x_l} \tag{5.2}$$

where V_s is the source (generator) RMS voltage, V_g is the RMS voltage of the connected grid (infinite) bus, δ is the rotor angle of the generator, x'_d is the transient reactance of the generator, and x_l is the line reactance. Note that in consistent with conventional power grid characteristics and for simplicity, the resistance of the generator and the transmission line can be neglected.

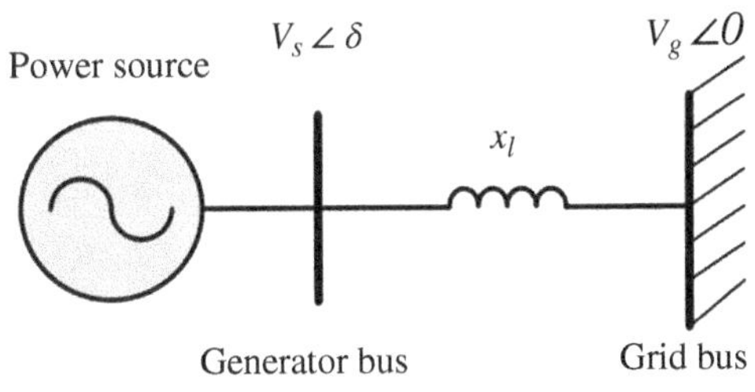

Figure 5.5 Single machine infinite bus system model.

From (5.1), assuming that the voltages V_s and V_g are constant, and $\cos \Delta\delta \cong 1$, $\sin \Delta\delta \cong \Delta\delta$, the output power deviation of the generator can be given by

$$\Delta P_e \cong \frac{V_s V_g \cos \delta_0}{x'_d + x_l} \Delta\delta \tag{5.3}$$

where the subscript 0 is used to denote the initial value of the equilibrium point, the δ_s and δ_g are the angles of V_g and V_∞, respectively; and

$$\Delta\delta = \delta_s - \delta_g \tag{5.4}$$

For small $\Delta\delta$, the ΔP mainly depends on the $\Delta\delta$. It means that the generator can determine the transferring real power P_e flows from itself to the grid by considering the phase (and therefore frequency) of its output voltage. The above relationship shows if the active power increases, the voltage angle must decrease, and vice versa. This relationship that is formulated in (5.5) allows us to establish a feedback loop in order to control generator's real power and frequency.

$$f - f_0 = -k_p(P - P_0) \tag{5.5}$$

The f_0 and P_0 are the nominal values (references) of frequency and active power, respectively. The ratio of frequency deviation (Δf) to change in output generated power (ΔP_s) is known as frequency-active power *droop* characteristic and can be expressed as

$$k_p \left(\text{Hz}/\text{pu MW} \right) = \frac{\Delta f}{\Delta P} \tag{5.6}$$

For example, a 5% droop means that a 5% deviation in nominal frequency (e.g., from 60 to 57 Hz) causes 100% change in output power. The interconnected generating units with different droop characteristics can jointly track the load change to restore the nominal system frequency. This is illustrated in Fig. 5.6, representing two units with different droop characteristics connected to a common load. Two generating units are operating at a unique nominal frequency with different output powers. Any change in the network load causes the units to decrease their speed, and the governors increase the outputs until they reach a new common operating frequency. As expressed in (5.7), the amount of produced power by each generating unit to compensate the network load change depends on the unit's droop characteristic.

$$\Delta P_i = \frac{\Delta f}{k_{pi}} \tag{5.7}$$

hence,

$$\frac{\Delta P_1}{\Delta P_2} = \frac{k_{p2}}{k_{p1}} \tag{5.8}$$

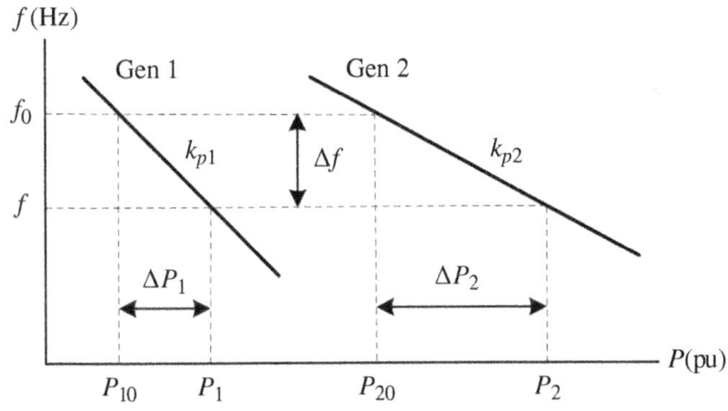

Figure 5.6 Load tracking by generators with different droops.

Similar relationship to (5.5) can be obtained for voltage and reactive power deviations:

$$v - v_0 = -k_q(Q - Q_0) \tag{5.9}$$

The k_q shows the voltage-reactive power droop characteristic. It is noteworthy that the described droop characteristics in (5.5) and (5.9) have been obtained for conventional power grids with inductive impedance ($X \gg R$) and a great amount of inertia.

$$X_l = R_l + jx_l; \quad x_l \gg R_l \tag{5.10}$$

In a conventional power system, immediately after a power imbalance due to a disturbance, the power is going to be balanced by natural response generators using rotating inertia in the system via the primary frequency control loop. From a control point of view, droop control is a decentralized proportional controller where the control gain (droop gain) determines the steady-state power distribution in the power grid. Since performance under droop control is satisfactory for large synchronous generators in existing conventional power systems, this technique has been also adapted to inverter-based DGs in MGs.

In MGs with low inertia RESs/DGs, there is no total significant inertia and if a load-generation imbalance occurs, the voltages of the power sources change. Moreover, voltage may be also triggered by the reactive power deviations. In fact, for MV and LV lines, with which MGs are working, the impedance is not dominantly inductive ($x_l \cong R_l$). This fact suggests a different formulation for the droop characteristics in MGs. Typical line impedance values for LV, MV, and high-voltage (HV) lines are given in Table 5.1 [21].

5.3.2 Droop Control in Inverter-based Distributed Generators

A simplified general representation for an inverter-based DG is shown in Fig. 5.7. A DG contains three basic elements: power source, DC interface, and

Table 5.1 Typical line impedance values [21].

Line type	*R* (Ω/km)	*X* (Ω/km)	*R*/*X* (pu)
LV	0.642	0.083	7.7
MV	0.161	0.190	0.85
HV	0.06	0.191	0.31

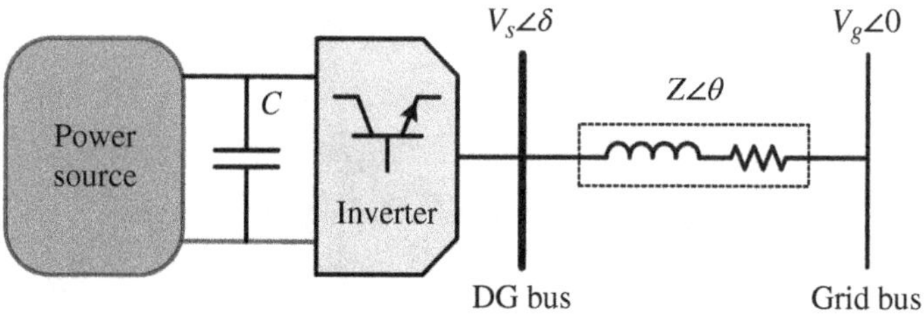

Figure 5.7 An inverter-based DG.

inverter. The DG couples to the MG through a power line. The output voltage and frequency, as well as real and reactive powers of the DG, can be controlled by using local feedbacks applied to the inverter.

In comparison to the conventional generators, the DGs such as natural gas and diesel generating units are very fast and can typically pick up load within 10–12 s from start-up and can serve full load just a few seconds thereafter. The DG can control the phase and magnitude of its output voltage V, and from the line reactance X, it can determine the transferring real power P and reactive power Q flows from itself to the grid.

Consider a simple inverter-based DG as shown in Fig. 5.7. The DG is connected to the grid via a line impedance $Z\angle\theta$. The complex power delivered to the grid bus is calculated as

$$S = V_g I^* = \frac{V_g V_s \angle(\theta - \delta)}{Z} - \frac{V_g^2 \angle\theta}{Z} \tag{5.11}$$

from which the active and reactive powers can be achieved as

$$P = \frac{V_g V_s}{Z} \cos(\theta - \delta) - \frac{V_g^2}{Z} \cos(\theta) \tag{5.12}$$

$$Q = \frac{V_g V_s}{Z} \sin(\theta - \delta) - \frac{V_g^2}{Z} \sin(\theta) \tag{5.13}$$

where $Z = R + jX$ is the interconnection line impedance angle and θ is the impedance angle. Considering $Ze^{j\theta} = R + jX = Z\cos(\theta) + jZ\sin(\theta)$, (5.12) and (5.13) can be rewritten as

$$P = \frac{V_s}{R^2 + X^2}[R(V_s - V_g \cos(\theta)) + XV_g \sin(\delta)] \tag{5.14}$$

$$Q = \frac{V_s}{R^2 + X^2}[-RV_g \sin(\delta) + X(V_s - V_g \cos(\theta))] \tag{5.15}$$

These equations show dependency of the inverter output voltage and power angle δ on the active and reactive powers.

5.3.2.1 Inductive Grid

The inductive component of the line impedance in HV power grids is typically much higher than the resistive one ($X \gg R$) as can be seen in Table 5.1. Therefore, the resistive component can be neglected. Furthermore, the power angle δ in such a grid is small, so it can be assumed that $\sin(\delta) \approx \delta$ and $\cos(\delta) \approx 1$. Thus, (5.14) and (5.15) can be rewritten as

$$P \approx \frac{V_s}{X} V_g \sin(\delta) \Rightarrow \delta \approx \frac{XP}{V_s V_g} \tag{5.16}$$

$$Q \approx \frac{V_s}{X}(V_s - V_g \cos(\delta)) \Rightarrow V_s - V_g \approx \frac{XQ}{V_s} \tag{5.17}$$

Equations (5.16) and (5.17) show that in the inductive MGs, similar to the conventional power grids, the output reactive power controls the source (inverter) voltage, and the active power controls the system frequency, independently. These strategies are known as *v/Q* and *f/P* controls, respectively. Therefore, similar to (5.5) and (5.9), the following droop control expressions can be written for the inductive MGs:

$$f - f_0 = -k_p(P - P_0) \tag{5.18}$$

$$v - v_0 = -k_q(Q - Q_0) \tag{5.19}$$

where f_0 and v_0 are, respectively, rated as normal frequency and voltage of the MG, the k_p and k_q are DG's droop characteristics, and $P - P_0$ and $Q - Q_0$ are the active and reactive power variations. According to (5.18) and (5.19), if a change happens in the frequency or voltage of an inverter, for any reason, the impact can be observed on the output active and reactive powers of the inverter. The amount of suitable frequency and voltage deviation can be achieved through the droop characteristics. The droop control as a local control provides a fast control action to keep the instantaneous balance between system production

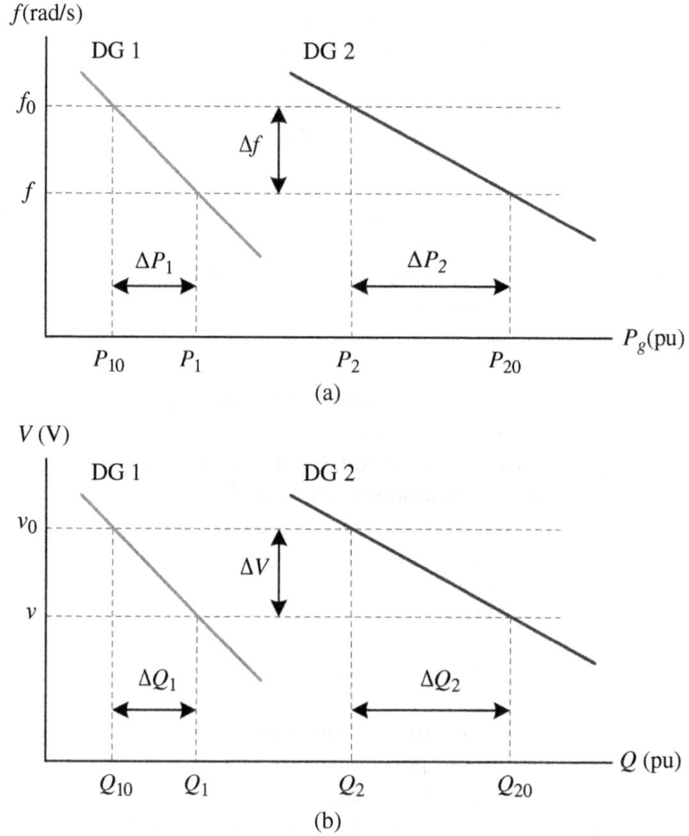

Figure 5.8 Droop characteristics for inverter-based DGs with inductive line impedance: (a) *f–p* droop, (b) *V–Q* droop.

and consumption [20, 22]. The given droop relationships in (5.18) and (5.19) can be graphically represented by the droop characteristics shown in Fig. 5.8.

5.3.2.2 Resistive Grid

Unlike an HV power grid, the LV grid impedance is mainly resistive ($X \ll R$), and thus, the inductive component can be neglected. This fact with assuming a small value for power angle δ leads to rewrite (5.14) and (5.15) as follows:

$$P \approx \frac{V_s}{R}(V_s - V_g \cos(\delta)) \Rightarrow V_s - V_g \approx \frac{RP}{V_s} \tag{5.20}$$

$$Q \approx -\frac{V_s}{R} V_g \sin(\delta) \Rightarrow \delta \approx -\frac{RQ}{V_s V_g} \tag{5.21}$$

Equations (5.20) and (5.21) show a strong linkage between reactive power and power angle, as well as between active power and voltage. Thus, in contrary to conventional power grids, in the resistive MGs, the *P/V* and *Q/f* droop control techniques are needed for the voltage and frequency controller synthesis, respectively. These relationships are expressed by the following droop controls:

$$f - f_0 = -k_q(Q - Q_0) \tag{5.22}$$

$$v - v_0 = -k_p(P - P_0) \tag{5.23}$$

5.3.2.3 General Case

In the general grids, both inductive and resistive components should be considered [23]. For this purpose, a rotation matrix T is required to transform the active and reactive powers (P, Q) into the rotational power components (P', Q') described as follows:

$$\begin{bmatrix} P' \\ Q' \end{bmatrix} = T \begin{bmatrix} P \\ Q \end{bmatrix} = \begin{bmatrix} \cos(\phi) & -\sin(\phi) \\ \sin(\phi) & \cos(\phi) \end{bmatrix} \begin{bmatrix} P \\ Q \end{bmatrix} \tag{5.24}$$

where $\phi = (\pi/2) - \theta$ and ϕ is the rotation angle of matrix T. Thus,

$$P' = \frac{X}{Z}P - \frac{R}{Z}Q \tag{5.25}$$

$$Q' = \frac{R}{Z}P + \frac{X}{Z}Q \tag{5.26}$$

Assumed that δ takes a small value, the application of the rotation matrix to (5.14) and (5.15) results in

$$P' \approx -\frac{V_s}{Z}V_g \sin(\delta) \Rightarrow \delta \approx -\frac{ZP'}{V_sV_g} \tag{5.27}$$

$$Q' \approx \frac{V_s}{X}(V_s - V_g \cos(\delta)) \Rightarrow V_s - V_g \approx \frac{ZQ'}{V_s} \tag{5.28}$$

where P' and Q' are the rotated components of P and Q according to (5.16) and (5.17). From (5.27) and (5.28), it can be seen that regulating the rotated active power P' and reactive power Q' controls the angle δ and voltage deviation Δv, respectively. Therefore, in a general case, the droop control equations can be written as follows:

$$f - f_0 = -k_p(P' - P'_0) = -k_p\frac{X}{Z}(P - P_0) + k_q\frac{R}{Z}(Q - Q_0) \tag{5.29}$$

$$v - v_0 = -k_q(Q' - Q'_0) = -k_p\frac{R}{Z}(P - P_0) + k_q\frac{X}{Z}(Q - Q_0) \tag{5.30}$$

Equations (5.29) and (5.30) show that by changing the values k_p and k_q in an MG, the system frequency and voltage amplitude can be controlled. This fact is examined in the following section.

5.3.2.4 General Case-based Voltage and Frequency Control

Defining the indices $K_R = R/X$, $K_f = -1/k_p$, and $K_v = -1/k_q$; having $\Delta f = f - f_0$ and $\Delta v = v - v_0$; and applying (5.18) and (5.19)– to (5.25) and (5.26), results in

$$P' = \frac{X}{Z}[K_f \Delta f + P_0 - K_R K_v \Delta v - K_R Q_0] \tag{5.31}$$

$$Q' = \frac{X}{Z}[K_R K_f \Delta f + K_R P_0 + K_v \Delta v + Q_0] \tag{5.32}$$

After some algebraic calculations on (5.31) and (5.32), following expressions are obtained:

$$\Delta f = \frac{1}{K_f}\left[\frac{Z}{X}P' - P_0\right] + \frac{K_R K_v}{K_f}\Delta v + \frac{K_R}{K_f}Q_0 \tag{5.33}$$

$$\Delta v = \frac{1}{K_v}\left[\frac{Z}{X}Q' - Q_0\right] + \frac{K_R K_f}{K_v}\Delta f + \frac{K_R}{K_v}P_0 \tag{5.34}$$

From (5.33), it is denoted that K_f affects the weighting coefficients of Δv and Q_0 in the second and third terms, respectively. Thus, to prevent an undesired effect of K_f on the Δv and Q_0 in the second and third terms of (5.33), K_f must be equal to unity. Similarly, in the second and third terms of (5.34), K_v should be fixed at unity. Then, from (5.33) and (5.34) one can obtain the following *general droop control* (GDC) relationships:

$$\Delta f = \frac{1}{K_f}\left[\frac{Z}{X}P' - P_0\right] + K_R K_v \Delta v + K_R Q_0 \tag{5.35}$$

$$\Delta v = \frac{1}{K_v}\left[\frac{Z}{X}Q' - Q_0\right] + K_R K_f \Delta f + K_R P_0 \tag{5.36}$$

Figure 5.9 shows an example of an inverter-based DG (working as a voltage source inverter (VSI) unit) for supplying a local load through a line. The LCL output filter has been added to prevent the resonance impact in the output network. Also, the LCL damps the distortion of output sinusoidal waveform and

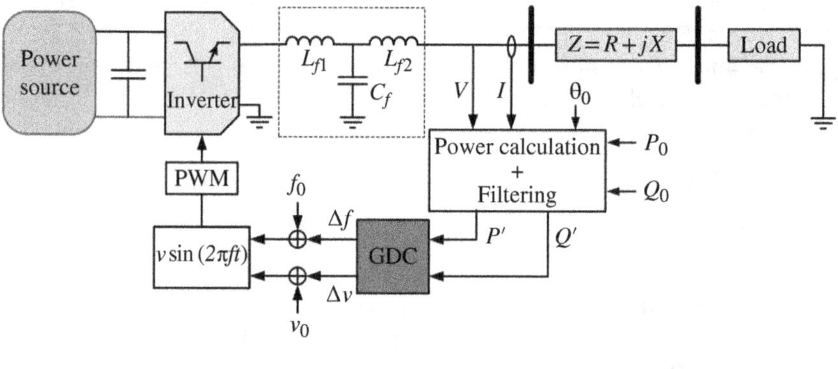

Figure 5.9 An example for GDC-based droop control.

reduces high-frequency harmonics caused by switching operations of the VSI. Therefore, it is used in the inverter output for the quality of output current and bus voltage in the context of weak grids [24]. The rated voltage and power are 220 V (rms) and 30 kVA, respectively. The system parameters are given in [11].

The instantaneous rotated active and reactive powers (P' and Q') can be calculated from the source (inverter) output voltage and current. Then, they are used by the GDC block, which is performed by (5.35) and (5.36) to provide corresponding voltage and frequency deviations. To illustrate the effectiveness of the GDC strategy, and to ensure that the developed droop control strategy is able to obtain a desirable result in inductive, resistive, and general types of MGs, some simulations have been done in states of $K_R = 0.1$ ($X = 10R$), $K_R = 1$ ($R = X$), and $K_R = 10$ ($R = 10X$).

In Fig. 5.10a, a severe scenario for load variations is considered so that before 0.6 s, the primary load is purely active and is fixed at about 0.7 pu. After 0.6 s, for next three steps, that is, 0.6, 0.8, and 1 s, active load is increased. Then, in other three steps at 1.2, 1.4, and 1.6 s, reactive load is increased while active load is kept constant. Active and reactive loads are simultaneously decreased to primary values at 1.8 s.

Finally, for better recognizing the GDC performance, in the final four steps, both active and reactive loads are changed together. The simulation results for different values of Z ($K_R = 0.1$, $K_R = 1$, and $K_R = 10$) are shown in Fig. 5.10b. This figure shows that the mentioned GDC method stabilizes the DG's frequency and voltage following the step load changes. It is also investigated that the developed control strategy is capable of maintaining frequency and voltage stability even in more serious conditions.

In the relevant literature, stability analysis of droop-controlled MGs has been traditionally carried out by means of detailed numerical small-signal analysis as well as extensive simulations and experimental studies aiming to characterize a range for the droop gains guaranteeing system stability. Some conditions on the droop gains to ensure stability of droop-controlled inverter-based MGs are given in [25].

5.3.3 Virtual Impedance Control

As mentioned, the performance of a drooped control system in an MG is highly dependent on the grid impedance parameters ($K_R = R/X$ index), especially in the general case. A small mismatching in the grid impedance estimation results in an inefficient power sharing among the droop-controlled DGs. Therefore, the application of droop control in all kind of MGs without using a sophisticated grid impedance estimation mechanism to calculate the line parameters or the rotation matrix T (5.24) may not be effective [11].

In order to solve this problem, over the years, several solutions are introduced to decrease the dependency of the conventional droop controller performance

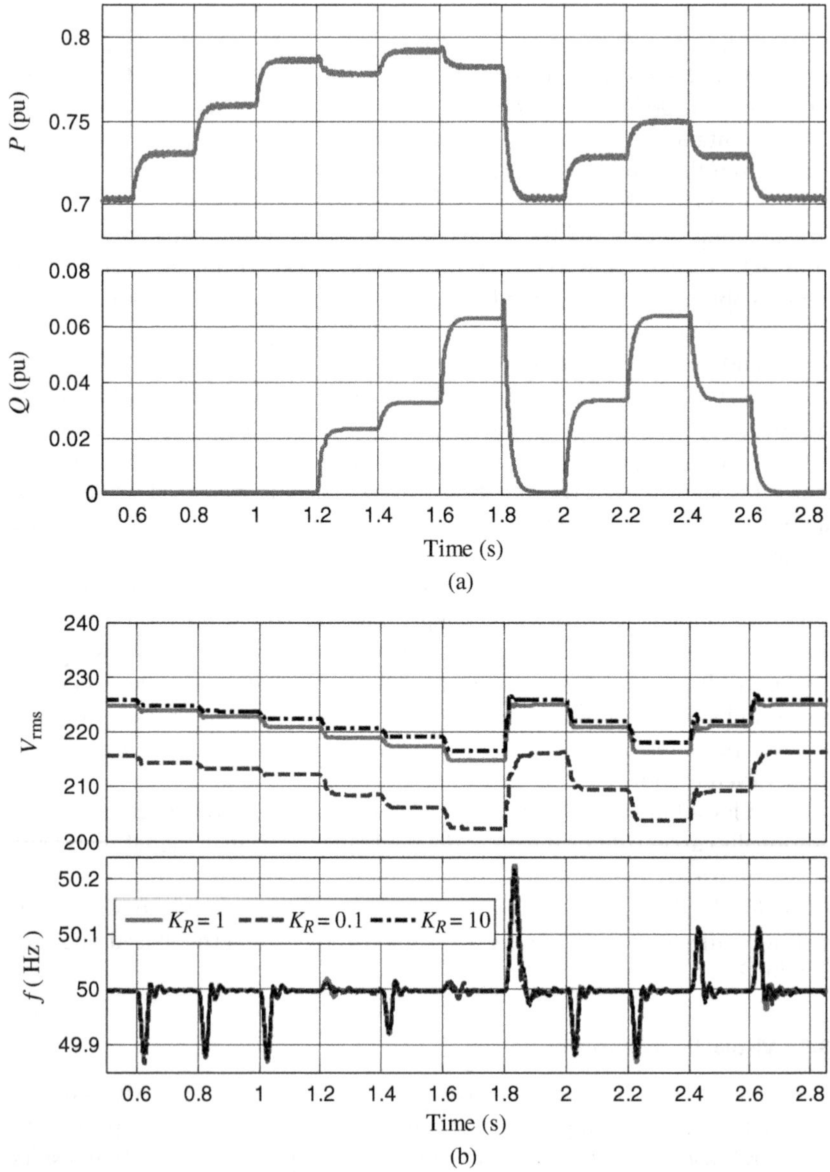

Figure 5.10 System response following a load disturbance: (a) load change pattern, (b) voltage and frequency for different K_R values.

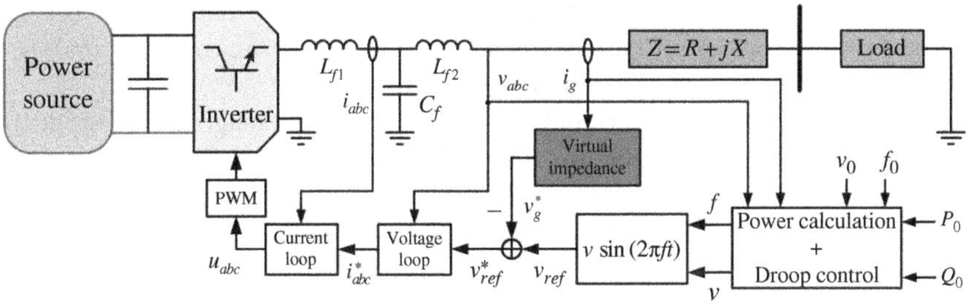

Figure 5.11 A power converter with virtual output impedance loop.

on the line impedance parameters. Using a large inductor between the power converter and the AC bus to make an inductive line impedance is one of these solutions. However, due to reduction in the overall efficiency and increase in the size and the costs, this solution does not provide an efficient methodology. Application of intelligent algorithms such as artificial neural network (ANN) and adaptive neurofuzzy system to estimate the line parameters and to cope the rotation matrix can be considered as another class of solutions.

A simple but effective solution is that updating the power converter control loop by virtually emulating the effect of an inductive element in the link impedance. This concept is known as ***virtual impedance*** and has been successfully implemented for regulating the power sharing among parallelized inverters and limiting over currents under grid disturbances [3, 26–28].

The virtual impedance updates the power converter output voltage reference as given in (5.37), where the new voltage reference (v_{ref}^*) is obtained by subtracting the virtual voltage drop across the virtual impedance from the reference value originally provided by the droop equations (v_{ref}) [29].

$$v_{ref}^* = v_{ref} - v_g^* = v_{ref} - Z_V i_g \tag{5.37}$$

The transfer function of virtual impedance block $Z_V(s)$ must be selected according to the nominal power of the converter. An example of the implementation of the virtual impedance concept in a grid-connected power converter (working with P–Q droop control method) is shown in Fig. 5.11.

5.4 Hierarchical Power Management and Control

Power management and control is one of the important control issues in an MG that hierarchically needs the contribution of all existing local, secondary, central, and global control levels, under supervision of the MGCC. As mentioned in Section 5.2, the task of MG supervision from the MGCC side (in central control level) is to manage the power and the energy between the MG sources and loads. Then the real and reactive powers must be shared among the DG/RES units (through the secondary and local controllers). So the MGCC must assign real and reactive power references and also other appropriate control signals to the DGs/RESs, ESSs, and controllable loads.

In comparison to a conventional power grid, power management system is more complex for MG application. The main reasons are (i) a large number of small DG units with different capacities and characteristics have to be managed and (ii) most of DGs/RESs are coupled to the grid with PE converters and so may have fast responses. Moreover, if conventional generators are also coupled, their voltage/angle stability may be affected negatively.

The MG central/emergency control should meet the safety, quality, reliability, and economy criteria, which govern the operation of the energy system. The safety is the first and the most important criterion. The consideration of the safety covers the staff personnel, the environment, and the property in every aspect of the MG operations. Quality is defined in terms of MG variables, such as frequency and voltage that must conform to certain standards to accommodate the requirements for proper operation of all connected loads to the MG system. Supply reliability does not mean a constant supply of power, but it means that any break in the supply of power is one that is agreed to and tolerated by both supplier and consumer sides. Minimizing the generation cost and losses motivates the economy criterion while mitigating the adverse impact of the MG operation on the environment.

So, according to the aforementioned criteria, some important tasks should be performed for an operating MG, such as (i) the real power balancing between the generation and load, (ii) the reactive power balancing for the voltage regulation, (iii) the optimum generation scheduling for limiting the cost and environment impact of the power generation, and (iv) the security of the MG network against credible contingencies. These tasks require protecting the MG network against failures of equipment, outage, or the main grid's faults.

5.4.1 Operation Layers and Control Functions

Similar to conventional power grids, to automate the operation of an MG, the system should rely on a high sophisticated integrated system for monitoring and control. Such an MG should have a multilayer structure according to the introduced control levels including numerous devices and elements. The bottom layer (layer 0) is the high-reliability PE devices and switchgear, which include automatic equipment such as local controllers, protective relays/circuits, breakers and automatic transformer tap changers. Layer 1 consists of remote-control cabinets mounted locally to the PE systems, switchgear, and provides facilities for actuator control, interlocking, and voltage and current measurement. Layer 2 includes the data concentrators/master remote terminal unit, which typically includes a man/machine interface giving the operator access to data produced by the lower layer equipment. The top layer

(layer 3) is the supervisory control and data acquisition (SCADA) system, which operates in the MGCC (in central/emergency control). Similar to the EMS in the conventional power grids [22], the SCADA system accepts telemetered values and displays them in a meaningful way to operators, usually via a one-line mimic diagram. The other main component of a SCADA system is an alarm management subsystem that automatically monitors all the inputs and informs the operators of abnormal conditions.

The EMS in the MGCC is intended to help the dispatchers in better monitoring and control of the MG. System dispatchers are required to make proper decisions on operational and outage scheduling on a daily basis. Moreover, they have to be always alerted and prepared to deal with contingencies that may arise occasionally. The EMS functions can be classified to *base functions*, *generation functions*, and *network functions* [30].

The required base functions of the EMS include the ability to acquire real-time data from monitoring equipment throughout the MG, and the processing of the raw data and the distribution of the processed data within the MGCC system in the central and emergency control level. So it consists of data acquisition, supervisory control, logical alarming, sequence of events function, historical database, automatic data collection, load shedding function, safety management, and so on. The generation functions involve the functions that are related to operational scheduling of the DGs such as load forecasting, UC, economic dispatch and power–frequency regulation, and interchange transaction scheduling with the distribution grid.

Network functions can be subdivided into *real-time* and *study* functions. The real-time functions are controlled by real-time sequence control that allows for a particular function(s) to be executed periodically or by a defined event manually. The network study functions essentially duplicate the real-time function and are used to study any number of "what if" situations. The functions that can be executed are topology processing (model update), state estimation, network parameter adaptation, dispatcher power flow, network sensitivity, security analysis, security dispatch, voltage control, and optimal power flow functions.

5.4.2 Timescale Analyzing and Implementation Constraints

The MG supervision and power management can be analyzed through various functions, which are classified in a timing scale as shown in Fig. 5.12. The short-term EM or power balancing corresponds to the local control level and includes (i) the real-time *balancing and power dispatching* among DGs and ESSs according to the storage level capacity and specific requirements/limitations of each unit, including available power from RESs, and (ii) the primary voltage and frequency regulation [31].

The medium-term EM includes (i) the adjustment of renewable energy production and load prediction (e.g., each half hour), (ii) the available storage

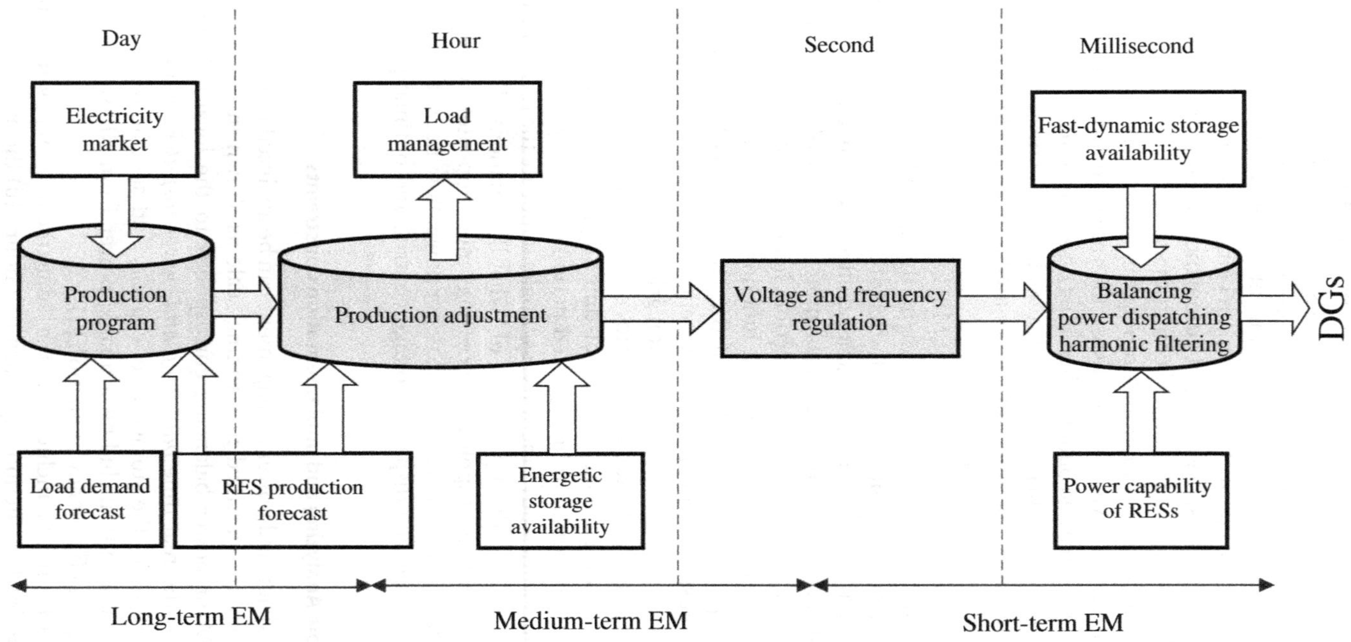

Figure 5.12 Timing classification of power control functions in the context of MG.

energy estimation, (iii) the correction of power set point of each source (e.g., each half hour), and (iv) the secondary regulation supply for the system ancillary services.

The long-term EM corresponds to central and secondary control levels and includes (i) the hourly *RES production forecast* including the time dependency of the power sources, environmental impacts, and cost of generation; (ii) the management of nonsensitive loads that may be disconnected/shed according to the emergency control plans; (iii) the maintenance intervals; and (iv) the provision of an appropriate level of power reserve capacity according to the electricity market and the load demand forecast.

In a conventional power grid, the long-term EM is implemented by the grid system operator and the short-term power balancing is implemented into the generators. For supplying these functions, DGs must be coordinated either in a grid-connected operation mode or an islanded operation mode of the MG. Two types of system managements exist [31, 32]: (i) The short-term power management by sensing electrical quantities (local measurements) is achieved with the knowledge of physical quantities at the point of common coupling (PCC) [33, 34] and a droop characteristic control; (ii) the long- and medium-term EM require a communication network to exchange information and control signals [35, 36]. In the following subsections, both control schemes are presented in detail.

5.4.2.1 Power Management Using Local Measurements

In a conventional power system, all generators must respond in a coordinated manner to a variable load in real time: the global balance of power must be ensured at the same time as the main parameters (voltage, frequency) must be controlled. In many power systems, coordination of all generators has been implemented through the measurement of the system frequency and voltage [33, 34]. With these data, a droop characteristic control (described in Section 5.3) performs the coordination of generators with f/P and v/Q power droop characteristics in order to share the adjustment of the total demand for real and reactive powers. In an MG, this function is mainly provided by the speed governors and voltage regulators of the microturbines (MTs) and diesel generators or by proper controlling of inverters in the inverter-based DGs (Section 5.3).

As mentioned, a droop characteristic control performs the coordination of DG's local controllers with an f/P and/or v/Q power droop characteristics. For example, according to the f/P droop characteristic, when the frequency decreases, the characteristic modifies the power reference in order to increase the generated real power. The droop control can be easily used in local controllers to set real and reactive power references for a *power dispatching control scheme*.

The main advantage of this method is its simple hardware implementations, since the development of central supervision devices is unnecessary. Moreover, this local implementation enables a very fast response and then a good adequacy for frequency control and voltage regulation. This organization works in an autonomous way and sometimes is called *noninteractive* since it is not coordinated with a higher control center. The main disadvantage is the fact that an optimization function of the MG cannot be designed with accuracy, since no sufficient information is known from operating points of other DGs.

For the reliability and stability of an MG, a v/Q droop control is also important. Without this droop, a small error in voltage set points exceeds the ratings of DGs due to circulating current. If the generated reactive power by a DG becomes more capacitive, the local voltage set point is reduced. Conversely, more inductive reactive power increases the voltage.

Each inverter-based DGs must have a droop control loop to share active and reactive powers among DG units and to improve the system performance and stability, and to regulate the output frequency and voltage. Using the concept of droop control, the *power dispatching strategy* mechanism is represented in Fig. 5.13. The magnitude and frequency of the output voltage directly influence generated currents through the droops. The P–Q mode and the droop control are implemented in the local control level.

The main advantage of this method is its simple hardware implementation, since the development of central supervision devices is unnecessary. Moreover, this local implementation enables a very fast response for frequency and voltage control loops. For the performed noninteractive power control scheme, the power references P_{ref_0} and Q_{ref_0} (Fig. 5.13) are constant, and the local controllers including droop and P–Q mode controls are activated in each DG.

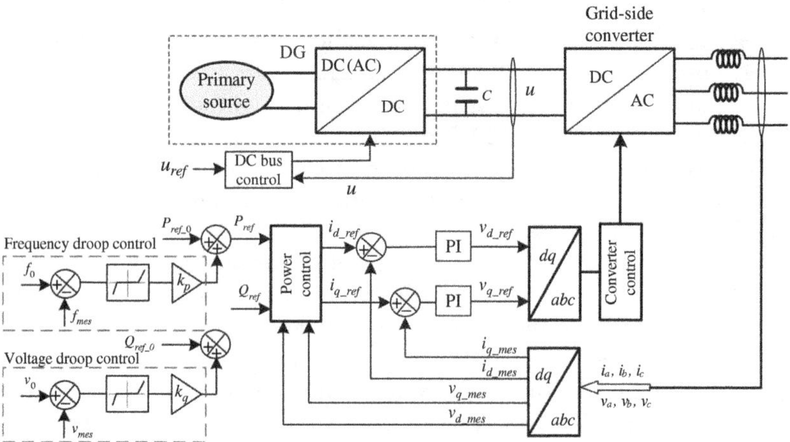

Figure 5.13 Droop controllers for the power dispatching strategy.

5.4.2.2 Power Management Using Communication Network

If a communication network can be used, then power references (P_{ref_0} and Q_{ref_0} in Fig. 5.13) can be changed to implement an EM of each DG in order to optimize the overall operation of the MG. This *interactive* control can be implemented with slower dynamics for changing of power references concerning the time delays of the communication network. Therefore, this scheme is suitable for the long-term EM in MGs. Control by signal communication enables information exchange between local, secondary, central/emergency (in MGCC), and global (in distribution network) control levels [32]. The mentioned system structure implies either a centralized control achieved by MGCC via a centralized supervision function or a decentralized control, which uses the results of negotiations between the DGs in local control level. This structure could be realized using a multiagent system (MAS) [35, 36].

The DNO or market operator is intended for an area in which more than one MG exists. It does not belong to an MG, but it operates in the distribution network as a global controller. The main interface between the DNO and the MG is the MGCC. The MGCC plays different roles ranging from the maximization of the MG value to the coordination of local controllers. In a centralized operation, each local controller receives set points from the corresponding MGCC. In this context, the MGCC corresponds to a virtual utility or an aggregator at the decentralized step. In a decentralized operation, each local controller makes decisions locally.

5.4.2.2.1 *Centralized Control* The centralized approach suggests that a central processing unit collects all the measurement and decides next actions for a coordinated operation inside the MG, as shown in Fig. 5.14. The goal is to coordinate and to schedule generators and controllable loads in order to maximize the revenues from energy market participations.

The centralized power management/control architecture consists of a central controller provided with the relevant information of every DG and load within an MG through the MGCC and the network itself, as well as the information from forecasting systems (e.g., local loads, wind speed, solar irradiance) in order to determine an appropriate UC and dispatch of the resources according to the selected objectives [1].

The MGCC is responsible for a real-time provision of set points for the DGs and controllable loads. The controller and communication infrastructure are constructed to allow optimal operation of the whole MG system. Real-time information of the production and consumption in the MG is collected at the central controller via a fast and reliable communication system. An advanced information and communication infrastructure is required, but an optimal operation economically and reliably can then be obtained. The central controller monitors the system frequency, real-time production/consumption in the MG, and energy and ancillary service market information and calculates

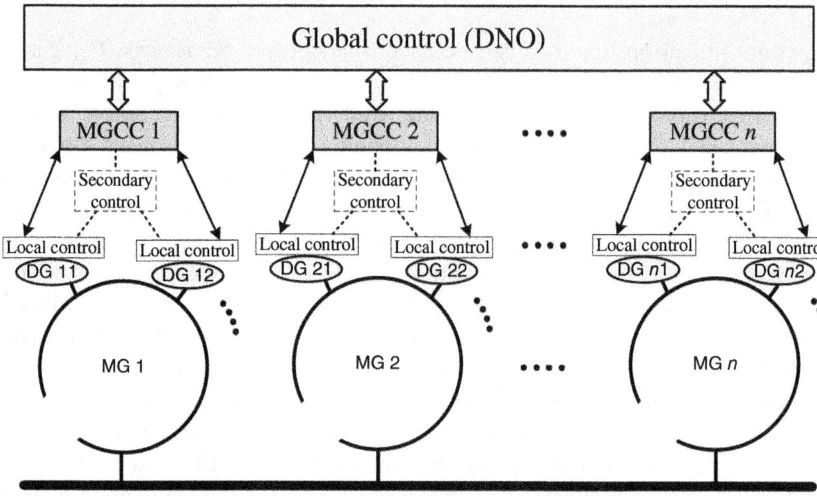

Figure 5.14 Centralized control of an MG in an interactive control framework.

and communicates the appropriate set points for each individual DG and responsive load based on the available forecast of market prices and the availability of microsources and responsive loads.

With a centralized control method, the MGCC takes into account the existing constraints in order to perform an optimization operating concerning the dispatched DGs and loads operating points, market prices, network security constrains, and demand and/or renewable production forecasts. This optimization is achieved by receiving the information from local controllers and returning appropriate reference signals.

In other words, the MGCC calculates the control reference and the necessary data for the optimization, such as power references for generators and dispatched DGs/RESs, set points for the noncritical loads, and market prices for the next optimization period. Therefore, the local controllers can adjust their own operating points by using the provided reference signals from the MGCC. Based on management tasks, the MGCC is built by considering a formulation of constraints, an implantation of economic optimizations and an interface with local controllers.

Figure 5.15 shows the structure of GMs' EMS layer in the MGCC. The constraints depend on weather forecasts, which have an impact on the daily load profiles and the energy potential of the intermittent renewable power. The operating cost of MG generators is presented in the constraints and also the environment impacts (by taking into account the generators using fossil energy, the efficiency of DGs, etc.). On the technical level, the ancillary services for the whole MG are quantified and dispatched to the different DGs. Moreover, the function mode of every DG is also specified. From available information

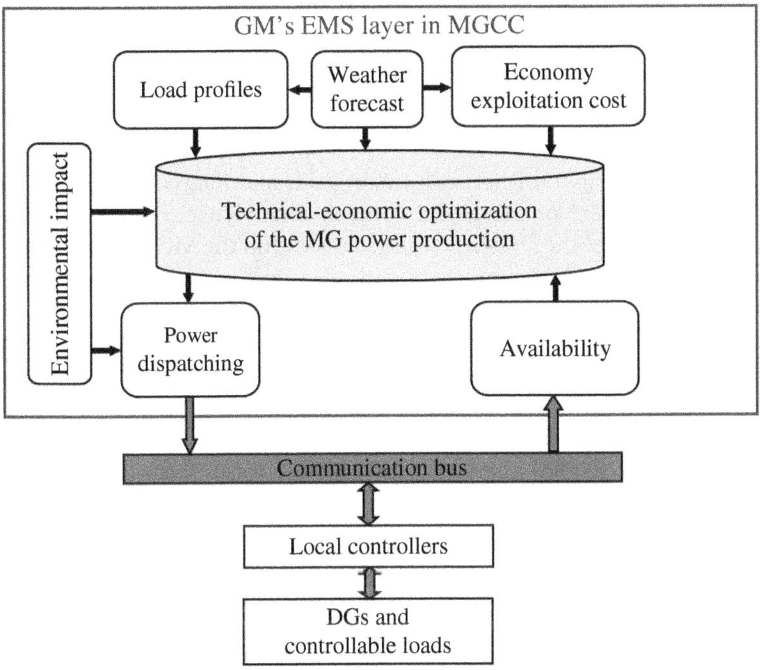

Figure 5.15 Architecture of EMS layer of an MG.

(frequency, AC bus voltage, storage level, etc.), these algorithms make the necessary decisions for the proper operation of the generators. The interface in the MGCC enables the real-time generation of the references for the DGs according to the selected function mode. It also shapes the DGs information about their availability (set point, storage level, production potential, etc.).

Figure 5.15 illustrates the information exchange path in a central supervision and indicates that a two-way communication bus between the MGCC and local controllers is required for transferring the relevant control and information signals. The communication can be performed through electric cables, optic cable with photoelectric devices, telephone lines, power line carriers, wireless transmission, intranet, or the internet network.

5.4.2.2.2 Decentralized Control The decentralized approach suggests to use the controllers installed in distributed nodes forming a distributed control system. This is similar to the case of local controls for the frequency and voltage. The controllers are tuned with predefined droop characteristics, so they can react to system frequency and voltage variations, autonomously.

A decentralized power management in the global control level intends to solve the EM problem of an MG while providing the highest possible autonomy for different DGs/RESs and loads. Although this approach can still use a hierarchical structure for data exchange, the required decisions on the control variables are made locally. The autonomy is achieved by using a hierarchical structure with all control levels: DNO (global control), MGCC (central control), and local controllers. The DNO is responsible for the interaction of the MG with the distribution network (main grid) and interconnecting MGs and is, therefore, a part of the global control. The MGCC coordinates the aggregated operation of the DGs/RESs and loads within the MG and is responsible for their reliable and economical operation, as well as for interaction with the main grid. Finally, the local controllers control DG units within the MG, or aggregation of them, interacting with central control (MGCC) through the secondary control level and trying to achieve local and global objectives. In a decentralized architecture, a local controller can communicate with the MGCC and other local controllers, through the secondary control, in order to share knowledge, request/offer a service, communicate expectations, and exchange any other information relevant to the operation of the MG [1].

For implementing decentralized control concept, one may use the intelligent MAS approach [22, 35, 37, 38]. The local DGs/RESs besides delivering power to the grid also have other tasks: producing heat for local installations, keeping the voltage locally at a certain level or providing a backup system for local critical loads in case of a failure of the main system [39, 40]. These tasks suggest the importance of the distributed control and autonomous operation.

5.5 Design Example

Consider again the MG case study shown in Fig. 3.11. Here, it is desirable to design an MGCC for power management in the mentioned MG, while working in the islanded operation mode. Obviously, the design of MGCC depends on the capacity of generators to participate in the MG power management. In the given MG, there are three different types of generators. The availability of electrical energy is ensured by a 30 kW gas MT, which is used for long-term EM. Renewable energy-based electrical production is considered by a 17.3 kW (peak power) PV generator. The third power source is a supercapacitor (SC) storage system, which is used only for short-term power balancing. Moreover, SCs have short time response constants and so are able to immediately provide the amount of required power to balance the system following disturbances and/or significant load changes. All generators are grid-connected with PE converters. Thus, the grid-inverter control is an important concern for the MG operation.

In the islanded operation mode, the gas MT is chosen as the single master generator in the MG system because it is the dominant source of energy generation. Hence, an indirect isochronous speed control mode is implemented by the use of the *grid-forming control scheme* as described in Section 3.4.2. Similar to the given description in Chapter 1 for the grid-connected DGs, the grid-side inverter controls the MG voltage with predefined voltage and frequency values. In case of a load increase, the grid currents will be extracted from the DC bus. Hence, the machine torque is adapted to maintain the DC bus voltage, constantly. In consequence, more fuel is injected to maintain the MT speed constant.

In the grid-connected operation mode, the voltage is imposed by the distribution grid. Then, the MT has to work in a droop speed control mode. Here, a communication bus is used in order to share this control function among all generators. Therefore, a *power dispatching control scheme* (Section 3.4.2) is applied for the gas MT in order to supply the provided real and reactive power set points.

In order to maximize the use of renewable energy, the PV generator is working in *grid-following control scheme* as described in Section 3.4.2, and all converted power is sent to the MG. Moreover, in order to facilitate the MG management, the knowledge of the produced power is available via the communication bus. To implement the short-term power balancing into the SC control system, a power dispatching control scheme has been used. The control scheme is also set the prescribed real and reactive powers for the MT.

The MGCC measures the MG state variables and dispatches orders to DGs by using the communication bus. Local controllers of the MT and SC units receive power set points from the central controller. At the same time, they send various information, such as the sensed power production of the unit. Hence, the MGCC has to manage the MT and SC units in order to control the power among the MG. Two cases of study are tackled for the grid-connected and islanded operation modes. For this organization, the gas MT has to sense electrical quantities and moreover send information signals to the MGCC.

5.5.1 Power Dispatching

The framework of the power dispatching system has been explained in Section 3.5.2. In order to define the operating objective of each DG unit, the following considerations must be taken into account. Firstly, the MG voltage in the islanded operation mode must be controlled by a controllable source. Therefore, the MT has been chosen for the control of this voltage in horizon of a long-term EM. The SC unit is also a controllable source and can be used to perform the short-time power balancing. It is noteworthy that the MG voltage cannot be controlled any longer when there is no more SC stored energy.

Secondly, by receiving the information given by the DG units, the MGCC calculates the set point of the SC, since it manages the power in a short-time range. By setting a correct power reference for SC unit, the operation of the MT can be well stabilized. In consequence, the gas injection is expected to be more continuous, which increases the lifetime of the MT unit and the economic consideration for the gas consumption. Therefore, the controls of DG units can be performed as discussed below:

1) The PV is operated in maximum power point tracking (MPPT) mode in order to generate power as much as possible, since its primary source (the sun irradiation) is free. Therefore, the power production from PV unit is intermittent according to the climatic conditions. The information of the produced power is transmitted to the MGCC.
2) The MT is controlled with a *grid-forming control scheme* to regulate the MG voltages. Although the power dispatching function of the MGCC calculates the power references for the MT, those references are not communicated to this generating unit.
3) The power dispatching function of the MGCC calculates the power reference for the SC unit to provide regulation power for reducing the MG power fluctuations.

The power dispatching function calculates the real power reference for the SC and MT in the islanded operation mode. The overall scheme of the power management and the relevant control framework to provide MT and SC reference set points (P_{mt_ref}, P_{sc_ref}) from the MGCC (P_{mg_ref}) are plotted in Fig. 3.16.

5.5.2 Hardware-In-the-Loop Test Results

Some simulation results for the performed control system are given in Section 3.5.3. Here, the objective is to validate the proposed MGCC by hardware-in-the-loop (HIL) laboratory tests. The HIL methodology is well explained in Section 3.6.1. The performed HIL framework is presented in Fig. 5.16. Test devices are divided into two classes: MG devices (DG units, loads, etc.) and control devices (local controllers, MGCC, etc.) [31].

Three types of devices are used: (i) Real devices, including the SC unit and 1.2 kW passive load. This load represents the noncritical load blocks. During the test, a load step is applied by switching the load blocks; (ii) the virtual devices, including the MT generator, the PV unit, and 6 kW passive load, are simulated under a real-time simulator (RTS). This passive load represents the critical load blocks and must remain connected to the MG during the test. The MT unit controls the MG voltage and the PV unit provides an intermittent power according to the sensed irradiance and the sensed temperature. (iii) A bidirectional power amplifier, which is used for the interface with the virtual RTS bus and the real MG bus. The amplifier imposes the voltage for the MG and senses the feedback current to the RTS system. Thus, the fictitious devices react with this feedback current.

The control devices are composed of four parts: (i) the MGCC is implemented in a digital signal processor (DSP) (as the SCADA system). It communicates

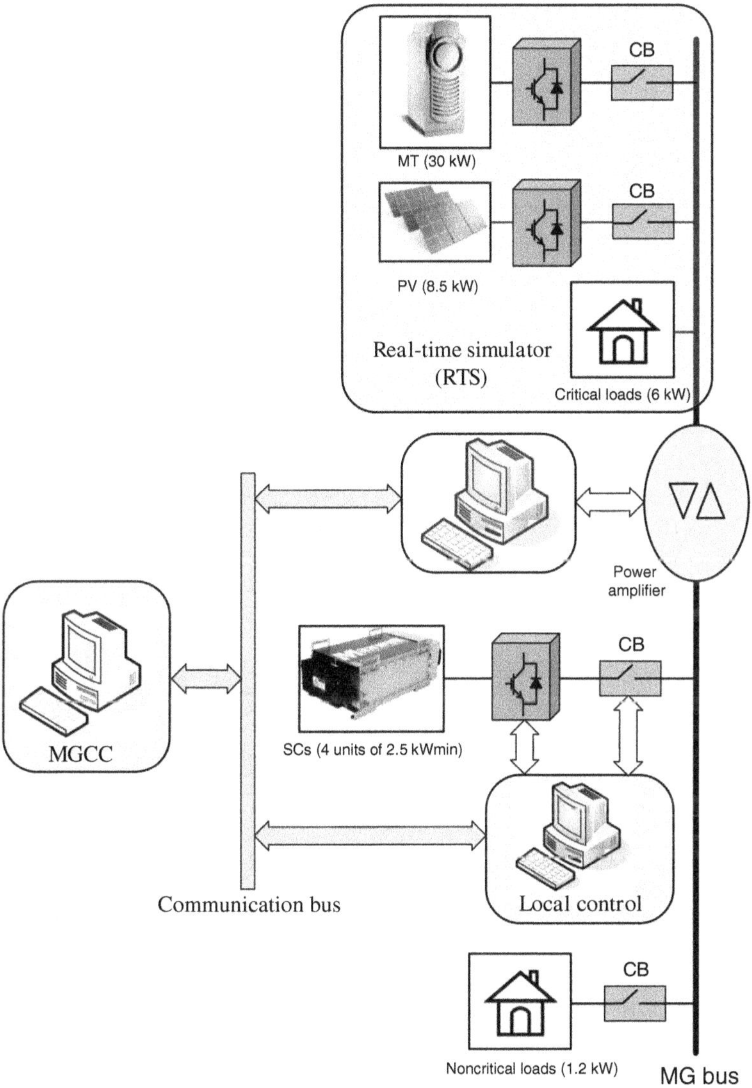

Figure 5.16 HIL test setup.

with the local controllers; (ii) another DSP is used to implement the local controller of the SC unit; (iii) the local controllers of the MT unit and the PV unit are integrated into the RTS software; and (iv) the communication bus exchanges the signals between the MGCC and local controllers. In this work, the secondary control is not activated and the electrical cables are used to build the communication bus.

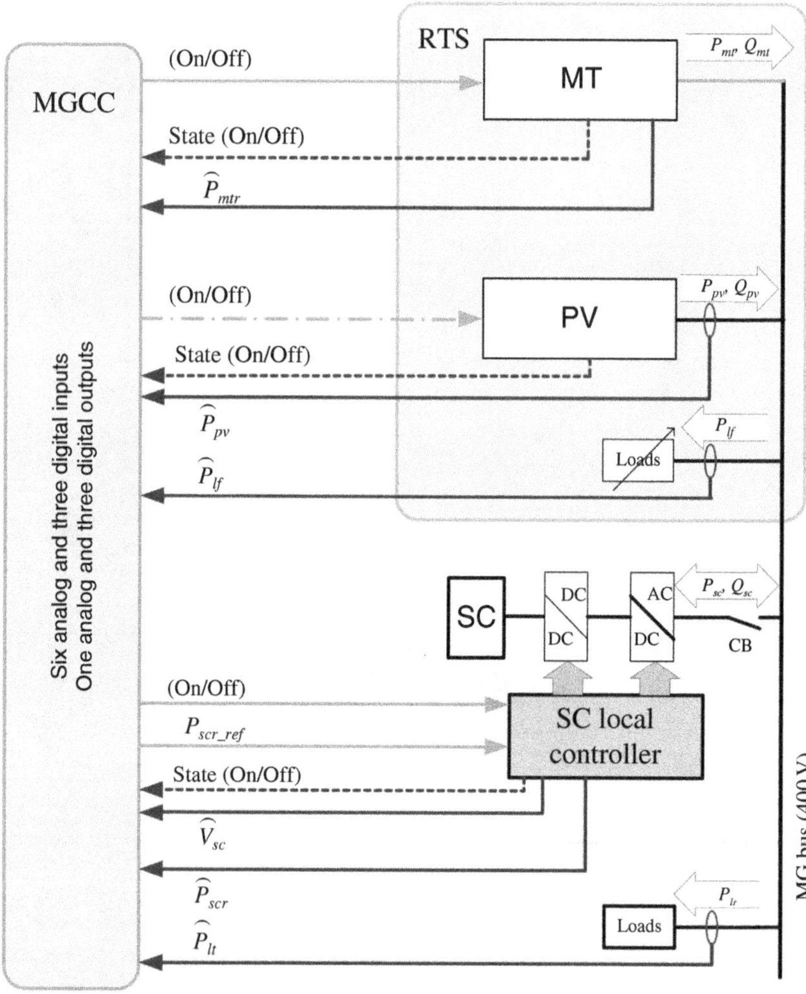

Figure 5.17 Signals communication.

The detail of the communication signals is presented at the Fig. 5.17. The PV generator, SC unit, and noncritical loads can be connected/disconnected to/from the MG by closing/opening the relevant CBs. The MGCC sends the On/Off state signal to the local controller of the MT unit to start/stop its operation. The MT returns its On/Off state as a confirmation, and in the meantime sends the sensed power to the MGCC. The MGCC exchanges the On/Off state with the PV local controller, and the PV returns the sensed power to the MGCC.

The MGCC sends the On/Off state signal to the SC local controller to start/stop its operation. Secondly, the MGCC sends the control signal for the exchanged real power with the SC unit. The SC returns its sensed real power as information. In the meantime, the local controller sends the SC terminal voltage to the MGCC as the information of the storage level or state of charge (SOC). Besides the signals mentioned above, the MGCC also must get the information of the loads, such as the power P_{lt} from the real load device and the power P_{lf} from the fictitious load device into the RTS system. The total local load power P_{chi} is the sum of both powers.

5.5.3 Test Procedure

The MG bus voltage is fixed to 400 V by the MT, and 6 kW critical loads remain connected and have to be supplied all the time. In the performed HIL test results, the CH represents an acronym for the local loads composed of 6 kW fictitious critical loads and of 1.2 kW real noncritical loads, and the CH On/Off is the switching signal for the 1.2 kW noncritical loads. Four different test scenarios are performed consecutively in order to make comparisons and validate the implementation of the proposed strategies by the MGCC.

5.5.3.1 Test of the Micro-turbine Unit: Zone 1 (MT)

The objective is to provide a basement for comparisons. During this test, the MT generator and the passive load blocks are connected to the MG. The test begins at time t_0 with the MT and 6 kW critical loads. At time t_1, we connect the 1.2 kW noncritical loads and at time t_2 we disconnect it. The total load power (power of critical loads and noncritical loads) change is shown in Fig. 5.18. When the test begins at t_0, only the 6 kW critical loads are connected to the MG. Between interval t_1–t_2, a load step is created by the connection and the disconnection of the 1.2 kW noncritical loads.

The MT response to the load changes is shown in Fig. 5.19. In Zone 1 of the Fig. 5.19a, the real power, which is generated by the MT, is clearly shown.

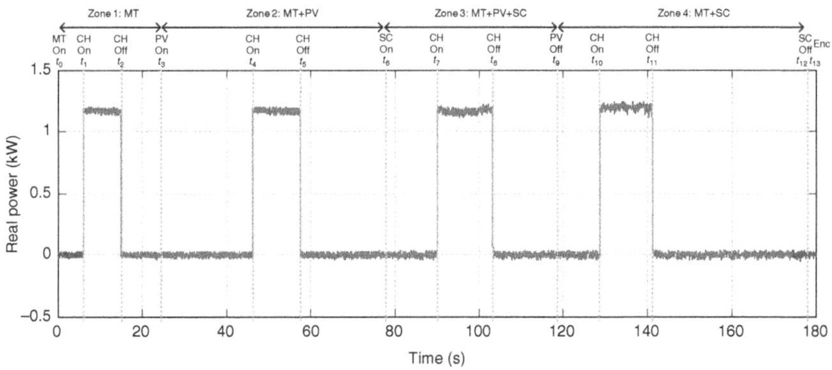

Figure 5.18 The MG noncritical loads (real) change pattern.

Since the MT is the only power generation unit, its power is totally matched with the load power. The shaft speed and the gas mass flow rate of the MT are visualized in Figs 5.19b,c. It can be seen that during the load transient, the shaft speed and the injection of the gas must vary very fast in order to follow the power production. This is not a benefit for the MT unit. The DC bus terminal voltage of the MT and its rezoomed view are shown in Figs 5.19d,e. This voltage is well controlled during the test. Two peaks appear at time t_1 and t_2 due to the sudden load transients.

5.5.3.2 Impact of the PV Unit: Zone 2 (MT + PV)

The PV unit is connected to the MG at time t_3, in order to study the influence of the renewable energy generation. The PV unit generates an intermittent power at time t_3 as shown in Fig. 5.20. 1.2 kW load step is performed between times t_4 and t_5 (Fig. 5.18). The MT compensates the power difference between the PV generation and the load consumption (Fig. 5.19b). Even if the load consumption is constant during a certain interval, the MT generation varies because of the intermittent power generation of PV.

5.5.3.3 Contribution of Supercapacitor: Zone 3 (MT + PV + SC)

The objective of this test scenario is to demonstrate the contribution of the SC unit. The SC unit is connected to the MG at time t_6 as shown in Fig. 5.21a. In this plot, "*0*" and "*1*" represent "*Off*" and "*On*" states, respectively. 1.2 kW load step is performed between times t_7 and t_8 (Fig. 5.18). The PV unit continues to generate the intermittent power (Fig. 5.20). In this case, the SC unit manages the power quickly as shown in Fig. 5.21b. It can be seen that more power is exchanged with the SC, when a fluctuation from the PEs is injected to the MG. This noise influences the balance of the DC bus voltage of the MT (Fig. 5.19e). But the peaks due to the load transient are now smaller (t_7 and t_8 of the Fig. 5.19d) than in the case without the SC unit (t_1, t_2, t_4, and t_5 of the Fig. 5.19d). The SC terminal voltage, which is recorded in Fig. 5.21c, varies according to the exchanged power with the MG. The MT power generation is better stabilized (Zone 3 of Fig. 5.19a) than the case without SC unit (Zones 1 and 2). Therefore, the shaft speed (Fig. 5.19d) and the gas mass flow rate (Fig. 5.19c) of the MT are also stabilized, thanks to the contribution of the SC unit.

5.5.3.4 Operating During the Night: Zone 4 (MT + SC)

The PV unit is disconnected from the MG at time t_9 (Fig. 5.20), the moment when it was generating about 4 kW. The only disturbance to the MG is the local loads. Therefore, this test can highlight the performance of the MT, which manages to control the power in a long-time range. 1.2 kW load step is performed between times t_{10} and t_{11}.

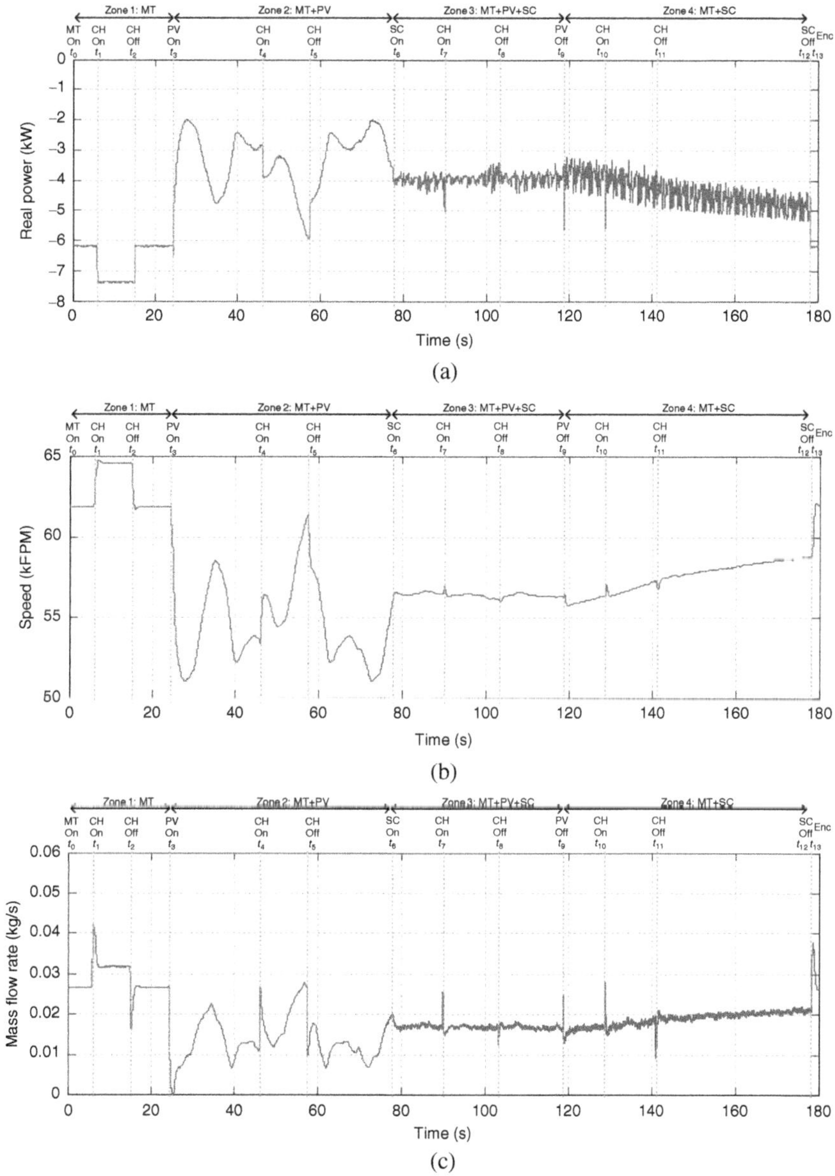

Figure 5.19 The MT (fictitious) response: (a) power, (b) shaft speed, (c) gas mass flow rate, (d) DC bus terminal voltage, and (e) a zoomed view of (d).

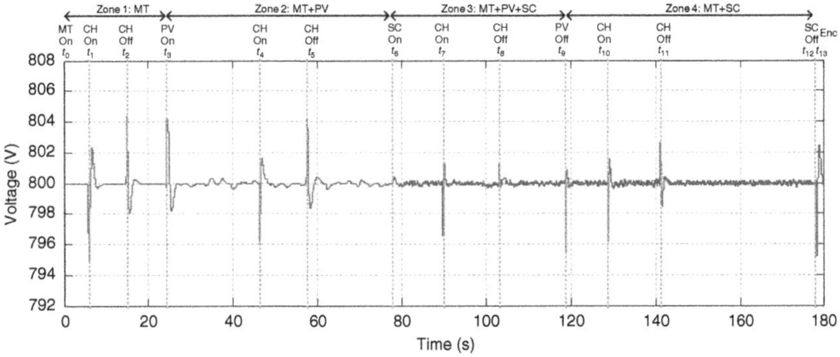

(d)

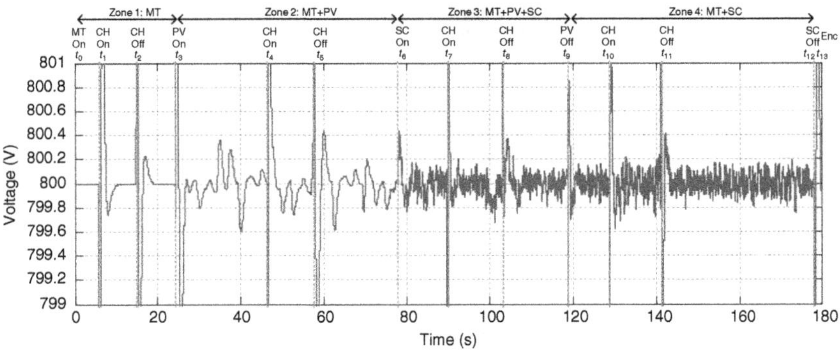

(e)

Figure 5.19 (*Continued*)

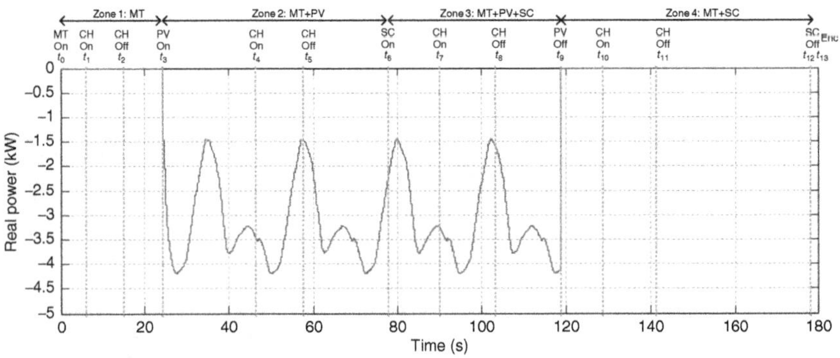

Figure 5.20 PV power (fictitious).

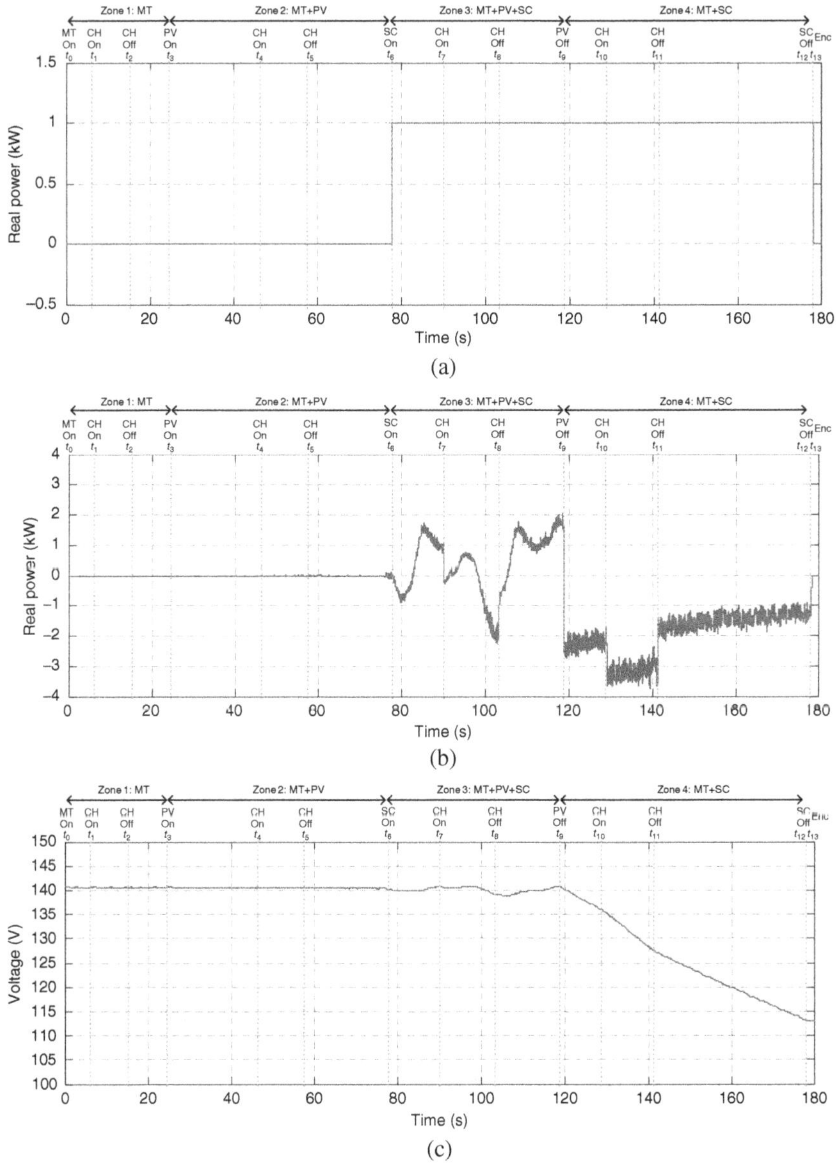

Figure 5.21 The SC unit response (real): (a) On/Off state, (b) exchanged power, and (c) terminal voltage.

As described in Fig. 5.21b, after the disconnection of the PV unit at time t_9, the SC unit compensates this 4 kW power variation immediately. Therefore, a fluctuation is injected to the MG (interval t_9–t_{12} in Fig. 5.21b) and influences the MT real power at the MG connection point (interval t_9–t_{12} in Fig. 5.20a). It is shown in Fig. 5.21c, that the SC terminal voltage decreases rapidly owing to its approximate 2 kW power generation. The MT power generation remains stabilized (Zone 4 in Fig. 5.19a) during variations of local loads (time t_{10} and t_{11} in Fig. 5.18). This test demonstrates the performance of the MT, which controls the power in a long-time range (interval t_9–t_{12}). The shaft speed (Fig. 5.19d) and the gas mass flow rate (Fig. 5.19c) are always stabilized with the contribution of the SC unit. Between t_{12} and t_{13}, the SC unit is unloaded (Fig. 5.21c) and can be seen that the real power of the MT unit is increased (Fig. 5.19a).

The test results justify the proposed MGCC design for the contribution of the SC unit. If the MG operates without the SC unit, the turbine must compensate all power fluctuations within the MG alone. The fluctuations can be either the load variations or the intermittent power generated by the RESs, even if a transient load step occurs. With the association of the SC unit, the operation of the MT is improved since power variations have been decreased.

5.6 Summary

This chapter presents an overview of the control levels and existing control technologies and challenges in MGs. The MG's hierarchical control structure is determined by four control levels: local, secondary, central/emergency, and global controls. The local control deals with each DG's internal voltage/frequency and current control loops, while the secondary control covers higher level control loops such as the voltage and frequency control among a bus connecting a set of DGs or MGs, following a load variation. The central/emergency control operates as an MG EMS, and it is responsible for emergency control and overall protection schemes. Finally, the global control supervises the power flow between the given MG, other interconnecting MGs, and the main grid, in an economic optimal operation.

This chapter describes the hierarchical control systems applied to MGs and discusses the classification of the control features in different control levels. Then, droop characteristics in inverter-based DGs and virtual impedance concept were addressed. Finally, the role of control levels in MGs' power management and control was explained, and a discussion supplemented by a real-time simulation for MGs' power control were given.

References

1 IEEE-PES Task Force on Microgrid Control (2014) Trends in microgrid control. *IEEE Transactions on Smart Grid*, **5** (4), 1905–1919.

2 Bidram, A. and Davoudi, A. (2012) Hierarchical structure of microgrids control system. *IEEE Transactions on Smart Grid*, **3** (4), 1963–1976.

3 Rocabert, J., Luna, A., Blaabjerg, F., and Rodriguez, P. (2012) Control of power converters in AC microgrids. *IEEE Transactions on Power Electronics*, **27** (11), 4734–4749.

4 Majumder, R. (2013) Some aspects of stability in microgrids. *IEEE Transactions on Power Systems*, **28** (3), 3243–3252.

5 Bevrani, H., Watanabe, M., and Mitani, Y. (2012) Microgrid controls, in *Standard Handbook for Electrical Engineers*, 16th edn (ed. H. Wayne Beaty), McGraw-Hill, pp. 160–176, Section 16.9.

6 Braun, M. and Notholt-Vergara, A. (2008) Advanced architectures and control concepts for more microgrids – inverter performance with regard to ancillary services and fault-ride-through capabilities [Online]. Available at http://www.microgrids.eu/documents/esr.pdf.

7 Bevrani, H., Watanabe, M., and Mitani, Y. (2014) *Power System Monitoring and Control*, IEEE-Wiley Press, New York.

8 Bevrani, H. (2014) Frequency control in microgrids, in *Robust Power System Frequency Control*, 2nd edn, Springer, Chapter 11.

9 Kundur, P. (1994) *Power System Stability and Control*, McGraw-Hill, New York.

10 Prodanović, M. and Green, T.C. (2006) High-quality power generation through distributed control of a power park microgrid. *IEEE Transactions on Industrial Electronics*, **53**, 1471–1482.

11 Bevrani, H. and Shokoohi, S. (2013) An intelligent droop control for simultaneous voltage and frequency regulation in islanded microgrids. *IEEE Transactions on Smart Grid*, **4** (3), 1505–1513.

12 Bevrani, H., Habibi, F., Babahajyani, P. *et al.* (2012) Intelligent frequency control in an AC microgrid: on-line PSO-based fuzzy tuning approach. *IEEE Transactions on Smart Grid*, **3** (4), 1935–1944.

13 Bevrani, H. and Hiyama, T. (2011) Neural network based AGC design, in *Intelligent Automatic Generation Control*, CRC Press, New York, pp. 95–122, Chapter 5.

14 Moreira, C.C.L. (2008) Identification and development of microgrids emergency control procedures. Ph.D. thesis. University of Porto.

15 Kapetanovic, T., Buchholz, B.M., Buchholz, B., and Buehner, V. (2008) Provision of ancillary services by dispersed generation and demand side

response: needs, barriers and solutions. *Elektrotechnik und Informationstechnik*, **125** (12), 452–459.
16 Dell, R.M. and Rand, D.A.J. (2001) Energy storage: a key technology for global energy sustainability. *Journal of Power Sources*, **100** (1/2), 2–17.
17 Jiayi, H., Chuanwen, J., and Rong, X. (2008) A review on distributed energy resources and microgrid. *Renewable and Sustainable Energy Reviews*, **12** (9), 2472–2483.
18 Fei, W., Duarte, J.L., and Hendrix, M.A.M. (2011) Grid-interfacing converter systems with enhanced voltage quality for microgrid application: Concept and implementation. *IEEE Transactions on Power Electronics*, **26** (12), 3501–3513.
19 Mercier, P., Cherkaoui, R., and Oudalov, A. (2009) Optimizing a battery energy storage system for frequency control application in an isolated power system. *IEEE Transactions on Power Systems*, **24** (3), 1469–1477.
20 Bevrani, H. (2014) Frequency response characteristics and dynamic performance, in *Robust Power System Frequency Control*, 2nd edn, Springer, pp. 49–69, Chapter 3.
21 Engler, A. (2005) Applicability of droops in low voltage grids. *International Journal of Distributed Energy Resources*, **1** (1), Technology and Science Publisher, Germany, Kassel.
22 Bevrani, H. and Hiyama, T. (2011) *Intelligent Automatic Generation Control*, CRC Press.
23 Guerrero, J.M., de Vicuña, L.G., Matas, J. *et al.* (2004) A wireless controller to enhance dynamic performance of parallel inverters in distributed generation systems. *IEEE Transactions on Power Electron*, **19**, 1205–1213.
24 Liserre, M., Teodorescu, R., and Blaabjerg, F. (2006) Stability of photovoltaic and wind turbine grid-connected inverters for a large set of grid impedance values. *IEEE Transactions on Power Electronics*, **21**, 263–272.
25 Schiffer, J., Ortega, R., Astolfi, A. *et al.* (2014) Conditions for stability of droop-controlled inverter-based microgrids. *Automatica*, **50**, 2457–2469.
26 Chiang, S.J., Yen, C.Y., and Chang, K.T. (2001) A multimodule parallelable series-connected PWM voltage regulator. *IEEE Transactions on Industrial Electronics*, **48** (3), 506–516.
27 Xiaoxiao, Y., Khambadkone, A.M., Huanhuan, W., and Terence, S. (2010) Control of parallel-connected power converters for low-voltage microgrid-part I: a hybrid control architecture. *IEEE Transactions on Power Electronics*, **25** (12), 2962–2970.
28 Vilathgamuwa, D.M., Poh Chiang, L., and Li, Y. (2006) Protection of microgrids during utility voltage sags. *IEEE Transactions on Industrial Electronics*, **53** (5), 1427–1436.
29 Guerrero, J.M., Hang, L., and Uceda, J. (2008) Control of distributed uninterruptible power supply systems. *IEEE Transactions on Industrial Electronics*, **55** (8), 2845–2859.

30 Lu, D. (2010) Design and control of a PV active generator with integrated energy storages: application to the aggregation of producers and consumers in an urban micro smart grid. PhD thesis. Ecole Centrale de Lille, France.

31 Li, P. (2010) Formalisme pour la Supervision des Systèmes Hybrides Multi-Sources de Générateurs d'Energie Répartie: Application à la Gestion d'un Micro Réseau. PhD thesis. Ecole Centrale de Lille, France.

32 Katiraei, F., Iravani, M.R., Hatziargyriou, N., and Dimeas, A. (2008) Microgrids management. *IEEE Power and Energy Magazine*, **6**, 54–65.

33 Lasseter, R., Akhil, A., Marnay, C., Stephens, J., Dagle, J., Guttromson, R., et al. (2002) Integration of distributed energy resources – The MicroGrid concept. CERTS MicroGrid Review, April 2002.

34 Katiraei, F. and Iravani, M.R. (2006) Power management strategies for a microgrid with multiple distributed generation units. *IEEE Transactions on Power Systems*, **21** (4), 1821–1831.

35 Dimeas, A. and Hatziargyriou, N.D. (2005) Operation of a multi-agent system for microgrid control. *IEEE Transactions on Power Systems*, **20** (3), 1447–1455.

36 Dimeas, A., Hatziargyriou, N.D. (2007) Agent based control for microgrids. IEEE Power Engineering Society General Meeting, Tampa, USA, June.

37 Bevrani, H., Daneshfar, F., and Hiyama, T. (2012) A new intelligent agent-based AGC design with real-time application. *IEEE Transactions on Systems, Man, and Cybernetics, Part C*, **42** (6), 992–1002.

38 Daneshfar, F. and Bevrani, H. (2010) Load–frequency control: a GA-based multi-agent reinforcement learning. *IET Generation, Transmission & Distribution*, **4** (1), 13–26.

39 Hatziargyriou, N., Dimeas, A., Tsikalakis, A.G., et al. (2005) *Management of Microgrids in Market Environment*. International Conference Future Power System, pp. 1–7.

40 Dimeas, A. and Hatziargyriou, N.D. (2004) A multiagent system for microgrids. *IEEE Power Energy Society General Meeting*, **1**, 55–58.

6

DC Microgrid Control

While most of the microgrids (MGs) adopt AC distribution and conventional power systems, DC MGs are proposed for useful connection with DC output type sources such as photovoltaics (PVs), fuel cells, and batteries. Moreover, if loads in the system are supplied with DC power, the conversion losses from sources to loads are reduced compared with the AC MG.

In recent years, DC MGs have gained popularity because DC output sources can be interconnected without AC/DC conversion, which increases the total system efficiency. Moreover, high-quality power can be supplied continuously when voltage sags or blackouts occur in utility grids.

In this chapter, the "low-voltage bipolar-type DC MG" is proposed, which can supply super high-quality power with a three-wire DC distribution line. In this chapter, a DC MG for a residential area is presented. In this area, each house has a cogeneration system (CGS) such as micro-combined heat and power (micro-CHP) unit or gas engine and fuel cell. The output electric power is shared among the houses, and the total power can be controlled by changing the running number of CGSs/micro-CHPs. Supercapacitors, electric double layer capacitors (EDLC), are chosen as main energy storage devices.

Then, the "low-voltage bipolar-type DC MG" is modified by adding another energy storage unit to the experimental system. A gain-scheduling control with a simple droop control has been proposed for parallel operations; some of them can be applied for both AC and DC MGs. To confirm the fundamental characteristics and system operations, laboratory-scale systems are also performed.

6.1 Introduction

In recent years, concerns about energy and environmental problems have been increasing, such as greenhouse gas, growth of energy demand, and depletion of energy resources. Against the background of these problems, a large number of distributed generations (DGs) including renewable energy sources (RESs)

Microgrid Dynamics and Control, First Edition. Hassan Bevrani, Bruno Francois, and Toshifumi Ise.

are being installed into power systems. It is well known that if many DGs are installed into a utility grid, they can cause problems such as voltage rise and protection issues. To solve these problems, new conceptual electric power systems, MGs, were proposed. In the last two decades, numerous research works and projects had been done all over the world [1–10]. For example, in Japan, MG projects are promoted by several private companies and NEDO (The New Energy and Industrial Technology Development Organization). Four popular NEDO's projects (Aichi Expo, Kyotango, Hachinohe, and Sendai) were undertaken from 2003, and the details are reported in [9, 10]. The NEDO also promoted other MG projects in Asian countries to research under several conditions. After the 2011 Tohoku earthquake and tsunami, the trends have accelerated even more in Japan.

Including those projects, most MGs adopt AC distribution as well as conventional power systems. In this case, DC output type sources such as PV system, fuel cell, and energy storages (e.g., Li-ion secondary battery, supercapacitor) need inverters. In addition, some gas engine cogenerations and wind turbines also need inverters because the output voltages and the frequencies are different from those of the utility grids. Therefore, DC distribution type MGs (DC MGs) were also developed in order to reduce conversion losses from the sources to loads [11–15]. Apart from a reduction in AC/DC conversions losses, DC MGs can supply continuous high-quality power when voltage sags or blackouts occur in utility grids. For instance, DC power supplies are commonly used in telecommunication buildings and Internet data centers where high-quality power is needed [16].

The advantages of DC MGs can be summarized as follows:

1) The system efficiency becomes higher because of the reduction of conversion losses of inverters between DC output sources and loads [15].
2) There is no need to consider synchronization with the utility grid and reactive power management.
3) When a blackout or voltage sag occurs in the utility grid, it does not affect the DC bus voltage of DC MG directly due to the existing stored energy of the DC capacitor and the voltage control of AC/DC converter. Therefore, DGs in DC systems are not easily tripped against these disturbances. In other words, DC MG already has fault ride through capability in its own.

On the other hand, there are some drawbacks to put DC MGs to practical use as follows:

1) Private DC distribution lines should be constructed for DC MGs.
2) Since there is no zero cross point of voltage in DC systems, protection is more difficult compared to the AC systems.
3) The loads adapted for DC power supply are required for high system efficiency.

As an instance of DC MGs, the system described in [14] adopted DC 380 V as DC bus voltage. The system has PV systems (2 × 10 kW), wind generator systems (10 + 2 kW), and storage battery (97 kW), but there are no controllable DGs such as gas engine or fuel cell. The system is normally operated in islanding mode. When the storage energy becomes low, the system supplies power from the utility grid and charges the battery. It is also unique that the system can be changed into AC MG by the switches in order to compare DC system with AC system.

On the other hand, high-quality power is essential for some customers such as banks, hospitals, and semiconductor factories because the downtime related to voltage sag or blackout is a great concern. Besides, high-quality power is also requested in our dependable society. Security of electric power is becoming more important in our daily life.

In this chapter, to achieve high efficiency and high-quality power supply, a "low-voltage bipolar-type DC MG" is developed [16]. In this system, DC power is distributed through three-wire lines, and it is converted into the required AC or DC voltages by load-side converters. When blackout or voltage sag occurs in the utility grid, the DC MG can supply a stable high-quality power, while inverters of DGs in AC MGs should be tripped unless they have fault ride through capability.

Here, a DC MG for a residential area is presented based on the DC MG concept [17]. Each house has a CGS such as micro-CHP unit or gas engine or fuel cell systems. The electric power from CGSs is shared among the houses with DC distribution lines, and the total power can be controlled by changing the number of running CGSs. As main energy storage devices, supercapacitors can be chosen in spite of their low energy density. To confirm the fundamental characteristics and system operation, some experiments are done with a laboratory-scale system. The results show that the proposed system could supply high-quality power under several situations. In addition, smooth disconnection and reconnection with the utility grid were also demonstrated.

In the mentioned experiments, one energy storage unit with a DC/DC converter is used to sustain DC bus voltage when the system is in intentional islanding operation. However, it is desired that the system should have two or more energy storage units for redundancy. Therefore, in continuation, another energy storage unit is added in the experimental system. Several types of droop controls have been proposed for parallel operations, some of which were applied for AC/DC MGs [16–21]. Here, an adopted gain-scheduling control technique is used as droop controls for sharing the outputs. In this case, although output power sharing of two energy storage units is achieved, it is also found that the storage energy control is necessary for the previous system operation.

In the updated experimental system, EDLCs are used as the energy storage units, and each DC/DC converter of EDLC controls the DC distribution voltage when the system is operated under an intentional islanding mode.

The experimental results show that the DC distribution voltages are within 340 V ± 5%, and the energy ratios of the storage units are approximately equal, implying that DC voltage regulation and stored energy balancing control can be realized simultaneously by applying the proposed control. Sections 6.2–6.6 are discussed in detail in [15, 17, 22–25].

This chapter is organized as follows. Section 6.2 shows a concrete configuration of the DC MG for a residential area and explains the characteristics and operation methods. Section 6.3 describes the configuration and features of low-voltage bipolar-type DC MG. Section 6.4 evaluates the stability of the given DC MG. Experimental study and results are explained in Section 6.5. Section 6.6 proposes a voltage control system for the updated version of the DC MG. Sections 6.7 and 6.8 show the simulations and the experimental results, which demonstrated the high-quality power supply of the proposed control system. Finally, the outcome is summarized in Section 6.9.

6.2 DC Microgrid for a Residential Area

6.2.1 System Configuration and Operation

As a concrete system, a DC MG for a residential area is considered as shown in Fig. 6.1. There are around 50–100 houses in the system, and each house has a CGS, such as micro-CHP unit and gas engine. The CGSs are connected to DC distribution line (three-wire, ±170 V), and the electric power is shared among the houses. From this configuration, it is expected that the total CGSs operation period increases and leads to effective utilization of primary energy [20]. In order to keep high efficiency, those CGSs should not be operated by a partial load condition but operated by a start/stop control. Cogenerated hot water is used in each house or shared with other houses. This system is connected to the utility grid by a rectifier. At the load side, various forms of electric power (AC 100 V, DC 48 V, etc.) can be obtained by the converters. Despite their low energy density, EDLCs are used as the main energy storage unit because of the fast response, safeness (especially compared with Li-ion battery), easy measurement of the stored energy, and no toxicity of the inner materials.

In a verification test of the energy storage system using an EDLC unit, the voltage and maximum energy were 500 V and 5 MJ, respectively [25]. Hence, EDLC is considered a viable energy storage system in a small grid. In addition, EDLCs can handle many charge–discharge cycles and therefore have a low cost per cycle considering their long life. The disadvantage of EDLCs is their low energy density. If a large energy capacity is needed for an MG, a relatively large EDLC is required. However, a large energy capacity is not necessary for the proposed DC MG because the micro-CHP units are operated to prevent overcharge/discharge of the EDLCs as described in the following section.

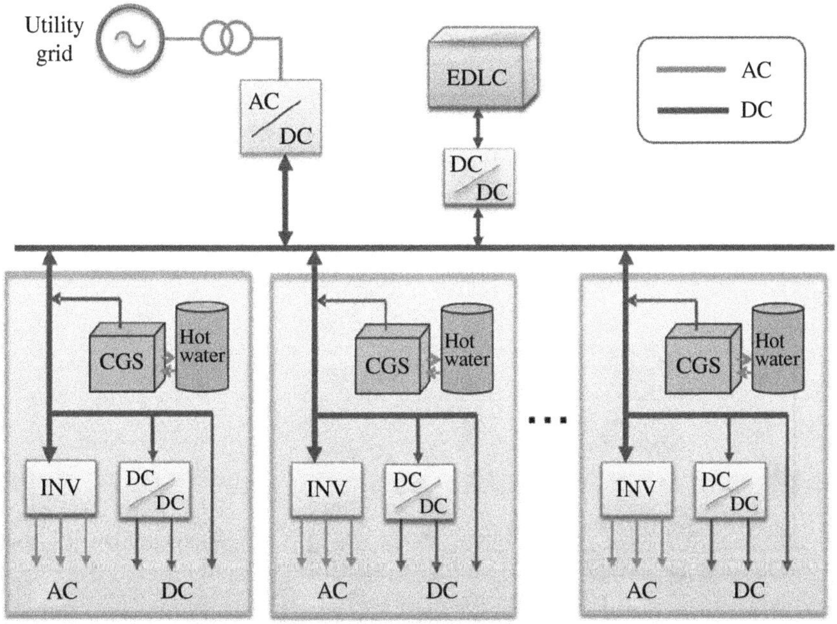

Figure 6.1 System configuration of the DC MG for residential area.

The total generated power is controlled by changing the number of running CGS/micro-CHP units. When the system is connected to the utility grid, the deficient power is compensated from the utility grid as shown in Fig. 6.2. This state is called *interconnected operation*. In this operation, the rectifier controls the DC distribution voltage, and the supervising computer changes the number of the running CGSs so that the power from the utility grid is within the contract demand. In other words, the generated power from CGSs does not flow to the utility grid in the interconnected operation.

When the system disconnects from the utility grid, the surplus or deficient power is compensated by the EDLC as shown in Fig. 6.3. This state is called *intentional islanding operation*. In this operation, the converter of the EDLC controls the DC distribution voltage, and the number of running CGSs is determined not only by the load consumption but also by the stored energy of EDLCs. When the stored energy exceeds the maximum limit, the system stops one of the operating CGSs. Then, the total output of CGSs becomes less than the load consumption, and the EDLC discharges until the stored energy becomes less than the minimum limit. On the contrary, when the stored energy is less than the minimum limit, the system starts CGSs. Then, the total output of CGSs becomes more than the load consumption, and the EDLC charges until the stored energy becomes more than the maximum limit. These two modes are repeated alternately in the intentional islanding operation.

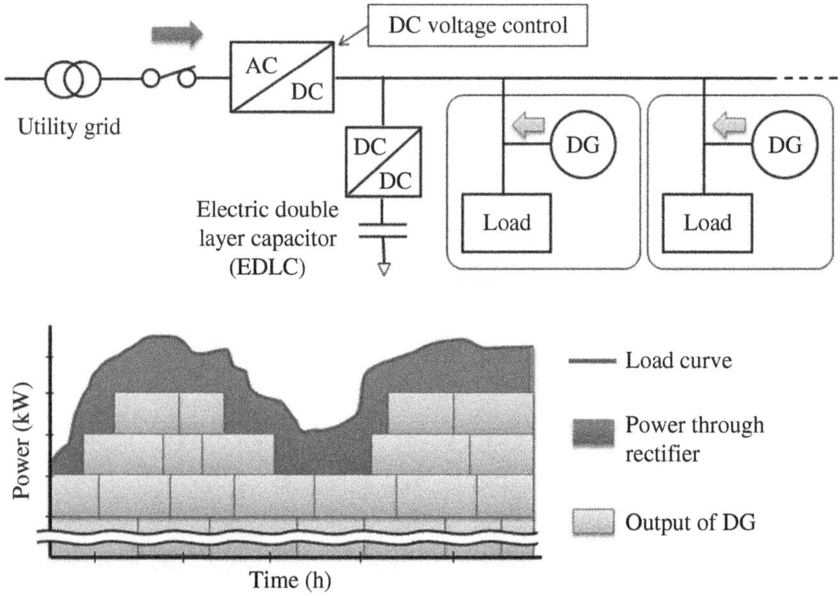

Figure 6.2 Interconnected operation.

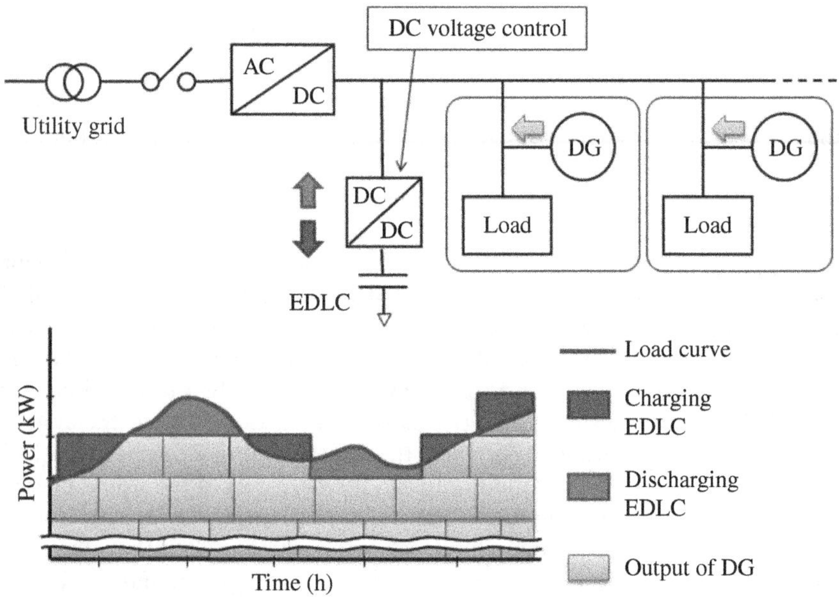

Figure 6.3 Intentional islanding operation.

This system should choose the CGSs that can start up for a few minutes, so that energy storage does not need a large capacity. Therefore, EDLCs can be used as main energy storage unit in this system. The output capacity of the EDLC and its converter are designed to compensate the assumed maximum load variation. However, if larger load variations occur in the system, the output of EDLC could reach the maximum charge or discharge level. If the charge is limited, the supervising computer reduces the number of required CGSs. If the discharge is limited, the supervising computer stops power supplying to specific loads that do not need high-quality power and runs CGSs to maintain the EDLC output within the allowed limits.

6.2.2 Voltage Clamp Control

When the system is in the interconnected operation, the EDLC does not charge or discharge unless the DC distribution voltage exceeds the limited range. We propose a voltage clamp control by the EDLC. When the DC distribution voltage exceeds the upper limit (360 V), the DC/DC converter of the EDLC is operated to clamp the DC voltage at the specified level. This clamp control contributes to prevent overvoltage of the devices connected to the DC line. The converter of the EDLC is also operated to clamp at the lower limit (320 V) when the DC voltage decreases to the lower limit; for example, the current from the utility grid is limited by the capacity of the rectifier. In addition, this clamp control assists the disconnection and reconnection process (as described in the following section).

6.2.3 Disconnection/Reconnection from/to the Utility Grid

Disconnection and reconnection procedures are performed by the voltage clamp control. Figure 6.4a shows the flowchart of the disconnection procedure. When a problem from the utility grid is detected, the system stops the grid interface rectifier. Then, the DC distribution voltage decreases, and the converter of the EDLC clamps the voltage at the lower limit (320 V). Then, the converter of the EDLC changes its operation from the clamp control to the DC voltage control. Then, the voltage reference is gradually changed from the lower clamping limit to the nominal DC distribution voltage (340 V).

Without the clamp control, the voltage control has to be moved from the rectifier to the converter of the EDLC immediately. However, a delay appears in the communication network. Therefore, this clamp control plays an important role to complete the disconnection properly.

Figure 6.4b shows the flowchart of the reconnection procedure, which is opposite to the disconnection process. After the utility grid is recovered, the controller of the rectifier detects the phase of the utility grid by phase-locked loop (PLL). Then, the system changes the operation of EDLC from voltage control to clamp control, and DC distribution voltage decreases to the lower limit (320 V). After that, the system starts the rectifier, and the voltage reference is gradually changed from the lower clamping limit to the nominal

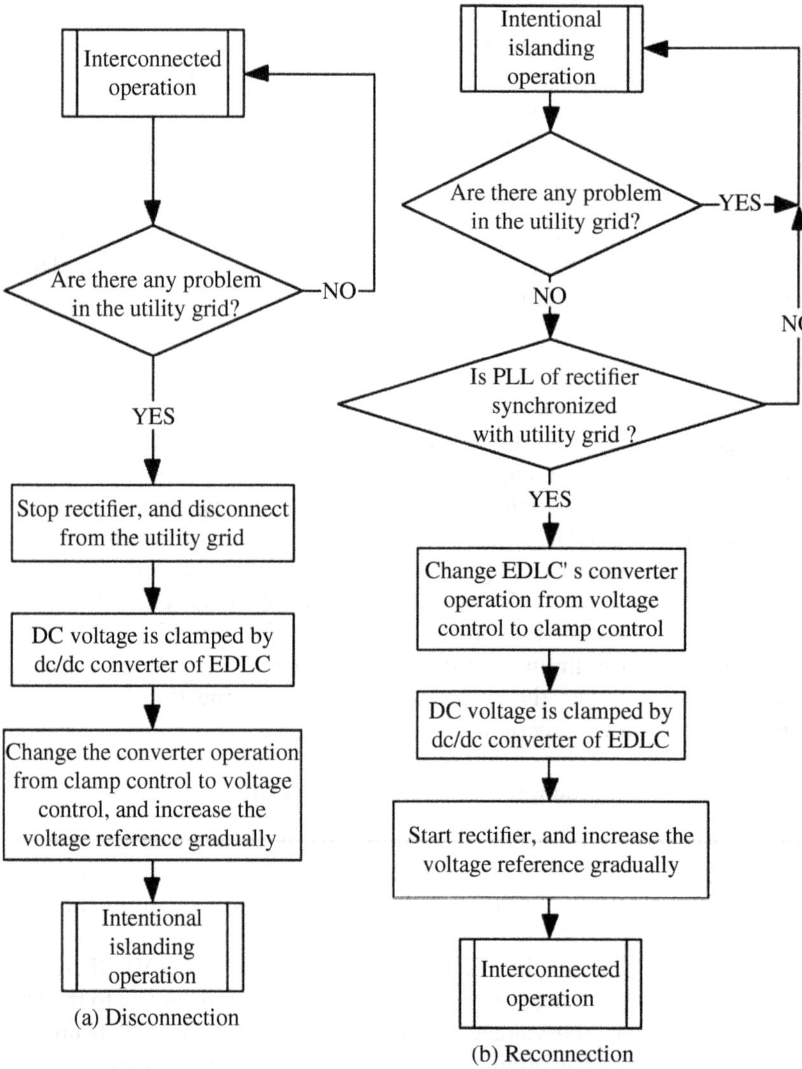

Figure 6.4 Flowcharts of disconnection (a) and reconnection (b).

voltage. In the performed experimental test described later, it depends on the voltage drop level and the period whether the problem occurs in the system or not. If the voltage decreases less than 30%, the system detects a problem in the utility grid. If the voltage drop is between 30% and 80%, and the duration is more than 1 s, the system also detects a problem in the utility grid. In this case, this system does not disconnect from the utility grid if voltage sag occurs because the duration of most voltage sags is usually within 0.5 s.

6.3 Low-voltage Bipolar-type DC Microgrid

Figure 6.5 shows a conceptual realization of the low-voltage bipolar-type DC MG. The utility grid voltage, 6.6 kV, is converted into DC 340 V by a transformer and a rectifier. It is a characteristic of the system to adopt three-wire DC distribution, which consists of +170 V line, neutral line, and −170 V line. The three-wire composition enables a low voltage to ground, and one of the single-phase 100 V output lines becomes a grounded neutral line according to an existing Japanese standard. In addition, load-side DC–DC converters can choose the source voltage from 340, +170, or −170 V. Moreover, if one wire snaps out, it is possible to supply power by other two lines and an auxiliary converter.

When there are DC/DC converters for loads and the source voltages of them are either +170 or −170 V, DC voltage balance control is essential [18]. Hence, a voltage balancer is placed near a rectifier to balance positive and negative voltages. It is also possible to connect the voltage balancer near load side.

As energy storage units, a secondary battery and a supercapacitor (EDLC) are connected to DC distribution line. A PV system and a gas engine cogeneration are also connected through DC/DC converter and AC/DC converter, respectively. At the load side, DC power is converted into required AC or DC voltages by each converter. Characteristics of the system are summarized as follows:

1) Three-wire bipolar DC distribution contributes to lowering the distribution voltage to ground. In addition, various source voltages are available (from 340, +170, or −170 V).
2) The distribution of the load-side converters contributes to provide a super high-quality power supplying. For instance, even if a short circuit occurs at one load side, it does not affect the other loads.
3) Various forms of electric power such as single-phase 100 V, three-phase 200 V, and DC 48 V can be obtained. These converters are transformerless; therefore, it contributes to downsizing and high efficiency.
4) DC distribution system is suitable for DC output type DGs and energy storage. If DC power can be supplied to loads directly or through DC/DC converters, the system efficiency becomes high.
5) When an accident occurs in the utility grid, this system can be disconnected from the utility grid seamlessly and supply electric power continuously. The reconnection to the utility grid is also smooth.
6) When a temporary overload occurs at a single-phase load, electric power can be shared between load-side converters by using additional electric power lines.

It is also possible to form DC loop configuration at DC distribution part [19] and share power between other DC MG systems.

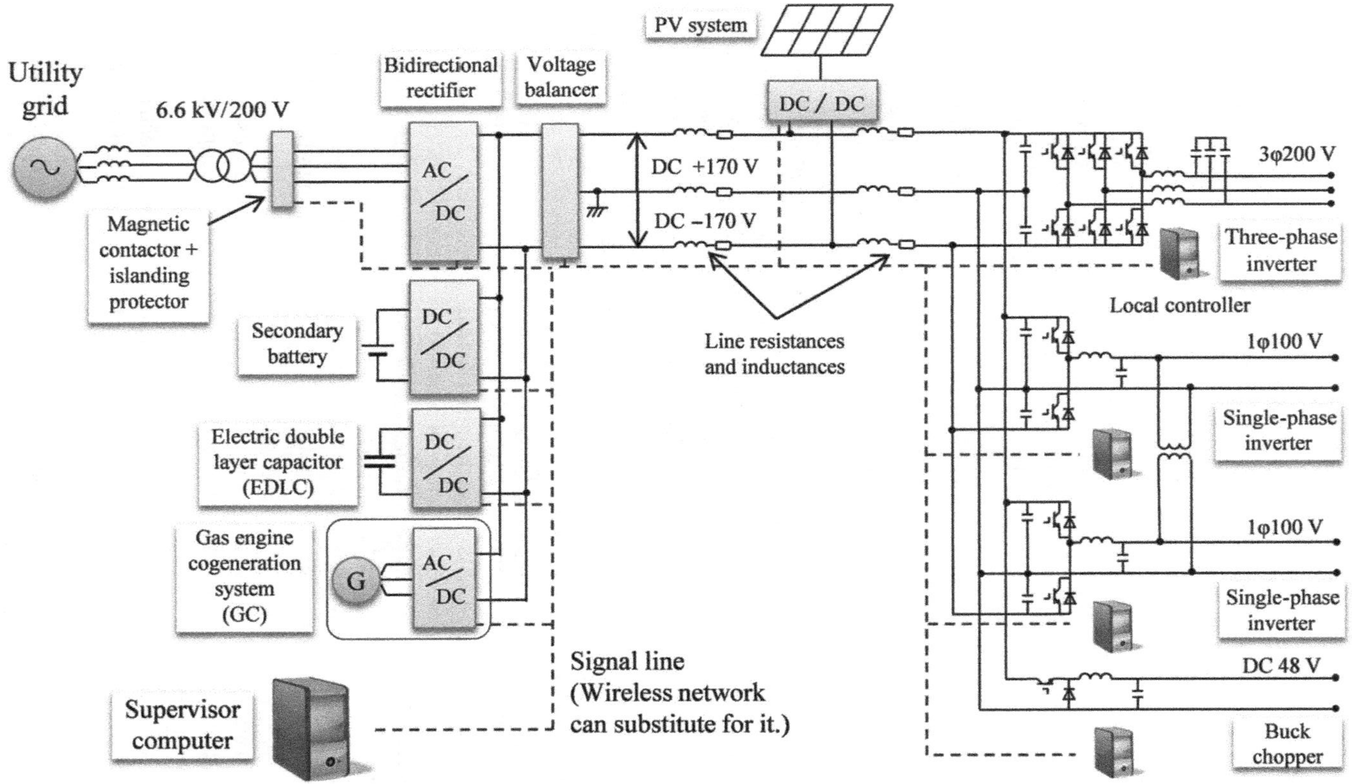

Figure 6.5 A low-voltage bipolar-type DC MG.

6.4 Stability Evaluation

It is well known that negative incremental impedance of converters can cause stability problems in DC power supply system [21, 26, 27] because those converters are operated like constant power load. Figure 6.6 shows a *V–I* curve of constant power load. The negative incremental impedance r' can cause a voltage oscillation on the distribution line. Figure 6.7 shows a simplified DC distribution circuit. If the source voltage is changed instantaneously, the current and voltage dynamics can be represented by the following equations:

$$\Delta v = r'\Delta i \tag{6.1}$$

$$\Delta V_s = R\Delta i_L + L\frac{d\Delta i_L}{dt} + \Delta v \tag{6.2}$$

$$\Delta i_L = C\frac{d\Delta v}{dt} + \Delta i \tag{6.3}$$

where ΔV_s, ΔI_L, and Δv are the instantaneous variations of source voltage, inductor current, and load voltage, respectively.

From (6.1)–(6.3), the stability condition can be calculated by the Hurwitz criterion. For example, when the parameters are set to be $L = 0.1$ (mH), $C = 10$ (mF), and $V_S = 340$ $(= 170 + 170)$ (V), the stability condition can be drawn as shown in Fig. 6.8. The stable condition is in the shaded area. There are two unstable areas. The upper one is an area where the power cannot be supplied to the load because of the voltage drop of resistance R. The lower one is an area where the resonance is caused by inductance L and the

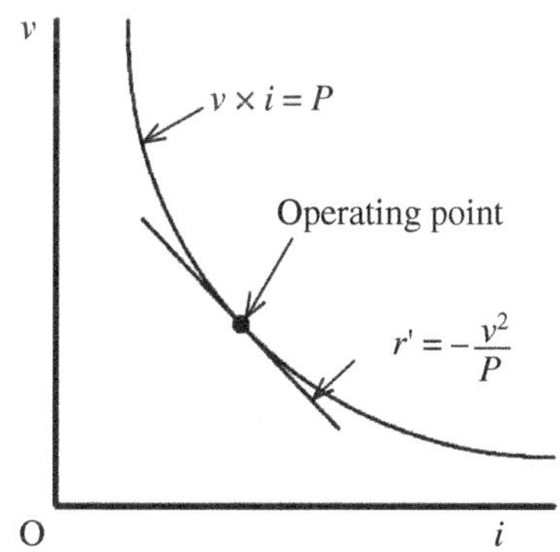

Figure 6.6 *V–I* curve of constant power load.

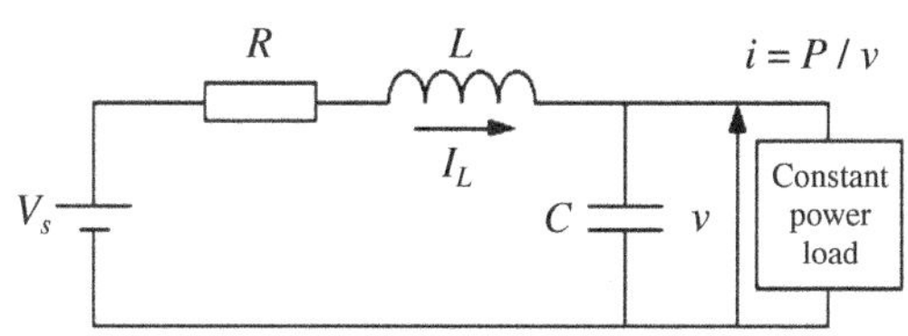

Figure 6.7 Simplified DC distribution circuit.

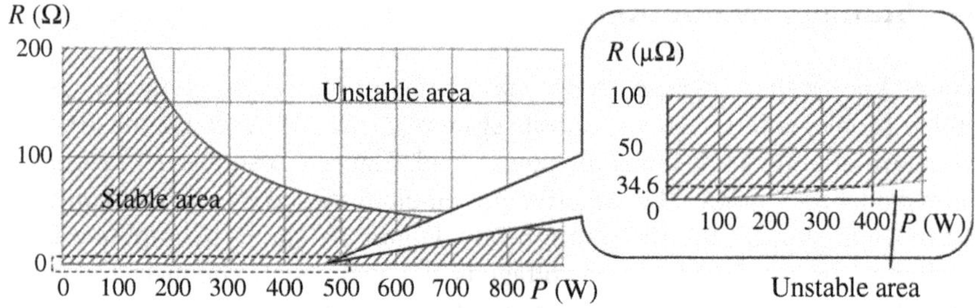

Figure 6.8 Stability condition of the simplified circuit.

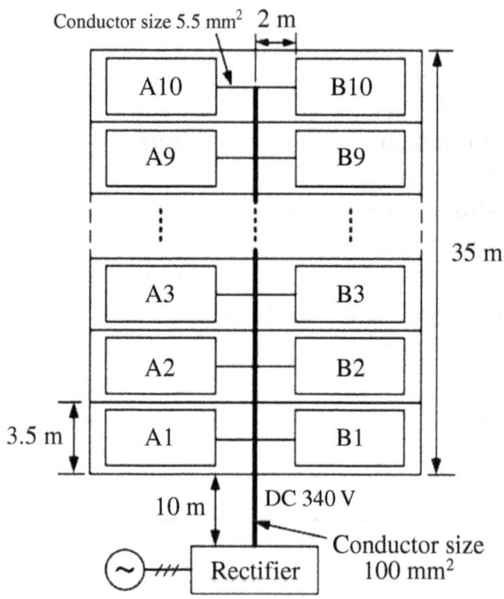

Figure 6.9 Target model in this study.

smoothing capacitor C. Oscillation in an upper unstable area does not happen in a practical system because proper conductor will be selected based on the system capacity. On the other hand, there is a possibility that the oscillation in a lower unstable region happens when the resistance R or the capacitance C is small or the inductance L is large. Here, the stability of the proposed DC MG is examined by computer simulations (MATLAB/Simulink).

Figure 6.9 shows the target model, an apartment building of 10 floors with 2 houses per floor. The rectifier is controlled to keep the distribution voltage constant (DC 340 V). Figure 6.10 shows the simulation circuit. The main parameters are given in Table 6.1. The line resistances and line inductances are calculated based on the impedance of unit length (1 km) of the CV cable. Each house is assumed to be a constant power load, and includes inverter and CGS. Figure 6.11 shows the circuit and control block of the rectifier. The rectifier is

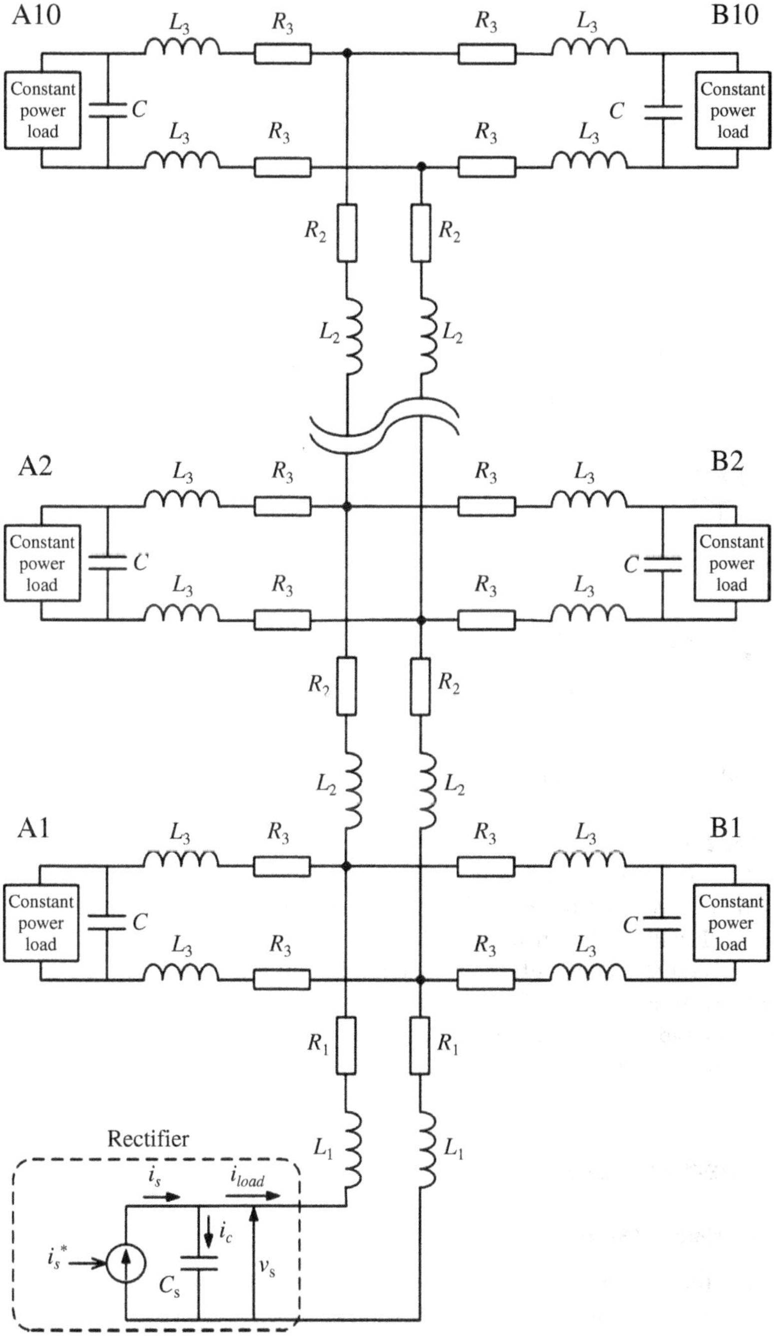

Figure 6.10 Simulation circuit.

Table 6.1 Main parameters of the simulation system.

R_1 (mΩ)	2.75	C_s (mF)	40
L_1 (μH)	3.3	C (mF)	5
R_2 (mΩ)	0.82	V_s^* (V)	340
L_2 (μH)	0.98	K_{vs}	2
R_3 (mΩ)	8.5	T_{im}	0.002
L_3 (μH)	0.77		

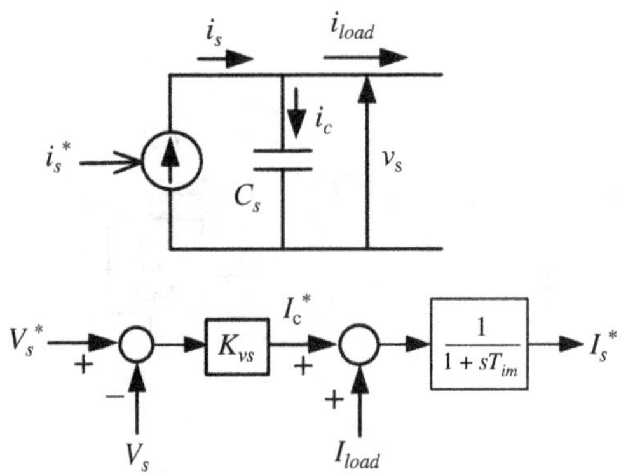

Figure 6.11 Equivalent circuit and control block of rectifier.

represented by a current source and a smoothing capacitor. The current minor loop of the control block is simulated as a first-order delay. The time constant T_{im} is set at 2 ms.

From the simulation results, voltage oscillation did not occur even when the line resistances were set to be smaller than 1/200 of the original value shown in Table 6.1. It will not be unstable in the case that line resistances become so small, because such a distribution line will not be selected from an economical point of view. In addition, it is confirmed that the oscillation does not occur under considerable situations such as the looped distribution line and power sharing among houses.

6.5 Experimental Study and Results

6.5.1 Experimental System

To examine the fundamental characteristics and the proposed system operations, a laboratory-scale experimental system is constructed. The circuit and main parameters are shown in Fig. 6.12 and Table 6.2. It is assumed that there

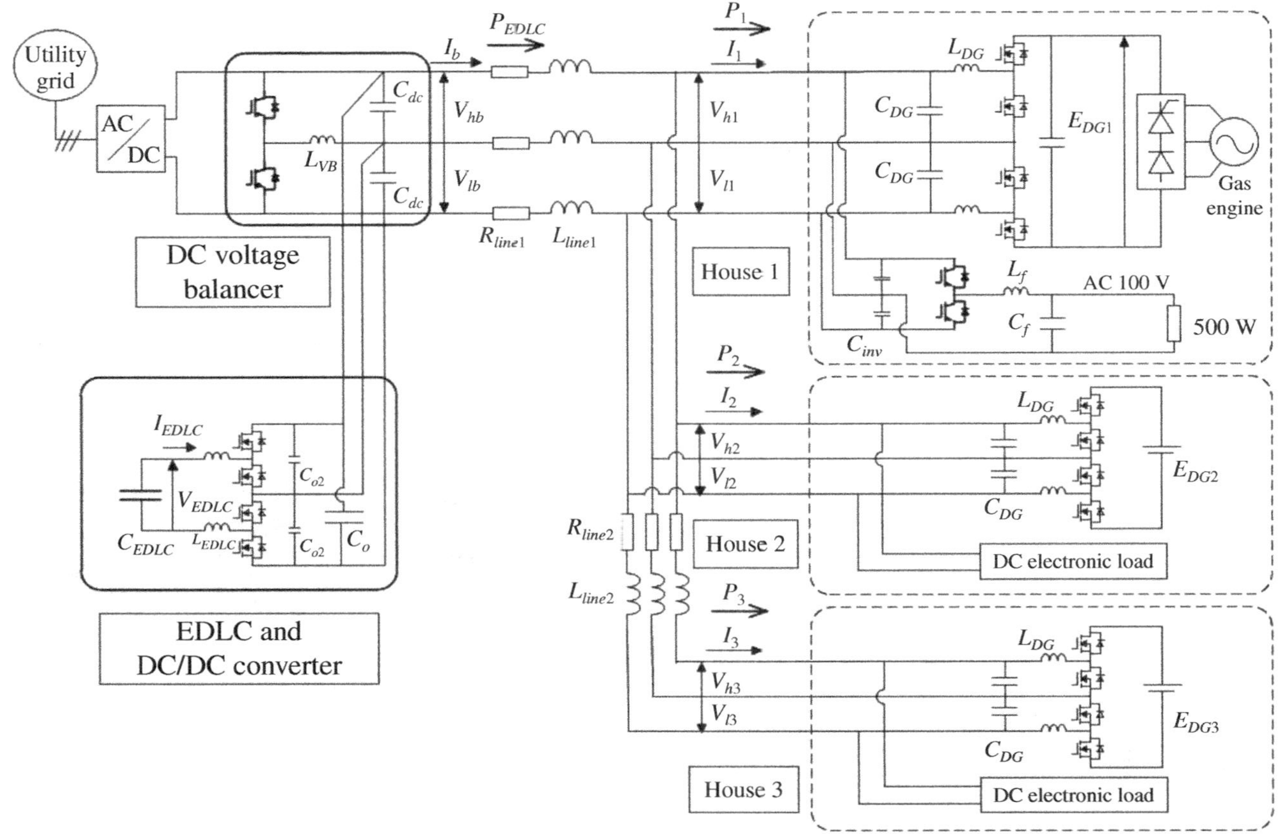

Figure 6.12 Circuit of the experimental system.

Table 6.2 Main parameters of the experimental system.

C_{DC} (μF)	9400	E_{DG1} (V)	390–400
C_o (μF)	1500	$E_{DG2,3}$ (V)	400
C_{o2} (μF)	220	C_{inv} (μF)	12,000
L_{VB} (mH)	4	C_{VB} (μF)	4700
R_{line1}	0.5 Ω	L_f	2 mH
L_{line1}	30 μH	C_f	18.3 μF
R_{line2}	1 Ω	C_{EDLC}	18 F
L_{line2}	30 μH	L_{EDLC}	7 mH
L_{DG}	5 mH	C_{DG}	220 μF

are three households, and each house has a gas engine cogeneration (GEC). Figure 6.13 shows the system, and Fig. 6.14 shows a GEC used in house 1. A commercial GEC with the rated capacity of 1 kW is used. Figure 6.15 shows the inside configuration of the GEC. The generator outputs AC 340 V, 307.5 Hz, and the AC power is converted into DC 390–400 V by a thyristor–diode rectifier. Normally, the DC power is converted into single-phase AC 200 V and flowed to

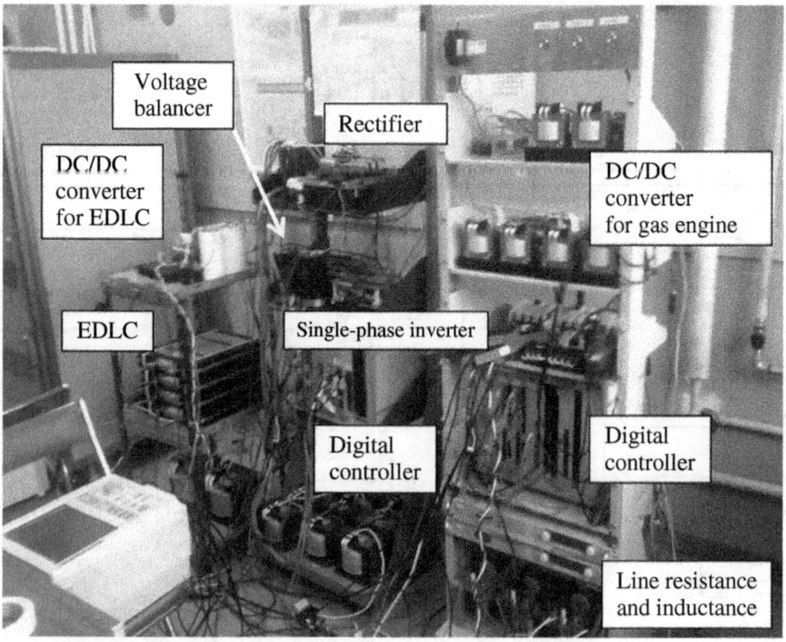

Figure 6.13 Laboratory-scale DC MG experimental system.

Figure 6.14 Gas engine unit and hot water tank.

Figure 6.15 DC power output.

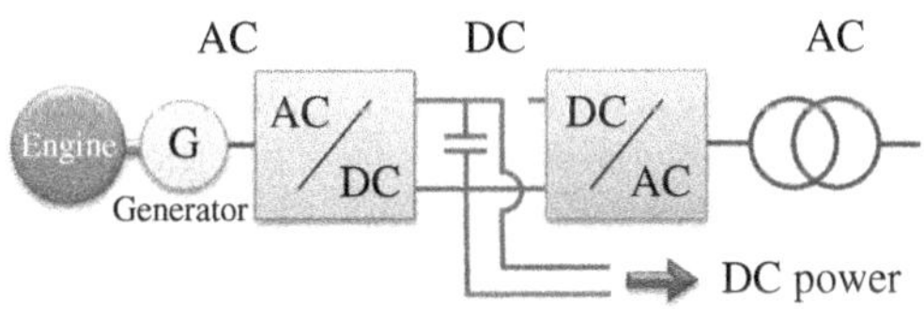

the utility grid, but here it is modified to take the DC power directly. In the other two houses, a DC power supply with constant output DC 400 V is substituted for the real GEC.

Since the voltage of the gas engine (400 V) is different from the distribution voltage (340 V), buck choppers are connected between GECs and DC distribution lines. The circuit is designed to be symmetric against the neutral line, because this converter is supposed to be worked as a voltage balancer under an appropriate control mechanism. For switching devices of this converter, concerning the converter efficiency, super junction MOS-FETs are adopted. The output power is changed gradually from 0 to 1 kW or from 1 to 0 kW for 1 s.

In house 1, DC power is converted into single-phase 100 V by a half-bridge inverter, and voltage feedback control with a current minor loop is used. In houses 2 and 3, DC power is supplied to each electronic load directly, which substituted for an inverter and loads.

The connected rectifier to the utility grid is controlled to keep the DC voltage constant (340 V) in the interconnected operation. The circuit is a conventional two-level voltage source converter, and the current control is adopted based on *dq* decoupling control. The current reference is calculated from the DC voltage reference and the feedback value. The control time constant of the voltage control is set at 15 ms. To balance positive voltage (+170 V) and negative voltage (−170 V), a voltage balancer is connected at the DC side of the rectifier as shown in Fig. 6.5.

As energy storage, EDLC is connected through DC/DC converter. The rated voltage is 160 V and the capacitance of one EDLC is 4.5 F. Four EDLCs are connected directly in parallel without any additional circuits, so the total capacitance is 18 F. Two digital signal processors – DSPs are used for this system. The sampling frequency is 10 kHz. One DSP is used for rectifier, DC voltage balancer, single-phase inverter, and converter of EDLC. Another DSP is employed for three DC/DC converters of GEC system. Those DSPs can communicate with each other.

6.5.2 Voltage Sag of the Utility Grid

The experimental results of the voltage sag in the utility grid are shown in Fig. 6.16. The voltage sag is simulated by a multipurpose power supply and is programmed to decrease 50% for 0.5 s. When the voltage sag occurred, the DC voltage is constantly controlled by the rectifier, and the current on the AC side of the rectifier is increased to keep the power from the utility grid. As a result, the DC voltage fluctuation is almost negligible and the power supplied to all loads. Voltage drops appeared, but the line resistances and inductances did not affect the system operation.

It is assumed that the distance between the rectifier and houses 1 and 2 is 100 m, and single conductor cable (5.5 mm^2) is used as the DC distribution line. To simulate the influence, resistors (0.5 Ω) and reactors (30 μH) have been placed. It is also assumed that single conductor cable (2.5 mm^2, 100 m) is used between houses 1 and 2 and house 3, and resistors (1 Ω) and reactors (30 μH) have been placed. Various experiments are carried out with the experiment system such as load variation, GEC operation, and short circuit at a load. [6]. Here, three kinds of experimental results are presented: voltage sag of the utility grid, disconnection procedure, and reconnection procedure. Table 6.3 shows the loads and GECs conditions in each experiment.

6.5.3 Disconnection/Reconnection from/to the Utility Grid

Figure 6.17 shows the experimental results of the disconnection procedure. At the initial condition, the system is in the interconnected operation. When the system detects that the voltage of the utility grid is lower than 30% of the nominal voltage, the rectifier stops. Then, the DC distribution voltage decreases, and the converter of EDLC clamps it at a lower limit (320 V). Following that, the converter control is changed from the clamp control to the DC voltage control.

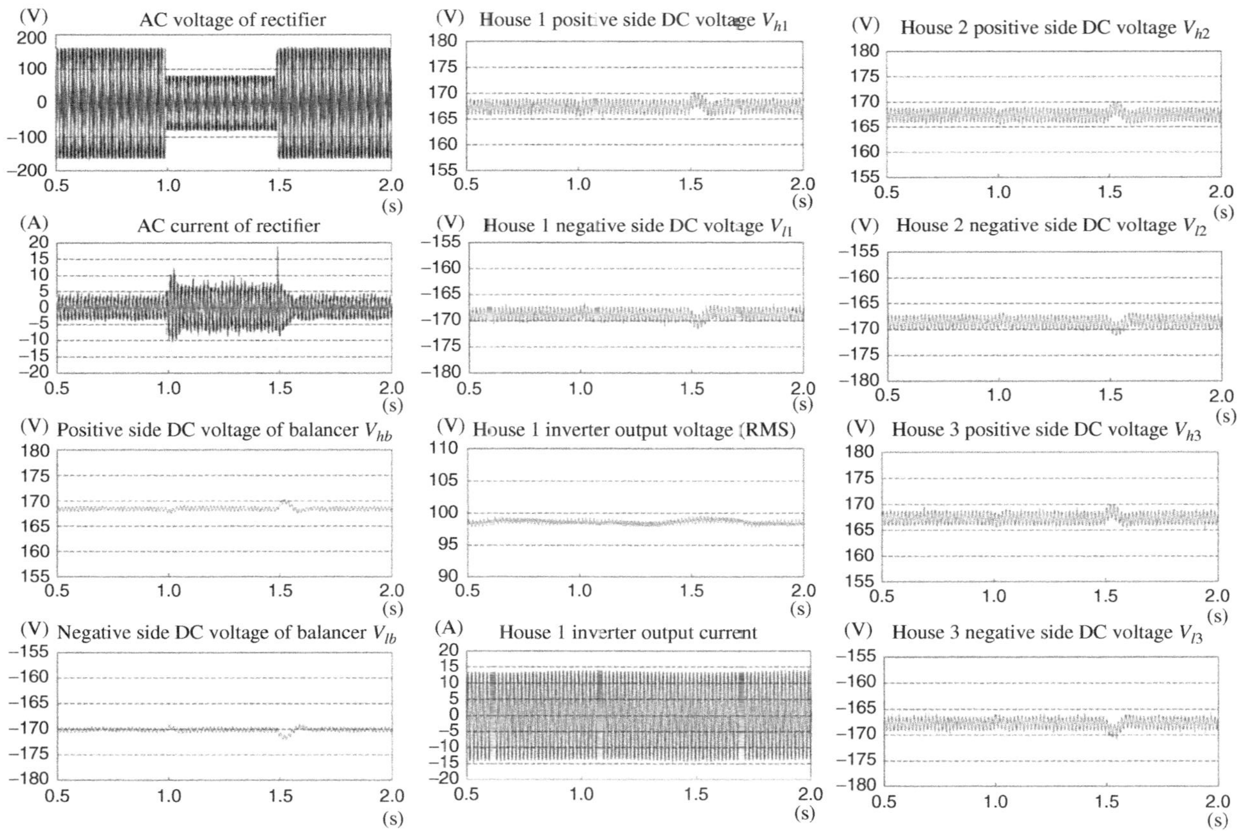

Figure 6.16 Experimental results of a voltage sag (50%, 0.5 s).

Table 6.3 Condition of each experiment.

	House 1		House 2		House 3	
	Load (W)	GEC (W)	Load (W)	GEC (W)	Load (W)	GEC (W)
A. Voltage sag	1000	1000	700	0	1000	1000
B. Disconnection	1000	0	700	0	0	0
C. Reconnection	1000	0	700	0	0	0

Finally, the voltage is gradually increased to 340 V for 1 s, and the system is in the intentional islanding operation. In this period, the RMS value of the single-phase inverter voltage, which is shown as "House 1 Inverter Output Voltage (RMS)," is not affected, and the smooth disconnection is confirmed. It is also shown that the line resistances and inductances do not affect the disconnection process.

Figure 6.18 shows the experimental results of reconnection procedure. When the system detects the utility grid is recovering from a problem, the controller of the rectifier monitors the phase of the utility grid by the PLL circuit. Then, the EDLC converter changes the control task from the DC voltage control to the clamp control. In the next step, the DC voltage decreases to the lower clamp limit, and the rectifier starts and controls the DC voltage. Finally, the voltage gradually increases to 340 V for 1 s, and the system is in the interconnected operation. In this period, the RMS value of the single-phase inverter voltage is not affected as well, and the smooth reconnection can be also seen. In addition, it is also confirmed that there are no impacts from the line resistances and inductances.

In the present work, it is shown that the disconnection and reconnection results under the conditions of turning off/on of the GEC are smooth enough.

6.6 A Voltage Control Approach

6.6.1 Case Study and Voltage Control System

The energy storage unit does not need a large capacity if suitable micro-CHP units with start up time of few minutes are chosen. For instance, one of the commercial micro gas engine CGSs can start up in about 5 min. Most of the heat in the starting-up period is utilized in heating the catalyst for the exhaust gas. Therefore, it is feasible that the gas engine CGS will be able to start in a few minutes every time, after improving the exhauster for quick start-up. In this case, the energy of the storage system is assumed to be less than 10 kWh and the charging/discharging time to be less than one half-hour, although the

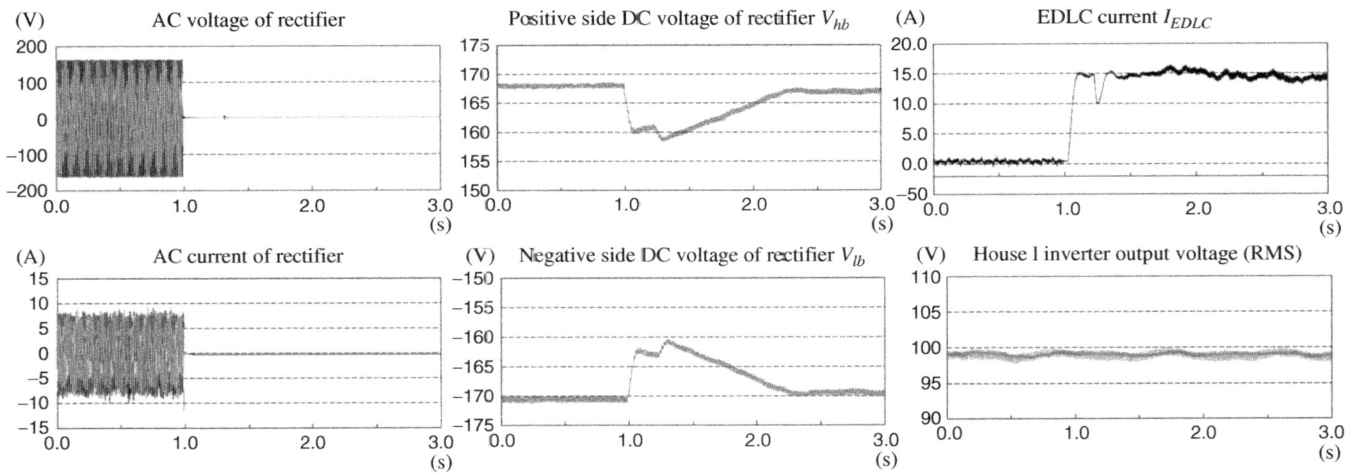

Figure 6.17 Experimental results of disconnection from the utility grid.

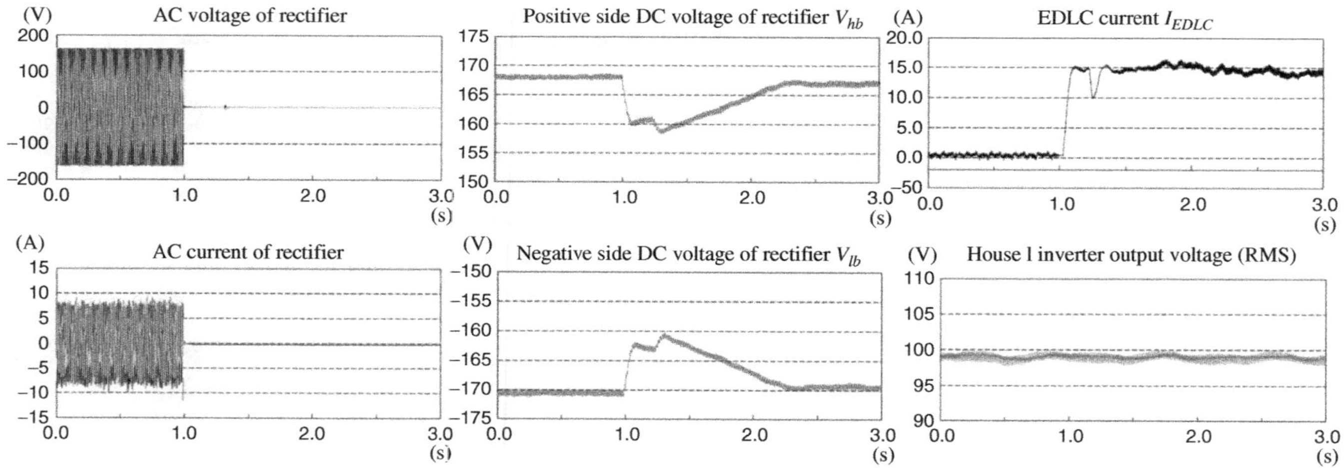

Figure 6.18 Experimental results of reconnection from the utility grid.

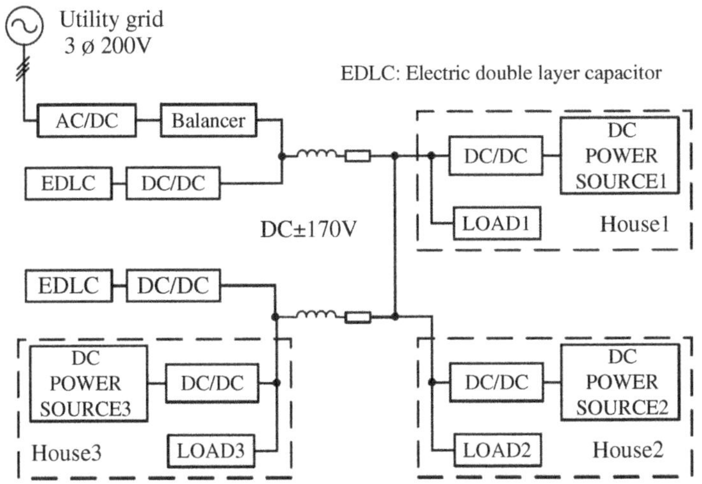

Figure 6.19 Configuration of the case study in the laboratory.

parameters depend on the specification of each system in practice. Therefore, EDLC can be a candidate of the main energy storage unit in this system [28].

Figure 6.19 shows an updated configuration of the experimental laboratory system for the given DC MG in Fig. 6.1. It is assumed that there are three households, and each house has a DC power source such as a fuel cell. The feature of the DC system is to adopt a three-wire DC distribution, which consists of a +170 V line, a neutral line, and a −170 V line. In a three-wire composition, the voltage relative to ground becomes low compared with that in a two-wire system, where one of the wires is grounded, and when half bridge inverters are adopted for it, one of the single-phase 100-V output lines can also become a grounded neutral line as long as abiding to the available Japanese standard [23]. In addition, load-side DC/DC converters can choose the source voltage from +340, +170, or −170 V. A rectifier connected to the utility grid is controlled to maintain the DC voltage constant (340 V) in the interconnected operation. To balance the positive voltage (+170 V) with a negative voltage (−170 V), a voltage balancer is connected at the DC side of the rectifier. Two EDLC banks are connected through a DC/DC converter as energy storage system.

As mentioned in the previous section, the DC distribution voltage is normally controlled by a grid-connected rectifier in the interconnected operation. The DC/DC converters of the storage systems are controlled to maintain the DC MG voltage within a specified range [23]. In intentional islanding operation, the DC/DC converters of the storage systems need to maintain the DC distribution voltage. Therefore, if the DC MG has two or more energy storage units and those converters can be operated in parallel, it contributes to the voltage regulation and system redundancy. Droop control is a well-known method for voltage control when two or more converters are used. In general, the droop controller detects the output power or current as a feedback parameter, and the deviation of DC voltage is controlled in proportion to the output power. However, if the converters are connected to energy storage units, the controller should consider not only the output power balance but also the stored energy. In particular, stored energy balance is important to carry out the operation described in the previous section because micro-CHP units have to start or stop frequently under an unbalanced condition of the stored energy. Therefore, a simple control method combining gain-scheduling and droop control accomplishes voltage regulation, load sharing, and energy balance simultaneously.

6.6.2 Energy Storage System Control

Figure 6.20 shows the circuit and proposed control diagram of a DC/DC converter for EDLC banks. We designed the circuit to be symmetric with respect to the neutral line because the converter is supposed to operate as a voltage balancer under the appropriate control.

6.6.2.1 Gain-scheduling Control

When DC voltage is controlled by several converters, it is difficult to achieve a good voltage regulation and sharing simultaneously. As illustrated in Fig. 6.21,

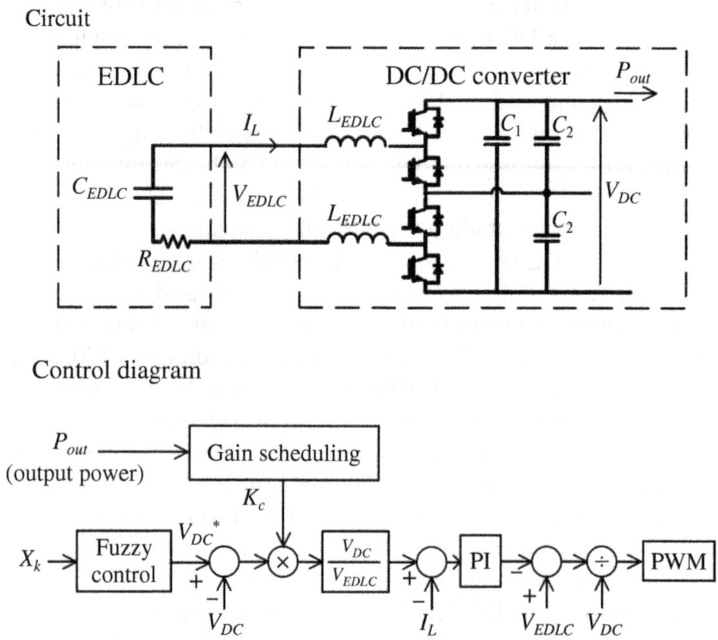

Figure 6.20 Control for DC/DC converter for energy storage.

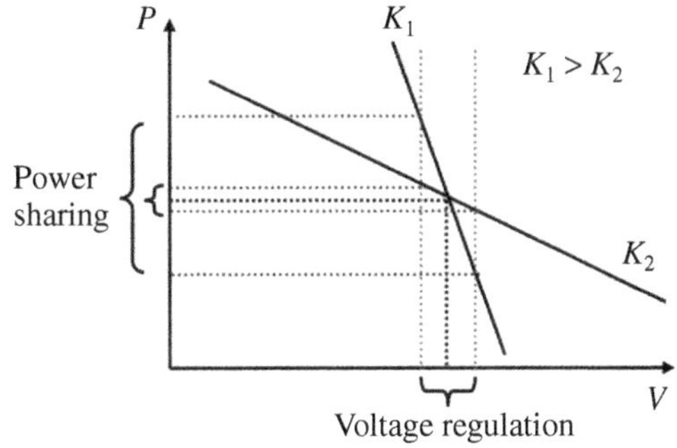

Figure 6.21 Droop control feature.

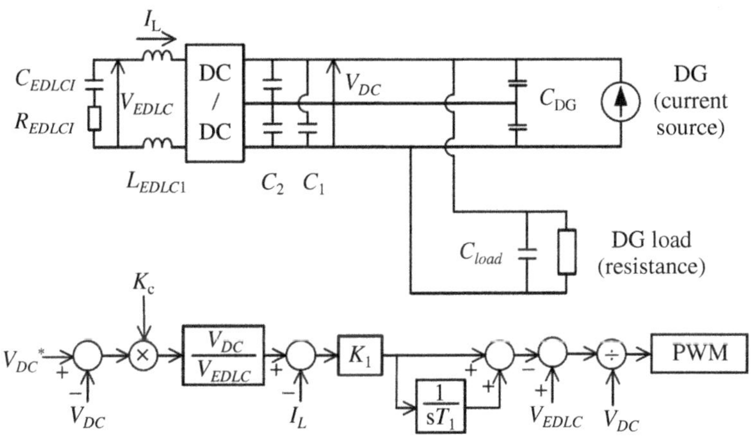

Figure 6.22 Circuit and control diagrams to obtain the relation between the steady-state error and the gain K_c.

better voltage regulation requires a higher DC gain, but it can decrease load sharing. Therefore, a gain-scheduling control is adopted to change the gain K_c according to the output power in order to obtain better voltage regulation and load sharing simultaneously [29].

The steady-state DC voltage error is examined when the gain K_c or the output power is changed by numerical simulations. Figure 6.22 and Table 6.4 show the circuit, control diagram, and parameters for the simulation. The rated output voltage and capacity were 340 V and 3 kW, respectively.

Results indicate that the steady-state error of the DC voltage becomes larger when the load is heavier or the gain K_c is smaller (Fig. 6.23). When the variation in the DC voltage is permitted within 2%, we can obtain the relation between the gain K_c and output power (pu) as shown in Fig. 6.23. The gain K_c can be expressed by a linear function of the output power. Similarly, we can obtain the relation between the gain and input power when the voltage variation is

Table 6.4 Parameters.

C_{EDLC1} (F)	18	C_1 (μF)	1500
R_{EDLC1} (Ω)	0.16	C_2 (μF)	220
L_{EDLC1} (mH)	7	C_{DG} (μF)	220
K_1	14	C_{load} (μF)	1500

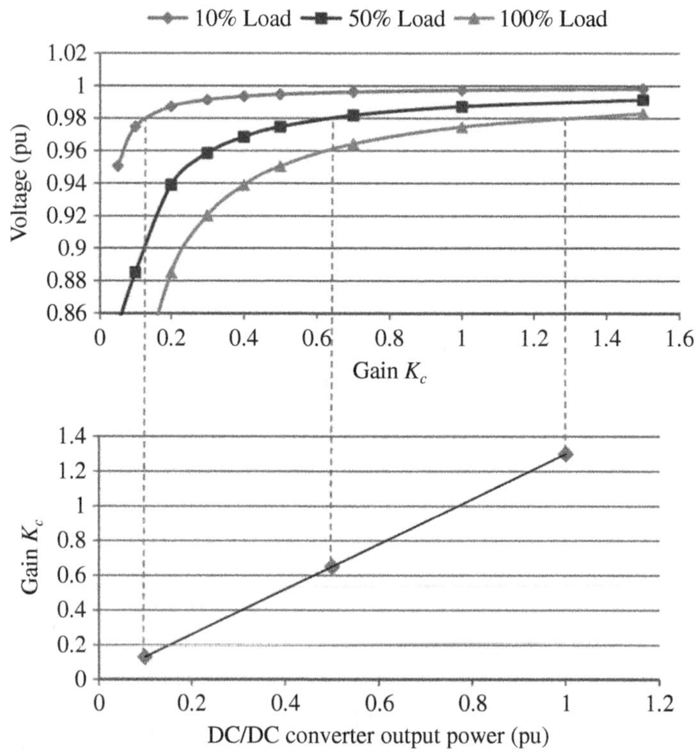

Figure 6.23 Voltage–gain and gain–output power characteristics of the DC/DC converter (voltage variation 2%).

permitted within 2%. Figure 6.24 shows the results, expressing it with a linear function. From the results, the equation between the output power (pu) P_{out} and the gain K_c can be obtained as follows:

$$K_c = \max(1.3|P_{out}|,\ 0.1) \tag{6.4}$$

where the input power is expressed by a negative number, and the minimum value of K_c is determined as 0.1. This gain-scheduling control allows a better load sharing and voltage regulation simultaneously. However, a stored energy

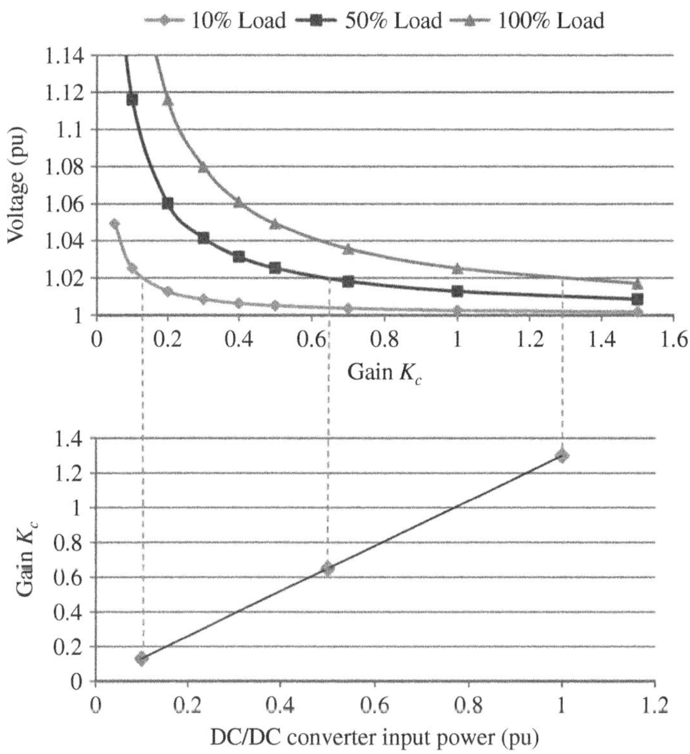

Figure 6.24 Voltage–gain and gain–input power characteristics of the DC/DC converter (voltage variation 2%).

control is also needed for the converter energy storage unit device to prevent a surplus or deficit of the stored energy.

6.6.2.2 Droop Control

A DC voltage control is proposed, which uses a droop control to change the DC voltage reference for balancing the stored energy. Figure 6.25 shows the

Figure 6.25 Droop control to obtain the voltage reference.

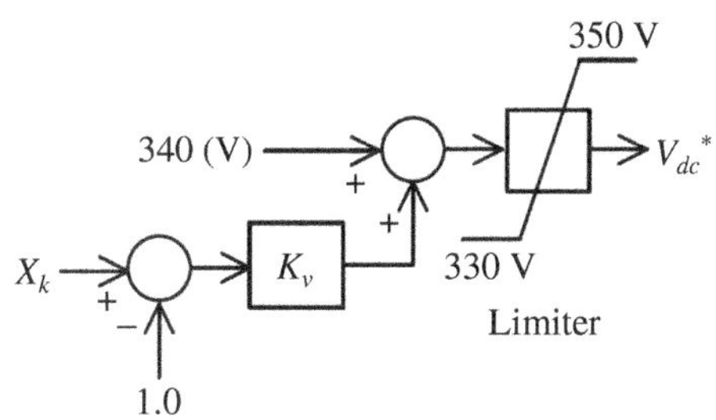

structure of the proposed droop control. The input X_k is the ratio of the stored energy and the average of all stored energies as shown below:

$$X_k = W_k \bigg/ \left\{ \frac{1}{n-1} \times \left(\sum_{i=1}^{n} W_i - W_k \right) \right\} \tag{6.5}$$

where W_k is the charged energy of the kth EDLC bank, which is described as follows:

$$\begin{aligned} W_k &= \left(\frac{1}{2}C_k {v_k}^2 - \frac{1}{2}C_k {V_{min}}^2\right) \bigg/ \left(\frac{1}{2}C_k {V_{max}}^2 - \frac{1}{2}C_k {V_{min}}^2\right) \\ &= \left({v_k}^2 - {V_{min}}^2\right) / \left({V_{max}}^2 - {V_{min}}^2\right) \end{aligned} \tag{6.6}$$

where n is the total number of EDLC banks. C_k and v_k are the capacitance and voltage of the kth EDLC bank, respectively. V_{max} and V_{min} are the maximum and minimum operation voltage of EDLC banks, respectively. To obtain the DC voltage reference V_{DC}^*, X_k is initially calculated from (6.5) and (6.6).

It is assumed that the DC distribution voltage has a tolerance of ±5% with reference to AC systems. In this case, the upper and lower references should be within 3% because the gain-scheduling control includes a tolerance of ±2%. Therefore, the upper and lower limits are selected as 350 and 330 V.

6.7 Simulation Results

To confirm the performance of the proposed method, some simulations are conducted in the PSCAD/EMTDC environment. The circuit and the main parameters are shown in Fig. 6.26 and Table 6.5, respectively. The configuration and the parameters are based on the experimental system shown in Fig. 6.19. Resistances and current sources are used as the loads and the DGs, respectively. The resistances are set to 115.6 Ω and consume 1 kW at 340 V. The output power of the current source is changed gradually from 0 to 1 kW or from 1 to 0 kW during 1 s.

Two EDLCs are connected through DC/DC converters: EDLC 1 (rated voltage, 160 V, and rated capacitance, 18 F) and EDLC 2 (rated voltage, 216 V, and rated capacitance, 18.75 F). The minimum voltage of both EDLCs is set to be 100 V. Therefore, EDLC 1 and EDLC 2 can store about 140 and 344 kJ, respectively.

Figure 6.27 shows the on/off events of the loads and DGs in the simulation. All loads and DGs are turned on in the first half, and the loads and DGs are turned off in the last half. The initial voltages of EDLC 1 and EDLC 2 are 125 and 200 V, respectively. Therefore, the initial stored energy ratio (W_2/W_1) is about 2.

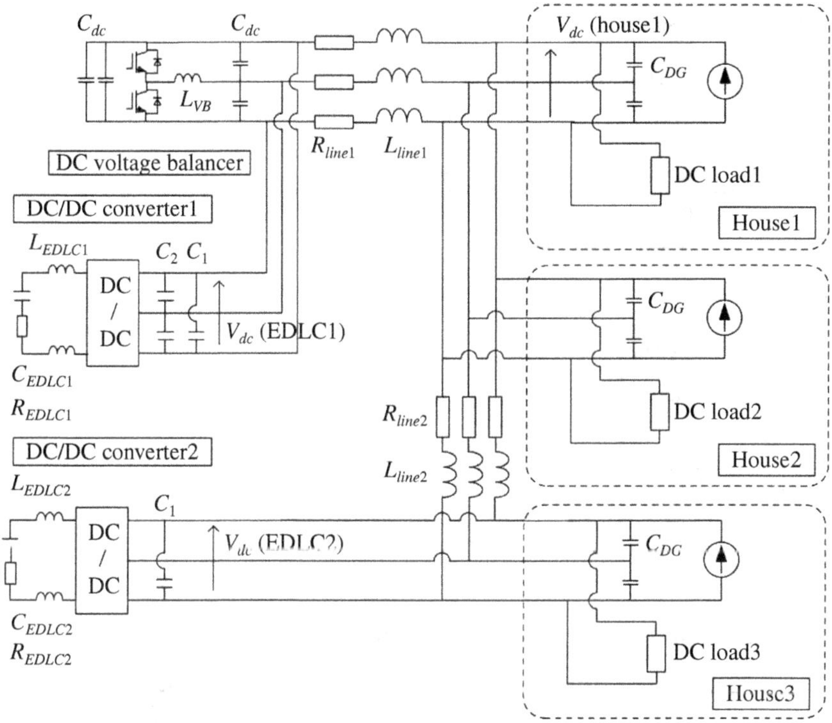

Figure 6.26 Simulation circuit.

Table 6.5 Main parameters.

C_{EDLC1} (F)	18	C_{EDLC2} (F)	18.75
R_{EDLC1} (Ω)	0.16	R_{EDLC2} (Ω)	0.11
L_{EDLC1} (mH)	7	L_{EDLC2} (mH)	5
R_{line1} (Ω)	0.5	R_{line2} (Ω)	1.0
L_{line1} (μH)	30	L_{line2} (μH)	30
C_{DC} (μF)	4700	C_{DG} (μF)	220
C_1 (μF)	1500	C_2 (μF)	220
L_{VB} (mH)	5		

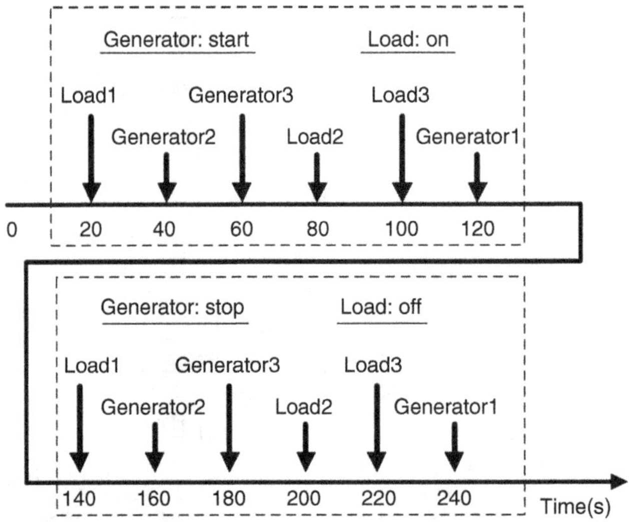

Figure 6.27 Events pattern for the performed simulations (initial condition $W_2/W_1 \approx 2$).

6.7.1 Simulation Results for the Gain-scheduling Control

Figure 6.28 shows the simulation results (the DC distribution voltage, the stored energy ratio, the current from EDLC, and the terminal voltage of EDLC) when only the gain-scheduling control technique is used. The maximum voltage of V_{DC} (EDLC1) and V_{DC} (EDLC2) is 346.7 V, which is about 340 V+2%, and the minimum voltage is 332.9 V, which is about 340 V−2%. Therefore, the distribution voltages at the output converter of the EDLCs are within 340 V±2%. However, the stored energy of EDLCs is not balanced, and the maximum of the ratio (W_2/W_1) reaches 4.2.

6.7.2 Simulation Results for Droop Control

To demonstrate the role of the proposed droop control (shown in Fig. 6.25), some simulations are conducted in the same condition. Figure 6.29 shows the simulation results when the droop control is used, and K_v is set to 10. The stored energy ratio was about 1. Figure 6.30 shows the simulation results when K_v is set to 50. The stored energy ratio was also 1, for about 80 s, which is almost half of the time of the previous test (Fig. 6.29). In this case, the maximum and minimum voltages of the output converters of the EDLCs are 349.9 and 331.6 V, respectively.

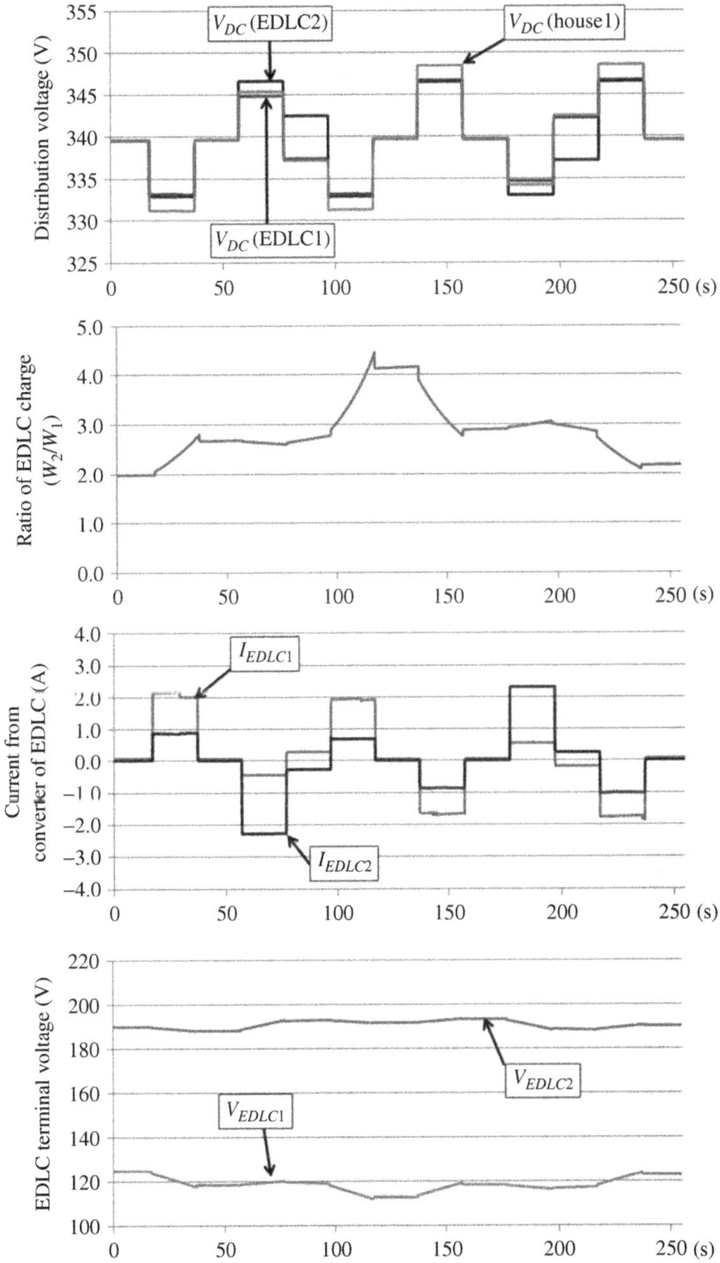

Figure 6.28 Simulation results for gain-scheduling control only (initial condition $W_2/W_1 \approx 2$).

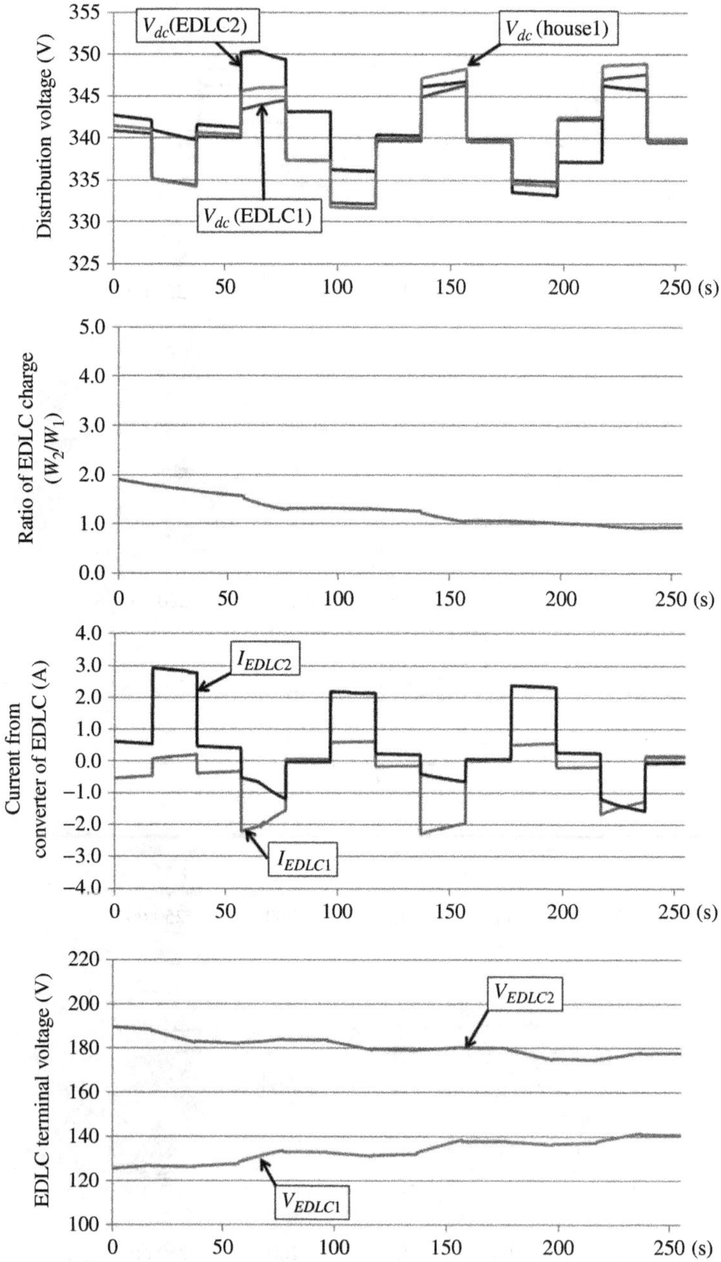

Figure 6.29 Simulation results for gain-scheduling control and droop control with $K_v = 10$ (initial condition $W_2/W_1 \approx 2$).

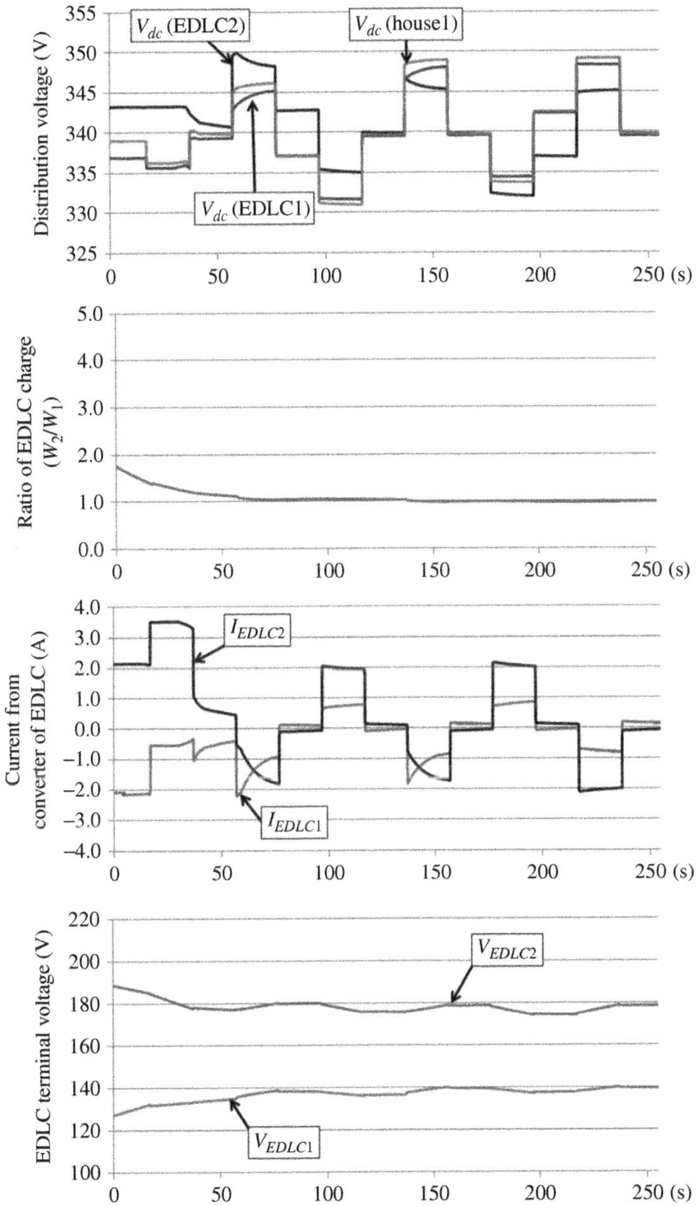

Figure 6.30 Simulation results for gain-scheduling control and droop control with $K_v = 50$ (initial condition $W_2/W_1 \approx 2$).

6.8 Experimental Results

Two experiments were conducted to confirm the performance of the proposed method in practice. The circuit and the main parameters are shown in Fig. 6.31 and Table 6.6, respectively. The system configuration and parameters are almost the same as that of the simulation circuit. In house 3, a commercial gas engine cogeneration (GEC) is used, and the rated capacity is 1 kW. The

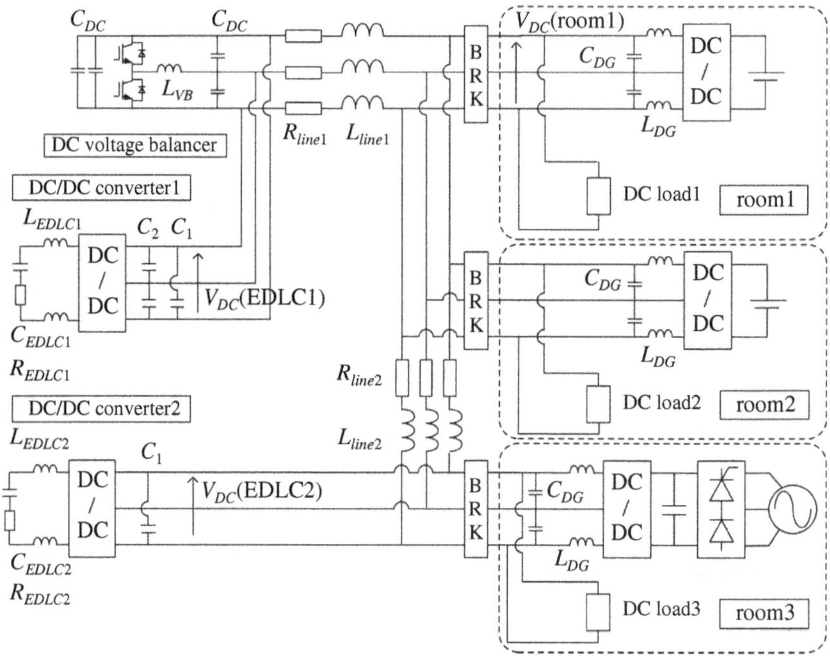

Figure 6.31 Circuit of experimental system.

Table 6.6 Main experiment parameters.

C_{EDLC1} (F)	18	C_{EDLC2} (F)	18.75
R_{EDLC1} (Ω)	0.16	R_{EDLC2} (Ω)	0.11
L_{EDLC1} (mH)	7	L_{EDLC2} (mH)	5
R_{line1} (Ω)	0.5	R_{line2} (Ω)	1.0
L_{line1} (μH)	30	L_{line2} (μH)	30
C_{DC} (μF)	4700	C_{DG} (μF)	220
C_1 (μF)	1500	C_2 (μF)	220
L_{VB} (mH)	5	L_{DG} (mH)	5

generator outputs AC 340 V, 307.5 Hz, and the AC power is converted into a DC 390–400 V by a thyristor–diode rectifier. In general, the DC power is converted to a single-phase AC 200 V and is supplied to the utility grid, but it is modified to accept the DC power directly. In the other two houses, the DC power supplies with 400 V DC constant outputs are used instead of the real GEC.

Buck choppers are connected between GECs and the DC distribution lines because the DC voltage of the gas engine is different from the distribution voltage. The output power is changed gradually from 0 to 1 kW or from 1 to 0 kW for 1 s. The ratings of two EDLCs are as follows: EDLC1 (rated voltage, 160 V, and rated capacitance, 18 F) and EDLC2 (rated voltage, 216 V, and rated capacitance, 18.75 F). The minimum voltage of both EDLCs is also set to be 100 V. For the DC loads, three electronic loads are used under constant resistance mode with a rated capacity of 1 kW for each.

6.8.1 Case I

The on/off events of the loads and DGs in case I are the same as the previous simulation shown in Fig. 6.27. The initial voltages of EDLC 1 and EDLC 2 are 125 and 200 V, respectively. Then, the initial stored energy ratio (W_2/W_1) is around 2. Figure 6.32 shows the experimental results when only the gain-scheduling control technique is used. The DC distribution voltage is controlled within 340 V ± 5%, but the stored energy of EDLCs is not balanced and the peak ratio is around 3.5. The energies of both EDLCs are balanced at around 100 s.

The experimental result is almost the same as the simulation result. However, the distribution voltage V_{DC} (EDLC 2) and the current I_{EDLC2} in Fig. 6.32 are higher than the simulation results in Fig. 6.28 during the first 120 s. It is conjectured that the feedback value of V_{DC} (EDLC 2) is a little lower than the real values in the experiment, which is confirmed by numerical simulations.

6.8.2 Case II

In this experiment, load 1 and the output of DG 3 are turned on from 0 to 1 kW at 5 s. Other loads and DGs are turned off through the experiment. The initial voltages of EDLC 1 and EDLC 2 are 130 and 180 V, respectively. Therefore, the initial stored energy ratio (W_2/W_1) is around 1.3. Figure 6.33 shows the experimental result when only the gain-scheduling control technique is used.

If there is no line resistance between house 1 and house 3, the distribution voltage V_{DC} (house 1) is equivalent to V_{DC} (EDLC 2) after the event. In this case, the output power of DG 3 is supplied to load 1 directly, and both EDLCs are not outputted their power ideally. However, there are line resistances R_2 ($1\ \Omega \times 2$) between house 1 and house 3 in the experimental system, and the distribution voltage V_{DC} (house 1) and V_{DC} (EDLC 1) are lower than V_{DC} (EDLC 2). To keep

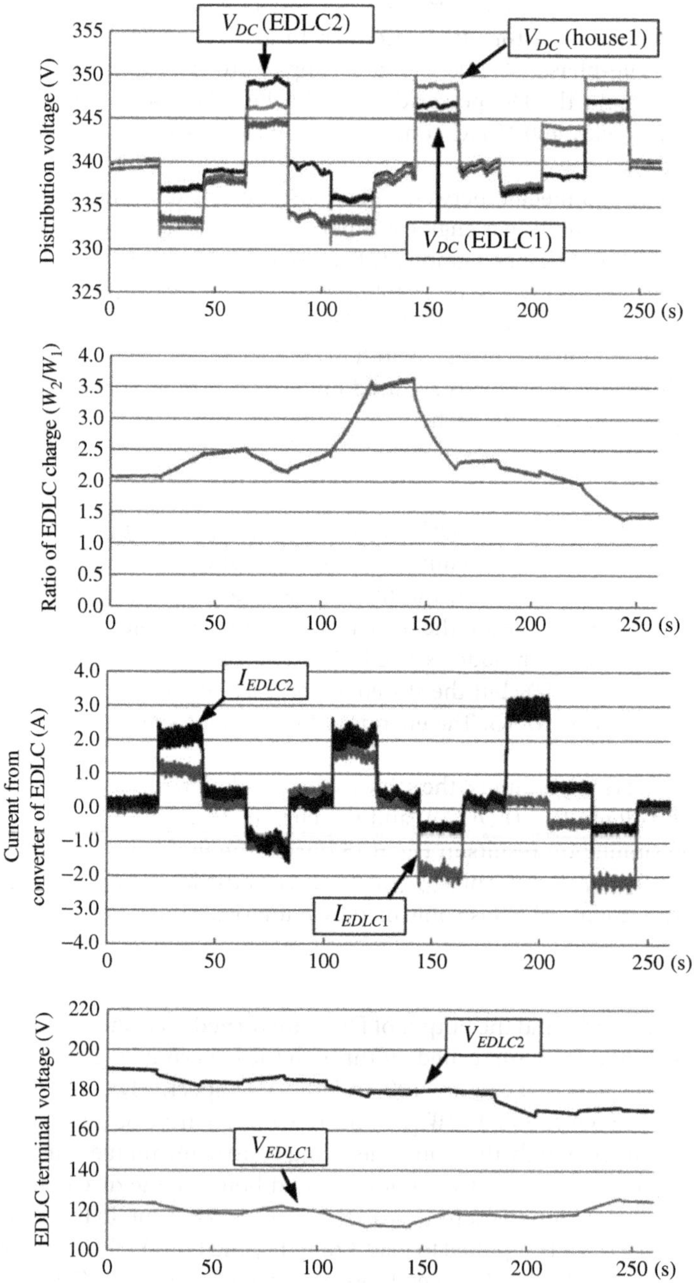

Figure 6.32 Experimental results of case I (gain-scheduling control only).

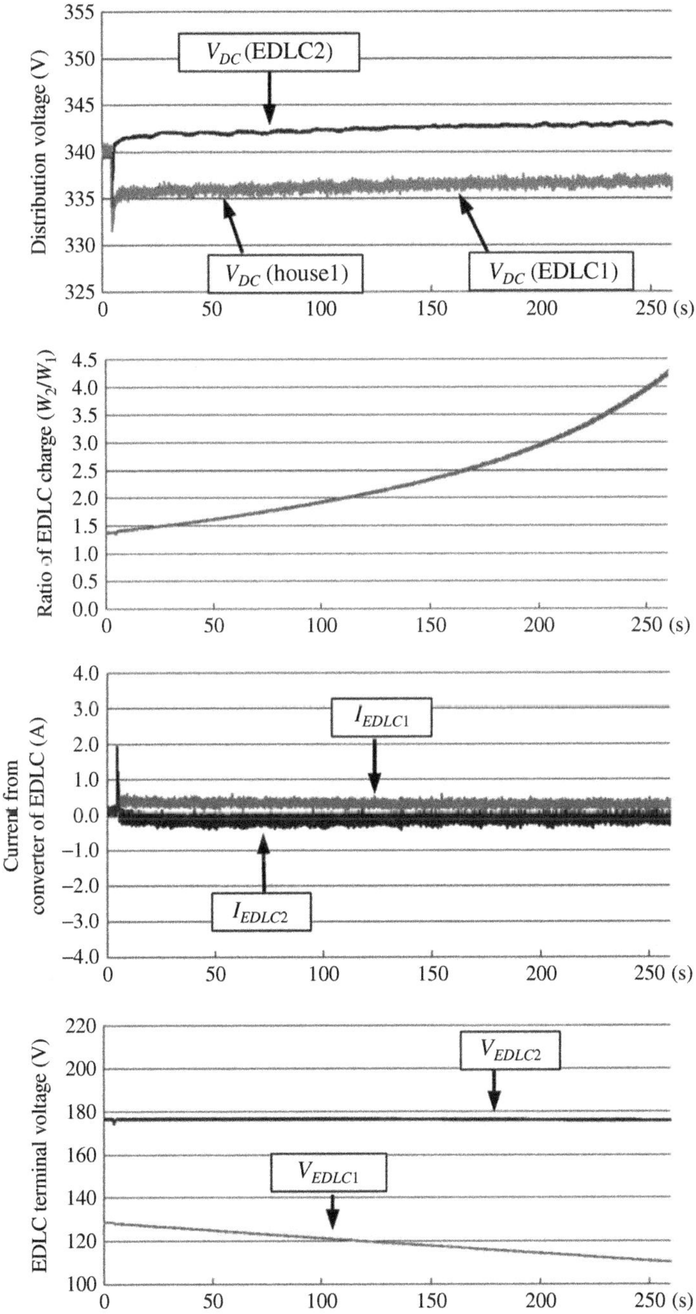

Figure 6.33 Experimental results of case II (gain-scheduling control only).

the distribution voltage, EDLC 1 supply load 1, and EDLC 2 is charged from DG3. Therefore, the stored energy increases.

6.9 Summary

This chapter addresses the DC MG structure and operation. To show their characteristics and for a smooth introduction of numerous DGs to achieve high efficiency and high-quality power supply, a low-voltage bipolar-type DC MG is developed. A DC MG system for residential area where each house has a CGS, and the power is shared among the houses by DC distribution. To confirm the fundamental characteristics and operation, a laboratory-scale experimental system is constructed. Several test scenarios are carried out such as a sudden load variation, short circuit at a load, interconnected operation, intentional islanding operation, and supplying commercial home appliances. The results of voltage sag in the utility grid, disconnection and reconnection procedures are presented. The results demonstrate the system could supply high-quality power to the loads in those conditions.

Then, a DC distribution voltage control system for the DC/DC converters with an energy storage unit is designed. The proposed control combines a gain-scheduling technique with droop control. The experimental results indicate that DC voltage regulation and stored energy balancing control are performed simultaneously.

References

1 Hatziargyriou, N. (2008) Microgrids, the key to unlock distributed energy resources? *IEEE Power and Energy Magazine*, **6** (3), 26–29.

2 Barnes, M., Ventakaramanan, G., Kondoh, J., Lasseter, R., Asano, H., Hatziargyriou, N., Oyarzabal, J., and Green, T. (2007) *Real-World Micro-Grids – An Overview*. IEEE International Conference on System of Systems Engineering (SoSE'07), pp. 1–8.

3 Nikkhajoei, H. and Lasseter, R.H. (2009) Distributed generation interface to the CERTS microgrid. *IEEE Transactions on Power Delivery*, **24** (3), 1598–1608.

4 Vasquez, J.C., Guerrero, J.M., Luna, A. *et al.* (2009) Adaptive droop control applied to voltage-source inverters operating in grid-connected and islanded modes. *IEEE Transactions on Industrial Electronics*, **56** (10), 4088–4096.

5 Barklund, E., Pogaku, N., Prodanovic, M. *et al.* (2008) Energy management in autonomous microgrid using stability-constrained droop control of inverters. *IEEE Transactions on Power Electronics*, **23** (5), 2346–2352.

6 Abdel-Rady, Y., Mohamed, I., and El-Saadany, E.F. (2008) Adaptive decentralized droop controller to preserve power sharing stability of paralleled inverters in distributed generation microgrids. *IEEE Transactions on Power Electronics*, **23** (6), 2806–2816.

7 Li, Y. and Kao, C. (2009) An accurate power control strategy for power-electronics-interfaced distributed generation units operating in a low-voltage multibus microgrid. *IEEE Transactions on Power Electronics*, **24** (12), 2977–2987.

8 Lawrence Berkeley National Laboratory, Microgrid Symposium Website, available at: http://der.lbl.gov/microgrid-symposiums.

9 Morozumi, S. (2007) *Micro-Grid Demonstration Projects in Japan*. The Fourth Power Conversion Conference, Japan, pp. 635–642.

10 Yamato, N., Fukui, A., and Hirose, K. (2007) *Effect of Breaking High Voltage Direct Current (HVDC) Circuit on Demonstrative Project on Power Supply Systems by Service Level in Sendai*. 29th International Telecommunications Energy Conference (INTELEC), pp. 46–51.

11 Ito, Y., Zhongqing, Y., and Akagi, H. (2004) *DC Microgrid Based Distribution Power Generation System*. IEEE The 4th International Power Electronics and Motion Control Conference (IPEMC), vol. 3, pp. 1740–1745.

12 Salomonsson, D., Soder, L., and Sannino, A. (2008) An adaptive control system for a DC microgrid for data centers. *IEEE Transactions on Industry Applications*, **44** (6), 1910–1917.

13 Guerrero, J.M., Vásquez, J.C., and Teodorescu,R. (2009) *Hierarchical Control of Droop-Controlled DC and AC Microgrids – A General Approach Towards Standardization*. IEEE Conference of Industrial Electronics (IECON), pp. 4305 4310.

14 Yukita, K., Shimizu, Y., Goto, Y., Yoda, M., Ueda, A., Ichiyanagi, K., Hirose, K., Takeda, T., Ota, T., Okui, Y., and Takabayashi, H. (2010) *Study of AC/DC Power Supply System with DGs using Parallel Processing Method*. The 2010 International Power Electronics Conference – ECCE Asia – (IPEC-Sapporo), 22A2-3, pp. 722–725.

15 Kakigano, H., Nomura, M., and Ise, T. (2010) *Loss Evaluation of DC Distribution for Residential Houses Compared with AC System*. The 2010 International Power Electronics Conference – ECCE Asia – (IPEC-Sapporo), 22A1-3, pp. 480–486.

16 Kakigano, H., Miura, Y., Ise, T., and Uchida, R. (2006) *DC Micro-grid for Super High Quality Distribution – System Configuration and Control of Distributed Generations and Energy Storage Devices*. 37th Annual IEEE Power Electronics Specialists Conference (PESC), Korea, pp. 3148–3154.

17 Kakigano, H., Miura, Y., Ise, T., Momose, T., and Hayakawa, H. (2008) *Fundamental Characteristics of DC Microgrid for Residential Houses with Cogeneration System in Each House. Proceedings of IEEE PESGM*, 08GM0500.

18 Kakigano, H., Miura, Y., Ise, T., and Uchida, R. (2007) *DC Voltage Control of the DC Micro-Grid for Super High Quality Distribution.* The Fourth Power Conversion Conference, Japan, pp. 518–525.

19 Saisho, M., Ise, T., and Tsuji, K. (2002) *DC Loop Type Quality Control Center for FRIENDS-System Configuration and Circuits of Power Factor Correctors.* Transmission and Distribution Conference and Exhibition 2002: Asia Pacific. IEEE/PES, vol. 3, pp. 2117–2122.

20 Hayashi, Y., Kawasaki, S., Funabashi, T., and Okuno, Y. (2007) *Power and Heat Interchange System Using Fuel Cells in Collective Housing*. The Fourth Power Conversion Conference, Japan, pp. 1207–1211.

21 Wildrick, C.M., Lee, F.C., Cho, B.H., and Choi, B. (1995) A method of defining the load impedance specification for a stable distributed power system. *IEEE Transactions on Power Electronics,* **10** (3), 280–285.

22 Kakigano, H., Miura, Y., Ise, T., Momose,T., and Hayakawa, H. (2008) *Fundamental Characteristics of DC Microgrid for Residential Houses with Cogeneration System in Each House.* IEEE Power & Energy Society 2008 General Meeting, 08GM0500.

23 Kakigano, H., Miura, Y., and Ise, T. (2010) Low-voltage bipolar-type DC microgrid for super high quality distribution. *IEEE Transactions on Power Electronics,* **25** (12), 3066–3075.

24 Kakigano, H., Miura, Y., and Ise, T. (2013) Distribution voltage control for DC Microgrid using fuzzy control and gain-scheduling technique. *IEEE Transactions on Power Electronics,* **28** (5), 2246–2258.

25 Kakigano, H., Hiraiwa, T., Fujiwara, H., Miura, Y., and Ise, T. (2012) An Analysis Method of a DC Microgrid Using Hardware-in-the-loop Simulation. The 13th IEEE Workshop on Control and Modeling for Power Electronics (COMPEL'12), Kyoto, Japan, 10–13 June.

26 Logue, D. L. and Krein, P.T. (2001) *Preventing Instability in DC Distribution Systems by Using Power Buffering*. Proceedings of IEEE PESC'01, vol. 1, pp. 33–37.

27 Feng, X., Liu, J., and Lee, F.C. (2002) Impedance specifications for stable DC distributed power systems. *IEEE Transactions on Power Electronics,* **17** (2), 157–162.

28 Muyeen, S.M., Takahashi, R., Murata, T., and Tamura, J. (2009) Integration of an energy capacitor system with a variable-speed wind generator. *IEEE Transactions on Energy Conversion,* **24** (3), 740–749.

29 Ye, Z., Boroyevich, D., Xing, K., and Lee, F. C. (1999) *Design of Parallel Sources in DC Distributed Power Systems by Using Gain-Scheduling Technique*. Proceedings of IEEE PESC, vol. 1, pp. 161–165.

7

Virtual Synchronous Generators: Dynamic Performance and Characteristics

Recent investigations indicate that relatively high integration of inverter-based distributed generators (DGs) and renewable energy sources (RESs) will have some impacts on power grid dynamics, frequency, and voltage regulation, as well as other control and operation issues. These impacts may increase at the penetration rates that are expected over the next several years. An important source of these impacts is the reduction of the overall inertia. Compared to conventional power systems with bulk power plants, microgrids (MGs) with DG/RES units have either small or no rotating mass and damping property. With growing penetration level of DGs/RESs, the impact of low inertia and damping effect on the grid dynamic performance and stability increases.

A solution toward stabilizing a grid/MG with numerous low-inertia DGs is to fortify the system with additional inertia, virtually. Virtual inertia can be established by using short-term energy storage together with a power electronics inverter/converter and a proper control mechanism in a system that is called virtual synchronous generator (VSG). The VSGs will then operate similarly to a synchronous generator (SG), exhibiting amount of inertia and damping properties of conventional synchronous machines for short time intervals. As a result, the virtual inertia concept may provide a basis for maintaining a large share of DGs/RESs in future grids without compromising the system stability.

As mentioned, the VSG simulates the dynamic behavior of an SG represented by its fundamental swing equation to create virtual inertia. Unlike a real synchronous machine, the parameters of VSG can be tuned to enhance the dynamic response of the system. From stability point of view, the power system would be significantly affected if the capacity of inverter-based DGs becomes larger and larger. The concept of VSG can be considered as a useful solution to improve the dynamic stability in the power grids including MGs.

In this chapter, first a basic VSG structure with the main components is explained. The VSG control performance is compared with the well-known droop control performance. Small-signal and state-space models are performed to study the frequency transient response and active power oscillation, respectively. An analytical approach to active power performance assessment

Microgrid Dynamics and Control, First Edition. Hassan Bevrani, Bruno Francois, and Toshifumi Ise.

in an MG with multiple VSGs is presented. Then, the stabilizing property of the VSG system is addressed, and a VSG-based oscillation damping approach is explained. Finally, a different VSG topology emulating the dynamic characteristics of a synchronous machine via modeling the SG properties is introduced.

7.1 Introduction

Compared to the conventional bulk power plants, in which the synchronous machine dominates, the DG/RES units have either very small or no rotating mass (which is the main source of inertia) and damping property. The intrinsic kinetic energy (rotor inertia) and damping property (due to mechanical friction and electrical losses in the stator, field, and damper windings) of the bulk SGs play a significant role in the grid stability.

With growing penetration level of DGs/RESs, the impact of low inertia and damping on the grid dynamic performance and stability increases. A solution toward stabilizing such a grid is to provide additional inertia, virtually. A virtual inertia can be established for DGs/RESs by using short-term energy storage together with a power inverter/converter and a proper control mechanism. This concept is known as VSG or *virtual synchronous machine* (VSM). The units will then operate similarly to an SG, exhibiting amount of inertia and damping properties of conventional synchronous machines for short time intervals. As a result, the virtual inertia concept may provide a basis for maintaining a large share of DGs/RESs in future grids without compromising the system stability. The VSG fundamentals and relevant main concepts are well discussed in [1].

The VSG consists of three fundamental components as shown in Fig. 7.1: energy storage, inverter, and a control mechanism. The VSG serves as an interface between a primary (usually DC) source and the grid. In the VSG control block, generally a dynamic function similar to the swing equation of the SGs

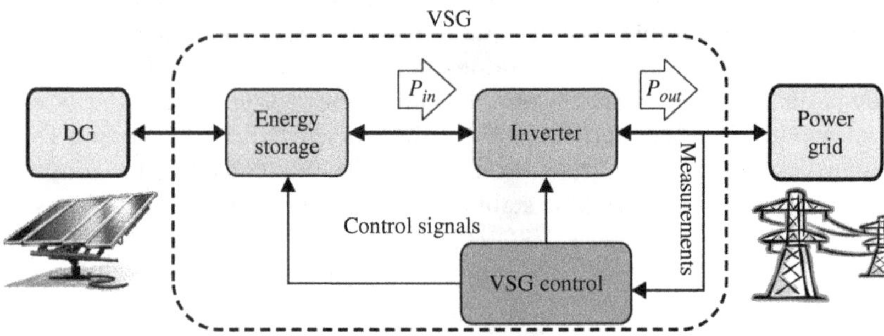

Figure 7.1 Conceptual structure of the VSG.

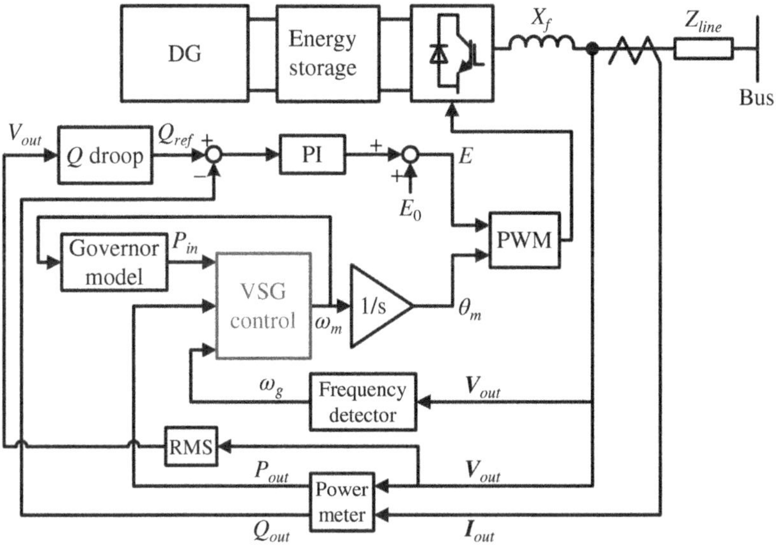

Figure 7.2 A general block diagram for the VSG system.

is embedded, which determines the output power based on the grid measurements and their rate of changes [2].

A basic block diagram for realizing the VSG system is shown in Fig. 7.2. In this scheme, a DG is connected to the power grid via a controlled inverter based on the VSG concept. An enhanced version of this VSG scheme is given in [3]. The well-known swing equation of SGs is used as the heart of the VSG system (VSG control block). The mentioned swing equation can be written as follows:

$$P_{in}(t) - P_{out}(t) = J\omega_m(t)\frac{d\omega_m(t)}{dt} + DP_{base}\frac{\omega_m(t) - \omega_g(t)}{\omega_0} \tag{7.1}$$

where P_{in} is the virtual shaft power determined by the governor, $P_{out}(t)$ is the measured output power, P_{base} is the power rating of DG, J is the virtual inertia, D is the virtual damping factor, ω_m is the virtual rotor angular frequency, ω_g is the angular frequency of the point where the voltage sensor is installed, and ω_0 is the nominal angular frequency.

The block "Governor model" in Fig. 7.2 is a ω-P droop controller, which can be represented as

$$P_{in}(t) = P_0 - k_p P_{base}\frac{\omega_m(t) - \omega_0}{\omega_0} \tag{7.2}$$

where P_0 is the normal value of active power and k_p is the droop coefficient (in per unit).

Let $k_p = (k_p P_{base})/\omega_0$, $D = (DP_{base})/\omega_0$, and eliminate $P_{in}(t)$ from (7.1) and (7.2), so that

$$P_0 - k_p(\omega_m(t) - \omega_0) - P_{out}(t) = J\omega_m(t)\frac{d\omega_m(t)}{dt} + D(\omega_m(t) - \omega_g(t)) \quad (7.3)$$

Here, $\omega_g(t)$, which is a measured parameter to provide the synchronous frequency for damping power calculation, is replaced by a constant value, that is, nominal frequency ω_0. Thus, the damping factor D becomes equivalent to droop coefficient k_p. This modification of the swing equation results in a simpler model, and as a result, the dedicated governor can even be removed. In this case, however, no damping effect caused by damper winding is emulated, which might result in larger output power oscillations.

The grid frequency is detected by a frequency detector block. The phase-locked loop (PLL) can be used as the frequency detector unit. The block diagram of a PLL unit is shown in Fig. 7.3a. The transformation from three-phase components to αβ-frame components is executed through the

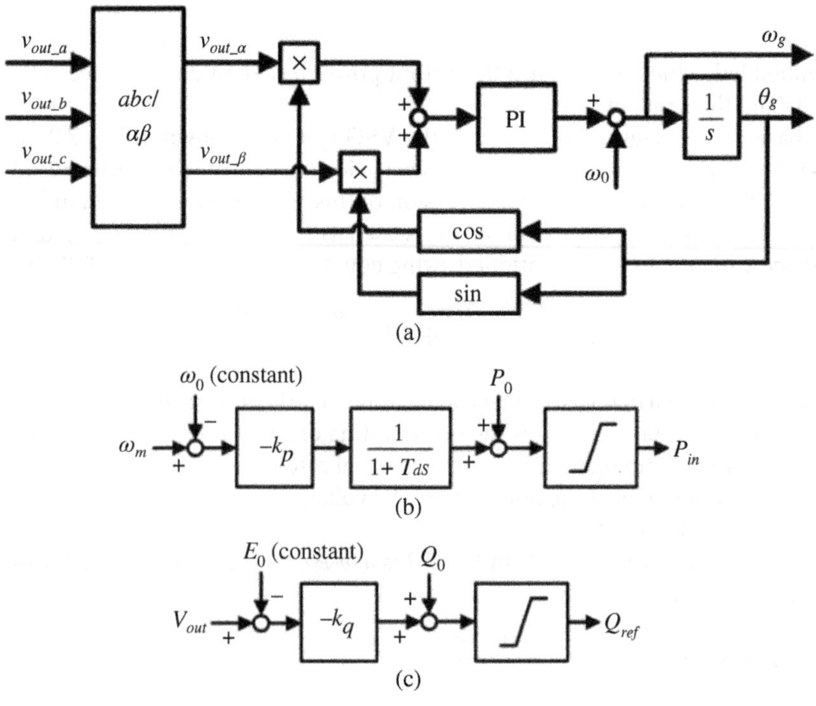

Figure 7.3 Detailed blocks of Fig. 7.2: (a) frequency detector block, (b) governor model block, and (c) Q droop block.

premultiplication by the transformation matrix $\boldsymbol{T}_{abc/\alpha\beta}$ defined as follows:

$$\boldsymbol{T}_{abc/\alpha\beta} = \sqrt{\frac{2}{3}} \cdot \begin{bmatrix} 1 & -\frac{1}{2} & -\frac{1}{2} \\ 0 & \frac{\sqrt{3}}{2} & -\frac{\sqrt{3}}{2} \end{bmatrix}$$

It should be pointed out that in the transient state, the frequency of output voltage may differ from the bus frequency, especially if the line impedance Z_{line} is large. However, measurement of bus voltage is usually difficult as the inverter may be installed far away from the ac bus of the MG. Normally, the influence of this compromise can be neglected as the MG line impedance Z_{line} is usually small.

In the block "Power Meter," the output active and reactive power P_{out} and Q_{out} are calculated as follows (for simplicity in presentation, the time index t is removed from the variables):

$$\begin{aligned} P_{out} &= v_{out\_a} i_{out\_a} + v_{out\_b} i_{out\_b} + v_{out\_c} i_{out\_c} \\ Q_{out} &= \frac{1}{\sqrt{3}} \left[i_{out\_a} \left(v_{out\_b} - v_{out\_c} \right) \right. \\ &\quad \left. + i_{out\_b} \left(v_{out\_c} - v_{out\_a} \right) + i_{out\_c} \left(v_{out\_a} - v_{out\_b} \right) \right] \end{aligned}$$

The line-to-line root mean square (RMS) value of output voltage is calculated in the block "RMS" as

$$V_{out} = \sqrt{v_{out\_a}^2 + v_{out\_b}^2 + v_{out\_c}^2}$$

The block "Governor Model" is an ω-P droop controller, as shown in Fig. 7.3b, where ω_0 is the nominal angular frequency, P_0 is the set value of active power, k_p is the ω-P droop coefficient, and T_d is the time constant of governor delay used to emulate the delayed mechanical response of a real governor of SG. The range of the output limiter for P_{in} is −0.05 to 1.05 pu. Figure 7.3b can be expressed as

$$P_{in} = P_0 - \frac{k_p}{1 + T_d s} (\omega_m - \omega_0)$$

The block "Q Droop" is a V-Q droop controller, as shown in Fig. 7.3c, where E_0 is the nominal voltage, Q_0 is the set value of reactive power, and k_q is the V-Q droop coefficient. The range of the output limiter for Q_{ref} is −1 to 1 pu. Figure 7.3c can be expressed as

$$Q_{ref} = Q_0 - k_q (V_{out} - E_0)$$

The "Governor Model" block creates the linear droop control law between active power and the frequency, and the "Q Droop" block creates the linear droop control law between reactive power and voltage. As this VSG control operates as a voltage source, it can be classified as a voltage-source-based grid-supporting control. Similarly to other voltage-source-based grid-supporting controls, VSG-control-based inverter can operate individually, in parallel, and even be connected to the power grid directly, and no change of control is needed during operation mode transitions. Therefore, the VSG control can be considered as an upgraded voltage-source-based grid-supporting control, which is able to provide inertia support.

A proportional–integral (PI) controller is used to track the output reactive power to the reference value Q_{ref} generated by the "Q Droop" controller. In order to diminish the influence from ripples in the measured output power, a 20 Hz first-order low-pass filter is applied for Q_{out} as shown in Fig. 7.2. As the output current is measured after the LC filter stage, the reactive power consumed by the LC filter is not included in Q_{out}. Therefore, no specific initial process is required for the reactive power PI controller. The output voltage of pulse width modulation (PWM) inverter is obtained from the sum of the PI controller output and the nominal voltage E_0. The function of this PI controller is similar to that of the automatic reactive power regulators of the SG, in which the electromotive force (emf) is controlled by regulating the field current. Therefore, the output voltage of PWM inverter can be considered as the virtual internal emf E of the VSG.

It is noteworthy that the inner current or voltage loop is not adopted in this control scheme, in order to make the filter inductor L_f contribute to the output impedance and be considered as the stator inductance of the VSG (note that the filter capacitor C_f is usually negligible at fundamental frequency). This stator inductance results in more inductive output impedance, which is especially important for active and reactive power decoupling in a low-voltage MG where line resistance is dominant. Nevertheless, output voltage is still regulated indirectly by the V-Q droop controller and the PI controller of reactive power.

Having the essential parameters, (7.1) can be solved by numerical calculation such as Runge–Kutta method. The procedure of calculating ω_m using Runge–Kutta method is given in Fig. 7.4. After calculating the momentary ω_m and passing it through an integrator, the virtual mechanical phase angle θ_m is produced. As shown in Fig. 7.2, the existing PI controller provides E signal, which is the voltage reference and can be used to determine the voltage magnitude and reactive power at the inverter terminal. The phase angle and the voltage reference are used to generate PWM pulses for the inverter.

Selecting the proper value for J and D would be a challenging issue. Mimicking a synchronous machine, J is the inertia characteristic given by $J = 2HS_0/\omega_0^2$, where H is the machine inertia constant, S_0 is the nominal apparent power of the machine, and ω_0 is the system frequency. The parameter H determines the time period during which the machine is able to supply the nominal load using the energy stored in the rotating mass. A system with a higher H, bigger time constant, provides a slower response but smaller frequency deviation after a disturbance. Although it depends on the machine size and power, for typical synchronous machines, H varies between 2 and 10 s [1].

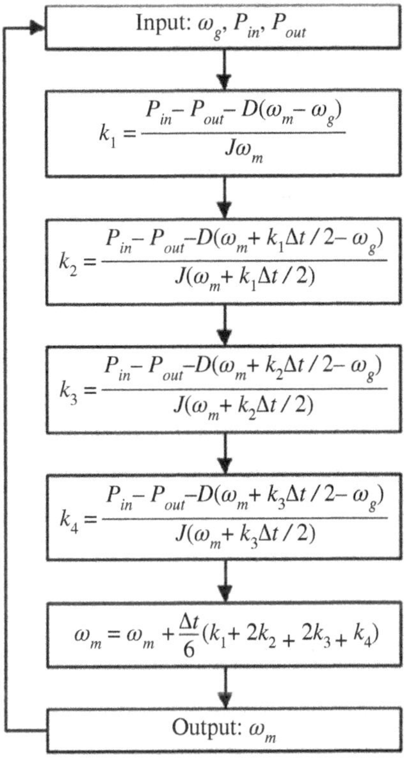

Figure 7.4 Calculating ω_m by Runge–Kutta method.

Concerning the frequency regulation, the J and D must be selected so that the VSG exchanges its maximum active power when the maximum specified frequency variation and rate of frequency change occur. The larger J and D means that more power will be either injected or absorbed for the same amount of frequency deviation and rate of frequency change. However, oppositely, large values of J and D with specific power rating result in a small frequency excursion.

As mentioned, increasing J results in higher amount of equivalent inertia for the VSG; however, there is a limitation, which is mainly imposed by inverter capacity and PLL accuracy. The inverter capacity does not have the overload capacity of a synchronous machine. Thus, a high derivative term causes bigger power overshoots during transient frequency deviations, and the inverter must sustain an important overload. The accuracy in frequency tracking depends on the performance of the implemented PLL. Therefore, the optimal value of derivative term in (7.1) can be obtained by a trade-off between the virtual inertia, the inverter overload capacity, and the PLL characteristics [5].

7.2 Virtual Synchronous Generator (VSG) and Droop Control

In this section, dynamic characteristics of VSG control and conventional droop control are compared to understand the difference in the mentioned two control methods caused by the presence of swing equation. For the sake of frequency response analysis, small-signal models of both methods are built, for both stand-alone and SG-connected modes, in order to understand the virtual inertia impacts. Although the idea of using a first-order lag unit in the droop control to emulate the virtual inertia is already mentioned in the literature, it is further developed here by adding a first-order lag unit to emulate the damping factor. State-space models of both methods are constructed to study oscillations of the output active power. Theoretical results are verified by performing simulations using the PSCAD/EMTDC software and laboratory experiments. It should be pointed out that although the proposed study is given for a specific VSG control scheme, the results should also be valid for other VSG control schemes with similar small-signal models.

7.2.1 Droop Control

A basic structure for the VSG control system is shown in Fig. 7.2. An analogous droop control system is shown in Fig. 7.5. The parameters and variables shown in this figure are the same as those explained in Fig. 7.2. Since the reactive power control is identical in both control methods, here, only the differences in active power control are studied.

By neglecting the time delay, the "*P* Droop" block in the droop control loop shown in Fig. 7.5 can be represented as follows:

$$\omega_m = -\frac{P_{out} - P_0}{k_p} + \omega_0 \tag{7.4}$$

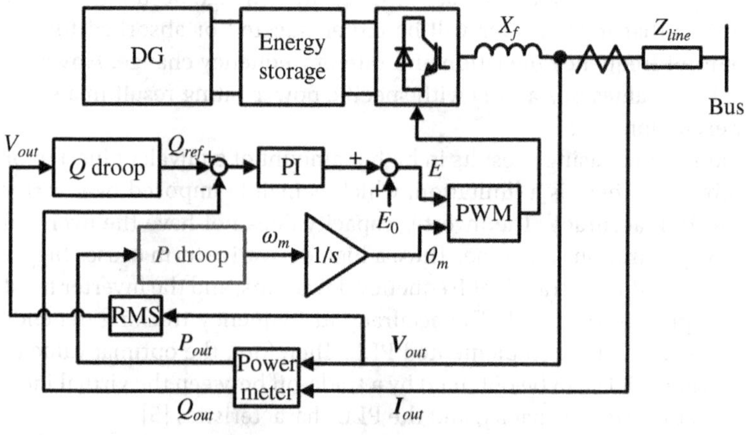

Figure 7.5 Droop control for inverter systems.

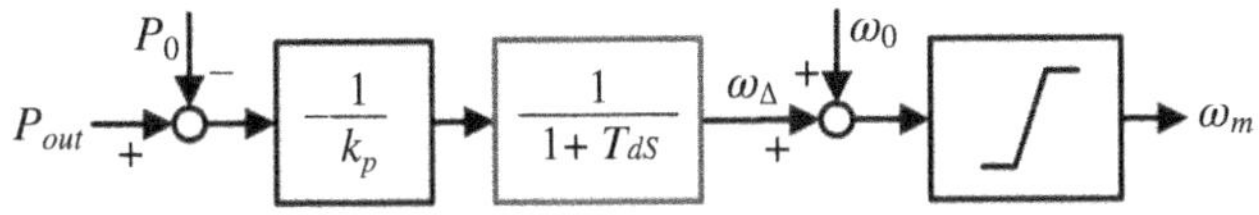

Figure 7.6 *P* droop control.

Let $J = 0, D = 0$, (7.3) is equivalent to (7.4). In other words, the droop control can be considered as a particular case of VSG control, where both inertia and damping factor are set to zero.

For droop control, similarly to the governor model, which is shown in Fig. 7.3a, a low-pass filter is usually added to the *P* droop control loop to filter the noises in the measured output power P_{out}, and to consider the physical time delay as shown in Fig. 7.6. Considering both delays as first-order lags with time constant T_d, the k_p must be replaced by $k_p/(1 + T_d s)$ for VSG control and by $k_p(1 + T_d s)$ for droop control in the small-signal models.

This difference is caused by the inverse of input and output in *P*-ω droop regulation. In VSG control, the input is the active power and the output is the frequency, while in droop control, it is the opposite, as illustrated in Fig. 7.6.

7.2.2 Transient Frequency Response

In this section, first, small-signal models of VSG and droop control are given for both stand-alone and SG-connected modes. Then, based on these models, step responses of the frequency change during a loading transition are calculated and compared with the corresponding simulation results. The system with the slower frequency change is preferable, because a lower df/dt indicates a larger inertia [6]. System with a smaller inertia is prone to exceed the df/dt relays threshold during a large loading transition, which may lead to unnecessary tripping or load shedding. Moreover, system with a larger inertia has a smaller maximum frequency excursion following a fault and during a loading transition when the secondary control is applied, as it is demonstrated in [4].

To simplify the model in order to focus on the active power control, the reactive power control of both control methods is inactivated in this study. Therefore, as shown in Figs 7.2 and 7.5, E is fixed at E_0.

7.2.2.1 Stand-Alone Mode

A stand-alone system is shown in Fig. 7.7a. The line between DG and load is considered as an inductive line, and its resistance can be neglected. This assumption is satisfied in both high-voltage (HV) and LV networks, specifically

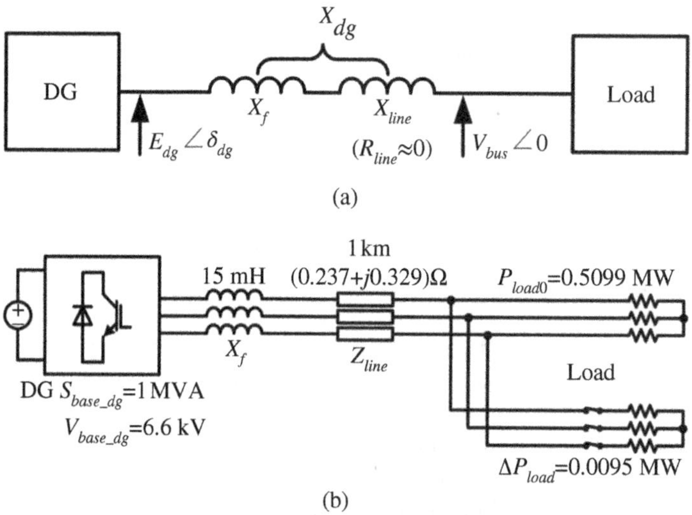

Figure 7.7 The stand-alone mode model: (a) single-line diagram and (b) simulation circuit.

when an inductive virtual output impedance is applied [7]. By calculating the power flow, (7.5) can be deduced as follows:

$$\Delta P_{out\_dg} = E_{dg} V_{bus} \cos \delta_{dg} \Delta\delta_{dg} / X_{dg} \tag{7.5}$$

where subscript "*dg*" indicates parameters related to DG, and $X_{dg} = X_f + X_{line}$ is the total output reactance of DG.

Knowing that $\Delta\delta_{dg} = (\Delta\omega_{m\_dg} - \Delta\omega_{bus})/s$, where ω_{bus} is the bus frequency, let $K_{dg} = (E_{dg} V_{bus} \cos \delta_{dg})/X_{dg}$, which is the synchronizing power coefficient of DG, so that (7.5) becomes

$$s\Delta P_{out\_dg} \approx K_{dg}(\Delta\omega_{m\_dg} - \Delta\omega_{bus}) \tag{7.6}$$

The small-signal model of (7.3) can be written as

$$\begin{aligned} -k_{p\_dg}\Delta\omega_{m\_dg} - \Delta P_{out\_dg} = {} & J_{dg}(\omega_{m\_dg} s\Delta\omega_{m\_dg} + s\Delta\omega_{m\_dg}^2) \\ & + D_{dg}(\Delta\omega_{m\_dg} - \Delta\omega_{g\_dg}) \end{aligned} \tag{7.7}$$

As the frequency deviation from the nominal frequency is relatively small, $\omega_{m\_dg} s\Delta\omega_{m\_dg} \approx \omega_0 s\Delta\omega_{m\_dg}$. Moreover, second-order perturbation terms can be neglected. Therefore, (7.7) becomes

$$-k_{p\_dg}\Delta\omega_{m\_dg} - \Delta P_{out\_dg} = J_{dg}\omega_0 s\Delta\omega_{m\_dg} + D_{dg}(\Delta\omega_{m\_dg} - \Delta\omega_{g\_dg}) \tag{7.8}$$

If line losses are neglected, $\Delta P_{out\_dg} = \Delta P_{load}$, where P_{load} is the active power consumed by the load. And if $X_f \gg X_{line}$, $\Delta\omega_{g\_dg} \approx \Delta\omega_{bus}$. In this case, by eliminating $\Delta\omega_{g\_dg}$ and $\Delta\omega_{bus}$ from (7.6) and (7.8), the following equation can be obtained:

$$\frac{\Delta\omega_{m\_dg}}{\Delta P_{load}} = -\frac{1 + (D_{dg}/K_{dg})s}{k_{p\_dg} + J_{dg}\omega_0 s} \tag{7.9}$$

Equation (7.9) is the transfer function of the frequency change over a small loading transition in the stand-alone mode for a VSG. Let $J_{dg} = 0$, $D_{dg} = 0$, and then the transfer function for a droop-control-based DG can be obtained as follows.

$$\frac{\Delta\omega_{m\_dg}}{\Delta P_{load}} = -\frac{1}{k_{p\_dg}} \tag{7.10}$$

Let $s = 0$ in (7.9), and then (7.9) is equivalent to (7.10). This implies that steady-state gain in (7.9) is determined by droop coefficient k_{p_dg} and is independent of the swing equation parameters. In other words, the steady state values of VSG control and droop control are the same if droop coefficients k_{p_dg} are set equally.

It is possible to calculate step responses of DG frequency change during a small loading transition through (7.9) and (7.10) with various parameters, and the results are shown in Fig. 7.8 (dotted lines) along with the corresponding simulation results (solid lines) obtained by using the PSCAD/EMTDC simulation software. Parameters used for both theoretical calculation and simulation are the same, as listed in Table A.1, and the line resistance is taken into consideration in the simulation, as shown in Fig. 7.7b.

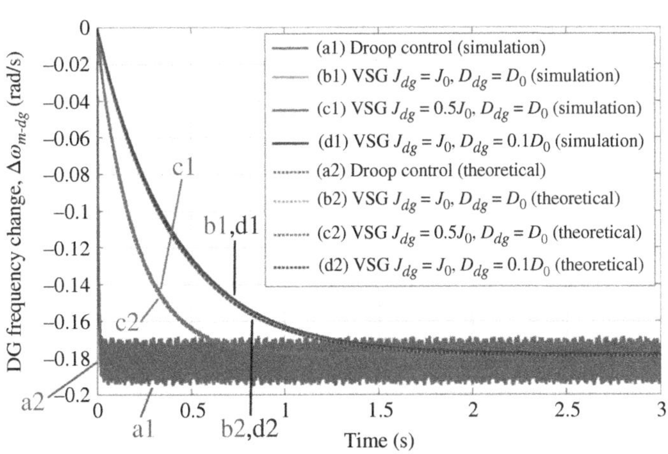

Figure 7.8 Step responses of DG frequency during a loading transition in stand-alone mode with various parameters.

As it is demonstrated in Fig. 7.8, in all cases, simulation results almost match with the corresponding theoretical results. Thus, it can be concluded that the small-signal models are verified. Ripples can be observed in the simulation result of droop control due to ripples in the measured output power ΔP_{out_dg}.

Normally, a filter should be applied to filter this ripple, and its effect is discussed later. To understand the dynamic response of a basic droop control without the filter, only a first-order filter with a small time constant (0.005 s) is applied in simulations and experimental tests, except that another value is specified.

It can be seen that the frequency of VSG control changes slowly, whereas that of droop control suffers a step change. This implies that DG with VSG control has an inertia, thanks to the swing equation, and that DG with a droop control has nearly no inertia. Moreover, a larger value of J_{dg} results in a slower frequency change, which indicates that inertia of the DG is determined by J_{dg}. On the other hand, D_{dg} has barely any influence on the dynamic response in this situation. Besides, steady states of all cases are the same. This verifies the previous conclusion that steady states only depend on k_{p_dg}.

7.2.2.2 Synchronous Generator (SG)-Connected Mode

In this section, an SG-connected system shown in Fig. 7.9a is studied to show the importance of inertia in a network with high penetration level of inverter-based DG. It is assumed that the penetration rate of DG in this network is up to 50%, and the rest of power generation comes from the SG. In this case, it is preferred to study the rotor frequency change of the SG rather than that of the DG because the SG is considered as the dominant one.

Neglecting the power losses,

$$\Delta P_{out\_dg} + \Delta P_{out\_sg} = \Delta P_{load} \tag{7.11}$$

where the subscript "*sg*" indicates parameters related to SG.

Let $\frac{\Delta P_{out\_dg}}{\Delta \omega_{bus}} = A$ and $\frac{\Delta P_{out\_sg}}{\Delta \omega_{bus}} = B$, so that (7.11) becomes

$$\Delta P_{out\_sg}(1 + A/B) = \Delta P_{load} \tag{7.12}$$

The same equation as (7.6) holds for SG.

$$s\Delta P_{out\_sg} \approx K_{sg}(\Delta \omega_{m\_sg} - \Delta \omega_{bus}) \tag{7.13}$$

Eliminate ΔP_{out_sg} from (7.12) and (7.13), so that

$$\frac{\Delta \omega_{m\_sg}}{\Delta P_{load}} = \frac{K_{sg} + Bs}{K_{sg}(A + B)} \tag{7.14}$$

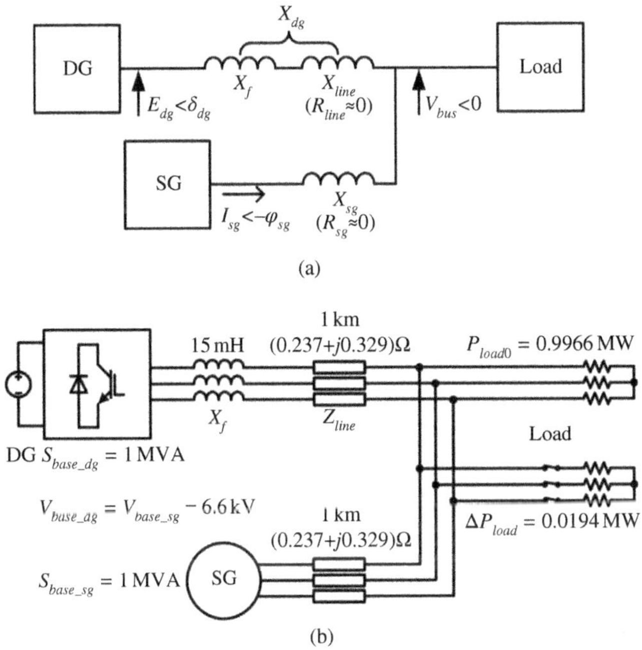

Figure 7.9 SG-connected mode model: (a) single-line diagram and (b) simulation circuit.

Now if A and B are known, the required transfer function can be obtained. Eliminate $\Delta\omega_{m\_dg}$ from (7.6) and (7.8), and then

$$A = \frac{\Delta P_{out\_dg}}{\Delta\omega_{bus}} = -\frac{k_{p\_dg} + J_{dg}\omega_0 s}{1 + \dfrac{k_{p\_dg} + D_{dg}}{K_{dg}} s + \dfrac{J_{dg}\omega_0}{K_{dg}} s^2} \tag{7.15}$$

Since SG has the same model as VSG, (7.15) is also true for SG.

$$B = \frac{\Delta P_{out\_sg}}{\Delta\omega_{bus}} = -\frac{k_{p\_sg} + J_{sg}\omega_0 s}{1 + \dfrac{k_{p\_sg} + D_{sg}}{K_{sg}} s + \dfrac{J_{sg}\omega_0}{K_{sg}} s^2} \tag{7.16}$$

Let $J_{dg} = 0, D_{dg} = 0$ in (7.15), and then A for droop control can be obtained as

$$A = \frac{\Delta P_{out\_dg}}{\Delta\omega_{bus}} = -\frac{k_{p\_dg}}{1 + (k_{p\_dg}/K_{dg})s} \tag{7.17}$$

However, for SG, K_{sg} and D_{sg} are not constant. They can be calculated from SG parameters and measurement data of V_{bus}, I_{sg} (output current of SG), and φ_{sg} (phase difference between V_{bus} and I_{sg}) [8]. Based on (7.14)–(7.17), it is possible to calculate the step responses of an SG frequency change during a small loading transition, and the results are shown in Fig. 7.10 along with the corresponding simulation results. (Note that for the calculation of B, (7.21) is used instead of (7.16), as explicated in the next part of this section.) Parameters of SG are listed in Table A.2, and parameters of DG are the same as those listed in Table A.1. The simulation circuit is shown in Fig. 7.9b, in which the line resistance is included.

As shown in Fig. 7.10, simulation results verify the obtained theoretical results with the small-signal model, although a slight delay during first 0.5 s is observed. It is probably because during simulation, the input disturbance ΔP_{load}, which is generated by switching on an additional load, is not an ideal

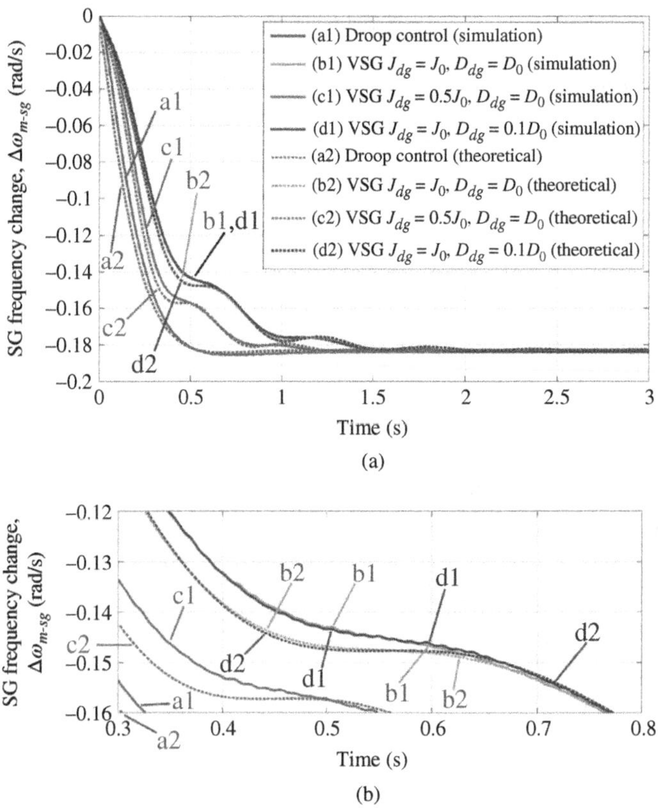

Figure 7.10 (a) Step responses of SG frequency during a loading transition in SG-connected mode with various parameters. (b) A zoomed view of (a).

step due to the existence of line reactance. It can also be noted that the amplitudes of oscillations in simulation results are slightly smaller than those in theoretical results, probably because simulation circuits include a line resistance, which provides an additional damping to the system.

Similar to the stand-alone mode, it is demonstrated that DG with VSG control can create a larger inertia for the system compared to DG with a droop control since df/dt of the former is slower. Further, a larger value of J_{dg} still increases the inertia of the system. In addition, a system with larger D_{dg} has slightly smaller oscillations. Similarly, parameters of swing equation have no influence on the steady state.

7.2.2.3 Effects of Delays in Governor and P Droop Controller

It is known that there is always a considerable delay in the mechanical governor of SG. In the previous research works on VSG, this delay is also imitated when (7.2) is applied [9], as illustrated in Fig. 7.3a. For a droop control, a low-pass filter is usually added into the loop as shown in Fig. 7.6.

Considering a time delay for VSG, in the stand-alone mode, (7.9) has to be modified as

$$\frac{\Delta\omega_{m\_dg}}{\Delta P_{load}} = -\frac{1+\left(T_{d\_dg}+\dfrac{D_{dg}}{K_{dg}}\right)s+\dfrac{D_{dg}T_{d\_dg}}{K_{dg}}s^2}{k_{p\_dg}+J_{dg}\omega_0 s+J_{dg}\omega_0 T_{d\_dg}s^2} \tag{7.18}$$

and for the droop control, (7.10) becomes

$$\frac{\Delta\omega_{m\_dg}}{\Delta P_{load}} = -\frac{1}{k_{p\_dg}+T_{d\_dg}k_{p\_dg}s} \tag{7.19}$$

Meanwhile, in the SG-connected mode, for VSG and SG, (7.15) and (7.16) have to be modified as

$$A = -\frac{k_{p_{dg}}+J_{dg}\omega_0 s+T_{d_{dg}}J_{dg}\omega_0 s^2}{1+\left(T_{d_{dg}}+\dfrac{k_{p_{dg}}+D_{dg}}{K_{dg}}\right)s+\dfrac{T_{d_{dg}}D_{dg}+J_{dg}\omega_0}{K_{dg}}s^2+\dfrac{T_{d_{dg}}J_{dg}\omega_0}{K_{dg}}s^3} \tag{7.20}$$

$$B = -\frac{k_{p\_sg}+J_{sg}\omega_0 s+T_{d\_sg}J_{sg}\omega_0 s^2}{1+\left(T_{d\_sg}+\dfrac{k_{p\_sg}+D_{sg}}{K_{sg}}\right)s+\dfrac{T_{d\_sg}D_{sg}+J_{sg}\omega_0}{K_{sg}}s^2+\dfrac{T_{d\_sg}J_{sg}\omega_0}{K_{sg}}s^3} \tag{7.21}$$

and for droop control, (7.17) becomes

$$A = -\frac{k_{p\_dg}+k_{p\_dg}T_{d\_dg}s}{1+(k_{p\_dg}/K_{dg})s+(k_{p\_dg}T_{d\_dg}/K_{dg})s^2} \tag{7.22}$$

Based on (7.18)–(7.22), frequency step responses for 0.1 s delay (T_{d_dg}) are recalculated and then compared with the cases without consideration delays, in [10].

7.2.2.4 Inertial Droop Control

It is interesting to compare (7.9) with (7.19), and (7.15) with (7.22). If T_{d_dg} is increased up to

$$T_{d\_dg} = J_{dg}\omega_0/k_{p\_dg} \tag{7.23}$$

the only difference between (7.9) and (7.19) or between (7.15) and (7.22) is that the equations of VSG control have terms coming from D_{dg}, whereas those of droop control do not. Therefore, by adding a first-order lag unit, with time constant T_{d_dg} specified by (7.23), into the P droop loop, the droop control can imitate the performance of VSG with a moment of inertia equal to J_{dg}.

Furthermore, if a first-order lead unit is also added into the P droop loop, as shown in Fig. 7.11, (7.19) and (7.22) become

$$\frac{\Delta\omega_{m\_dg}}{\Delta P_{load}} = -\frac{1 + T_{a\_dg}s}{k_{p\_dg} + T_{d\_dg}k_{p\_dg}s} \tag{7.24}$$

$$A = -\frac{k_{p\_dg} + k_{p\_dg}T_{d\_dg}s}{1 + (k_{p\_dg}/K_{dg} + T_{a\_dg})s + (k_{p\_dg}T_{d\_dg}/K_{dg})s^2} \tag{7.25}$$

respectively. If the time constant is set to

$$T_{a\_dg} = D_{dg}/K_{dg} \tag{7.26}$$

then, (7.24) is equivalent to (7.9), and (7.25) is equivalent to (7.15). This conclusion is proved in Fig. 7.12 with both simulation results and calculated theoretical results. By adding a first-order lag unit with time constant specified by (7.23), droop control generates a similar step response to VSG control, with slightly larger oscillations due to lack of damping factor. Additionally, when a first-order lead unit with time constant specified by (7.26) is also added, droop control leads to the exactly same step response as VSG control in theoretical results. Although in simulation results, there is still a slight difference between the result of droop control with specified lead–lag unit (c1) and that of VSG (d1), probably due to the presence of line resistance, the damping effect of lead unit is confirmed by comparing lines (b1) and (c1). This implies that

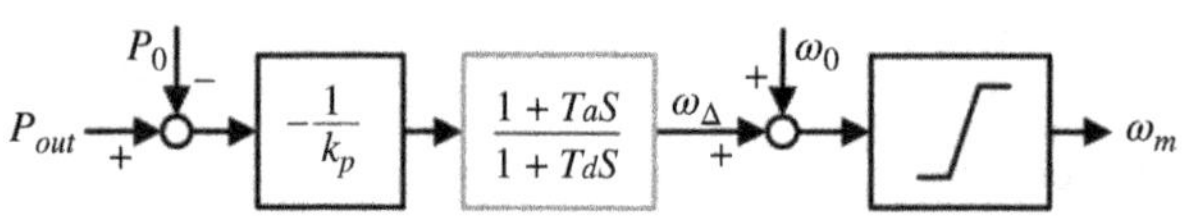

Figure 7.11 P Droop control with a first-order lead–lag unit.

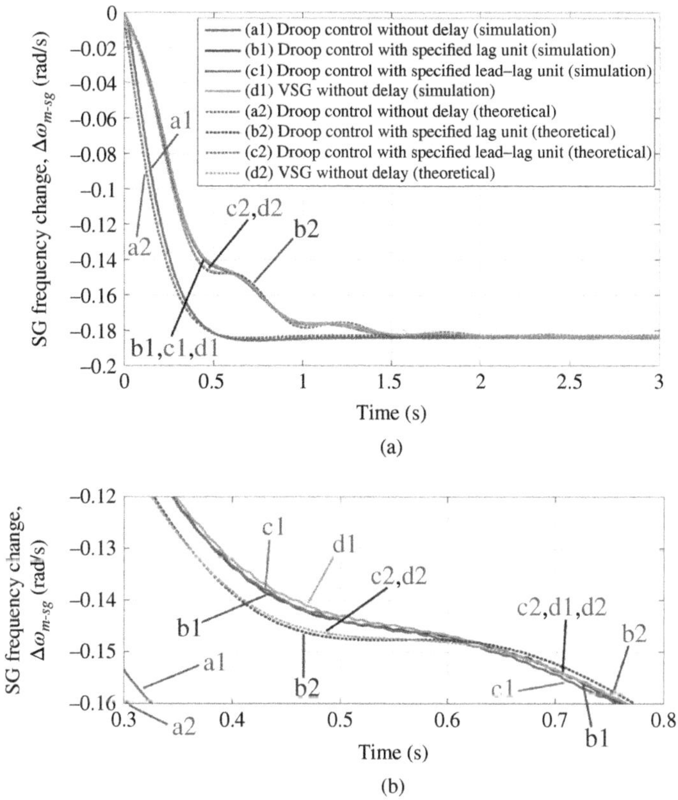

Figure 7.12 (a) Step responses of SG frequency during a loading transition in SG-connected mode with specified lag or lead–lag unit in droop control. (b) Zoom in of (a).

during small disturbances, VSG can be approximated by droop control, with a first-order lag unit to simulate inertia and with a first-order lead unit to simulate damping factor.

7.2.3 Active Power Response

To compare the oscillation of output active power in both VSG and droop control systems, the closed-loop state-space models of SG-connected mode are built, by using the small-signal relations developed in Section 7.2.2:

$$\begin{cases} \dot{\boldsymbol{x}} = \boldsymbol{A}\boldsymbol{x} + \boldsymbol{B}\boldsymbol{w} \\ \boldsymbol{y} = \boldsymbol{C}\boldsymbol{x} + \boldsymbol{D}\boldsymbol{w} \end{cases} \tag{7.27}$$

where

$$w = [\Delta P_{load}] \tag{7.28}$$

$$y = \left[\Delta\omega_{m\_sg}\ \Delta\omega_{m\_dg}\ \Delta P_{out\_sg}\ \Delta P_{out\_dg}\right]^T \tag{7.29}$$

The state vector and matrices $\boldsymbol{A}$, $\boldsymbol{B}$, $\boldsymbol{C}$, and $\boldsymbol{D}$ for the closed-loop VSG control system are expressed as

$$\boldsymbol{x} = \begin{bmatrix} \Delta\omega_{m_{sg}} + \dfrac{D_{sg}}{J_{sg}\omega_0\left(K_{sg}+K_{dg}\right)}\Delta P_{load} \\ \Delta\omega_{m_{dg}} + \dfrac{D_{dg}}{J_{dg}\omega_0\left(K_{sg}+K_{dg}\right)}\Delta P_{load} \\ \Delta\delta_{sg} - \dfrac{1}{K_{sg}+K_{dg}}\Delta P_{load} \\ \Delta P_{in_{sg}} \\ \Delta P_{in_{dg}} \end{bmatrix}', \quad \boldsymbol{C} = \begin{bmatrix} 1 & 0 & 0 & 0 & 0 \\ 0 & 1 & 0 & 0 & 0 \\ 0 & 0 & K_{sg} & 0 & 0 \\ 0 & 0 & -K_{sg} & 0 & 0 \end{bmatrix},$$

$$\boldsymbol{D} = \begin{bmatrix} -\dfrac{D_{sg}}{J_{sg}\omega_0\left(K_{sg}+K_{dg}\right)} \\ -\dfrac{D_{dg}}{J_{dg}\omega_0\left(K_{sg}+K_{dg}\right)} \\ \dfrac{K_{sg}}{K_{sg}+K_{dg}} \\ \dfrac{K_{dg}}{K_{sg}+K_{dg}} \end{bmatrix}$$

$$\boldsymbol{A} = \begin{bmatrix} -\dfrac{D_{sg}K_{dg}}{J_{sg}\omega_0\left(K_{sg}+K_{dg}\right)} & \dfrac{D_{sg}K_{dg}}{J_{sg}\omega_0\left(K_{sg}+K_{dg}\right)} & -\dfrac{K_{sg}}{J_{sg}\omega_0} & \dfrac{1}{J_{sg}\omega_0} & 0 \\ \dfrac{D_{dg}K_{sg}}{J_{dg}\omega_0\left(K_{sg}+K_{dg}\right)} & -\dfrac{D_{dg}K_{sg}}{J_{dg}\omega_0\left(K_{sg}+K_{dg}\right)} & \dfrac{K_{sg}}{J_{dg}\omega_0} & 0 & \dfrac{1}{J_{dg}\omega_0} \\ \dfrac{K_{dg}}{K_{sg}+K_{dg}} & -\dfrac{K_{dg}}{K_{sg}+K_{dg}} & 0 & 0 & 0 \\ -\dfrac{k_{p\_sg}}{T_{d\_sg}} & 0 & 0 & -\dfrac{1}{T_{d\_sg}} & 0 \\ 0 & -\dfrac{k_{p\_dg}}{T_{d\_dg}} & 0 & 0 & -\dfrac{1}{T_{d\_dg}} \end{bmatrix},$$

$$\boldsymbol{B}=\begin{bmatrix} -\dfrac{K_{sg}}{J_{sg}\omega_0\left(K_{sg}+K_{dg}\right)}-\dfrac{D_{sg}D_{dg}K_{dg}}{J_{sg}J_{dg}\omega_0^2\left(K_{sg}+K_{dg}\right)^2}+\dfrac{D_{sg}^2K_{dg}}{J_{sg}^2\omega_0^2\left(K_{sg}+K_{dg}\right)^2} \\ -\dfrac{1}{J_{dg}\omega_0}+\dfrac{K_{sg}}{J_{dg}\omega_0\left(K_{sg}+K_{dg}\right)}-\dfrac{D_{sg}D_{dg}K_{sg}}{J_{sg}J_{dg}\omega_0^2\left(K_{sg}+K_{dg}\right)^2}+\dfrac{D_{dg}^2K_{sg}}{J_{dg}^2\omega_0^2\left(K_{sg}+K_{dg}\right)^2} \\ -\dfrac{D_{sg}K_{dg}}{J_{sg}\omega_0\left(K_{sg}+K_{dg}\right)^2}+\dfrac{D_{dg}K_{dg}}{J_{dg}\omega_0\left(K_{sg}+K_{dg}\right)^2} \\ \dfrac{k_{p\_sg}D_{sg}}{T_{d\_sg}J_{sg}\omega_0\left(K_{sg}+K_{dg}\right)} \\ \dfrac{k_{p\_dg}D_{dg}}{T_{d\_dg}J_{dg}\omega_0\left(K_{sg}+K_{dg}\right)} \end{bmatrix}$$

Similarly, the state-space model matrices for the closed-loop droop control system can be obtained as follows:

$$x=\begin{bmatrix} \Delta\omega_{m\_sg}+\dfrac{D_{sg}}{J_{sg}\omega_0\left(K_{sg}+K_{dg}\right)}\Delta P_{load} \\ \Delta\omega_{m\_dg} \\ \Delta\delta_{sg}-\dfrac{1}{K_{sg}+K_{dg}}\Delta P_{load} \\ \Delta P_{in\_sg} \end{bmatrix},\quad \boldsymbol{C}=\begin{bmatrix} 1 & 0 & 0 & 0 \\ 0 & 1 & 0 & 0 \\ 0 & 0 & K_{sg} & 0 \\ 0 & 0 & -K_{sg} & 0 \end{bmatrix},$$

$$\boldsymbol{D}=\begin{bmatrix} -\dfrac{D_{sg}}{J_{sg}\omega_0\left(K_{sg}+K_{dg}\right)} \\ 0 \\ \dfrac{K_{sg}}{K_{sg}+K_{dg}} \\ \dfrac{K_{dg}}{K_{sg}+K_{dg}} \end{bmatrix}$$

$$\boldsymbol{A}=\begin{bmatrix} -\dfrac{D_{sg}K_{dg}}{J_{sg}\omega_0\left(K_{sg}+K_{dg}\right)} & \dfrac{D_{sg}K_{dg}}{J_{sg}\omega_0\left(K_{sg}+K_{dg}\right)} & -\dfrac{K_{sg}}{J_{sg}\omega_0} & \dfrac{1}{J_{sg}\omega_0} \\ 0 & -\dfrac{1}{T_{d\_dg}} & \dfrac{K_{sg}}{k_{p\_dg}T_{d\_dg}} & 0 \\ \dfrac{K_{dg}}{K_{sg}+K_{dg}} & -\dfrac{K_{dg}}{K_{sg}+K_{dg}} & 0 & 0 \\ -\dfrac{k_{p\_sg}}{T_{d\_sg}} & 0 & 0 & -\dfrac{1}{T_{d\_sg}} \end{bmatrix},$$

$$
\boldsymbol{B} = \begin{bmatrix} -\dfrac{K_{sg}}{J_{sg}\omega_0\left(K_{sg}+K_{dg}\right)} + \dfrac{D_{sg}^2 K_{dg}}{J_{sg}^2\omega_0^2\left(K_{sg}+K_{dg}\right)^2} \\ -\dfrac{1}{k_{p\_dg}T_{d\_dg}} + \dfrac{K_{sg}}{k_{p\_dg}T_{d\_dg}\left(K_{sg}+K_{dg}\right)} \\ -\dfrac{D_{sg}K_{dg}}{J_{sg}\omega_0\left(K_{sg}+K_{dg}\right)^2} \\ \dfrac{k_{p\_sg}D_{sg}}{T_{d\_sg}J_{sg}\omega_0\left(K_{sg}+K_{dg}\right)} \end{bmatrix}
$$

It can be proved that the transfer function of the first output $\Delta\omega_{m\_sg}$ over the input ΔP_{load} deduced from this state-space model is the same as that described by (7.14).

Since oscillations in output variables, including DG output power ΔP_{out_dg}, are determined by the eigenvalues of state matrix $\boldsymbol{A}$, it is important to analyze how these eigenvalues vary according to different parameters. The relevant root locus plots are shown in Fig. 7.13. In each diagram, radial lines indicate damping ratio ζ, while circle lines indicate natural frequency ω_n.

Parameters used for calculation are the same as those listed in Tables A.1 and A.2. It is implied that VSG control is more oscillatory compared to droop control due to smaller ζ of the complex conjugate eigenvalues. Although oscillation in the output power is a common phenomenon in SG, it may cause overcurrent in DG and stop the inverter, because an inverter usually has weaker overload ability compared to an SG with the same rating. However, this problem can be solved by increasing damping factor D_{dg} and/or line reactance X_{dg}, as ζ of the complex conjugate eigenvalues increases in these cases as it is shown in Fig. 7.13. For example, a damping method based on modification of D_{dg} and considering line resistance is presented in Section 7.3. As for X_{dg}, it can be increased by the virtual impedance control, for the decoupling, the active and reactive power.

In addition, it is shown that J_{dg} affects the oscillation frequency ω_d, which is indicated by the distance between the eigenvalues and the real axis, but has barely any effect on the damping ratio. It is also shown that the increase in T_{d_dg} not only results in a higher df/dt, as discussed in the previous section, but also makes the system more oscillatory. Therefore, it is better not to simulate the governor delay in the VSG control.

The droop control may also become oscillatory if the lag time constant T_{d_dg} is large. In this case, the distribution of its eigenvalues becomes similar to that of VSG control. This verifies that the droop control can be used to simulate VSG by increasing T_{d_dg}. Moreover, for both control methods, no matter how parameters change, no right-half-plane pole is observed. This implies that

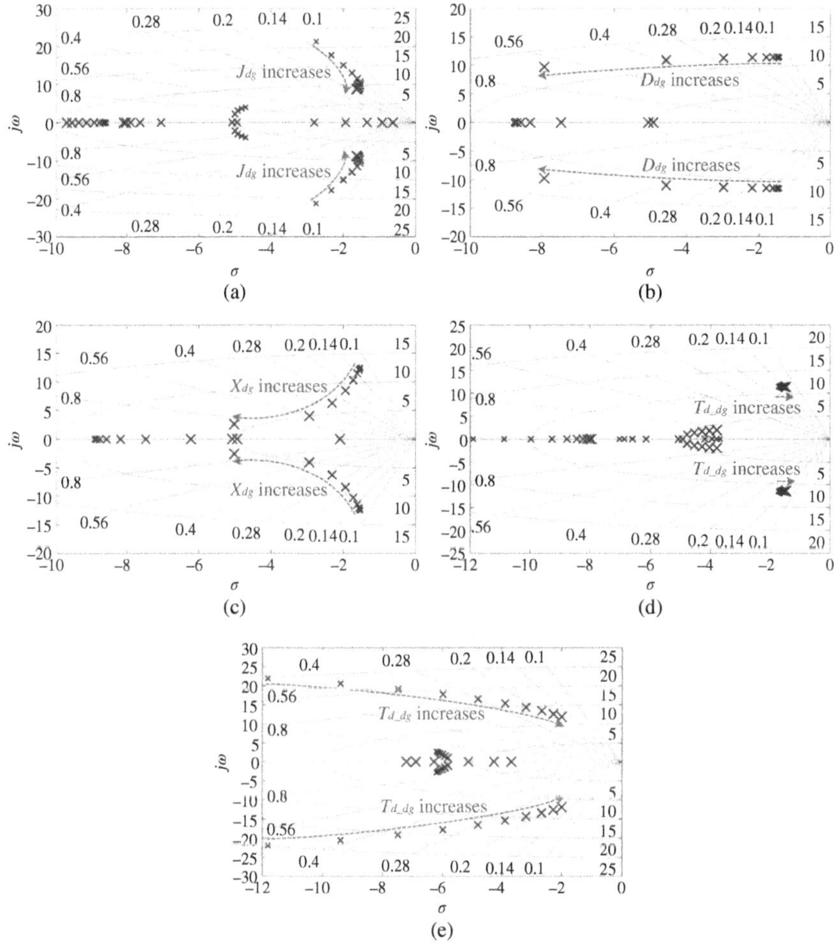

Figure 7.13 Eigenvalues when (a) J_{dg} of VSG varies from $1.5^{-4}J_0$ to 1.5^5J_0, (b) D_{dg} of VSG varies from $2^{-4}D_0$ to 2^5D_0, (c) X_{dg} of VSG varies from $2.5^{-4} \times 0.1374$ to $2.5^5 \times 0.1374$ pu, (d) T_{d_dg} of VSG varies from $1.09^{-4} \times 0.1$ to $1.09^5 \times 0.1$ s, and (e) T_{d_dg} of droop control varies from $1.25^{-4} \times 0.1$ to $1.25^5 \times 0.1$ s.

evaluated parameters do not influence the stability of active power control in SG-connected mode.

7.2.4 Experimental Results

Laboratory experiments are also executed to support the discussed points in Section 7.2.3. Experimental circuits of stand-alone and connected modes are shown in Fig. 7.14a,b, respectively. A three-phase power supply rectified by a

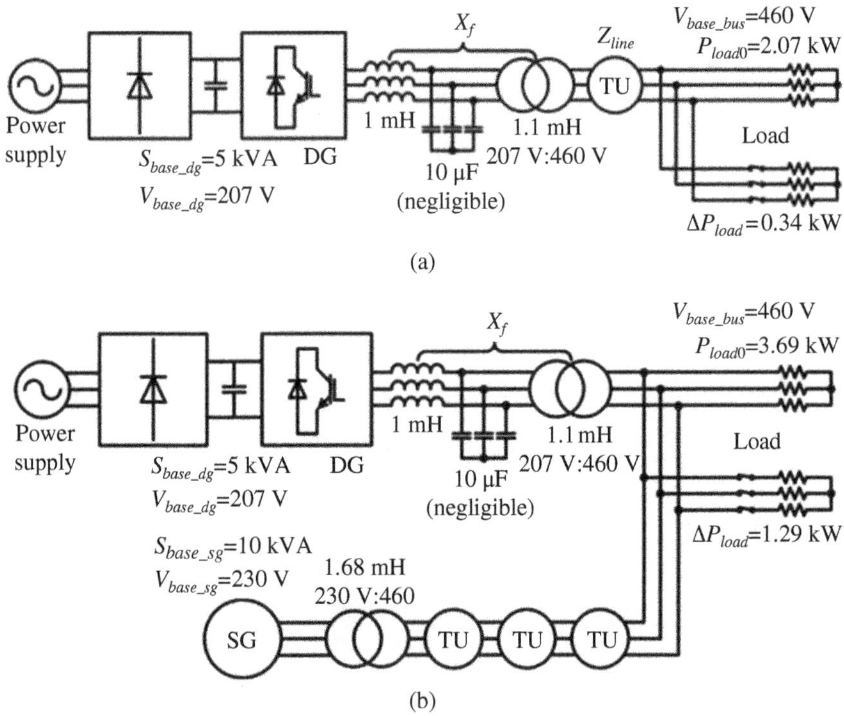

Figure 7.14 Experimental circuit of (a) stand-alone mode and (b) SG-connected mode.

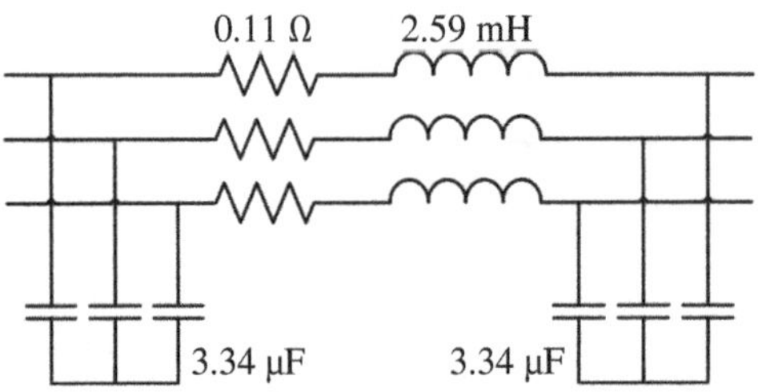

Figure 7.15 TU.

diode bridge is used to imitate the DC output of a DG. The block "transmission line (TU)" in these graphs is a Π circuit to simulate a section of 40 km HV TU, which is demonstrated in Fig. 7.15. Parameters of stand-alone and SG-connected modes are listed in Tables A.3 and A.4, and the results are shown in Figs 7.16 and 7.17. For a stand-alone mode, experimental results (solid lines) coincide with the corresponding theoretical results (dotted lines). This proves again the small-signal models and the conclusions discussed in Section 7.2.3.

For SG-connected mode, by comparing the experimental results, the same conclusions can be drawn. As shown in Fig. 7.17a, system with VSG control

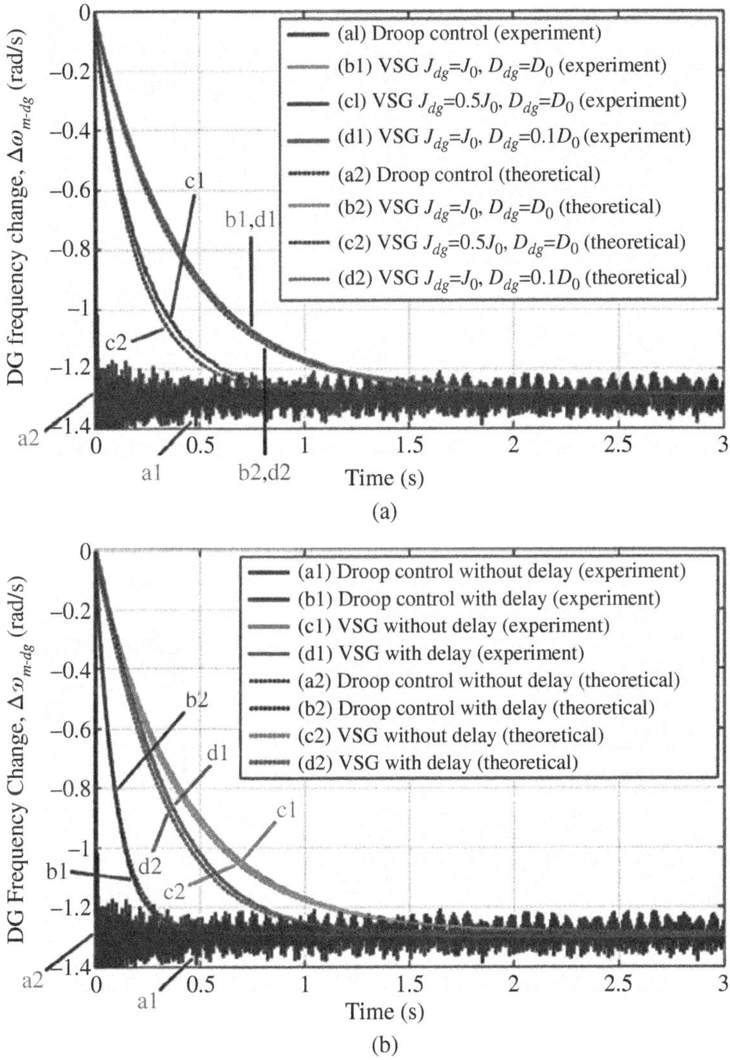

Figure 7.16 Experimental results of stand-alone mode to verify (a) effects of parameters and (b) effects of delays.

still results in slower frequency deviation than that with droop control, and the frequency changes faster when J_{dg} decreases. In Fig. 7.17b, it can still be noticed that the delay ($T_{d\_dg} = 0.1$ s) makes the frequency change faster in VSG control but makes the frequency deviation more slowly in droop control. Similarly, in Fig. 7.17c, with specified lag unit or lead–lag unit, the droop control has similar dynamic response to VSG. Although as shown in the experimental results, it is

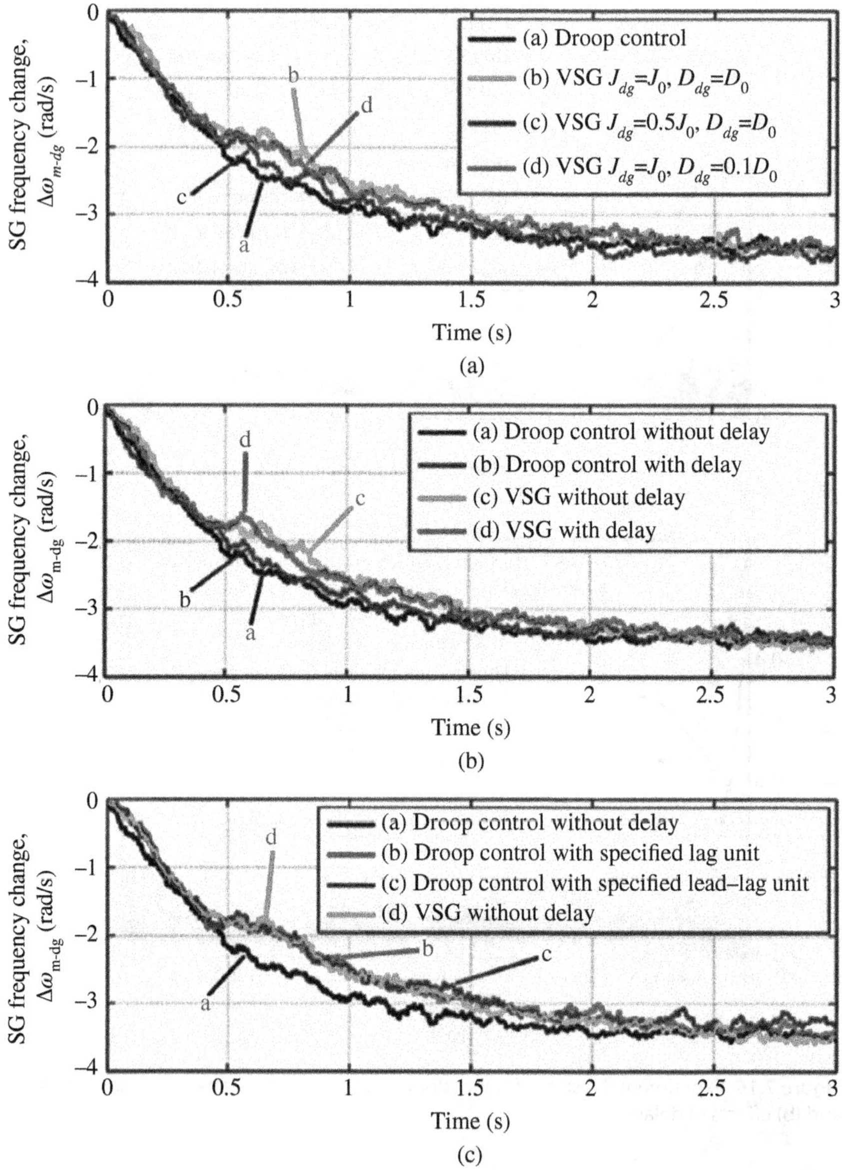

Figure 7.17 Experimental results of SG-connected mode to verify the effects of (a) parameters, (b) delays, and (c) inertial droop control.

difficult to confirm the difference in amplitude of oscillations due to noises and insufficient measurement resolution, simulation results are sufficient to confirm the claimed points.

7.3 Virtual Synchronous Generator-Based Oscillation Damping

In this section, using the linear control theory, the damping property of the VSG system is analyzed, and an oscillation damping approach is developed for DG-based power grids. The method is confirmed analytically and examined by using computer simulations as well as laboratory experiments.

7.3.1 Mathematical Formulation

Consider the VSG block diagram shown in Fig. 7.2, once again. To match the given discussion in this section with the given variables in Fig. 7.2, rename I_{out}, V_{out}, E, and Q_{ref} as I_{grid}, V_{grid}, V^* and Q^*, respectively. When the VSG inverter is connected to the infinite bus, active and reactive powers (P and Q) can be expressed as

$$P + jQ = V_{grid} I^*_{grid} = \frac{V_t V_{grid}}{X_L} \sin\delta + j\frac{V_t V_{grid}\cos\delta - V^2_{grid}}{X_L} \tag{7.30}$$

where "*" represents the conjugate operator, δ is the phase difference between V_t and V_{grid}, V_t is the inverter output voltage, and the V_{grid} is the grid voltage (at the infinite bus). By controlling the amplitude of V_t such that $V_t \cos\delta > V_{grid}$, the inverter supplies reactive power to the grid. Inversely, when $V_t\cos\delta < V_{grid}$, the inverter consumes reactive power. It can be seen that the voltage reference (V^*) is obtained through the PI controller by using the feedback of Q_{out}. The Q^* is the reference reactive power signal.

In the mentioned VSG topology (Fig. 7.2), the active power P and reactive power Q can be controlled by changing δ and V_t, respectively. However, as explained in Chapter 3 and illustrated in [11], for a considerable δ (over 30°) or for a large resistance in connecting line impedance $R+jX$, P is not only a function of δ. The R and X are resistance and inductance of the line between the inverter output and the grid bus, respectively. In the case of an existing considerable resistance R, instead of (7.30), P and Q can be reformulated as follows:

$$P = \frac{V_t V_{grid}(R\cos\delta + X\sin\delta) - RV^2_{grid}}{R^2 + X^2},$$
$$Q = \frac{V_t V_{grid}(X\cos\delta - R\sin\delta) - XV^2_{grid}}{R^2 + X^2} \tag{7.31}$$

Following a manipulation, (7.31) becomes (7.32).

$$P = \frac{V_t V_{grid}}{\sqrt{R^2 + X^2}} \sin\left(\delta + \tan^{-1}\frac{R}{X}\right) - \frac{R V_{grid}^2}{R^2 + X^2},$$
$$Q = \frac{V_t V_{grid}}{\sqrt{R^2 + X^2}} \cos\left(\delta + \tan^{-1}\frac{R}{X}\right) - \frac{X V_{grid}^2}{R^2 + X^2} \tag{7.32}$$

Thus, P contains *sine* function and provides a nonlinear swing equation. Additionally, active and reactive power controls are not independent. The δ and V_t affect both P and Q. This conflicting behavior makes the swing equation more complicated for the realization and control synthesis.

In order to solve this problem, a control function independent of V_t is required. Assume that P and Q are binary functions of δ and V_t, so total differentiation of $P(\delta, V_t)$ and $Q(\delta, V_t)$ can be expressed as follows:

$$dP = P_\delta d\delta + P_{Vt} dV_t, \quad dQ = Q_\delta d\delta + Q_{Vt} dV_t \tag{7.33}$$

where P_δ, P_{Vt}, Q_δ, and Q_{Vt} are partial differentials of P and Q. Following the setting of the initial power values at P_0 and Q_0, respectively, the P and Q can be expressed as (7.34) for a small change in δ and V_t.

$$P = P_0 + dP = P_0 + P_\delta d\delta + P_{Vt} dV_t,$$
$$Q = Q_0 + dQ = Q_0 + Q_\delta d\delta + Q_{Vt} dV_t \tag{7.34}$$

Then, the calculated small change $d\omega_m$ can be expressed as (7.35) by substituting (7.34) for P_{out} in (7.1), by assuming $DP_{base}/\omega_0 = D$,

$$d\omega_m = \frac{(P_{in} - P_0 - P_\delta d\delta) - P_{Vt} dV_t + D(\omega_g - \omega_m)}{J\omega_m} \tag{7.35}$$

The term $P_{Vt}\, dV_t$ shows the effect of the voltage change. In case of neglecting this term, the swing equation is no longer subjected to the voltage change. Therefore, the simplified basic swing equation can be updated as follows:

$$P_{in} - P_{out} + m(Q^* - Q_{out}) = J\omega_m \frac{d\omega_m}{dt} + D(\omega_g - \omega_m) \tag{7.36}$$

where

$$m = -\frac{P_{Vt}}{Q_{Vt}} \tag{7.37}$$

By substituting (7.34) for P_{out} and Q_{out} in (7.36), the calculated small change $d\omega_m$ is expressed as (7.38).

$$d\omega_m = \frac{1}{J\omega_m}\{P_{in} - P_0 - P_\delta d\delta + m(Q^* - Q_0 - Q_\delta dQ_\delta)$$
$$- (P_{Vt} dV_t + mQ_{Vt} dV_t) + D(\omega_g - \omega_m)\} \tag{7.38}$$

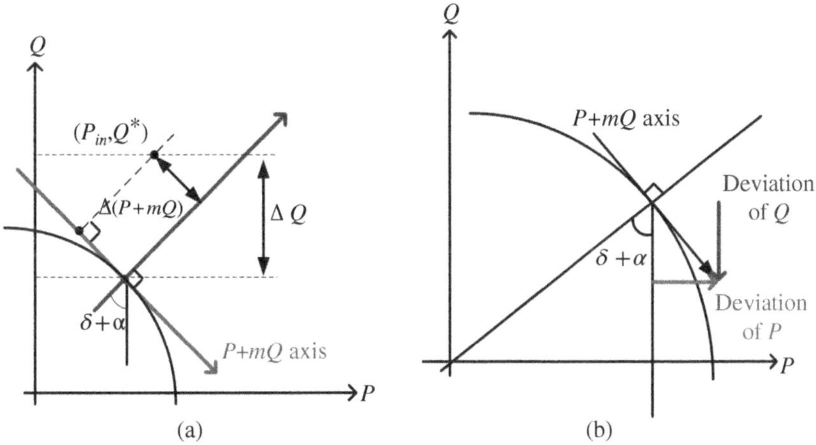

Figure 7.18 (a) Control of $P + mQ$ axis, (b) $P + mQ$ deviation for P and Q.

Considering (7.37), the term $(P_{Vt}\, dV_t + mQ_{Vt}\, dV_t)$ can be modified as follows:

$$P_{Vt}\,dV_t + mQ_{Vt}\,dV_t = P_{Vt}\,dV_t - \frac{P_{Vt}}{Q_{Vt}} Q_{Vt}\,dV_t = 0 \tag{7.39}$$

Equation (7.39) is graphically demonstrated in Fig. 7.18a. This means that Equation (7.36) is not subjected to the voltage change. By feed-backing $P_{out} + mQ_{out}$, the VSG control provides $P_{in} + mQ^*$. On the other hand, PI control for reactive power control produces Q^* from Q_{out}. As a result, P and Q follow their references properly. In addition, $P + mQ$ axis is vertical to radial direction. Thus, the VSG control for $P + mQ$ is not subjected to the voltage deviation and can be only controlled by the angle δ.

Here, m is the gradient of vector $(P_\delta\, d\delta,\, Q_\delta\, d\delta)$. The connection of P and Q is in a circular form; thus, the change directions of P and Q (which are invoked angle and voltage changes) are orthogonal [12]. Therefore, the ratio m can be calculated as follows:

$$\frac{Q_\delta\, d\delta}{P_\delta\, d\delta} = -\frac{P_{Vt}\,dV_t}{Q_{Vt}\,dV_t} = \frac{Q_\delta}{P_\delta} = -\frac{P_{Vt}}{Q_{Vt}} = m \tag{7.40}$$

When $d\delta = 0$ (δ is not changed), dP is equal to $P_{Vt}\, dV_t$ and dQ is equal to $Q_{Vt}\, dV_t$; thus, m is calculated via dividing $-dP$ by dQ. Conversely, when $dV_t = 0$ (V_t is not changed), dP is equal to $P_\delta\, d\delta$ and dQ is equal to $Q_\delta\, d\delta$. Thus, m is calculated via dividing dQ by dP.

The term $P_{in} - P_0 - P_\delta\, d\delta + m(Q_{in} - Q_0 - Q_\delta\, d\delta)$ in (7.38) can be rewritten as

$$P_{in} - P_0 + m\left(Q^* - Q_0\right) - \left(\frac{P_\delta^2 + Q_\delta^2}{P_\delta}\right) d\delta \tag{7.41}$$

Assuming P and Q as

$$P = A\sin(\delta + \alpha) - B, \quad Q = A\cos(\delta + \alpha) - C \tag{7.42}$$

where

$$A = \frac{V_t V_{grid}}{\sqrt{R^2 + X^2}}, \quad B = \frac{RV_{grid}^2}{R^2 + X^2}, \quad C = \frac{XV_{grid}^2}{R^2 + X^2}, \quad \alpha = \tan^{-1}\frac{R}{X} \tag{7.43}$$

and

$$\frac{P_\delta^2 + Q_\delta^2}{P_\delta} = \frac{A^2}{A\cos(\delta + \alpha)} = \frac{A}{\cos(\delta + \alpha)} \tag{7.44}$$

The change of δ causes a deviation in the tangent line direction as shown in Fig. 7.18b. Thus, the synchronizing coefficient P_δ under $P+mQ$ control is expressed as (7.45), by making a projection from the $P+mQ$ axis to P axis.

$$P_\delta = \frac{A}{\cos(\delta + \alpha)}\cos(\delta + \alpha) = A \tag{7.45}$$

It is noteworthy that the quantity of P deviation is equal to the quantity of $P+mQ$ deviation multiplied by $\cos(\delta+\alpha)$. By integrating (7.45), P_{out} is expressed as

$$P_{out} = A\delta + P_0 \tag{7.46}$$

On the other hand, (7.38) can be rewritten as given in (7.47) by making the projection in the same way.

$$d\omega_m = \frac{1}{J\omega_m}\left\{P_{in} - P_0 - P_\delta d\delta + D(\omega_g - \omega_m)\cos(\delta + \alpha)\right\} \tag{7.47}$$

The mQ_{in} and mQ_0 are neglected, $(P_\delta^2 + Q_\delta^2)/P_\delta$ and $D\Delta\omega_m$ become P_δ and $D\Delta\omega_m \cos(\delta+\alpha)$, respectively. Considering D=$D/\cos(\delta+\alpha)$, (7.47) will be changed to

$$d\omega_m = \frac{1}{J\omega_m}\left\{P_{in} - P_0 - P_\delta d\delta + D(\omega_g - \omega_m)\right\} \tag{7.48}$$

Therefore, by replacing $P_0 + P_\delta\, d\delta$ to P_{out}, (7.48) is returned to the temporal differentiation form of ω_m.

$$\frac{d\omega_m}{dt} = \frac{P_{in} - P_{out} + D(\omega_g - \omega_m)}{J\omega_m} \tag{7.49}$$

Assuming $J\omega_m$ as a constant term, (7.49) can be considered as a second-order linear differential equation.

7.3.2 Oscillation Damping Methodology

Considering $J\omega_m$ as a constant term, $\delta = \int(\omega_m - \omega_g)dt$, $\omega_g = 2\pi f_0$, and $P_0 = 0$; first- and second-order temporal differentiation of P are expressed as

$$\frac{dP}{dt} = A(\omega_m - \omega_g), \quad \frac{d^2P}{dt^2} = A\frac{d\omega_m}{dt} \tag{7.50}$$

By substituting (7.50) in (7.49) and labeling $J = J_m$ and $P_{out} = P$, (7.49) is updated in terms of P:

$$\frac{J_m}{A} \cdot \frac{d^2P}{dt^2} + \frac{D}{A} \cdot \frac{dP}{dt} + P = P_{in} \tag{7.51}$$

Using Laplace transformation, (7.51) can be rewritten as (7.52).

$$P(s) = P_{in} \frac{A/J_m}{\left\{s^2 + (D/J_m)s + A/J_m\right\} s} \tag{7.52}$$

Equation (7.52) has the same form as the well-known response of a typical second-order control system to a step input:

$$Y(s) = \frac{\omega_n^2}{\left(s^2 + 2\zeta\omega_n s + \omega_n^2\right) s} \tag{7.53}$$

where ω_n is the undamped natural frequency and ζ is the damping ratio of the system. The system response performance can be changed by tuning the damping ratio. Comparing (7.52) to (7.53), the ω_n and ζ are expressed as follows:

$$\omega_n = \sqrt{\frac{A}{J_m}}, \quad \zeta = \frac{D}{2\sqrt{AJ_m}} \tag{7.54}$$

Thus, via a proper realization of $D = 2\zeta\sqrt{AJ_m}$, it is possible to adequately damp the oscillation.

Therefore, following the representation of the VSG control core as (7.36) instead of (7.1), the VSG control for active power is linearized independently from the voltage variation. Finally, to propose an effective oscillation methodology, first the gradient m should be obtained for the mentioned control function. The algorithm for estimating the m value is summarized as Algorithm 1.

Algorithm 1 ***m* estimation**

1. Set an initial value for m (here, −0.1).
2. By temporal differentiating P, Q, δ, and V_t, calculate dP/dt, dQ/dt, $d\delta/dt$, and dV_t/dt.
3. Compare the temporal differentials to the previous values. When only the sign of one of $d\delta/dt$ and dV_t/dt is changed, go to step 4. Otherwise, go to step 6.
4. In steady state, if $dP/dt = 0$ or $dQ/dt = 0$, m is not estimable, and go to step 6. Otherwise, go to step 5.
5. Renew m. When the sign of $d\delta/dt$ is changed, set $m = -(dP/dt)/(dQ/dt)$. When the sign of dV_t/dt is changed, set $m = (dQ/dt)/(dP/dt)$. If m is positive, stop renewing m. Then, go to step 6.
6. Use low-pass filter in order to smooth the m output. Then, go back to step 2.

The P, Q, δ, and V_t are measured in each sampling time step. The m can be calculated from (7.40) in each iteration. Additionally, by considering (7.33), it is clear that

$$\text{If } d\delta/dt = 0, \text{ then } m = -\frac{P_{Vt}}{Q_{Vt}} = -\frac{dP}{dQ} = -\frac{dP}{dt}\Big/\frac{dQ}{dt};$$

$$\text{and if } dV_t/dt = 0, \text{ then } m = \frac{Q_\delta}{P_\delta} = \frac{dQ}{dP} = \frac{dQ}{dt}\Big/\frac{dP}{dt} \tag{7.55}$$

In the VSG system, since the δ is an unknown value, $(\omega_m - \omega_g)$ is used instead of $d\delta/dt$. If $dP/dt = 0$ or $dQ/dt = 0$, m is inestimable. Thus, if these values are under a given threshold value, the calculation is skipped. In the performed simulation, m is a negative value. Moreover, since the m is calculated at the point that δ and V_t are extrema, the m is a discrete and variable parameter. To smooth the m behavior, a low-pass filter is used.

For linearizing purpose, here, the damping factor D is modified to $D/\cos(\delta + \alpha)$. Thus, considering $m = -\tan(\delta + \alpha)$, the damping factor can be calculated as follows:

$$\frac{D}{\cos(\delta + \alpha)} = D\sqrt{1 + \tan^2(\delta + \alpha)} = D\sqrt{m^2 + 1} \tag{7.56}$$

After obtaining m, in order to analyze the system oscillation, according to the standard second-order system in linear control theory as explained in (7.53) and (7.54), the coefficient A should also be determined. The developed algorithm for calculating A is as follows:

Algorithm 2 Calculation of A

---

1. Set the initial value of A such that the D becomes the same as in the conventional control mode (D_0).
2. By temporal differentiation of P, δ, and V_t, calculate dP/dt, $d\delta/dt$, and dV_t/dt.
3. Compare the temporal differential to the previous values. When the sign of dV_t/dt is changed and $d\delta/dt$ is not, go to step 4. Otherwise, go back to step 2.
4. If $d\delta/dt = 0$, A is not estimable and go back to step 2. Otherwise, A is calculated as $(dP/dt)/(\omega_m - \omega_g)$ according to (7.50).
5. Calculate the D using (7.54). If $D < D_0$, reset the D at D_0, and go back to step 2.

The J_m is initially set at (moment of inertia) × (rated angular speed). The calculation of step 2 in Algorithm 2 is the same as step 2 for the m estimation Algorithm 1. If $d\delta/dt$ is under the given threshold value, the calculation is skipped in step 4. To reduce the oscillation, the D is compelled to be larger than D_0. Additionally, the damping factor has to be calculated as (7.56). Therefore, (7.56) can be updated as follows:

$$\frac{D}{\cos(\delta + \alpha)} = 2\zeta\sqrt{AJ_m(m^2 + 1)} \tag{7.57}$$

7.3.3 Simulation Results

To verify the proposed oscillation damping methodology, a simulation using PSCAD/EMTDC software is performed. The results of conventional VSG damping control and the developed approach are compared.

7.3.3.1 Conventional Virtual Synchronous Generator Control

To affirm the difficulty of control in the case of $\sin\delta \neq \delta$, a simulation is carried out. The impedance with a resistance part is inserted between the VSG and the grid. Then, the VSG is driven in the grid-connected mode. The model of the simulated system is shown in Fig. 7.2, and the simulation parameters are shown in Table A.5. Equation (7.1) is used to calculate the ω_m as explained in Fig. 7.4. For the constant active power of 1.0 MW, the reactive power reference is changed from 0 to −0.5 Mvar at 2 s. Here, "−" sign for reactive power shows the leading property. The system response is shown in Fig. 7.19.

After changing the reactive power reference, an undamped oscillation appears in the system response. This phenomenon may occur due to the mentioned conflict between active and reactive power controls. Under this condition, the V_t deviation cancels the damping factor effect in (7.35). Thus, P and Q are not able to converge to the reference values.

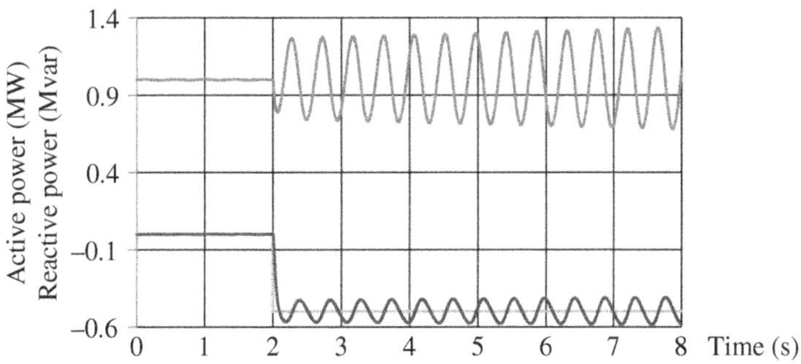

Figure 7.19 System response for conventional VSG control: P (up signal), Q^* (light down signal), and Q (dark down signal).

7.3.3.2 The Proposed Damping Approach

To verify the proposed damping control scheme, a simulation is carried out. The condition of the simulation is the same as before. Reactive power reference is changed from 0 to −0.5 Mvar at 2 s. By applying (7.36) and (7.56), the VSG damping control becomes independent of the voltage deviation. The thresholds for dP/dt and dQ/dt are set at 0.3 pu, and the time constant of low-pass filter is fixed at 1 s. System response and corresponding m change are shown in Fig. 7.20.

After finding $m = -0.36$, for the problem at hand ($D_0 = 0.045$), D is minimized by using the developed algorithms. By applying the algorithm, the oscillation is controlled by the value of ζ. The scenarios of $D = 0.045$ (constant), $\zeta = 0.707$, and $\zeta = 1.5$ are considered. The thresholds for dP/dt and $d\delta/dt$ are set at 0.02 pu and 0.02 rad/s, respectively. For 0 Mvar reactive power, the active power reference (P_{ref}) is changed from 1 to 1.5 MW. The simulation results are shown in Fig. 7.21.

In comparison with Fig. 7.19, the oscillation is effectively damped. The dP/dt and dQ/dt are under the assigned threshold at 3.5 s, the control system ceases updating the m value, and m converges to −0.36. As a result, linearizing and decoupling are succeeded.

In the scenario of constant D, the active power oscillates following the reference change, because the VSG emulates the SG properties. By applying the proposed damping approach, the output wave form varies along the ideal response

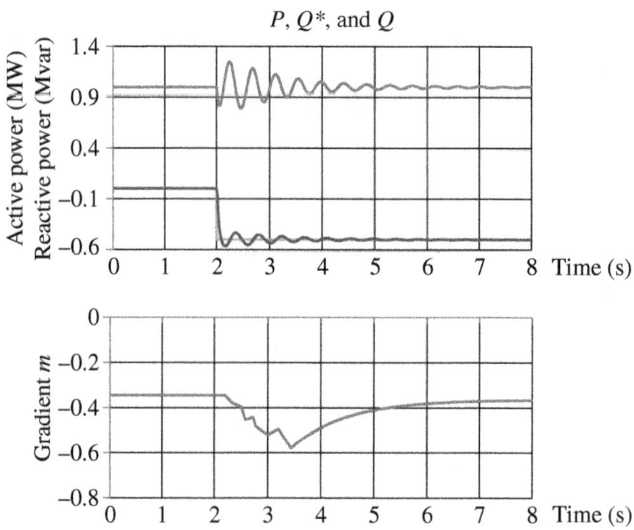

Figure 7.20 System response using proposed control method: P (up signal), P^* (light down signal), and Q (dark down signal).

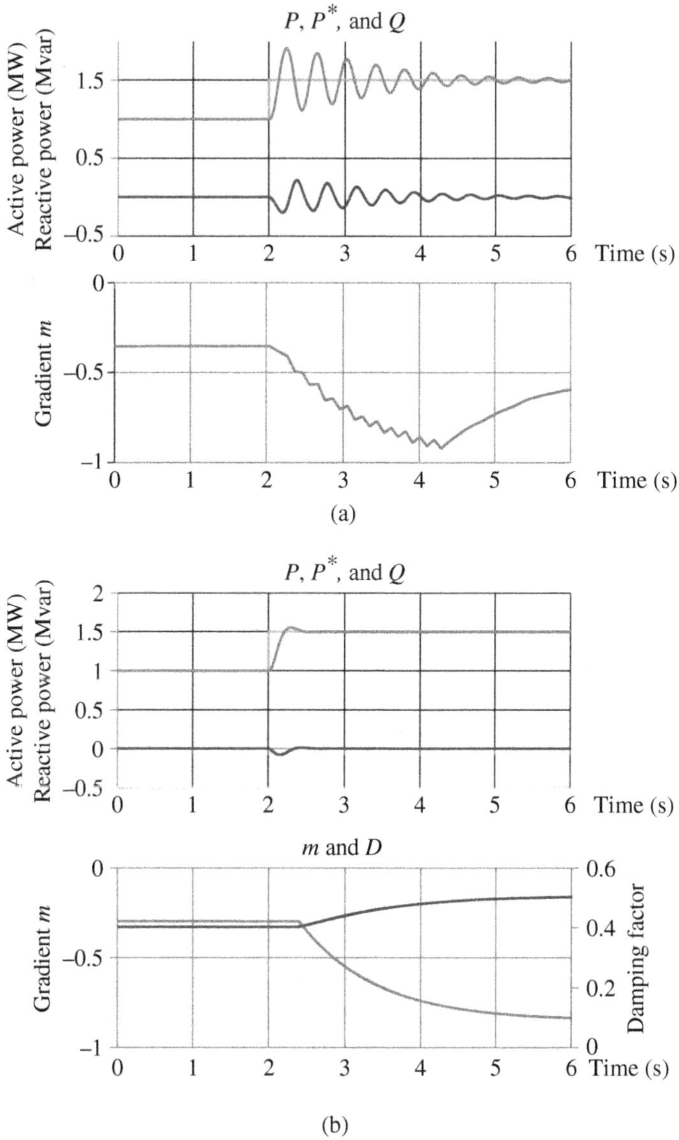

Figure 7.21 System response using the proposed damping approach: (a) $D = 0.045$ (constant), (b) $\zeta = 0.707$, and (c) $\zeta = 1.5$.

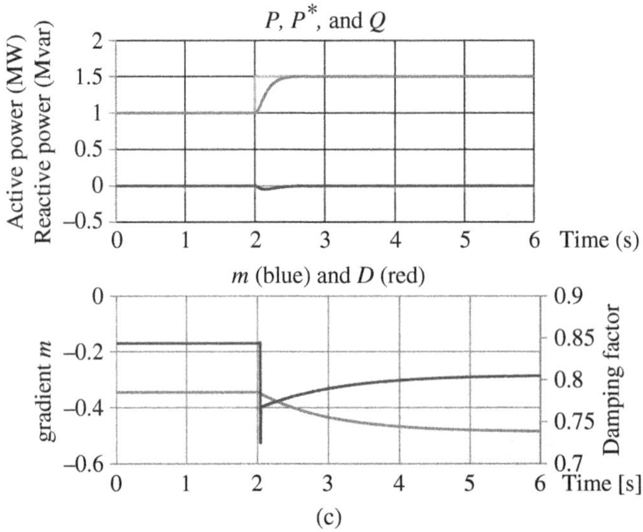

Figure 7.21 (Continued)

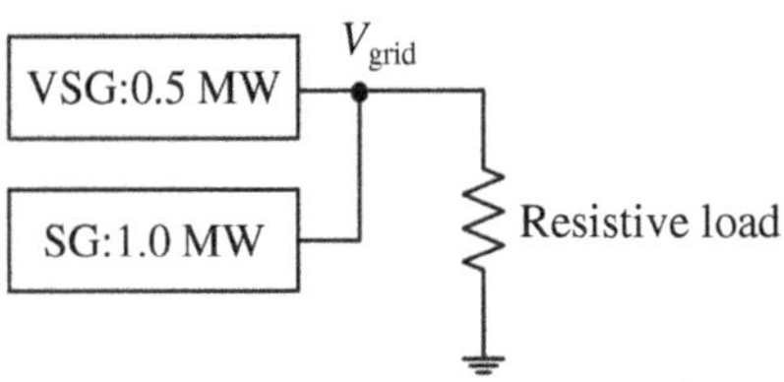

Figure 7.22 Parallel running system of VSG and SG.

form. For $\zeta = 0.707$, the response is fast with a small overshoot. With $\zeta = 1.5$, there is no overshoot, but the response is slower. As a result, the oscillation is damped adequately in both scenarios.

To verify the damping ability in the intentional islanding mode, a simulation for parallel running of VSG and SG is carried out. The model of the simulated system is shown in Fig. 7.22, and the simulation parameters of SG are the same as the given parameters in [13]. In addition, the VSG uses the governor model (Fig. 7.2) in intentional islanding operation. Base power of VSG and SG is set at 0.5 and 1.0 MW, respectively. The damping ability is verified in the case that the load consumption is changed from 1.5 to 0.75 MW. The simulation results are shown in Fig. 7.23.

The P_{SG} and P_{VSG} are the output active power of SG and VSG, respectively. The ω_{SG} is the rotor speed of SG. There is a crosscurrent between SG and VSG (Fig. 7.23a). The crosscurrent is suppressed by applying the proposed damping control. However, the oscillation in the first cycle still remains.

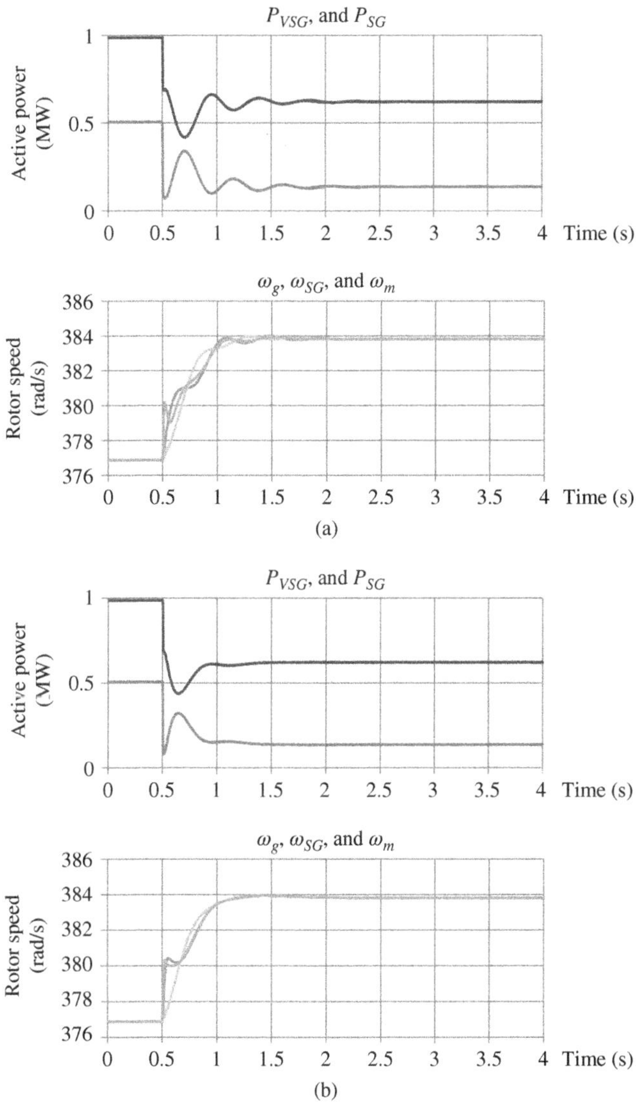

Figure 7.23 Power response of VSG and SG: (a) $D = 0.045$ (constant) and (b) $\zeta = 1.5$.

7.3.4 Experimental Results

To show the effectiveness of the proposed damping methodology, an experimental study for a system configuration, which is shown in Fig. 7.24a, is carried out. The experimental system consists of a VSG controlled by a digital signal processor (DSP). Base power and voltage are 10 kW and 207 V, respectively.

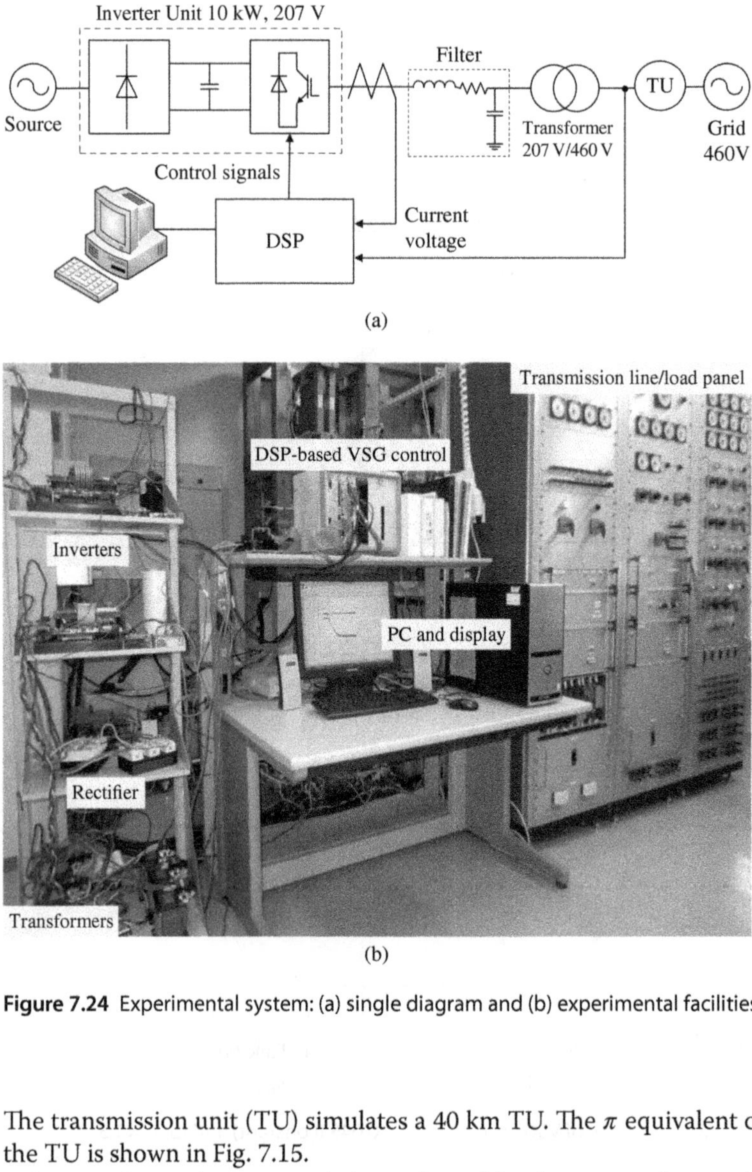

Figure 7.24 Experimental system: (a) single diagram and (b) experimental facilities.

The transmission unit (TU) simulates a 40 km TU. The π equivalent circuit of the TU is shown in Fig. 7.15.

An experimental test scenario is conducted for the proposed control method described in Section 7.3.2. VSG parameters are shown in Table 7.1. For 0 kvar reactive power, the active power reference is changed from 1 to 3 kW. The thresholds for m and A are set at 0.01 pu and 0.1 rad/s, respectively. The scenarios of $D = 0.045$ (constant), $\zeta = 0.707$, and $\zeta = 1.5$ are considered again. The experimental results are shown in Fig. 7.25.

Table 7.1 VSG parameters.

Base power P_{base} (kW)	10
Base voltage V (V)	207
Per unit inertia constant H (s)	4
Moment of inertia J (kg m^2)	0.563
Damping factor D	0.045
Interconnecting reactance X_L (%)	7.0
PI control gain k_p	0.0025
PI control time constant k_i (s)	0.015

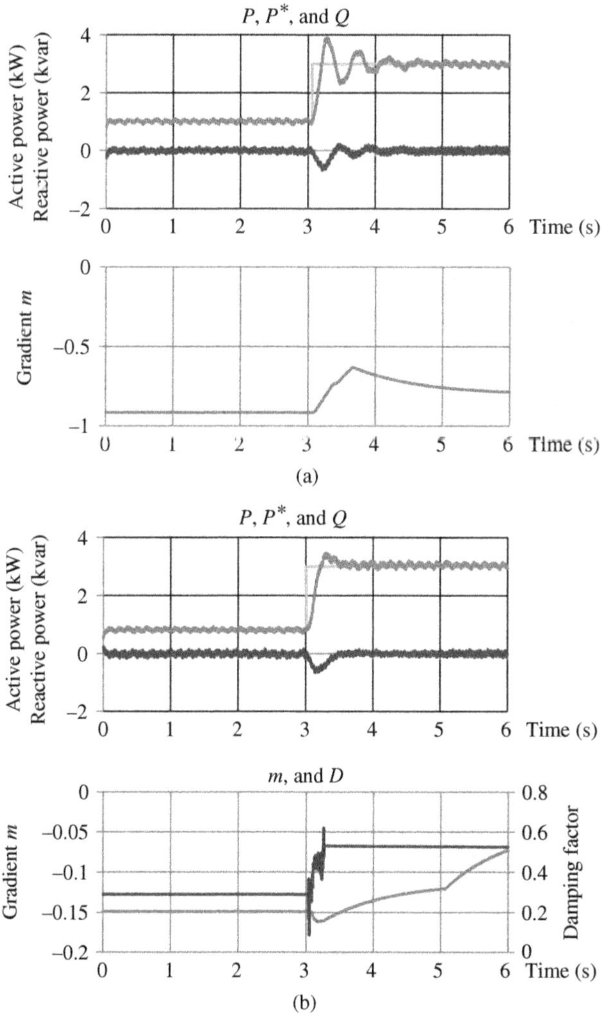

Figure 7.25 The experimental results of the proposed approach: (a) $D = 0.045$ (constant), (b) $\zeta = 0.707$, and (c) $\zeta = 1.5$.

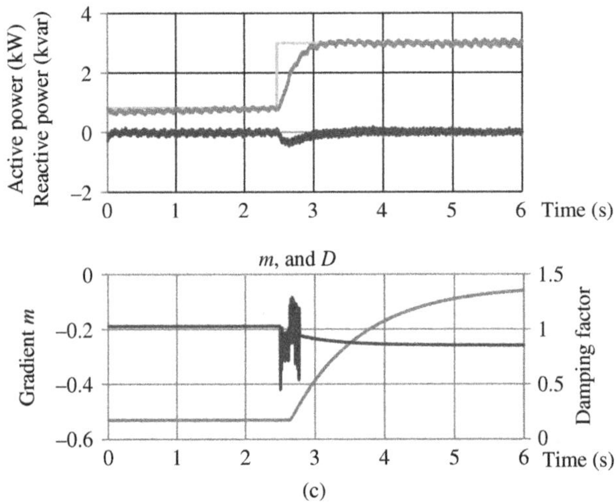

Figure 7.25 (*Continued*)

Using the proposed damping approach, the oscillation by emulating the SG inertia and damping characteristics is significantly damped. Similar to the simulation results, the response for scenario of $\zeta = 0.707$ shows an overshoot. However, the response is quicker than that for the other scenario with $\zeta = 1.5$.

7.4 A Virtual Synchronous Generator Scheme with Emulating More Synchronous Generator Characteristics

The SGs can be synchronized to the grid by synchronizing power and can operate in both connected and islanded modes. When SGs are disconnected from the grid, they can continue to operate without changing the local control logic [14]. Since the VSG philosophy attempts to reproduce the characteristics of the actual SGs on inverters, so using the VSG control, the inverter can operate in both mentioned operating modes. Various methods and topologies have been proposed for the VSG control [5]. In the described VSG schemes in the previous sections, the VSGs are constructed on the voltage-controlled inverters.

However, most of conventional DG inverters are controlled by current controllers. The current controllers can be synchronized with the grid by a PLL and control the inverter currents according to current references. The current controller is using a premise that the inverter is connected to the grid, so it cannot operate in the isolated grid. The conventional method for the inverters in isolated grid is constant voltage and constant frequency (CVCF) control. The output frequency and voltage RMS of the CVCF inverters are constant. The CVCF control cannot operate in the grid-connected operation mode since it cannot be synchronized with the grid. The DGs are required to continue to operate in various conditions, and the system must continue to supply power to critical loads without interruption due to a grid fault.

The DGs must withstand a grid failure, that is, fault ride through (FRT) property. It is difficult to reduce the rush current due to the grid fault for the voltage-controlled inverters since the inverter currents are not controlled directly. It is also difficult for the current-controlled inverters with PLL to have the FRT capability since there is no evidence that the PLL behaves precisely following a grid fault. Here, a current-control-based VSG without PLL is proposed. The control system for the given VSG topology shown in Fig. 7.2 was working based on simulation of the basic SG swing equation in (7.1), while, in the present VSG scheme, more SG properties including governor, rotor, damper, and automatic voltage regulator (AVR) are emulated to improve the system performance.

7.4.1 Emulating Synchronous Generator Characteristics

Here, the proposed VSG control is based on a current-control-type inverter. The proposed VSG scheme is shown in Fig. 7.26. It consists of an impedance model, a governor model, an AVR model, a rotor model, and a damper model. The L_F and R_F are reactance and resistance, respectively, of the inverter filter reactor. The grid in Fig. 7.26 represents an aggregated model for the rest of the network including generators, TUs, and loads.

7.4.1.1 Impedance

The inverter current references of the VSG are calculated by using impedance model. The impedance model is based on the algebraic approximation of Park's equation [15]. Since the inverter current controller cannot respond fast enough to simulate transient and subtransient impedance components, it simulates the impedance of an SG without transient and subtransient impedances.

The generator is assumed to be a cylindrical type, where the reactances of direct axis and quadrature axis are equal. Figure 7.27 shows the impedance model of the VSG and its phasor diagram. Here, the synchronous impedance of the generator is denoted by $r+jx$. The emf induced by the field windings is denoted by E_q. The phase difference between the rotor and the grid is denoted by δ, which is known as load angle [16].

The q axis is chosen in parallel with E_q, while the d axis is taken at a right-side angle with it. The grid voltage is denoted by V_g, and its projections to d and q axes are denoted by V_{gd} and V_{gq}. The generator currents and its projections to d–q axes are denoted by I^*, I_d^*, I_q^*, respectively. Then, one can write

$$\begin{aligned}\dot{E}_q - \dot{V}_g &= (E_q - V_{gq}) - jV_{gd} = (r+jx)(I_q^* + jI_d^*)\\ &= (rI_q^* - xI_d^*) + j(rI_d^* + xI_q^*)\end{aligned} \tag{7.58}$$

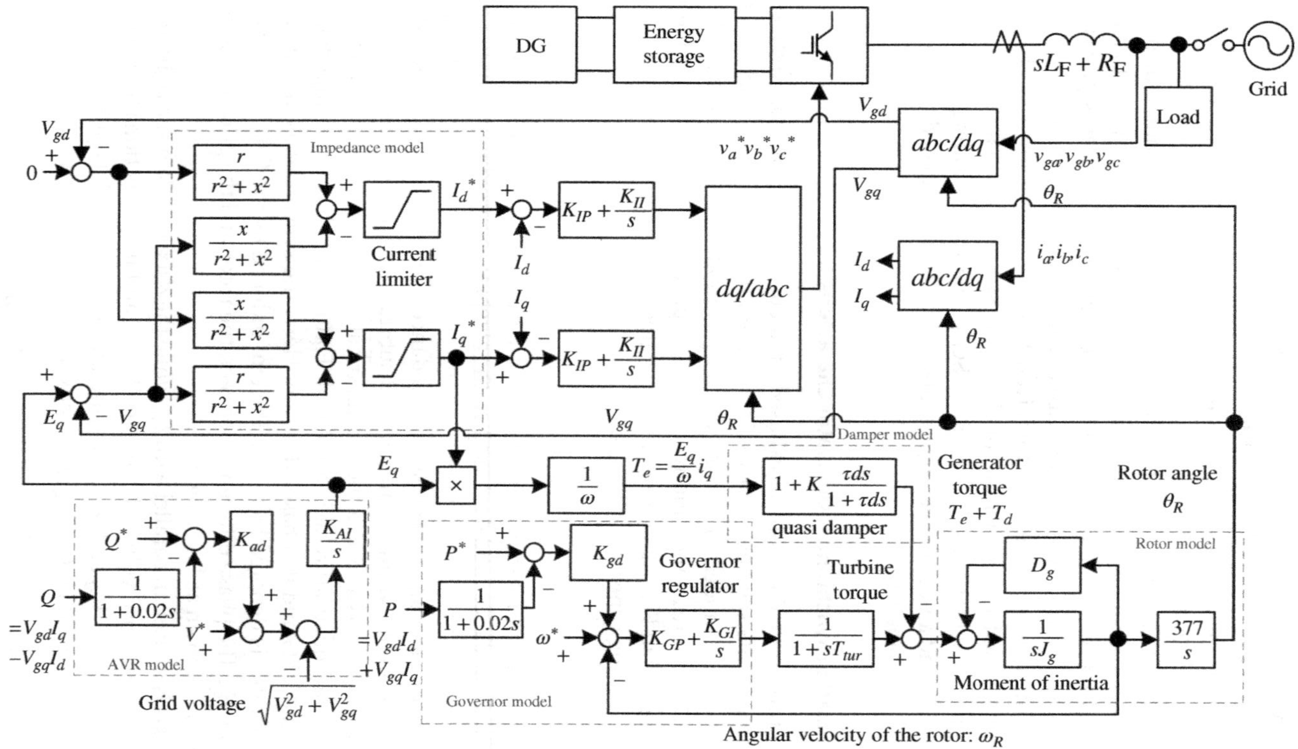

Figure 7.26 VSG control block diagram.

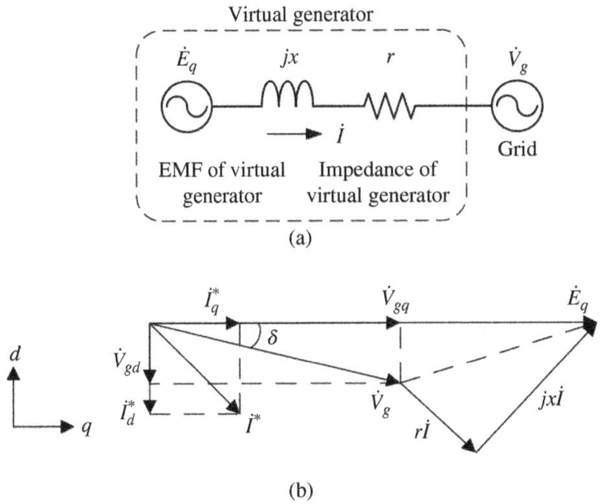

Figure 7.27 VSG: (a) impedances model and (b) phasor diagram.

From (7.58), the current references fed to the inverter are taken as

$$\begin{bmatrix} I_q^* \\ I_d^* \end{bmatrix} = \frac{1}{r^2 + x^2} \begin{bmatrix} r & x \\ -x & r \end{bmatrix} \begin{bmatrix} E_q - V_{gq} \\ -V_{gd} \end{bmatrix} \tag{7.59}$$

This equation is realized in the "Impedance model" block of Fig. 7.26, where the signals represent the instantaneous values. The x of the impedance model corresponds to reactance of actual SGs. The VSG impedance x should be set to approximately transient impedance $x_d{}'$ of other SGs in order to share loads among generators.

7.4.1.2 Governor and Automatic Voltage Regulator

The VSG model must simulate governor and an AVR in order to share the existing loads with other generators. The governor model is shown in Fig. 7.26. A PI controller is used as the governor regulator. The turbine model consists of a first-order system. The governor uses a droop characteristic. The difference between active power P and its reference P^* is fed to the input of the regulator via the gain K_{gd}. The parameter K_{gd} is referred to as the speed regulation.

The AVR model is also illustrated in Fig. 7.26. A simple integral controller is used as the AVR. It simulates the AVR function and field winding characteristic of the actual SGs. The difference between the reactive power Q and its reference Q^* is also fed to the input of the regulator via the gain K_{ad}. The AVR also uses the droop characteristic.

7.4.1.3 Rotor

The phase of d–q transformation is calculated by using the rotor model. Assuming no winding losses, the generator electric torque T_e is given by

$$T_e = E_q i_q / \omega \tag{7.60}$$

The rotor dynamics of the virtual generator is given by

$$J\frac{d\omega_R}{dt} = T_p - T_e - T_d - D_g\omega_R \tag{7.61}$$

where T_p and D_g denote the torque and mechanical damping of the prime mover, which is proportional to the rotor speed, and T_d is the damping torque. The moment of inertia of the generator and the prime mover are merged as J. The J should be set to the same level with other generators because of well-balanced load sharing in transient condition. The rotor angle θ_R can be presented as follows:

$$\theta_R = \int \omega_R dt + \theta_{R0} \tag{7.62}$$

where θ_{R0} is the initial value of θ_R; θ_R is the phase of E_q as shown in Fig. 7.27.

7.4.1.4 Synchronizing Power

In the real SGs, if the load angle δ changes, load angle returns back to the original value by the existing synchronizing power. The proposed VSG can also be synchronized to the grid since the VSG controller has a synchronizing power. In this section, the grid characteristics are expressed in a block diagram in order to analyze the VSG synchronizing power. The relation between the grid voltage V_g and the rotor angle θ_R is obtained. Grid phase θ_g is the phase of V_g as shown in Fig. 7.27 and is given by

$$\theta_g = \int \omega_g dt + \theta_{g0} \tag{7.63}$$

where ω_g is the grid frequency and θ_{g0} is the initial value of θ_g. As shown in Fig. 7.27b, the δ is the difference between θ_R and the grid phase θ_g and is given by

$$\delta = \theta_R - \theta_g \tag{7.64}$$

From Fig. 7.27b, the relation between δ and V_g is known as

$$\mathbf{V}_g = jV_{gd} + V_{gq} \tag{7.65}$$

where $V_{gd} = -V_g \sin\delta$, $V_{gq} = V_g \cos\delta$.

During VSG startup, the θ_{R0} is set to be equal to the θ_{g0}. In steady state, V_g is equal to V_{g0} $(=jV_{gd0} + V_{gq0})$. Considering $\delta = \delta_0 + \Delta\delta$, the grid voltage can be presented as follows:

$$V_g = V_{g0} + \Delta V_g = j(V_{gd0} + \Delta V_{gd}) + (V_{gq0} + \Delta V_{gq}) \tag{7.66}$$

where

$$\Delta V_{gd} = \frac{\partial}{\partial \delta}(-V_g \sin \delta)\Delta\delta = -V_g \cos \delta_0 \cdot \Delta\delta = -V_{gq0} \cdot \Delta\delta \tag{7.67}$$

$$\Delta V_{gq} = \frac{\partial}{\partial \delta}(V_g \cos \delta)\Delta\delta = -V_g \sin \delta_0 \cdot \Delta\delta = V_{gd0} \cdot \Delta\delta \tag{7.68}$$

Figure 7.28 realizes Equations (7.64)–(7.68) in the form of a block diagram. This figure shows the characteristics of grid voltages V_{gd} and V_{gq}. Figure 7.26 is redrawn in Fig. 7.29 in which the inverter, filter, and grid are represented by their block diagrams. The arrows in Fig. 7.29 indicate the feedback loops of the rotor angle θ_R. The solid red lines indicate the positive polarity of the signal flows, while the dashed blue line indicates the reverse. If the change of the rotor angle θ_R is positive, then the generator torque increases, the angular velocity decreases, and finally, the rotor angle θ_R decreases. In the same way, if the deviation of θ_R is negative, the generator torque decreases and the deviation returns back to the equilibrium. This system consists of negative feedback of $\theta_{R,}$ and it can be synchronized with the grid, which is called synchronizing power. When conventional inverters with PLL are operating while a voltage sag occurs, they are required to adjust the PLL parameters to detect the exact grid phase. The proposed VSG can also operate in the voltage sag condition without needing a PLL.

After disconnection from the grid, the voltage V_g is changed from grid voltage to load voltage, and VSG can continue to operate without changing controller. The VSG can be reconnected to the grid by using synchroscope according to the same producer as conventional SGs.

7.4.1.5 Damping

In actual SGs, the damper winding has a significant role in stabilizing the rotor swing. In the developed VSG scheme, the damping torque, which is generated by damper windings of an actual SG, is simulated. For a small τ_d in the

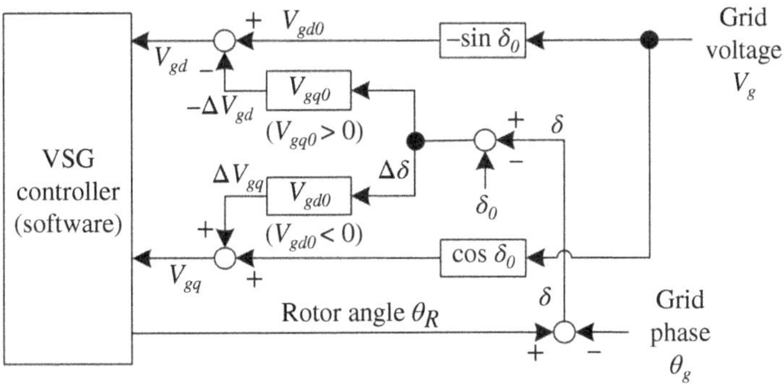

Figure 7.28 Characteristics of grid voltage V_{gd} and V_{gq}.

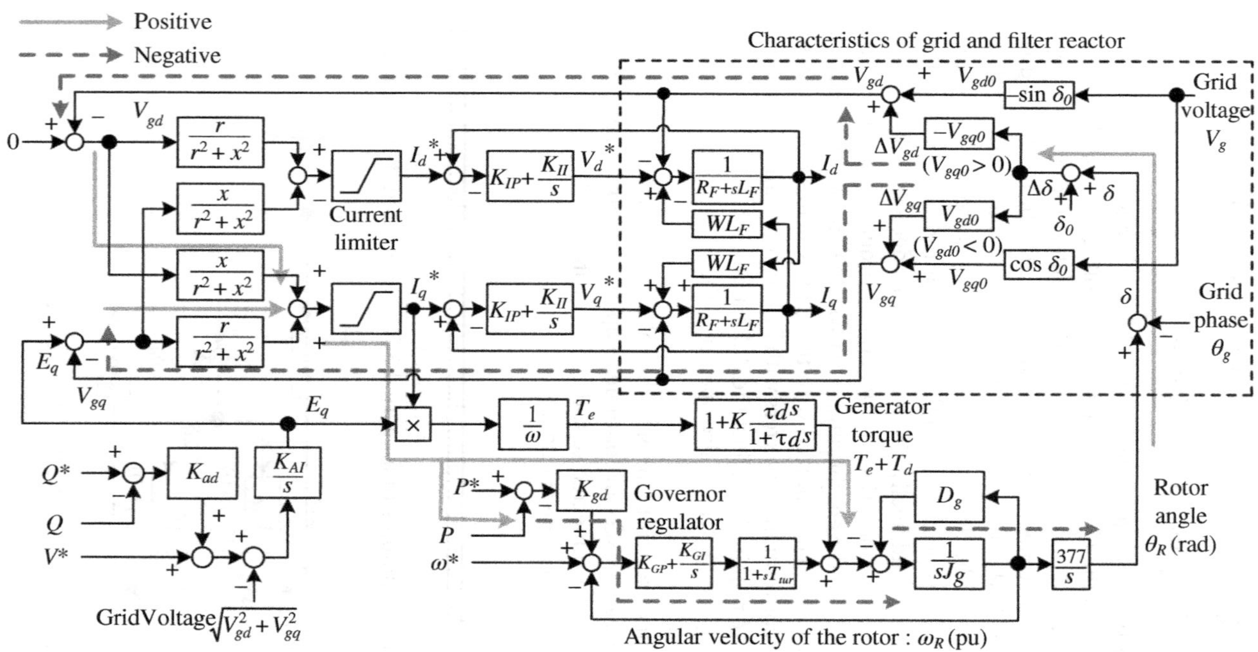

Figure 7.29 The VSG control and grid voltage characteristics.

first-order delay element of damper model, damping torque T_d can be given by the following equation:

$$T_d = K\tau_d \frac{d}{dt} T_e \tag{7.69}$$

To simplify the analysis, the model is approximated by setting $E_q = 1.0$ pu and $V_{gq0} = 1.0$ pu. From Fig. 7.29, T_e and T_d are given by (7.70) and (7.71) using (7.67) and (7.69).

$$T_e = -E_q \frac{x}{r^2 + x^2}(V_{gd0} + \Delta V_{gd}) = -\frac{x}{r^2 + x^2}(V_{gd0} - \Delta\delta) \tag{7.70}$$

$$T_d = K\tau_d \frac{dT_e}{dt} = K\tau_d \frac{x}{r^2 + x^2}\frac{d\Delta\delta}{dt} = D\frac{d\Delta\delta}{dt} \tag{7.71}$$

where D is

$$D = K\tau_d \frac{x}{r^2 + x^2} \tag{7.72}$$

The damping power generated by the damper winding of an actual SG can be given by [17]

$$P_{damp} = P_D \frac{d\Delta\delta}{dt} \tag{7.73}$$

where P_D is called damping factor and is expressed as follows:

$$P_D = V^2 \left\{ \left(\frac{1}{X''_d} - \frac{1}{X'_d} \right) T''_d \sin^2 \delta_0 + \left(\frac{1}{X''_q} - \frac{1}{X'_q} \right) T''_q \cos^2 \delta_0 \right\} \tag{7.74}$$

Assuming that the rotor speed ω_R is 1.0 pu, the damping torque T_d of (7.71) is equal to damping power P_{damp} of (7.73). It proves that the damping torque T_d in VSG simulates the power generated by damper winding of an actual SG. Thus,

$$P_D = D = K\tau_d \frac{x}{r^2 + x^2} \tag{7.75}$$

7.4.2 Stability Analysis and Parameters Design

It is necessary to analyze the stability of the introduced VSG scheme in order to confirm its feasibility and to provide a method for parameter tuning.

7.4.2.1 Islanded Operation

In the islanded operation, the output active power P is determined by the connected loads to the inverter. In this section, the control stability is analyzed in an islanded operation in order to design the PI control gain of the governor model. Assuming that the rotor speed ω_R is 1 pu, the electrical torque T_e is equal to the active power P. Therefore, the system block diagram in the islanded operation can be simplified as shown in Fig. 7.30. In Fig. 7.30a, P^* and ω^* are omitted. The transfer function between P and ω_R can be obtained from Fig. 7.30 as follows:

$$\frac{\omega_R}{P} = -\frac{(K\tau_d T_{tur}s^3 + (T_{tur} + K\tau_d)s^2 + (K_{gd}K_{GP} + 1)s + K_{gd}K_{GI})}{(JT_{tur}s^3 + (D_g T_{tur} + J)s^2 + (D_g + K_{GP})s + K_{GI})} \tag{7.76}$$

Using the Routh stability criterion, the stability condition for the aforementioned transfer function is given by

$$\frac{(D_g T_{tur} + J)(D_g + K_{GP})}{JT_{tur}} > K_{GI} \tag{7.77}$$

7.4.2.2 Grid-connected Operation

In the grid-connected operation, the rotor speed ω_R is determined by the grid frequency. To simplify the analysis, the model is approximated by assuming E_q as a constant parameter, $\omega_R = 1.0$ pu, and $V_{gq0} = 1.0$ pu. The relevant feedback loop shown in Fig. 7.29 is also neglected (for the case of $r = 0$). The simplified block diagram in the connected operation is shown in Fig. 7.30b. In this figure, the P^* and ω^* are removed and the K_{plant} is

$$K_{plant} = \frac{377x}{r^2 + x^2} \tag{7.78}$$

From Fig. 7.30b, the transfer function between P and ω_R can be obtained as follows:

$$\frac{P}{P^*} = \frac{K_{gd}(sK_{GP} + K_{GI})K_{plant}}{\begin{array}{c}(T_{tur}Js^4 + T_{tur}(K_{plant}K\tau_d + D_g + J)s^3 \\ + \left(T_{tur}K_{plant} + K_{plant}K\tau_d + D_g\right)s^2 \\ + \left(K_{GP}K_{gd} + 1\right)K_{plant}s + K_{GI}K_{gd}K_{plant})\end{array}} \tag{7.79}$$

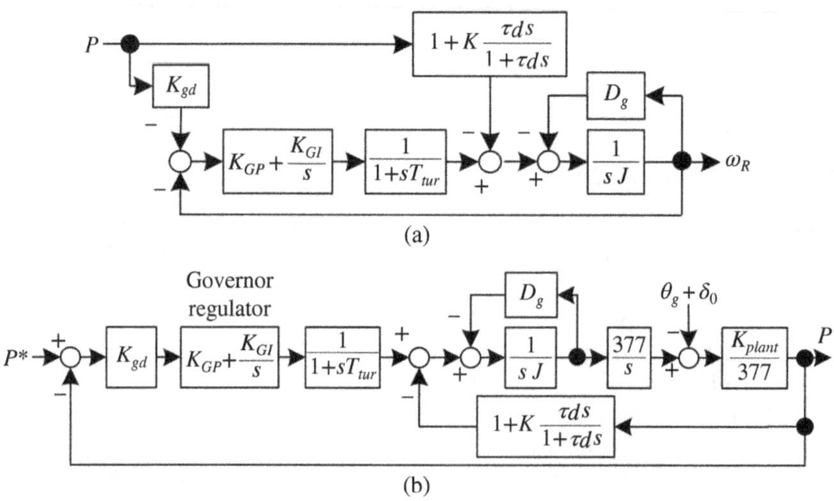

Figure 7.30 Simplified control loops: (a) from P to ω_R in an isolated grid and (b) from P^* to P in a grid-connected operation.

Using the Routh stability criterion, the stability condition for the aforementioned transfer function is achieved.

$$\tau_d K > \frac{-\left(T_{tur}K_{plant} + 2D_g + J\right) + \sqrt{\left(T_{tur}K_{plant} - J\right)^2 + 4JK_{plant}\left(1 + K_{GP}K_{gd}\right)}}{2K_{plant}} \tag{7.80}$$

The τ_d to remove the noise from the inverter switching must be enough small to neglect the first-order delay element of the damper model. As J increases, damping effect decreases, and the damping factor K must be set to a larger value. For example, using (7.72) for the damping factor of $D = 0.2$, K should be set at 10.

7.5 Active Power Performance Analysis in a Microgrid with Multiple Virtual Synchronous Generators

7.5.1 Closed-Loop State-Space Model

In this section, an islanded MG, which consists of two DGs using VSG control, is studied, as shown in Fig. 7.31. The DGs are connected to a common ac bus via distribution lines to supply the loads inside the MG. Note that the capacitor of the DG output LC filter shown in Fig. 7.2 is neglected, as its susceptance is usually negligible at fundamental frequency. In order to understand the causes of active power oscillation and to find proper solutions, a state-space model for the closed-loop active power control of the MG (Fig. 7.31) can be obtained as given in (7.27), (7.81)–(7.83) with a deduction process same as that shown in Section 7.2. Here, the governor first-order lag unit (delay) is removed by setting $T_d = 0$. Therefore, the rank of the state-space model is reduced to 3.

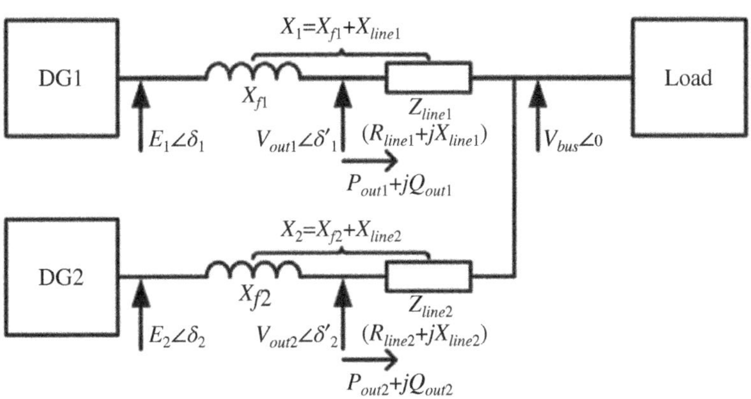

Figure 7.31 Structure of an MG composed of two DGs in islanded mode.

Here, the state-space model variables are

$$\boldsymbol{w} = \left[\Delta P_{load}\ \ \Delta P_{0\_1}\ \ \Delta P_{0\_2}\right]^T \tag{7.81}$$

$$\boldsymbol{y} = \left[\Delta \omega_{m1}\ \ \Delta \omega_{m2}\ \ \Delta P_{out1}\ \ \Delta P_{out2}\right]^T \tag{7.82}$$

$$\boldsymbol{x} = \begin{bmatrix} \Delta \omega_{m1} + \dfrac{D_1}{J_1\omega_0\left(K_1+K_2\right)}\Delta P_{load} \\ \Delta \omega_{m2} + \dfrac{D_2}{J_2\omega_0\left(K_1+K_2\right)}\Delta P_{load} \\ \Delta\delta_1 - \dfrac{1}{K_1+K_2}\Delta P_{load} \end{bmatrix} \tag{7.83}$$

where $\boldsymbol{A}$, $\boldsymbol{B}$, $\boldsymbol{C}$, and $\boldsymbol{D}$ are obtained as follows:

$$\boldsymbol{A} = \begin{bmatrix} -\dfrac{D_1K_2}{J_1\omega_0(K_1+K_2)} - \dfrac{k_{p1}}{J_1\omega_0} & \dfrac{D_1K_2}{J_1\omega_0(K_1+K_2)} & -\dfrac{K_1}{J_1\omega_0} \\ \dfrac{D_2K_1}{J_2\omega_0(K_1+K_2)} & -\dfrac{D_2K_1}{J_2\omega_0(K_1+K_2)} - \dfrac{k_{p2}}{J_2\omega_0} & \dfrac{K_1}{J_2\omega_0} \\ \dfrac{K_2}{K_1+K_2} & -\dfrac{K_2}{K_1+K_2} & 0 \end{bmatrix}$$

$$\boldsymbol{B} = \begin{bmatrix} -\dfrac{K_1}{J_1\omega_0(K_1+K_2)} - \dfrac{D_1D_2K_2}{J_1J_2\omega_0^2(K_1+K_2)^2} + \dfrac{D_1^2K_2}{J_1^2\omega_0^2(K_1+K_2)^2} + \dfrac{D_1k_{p1}}{J_1^2\omega_0^2(K_1+K_2)} & \dfrac{1}{J_1\omega_0} & 0 \\ -\dfrac{1}{J_2\omega_0} + \dfrac{K_1}{J_2\omega_0(K_1+K_2)} - \dfrac{D_1D_2K_1}{J_1J_2\omega_0^2(K_1+K_2)^2} + \dfrac{D_2^2K_1}{J_2^2\omega_0^2(K_1+K_2)^2} + \dfrac{D_2k_{p2}}{J_2^2\omega_0^2(K_1+K_2)} & 0 & \dfrac{1}{J_2\omega_0} \\ -\dfrac{D_1K_2}{J_1\omega_0(K_1+K_2)^2} + \dfrac{D_2K_2}{J_2\omega_0(K_1+K_2)^2} & 0 & 0 \end{bmatrix}$$

$$\boldsymbol{C} = \begin{bmatrix} 1 & 0 & 0 \\ 0 & 1 & 0 \\ 0 & 0 & K_1 \\ 0 & 0 & -K_1 \end{bmatrix}, \quad \boldsymbol{D} = \begin{bmatrix} -\dfrac{D_1}{J_1\omega_0(K_1+K_2)} & 0 & 0 \\ -\dfrac{D_2}{J_2\omega_0(K_1+K_2)} & 0 & 0 \\ \dfrac{K_1}{K_1+K_2} & 0 & 0 \\ \dfrac{K_2}{K_1+K_2} & 0 & 0 \end{bmatrix}$$

where the subscript $i = 1,2$ indicates the ith DG. Here, $K_i = \left(E_iV_{bus}\cos\,\delta_i\right)/X_i$, and $X_i \approx X_{f\ i} + X_{line\ i}$, as it is discussed in Section 7.2. Analyses of transient active power control performance in the following parts of this section are based on this model, as it describes the transient performance of variables in $\boldsymbol{y}$ after a given disturbance $\boldsymbol{w}$.

7.5.2 Oscillation Damping

It is a known conclusion in the control theory that the poles of transfer function of $Y_j(s)/W_k(s)$ are available in the eigenvalues of $\boldsymbol{A}$. Therefore, the eigenvalues study gives some clues to damping methods for oscillations in $\Delta P_{out\ i}$. The loci of eigenvalues of $\boldsymbol{A}$ with a variation of D_1 or X_1 are shown in Fig. 7.32. Nominal parameters for the eigenvalue loci plots are listed in Table 7.2.

In the eigenvalue loci plots, radial dash lines indicate damping ratio ζ, and circle dash lines indicate undamped natural frequency ω_n. As it is shown in Fig. 7.32, damping ratio of the complex conjugate eigenvalues increases if the damping factor D_i and the output reactance impedance X_i are increased. It should be pointed out that increasing X_i causes a decrease in damped natural frequency ω_d, which is indicated by the distance between eigenvalue and the real axis. This may result in longer settling time compared with the method of increasing D_i. However, the approach of increasing output reactance has other three merits: (i) the state-space model is obtained under the assumption that the output impedance of DGs is inductive, as it is discussed in Section 7.2.

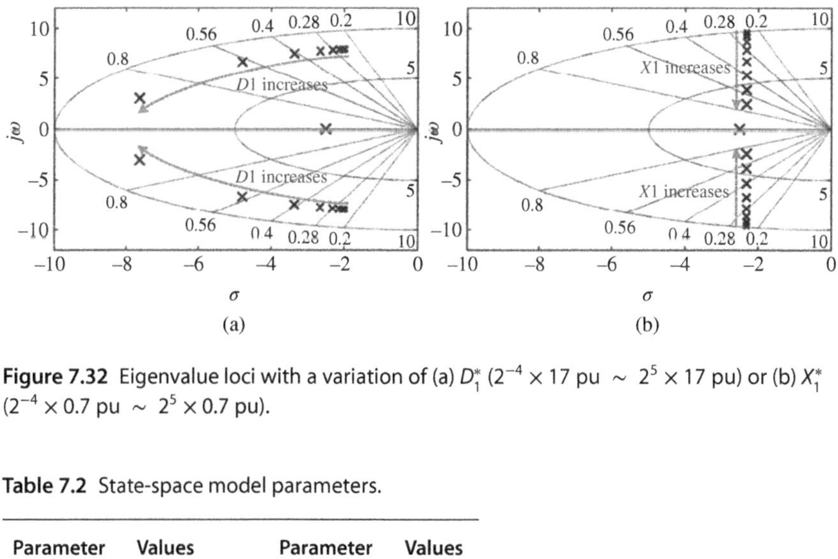

Figure 7.32 Eigenvalue loci with a variation of (a) D_1^* ($2^{-4} \times 17$ pu ~ $2^5 \times 17$ pu) or (b) X_1^* ($2^{-4} \times 0.7$ pu ~ $2^5 \times 0.7$ pu).

Table 7.2 State-space model parameters.

Parameter	Values	Parameter	Values
S_{base1}	10 kVA	M_i^*	8 s
S_{base2}	5 kVA	D_i^*	17 pu
E_0	200 V	$k_{p\ i}^*$	20 pu
ω_0	376.99 rad/s	X_i^*	0.7 pu
$E_i = V_{bus}$	200 V	$\cos\ \theta_i$	≈1

This assumption is less valid if X_i is small, especially in an LV MG in which line impedances are mainly resistive [11]. If this assumption is not valid, the active power and reactive power control cannot be decoupled correctly, and the system may become more oscillatory and even unstable. (ii) To share transient active power properly, output reactance of each DG should be designed equally in per unit value, as it is discussed in the next part of this section. Therefore, the problem of oscillation and that of transient active power sharing can be solved simultaneously by proper stator reactance design. (iii) Moreover, the influence of output reactance mismatch on transient active power sharing becomes smaller if output reactance of DGs is increased, owing to decreased relative errors.

7.5.3 Transient Active Power Sharing

If the disturbance comes from a loading transition, the disturbance vector can be written as $\boldsymbol{w} = [\Delta P_{load}\ \ 0\ \ 0]^T$. The output vector at $t = 0^+$ can be calculated as follows:

$$\boldsymbol{y}(0^+) = \boldsymbol{D}\boldsymbol{w} \tag{7.84}$$

From (7.84), the transient active power sharing at $t = 0^+$ is obtained as

$$\Delta P_{out1}(0^+) = \frac{K_1}{K_1 + K_2}\Delta P_{load},\quad \Delta P_{out2}(0^+) = \frac{K_2}{K_1 + K_2}\Delta P_{load} \tag{7.85}$$

The output vector at steady state can be calculated as

$$\boldsymbol{y}(\infty) = (\boldsymbol{D} - \boldsymbol{C}\boldsymbol{A}^{-1}\boldsymbol{B})\boldsymbol{w} \tag{7.86}$$

From (7.86), the steady-state active power sharing is obtained as

$$\Delta P_{out1}(\infty) = \frac{k_{p1}}{k_{p1} + k_{p2}}\Delta P_{load},\quad \Delta P_{out2}(\infty) = \frac{k_{p2}}{k_{p1} + k_{p2}}\Delta P_{load} \tag{7.87}$$

The transient power sharing should be equivalent to the steady-state power sharing, in order to avoid unnecessary transient oscillation. Therefore,

$$\frac{k_{p1}}{k_{p2}} = \frac{K_1}{K_2} = \frac{\left(E_1 V_{bus}\cos\delta_1\right)/X_1}{\left(E_2 V_{bus}\cos\delta_2\right)/X_2} \tag{7.88}$$

Knowing that $k^*_{p\ i} = (k_{p\ i}\omega_0)/S_{base\ i}$, $X^*_i = (X_i S_{base\ i})/E_0^2$, and supposing $E_i \approx V_{bus} \approx E_0$, $\cos\delta_i \approx 1$, thus

$$\frac{X^*_1}{X^*_2} = \frac{k^*_{p2}}{k^*_{p1}} \tag{7.89}$$

As mentioned in Section 7.2, $k^*_{p\ i}$ is designed equally so that the active power is shared according to power ratings of DGs in steady state [11]. Therefore, $X^*_1 = X^*_2$, which shows the condition for appropriate transient power sharing at

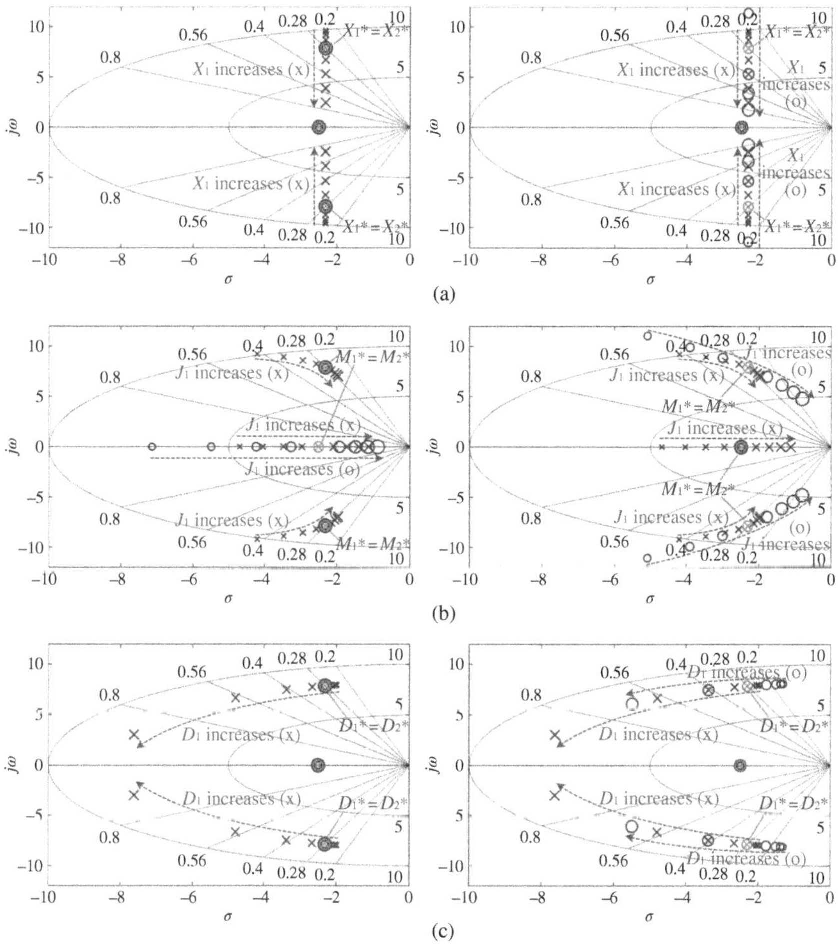

Figure 7.33 Poles and zeros of $\frac{\Delta P_{out1}(s)}{\Delta P_{load}(s)}$ in the left column and $\frac{\Delta P_{out2}(s)}{\Delta P_{load}(s)}$ in the right column with a variation of (a) X_1^* ($2^{-4} \times 0.7$ pu ~ $2^5 \times 0.7$ pu), (b) M_1^* ($1.3^{-4} \times 8$ s ~ $1.3^5 \times 8$ s), and (c) D_1^* ($2^{-4} \times 17$ pu ~ $2^5 \times 17$ pu).

$t = 0^+$. This conclusion can be verified by evaluating the poles and zeros of $\frac{\Delta P_{out1}(s)}{\Delta P_{load}(s)}$ and $\frac{\Delta P_{out2}(s)}{\Delta P_{load}(s)}$, as demonstrated in Fig. 7.33a. $\frac{\Delta P_{out1}(s)}{\Delta P_{load}(s)}$ and $\frac{\Delta P_{out2}(s)}{\Delta P_{load}(s)}$ can be obtained as the elements $G_{31}(s)$ and $G_{41}(s)$ of the matrix $\boldsymbol{G}(s)$ shown in (7.90), respectively.

$$\boldsymbol{G}(s) = \boldsymbol{C}(s\boldsymbol{I} - \boldsymbol{A})^{-1}\boldsymbol{B} + \boldsymbol{D} \tag{7.90}$$

Figure 7.33a shows that if $X_1^* = X_2^*$ holds, all poles of $\frac{\Delta P_{out1}(s)}{\Delta P_{load}(s)}$ and $\frac{\Delta P_{out2}(s)}{\Delta P_{load}(s)}$ are canceled by zeros, and if it does not hold, this cancelation does not happen for the complex conjugate poles. The cancelation of all poles implies a desirable step change of ΔP_{out1} and ΔP_{out2} directly to their respective steady-state values without any oscillation during a loading transition.

However, $X_1^* = X_2^*$ can only guarantee appropriate power sharing at $t = 0^+$. The transient state during the first several seconds after $t = 0^+$ depends on the swing equation parameters, that is, moment of inertia and damping factor. As shown in Fig. 7.33b,c, if the inertia constant M_i^* and the per unit damping ratio D_i^* are not designed equally, the cancelation of poles and zeros does not happen. Therefore, to obtain ideal transient active power sharing, (7.91) should also be guaranteed.

$$M_1^* = M_2^*, \quad D_1^* = D_2^* \tag{7.91}$$

Equation (7.91) is not difficult to realize since M_i^* and D_i^* are virtual parameters that can be easily changed in the control program. As for $X_1^* = X_2^*$, a control method to adjust stator reactance is presented in Section 8.2 and [3].

7.6 Summary

In this chapter, the most important characteristics of a typical VSG system were discussed. The VSG dynamic performance in islanded and connected modes was described. Then the chapter was focused on the potential role of VSGs in the grid oscillation damping. The small-signal models and state-space models of VSG control system and droop control system were presented, and their dynamic responses compared. A different VSG structure for emulating a more complete set of SG properties was introduced, and its feasibility and stability were investigated. Finally, an analytical approach for active power performance assessment in an MG with multiple VSGs was presented. All given results are examined by simulations and laboratory experiments. In the next chapter, several VSG-based MG stabilization and regulation methodologies are addressed.

References

1 Bevrani, H. (2014) *Robust Power System Frequency Control*, 2nd edn, Springer.

2 Ise, T. and Bevrani, H. (2016, Chapter 12) Virtual synchronous generators and their applications in microgrids, in *Integration of Distributed*

Energy Resources in Power Systems, 1st edn (ed. T. Funabashi), Elsevier, pp. 282–294.

3 J. Liu, Y. Miura, H. Bevrani, and T. Ise, Enhanced virtual synchronous generator control for parallel inverters in a microgrid, Submitted to the *IEEE Transactions on Smart Grid*, In press, DOI: 10.1109/TSG.2016.2521405.

4 Guan, M., Pan, W., Zhang, J. *et al.* (2015) Synchronous generator emulation control strategy for voltage source converter (VSC) stations. *IEEE Transactions on Power Systems*, **30** (6), 3093–3101. DOI: 10.1109/TPWRS.2014.2384498

5 Bevrani, H., Ise, T., and Miura, Y. (2014) Virtual synchronous generators: a survey and new perspectives. *International Journal of Electrical Power and Energy Systems*, **54**, 244–254.

6 Delille, G., Francois, B., and Malarange, G. (2012) Dynamic frequency control support by energy storage to reduce the impact of wind and solar generation on isolated power system's inertia. *IEEE Transactions on Sustainable Energy*, **3** (4), 931–939.

7 Vasquez, J.C., Guerrero, J.M., Savaghebi, M. *et al.* (2013) Modeling, analysis, and design of stationary-reference-frame droop-controlled parallel three-phase voltage source inverters. *IEEE Transactions on Industrial Electronics*, **60** (4), 1271–1280.

8 Machowski, J., Bialek, J., and Bumby, J.R. (1997) *Power System Dynamics and Stability*, John Wiley & Sons, New York, pp. 141–182.

9 Alipoor, J., Miura, Y., and Ise, T. (2015) Power system stabilization using virtual synchronous generator with alternating moment of inertia. *IEEE Journal of Emerging and Selected Topics in Power Electronics*, **3** (2), 451–458.

10 Liu, J., Miura, Y., and Ise, T. (2016) Comparison virtual synchronous generator and droop control in inverter-based distributed generators. *IEEE Transactions on Power Electronics*, **31** (5), 3600–3611. DOI: 10.1109/TPEL.2015.2465852

11 Bevrani, H. and Shokoohi, S. (2013) An intelligent droop control for simultaneous voltage and frequency regulation in islanded microgrids. *IEEE Transactions on Smart Grid*, **4** (3), 1505–1513.

12 Shintai, T., Miura, Y., and Ise, T. (2014) Oscillation damping of a distributed generator using a virtual synchronous generator. *IEEE Transactions on Power Delivery*, **29** (2), 668–676.

13 Sakimoto, K., Miura, Y., and Ise, T. (2011), *Stabilization of a Power System with a Distributed Generators by a Virtual Synchronous Generator Function*. 8th IEEE International Conference on Power Electronics–ECCE Asia, pp. 1498–1505.

14 Bevrani, H., Watanabe, M., and Mitani, Y. (2014) *Power System Monitoring and Control*, IEEE-Wiley Press, New York.

15 Hirase, Y., Abe, K., Sugimoto, K., and Shindo, Y. (2013) A grid-connected inverter with virtual synchronous generator model of algebraic type. *Electrical Engineering in Japan*, **184** (4), 10–21.

16 Kundur, P. (1993) *Power System Stability and Control*, McGraw-Hill.

17 The Institute of Electrical Engineering of Japan (IEEJ) (2013) *Electrical Engineering Handbook*, 7th edn, IEEJ, Chapter 15 Synchronous machine, Induction machine.

8

Virtual Inertia-based Stability and Regulation Support

Compared to the conventional power systems with synchronous machines, the microgrids (MG) with distributed generators (DGs) have either very small or no inertia, which is the main source of stability. With growing penetration level of DGs, the negative impact of low inertia and damping effect on the MG dynamic performance and stability increases. A solution toward the MG stability enhancement is to provide virtual inertia, by using the concept of virtual synchronous generator (VSG). The VSG simulates the dynamic behavior of a synchronous generator (SG) represented by its fundamental swing equation to create virtual inertia. Unlike a real synchronous machine, the parameters of VSG can be controlled to enhance the dynamic response of the system.

In this chapter, several VSG control schemes are introduced to support the MG stability and regulation performance. First, an enhanced VSG control is proposed to achieve oscillation damping and proper transient active power sharing by the virtual stator reactance adjustment and common AC bus voltage estimation. Then, for a parallel operation of VSG and SG, to prevent unbalanced SG current and inverter overcurrent, a new control mechanism is developed.

An alternating inertia concept is then introduced to enhance the VSG control performance. This scheme adopts the suitable value of the moment of inertia of the VSG, considering the virtual angular velocity and its rate of change during oscillations. The VSG support in case of the serious voltage sags is also assessed in this chapter. Finally, the impacts of a VSG topology (without phase-locked loop unit) on the MG stability are examined. In this chapter, all given control approaches and synthesis methodologies are verified analytically and examined through computer simulations as well as laboratory experiments.

8.1 Introduction

Recently, the MG concept provides a quite appealing alternative for overcoming the challenges of integrating renewable energy sources (RESs) and DGs

Microgrid Dynamics and Control, First Edition. Hassan Bevrani, Bruno Francois, and Toshifumi Ise.

into power systems. However, in order to allow seamless deployment of MGs, the dynamic stability and control challenges should be solved. Current efforts are mostly being put into the design of more effective control strategies in different control levels and special protection schemes that ensure reliable, secure, and economical operation of MGs in either grid-connected or islanded operation mode. A general overview of the existing MG control technologies and remaining challenges is addressed in [1–4]. Distributed-inverter-based power sources are the main blocks of an MG. However, if the applied inverters are only equipped with simple and conventional current controls, they may not only support the grid stability but also provide negative dynamic impacts. Therefore, in the MGs without existing large commercial power plants, the voltage and frequency may not be easily maintained within the allowed variation bounds [5]. Moreover, in terms of introduction expenses, it is not practical to replace all existing fossil-fuel-based power to inverters with storage batteries. Therefore, DGs must be combined with existing SGs without additional changes. This fact has hampered the widespread use of renewable energy. Furthermore, since the electrical power generated from renewable energy often fluctuates over a short time (cycle), stabilizing the frequency of the system is difficult. A similar concern arises when a greatly varying load is connected to an MG. It is reported that a larger amount of inverters with conventional inverter-control methods cannot provide electrical power stably compared to the rotation-type SGs because whole MG systems have lower inertial forces compared to typical commercial grids [6, 7]. This problem becomes conspicuous in isolated islands and small communities. If system destabilization is caused by specific load disturbances/faults, it may be sufficient to add a stabilizing/protection mechanism to the faulted loads. However, if numerous interspersed DGs cause the destabilization, it is difficult to countermeasure it in each DG. Therefore, we need some sort of autonomous controls without monitoring the state of each DG.

So far, various types of control systems that enable an inverter to be operated as an SG have been studied [8–19]. These systems are known as VSG and enable to work in both stand-alone and parallel operations with other SGs. The VSG simulates the dynamic behavior of an SG represented by its fundamental swing equation to create virtual inertia. Unlike a real synchronous machine, the parameters of VSG can be tuned to enhance the dynamic response of the system. Therefore, the concept of VSG can be considered as a useful solution to improve the dynamic stability in the MG systems [19]. However, system instability can be exacerbated by resonance among SGs and DGs (loads). Inverters, which have the characteristics of an SG, can also cause resonance among other SGs and DGs (loads). Therefore, in order to achieve stable parallel operation of multiple SGs, including VSG-controlled inverters, VSGs should be designed in a manner that allows them to easily avoid the aforementioned problems, using a minimal number of required parameters. In the simple VSG controls that have been introduced in Chapter 7, the essential elements of the dynamic characteristics of an SG are applied. Thus, the mentioned VSGs can effectively improve the MG stability and control performance.

As mentioned in Chapters 4 and 5, understanding the dynamics and using appropriate analytic methodologies are significant issues for MG stability analysis and control synthesis. Despite the small scale of an MG, it has many of the complexities of a large-scale conventional power system. In conventional power grids, stability analysis is well established with standard models of SGs, governors, and excitation systems of varying orders that are known to capture the important modes for particular classes of problems. This does not yet exist for the MGs and may be hard to achieve because of the wide range of power technologies that might be deployed [12]. In the literature [5], the need for automatic voltage regulators (AVRs) and power system stabilizers (PSSs) is described. In cases in which an SG is connected to an infinite bus, various similar methods have also been proposed. However, all these methods require adding a stabilizer outside the SG.

In this chapter, several VSG control schemes are introduced to support the MG stability and regulation performance. The chapter is organized as follows: Section 8.2 addresses an enhanced VSG control to achieve oscillation damping and proper transient active power sharing by the virtual stator reactance adjustment and common AC bus voltage estimation. Then, for a parallel operation of VSG and SG, to prevent unbalanced SG current and inverter overcurrent, a new control mechanism is developed in Section 8.3. The alternating inertia concept is introduced in Section 8.4 to enhance the VSG control performance. This scheme adopts the suitable value of the moment of inertia of the VSG, considering the virtual angular velocity and its rate of change during oscillations. The VSG support for the serious voltage sags is presented in Section 8.5. In Section 8.6, the impacts of a VSG topology (without phase-locked loop unit) on the MG stability are examined, and finally, the chapter is summarized in Section 8.7. All given control approaches and synthesis methodologies are verified analytically and examined through computer simulations as well as laboratory experiments.

8.2 An Enhanced Virtual Synchronous Generator Control Scheme

In this section, the basic VSG control mechanism shown in Fig. 7.2 is improved [19, 20]. The proposed enhanced VSG control scheme is shown in Fig. 8.1. Compared with the basic VSG control, two major modifications are made, that is, the stator impedance adjuster and the bus voltage estimator, as shown in Fig. 8.2a and b, respectively.

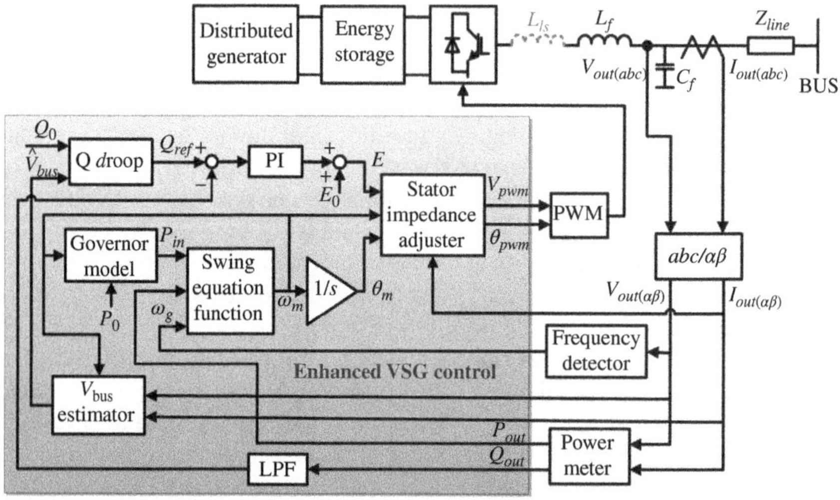

Figure 8.1 Block diagram of the proposed enhanced VSG control.

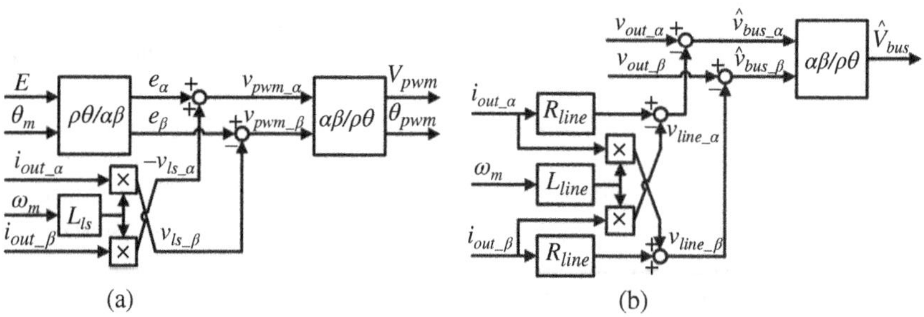

Figure 8.2 Block diagram of (a) the "stator impedance adjuster" and (b) the "V_{bus} Estimator" blocks of the enhanced VSG control.

8.2.1 Proposed Virtual Synchronous Generator Control Scheme

In order to facilitate the VSG control algorithm in the stator impedance adjuster and the bus voltage estimator, the measured three-phase output voltage and current are converted to the $\alpha\beta$ frame. As a result, the calculation in the power meter should be based on the αβ-frame components instead of three-phase components. Therefore, the given equations for P_{out} and Q_{out} in Section 7.1 should be replaced by (8.1) and (8.2).

$$P_{out} = v_{out\_\alpha} i_{out\_\alpha} + v_{out\_\beta} i_{out\_\beta} \tag{8.1}$$

$$Q_{out} = -v_{out\_\alpha} i_{out\_\beta} + v_{out\_\beta} i_{out\_\alpha} \tag{8.2}$$

The function of stator impedance adjuster shown in Fig. 8.2a is to adjust the output reactance of the DG freely. It is operating as a virtual impedance controller. The transformation from the $\rho\theta$ frame to the $\alpha\beta$ frame and the inverse transformation used in Fig. 8.2a are defined in (8.3) and (8.4).

$$\begin{bmatrix}\alpha\\ \beta\end{bmatrix} = \begin{bmatrix}\rho\cos\theta\\ \rho\sin\theta\end{bmatrix} \tag{8.3}$$

$$\begin{bmatrix}\rho\\ \theta\end{bmatrix} = \begin{bmatrix}\sqrt{\alpha^2+\beta^2}\\ \tan^{-1}\dfrac{\beta}{\alpha}\end{bmatrix} \tag{8.4}$$

The virtual stator inductor is realized by multiplying output current by the virtual stator inductor L_{ls} in the $\alpha\beta$ frame. It will be more accurate if inductor current through L_f is used. As the current flowing into C_f at fundamental frequency is less than few percent of the inductor current, using output current instead of inductor current does not affect the performance of the control scheme.

Based on the given analyses in Section 7.5, tuning method of virtual stator inductor L_{ls} is suggested to set total output reactance X_i^* for both DGs to same large per unit value, as shown in (8.5). This approach increases active power damping ratio and shares transient load without oscillation. The target value is proposed to be 0.7 pu because it is a typical value for the total direct-axis transient reactance X_d' of a real SG.

$$X_i^* = \frac{S_{base\ i}\omega_{m\ i}(L_{ls\ i} + L_{f\ i} + L_{line\ i})}{E_0^2} = 0.7\ \text{pu} \tag{8.5}$$

Here, the $L_{f\ i}$ and $Z_{line\ i}$ $(R_{line\ i} + jL_{line\ i})$ are considered as known parameters. As the scale of MG is usually small, the line distance can easily be measured or fed by the planner. Even if it is not the case, several online measurement or intelligent tuning methods for $Z_{line\ i}$ are available in [4, 21].

With the proposed design of stator impedance adjustment, oscillation in a VSG-control-based MG should be almost eliminated during a loading transition in the islanded operation mode. Particularly, transition from grid-connected operation mode to islanded operation mode can also be considered as a loading transition; therefore, the oscillation during an islanding event should be eliminated with the proposed control strategy, as it is proved by the simulation results given in the next section. As for other disturbances in islanded operation mode during the change of power sharing ratio, such as change of active power set value of DG(s) and connection/disconnection of DG(s), oscillation cannot be completely eliminated with the proposed solution but can still be damped owing to the increased total output reactance.

The principle of the bus voltage estimator shown in Fig. 8.2b is similar to that of the stator impedance adjuster shown in Fig. 8.2a. By calculating the line voltage drop in $\alpha\beta$ frame using measured output current and line impedance data, the bus voltage can be estimated from the difference of output voltage and calculated line voltage drop. Since the line-to-line RMS value of estimated

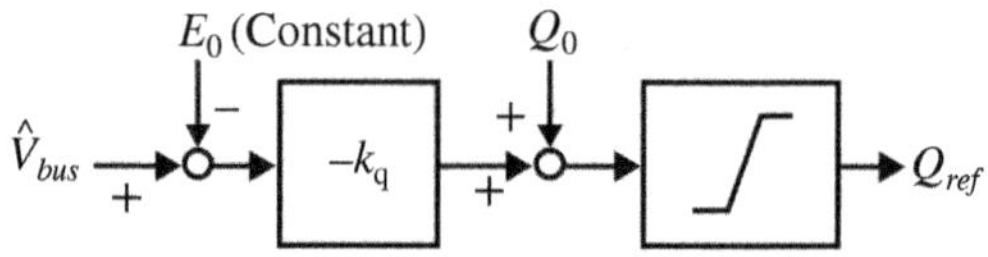

Figure 8.3 Block diagram of the "Q droop" block of enhanced VSG control.

bus voltage $\hat{V}_{bus}$ for each DG should be approximately equal, as it is discussed later, accurate reactive power sharing can be obtained by using estimated bus voltages as the input references of "Q droop" instead of respective output voltages of DGs. Although, the principle of the presented bus voltage estimator is not new, the idea of using this estimator to realize communicationless accurate reactive power sharing can be considered as a contribution to the present VSG control scheme.

The modified Q droop controller is shown in Fig. 8.3 and can be expressed as

$$Q_{ref} = Q_0 - k_q(\hat{V}_{bus} - E_0) \tag{8.6}$$

However, if there is an estimation error in $\hat{V}_{bus}$, it will cause a reactive power sharing error. Supposing $\hat{V}^*_{bus1} = V^*_{bus} + \Delta\hat{V}^*_1$ and $\hat{V}^*_{bus2} = V^*_{bus} + \Delta\hat{V}^*_2$, substitute them into (8.6),

$$Q^*_{ref1} - Q^*_{ref2} = -k^*_q(\Delta\hat{V}^*_1 - \Delta\hat{V}^*_2) \tag{8.7}$$

In steady state, as $Q^*_{ref\ i} = Q^*_{out\ i}$,

$$Q^*_{out1} - Q^*_{out2} = -k^*_q(\Delta\hat{V}^*_1 - \Delta\hat{V}^*_2) \tag{8.8}$$

Therefore, the reactive power sharing error caused by estimation errors is determined by the V–Q droop gain k^*_q. The design of k^*_q is a well-known trade-off between voltage deviation and reactive power control accuracy. Considering the probable ripples in the measured RMS value of $\hat{V}_{bus}$, k^*_q is recommended to be 5 pu for the present example.

It is noteworthy that the increased output reactance by adding the virtual stator inductor L_{ls} causes a decrease in the reactive control plant gain. The reactive output power can be calculated as [20]:

$$Q_{out\ i} = \frac{E_i(E_i - V_{bus}\cos\delta_i)}{X_i} \tag{8.9}$$

The small-signal model of (8.9) is

$$\Delta Q_{out\ i} = \frac{(E_i - V_{bus}\cos\delta_i)\Delta E_i + E_i(\Delta E_i - \Delta V_{bus}\cos\delta_i)}{X_i} \tag{8.10}$$

As $\cos\delta_i \approx 1$, $E_i \approx V_{bus} \approx E_0$, (8.10) can be simplified as

$$\Delta Q_{out\ i} = \frac{E_0(\Delta E_i - \Delta V_{bus})}{X_i} \tag{8.11}$$

With (8.11), the small-signal model of the reactive power control part is illustrated in Fig. 8.4a. Therefore, in the enhanced VSG control where X_i is

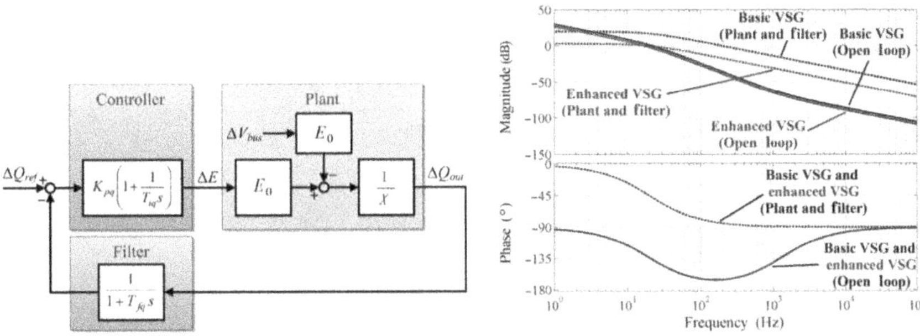

(a) (b)

Figure 8.4 Reactive power control loop: (a) small-signal model and (b) bode plot.

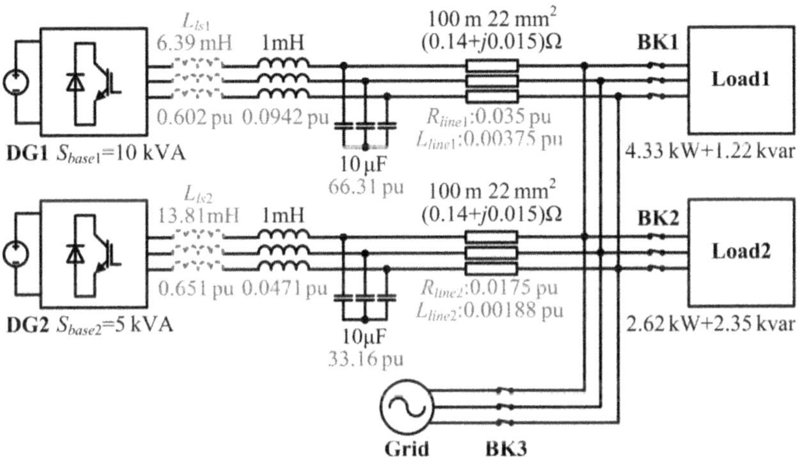

Figure 8.5 Simulation circuit.

largely increased, in order to keep the bandwidth of the reactive power control loop at 20 Hz, the gain of PI controller should be increased to compensate the decreased plant gain, as illustrated in Fig. 8.4b. The parameters used to plot Fig. 8.4b are related to DG1, which are shown in Fig. 8.5 and Table 8.1. The 20 Hz bandwidth is relatively low compared with control methods working on instantaneous value; however, it is fast enough to track the reactive power and regulate the output voltage as it is demonstrated in the simulation and experimental results.

8.2.2 Simulation Results

Some simulations are performed in PSCAD/EMTDC environment to verify the effectiveness of the proposed enhanced VSG control scheme. The MG case study is shown in Fig. 8.5. Impedances of output filters and distribution

Table 8.1 Simulation parameters.

Common parameters			
Parameter	**Values**	**Parameter**	**Values**
S_{base1} (kVA)	10	M_i^* (s)	8
S_{base2} (kVA)	5	D_i^* (pu)	17
$E_0 = V_{grid}$ (V)	200	$k_{p\ i}^*$ (pu)	20
$\omega_0 = \omega_{grid}$ (rad/s)	376.99	$k_{q\ i}^*$ (pu)	5
$P_{0\ i}^*$ (pu)	1	$T_{fq\ i}$ (s)	7.96×10^{-3}
$Q_{0\ i}^*$ (pu)	0		
The basic VSG control			
$K_{pq\ i}^*$ (pu)	0.0025	$T_{iq\ i}$ (s)	1.25×10^{-4}
The enhanced VSG control			
$K_{pq\ i}^*$ (pu)	0.0125	$T_{iq\ i}$ (s)	1.25×10^{-4}

Table 8.2 Simulation sequence.

Time	Grid	$P_{0\_1}^*$ (pu)	$P_{0\_2}^*$ (pu)	Load
$t < 21$ s	Connected	1	1	Load1
$21\text{ s} \le t < 24$ s	Disconnected	–	–	–
$24\text{ s} \le t < 27$ s	–	–	–	Load1 + 2
$27\text{ s} \le t < 30$ s	–	–	0.6	–

lines of each DG differ in per unit values. Other main parameters are listed in Table 8.1, and the sequence of simulation is shown in Table 8.2. Events of islanding from main grid, loading transition, and intentional active power sharing change are simulated at 21, 24, and 27 s, respectively. The simulation results are shown in Fig. 8.6.

As it is illustrated in Fig. 8.6a, when the MG is islanded at 21 s, and when load 2 is connected at 24 s, oscillation can be observed in active power when the basic VSG control is applied for both DGs. This oscillation is almost eliminated by applying the proposed stator impedance adjuster as shown in Fig. 8.6b and c. As the disturbance at 27 s is caused by change of active power set value of DG1, which is not a loading transition, active power oscillation cannot be eliminated in this case. However, the proposed stator impedance adjuster increases the damping ratio; therefore, the overshoots in Fig. 8.6b and c are smaller than that in Fig. 8.6a. Meanwhile, the oscillation periods become longer, because the damped natural frequencies are decreased as it is discussed in Section 7.5.2.

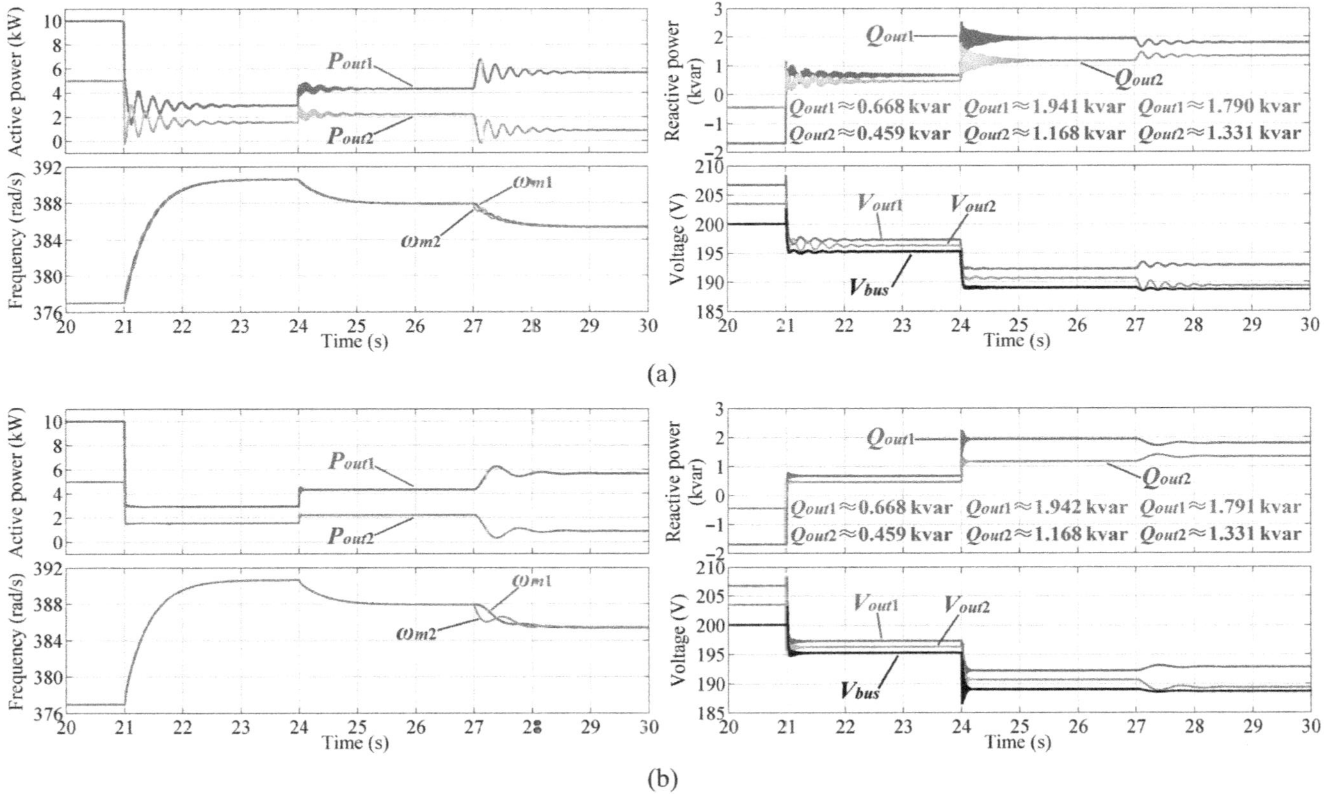

Figure 8.6 Simulation results when both DGs are controlled by (a) the basic VSG control, (b) the basic VSG control with the proposed stator impedance adjuster, and (c) the complete proposed enhanced VSG control.

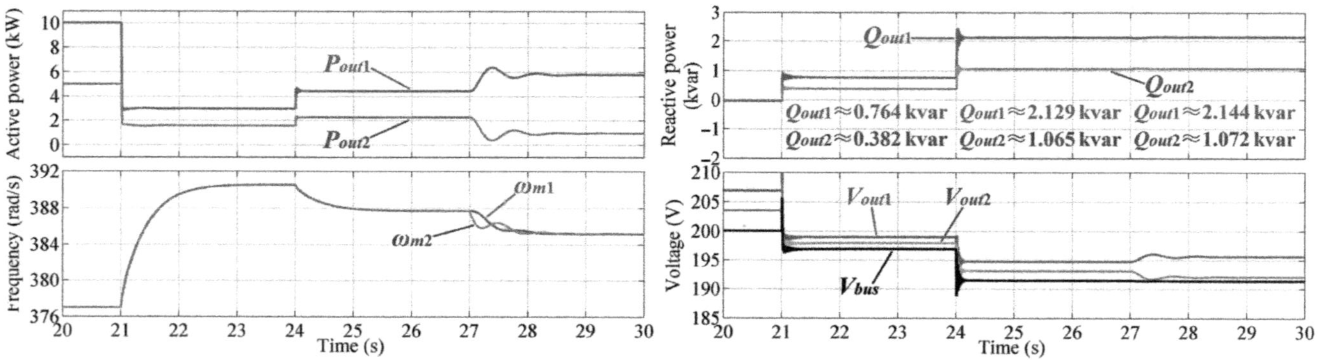

(c)

Figure 8.6 *(Continued)*

Note that the rate of change of frequency remains the same in all cases, which suggests that the proposed enhanced VSG control has no influence on the inertia support feature of VSG control.

Moreover, as shown in Fig. 8.6a and b, reactive power is not shared properly in the islanded operation mode and is not controlled at set value in the grid-connected operation mode, due to the voltage drop through the line impedance. Besides, reactive power control is not independent from active power control, as a change of set value of active power at 27 s also causes a change of reactive power sharing. These problems are all solved in the developed enhanced VSG control by applying the bus voltage estimator, as shown in Fig. 8.6c. It is also noteworthy that the steady-state deviations of DG voltage and bus voltage become smaller when the enhanced VSG control is applied.

Figure 8.7 illustrates the dynamic performance of reactive power and voltage during the loading transition at 24 s. Although the virtual internal emf E_1

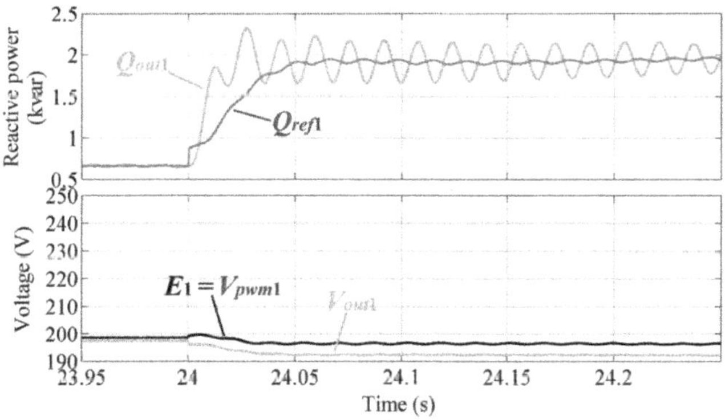

(a)

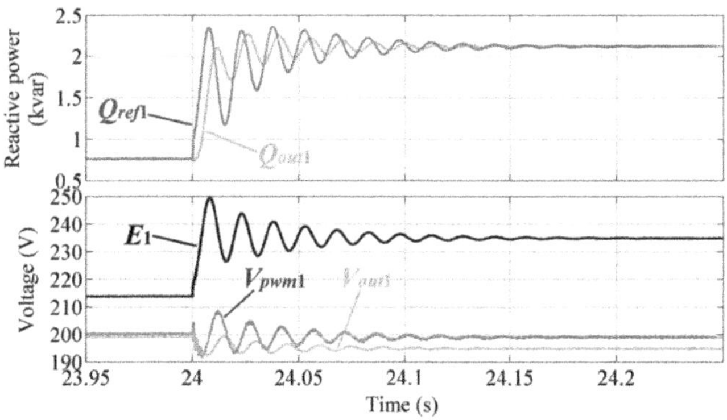

(b)

Figure 8.7 Zoom-in simulation results of reactive power and voltage of DG1 at 24 s: (a) the basic VSG control and (b) the proposed enhanced VSG control.

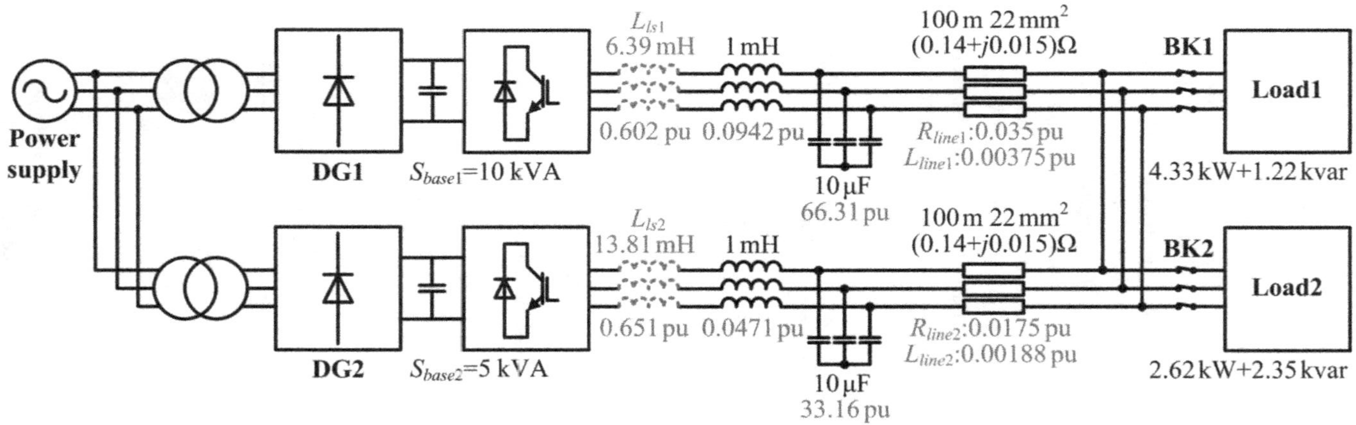

Figure 8.8 Experiment circuit.

becomes much higher in the proposed enhanced VSG control owing to large voltage drop on the virtual stator inductance L_{ls}, the maximum voltage sag of PWM inverter reference V_{pwm1} and output voltage V_{out1} are kept within the same level as the basic VSG control. This implies that the voltage drop on L_{ls} is compensated well by the reactive power PI controller. Besides, although in the enhanced VSG control, the voltages become slightly oscillatory, the reactive power oscillations converge within 0.1 s.

8.2.3 Experimental Results

Experiments are executed in an islanded MG, at which the circuit is the same as that of simulation shown in Fig. 8.5, except that instead of DC sources, AC supply rectified by diode bridge is used to imitate the DC output of DGs, and the breaker BK3 is opened, as shown in Fig. 8.8. The experiment setup is the same as those shown in Section 7.2.4. Experiment sequence is described in Table 8.3. Control parameters are the same as those listed in Table 8.2, and the experimental results are shown in Fig. 8.9.

Experimental results verify again the effectiveness of the proposed enhanced VSG control. First, by comparing Fig. 8.9a and b with Fig. 8.9c, it can be observed that with the proposed stator impedance adjuster, the oscillation due to loading transition at 0.5 s is eliminated, and the oscillation due to change of set value of acitive power at 3.0 s is damped. It implies that the VSG control with the proposed stator impedance adjuster is able to track the loading transition rapidly and accurately without oscillation; meanwhile, the inertia support of the basic VSG control is kept. Even when an oscillation occurs, the overshoot is suppressed owing to increased system damping.

Furthermore, by comparing Fig. 8.9c with Fig. 8.9a and b, it can be concluded that by applying the proposed bus voltage estimator, the reactive power is shared according to the power rating ratio and is immune to active power sharing change and line impedance mismatch in per unit values. Although ripples in RMS value of output voltage can be observed due to a slight load unbalance, the reactive power is controlled well when the enhanced VSG control is applied.

Table 8.3 Experiment sequence.

Time	$P^*_{0\_1}$ (pu)	$P^*_{0\_2}$ (pu)	Load
$t < 0.5$ s	1	1	Load1
$0.5\ \text{s} \leq t < 3$ s	–	–	Load1 + 2
$3\ \text{s} \leq t < 5$ s	–	0.6	–

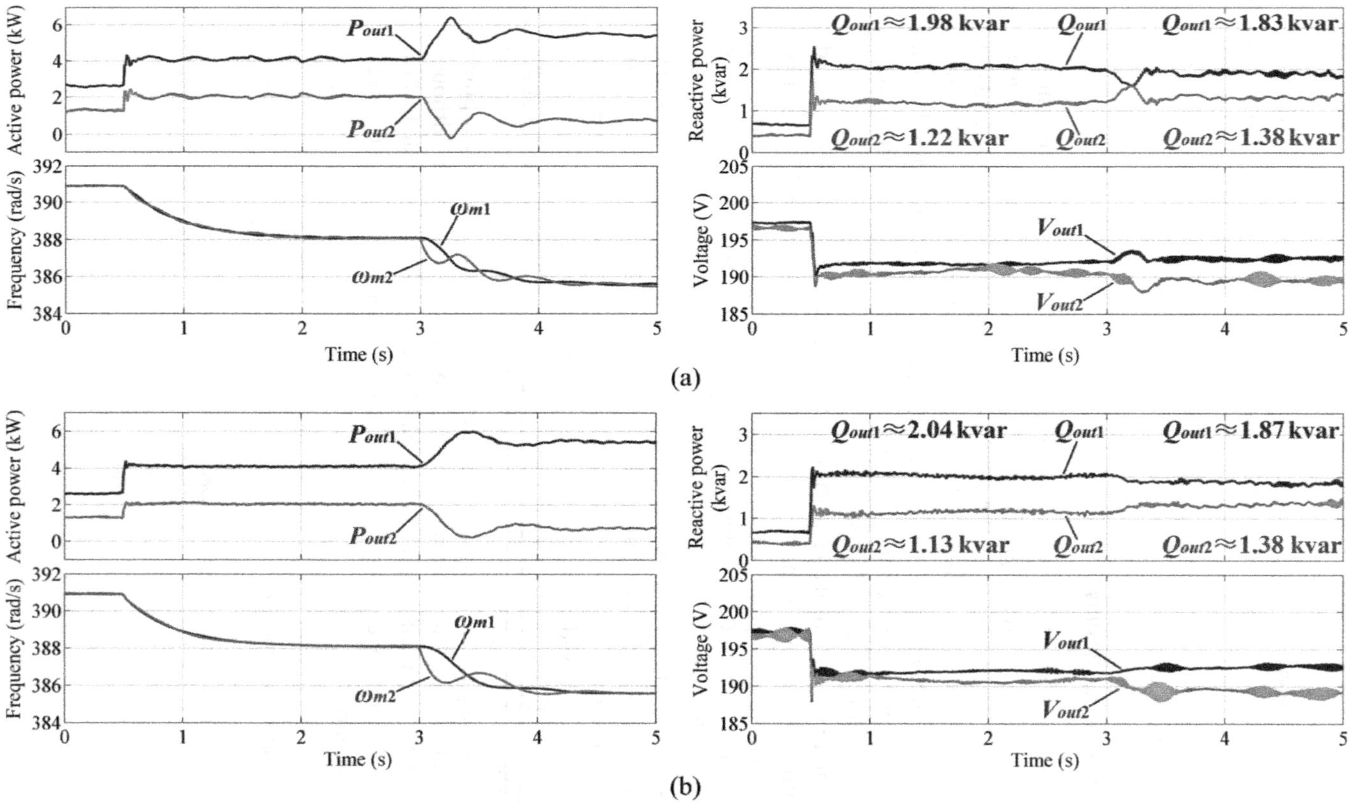

Figure 8.9 Experimental results when both DGs are controlled by (a) the basic VSG control, (b) the basic VSG control with the proposed stator impedance adjuster, and (c) the complete proposed enhanced VSG control.

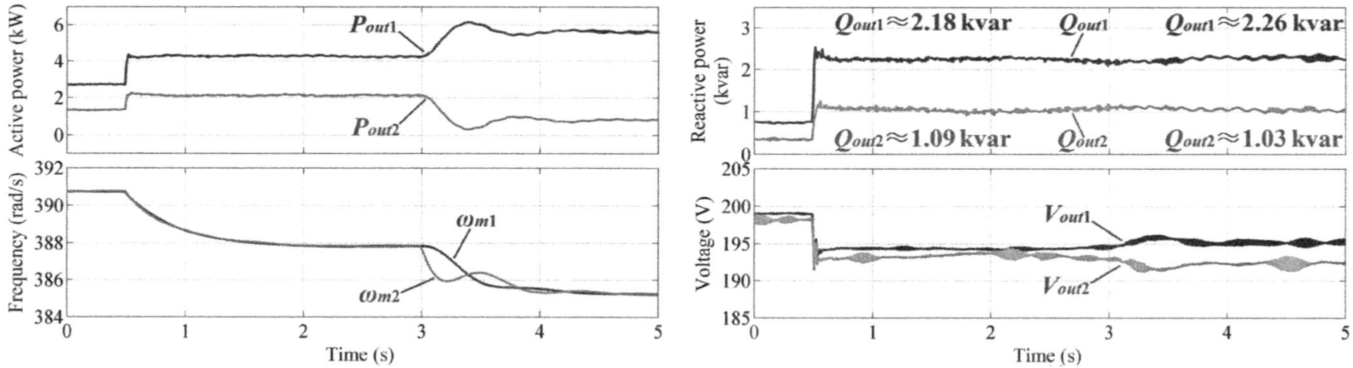

(c)

Figure 8.9 (*Continued*)

8.3 Virtual Synchronous Generator Control in Parallel Operation with Synchronous Generator

In an islanded MG where the main grid is not available, a dispatchable small SG using a diesel engine or a gas engine is usually used as the main power supply. Meanwhile, to save the fuel consumption, an inverter-interfaced DG using RESs, for example, photovoltaic, can be used as the secondary supply. Therefore, the control method of the inverter-interfaced DG for parallel operation with an SG should be established.

However, the dynamic performance of a small SG is usually poor due to relatively smaller moment of inertia. As a result, large rotor speed deviation during loading transition can be observed even for single SG operation. A method using back-to-back converter and electric double-layer capacitor (EDLC) is proposed to address this issue [22]; however, additional back-to-back converter is needed for SG. Another issue for a small SG is the operation under three-phase unbalanced loading condition. As three-phase unbalanced current will heat the SG and give rise to torsional stresses, the negative sequence current from the SG should be prevented. In this section, these two issues are addressed by the control design of the inverter-interfaced DG parallel to the SG. Parallel operation of a SG and an inverter-based DG is studied in [23, 24]. However, a motor-generator set is used in these works, in which the response of governor is faster than that in a diesel or gas engine. Moreover, operation under unbalanced loading condition is not discussed in these works.

As it is demonstrated in Chapter 7, VSG control is able to improve the transient performance of rotor speed of the SG paralleled with it. Therefore, in this section, as continuation of Section 7.5, a modified VSG control scheme is proposed for the parallel operation of an SG and an inverter-based DG. It is shown that the transient rotor speed deviation of a small SG is improved by the parallel operation of a VSG-control-based inverter. Moreover, a modification of VSG control based on double-decoupled synchronous reference frame (DDSRF) decomposition, SG negative-sequence current compensation, and modified calculation methods for power and voltage is proposed, in order to deal with the unbalanced loading condition and to compensate the negative-sequence current of the SG. Although a VSG control method using DDSRF is proposed in [13] for the normal operation of a single-phase inverter, the idea to use it for unbalanced three-phase operation is a new challenge. Transient virtual stator impedance proposed in [24] is also introduced in the control scheme to limit the overcurrent during a large loading transition. Tuning methods of main control parameters are discussed based on eigenvalue analyses and simulation results executed in PSCAD/EMTDC. The overall performance of the proposed modified VSG control is verified by simulation results, which demonstrate that the issues of SG rotor speed deviation, unbalanced SG current, and overcurrent limiting are properly addressed with the proposed control method.

8.3.1 System Description

An islanded MG composed of an SG and an inverter-based DG is shown in Fig. 8.10a. An LC filter is installed at the output terminal of the inverter to filter the higher order harmonics, and an additional reactor L_{ad_sg} is installed at the output terminal of the SG, whose function is discussed in Section 3.3. In the present application, it is assumed that the generators and the AC bus are installed in the same power plant. As a result, the short distribution lines (10 m with cross section area of 8 mm^2) between the generators and the AC bus are negligible. Three-phase and single-phase loads representing the residential consumptions are installed at a distance and are connected to the AC bus through a distribution line of 100 m with a cross section area of 22 mm^2.

A round-rotor SG with a primary mover of gas engine is discussed in the present work, the main parameters of which are listed in Table 8.4. The control system of the SG is shown in Fig. 8.10b, which is composed of a governor to adjust the shaft power and an AVR to regulate the exciting voltage [5]. Droop coefficients $k^*_{p\_sg}$ and $k^*_{q\_sg}$ and set value of active and reactive power $P^*_{0\_sg}$ and $Q^*_{0\_sg}$ are designed equally to those of DG, in order to share the active and reactive power according to generator power rating ratio. Here, subscript "*sg*" indicates parameters related to SG. A first-order lag unit with time constant T_{d_sg} is presented in the governor to indicate the response delay of the mechanical system of the governor.

As can be seen in Table 8.4, the inertia constant M^*_{sg} is very small (0.16 s) compared with large power rating SG of MVAs (usually several seconds), and the governor response is very slow in view of the fact that T_{d_sg} is 1 s. This implies that during a loading transition, the shaft power cannot be adjusted immediately to catch up the load power, and the kinetic energy stored in the rotating mass of the SG is not sufficient to compensate the transient power shortage/surplus. As a result, a large deviation of rotor speed can be observed, as shown in the simulation results of a single SG operation (Fig. 8.11a). For the

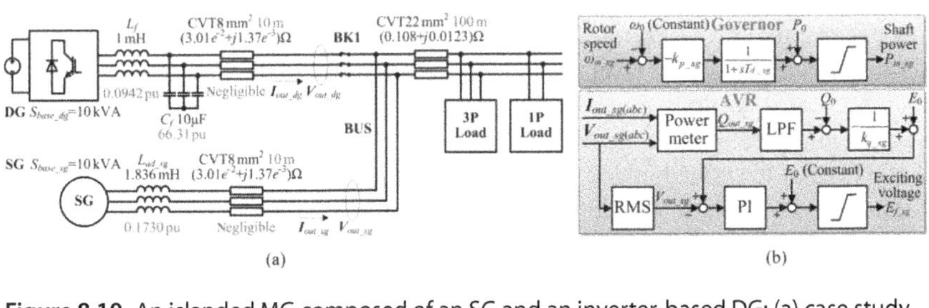

Figure 8.10 An islanded MG composed of an SG and an inverter-based DG: (a) case study and (b) SG's control system.

Table 8.4 SG parameters.

Parameter	Value	Parameter	Value
Rated voltage E_0 (V)	200	Set value of active power $P^*_{0\_sg}$ (pu)	1
Rated power S_{base_sg} (kVA)	10	Set value of reactive power $Q^*_{0\_sg}$ (pu)	0
Nominal frequency ω_0 (rad/s)	376.99	$\omega - P$ droop coefficient $k^*_{p\_sg}$ (pu)	20
Inertia constant M^*_{sg} (s)	0.16	$V - Q$ droop coefficient $k^*_{p\_sg}$ (pu)	5
AVR PI K^*_{pAVR} (pu)	20	Governor time constant T_{d_sg} (s)	1
AVR PI T_{iAVR} (s)	0.025	AVR LPF cut-off frequency (Hz)	20
Impedance model			
$X^*_d = X^*_q$ (pu)	0.219	T'_{do} (s)	6.55
$X'^*_d = X'^*_q$ (pu)	0.027	T''_{do} (s)	0.039
$X''^*_d = X''^*_q$ (pu)	0.01	T'_{qo} (s)	0.85
		T''_{qo} (s)	0.071

sake of comparison, the SG rotor speed ω_{m_sg} shown in simulation results is multiplied by the number of pairs of poles.

In Fig. 8.11a, SG is initially operated with a three-phase load of 1 kW, 0.5 kvar. At 20 s, a single-phase load of 4.8 kW, 2.1 kvar is connected, which leads to a large rotor speed drop down to around 330 rad/s. Besides, large ripples can be observed in the steady-state output voltage, rotor frequency, and active and reactive power of SG due to the three-phase unbalance introduced by the single-phase load. This is verified by the steady-state three-phase output current waveforms of the SG shown in Fig. 8.11b, in which a large three-phase unbalance can be observed.

8.3.2 The Proposed Modified Virtual Synchronous Generator Control Scheme

The control scheme of the proposed modified VSG control is shown in Fig. 8.12. The measured three-phase DG and SG output voltage and current are firstly transformed into $\alpha\beta$ frame and then decomposed into positive- and negative-sequence dq frame using the DDSRF algorithm shown in Fig. 8.13a [13]. For simplification, the $\boldsymbol{V}_{\alpha\beta}$ in Fig. 8.13a signifies both $\boldsymbol{V}_{out\_sg(\alpha\beta)}$ and $\boldsymbol{V}_{out\_dg(\alpha\beta)}$ and so does $\boldsymbol{I}_{\alpha\beta}$, as this algorithm is applied for both SG and DG output voltage and current. Here, subscript "dg" indicates parameters related to DG. The transformation matrices $\boldsymbol{T}^{\pm} = \boldsymbol{T}(\pm\theta_g)$ and $\boldsymbol{T}^{\pm 2} = \boldsymbol{T}(\pm 2\theta_g)$ are defined as (8.12), and the low-pass filter is a first-order filter with a cut-off frequency at 40 Hz.

$$\boldsymbol{T}(\theta_g) = \begin{bmatrix} \cos\theta_g & \sin\theta_g \\ -\sin\theta_g & \cos\theta_g \end{bmatrix} \tag{8.12}$$

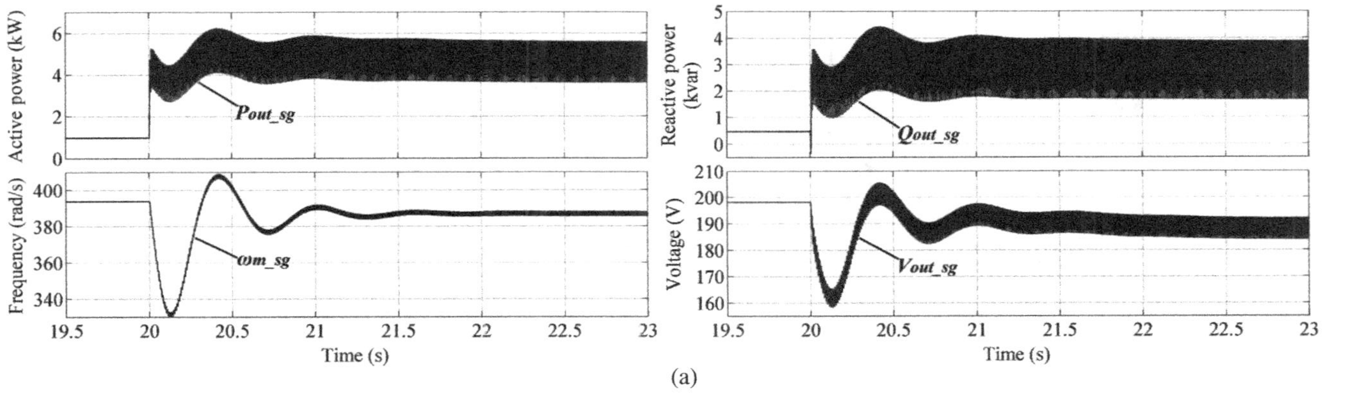

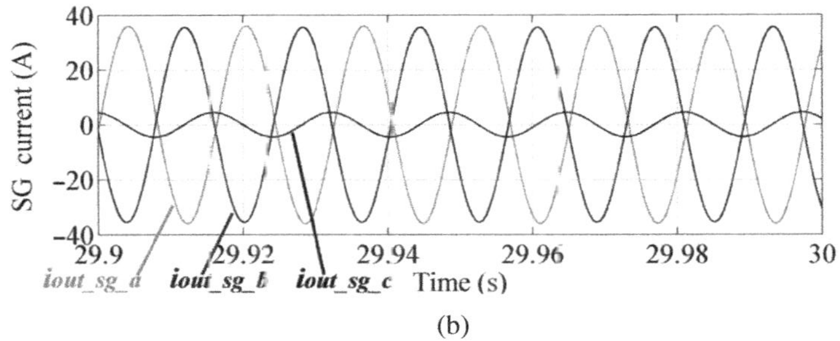

Figure 8.11 Single SG operation: (a) active power and frequency as well as reactive power and voltage and (b) steady-state SG current waveforms.

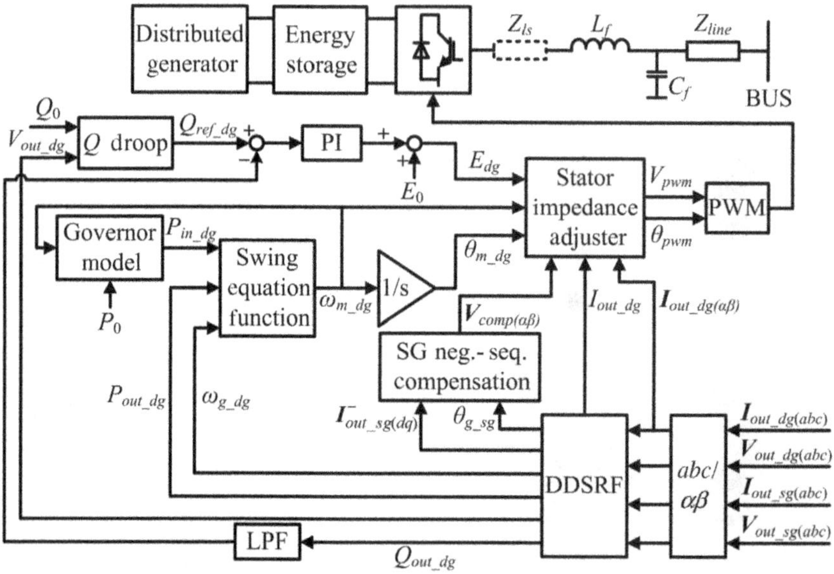

Figure 8.12 Block diagram of the proposed modified VSG control system.

The DG output active and reactive power P_{out_dg} and Q_{out_dg} and line-to-line RMS value of output voltage and current V_{out_dg} and I_{out_dg} used for the rest part of control are calculated based on only the positive-sequence components, as shown in (8.13)–(8.16). If $\alpha\beta$ frame or three-phase components are used directly for these calculations, that is, using (8.1) and (8.2) for the calculations of the output active and reactive power and calculation of the line-to-line RMS value of output voltage (using the equation given in section 7.1), negative-sequence components will be included in these calculated variables during unbalanced loading condition. These negative-sequence components present large ripples with oscillation frequency. Thus, this consequently results in oscillations in output voltage and frequency through the control algorithm in the swing equation function and the Q droop and Q PI controller. Therefore, calculations based on only the positive-sequence components shown in (8.13)–(8.16) are quite important for the present application.

$$P_{out\_dg} = v^+_{out\_dg\_d} i^+_{out\_dg\_d} + v^+_{out\_dg\_q} i^+_{out\_dg\_q} \tag{8.13}$$

$$Q_{out\_dg} = -v^+_{out\_dg\_d} i^+_{out\_dg\_q} + v^+_{out\_dg\_q} i^+_{out\_dg\_d} \tag{8.14}$$

$$V_{out\_dg} = \sqrt{(v^+_{out\_dg\_d})^2 + (v^+_{out\_dg\_q})^2} \tag{8.15}$$

$$I_{out\_dg} = \sqrt{(i^+_{out\_dg\_d})^2 + (i^+_{out\_dg\_q})^2} \tag{8.16}$$

The SG negative-sequence current compensator is shown in Fig. 8.13b. Conventional PI control is used to eliminate the SG negative-sequence current. The output $V_{comp(\alpha\beta)}$ is a counter-rotating voltage vector, which is

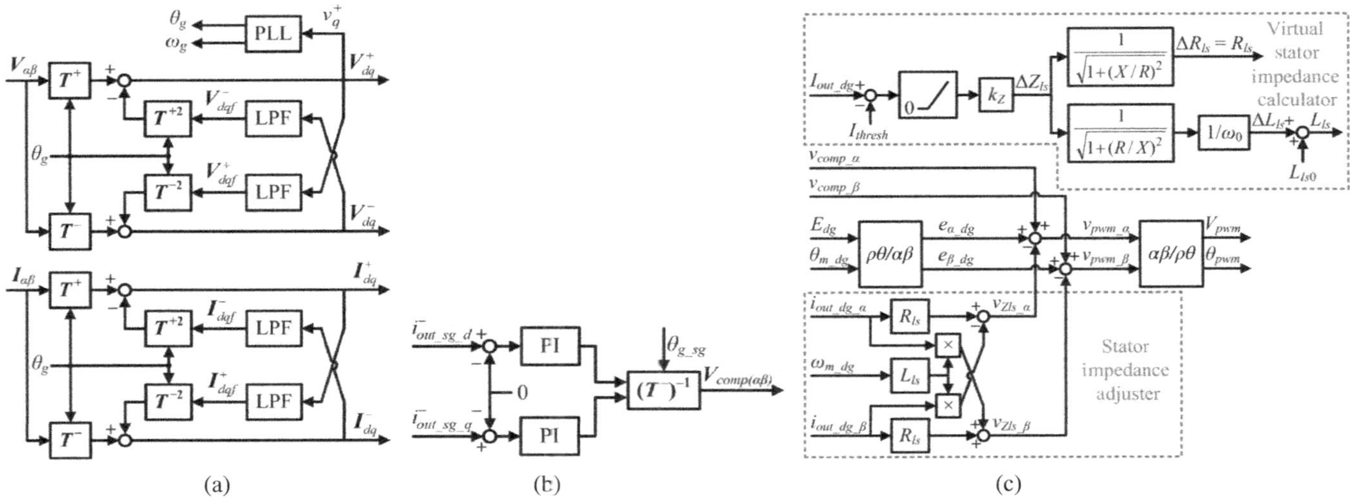

Figure 8.13 Detail of existing blocks in the modified VSG control system: (a) "DDSRF" block, (b) "SG Neg.-Seq. Compensation" block, and (c) "stator impedance adjuster" block.

added to the forward-rotating output voltage reference for the PWM inverter in Fig. 8.13c.

The stator impedance adjuster shown in Fig. 8.13c is used to set the total output impedance to a desired value. The virtual stator impedance is composed of two parts: the constant virtual stator inductance L_{ls0} and the transient virtual stator impedance $\Delta Z_{ls} = \Delta R_{ls} + \omega_0 \Delta L_{ls}$. The constant virtual stator inductance L_{ls0} follows the same principle as that of the virtual stator inductance L_{ls} discussed in Section 8.2, and its tuning method is discussed in Section 8.3.3. The transient virtual stator impedance ΔZ_{ls} is proposed in [24] to limit the DG overcurrent during a large loading transition. The transient virtual stator impedance only functions when the line-to-line RMS value of output current I_{out_dg} exceeds the predefined threshold value I_{thresh}. As I_{thresh} is normally set to 1 pu, this implies an overcurrent condition. The transient virtual stator impedance is designed proportionally to the amount of overcurrent. Tuning methods of the proportional gain k_Z and the reactance/resistance ratio X/R are discussed in Section 8.3.3.

It is noteworthy that the bus voltage estimator proposed in Section 8.2 is omitted in the proposed modified VSG control, as the output voltages of the DG and the SG are measured in the proximity of the common AC bus, as shown in Fig. 8.10a. Therefore, based on the discussion in Section 8.2, accurate steady-state reactive power sharing is guaranteed even without common AC bus voltage estimation.

8.3.3 Parameter Tuning Methods

The control parameters used in the proposed modified VSG control are listed in Table 8.5. As mentioned previously, droop coefficients $k^*_{p\_dg}$ and $k^*_{q\_dg}$, and the set values of active and reactive power $P^*_{0\_dg}$ and $Q^*_{0\_dg}$ are designed equally to those of SG for active and reactive sharing. Tuning methods of other main parameters are discussed in this section. The eigenvalue analyses and simulations given in this section are based on the system shown in Fig. 8.10a, and parameters are listed in Tables 8.4 and 8.5. It is noteworthy that the transient virtual stator impedance ΔZ_{ls} is set to 0 in the eigenvalue analyses and the performed simulations.

8.3.3.1 Swing Equation Parameters

As it is analyzed in Section 7.5.3, to share the transient active power properly, the swing equation parameters of VSG should be set to the same per unit value as those of SG. That is to say, M^*_{dg} should be set to 0.16 s. As for damping factor, from [25], the average damping factor of an SG can be calculated through (8.17).

$$D_{av\_sg} \approx \frac{E_0^2}{2}\left[\frac{T''_d X'_d (X'_d - X''_d)}{X''_d (X'_d + X_{ad\_sg})^2} + \frac{T''_q X'_q (X'_q - X''_q)}{X''_q (X'_q + X_{ad\_sg})^2}\right] \tag{8.17}$$

Table 8.5 DG control parameters.

Parameter	Values	Parameter	Values
E_0 (V)	200	$P^*_{0\_dg}$ (pu)	1
S_{base_dg} (kVA)	10	$Q^*_{0\_dg}$	0
ω_0 (rad/s)	376.99	$k^*_{p\_dg}$ (pu)	20
M^*_{dg} (s)	0.16	$k^*_{q\_dg}$ (pu)	5
D^*_{dg} (pu)	8.7	T_{d_dg} (s)	0
L_{ls0} (mH)	1.122	k^*_Z (pu)	0.69
L_{ad_sg} (mH)	1.836	X/R	5
PI controller for reactive power			
K^*_{pq} (pu)	0.05	T_{iq} (s)	1.25×10^{-2}
PI controller for SG Neg.-Seq. compensation			
$K^*_{p\_neg}$ (pu)	0.1	T_{i_neg} (s)	0.01
Cut-off frequency of LPF for Q_{out_dg} (Hz)			20

From parameters listed in Table 8.5 and Fig. 8.10a, $D^*_{av\_sg}$ is 8.7 pu. Therefore, D^*_{dg} should be set to 8.7 pu.

8.3.3.2 Governor Delay

As it is discussed in Chapter 7, the governor delay T_{d_dg} reduces the total inertia of MGs and amplifies oscillations. As a result, it is removed in Section 8.2. In [23], it is suggested to set T_{d_dg} to the same value as T_{d_sg}, that is, 1 s, in order to share the transient active power properly. However, based on the steady-state model presented in Chapter 7, if T_{d_dg} is set to 1 s, a very oscillatory pair of conjugated eigenvalues could be observed near the origin, with a damping ratio ζ less than 0.14, as depicted in Fig. 8.14. These oscillatory eigenvalues indicate large oscillations in output variables explained in Section 7.3, including the output active power and the frequency of both the DG and the SG. These eigenvalues become a fast nonconjugated eigenvalue if T_{d_dg} is set to 0 s. To verify this point, simulation results with both $T_{d\_dg} = 1$ s and $T_{d\_dg} = 0$ s during a three-phase loading transient from 2 kW, 1 kvar to 15 kW, 3 kvar are shown in Fig. 8.15a and b, respectively. It is clear that in the case of $T_{d\_dg} = 1$ s, large oscillations with a period around 1 s can be observed in all presented parameters, which coincide with the oscillatory eigenvalues near the origin shown in Fig. 8.14. These oscillations result in much larger frequency and voltage deviations and poorer active and reactive power transient performances compared to the case of $T_{d\_dg} = 0$ s. Therefore, the governor delay T_{d_dg} is set to 0 s in the present work.

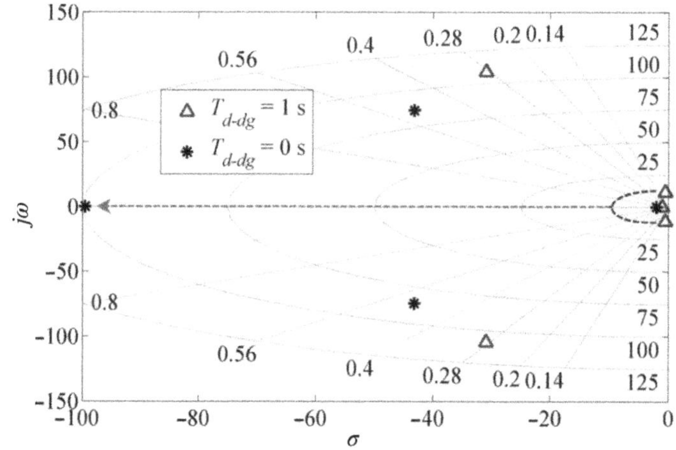

Figure 8.14 Eigenvalues with a variation of T_{d_dg}.

8.3.3.3 Constant Virtual Stator Reactance

From the principle of a round-rotor SG, the synchronizing power coefficient of the SG shown in Fig. 8.10a can be calculated through (8.18) [25].

$$K_{sg} \approx \frac{E'_q V_{bus}}{X'_d + X_{ad\_sg}} \cos \delta'_{sg} \tag{8.18}$$

where δ'_{sg} is the transient power angle of SG.

As it is discussed in Section 7.5.3, the per unit value of synchronizing power coefficient of the DG and the SG should be designed equally to provide a proper immediate active power sharing after a loading transition. Defining $X_{sg} = X'_d + X_{ad\_sg}$ and supposing $E'_q \approx E_{dg}$, $\cos \delta'_{sg} \approx \cos \delta_{dg}$, by comparing the synchronizing power coefficient of DG in Section 7.2.2.1 with (8.18), the key is to guarantee $X^*_{sg} = X^*_{dg}$. Knowing that $X_{dg} = X_f + X_{ls0}$, and X'_d of the SG and X_f of the DG are fixed, $X^*_{sg} = X^*_{dg}$ can be guaranteed by tuning X_{ls0} and X_{ad_sg}.

It should be pointed out that a small value of $X^* = X^*_{sg} = X^*_{dg}$ leads to an unstable system, as demonstrated in Fig. 8.16a by eigenvalue analyses based on the steady-state model presented in Section 7.2. As X^* gets small, the damped natural frequency of a pair of conjugated eigenvalues moves toward the synchronous frequency, which implies a probable resonance near synchronous frequency. The simulation results of the synchronization process shown in Fig. 8.16b verify this point. In these simulations, the SG operates with a three-phase load of 2 kW, 1 kvar, and the DG is connected at around 14 s. In the case where $X_{ls0} = 0$ and X_{ad_sg} is tuned to make $X^* = X^*_{sg} = X^*_{dg} = 0.0942$ pu, a divergent oscillation occurs after the connection of the DG. The system becomes unstable, and the DG is finally tripped due to overcurrent. This issue can be solved by increasing X^* up to 0.15 pu, and the performance is even

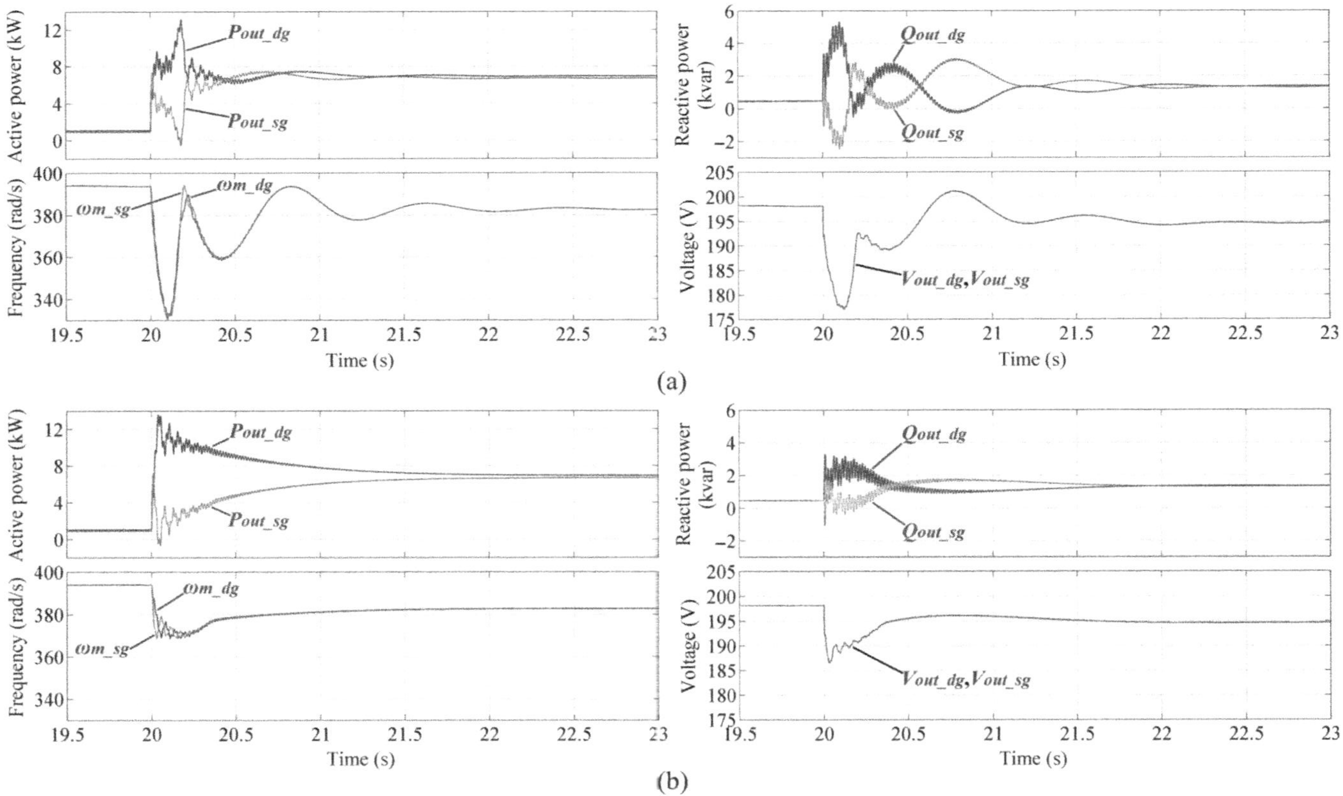

Figure 8.15 Simulation results for the governor delay: (a) $T_{d\_dg} = 1$ s and (b) $T_{d\_dg} = 0$ s.

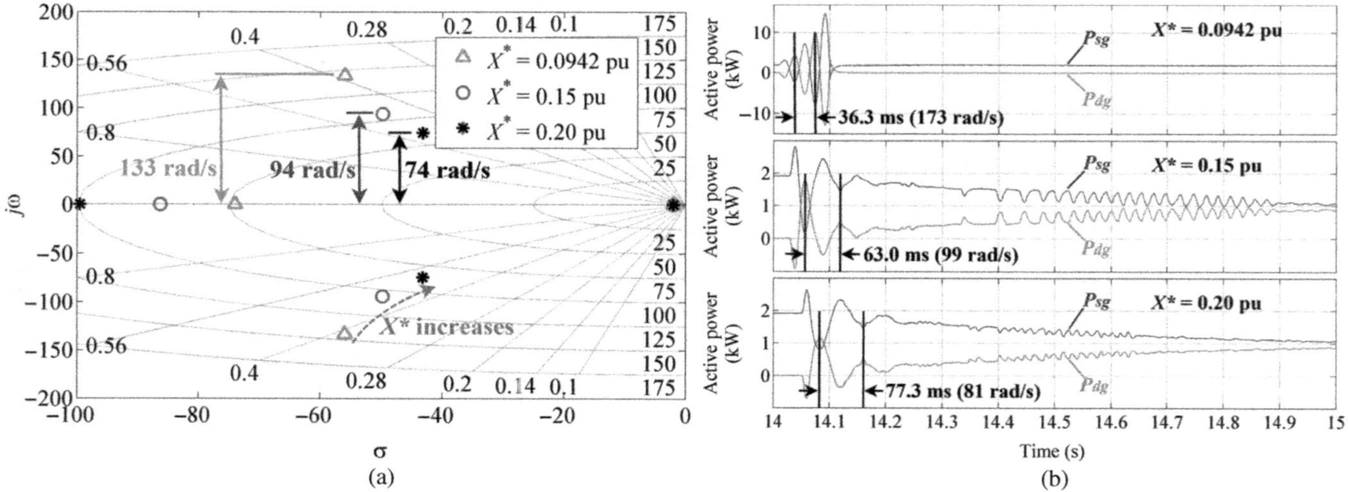

Figure 8.16 Analysis of different values of total output reactance X^*: (a) eigenvalues and (b) simulation results.

better in the case of 0.20 pu, as it is shown in Fig. 8.16b. This is because the damped natural frequency of the conjugated eigenvalues becomes smaller, thus difficult to be excited by the disturbances at synchronous frequency. It is noteworthy that the oscillation frequencies in the simulation results coincide with the damped natural frequencies in eigenvalue analyses.

As a result, $X^* = X^*_{sg} = X^*_{dg} = 0.20$ pu is suggested for the present application. For the DG, the constant virtual stator inductance L_{ls0} is set to 1.122 mH (X^*_{ls0} = 0.1057 pu), and for the SG, an additional inductor L_{ad_sg} of 1.836 mH ($X^*_{ad\_sg}$ = 0.1730 pu) is added to the output of the SG as shown in Fig. 8.10a.

8.3.3.4 Transient Virtual Stator Impedance for Current Limiting

The voltage drop through the transient virtual stator impedance should equal the nominal voltage during a three-phase fault, as given by (8.19) [24].

$$E_0^* = I^*_{max} \Delta Z^*_{ls} \tag{8.19}$$

where I^*_{max} is the desired maximum RMS value of phase current in per unit value. From Fig. 8.13c and (8.19), the proportional gain k_Z can be calculated through (8.20).

$$k_Z^* = \frac{\Delta Z^*_{ls}}{(I^*_{max} - I^*_{thresh})} = \frac{E_0^*}{I^*_{max}(I^*_{max} - I^*_{thresh})} \tag{8.20}$$

As I^*_{thresh} is set to 1 pu and $E_0^* = 1$ pu, based on (8.20), k_Z^* can be determined by a given I^*_{max}.

To choose an appropriate value for I^*_{max}, a series of simulations are executed, and the results are shown in Fig. 8.17. In these simulations, loading transitions

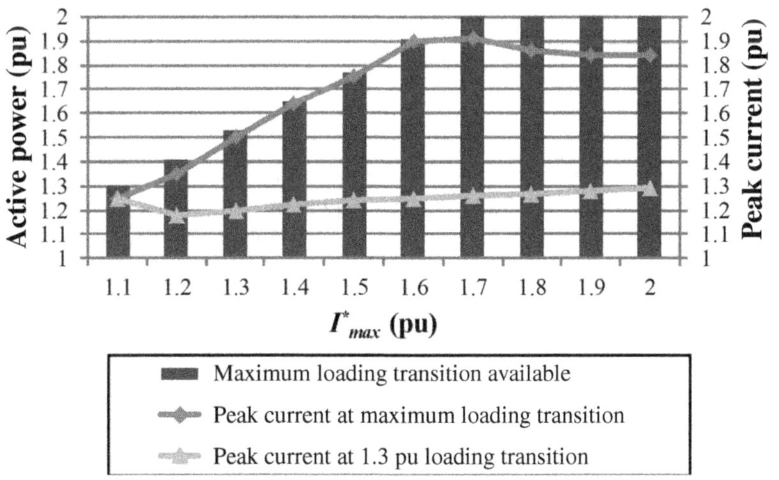

Figure 8.17 Simulation results for tuning the proportional gain k_Z of transient virtual stator impedance.

from 0 pu to a specified active power loading are executed during the parallel operation of SG and DG. As the sum of power ratings of DG and SG is 2 pu (20 kVA), loading transitions up to 2 pu are tested.

As it is illustrated in Fig. 8.17, if I^*_{max} is chosen to be less than 1.7 pu, the MG becomes unstable during a 2 pu active power loading transition. For $I^*_{max} \geq$ 1.7 pu, larger value of I^*_{max} leads to smaller peak current during a 2 pu active power loading transition, but larger peak current during a 1.3 pu active power loading transition. A compromise can be reached when $I^*_{max} = 1.8$ pu; therefore, according to (8.20), k^*_Z is set to 0.69 pu for the present application.

Another issue is the design of X/R ratio for the transient virtual stator impedance. A series of simulations are executed during a three-phase loading transition from 2 kW, 1 kvar to 15 kW, 3 kvar, with $k^*_Z = 0.69$ pu in all cases and X/R is set to 1, 5, and 20. The simulation results are shown in Fig. 8.18. It can be observed that the peak value of DG output active power is limited better with more resistive transient virtual stator impedance owing to the damping feature of resistance. However, resistive transient virtual stator impedance may deteriorate the transient reactive power and lead to larger voltage deviation, as it is illustrated in the right column of Fig. 8.18. Therefore, a trade-off should be made between the active power limiting and the transient performances of reactive power and voltage. In the present work, X/R is set to 5 to achieve a balanced overall performance.

8.3.4 Simulation Results

Comparative simulations are executed in PSCAD/EMTDC environment, in order to verify the proposed modification on the VSG control. The results of the first simulation group are shown in Figs 8.19 and 8.20, whereas the control schemes are listed in Table 8.6. In these simulations, SG and DG operate in parallel with an initial three-phase load of 2 kW, 1 kvar. At 20 s, a single-phase load of 9.6 kW, 4.2 kvar is connected. In comparison of the case of single SG operation discussed in Section 8.3.1, all loads are doubled.

As it is shown in Fig. 8.19a, without the proposed modification, slight ripples can be observed in the SG active and reactive power after the single-phase load is connected, due to a three-phase unbalance. This unbalance can be further confirmed in the steady-state SG current waveforms shown in Fig. 8.20a. Moreover, due to the ripples in calculated active and reactive power, ripples can be observed in frequency and voltage of the DG, as well as those of the SG. These problems can be solved by the proposed DDSRF decomposition, the modified power and voltage calculations based on only positive-sequence components shown in (8.13)–(8.15), and the proposed SG negative-sequence current compensator, as shown in Fig. 8.19b. The steady-state current waveforms of this case are shown in Fig. 8.20b, in which no three-phase unbalance is observed in the SG current.

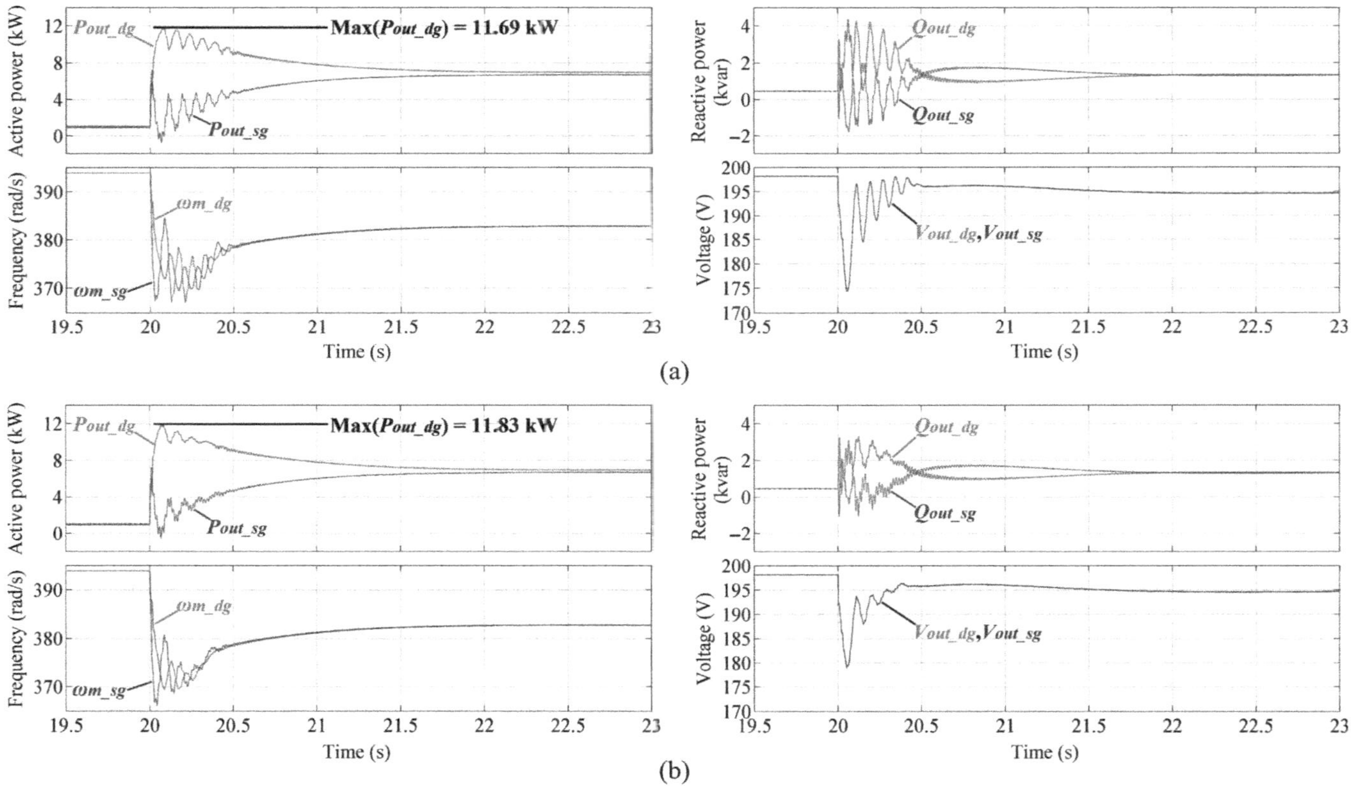

Figure 8.18 Simulation results for tuning the X/R ratio of transient virtual stator impedance when (a) $X/R = 1$, (b) $X/R = 5$, and (c) $X/R = 20$.

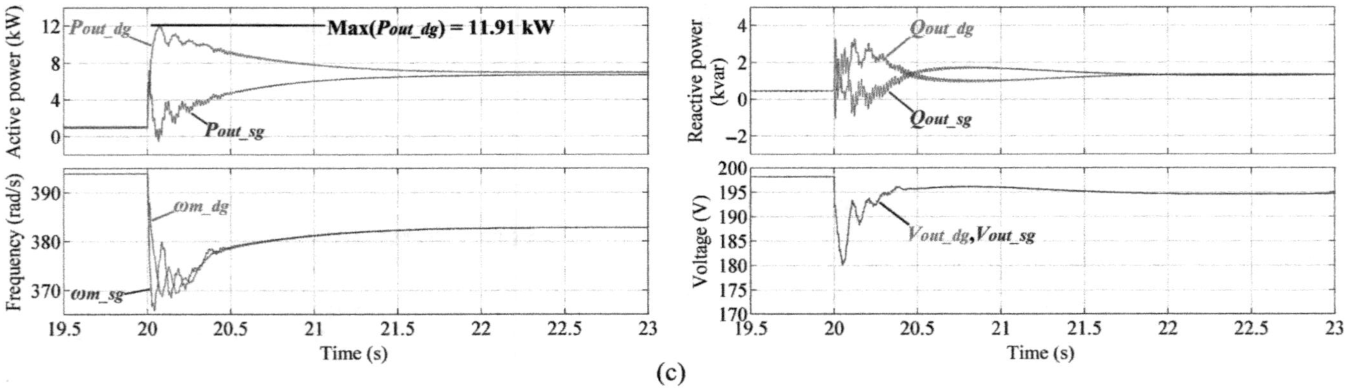

Figure 8.18 (*Continued*)

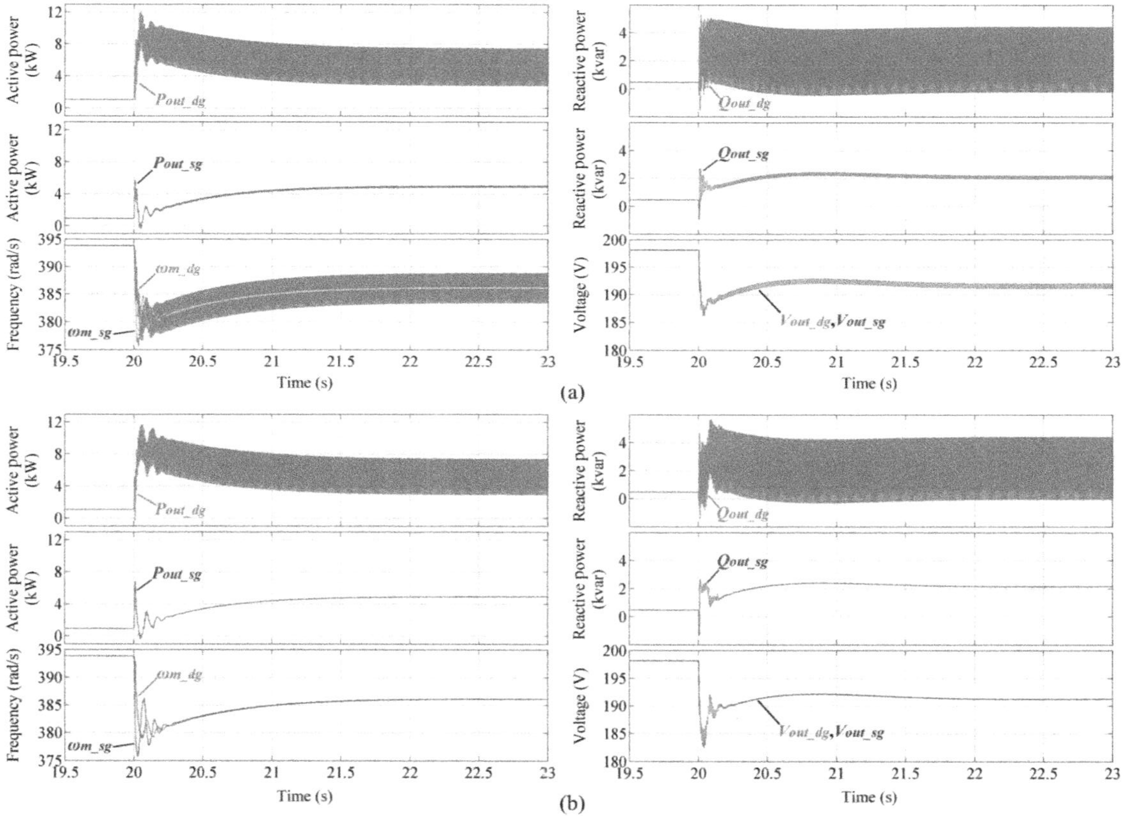

Figure 8.19 Simulation results under unbalanced loading condition: (a) Case A and (b) Case C (see Table 8.7).

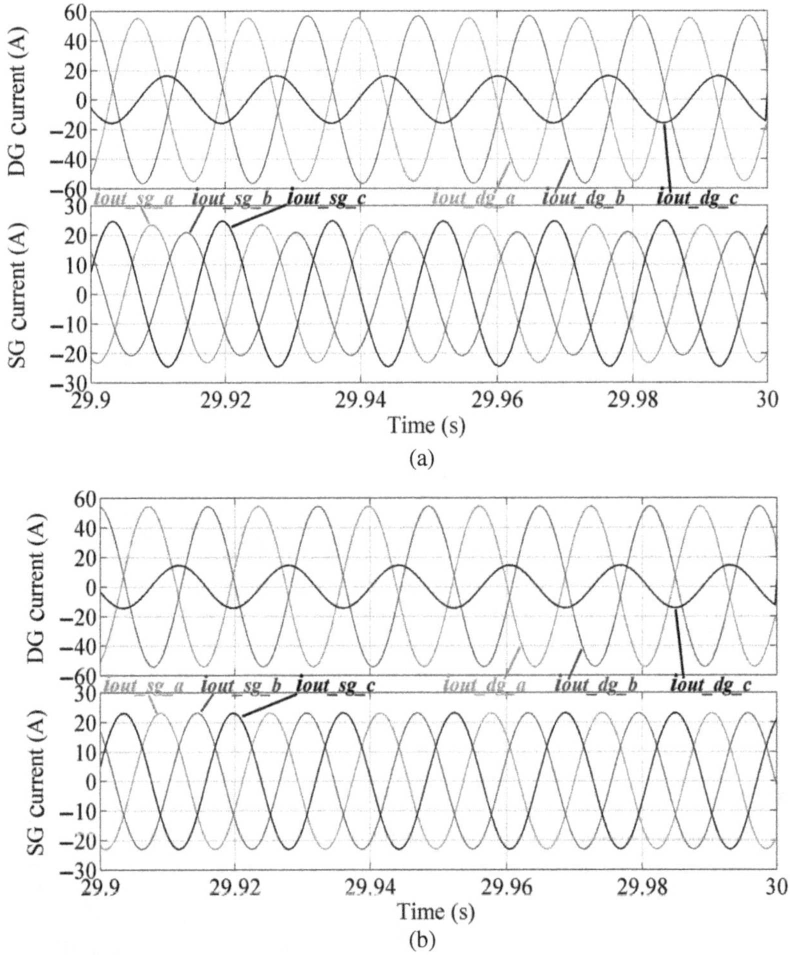

Figure 8.20 Simulation results of steady-state SG and DG current waveforms under unbalanced loading condition: (a) Case A and (b) Case C (see Table 8.7).

Table 8.6 Control schemes of simulations.

Control part	Case A	Case B	Case C
Enhanced VSG control (Section 8.2)	✓	✓	✓
DDSRF with new power and voltage calculation	×	✓	✓
SG neg.-seq. compensation	×	✓	✓
Transient virtual stator impedance	×	×	✓

Furthermore, by comparing the simulation results of parallel operation of SG and VSG with the proposed modified VSG control shown in Fig. 8.19b with those of the single SG operation shown in Fig. 8.11a, it can be observed that the SG rotor speed deviation is improved tremendously by the presence of the VSG. The 4.8 kW, 2.1 kvar single-phase load causes a transient rotor speed drop down to around 330 rad/s as shown in Fig. 8.11a, but when a doubled loading transition occurs during the parallel operation of SG and VSG, the minimum frequency is still higher than 375 rad/s. Moreover, by comparing the SG current waveforms shown in Fig. 8.11b with those in Fig. 8.20b, it is clear that the unbalanced SG current is compensated properly by the DG equipped with the proposed modified VSG control.

8.4 Alternating Inertia-based Virtual Synchronous Generator Control

The quantities of the VSG, such as its output frequency and power, oscillate after a change or disturbance similar to those of an SG. However, the transient condition tolerance of an inverter-based DG is much less than that of a real SG. On the other hand, VSG control has an advantage in that its swing equation parameters can be adopted in real time to obtain a faster and more stable operation. In this concept, the value of the virtual moment of inertia is changed based on the relative virtual angular velocity (the difference between virtual mechanical velocity generated by the VSG and grid angular frequency) and its rate of change [26]. Therefore, it is called *alternating inertia* scheme. This scheme removes the oscillations and, thereby, increases the reliability of the VSG unit against parameter changes and disturbances. This section goes into detail on the alternating inertia control with the objective of clarifying its damping and stabilizing effect. The damping effect is investigated by the transient energy analysis, and its stabilizing performance on the nearby machines is investigated by simulations. Finally, the performance is verified by experiments on a laboratory-scale DSP-based controlled inverter.

8.4.1 Control Strategy

Consider the power angle curve shown in Fig. 8.21. After a change in the MG, for example, a change in the prime mover power from P_{in0} to P_{in1}, the operating point moves along the power curve, from point "a" to "c" and then from "c" to "a." One cycle of the oscillation consists of four segments indicated as ①–④ in Fig. 8.21. The machine condition during each segment of an oscillation cycle is summarized in Table 8.7. During each segment, the sign of the $d\omega_m/dt$ together with the sign of the relative angular velocity $\Delta\omega$ defines the acceleration or deceleration. For example, in segment ③ of Fig. 8.21, during

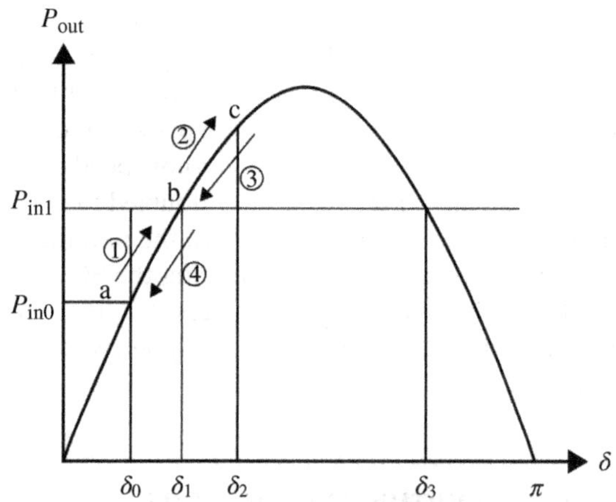

Figure 8.21 Power angle curve of a typical SG.

transition from points "c" to "b," both $d\omega_m/dt$ and $\Delta\omega$ are negative and act in the same direction; therefore, it is an acceleration period, whereas when they have opposite signs as segment ④, it is a deceleration period.

The objective is to damp frequency and power oscillation quickly by controlling the acceleration and deceleration term. The derivative of angular velocity, $d\omega_m/dt$, indicates the rate of acceleration or deceleration. Considering (7.1), it is observed that this rate has a reverse relation to the moment of inertia, J. Based on this fact, one can select a large value of J during acceleration phases ("a" to "b" and "c" to "b") to reduce the acceleration and a small value of J during deceleration phases ("b" to "c" and "b" to "a") to boost the deceleration. The large moment of inertia J_{big} and the small one J_{small} can be chosen within a wide range, depending on the rated power so that the difference between J_{big} and J_{small} determines the damped power in each half-cycle of oscillation by alternating inertia. The value of J_{big} can be equal to the normal value of J (calculated by $J = 2HS_0/\omega_0^2$). However, applying a considerably larger value than the normal J will result in a smaller frequency excursion at the first quarter cycle (segment ①) but a sluggish response. The value of J_{small} determines the transient of the second quarter cycle of oscillation (segment ②). It will be shown that a very small value of J_{small} (1% of J_{big}) will result in a quick damping effect.

Table 8.7 is fully matched with the concept of bang–bang control strategy [26]. During each cycle of oscillations, the value of J is switched four times. Each switching occurs at the points that the sign of either $\Delta\omega$ or $d\omega_m/dt$ varies. Before the disturbance, the VSG operates with the normal value of J. When the disturbance occurs, the transition from "a" to "b" starts with $\Delta\omega > 0$ and $d\omega_m/dt > 0$. In this condition, the J_{big} is adopted. At the end of the first quarter cycle, that is, point "b," the sign of $d\omega_m/dt$ changes. It means that the small value for J is adopted at this point. At point "c," the sign of $\Delta\omega$ changes, and

Table 8.7 Machine modes during oscillation.

Segment		$\Delta\omega$	$d\omega_m/dt$	Mode	Alternating J
①	a → b	$\Delta\omega > 0$	$d\omega_m/dt > 0$	Accelerating	Big value of J
②	b → c	$\Delta\omega > 0$	$d\omega_m/dt < 0$	Decelerating	Small value of J
③	c → b	$\Delta\omega < 0$	$d\omega_m/dt < 0$	Accelerating	Big value of J
④	b → a	$\Delta\omega < 0$	$d\omega_m/dt > 0$	Decelerating	Small value of J

J retrieves its large value. It will be the end of the first half cycle. During the second half cycle, the value of J is switched to the J_{small} at point "b," and again at the end of one cycle at point "a," J_{big} is adopted. This procedure is repeated for each cycle of oscillation until the transients are suppressed and $\Delta\omega$ equals zero at the new equilibrium point (point "b"). A threshold for $\Delta\omega$ can be applied to avoid the chattering of J during normal operation. However, here, the threshold is set to zero.

Figure 8.22 shows the simulation results of this concept. Simulations are performed on the system model of Fig. 7.2 with the parameters of $P_{base} = 50$ kW, $f_{base} = 60$ Hz, $R_L = 12.5\%$, $X_L = 33.0\%$, $X_F = 42.4\%$, and $D = 17$ pu at the base kW and the base voltage of 0.2 kV. The system is subjected to increase in the VSG power reference in two steps of 70% and 30% at $t = 2$ s and $t = 8$ s, respectively. Figure 8.22a shows the output power and frequency of the VSG with the fixed value of $J = 6$ kg m^2. To show the effectiveness of the proposed idea, simulations are carried out on a weak system that VSG with fixed J is not able to track the grid frequency and keep up synchronism after the second step of power increase. Then, the control scheme of the VSG is changed to alternating inertia control, and the same scenario is applied. As shown in Fig. 8.22b, alternating inertia selects the values of J out of $J_{big} = 6$ kg m^2 and $J_{small} = 1$ kg m^2 (the value for J_{small} is decided by experience; any value that is smaller than J_{big} can be applied. However, the difference between J_{big} and J_{small} determines the suppressed energy at each half cycle and, thereby, affects the speed of oscillation damping). This process does not only stabilize the system but also suppresses the frequency and power oscillations effectively. In Fig. 8.22b, it is observed that the value of inertia is chattering during steady-state operation. This chattering appears because of the slight variation of the angular velocity around the equilibrium point. However, the chattering can be removed by setting a threshold for the $\Delta\omega$ in the alternating inertia algorithm. It is carried out by replacing the $\Delta\omega > 0$ and $\Delta\omega < 0$ in Table 8.8 with $\Delta\omega > \varepsilon$ and $\Delta\omega < \varepsilon$, respectively, where ε is a negligible positive value.

Damping factor is an important term that defines the VSG response. An inappropriate value of damping factor may result in a high magnitude of oscillation or a sluggish response. The alternating inertia concept allows the VSG system

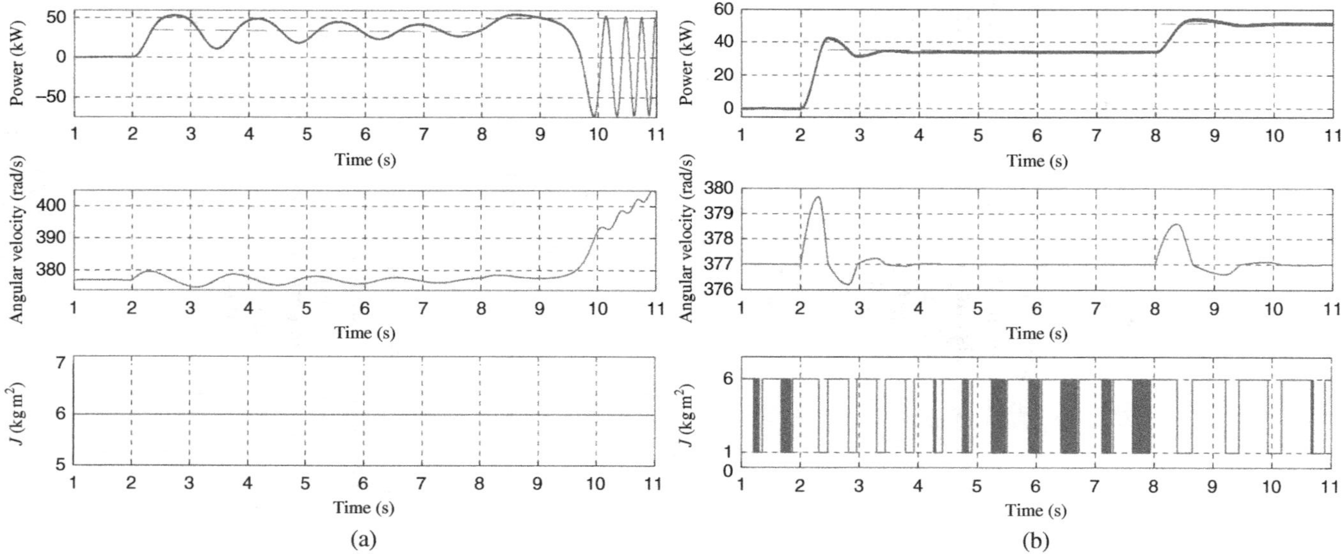

Figure 8.22 Output power, virtual angular velocity, and virtual moment of inertia of VSG with $J = 6$ kg m^2 and $D = 17$ pu: (a) fixed J and (b) alternating J.

Table 8.8 Specifications of the experimental system.

Base power (kVA)	10
Base frequency (Hz)	60
Base voltage (V)	207
Switching frequency (kHz)	14
DC-link voltage (V)	320
Filter stray resistance (%)	0.23
Filter inductive reactance (%)	8.8
Filter capacitor VAR (%)	1.62
Resonance frequency of LC filter (kHz)	1.59
Transformer reactance (%)	9.68

to exert a suitable time constant in each phase of oscillation; therefore, the importance of the damping factor in the behavior of the VSG system is reduced considerably [26].

8.4.2 Stability Analysis

Transient stability concerns the stability of the rotor angle of SGs (voltage angle in the case of VSG) after a significant disturbance. Having the advantage of not solving the nonlinear differential equations, Lyapunov direct method has become the center of attention for transient stability analysis. Consider a system expressed by a set of nonlinear differential equations of the form $\boldsymbol{x} = \boldsymbol{F}(\boldsymbol{x})$, $\boldsymbol{x}$ being a vector of state variables. The point $\hat{\boldsymbol{x}}$ for which $\boldsymbol{F}(\hat{\boldsymbol{x}}) = 0$ is the equilibrium point of the system in a state space. The solution of system state equations from initial point to the equilibrium point forms the system trajectory. The trajectory of an asymptotically stable system converges at the equilibrium point as time approaches infinity.

Based on the Lyapunov stability theorem, point $\hat{\boldsymbol{x}}$ is asymptotically stable if a continuous differentiable function $V(\boldsymbol{x})$ exists and $V(\boldsymbol{x}) \leq 0$. In the other words, the rate of change of $V(\boldsymbol{x})$ along the system trajectory should be negative. The $V(\boldsymbol{x})$ is declining during time as state variables converge to the equilibrium point that is a minimum stationary point. Whereas for an unstable initial point, the value of $V(\boldsymbol{x})$ rises, and the trajectory diverges on the state variable plane.

Finding a candidate Lyapunov function is the next step of stability analysis. Reference [25] calculated the energy function through removing the damping factor, multiplying the swing equation by $\Delta\omega$, and integrating the product from the first equilibrium point, that is, point "b" in Fig. 8.21, with δ_1 and $\Delta\omega = 0$, to any point on the system transient trajectory. The resultant expression is as follows:

$$V = E_k + E_p = \frac{1}{2}\omega_0 J\Delta\omega^2 - [P_{in}(\delta - \delta_1) + b(\cos\delta - \cos\delta_1)] \quad (8.21)$$

where V is the system transient energy after a change or disturbance, b is the amplitude of the power angle curve, and ω_0 is the system frequency. The transient energy is the energy required to move the system states from a pre-disturbance operating point to the new stable operating point. This term can be considered as an energy gap. Depending on the form of change or disturbance, the transient energy may be needed to be injected or extracted. The transition of the system states to the new stable point (filling the energy gap) includes oscillation that may result in instability (transient instability evaluated by equal-area criterion). Besides, the transients involve power oscillation that is dissipated in resistive elements. It means that the high-magnitude and long-duration oscillation results in power loss in the system. In the case of a VSG with energy storage, the dissipated power comes from the storage unit. It can be concluded that if the transient oscillation is removed, not only stabilization is achieved, but also the energy dissipation is prevented. The first term of (8.21), which is denoted as E_k, is the kinetic energy of the rotor of SG and virtual kinetic energy for the VSG system. The other term, E_p, is the potential energy that is stored and released electromagnetically during the interaction between the electromagnetic fields of the rotor and stator of the machine. It is proved that V satisfies the Lyapunov function criteria: (i) it has stationary points at the equilibrium points (the gradient of V is equal to zero at the stationary points); (ii) it is positive definite in the vicinity of one of the equilibrium points (its Hessian matrix is positive definite); and (iii) its derivative is not positive [25]. In the case of VSG with alternating moment of inertia, a presumption of $J > 0$ is needed. Moreover, the derivative of the V with alternating inertia should be checked to be negative.

When oscillation starts at point "a" of Fig. 8.21, $\Delta\omega$ is zero, and $\delta - \delta_1$ is maximum. Therefore, $E_k = 0$ and E_p is maximum. During the transition from point "a" to "b" with a large value of J, E_k is increasing, and E_p is decreasing as $\Delta\omega$ increases and $\delta - \delta_1$ decreases. All system transient energy is converted into the kinetic form at point "b" with maximum $\Delta\omega$ and J_{big}. At this point, the change in the moment of inertia to the small value is applied. Therefore, the system transient energy will be decreased to a smaller-value kinetic form with the same $\Delta\omega$ but J_{small}. Now this energy will be converted into the potential form during the transition from point "b" to "c" as $\Delta\omega$ decreases and $\delta - \delta_1$ increases. Because the total energy has decreased, the amplitude of oscillation will be reduced. In other words, $\delta_2 - \delta_1$ will be much smaller than $\delta_0 - \delta_1$. J adopts its large value at point "c." However, all of the system transient energy is in potential form at this point as $\Delta\omega = 0$ and $\delta - \delta_1$ has its maximum value. Therefore, an increase in J does not increase the system energy level based on (8.21). This process occurs in each half cycle until $\Delta\omega$ becomes less than a desired threshold.

To see the damping effect of the alternating moment of inertia scheme, the third criterion of the Lyapunov function is considered. This criterion demands that the derivative of the energy function is negative. Thus, the system transient energy declines for the time until the system state variables are settled at the equilibrium point. For a VSG with variable J, by calculating the derivatives of E_k and E_p of (8.21) separately and considering the swing equation, the derivative of V is expressed as

$$\frac{dV}{dt} = \frac{\omega_0}{2}\Delta\omega^2\frac{dJ}{dt} - D\Delta\omega^2 \tag{8.22}$$

This expression must be negative to have decay in the system transient energy during oscillation. The term $-D\Delta\omega^2$ is obviously negative for $D > 0$. Because J varies discontinuously, dJ/dt is approximated by its average value at the points at which J is switched. At points "a" and "c" at which J is varied from J_{small} to J_{big}, the variation of J is positive ($\Delta J = J_{big} - J_{small}$). However, dJ/dt is multiplied by $\Delta\omega^2$ in (8.22), and $\Delta\omega$ is zero at these points. It means that the variation of J does not change the transient energy at these points. Inversely, at point "b," at which J is varied from J_{big} to J_{small}, ΔJ is calculated as $J_{small} - J_{big}$, and $\Delta\omega$ has its maximum value. Therefore, the term dJ/dt is effective at this point, and it is estimated for each half cycle as follows:

$$\frac{dJ}{dt} \approx \frac{\Delta J}{\Delta t} = \frac{J_{small} - J_{big}}{0.5T} \tag{8.23}$$

Assuming a zero damping factor D, (8.22) can be rewritten as

$$\frac{dV}{dt} = \frac{\omega_0}{2}\Delta\omega^2\frac{J_{small} - J_{big}}{0.5T} < 0 \tag{8.24}$$

Equation (8.24) shows that an additional damping effect is imposed in each half cycle by varying the value of the moment of inertia. This damping acts directly on the transient energy and cuts it to a desired level decided by the difference in the values of J. Figure 8.23 illustrates the performance of the alternating inertia method in transient energy suppression. Some simulations are performed on the system model of Fig. 7.2. Figure 8.23a contains the system transient energy trajectory for a VSG with the fixed moment of inertia of 6 kg m^2 after a step increase of 1 pu (at 50 kW base power) in power reference.

It is observed that the system transient energy declines by the damping factor as the state variables converge to the equilibrium point. When the alternating inertia scheme is applied, the transient energy drops to a desired level by applying the small J of 3 kg m^2 as shown in Fig. 8.23b. By adopting a tiny J_{small}, fast decay of transient energy can be achieved, which bypasses the state variables to the stable stationary point in the first half cycle.

There is another stationary point for the swing equation, with $\Delta\omega = 0$ and $\delta = \pi - \delta_1$ (that equals δ_3 in Fig. 8.21). The value of the energy function at this point is the critical transient energy. For an SG and VSG with fixed inertia, the value of energy function at the initial point of oscillation must be smaller than

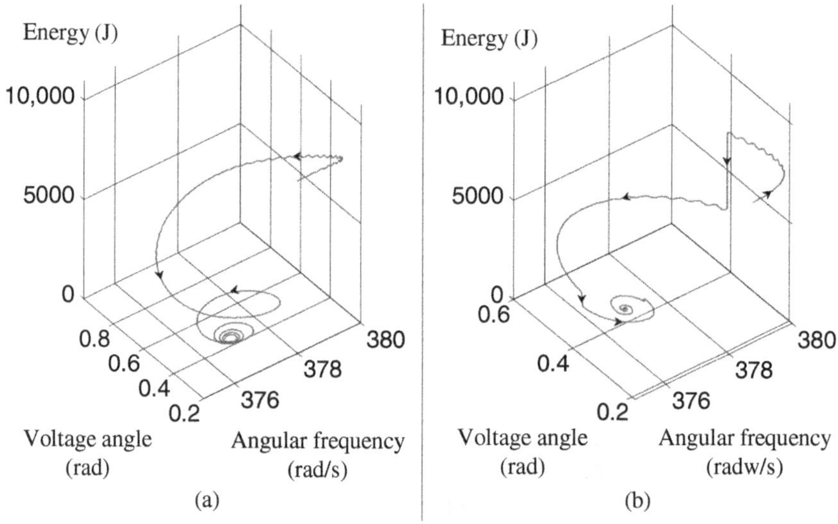

Figure 8.23 Transient energy trajectory after a step change in power reference of VSG with: (a) fixed moment of inertia, and (b) alternating inertia.

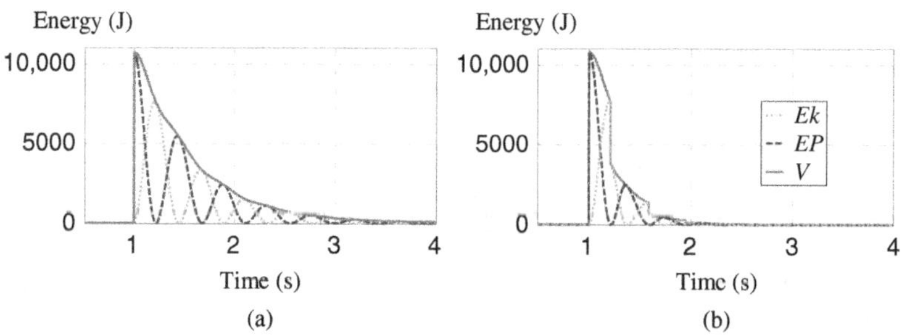

Figure 8.24 The kinetic energy (E_k), potential energy (E_P), and total transient energy (V) waveforms after a step increase in power reference of VSG with: (a) fixed moment of inertia and (b) alternating inertia.

the critical value. However, in the case of VSG with alternating inertia scheme, any initial energy gap can be reduced to nearly zero value at the end of the first quarter cycle; therefore, the critical transient energy and thereby transient stability area will always have their maximum value, that is, $2b - \pi P_{in}$ from (8.21). Other corollaries regarding the transient stability area and the critical clearance time that can be inferred from the alternating inertia concept are forgone because of prolixity. The system kinetic, potential, and total energy under the same simulation condition shown in Fig. 8.23 are plotted in Fig. 8.24.

8.4.3 Effect of Alternating Inertia on Dissipated Energy

In the case of a real SG, during oscillation, the kinetic energy preserved in the rotor mass will be converted into the potential energy accumulated in the magnetic field of SG, and then in reverse, until the oscillation is damped and system states are stabilized at the equilibrium point. For the VSG system, the kinetic energy of the rotor and the potential energy are virtual quantities, and the VSG scheme controls the inverter output similar to the output of an SG. For example, when $\Delta\omega$ is not zero, the VSG imagines that there is a specific value of kinetic energy considering the value of J and oscillates the output power of the inverter (by changing the voltage angle) based on the swing equation of an SG that expressed in (7.1). When the alternating inertia switches the value of J to J_{small} at the maximum value of $\Delta\omega$, the VSG scheme just imagines that a smaller value of kinetic energy is available rather than its large value with the J_{big}. In other words, the kinetic energy that is removed by alternating inertia is imaginary and is used by VSG algorithm to control the output of VSG and does not exist in reality. Thus, for quick convergence of the states of the system, the energy value can be varied at the proper time without concerning where the energy goes to.

Another issue is the effect of alternating inertia on the dissipated energy. The dissipated energy is different from the transient energy (energy gap) and refers to the energy that is dissipated in the system during oscillations of power and current. Hereafter, it is shown that alternating inertia reduces the energy dissipated in the system by monitoring the DC-link power and energy. Consider the VSG system of Fig 7.2. The system parameters are as follows: $P_{base} = 50$ kW, $f_{base} = 60$ Hz, $R_L = 12.5\%$, $X_L = 33.0\%$, $X_F = 42.4\%$, and $D = 17$ pu at the base kW and the base voltage of 0.2 kV. The energy storage is a 5 mF capacitor. At $t = 1$ s, the VSG power command has been changed from 0 to 1 pu. The DC-link power is injected to the grid through inverter as shown in Fig. 8.25 under two conditions: (i) VSG with fixed inertia ($J = 6$ kg m^2) and (ii) VSG with alternating inertia ($J_{small} = 1$ and $J_{big} = 6$ kg m^2). The alternating inertia removed the oscillation of DC-link power. It is shown in [26] that although the difference is not considerable, the DC-link energy of VSG with alternating inertia is slightly less than the case with fixed inertia.

8.4.4 Grid Stability Improvement

8.4.4.1 Virtual Synchronous Generator in Parallel with Synchronous Generator

In MG applications, an inverter-based DG works in parallel with other DGs that may include SGs. Consider the islanded MG of Fig. 8.26. The VSG block has the control scheme shown in Fig. 7.2 with the output filter reactance of 9.7%. The objective is to assess the effect of the alternating inertia on the stability of the paralleled SG. To clarify the effectiveness of the idea, the capacity of the VSG unit is assumed to be 20% of the SG that is insignificant. A symmetrical

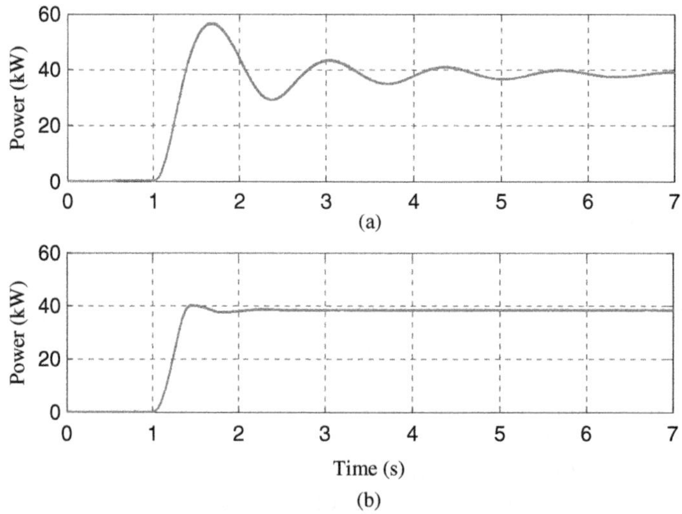

Figure 8.25 DC-link power of VSG subjected to a step change in the power reference: (a) VSG with fixed inertia and (b) VSG with alternating inertia.

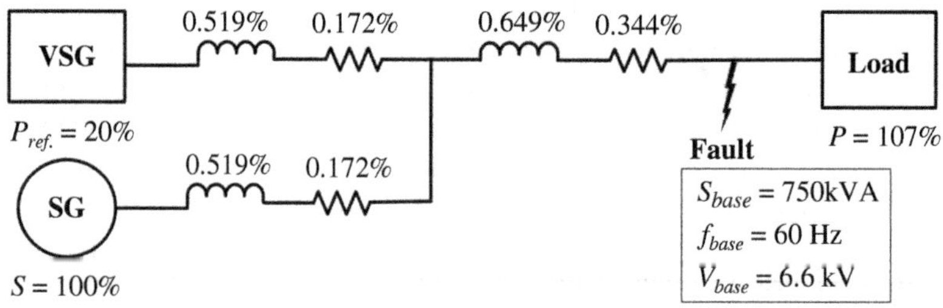

Figure 8.26 VSG unit in parallel with the SG in MG.

three lines to ground fault is occurred at the load point at $t = 0.2$ s and lasted for 0.3 s. In this condition, the system comprising the VSG with the fixed value of moment of inertia $J = 8.445$ kg m^2 is not able to recover from the fault as shown in Fig. 8.27a. The same scenario is applied to the system with alternating inertia of $J_{big} = 8.445$ and $J_{small} = 0.0844$ kg m^2. The waveforms of power, SG rotor angle, and angular frequency are shown in Fig. 8.27b. As it can be seen, the alternating inertia scheme improves the stability of the adjacent machine by the extra damping effect directly imposed on the transient energy.

8.4.4.2 Virtual Synchronous Generator as an Interface between the Synchronous Generator and the Grid

Another configuration is shown in Fig. 8.28a. An SG is connected to the grid/MG through a VSG unit. The prime mover of the SG can be considered

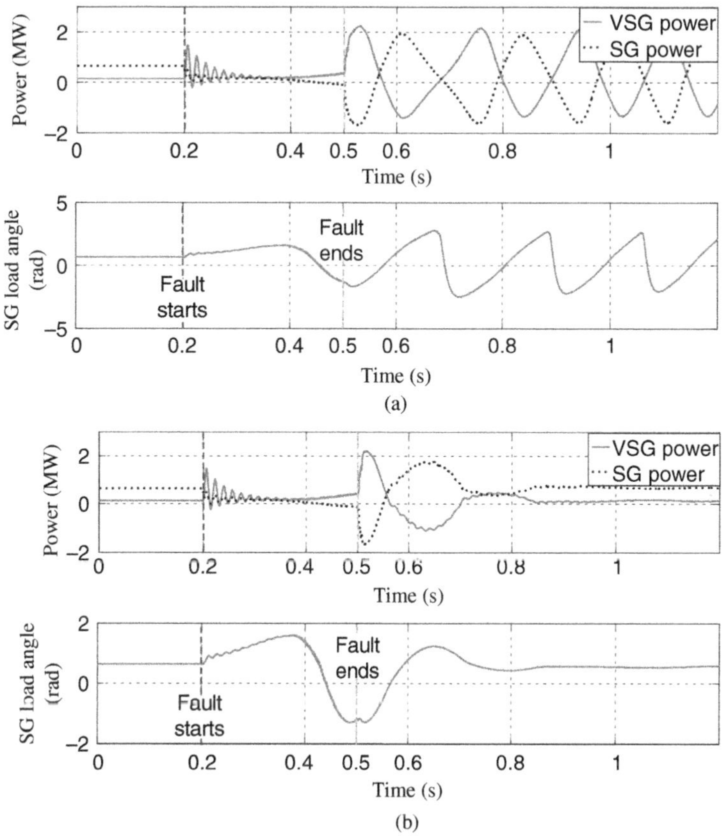

Figure 8.27 VSG and SG powers and SG rotor angle waveforms using (a) fixed moment of inertia and $D = 17$ pu and (b) alternating inertia control and $D = 0$ pu.

as a gas or diesel engine, and an inverter interface is required to correct the generated power to be injected to the grid. If the VSG unit is not robust enough, the disturbances from grid/MG will affect the stable operation of the SG. To assess the effect of the alternating inertia control on the stability of such systems, a symmetrical three-phase voltage sag with 10% remained voltage magnitude and the duration of 0.2 s is applied from grid side, and the performance of the system is monitored. The reference power and damping factor of the VSG are 1 and 17 pu, respectively, and a fixed inertia factor equal to 5 kg m^2 is applied. Figure 8.28b shows that the SG rotor angle and DC-link voltage are considerably affected by the grid voltage sag. The high-peak transient of DC-link voltage is mainly because of the oscillation of the VSG output power. The same scenario is applied to the system with the alternating inertia control with $J_{big} = 5$ and $J_{small} = 0.05$ kg m^2. To discriminate

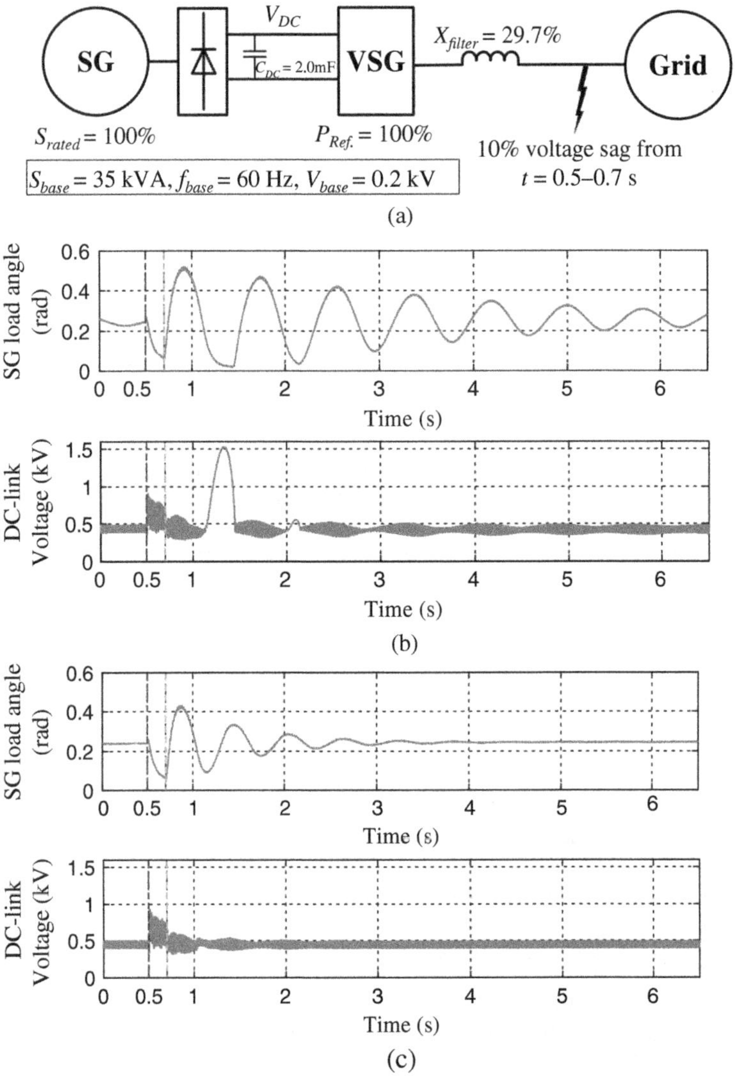

Figure 8.28 SG connected to the grid via VSG unit: (a) system configuration, (b) for VSG with fixed moment of inertia, and (c) for VSG with alternating inertia.

the stabilizing effect of alternating inertia as the only stabilizing effect in the system, the damping factor D is set to zero. It should be mentioned that the system with fixed inertia and a zero damping factor is unable to recover from much milder faults (higher remained voltage and shorter duration). As it is observed in Fig. 8.28c, the oscillation is suppressed by the alternating inertia scheme, and the severe transient of DC-link voltage is also eliminated.

8.4.5 Experimental Results

The damping effect of alternating inertia is verified by applying it to a laboratory-scale test system. The overall system configuration is depicted in Fig. 7.24, and the main parameters of the system are presented in Table 8.8. Initially, the VSG with the constant moment of inertia $J = 0.563$ kg m^2 is subjected to a step change of 3 kW in the power reference. This value of J is calculated by assuming the inertia constant $H = 8$ s at rated power and frequency. It is shown that for this test scenario, the VSG with fixed moment of inertia is not stable [26].

Then, the VSG control is changed to the alternating inertia scheme with $J_{big} = 0.563$ kg m^2 and $J_{small} = 0.1$ kg m^2, and the step power change of 4.5 kW is applied. The result is shown in Fig. 8.29. It is observed that the VSG follows the

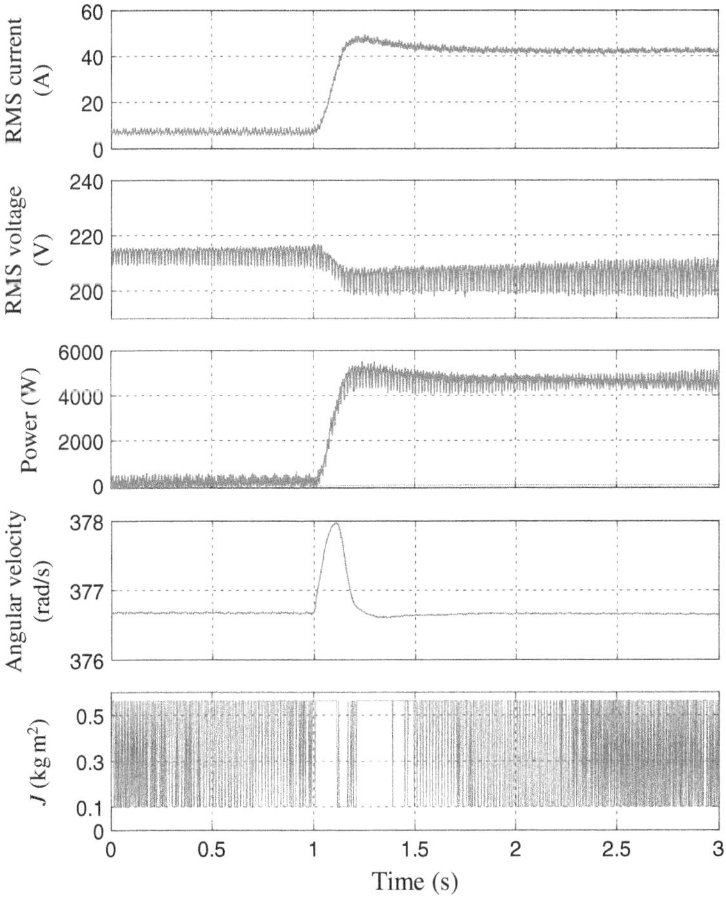

Figure 8.29 RMS current, RMS voltage, output power, virtual angular velocity, and virtual moment of inertia of VSG with alternating J and $D = 17$ pu after a power command of 4.5 kW.

power command without oscillations. It can be concluded that the VSG with alternating inertia can be loaded reliably at power levels close to its rating and can also ride-through severer disturbances compared to the disturbances that VSG with fixed moment of inertia can ride-through them. The effectiveness of the alternating inertia in the smooth transition of current level and reduction of the voltage ripples at the VSG terminal is obvious in this figure.

8.5 Voltage Sag Ride-through Enhancement Using Virtual Synchronous Generator

Faults on transmission lines cause voltage drops in several points in the power system, which affect the electrical equipment. Voltage sags (drops) are classified into several types based on the fault type. A three-phase fault on power line produces symmetrical voltage sag, while other fault types cause various unsymmetrical sags. Several works addressed the effect of inverter-based DG units on voltage sags in power system focusing on compensation effect of DGs [27, 28]. Reference [29] compared the performance of two control strategies in voltage sag ride-through improvement of a single-phase converter-connected DG.

Since the VSG is inherently a power-electronics-based unit [30], it is extremely sensitive to the grid side faults and disturbances. Consequently, before practical usage, their operation must be evaluated under disturbances conditions. In this section, the VSG is tested under symmetrical and unsymmetrical voltage sags, and the influence of the characteristics of voltage sags on the VSG transient current is investigated. To verify the results by equations, current trajectory analysis in phase plane is introduced. The trajectory of system state variable (VSG current) is monitored in phase plane, and its equations are extracted during and after the sag. The effects of the characteristic of voltage sags can be observed clearly by this method. High-magnitude transient current is the major hazardous consequence of voltage sags on inverter-based DGs. To limit the overcurrent, three additional controllers of voltage amplitude control, output power control, and alternating inertia control are embedded in the model and tested. Experiments are performed on a 10 kVA VSG-controlled inverter, and the results show that the additional controllers enhance the voltage sag ride-though capability of the VSG system.

8.5.1 Virtual Synchronous Generator Subjected to Voltage Sags

The basic VSG model (Fig. 7.2) is simulated by PSCAD/EMTDC software. The parameters of the system are as follows: $S_{base} = 35$ kVA, $f_{base} = 60$ Hz, $X_F = 42.4\%$, $J = 5.63$ kg m^2, and $D = 17$ pu. Voltage sags are applied to the VSG, and its transient currents are investigated. Figure 8.30a shows the current waveforms of VSG subjected to voltage sag for three-phase short

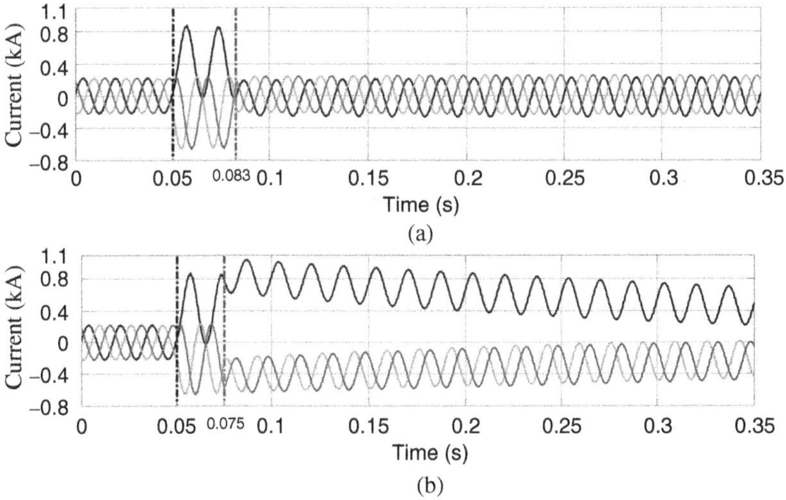

Figure 8.30 Currents of the VSG subjected to the applied symmetrical voltage sag with (a) the duration of 2 cycles and $h = 0.1$ and (b) the duration of 1.5 cycles and $h = 0.1$.

circuit (symmetrical sag) with the duration of two cycles. Since this is the lightest characteristic of voltage sag, overcurrent occurs only during voltage sag, whereas, as shown in Fig. 8.30b, for the mentioned voltage sag with the duration of 1.5 cycles, current oscillations appeared after voltage recovery.

The point achieved by simulations is that there are two sorts of transients that should be considered: one during the voltage sag and the other after voltage recovery. In the next part, the effect of the voltage sag characteristics will be clarified analytically.

8.5.2 State Variable Analysis in Phase Plane

The disparity of the VSG responses to various sags and their characteristics can be illustrated by VSG state variable (current) analysis in the phase plane. In addition, the origin of the transients after voltage recovery can be explained by this analysis. Inverter output current during the applied symmetrical voltage sag is calculated by integrating the voltage in polar coordinates in stationary frame (the inverter current before voltage sag is not considered). To obtain the voltage equation in polar form, three-phase gird voltage is transferred into the stationary dq frame and expressed in the form of

$$v(t) = v_d(t) + jv_q(t) = V_m e^{j(\omega_{grid}t+\theta_0-\pi/2)} \tag{8.25}$$

where V_m is the voltage vector magnitude in polar plane. This voltage vector rotates with the synchronous frequency on the stationary phase plane with the initial angle of $\theta_0 - \pi/2$. Using this transformation, the voltage equation of sag type A in the stationary polar coordinates is expressed as

$$v_{sag}(t) = hV_m e^{j(\theta_0 - \pi/2)}\ e^{j\omega_{grid} t} \tag{8.26}$$

where hV_m is the remaining voltage amplitude and θ_0 is the initial point-on-wave. As mentioned before, h is the voltage sag intensity parameter varying from 0 (remaining voltage magnitude) to 1 (normal voltage). The fault current is calculated by integrating the voltage difference as

$$i_{sag}(t) = \frac{1}{L}\int_0^{t_{sag}} (1-h)V_m e^{j(\omega_{grid} t + \theta_0 - \pi/2)}\ dt \tag{8.27}$$

where i_{sag}, L, and t_{sag} are the fault current in the stationary reference frame, the interconnecting inductance, and the sag ending time, respectively. Solving (8.27) and transferring to the synchronous reference frame yields

$$i_{sag}^{synch}(t) = (1-h)\frac{V_m}{\omega_{grid} L} e^{j(\theta_0 - \pi)}(1 - e^{-j\omega_{grid} t}) \tag{8.28}$$

Based on (8.28), during voltage sag, current vector is circulating in the synchronous dq frame phase plane with fixed radius as shown in Fig. 8.31. The damping terms are neglected in this figure. The radius of the circle, $(1-h)V_m/L\omega_{grid}$, determines the oscillation magnitude during voltage sag, and it is independent of the θ_0 obviously.

If the duration of voltage sag is a multiple of full cycles, the state variable position at the sag ending moment will reach to its normal point (presag point) after passing through circles in phase plane. When the voltage recovered

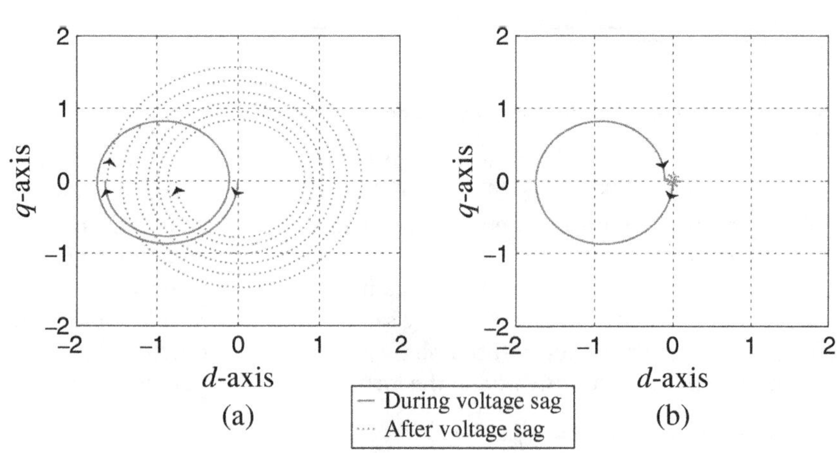

Figure 8.31 VSG current trajectory in phase plane during (solid line) and after (dotted line) voltage sag with $h = 0.1$: (a) voltage sag with the duration of 1.5 cycles and (b) voltage sag with the duration of 1 cycle.

from the sag, the state variable has to move to its normal point, and since it is extremely close to it, minimum oscillations occur after voltage sag. Whereas if the duration is half a cycle more than any number of full cycles, the state variable position has maximum distance to its normal point at the sag ending moment based on (8.28). It means that transient current has its maximum magnitude. The transient current must settle down on the origin of coordinate (current before fault is neglected) after voltage recovery. The second transient can be expressed as

$$i_{after\ sag}^{synch}(t) = 2(1-h)\frac{V_m}{\omega_{grid}L}e^{j(\theta_0-\pi)}e^{-j\omega t} \tag{8.29}$$

Equation (8.29) is a circular trajectory with the center coinciding with the origin. The radius of trajectory after voltage sag has its maximum possible value proving that the assumed duration is the severest one. Figure 8.31 includes the current paths in phase plane. In this Section, d axis is taken as the reference axis, and phase "a" and q axis lead d axis by θ_0 and $\pi/2$, respectively. Figure 8.31a is related to the sag with the duration of 1.5 cycles and $h=0.1$. At the sag ending point, the state variable goes through circles with a large radius (dotted line) that causes severe current transient. As shown in Fig. 8.31b, when the duration is one full cycle, there is a small distance between the position of the variable after and before the sag, and slight transient occurs.

It is noteworthy that the transient during voltage sag is not affected by the sag duration. However, the sag magnitude, parameter h, determines the transient severity during voltage sag. If the voltage sag is shallow (h close to unity) and/or if the sag lasts for a relatively long time, the state variable transient during voltage sag will settle down at the center of the solid-line circles before voltage sag ends. Afterward, when voltage amplitude is recovered, the state variable has to move to its normal point from the center of the first transient circles. It means that the oscillations will have identical amplitude regardless of the duration of the voltage sag.

8.5.3 Voltage Sag Ride-through Enhancement

Three approaches are added to the VSG system to limit the overcurrent during and after voltage sag. In comparison with Figure 7.2, Figure 8.32 shows an updated VSG control scheme. Subsystem A calculates the RMS value of grid voltage and uses it as the reference voltage for inverter output. By this control, when grid voltage drops during voltage sag, inverter output voltage will also be reduced to prevent overcurrent. The RMS voltage is calculated as follows:

$$V_{RMS} = \sqrt{V_{grid\ d}^2 + V_{grid\ q}^2} \tag{8.30}$$

During unsymmetrical voltage sags, the RMS voltage oscillates. A first-order low-pass filter (LPF) is used to remove the oscillation.

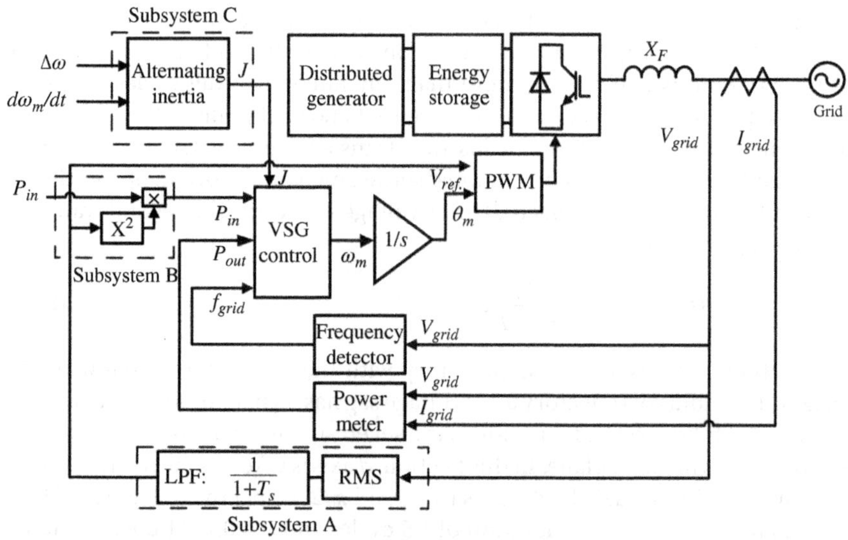

Figure 8.32 An updated VSG control scheme for voltage sag ride-through enhancement.

Since inverter output power is proportional to the product of the inverter and grid voltages, once the output voltage follows the grid voltage, inverter output power must be limited proportional to the square of the grid voltage based on the power transfer equation, that is,

$$\begin{aligned} P &= P_{max} \sin\delta \\ &= \frac{V_{vsg} V_{grid}}{X} \sin\delta \end{aligned} \tag{8.31}$$

where δ is the voltage angle difference between the VSG and the grid. Subsystem B calculates the output power reference of VSG proportional to the square of the grid voltage.

The alternating inertia control (Subsystem C) is used to quickly suppress the oscillation after voltage recovery. The disturbance in this case is different from the power reference change shown in Fig. 8.21. In the case of voltage sag, the power angle curve follows a new curve in which the maximum transferable power decreases due to voltage drop based on (8.31). In this condition, the power angle relation follows the dotted line curve shown in Fig. 8.33, and load angle moves to the point δ_0. After the fault clearance, operating point moves along the original power angle curve, and load angle oscillates around the equilibrium point δ_1. The machine characteristics during each phase of an oscillation cycle are summarized in Table 8.8. Here, to suppress the oscillation, the alternating inertia control selects a large value of J during acceleration phases ("a" to "b" and "c" to "b") to reduce the acceleration, and a small value of J during

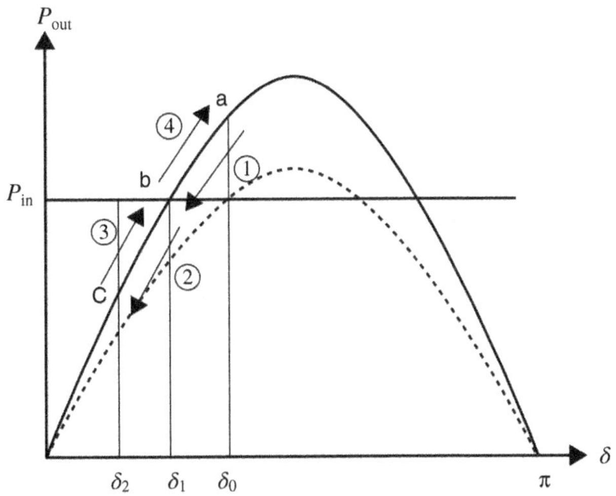

Figure 8.33 Power angle curve of an SG subjected to a fault. When a fault occurs, operating point moves on the dotted line and reaches to the point δ_0. After fault clearance, it returns to the original curve and oscillates around the equilibrium point δ_1.

deceleration phases ("b" to "c" and "b" to "a") to boost the deceleration. Using the transient energy function analysis, it is proved that the proposed algorithm brings in a damping effect that suppresses the transient quickly [31]. Subsystem C adopts the value of the virtual inertia based on the stated algorithm summarized in Table 8.7.

The simulations discussed in Section 5.1 are repeated for the VSG with the voltage amplitude, output power, and alternating inertia controls under the severest characteristics condition of voltage sags. To let the voltage amplitude control be quick enough, the LPF of the voltage reference loop of Fig. 8.32 is not included during the simulations. Figure 8.34 shows the current of the VSG subjected to a critical symmetrical sag. The overcurrent that appeared during and after voltage sag, shown in Fig. 8.30b, is eliminated effectively by the additional controllers. The current position in phase plane is moved from its normal position toward the origin of the phase plane. This small distance from the normal position caused a slight transient after voltage recovery that is damped slowly.

8.5.4 Simulation Results

A simulation model similar to the experiment system discussed in the previous section is built in PSCAD/EMTDC. The system configuration is shown in Fig. 8.35, and the system parameters are given in Table 8.9. In the simulation, the transformer is replaced by its equivalent reactance, X_T, and all system operated

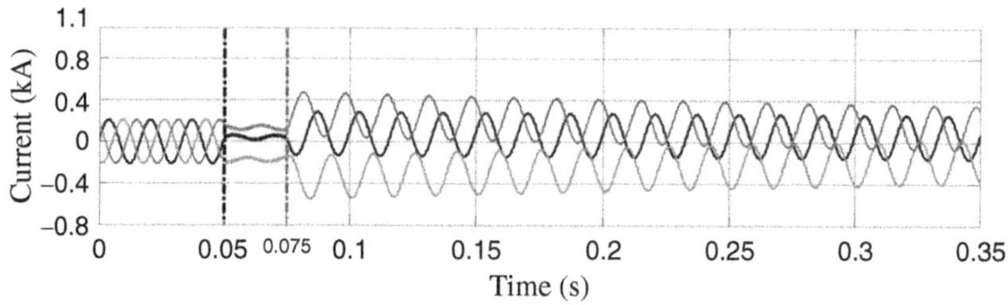

Figure 8.34 Currents of the VSG with the voltage amplitude, output power, and alternating inertia controls, subjected to a symmetrical voltage sag with the duration of 1.5 cycles and $h = 0.1$ (the severest case).

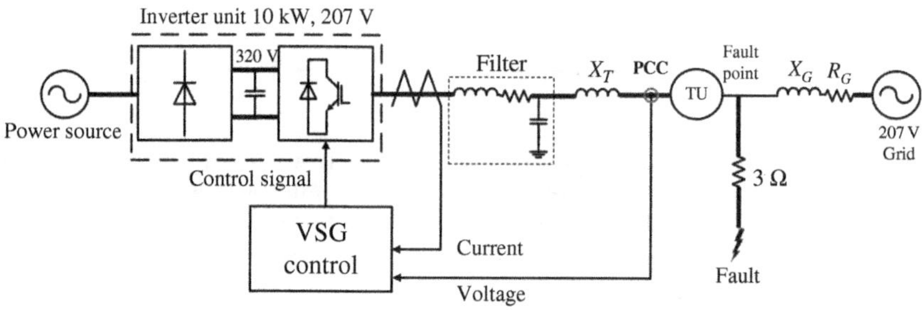

Figure 8.35 Simulation system.

Table 8.9 The specifications of the simulation system.

Base power (kVA)	10
Base frequency (Hz)	60
Base voltage (V)	207
Switching frequency (kHz)	14
DC-link voltage (V)	320
DC-link capacitor (mF)	4.7
Filter stray resistance (%)	0.23
Filter inductive reactance (%)	8.8
Filter capacitor VAR (%)	1.62
Resonance frequency of LC filter (kHz)	1.59
X_T' (%)	9.68
Grid inductive reactance, X_G (%)	44
Grid resistance, R_G (%)	46
Damping factor (pu)	17
Moment of inertia J (kg m^2)	0.563
J_{big} (kg m^2)	0.563
J_{small} (kg m^2)	0.1

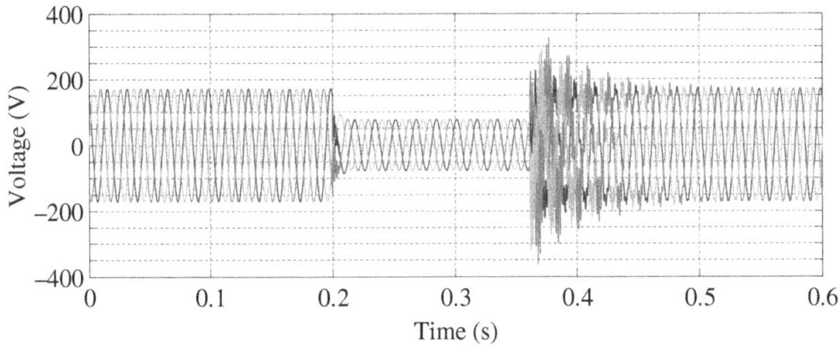

Figure 8.36 Symmetrical voltage sag at PCC of simulated system due to the three-phase fault.

at 207 V nominal voltage. Furthermore, the fault path resistance is chosen to be 3 Ω to have deep voltage sags. To see the voltage profile during fault, the VSG was disconnected from the point of common coupling (PCC), and three-phase fault established for 10 cycles of system frequency. The phase-to-neutral volt ages of PCC affected by the fault are shown in Fig. 8.36. The remaining voltage in this case is 45% of the rated voltage ($h = 0.45$). However, the voltage drop in the presence of VSG is less due to the compensation by VSG.

In the next step, the VSG system without the additional controller is connected, and a fault occurs while the VSG injects 2.6 kW power to the grid. The RMS voltage at PCC, VSG current, and the DC-link voltage waveforms are presented in Fig. 8.37. In this simulation, the RMS voltage is calculated by (8.30) with a smoothing time constant of 3 ms. As it is observed, the RMS current increased immediately up to 70 A when voltage drop at PCC occurred. Moreover, when the voltage magnitude is recovered, the current transient moved to a negative level (reverse current) and increased the DC-link voltage. The transient during voltage sag may result in the overcurrent failure, and the transient after voltage recovery may result in the DC-link overvoltage failure.

The waveforms of the VSG angular velocity, power reference calculated by the governor, VSG output active power, and VSG output reactive power are shown in Fig. 8.38. During transients, the governor calculates the power reference opposite to the frequency oscillation to fix the VSG frequency at system nominal frequency. The active power of VSG has a waveform similar to the I_d shown in Fig. 8.37. The reverse power that increases the DC-link voltage is observed more clearly in the power waveform of Fig. 8.38. The reactive power transient is similar to the transient of I_q shown in Fig. 8.37.

To mitigate the possible failures, the proposed voltage amplitude control and output power control are added to limit the overcurrent current during voltage sag, and the alternating inertia control is added to eliminate the after-sag transients and prevent DC-link overvoltage. The RMS voltage at PCC, VSG current,

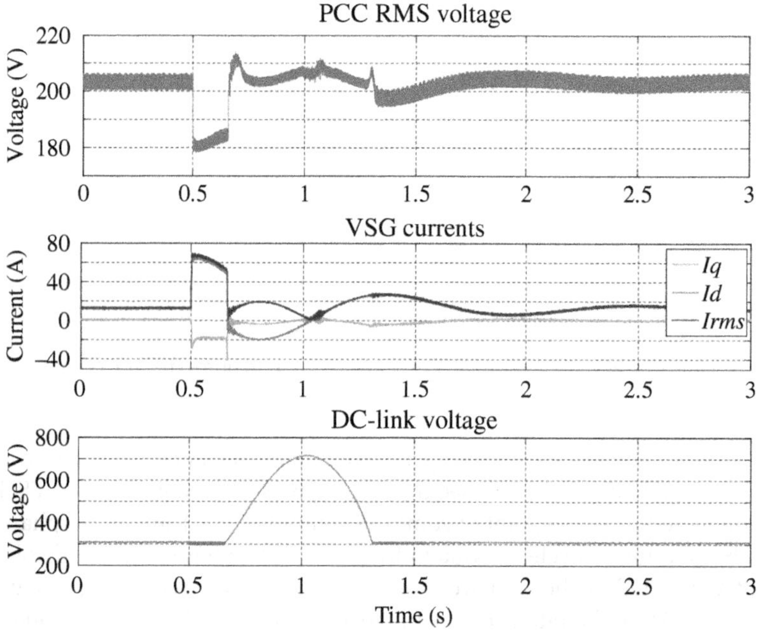

Figure 8.37 PCC RMS voltage, VSG currents, and DC-link voltage of the system with VSG without the additional controllers, affected by voltage sag.

and the DC-link voltage waveforms are shown in Fig. 8.39. The PCC voltage sag is deeper because the current injected by VSG is reduced by additional controllers. With the voltage amplitude and output power controls, although the transient current peak at the sag starting moment is reduced to some extent, however, because of the delay in RMS voltage calculation, the voltage amplitude control is not able to apply the voltage reference equal to the PCC voltage promptly, and the overcurrent prevention performance is limited by this delay. During the voltage sag, since the voltage reference followed the grid voltage, the overcurrent reduced with a steep ramp. The DC-link voltage rise is reduced by around 190 V via the alternating inertia control. However, by setting a smaller value for J_{small}, this controller is more effective.

The VSG angular velocity, power reference calculated by the governor, VSG output active power, and VSG output reactive power are shown in Fig. 8.40. The alternating inertia control suppressed the VSG angular velocity quickly. By the output power control, the power reference calculated by governor is reduced at the sag starting moment proportional to the square of the PCC voltage. However, the power reference is increased after a short time to regulate the frequency. Similarly to I_d, the active power transient during voltage sag is reduced by the additional controllers. After voltage recovery, the transient of output

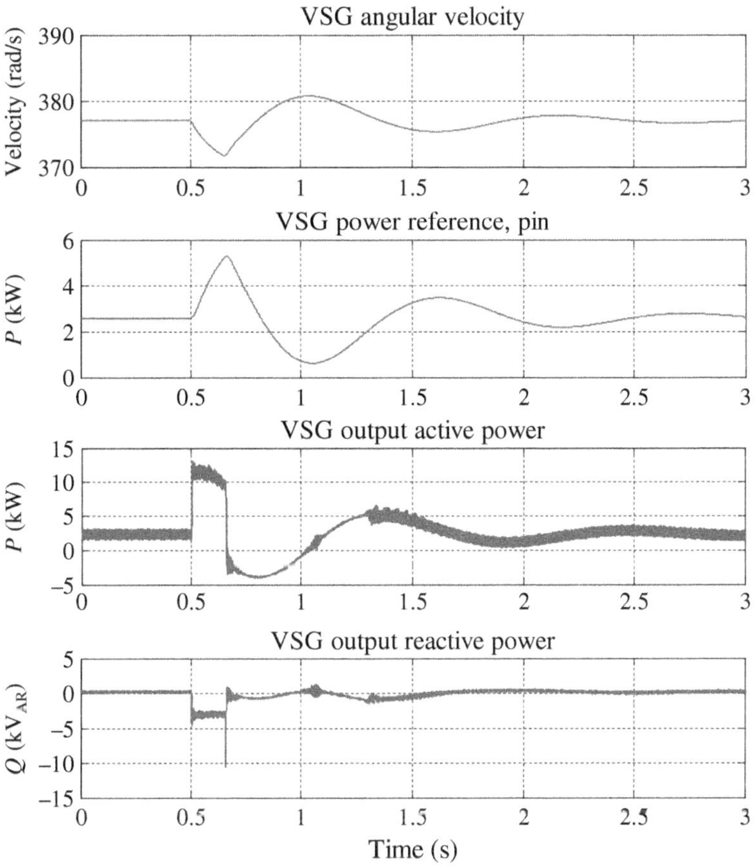

Figure 8.38 VSG angular velocity, power reference calculated by the governor, VSG output active power, and VSG output reactive power of the system with VSG without the additional controllers, affected by voltage sag.

active power had the negative amplitude almost equal to the curve shown in Fig. 8.38. However, this transient is suppressed quickly by the alternating inertia control as shown in Fig. 8.39. Therefore, the reverse energy and, consequently, the DC-link voltage rise are reduced. As mentioned before, the delay in sensing, calculating, and filtering the PCC voltage to calculate the VSG voltage reference is an obstacle to the effectiveness of the voltage amplitude control. To evaluate the effect of this delay, some simulations are performed in [26].

8.5.5 Experimental Results

A laboratory-scale system is used to investigate the voltage sag ride-through performance of the VSG. The overall system configuration is depicted in

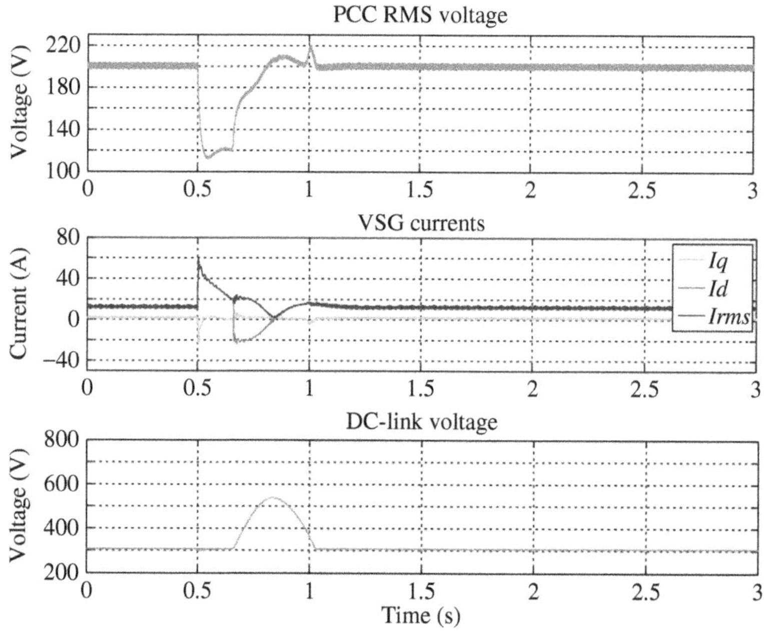

Figure 8.39 PCC RMS voltage, VSG currents, and DC-link voltage of the system with VSG with the additional controllers, affected by voltage sag.

Fig. 8.41, and the main parameters of the system are presented in Table 8.10. Due to the strict overcurrent limitation of the inverter unit, when a deep voltage sag occurs, the VSG unit is stopped by its protection system. Therefore, mild voltage sags with magnitudes larger than 90% with the duration of 10 cycles are tested on this system, and voltage sag ride-through performance of the VSG unit is evaluated. For light voltage sags, since the state variable position in phase plane at the sag ending moment does not change considerably by the characteristics, the characteristics do not affect the severity of transient oscillation. The voltage amplitude control in experiments has an LPF with 10 Hz cut-off frequency ($T = 0.0159$ s in Fig. 8.32).

The voltage sag shown in Fig. 8.42 appeared due to three-phase short circuit at the fault point and measured at the PCC indicated in Fig. 8.41. The voltage magnitude during voltage sag is 92% of the normal value. The fault is controlled by switching thyristors in the fault path. The initial point-on-wave can be adjusted by a phase detection algorithm in the DSP unit. Since the exact turn-off point of the thyristors cannot be controlled by the gate signal, the duration of the voltage sag will have utmost a half-cycle error. However, as mentioned before, the duration and initial point-on-wave are not effective for

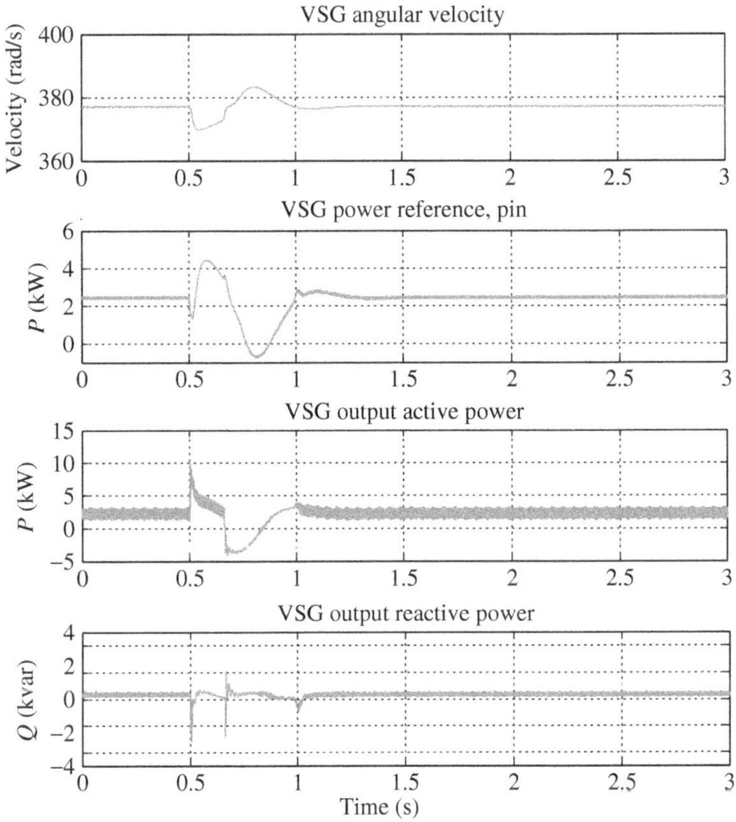

Figure 8.40 VSG angular velocity, power reference calculated by the governor, VSG output active power, and VSG output reactive power of the system with VSG with the additional controllers, affected by voltage sag.

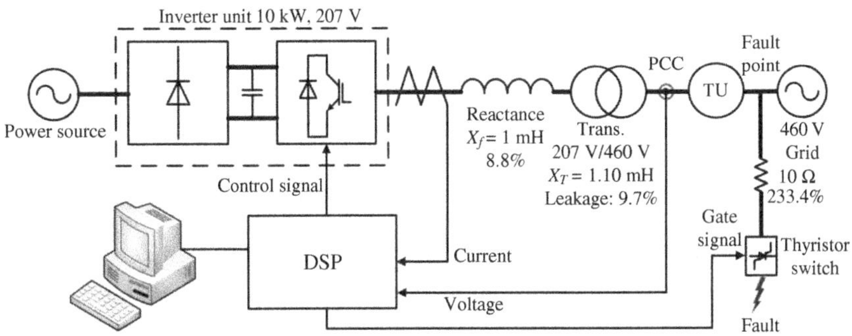

Figure 8.41 Experimental system.

Table 8.10 The specifications of the simulation system.

Base power (kVA)	10
Base voltage (V)	207
Base frequency (Hz)	60
Switching frequency (kHz)	14
DC-link voltage (V)	320
DC-link capacitor (mF)	4.7
Filter stray resistance (%)	0.23
Filter inductive reactance (%)	8.8
Filter capacitor VAR (%)	1.62
Resonance frequency of LC filter (kHz)	1.59
Transformer reactance (%)	9.68
Damping factor (pu)	17
Moment of inertia J (kg m^2)	0.563
J_{big} (kg m^2)	0.563
J_{small} (kg m^2)	0.1

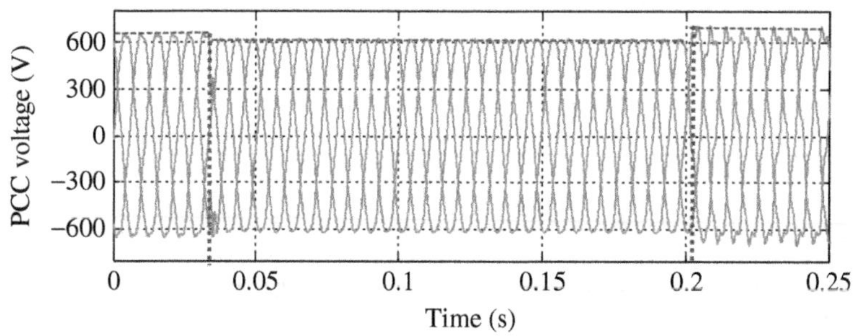

Figure 8.42 Symmetrical voltage sag at PCC due to symmetrical three-phase fault.

shallow voltage sags. Therefore, the effects of the characteristics of voltage sags are not investigated in the experiments.

To assess the performance of the additional controls, first, the VSG unit without the controllers is subjected to the voltage sag presented in Fig. 8.42. As stated before, two transient states occur: the transient during voltage sag and the one after voltage recovery. To see the importance of the transient after voltage recovery, the VSG is subjected to the voltage sag while injecting 1 kW power to the grid, and its current in synchronous dq coordinates is monitored. Figure 8.43a shows the currents and DC-link voltage in this condition. The oscillations after voltage recovery caused power and current oscillation, which reached a negative value. This reverse current increased the DC-link voltage

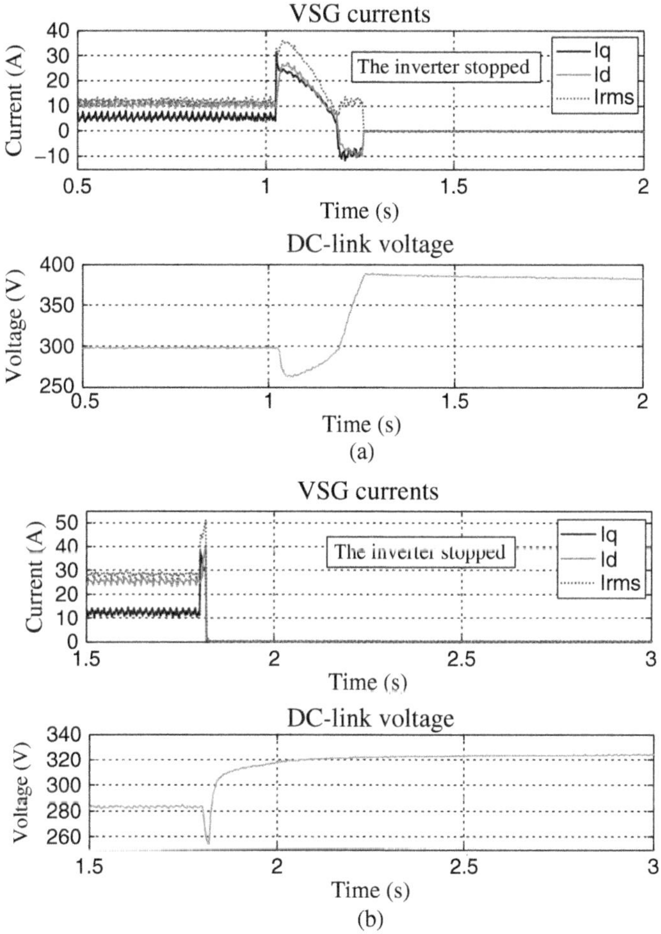

Figure 8.43 Currents and DC-link voltage of the VSG without additional controller subjected to the applied voltage sag, with (a) 1 kW output power, and (b) 2.6 kW output power.

and resulted in failure. Using an energy storage unit with high rate of charging or a larger DC-link capacitor can prevent the DC-link overvoltage failure. However, the objective is to prevent the failure of the existing system by a proper control scheme. It is observed later that alternating inertia control will prevent this kind of failure.

During the voltage sag, the VSG current is increased due to the voltage difference between the VSG and the grid. Since the voltage sag amplitude (the remained voltage) is high in this research (which means a mild voltage sag), the initial current of the VSG affects the fault ride-through performance. When the VSG unit is loaded at 2.6 kW, the voltage sag type A is applied.

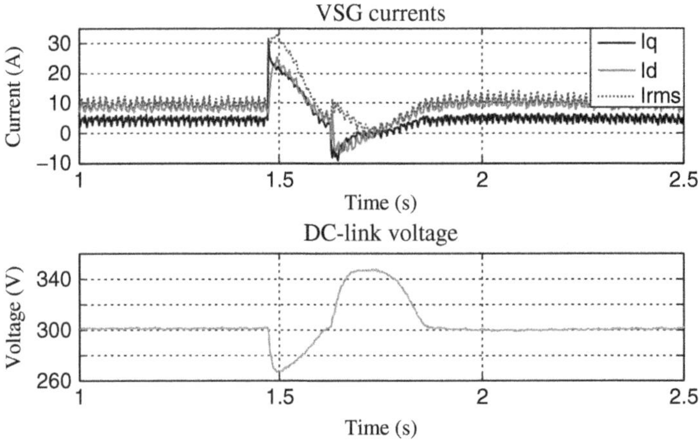

Figure 8.44 Currents and DC-link voltage of the VSG with 1 kW output power and with voltage amplitude, output power, and alternating inertia controller subjected to the applied voltage sag.

Fig. 8.43b shows that the VSG current increased sharply, and VSG is stopped. It can be concluded that for lower loading levels, the oscillations after fault recovery cause the failure, and for higher loading levels, the overcurrent during voltage sag results in failure. In the next experiment, all of the additional controllers are activated. First, the VSG with the low output power of 1 kW is subjected to the applied voltage sag. Figure 8.44 shows that the alternating inertia control suppresses the after-sag oscillations by imposing a damping effect. By quick suppression of the after-sag transient, the negative current (and the reverse power) is limited, and as a result, the increase in the DC-link voltage is limited effectively. The PCC RMS voltage, power reference calculated by governor, alternating inertia, angular velocity, VSG output active power, and VSG output reactive power for the experiment case with 1 kW output power reference and with voltage amplitude, output power, and alternating inertia controllers subjected to voltage sag type A are depicted in Fig. 8.45.

It is observed that when the voltage dropped, the power reference calculated by governor has been slightly reduced at first, because it is proportional to the square of RMS voltage by the output power control. However, it has been increased by the governor afterward to compensate for the frequency drop. It should be noted that the overcurrent at the sag starting moment has maximum value and should be limited. Accordingly, the reduction in the power reference at the beginning of voltage sag is effective enough to prevent the overcurrent failure in this case. Although the oscillation after voltage recovery is suppressed by alternating inertia, transient excursions are still observed in the active and reactive power waveforms. However, the magnitude of the transients is reduced, and the inverter is able to ride-through them.

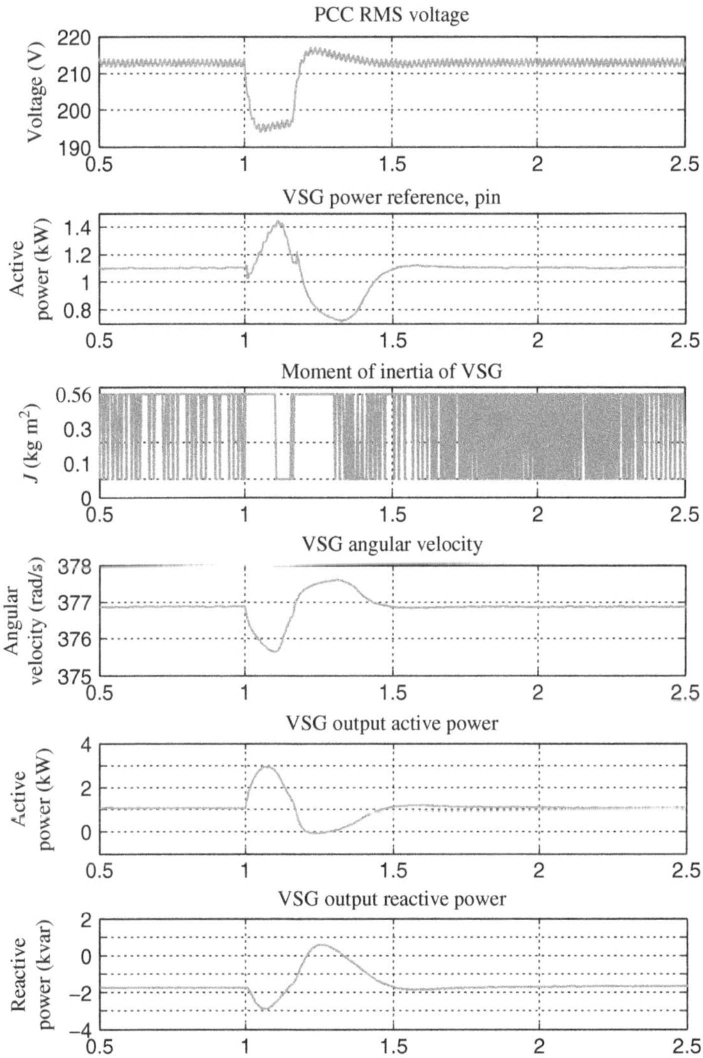

Figure 8.45 System response in the presence of a voltage sag.

8.6 Performance Evaluation of the Virtual Synchronous Generator with More Synchronous Generator Characteristics

In Section 7.4, a VSG scheme with emulating more SG characteristics is introduced (Fig. 7.26). The mentioned VSG does not use PLL and operates as a current control inverter. The proposed control scheme emulates impedance model, governor, AVR, damper model, and rotor model. By using synchronizing power, it can be synchronized with the grid without PLL. In this section, the control performance of the mentioned VSG scheme is evaluated via simulation and experimental tests.

8.6.1 System Configuration and Parameters

Here, in the impedance model of the VSG (Fig. 7.27a), the $r + jx$ is set to $0.2 + j0.4$ (pu). Time constant T_{tur} in the turbine model is set to 120 ms. The K_{gd} (Fig. 7.30a) should be set to the same level with the droop gain of other generator's governor. The K_{gd} set to 5%. P^* is controlled by host controller, considering the generator characteristics and states. The K_{ad} and K_{GI} are set to 5% and 100%, respectively. Here, J is set to 4 s since it is typically 2–10 s in actual SGs [32].

The proportional gain of the governor K_{GP} is designed about a reciprocal of droop gain K_{gd}, preferably. Therefore, K_{GP} is set to 20(=1/0.05). From (7.77), the integral gain K_{GI} must satisfy the following equation:

$$\frac{(1.0 \times 0.12 + 4.0)(1.0 + 20)}{4.0 \times 0.12} = 180.25 > K_{GI} \tag{8.32}$$

In this work, τ_d is set to 0.01, and J is set to 2, 4, and 6 s in the simulation and experimental system. From (8.32), the damping factor K must satisfy the following equations to operate stably:

$$\begin{cases} K > 3.09 & (J = 2) \\ K > 5.39 & (J = 4) \\ K > 7.27 & (J = 6) \end{cases} \tag{8.33}$$

In the simulation, the VSG is connected to the grid as shown in Fig. 8.46. The filter consists of reactors, capacitors, and a transformer. The capacitors do not only work as a filter but make the current control stable in an isolated operation. The VSG control parameters and other circuit parameters are listed in Tables 8.11 and 8.12, respectively. Load model is constant impedance model in the performed simulations.

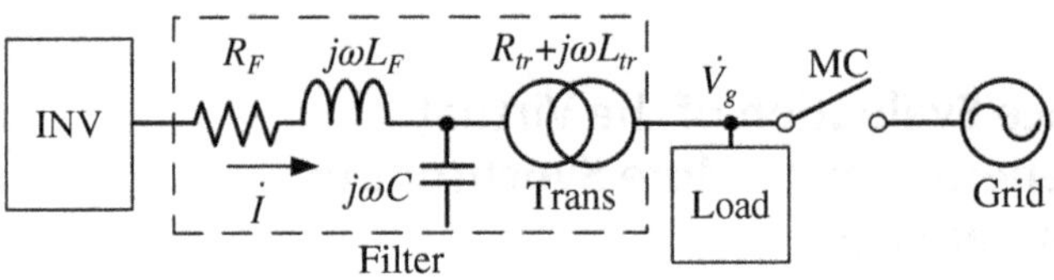

Figure 8.46 The circuit of the VSG connected to a grid.

Table 8.11 Control parameters.

Model	Parameter	Symbol	Values
AVR	Droop gain (pu)	K_{ad}	0.05
	Regulator gain (pu)	K_{fI}	20
Governor	Droop gain (pu)	K_{gd}	0.05
	Regulator gain (pu)	$K_{GP}+K_{GI}/s$	$20+100/s$
Impedance model	Resistance (pu)	r	0.2
	Reactance (pu)	x	0.4
Current control	Proportional gain (pu)	K_{IP}	1.49
	Integral gain (pu)	K_{II}	71.95
Damper	Damping gain (pu)	K	10
	Time constant	τ_d	0.01
Rotor	Moment of inertia (s)	J_g	4.0
	Damping (pu)	D_g	1.0
Power command	Active power (pu)	P^*	0.0
	Reactive power (pu)	Q^*	0.0

Table 8.12 Circuit parameters.

R_F (%)	0.22	ωL_F (%)	8.8
R_{tr} (%)	2.1	ωL_{tr} (%)	9.1%
C (μF)	10	Grid voltage (V)	460
Rated power (kW)	10	Load (kW)	5

8.6.2 Simulation Results

8.6.2.1 Grid-connected Operation

In the simulation, the VSG is connected to the grid, and an active power command P^* is changed from 0.0 to 0.5 pu in steps. Table 8.13 lists the parameters of J and K in each scenario. The parameters of scenarios 1-2 and 2-2 are equal. Figure 8.47 shows the simulation results in the grid-connected operation mode.

Resistance r in impedance model, which is neglected in stability analysis, contributes to damping effect. Therefore, in scenario 1-1 (Fig. 8.47a), which is an unstable condition in stability analysis, the VSG operates stably and near limit of stable. Figure 8.47b shows that the output power P is equal to the power reference P^* in steady state. VSG output power followed the power command P^* quickly, and rotor speed ω_R is converged after 0.5 s.

Table 8.13 Test scenarios and parameter in grid connected operation.

Scenarios	Moment of inertia J	Damping factor K	Stability condition by analysis
1-1		5	Unstable
1-2	4 s	10	Stable
1-3		15	Stable
2-1	2 s		Stable
2-2	4 s	10	Stable
2-3	6 s		Stable

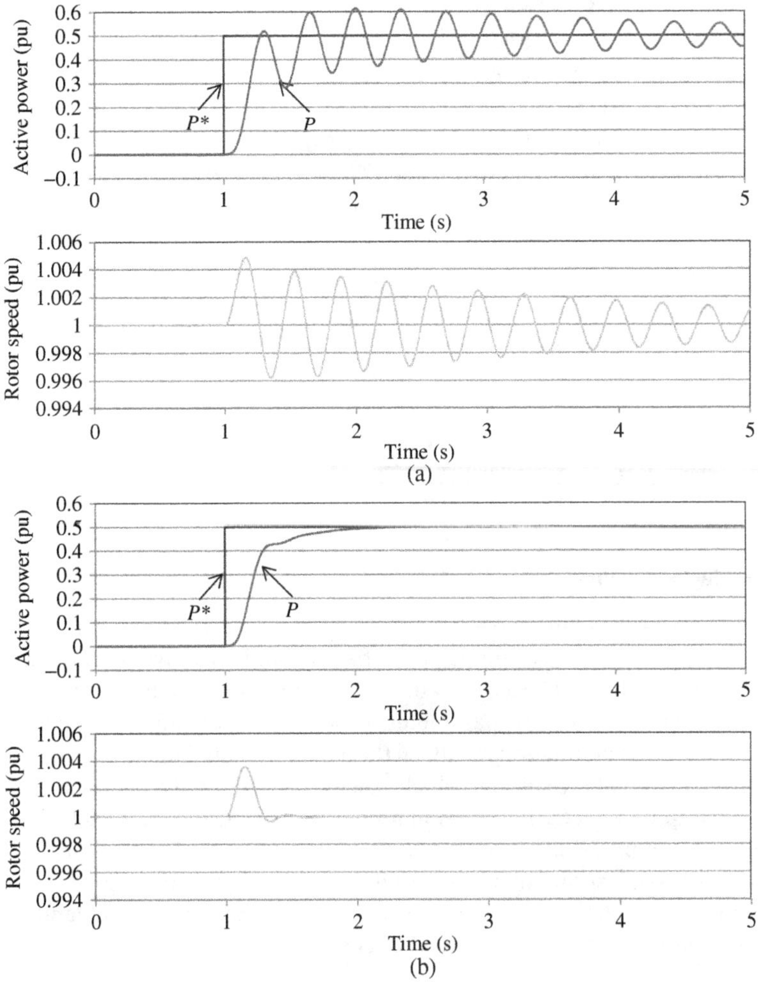

Figure 8.47 System response in grid-connected operation: (a) scenario 1-1 ($J = 4$ s, $K = 5$) and (b) scenarios 1-2 and 2-2 ($J = 4$ s, $K = 10$).

Table 8.14 Test cases and parameter in islanded operation.

Case	Power command P^* (pu)	Load	
		Active power (kW)	Power factor
3-1	0.0	0.0	1.0
3-2	0.0	5.0	1.0
3-3	0.5	0.0	1.0
3-4	0.5	5.0	1.0

8.6.2.2 Islanded Operation

In the simulation, the VSG is disconnected from the grid by opening the MC at 1 s. The control method of the inverter is not changed. Table 8.14 summarizes the simulation cases. Figure 8.48 shows the simulation results of gird disconnection. In all cases, the VSG can continue to operate from grid-connected to islanded operation mode smoothly. In the grid-connected operation mode, the output power of the VSG equal to the power command P^* and changed to load power after grid disconnection. In scenario 3-1 shown in Fig. 8.48a and in scenario 3-4, the rotor speed is not changed after grid disconnection since the power command P^* is consistent with the power of the load. In scenario 3-2 shown in Fig. 8.48b and in scenario 3-3, the rotor speed ω_R in steady state after the grid disconnection is not equal to 1 pu. It is caused by the droop characteristics of the governor. Relationship between power and rotor speed in steady state is given by (8.34) because of the droop characteristics of the governor

$$\omega^* - \omega = -K_{gd}(P^* - P) \tag{8.34}$$

In scenario 3-2, the power command P^* is set to 0.0 pu and the load power P is 5.0 kW (=0.50 pu). The rotor speed $\omega_R = 0.975$ is obtained by (8.34). It is consistent with the simulation results shown in Fig. 8.48b.

8.6.3 Experimental System

In the experimental system, the VSG is connected to a commercial grid as shown in Fig. 8.46. The experimental system is same as the simulations. The load is resistive, and the configuration of inverter system is shown in Fig. 8.49. A DSP is used for controller of inverter. The controller time step is set to 125 μs. All calculations for the VSG control including calculations of active power P and reactive power Q are carried out by the DSP within this time step. The control parameters of the VSG and the circuit parameters are same as those used in the simulations.

First, the inverter is started using conventional current control with PLL. Current commands I_d^* and I_q^* are set to 0.0, and the phase of dq transform is

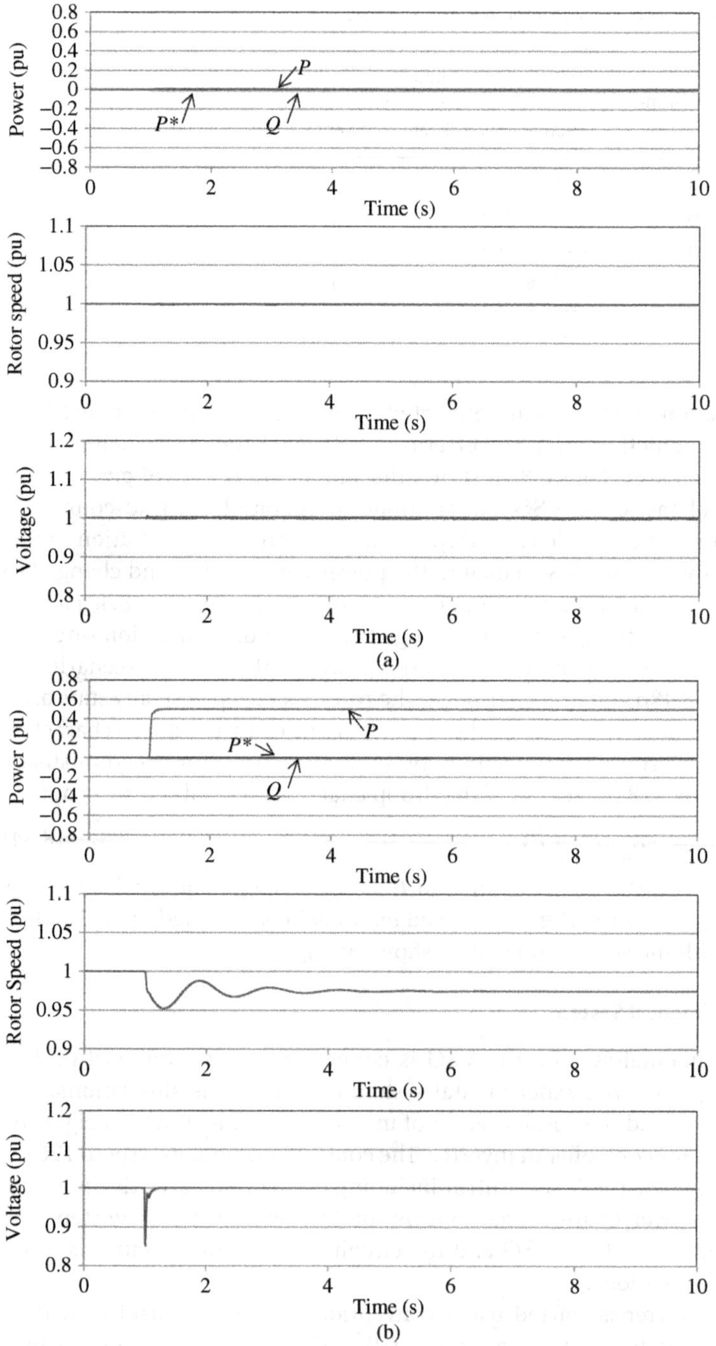

Figure 8.48 System response in islanded operation: (a) scenario 3-1 ($P^* = 0.0$ pu, load is 0.0 kW) and (b) scenario 3-2 ($P^* = 0.0$ pu, load is 5.0 kW).

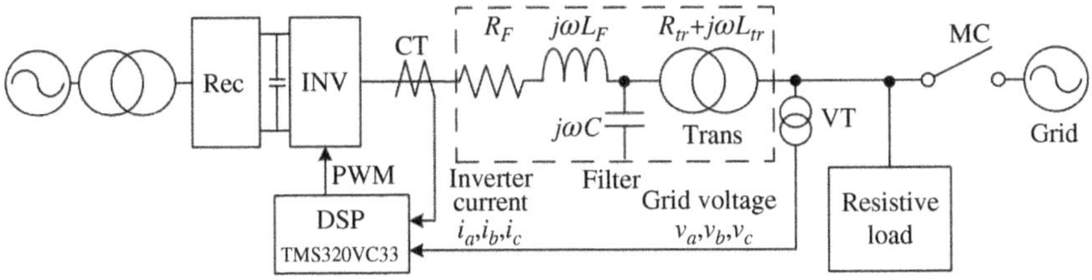

Figure 8.49 Experimental test system.

set to θ_g calculated by the PLL. Controller is switched from conventional current control to VSG control by switching I_d^* and I_q^* to the output of impedance model, and the phase of dq is transformed to the rotor phase θ_R. The initial parameter of θ_R is set to θ_g at the time of switching to the VSG control. After switching to VSG control, the controller does not use the PLL. The PLL is used for only synchronizing to the grid upon starting up the inverter.

8.6.3.1 Grid-connected Operation

In the experimental test, the VSG is connected to the grid and active power command P^* is changed from 0.0 to 0.5 pu in steps. J and K are the same as the simulation values listed in Table 8.13. The other parameters are listed in Table 8.12. Figure 8.50 shows the experimental results of grid-connected operation mode.

The experimental results of case 1-1 are shown in Fig. 8.50a. The VSG control is enabled at 1 s, and the VSG output power became 0 pu because the inverter is stopped by the inverter's protection against overcurrents at 1.05 s. Rotor speed fluctuates, since DSP continues to operate after the inverter stopped. The damping factor K in case 1-1 is smaller than 5.39, and it is an unstable condition. Therefore, the VSG control is unstable and cannot continue to operate. In the other cases, the VSG output power follows the power command P^* quickly, and the rotor speed ω_R is converged within 1 s. In case 2-1, the experimental result shown in Fig. 8.50b is consistent with the simulation result shown in Fig. 8.47b. In scenarios 1-2 and 1-3 shown in Fig. 8.50b and c, the difference is only in the damping factor K. These results show that the damper model provides the damping effect, and oscillations are decreased by increasing the damping factor K. In scenarios 2-1, 2-2, and 2-3 shown in Fig. 8.50d, b, and e, the difference is only in the inertia J. These experimental results show that the damping effect is decreased with increasing the inertia.

There is an error between the output power P and the power reference P^*. It is caused by the droop characteristics of the governor. In case 1-3, Fig. 8.50c shows that the rotor speed is 0.999 pu (=59.94 Hz) in steady state. Relationship between power and rotor speed in steady state is given by (8.34). The output

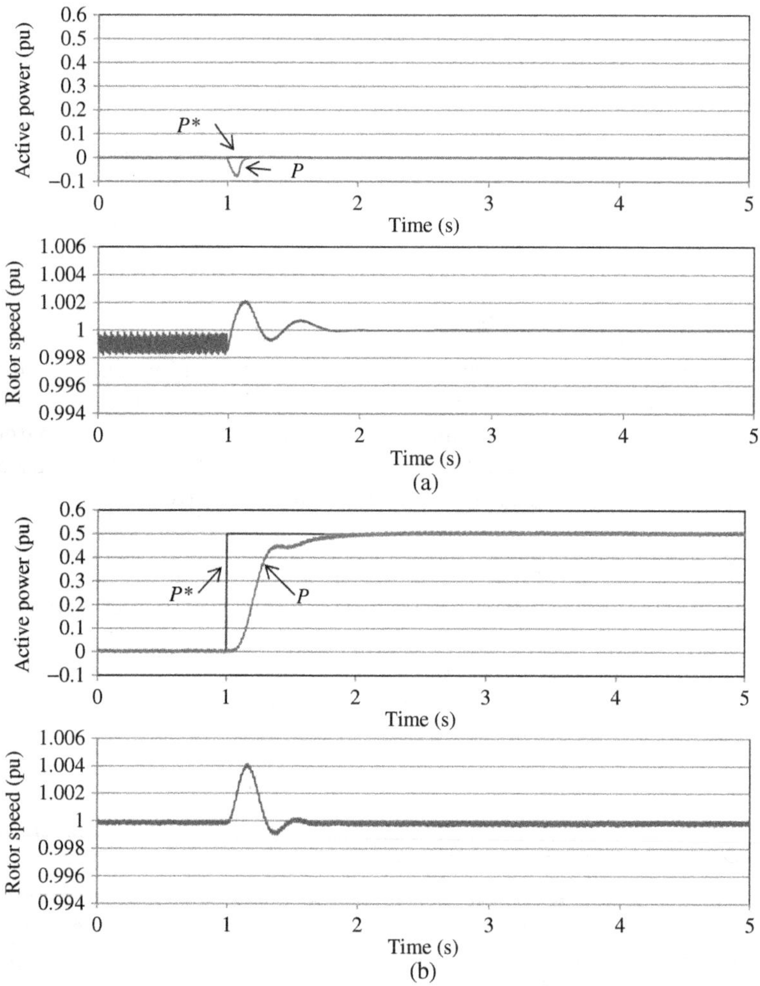

Figure 8.50 System response in grid-connected operation: (a) scenario 1-1 ($P^*=0.5$ pu, $J=4$ s, $K=5$), (b) scenarios 1-2 and 2-2 ($J=4$ s, $K=10$), (c) scenario 1-3 ($J=4$ s, $K=15$), (d) scenario 2-1 ($J=2$ s, $K=10$), and (e) scenario 2-3 ($J=6$ s, $K=10$).

power $P=0.02+P^*$ is obtained by (8.34), and it agrees with the experimental results shown in Fig. 8.50c.

8.6.3.2 Islanded Operation

Figure 8.51 shows the experimental results of grid disconnection corresponding to the scenarios shown in Table 8.14. In the experiments, the VSG is disconnected from the grid by opening the MC shown in Fig. 8.49 at 1 s. In all cases,

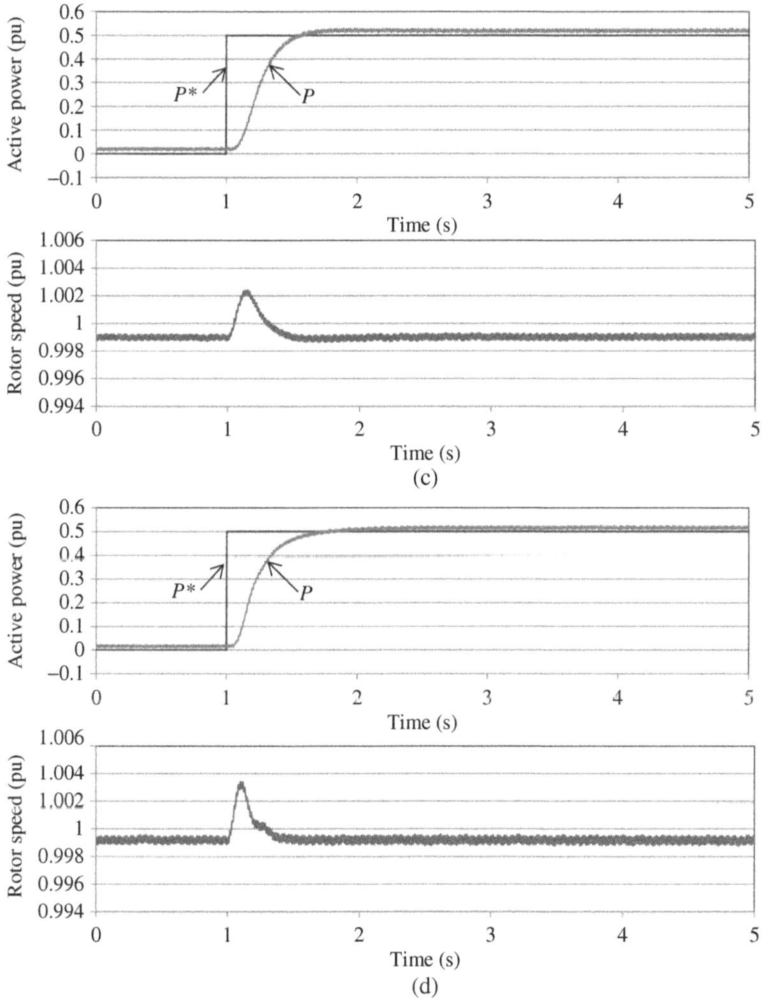

Figure 8.50 *(Continued)*

the VSG passes the transient state from the grid-connected to islanded operation mode, smoothly. The control logic of the inverter is not changed between two operation modes. After grid disconnection, output power of VSG (P) is larger than the load power. The difference power between VSG and load is due to losses of the transformer and lines. These losses are about 0.4 kW (=0.04 pu). In scenario 3-2, the power command P^{*} is set to 0.0 pu and the load power P is 5.4 kW (=0.54 pu) including transformer losses. In scenario 3-2, the rotor speed $\omega_R = 0.973$ pu is obtained by (8.34). It is consistent with the experimental

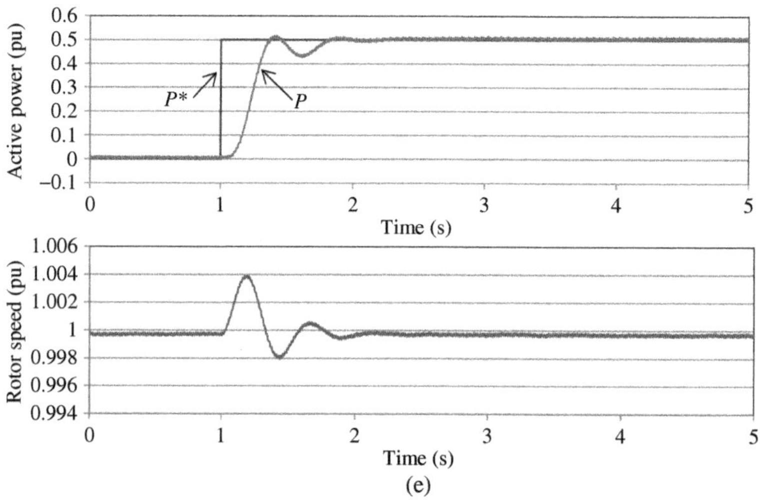

Figure 8.50 *(Continued)*

results shown in Fig. 8.51b. In scenarios 3-1 and 3-3, filtering effects by filter circuit are weakened because the load is zero after the grid disconnection.

8.7 Summary

In this chapter, several VSG control schemes for the sake of MG stabilizing and frequency/voltage regulation support are introduced. First, an enhanced VSG control is proposed to achieve oscillation damping and proper transient active power sharing by the virtual stator reactance adjustment and common AC bus voltage estimation. Then, for a parallel operation of VSG and SG, an effective control mechanism to prevent unbalanced SG current and inverter overcurrent is developed. The alternating inertia concept is used to enhance the VSG control performance. This scheme adopts the suitable value of the moment of inertia of the VSG, considering the virtual angular velocity and its rate of change during oscillations. The VSG support for the serious voltage sags is also presented.

Finally, the control performance of a current control inverter-based VSG without PLL is examined. The proposed VSG system emulates the impedance, governor dynamic, AVR unit, rotor, and damper models. System stability in both grid-connected and islanded operation modes is analyzed. All given control approaches and synthesis methodologies are verified analytically and examined through computer simulations as well as laboratory experiments.

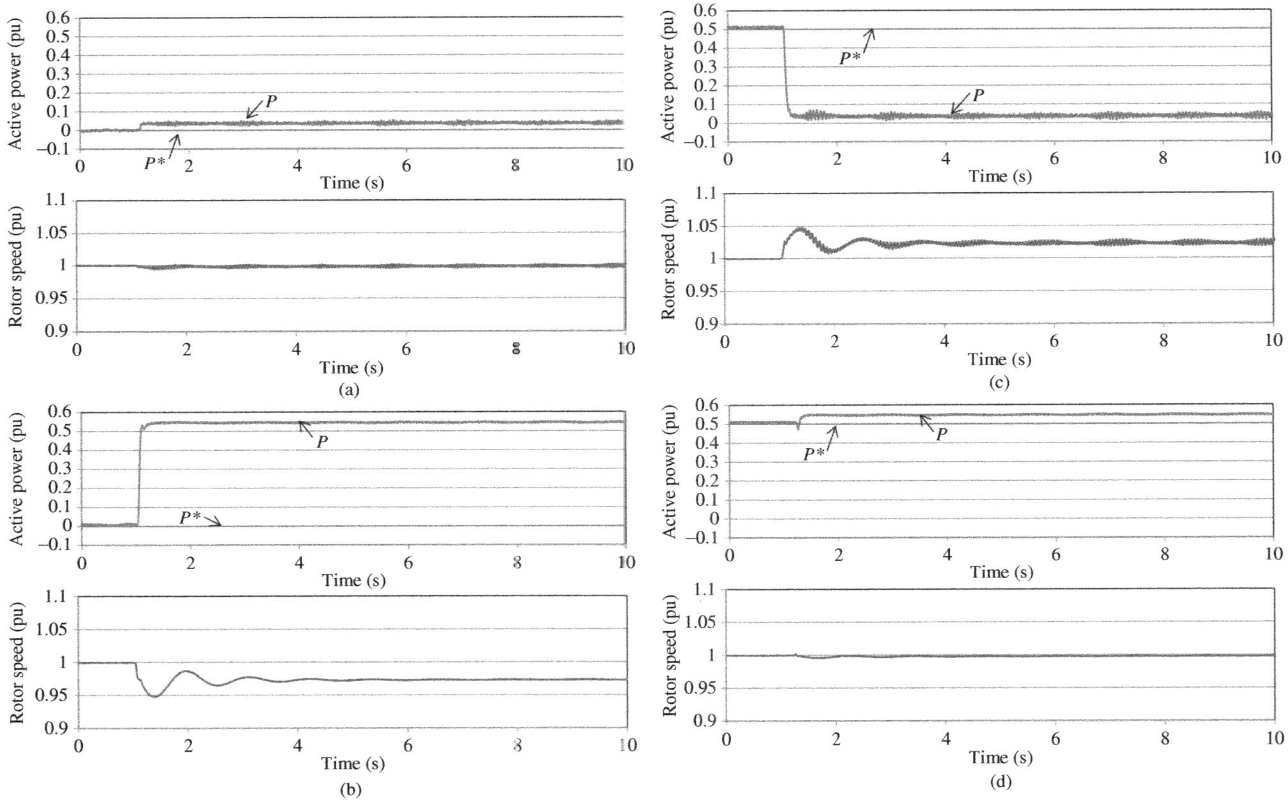

Figure 8.51 System response in islanded operation: (a) scenario 3-1 ($P^* = 0.0$ pu, load is 0.0 kW), (b) scenario 3-2 ($P^* = 0.0$ pu, load is 5.0 kW), (c) scenario 3-3 ($P^* = 0.5$ pu, load is 0.0 kW), and (d) scenario 3-4 ($P^* = 0.5$ pu, load is 5.0 kW).

References

1 IEEE-PES Task Force on Microgrid Control (2014) Trends in microgrid control. *IEEE Transactions on Smart Grid*, **5** (4), 1905–1919.

2 Bidram, A. and Davoudi, A. (2012) Hierarchical structure of microgrids control system. *IEEE Transactions on Smart Grid*, **3** (4), 1963–1976.

3 Bevrani, H., Watanabe, M., and Mitani, Y. (2012) Microgrid controls. *In Standard Handbook for Electrical Engineers*, 16th edn (ed. H. Wayne Beaty), McGraw-Hill, pp. 160–176, Section 16.9.

4 Rocabert, J., Luna, A., Blaabjerg, F., and Rodriguez, P. (2012) Control of power converters in AC microgrids. *IEEE Transactions on Power Electronics*, **27** (11), 4734–4749.

5 Bevrani, H., Watanabe, M., and Mitani, Y. (2014) *Power System Monitoring and Control*, Wiley-IEEE Press, Hoboken, NJ, USA.

6 Sakimoto, K., Miura, Y., Ise, T. (2011) *Stabilization of a Power System with a Distributed Generator by a Virtual Synchronous Generator Function.* Proceedings of IEEE Power Electronics and ECCE Asia (ICPE & ECCE), May, Jeju, Korea, pp. 1498–1505.

7 Bevrani, H., Ise, T., and Miura, Y. (2014) Virtual synchronous generators: a survey and new perspectives. *International Journal of Electrical Power & Energy Systems*, **54**, 244–254.

8 Zhong, Q.-C. (2011) Synchronverters: inverters that mimic synchronous generators. *IEEE Transactions on Industrial Electronics*, **58** (4), 1259–1267.

9 Chen, Y., Hesse, R., Turschner, D., and Beck, H.-P. (2011) *Improving the Grid Power Quality Using Virtual Synchronous Machines*, in International Conference on Power Engineering Energy and Electrical Drivers (POWERENG), May, pp. 1–6.

10 Chen, Y., Hesse, R., Turschner, D., and Beck, H.-P. (2012) *Comparison of Methods for Implementing Virtual Synchronous Machine on Inverters.* 28th–30th International Conference on Renewable Energies and Power Quality ICREPQ, March 2012, pp. 1–6.

11 Hirase, Y., Abe, K., Sugimoto, K., and Shindo, Y. (2013) A grid-connected inverter with virtual synchronous generator model of algebraic type. *Electrical Engineering in Japan*, **184** (4), 10–21.

12 Tang, X., Deng, W., and Qi, Z. (2014) Investigation of the dynamic stability of microgrid, *IEEE Transactions on Power Systems*, **29** (2), 698–706.

13 Hirase, Y., Noro, O., Yoshimura, E. *et al.* (2015) Virtual synchronous generator control with double decoupled synchronous reference frame for single-phase inverter. *IEEJ Journal of Industry Applications*, **4** (3), 143–151.

14 Liu, J., Miura, Y., and Ise, T. (2014) *Dynamic Characteristics and Stability Comparisons Between Virtual Synchronous Generator and Droop Control in Inverter-Based Distributed Generators*. Proceedings of International Power Electronics Conference (IPEC), Hiroshima, Japan, pp. 1536–1543.

15 Shintai, T., Miura, Y., and Ise, T. (2014) Oscillation damping of a distributed generator using a virtual synchronous generator. *IEEE Transactions of Power Delivery*, **29** (2), 668–676.

16 Sakimoto, K., Sugimoto, K., Shindo, Y., and Ise, T. (2015) Virtual synchronous generator without phase locked loop based on current controlled inverter and its parameter design. *IEEJ Transactions on Power and Energy*, **135** (7), 462–471.

17 Xiong, L., Zhuo, F., Wang, F. *et al.* (2016) Static synchronous generator model: a new perspective to investigate dynamic characteristics and stability issues of grid-tied PWM inverter. *IEEE Transactions on Power Electronics*, **31** (9), 6264–6280.

18 Hirase, Y., Sugimoto, K., Sakimoto, K., and Ise, T. (2016) Analysis of resonance in microgrids and effects of system frequency stabilization using a virtual synchronous generator. *IEEE Journal of Emerging and Selected Topics in Power Electronics*, **4** (4), 1287–1298.

19 Liu, J., Miura, Y., Bevrani, H., and Ise, T. , Enhanced virtual synchronous generator control for parallel inverters in a microgrid. *IEEE Transactions on Smart Grid*, (99). In press, DOI: 10.1109/TSG.2016.2521405

20 Liu, J. (2016) Studies on improving dynamic performance of microgrids by applying virtual synchronous generator control to distributed generators. PhD thesis. Osaka University, January.

21 Bevrani, H. and Shokoohi, S. (2013) An intelligent droop control for simultaneous voltage and frequency regulation in islanded microgrids. *IEEE Transactions on Smart Grid*, **4** (3), 1505–1513.

22 Banjo, Y., Miura, Y., Ise, T., and Shintai, T. (2015) *Enhanced Stand-Alone Operating Characteristics of an Engine Generator Interconnected through the Inverter using Virtual Synchronous Generator Control*. Proceedings of IEEE International Vacuum Electronics Conference, 1003–1010.

23 Paquette, A.D., Reno, M.J., Harley, R.G., and Divan, D.M. (2014) Sharing transient loads: causes of unequal transient load sharing in islanded microgrid operation. *IEEE Industry Applications Magazine*, **20** (2), 23–34.

24 Paquette, A.D. and Divan, D.M. (2015) Virtual impedance current limiting for inverters in microgrids with synchronous generators. *IEEE Industry Applications Magazine*, **51** (2), 1630–1638.

25 Machowski, J., Bialek, J., and Bumby, J.R. (1997) *Power System Dynamics and Stability*, John Wiley & Sons, New York, pp. 141–182.

26 Alipoor, J. (2015) Study on stabilization of power system with distributed generations using virtual synchronous generator. PhD thesis. Osaka University, January.

27 Macken, K.J.P., Bollen, M.H.J., and Belmans, R.J.M. (2004) Mitigation of voltage dips through distributed generation systems. *IEEE Transactions on Industry Applications*, **40** (6), 1686–1693.

28 Renders, B., De Gusseme, K., Ryckaert, W. *et al.* (2008) Distributed generation for mitigating voltage dips in low-voltage distribution grids. *IEEE Transactions on Power Delivery*, **23** (3), 1581–1588.

29 Renders, B., Ryckaert, W.R., De Gusseme, K. *et al.* (2007) Improving the voltage dip immunity of converter-connected distributed generation units. *Renewable Energy*, **33** (5), 1011–1018.

30 Ise, T. and Bevrani, H. (2016) Virtual synchronous generators and their applications in microgrids, in *Integration of Distributed Energy Resources in Power Systems*, 1st edn (ed. T. Funabashi), Elsevier, Chapter 12.

31 Alipoor, J., Miura, Y., and Ise, T. (2014) Power system stabilization using virtual synchronous generator with alternating moment of inertia. *IEEE Journal of Emerging and Selected Topics in Power Electronics*, **PP** (99), 1–8.

32 Bevrani, H. (2014) *Robust Power System Frequency Control*, 2nd edn, Springer.

9

Robust Microgrid Control Synthesis

This chapter addresses the application of robust control in microgrids (MGs). The robust frequency control in islanded MGs is emphasized. In an islanded MG with renewable sources, load change, wind power fluctuation, and solar irradiation power disturbance, as well as dynamical perturbation such as damping coefficient and inertia constants, can significantly influence the system frequency, and hence, the MG frequency control problem faces some new challenges. In response to these challenges, in this chapter, H_2, H_∞, and μ-synthesis robust control techniques are used to develop the MG frequency control loop. In the proposed control scheme, some microsources (diesel engine generators (DEGs), microturbine (MTs), and fuel cells (FC)) are assumed to be responsible for balancing the load and power in the MG system. The synthesized H_∞ and μ-controllers are examined on an MG test platform, and the controllers' robustness and performance are evaluated in the presence of various disturbances and parametric uncertainties. The results are compared with two well-tuned proportional–integral control methodologies.

The robust control theory is also used for optimal tuning of a virtual synchronous generator (VSG), which is introduced in Chapter 7. It is shown that the proposed robust control methodologies provide quite better closed-loop performance in comparison with conventional control techniques.

9.1 Introduction

Recently, remote off-grid MGs have been widely developed especially for rural and distant areas, where providing electrical energy from the main utility grid is costly and has destructive environmental effects. There are many real MGs installed for providing the electrical energy for distant areas [1–4]. On the other hand, MG control in an islanded operation mode is more difficult than in the grid-connected operation mode, because as mentioned in Chapter 5, voltage and frequency regulation of the MGs in grid-connected operation mode is initially supported by the main utility grid. In the islanded operation mode, some

Microgrid Dynamics and Control, First Edition. Hassan Bevrani, Bruno Francois, and Toshifumi Ise.

MG resources should compensate the fluctuation in load and generation (wind power and solar irradiation).

Low inertia, uncertainties, dynamic complexity, nonlinear structure, and intermittent nature of distributed energy resources (DERs) are the most important challenges in the MGs. Therefore, if a mismatch between the load and power generation occurs, the MG frequency and voltage deviation (from nominal value) is unavoidable; even it may lead to the MG blackout. Therefore, it seems that the robust control design strategies can be considered as a powerful solution to achieve robust stability (RS)/performance under the environmental condition and load deviation in the MGs [5, 6].

In the MGs, using distributed resources is essential. Distributed resources that mostly participate in the structure of an MG are wind turbine generators (WTGs), DEGs, photovoltaic (PV) units, FCs, battery energy storage (BES), and flywheel energy storage (FES) systems. Environmental and economic constraints are the main factors for choosing DGs/renewable energy sources (RESs) for an islanded MG system. Since the produced power by RESs such as PVs and WTGs depends on the weather condition, the RESs are not preferred to be used for secondary frequency control. The MT and DEG units usually supply electrical energy for the demand side to compensate the electrical energy deficiency. However, they have a slow response time and cannot handle MG control for sudden changes in load and power. So, for improving the MT or DEG control efforts, coordination with energy storage systems (ESSs) is required to quickly compensate the deviations [7–10].

As mentioned in Chapter 5, similar to the conventional power networks, a hierarchical control approach for MGs is proposed. Local/primary control, secondary control, central/emergency control, and global control are the MG control levels [10–12]. Robust control synthesis for secondary frequency regulation loop is the aim of the first control synthesis example in this chapter. The secondary control ensures that the set points of the MG are adapted to the optimality requirement of the MG, and the frequency and average voltage deviations are regulated toward zero after any change in load or supply [13]. Two main structures have been identified for secondary control in MGs: centralized and decentralized structures. The centralized scheme relies on the operation of the microgrid central controller (MGCC) in the central control level and the decentralized scheme allows the interaction of the various units within the MG [14, 15]. Generally, the centralized structures are more suitable for islanded MGs, while decentralized schemes are more suitable for grid-connected MGs, with multiple owners and fast-changing number of DG/RES units [14]. In the present work, a centralized approach is used.

In [16–19], some aspects of secondary control in MGs are investigated. In [16], a potential function is obtained for controllable units in the MG, and a central controller determines the set points, to optimize the overall performance of the MG. In comparison with [16], due to the possibility of failure in the communication links, [17] did not build secondary control in the MGCC, so failure in a communication link does not affect the other distributed units, and primary/secondary frequency controls are implemented as local controllers.

In [5, 6, 19–21], several methods are applied to the frequency control problem in MGs. In [5, 19], metaheuristic optimization algorithms such as Hopfield fuzzy neural network method and combined particle swarm optimization (PSO) with fuzzy logic are applied to regulate frequency deviation. In [20, 21], to enhance the frequency control performance and robustness in the presence of uncertainties, the mixed H_2/H_∞ and PSO-based mixed H_2/H_∞ are proposed for tuning proportional–integral derivative (PID) parameters. In [22], using a fuzzy-based PI control strategy, the stability of hybrid MG system, including MT, FC, and ESSs, is investigated.

The applications of some classical and intelligent control methods are also reported in the MG control literature [5–22]. Using traditional control methods, it is not easy to achieve a successful trade-off between nominal and robust performances (RPs). It is also difficult to guarantee simultaneous RS and RP for a wide range of disturbances and uncertainties using the heuristic and intelligent control strategies. However, robust control techniques due to the possibility of uncertainty formulation in the control synthesis procedure satisfy this objective effectively.

Because of considering physical constraints and uncertainties, the linear robust control techniques provide effective control synthesis methodologies for dynamical systems. But most linear robust control methods suggest complex state-feedback controllers, whose orders are not smaller than the order of controlled systems [23]. However, the small size of MGs in comparison with conventional large-scale power systems once again directs our attention to use these powerful synthesis methodologies for the MG control problems.

In the literature [15, 24–31], some studies on robust control applications for various MG systems are presented. In [15], a decentralized robust control strategy for an islanded MG is investigated. The MG consists of different DGs/RESs, and for each generation unit, a robust controller is designed. In [27], for frequency regulation in an autonomous MG, a μ-based robust controller is proposed. The μ-synthesis robust controller is designed via *D–K* iteration method. In [28, 29], more works on μ-synthesis robust methods are presented. In [30], an H_∞ robust controller for power sharing in both interconnected and islanded modes is proposed. Similar works for utilizing H_∞ robust control strategies on the MGs are presented in [31, 32]. In the mentioned [15, 24–31], the comparison between H_∞ and μ-synthesis applications in MG control is not given. The impacts of uncertainty modeling on the control performance are also not discussed.

This chapter focuses on the application of H_2, H_∞, and μ-synthesis robust controls for improving control performance in islanded MGs. The proposed robust approaches are flexible enough to include uncertainties in the MG model and control synthesis procedure. For H_2, H_∞, and μ-methods, linear fractional transformation (LFT) is used. In H_∞ control method, the parametric perturbation is lumped into one block as unstructured uncertainty, while in μ-synthesis method, structured uncertainty is used. The RS and RP of all controllers are investigated. In this chapter, the robust control theorem is also used for optimal tuning of an extended virtual synchronous generator (EVSG). Some parts of this chapter have already been published in [33–35].

The rest of this chapter is organized as follows: Section 9.2 describes the MG case study and the state-space dynamic model for the first robust control synthesis example. The H_∞ and μ-control theorems are briefly described in Section 9.3. The H_∞ method to obtain robust controller is investigated in Section 9.4. In Section 9.5, the μ-synthesis technique is applied. Controller order reduction and application results are presented in Section 9.6. In Section 9.7, MG control design problem is considered as a multivariable control synthesis problem and is solved using H_2 and H_∞ control techniques. Robust tuning of VSG parameters is addressed in Section 9.8, and, finally, Section 9.9 summarizes this chapter.

9.2 Case Study and State-Space Model

The present study focuses on the islanded AC MG including AC loads, WTG, PV, DEG, MT, FC, and energy storage devices such as FES system and BES system. Figure 9.1 illustrates a simplified MG configuration. As shown in Fig. 9.1, the distributed resources are connected to the AC bus by power electronic devices used for synchronization of AC sources such as DEG and WTG, and they are used to invert DC voltages into AC voltage in sources such as PV arrays and FCs. A converter is also considered for the BES system to convert AC into DC in charging mode and DC into AC in discharging mode.

The total power generation of distributed resources, for supplying the demand side, comprises the output power of DEG, MT, WTG, PV, FC, and exchange power of FES and BES.

$$P_{Load} = P_{DEG} + P_{MT} + P_{WTG} + P_{PV} + P_{FC} \pm P_{BES} \pm P_{FES} \tag{9.1}$$

Since the produced power by the RESs such as PVs and WTGs depends on the environmental condition, they are not commonly used for frequency regulation; so in the first robust control design example, for frequency control problem, MT, DEG, and FC are considered. In the secondary frequency control loop, the fluctuations in load, WTG, and PV output powers are compensated by decrease/increase in the DEG, MT, and FC output powers. An expression for changes in the MG resources associated with frequency regulation can be

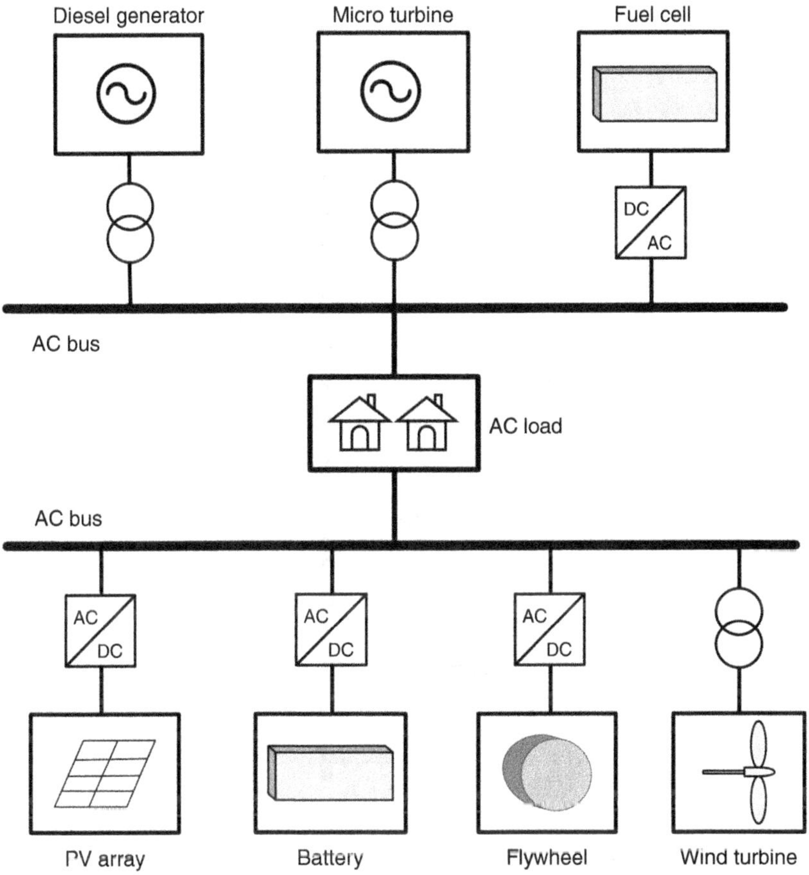

Figure 9.1 Simplified schematic of an islanded MG.

presented as follows:

$$\Delta P_{\text{Load}} + \Delta P_{\text{DEG}} + \Delta P_{\text{MT}} + \Delta P_{\text{FC}} + \Delta P_{\text{PV}} + \cdots$$
$$\Delta P_{WTG} + \Delta P_{BES} + \Delta P_{FES} = 0 \tag{9.2}$$

In [19, 32, 36], some simplified dynamical models for DGs/storage units are presented. Some distributed sources may have high-order dynamical frequency response models, but low-order dynamical models considered in this chapter are sufficient for investigating frequency control issue [19]. Figure 9.2 shows an MG dynamical frequency response model, and the relevant system parameters for a typical MG are represented in Table 9.1.

Linearized state-space model is a useful model representation for the application of robust control theory in the MG control synthesis. Using appropriate

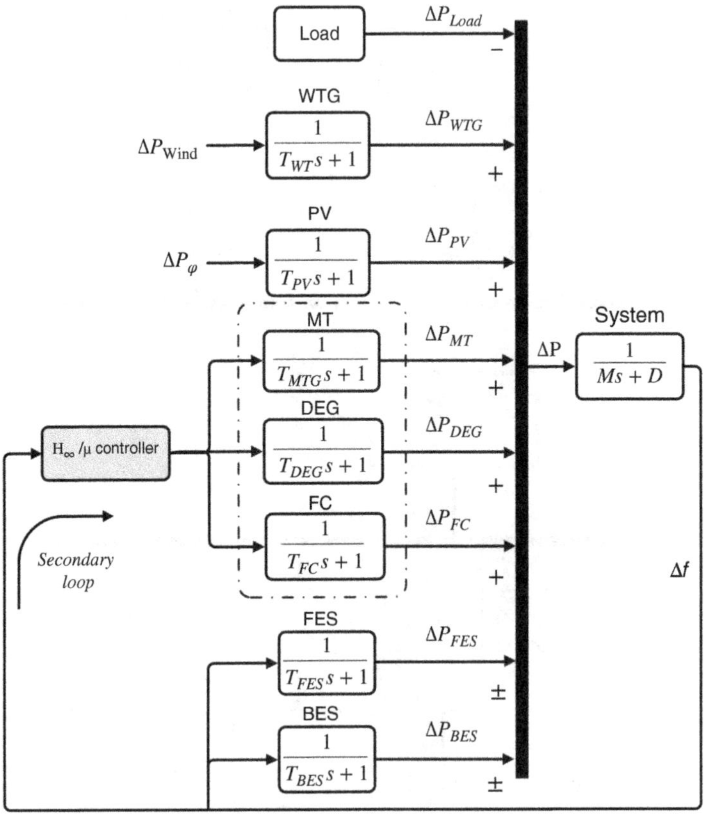

Figure 9.2 MG dynamical frequency response model.

Table 9.1 The parameters of frequency response model (Fig. 9.2).

Parameter	Values	Parameter	Values
D (pu/Hz)	0.012	T_{DEG} (s)	2
M (pu/s)	0.2	T_{MT} (s)	2
T_{FC} (s)	4	T_{WTG} (s)	1.5
T_{BESS} (s)	0.1	T_{PV} (s)	1.8
T_{FESS} (s)	0.1		

definitions and state variables, as given in (9.3)–(9.6), the linearized state-space realization of the MG system (Fig. 9.2) can be easily obtained in the form of (9.3) [36]. An elaborated expression of the MG state-space model is given in (9.7).

$$\dot{x} = Ax + B_1 w + B_2 u$$
$$y = Cx \tag{9.3}$$

Here, u is the control input signal, and state vector can be defined as follows:

$$x^T = [\Delta P_{WTG} \Delta P_{PV} \Delta P_{DEG} \Delta P_{FC} \Delta P_{MT} \Delta P_{BES} \Delta P_{FES} \Delta f] \tag{9.4}$$

$$w^T = [\Delta P_{Wind} \Delta P_{\varphi} \Delta P_{Load}] \tag{9.5}$$

$$y = \Delta f \tag{9.6}$$

$$\dot{x} = \begin{bmatrix} -1/T_{WTG} & 0 & 0 & 0 & 0 & 0 & 0 & 0 \\ 0 & -1/T_{PV} & 0 & 0 & 0 & 0 & 0 & 0 \\ 0 & 0 & -1/T_{DEG} & 0 & 0 & 0 & 0 & 0 \\ 0 & 0 & 0 & -1/T_{FC} & 0 & 0 & 0 & 0 \\ 0 & 0 & 0 & 0 & -1/T_{MT} & 0 & 0 & 0 \\ 0 & 0 & 0 & 0 & 0 & -1/T_{BES} & 0 & 1/T_{BES} \\ 0 & 0 & 0 & 0 & 0 & 0 & -1/T_{FES} & 1/T_{FES} \\ 0 & 0 & 0 & 0 & 0 & 0 & 0 & -2D/M \end{bmatrix}$$

$$\times \begin{bmatrix} \Delta P_{WTG} \\ \Delta P_{PV} \\ \Delta P_{DEG} \\ \Delta P_{FC} \\ \Delta P_{MT} \\ \Delta P_{BES} \\ \Delta P_{FES} \\ \Delta f \end{bmatrix} + \begin{bmatrix} 1/T_{WTG} & 0 & 0 \\ 0 & 1/T_{PV} & 0 \\ 0 & 0 & 0 \\ 0 & 0 & 0 \\ 0 & 0 & 0 \\ 0 & 0 & 0 \\ 0 & 0 & 0 \\ 0 & 0 & 2/M \end{bmatrix} \begin{bmatrix} \Delta P_{Wind} \\ \Delta P_{\varphi} \\ \Delta P_{Load} \end{bmatrix} + \begin{bmatrix} 0 \\ 0 \\ 1/T_{DEG} \\ 1/T_{FC} \\ 1/T_{MT} \\ 0 \\ 0 \end{bmatrix} u,$$

$$y = \begin{bmatrix} 0 & 0 & 0 & 0 & 0 & 0 & 0 & 0 \end{bmatrix} x \tag{9.7}$$

In the mentioned MG case study (Fig. 9.2), ΔP_{Wind}, ΔP_{φ}, and ΔP_{Load} are considered as MG disturbance signals and the M and D parameters are considered as uncertain parameters. More complete (higher order or nonlinear) models of the MG components for interested readers are given in [9, 18, 26, 30, 32, 37, 38].

9.3 H_∞ and Structured Singular Value (μ) Control Theorems

Here, the basics of H_∞ and structured singular value (μ) theorems are briefly given. Interested readers can find details and proofs in [36].

9.3.1 H_∞ Control Theory

Consider a linear time-invariant system $G(s)$ with the following state-space realization:

$$\begin{aligned} \dot{x} &= Ax + B_1 w + B_2 u \\ z &= C_1 x + D_{12} u \\ y &= C_2 x \end{aligned} \tag{9.8}$$

where x is the state variable vector, w is the disturbance and other external input vector, z is the controlled output vector, and y is the measured output vector.

It is assumed that (A, B_2, C_2) is stabilizable and detectable. The H_∞ control problem for the linear time-invariant system $G(s)$ with the state-space realization of (9.8) is to find a controller $K(s)$ such that the resulting closed-loop system is internally stable, and the ∞-norm from w to z is smaller than γ, a specified positive number, that is,

$$\|T_{wz}(s)\|_\infty < \gamma \tag{9.9}$$

Here, the $T_{wz}(s)$ is closed-loop transfer function from input w to output z. Using matrix representation, the K is a dynamic H_∞ controller, if and only if there exists a symmetric matrix $X > 0$ such that

$$\begin{bmatrix} A_{cl}^T X + X A_{cl} & X B_{cl} & C_{cl}^T \\ B_{cl}^T X & -\gamma I & D_{cl}^T \\ C_{cl} & D_{cl} & -\gamma I \end{bmatrix} < 0 \tag{9.10}$$

where $(\cdot)_{cl}$ shows the closed-loop components, and

$$\begin{aligned} &\mathrm{A_{cl} = A + B_2 K C_2,\ B_{cl} = B_1} \\ &\mathrm{A_{cl} = A + B_{12} K C_2,\ D_{cl} = 0} \end{aligned}$$

9.3.2 Structured Singular Value (μ) Control Theory

The structured singular value (μ) provides sufficient and necessary conditions to ensure the RS and RP of a dynamic system in the presence of structured uncertainties. The overall structured uncertainty block Δ, which can be represented in the general form of (9.11), includes the parametric uncertainties (δ_i)

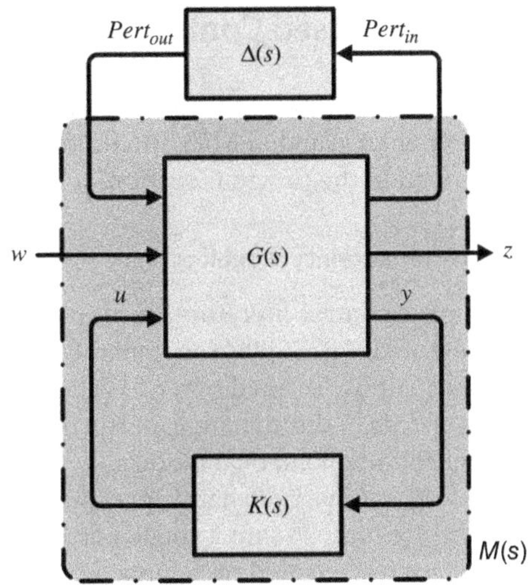

Figure 9.3 Standard M–Δ configuration for μ-synthesis.

and block uncertainties (Δ_i) that can be extracted from the system dynamics.

$$\Delta = \{diag[\delta_1 I_{r1}, \ldots, \delta_k I_{rk}, \Delta_1, \ldots, \Delta_f], \delta_i \in C, \Delta_j \in C^{k_i \times k_j}\} \tag{9.11}$$

The whole system can be rearranged in a standard configuration, which is known as $M-\Delta$ configuration, and it is shown in Fig. 9.3. The w, u, z, y, $pert_{in}$, and $pert_{out}$ are the exogenous inputs, control signals, control performance signals, measured outputs, input and output perturbations signals of the uncertain block, respectively.

Based on the μ-synthesis theorem, for a given $M-\Delta$ configuration (Fig. 9.3), the RS/performance is satisfied by controller $K(s)$ if and only if

$$\inf_{K} \sup_{\omega \in R} \mu[M(j\omega)] < 1 \tag{9.12}$$

where the μ-function is defined as

$$\mu_\Delta(M) = \frac{1}{\min_\Delta\{\overline{\sigma}(\Delta) : |I - M\Delta| = 0, \Delta \in \Delta\}} \tag{9.13}$$

The Δ is defined as (9.11). Considering the upper bound of μ, the robust synthesis problem (9.12) can be reduced to determine

$$\min_{K} \inf_{D} \sup_{\omega \in R} \overline{\sigma}[DM(j\omega)D^{-1}] \tag{9.14}$$

or equivalently

$$\min_{K,D} \|DM(G(j\omega), K(j\omega))D^{-1}\|_\infty \tag{9.15}$$

by iteratively solving for D and K. Here, D is a positive definite symmetric matrix with appropriate dimension, and $\overline{\sigma}(\cdot)$ denotes the maximum singular value of a matrix.

9.4 H_∞-Based Control Design

The aim of this section is to design an H_∞-based controller for MG frequency control in an islanded MG. The fundamentals of H_∞ control theory are briefly described in the previous section.

9.4.1 Uncertainty Modeling

In robust control literature, several definitions for uncertainty modeling are represented [23, 36], but in general, the uncertainty as dynamic perturbation can be mainly classified into two categories: "unmodeled dynamics" and "modeling errors." The dynamic perturbations represent the difference between actual and mathematical models.

For H_∞ synthesis, dynamic parametric perturbations considered for the MG system are lumped into a single perturbation block $\Delta(s)$. This uncertainty representation is referred to as "unstructured uncertainty" [23]. There are different methods for modeling unstructured uncertainty in robust control theory, such as additive perturbation, inverse additive perturbation, and input/output multiplicative perturbation. In this section, the output multiplicative perturbation method is used for uncertainty modeling.

Figure 9.4 shows the block diagram of the closed-loop configuration used for the H_∞ design procedure. The lumped uncertainty and selected weighting functions are shown. $P(s)$ denotes the actual (perturbed system) dynamics, while $G(s)$ represents the nominal model of the MG physical system without perturbation. As it can be seen from Fig. 9.4, the weighting functions $W_e(s)$, $W_u(s)$, $W_{d1}(s)$, $W_{d2}(s)$, and $W_{d3}(s)$ are chosen to improve the RS and RP of system and $w_{\{1-3\}}$, u, $z_{\{1-5\}}$ and y are the exogenous disturbance inputs, control signals, desired performance signals, and measured output, respectively. A proper selection of the weighting functions is demonstrated in (9.16)–(9.18).

$$W_e(s) = 0.01\frac{s^3 + 5s^2 + 10s + 60}{s^3 + 100s^2 + 15s + 3} \tag{9.16}$$

$$W_u(s) = \frac{2(s+1)}{0.01s + 9} \tag{9.17}$$

$$W_d = \begin{bmatrix} 0.01 & & \\ & 0.01 & \\ & & 0.01 \end{bmatrix} = 0.01 I_{3\times 3} \tag{9.18}$$

Considering ±50% deviations for D and M, Bode diagram of output multiplicative perturbed system $P(s)$ (with three disturbance inputs, one control input, and one measured output) is shown in Fig. 9.5. As mentioned, ΔP_{Wind}, ΔP_{φ}, and ΔP_{Load} are considered as MG disturbance signals and Δf is the measured system output. In Fig. 9.6, $T_{w_1y}(j\omega)$, $T_{w_2y}(j\omega)$, $T_{w_3y}(j\omega)$, $T_{uy}(j\omega)$ represent the system Bode diagrams from the MG input signals (ΔP_{Wind}, ΔP_{φ}, ΔP_{Load},

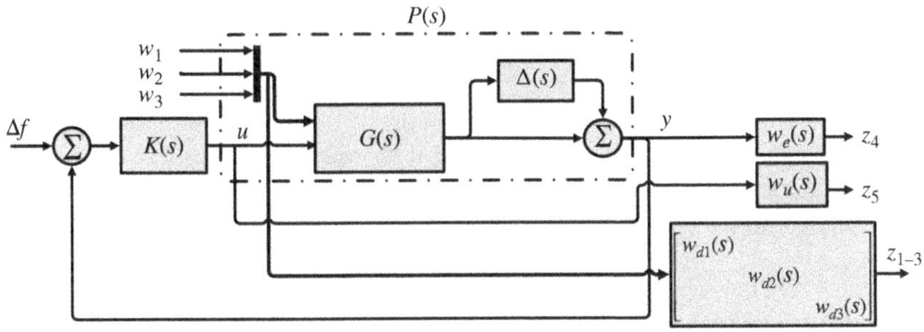

Figure 9.4 Closed-loop system structure with lumped multiplicative uncertainty.

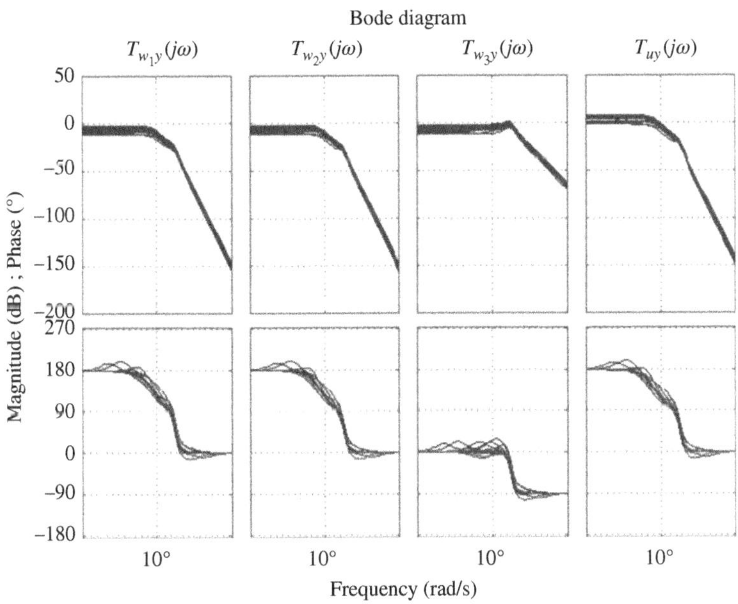

Figure 9.5 Bode diagram of the perturbed system *P*(*s*).

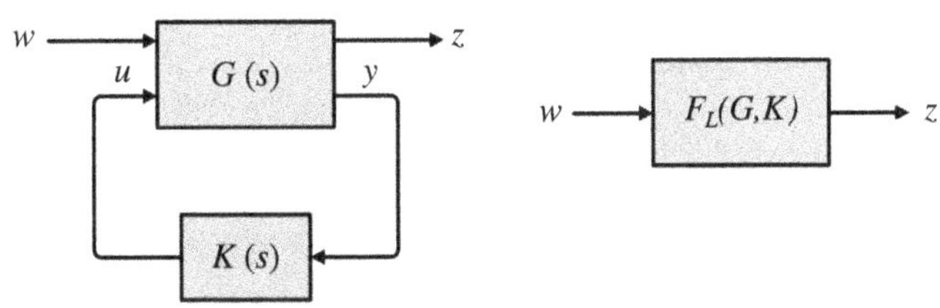

Figure 9.6 H_∞ standard LFT configuration.

and control command) to output signal (Δf). This plots show the frequency response behavior for the perturbed input–output subsystems.

9.4.2 H$_\infty$ Optimal Controller

The H$_\infty$ controller is an optimization control problem, which determines a feasible robust controller by minimizing the infinity-norm of an appropriate LFT, $F_L(G, K)$, as follows:

$$\|F_L(G, K)\|_\infty < 1 \tag{9.19}$$

$F_L(G, K)$ is the transfer function matrix of the nominal closed-loop system from the disturbance input signals to the controlled output signals, which can be described as transfer function of T_{wz} [28]. There is no analytic method for solving the aforementioned optimization problem. On the other hand, the solution for this minimization problem is not unique; so, it is usually sufficient to find a stabilizing controller K_∞ such that the H$_\infty$-norm of (9.19) tends to be below 1. Figure 9.6 shows the standard closed-loop LFT for the H$_\infty$ synthesis.

9.4.3 Closed-Loop Nominal Stability and Performance

The nominal stability (NS) is satisfied because the closed-loop system T_{wz} is internally stable for the designed K_∞. For evaluating the nominal performance (NP), the infinity-norm of both sensitivity function $S = (I + GK)^{-1}$ and complementary function KS should be less than a positive value. The NP criterion for the closed-loop system can be described by (9.20) [21, 24], where W_e and W_u are weighting functions represented in (9.8) and (9.9).

$$\left\| \begin{bmatrix} W_e(I + GK)^{-1} \\ W_P K(I + GK)^{-1} \end{bmatrix} \right\|_\infty < 1 \tag{9.20}$$

Figure 9.7 shows that the ∞-norm inequality of (9.20) is satisfied and is always less than 1. So, the closed-loop system successfully reduces the influence of the disturbance, and the required performance is fully achieved. In the present work, the *hinfsyn* command in MATLAB Robust Control Toolbox is used for solving the aforementioned inequality.

9.4.4 Closed-Loop Robust Stability and Performance

The RS is satisfied because the closed-loop system (T_{wz} transfer function) is internally stable for all possible plants $P = (I + \Delta(s))G(s)$ in which the block Δ is the uncertainty block. For analyzing the RP, the performance criterion given in (9.21) must be satisfied for all $P = (I + \Delta(s))G(s)$. In this section, T_{wz} function in the presence of 50% perturbations in D and M parameters is investigated. As shown in Fig. 9.8, the ∞-norm inequality represented in (9.21) is less than 1.

$$\left\| \begin{bmatrix} W_e(I + GP)^{-1} \\ W_P K(I + GP)^{-1} \end{bmatrix} \right\|_\infty < 1 \tag{9.21}$$

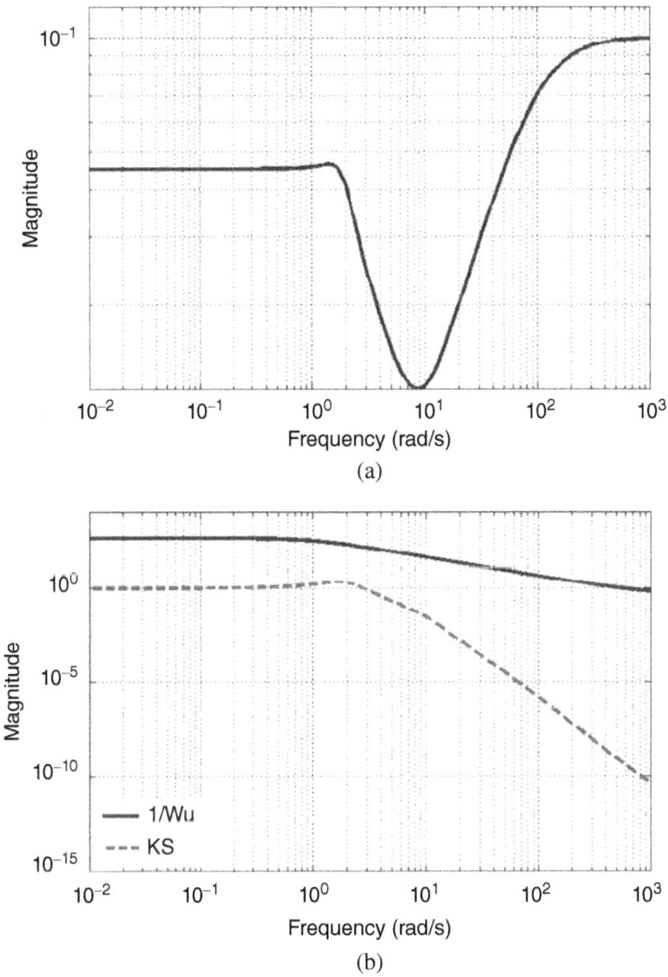

Figure 9.7 (a) S and (b) KS functions of nominal system.

Figure 9.8a shows that the inequality $\|[W_e(I + GP)^{-1}]\|_\infty < 1$ for $P = (I + \Delta(s))G(s)$ is always satisfied. Figure 9.8b shows that the inequality $\|[W_PK(I + GP)^{-1}]\|_\infty < 1$ is also satisfied, or $\|[K(I + GP)^{-1}]\|_\infty$ is less than $1/W_P$. Therefore, the RS and RP are simultaneously satisfied for the obtained controller K via a single ∞-norm inequality, which is represented in (9.21).

9.5 μ-Based Control Design

The basics of structured singular value (μ) theorem are briefly given in [36, 39].

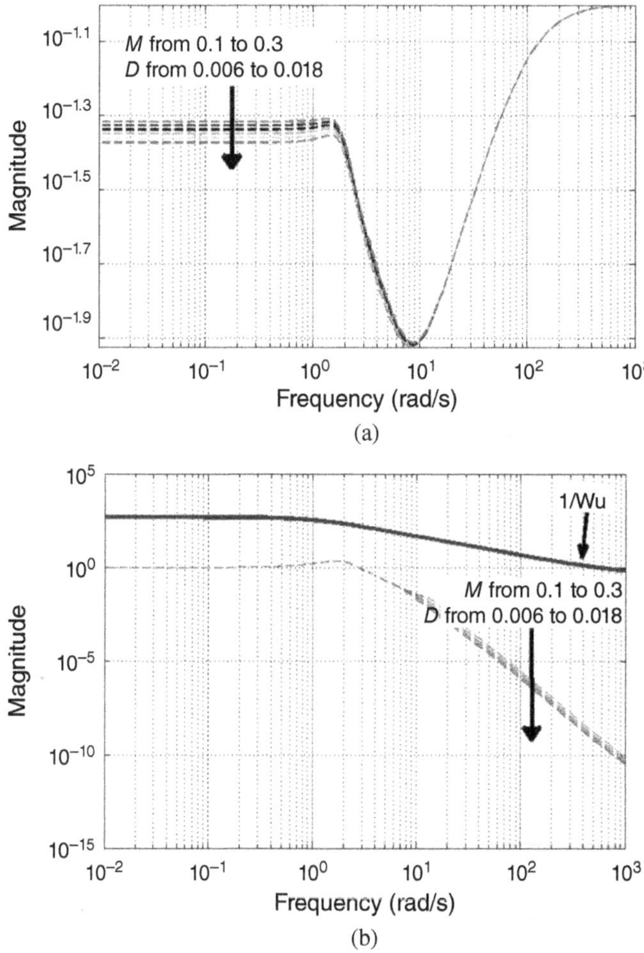

Figure 9.8 (a) S and (b) KS functions in the presence of perturbations (robust performance).

9.5.1 Uncertainty Modeling in μ-Synthesis

The uncertainties have already been modeled in a lumped block Δ, which refers to the "unstructured uncertainties." This type of uncertainty provides conservative results in the synthesis procedure. For solving this problem, the "structured uncertainty" should be used [21]. The structured uncertainty may include structured unmodeled dynamics and parametric perturbation.

In order to apply μ-synthesis theorem to the problem at hand, first, the structured uncertainty block Δ that is represented in the general form of (9.11) should be extracted from the MG dynamics, and then, the whole system can

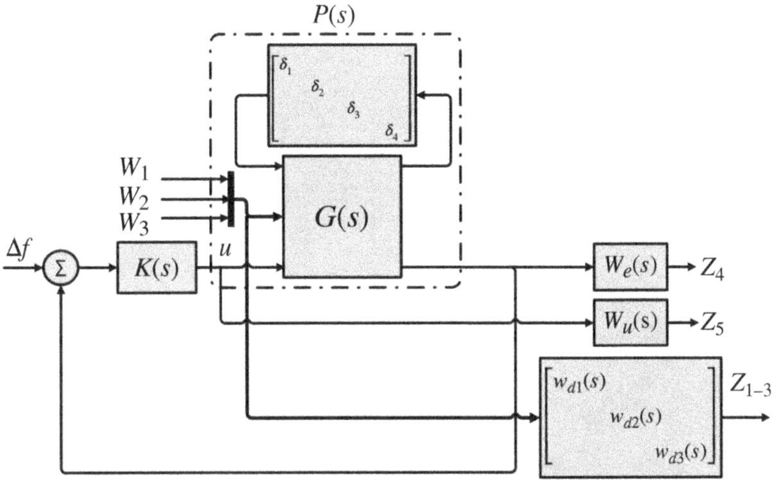

Figure 9.9 Closed-loop system diagram with structured diagonal uncertainty block.

be rearranged in a standard configuration of the upper LFT. Figure 9.9 shows the closed-loop block diagram of the MG system with a structured diagonal uncertainty. In the present work, the system has a 4×4 parametric diagonal uncertainty as shown in Fig. 9.9.

9.5.2 D–K Iteration

As has already been mentioned, the structured singular value (μ-based control) framework provides conditions to ensure the RP of a dynamic system. Considering the $M\text{–}\Delta$ configuration (Fig. 9.3), the w, u, z, y, $pert_{in}$, and $pert_{out}$ are the exogenous inputs, control signals, control performance signals, measured outputs, input and output perturbations signals of the uncertain block, respectively. In μ-synthesis, for satisfying the RP condition, $\|T_{wz}\|_\infty \leq 1$ must be satisfied for all $\Delta \in \Delta_P$, where Δ_P is defined as follows:

$$\Delta_p := \left\{ \begin{bmatrix} \Delta & 0 \\ 0 & \Delta_F \end{bmatrix} : \ \Delta \in R^{4\times4}, \ \Delta_F \in C^{4\times5} \right\}. \tag{9.22}$$

The first uncertainty block Δ of this structured matrix is a 4×4 diagonal matrix and corresponds to the perturbations used in the dynamical modeling of islanded MG system. The second block Δ_F is a fictitious uncertainty block that is introduced to represent the performance requirement in the M-Δ framework. To achieve RP, the stabilizing controller $K(s)$ should minimize.

$$\max_\omega \mu_{\Delta_p}(M(s)) \tag{9.23}$$

There is no standard analytical method to calculate the μ-optimal controller via the optimization problem given in (9.23). Therefore, a numerical method for

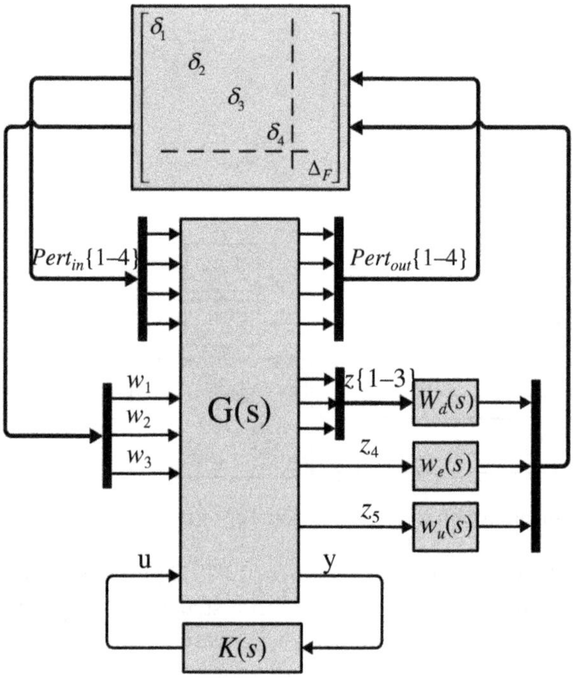

Figure 9.10 The closed-loop configuration for using *D–K* iteration method.

complex perturbation known as *D–K* iteration [36] is used to solve (9.23). The idea is to find a controller *K* such that (9.18) should be satisfied by alternating the minimization with respect to *K* or *D*, while fixing the other one.

$$\min_K(\min_{D(j\omega)}\|D(j\omega)M(K(j\omega))D^{-1}(j\omega)\|_\infty) \tag{9.24}$$

Figure 9.10 shows the simplified LFT configuration for implementing *D–K* iteration method in order to design robust μ-controller for the uncertain MG system using MATLAB toolbox. The iteration for designing μ-controller with *D–K* iteration method includes the following main steps:

1) *K*-Step: synthesize an H_∞ controller for the scaled problem $\min\|D(s)M(K)D^{-1}(s)\|_\infty$, while $D(s)$ is fixed.
2) *D*-Step: find $D(j\omega)$ to minimize $\overline{\sigma}(DM(K)D^{-1}(j\omega))$ over the given frequency range, while $M(K)$ is fixed.
3) Fit the magnitude of each element $D(j\omega)$ to a stable and minimum-phase transfer function $D(s)$. Then, go back to step *i*.

These iterations continue until the $\|D(s)M(K)D^{-1}(s)\|_\infty$ is less than 1 or until the ∞-norm no longer decreases. RP characteristics are extremely dependent on appropriate selection of weighting functions. To have a better comparison between H_∞ and μ-synthesis methods, the same weighting functions are considered in both synthesis approaches.

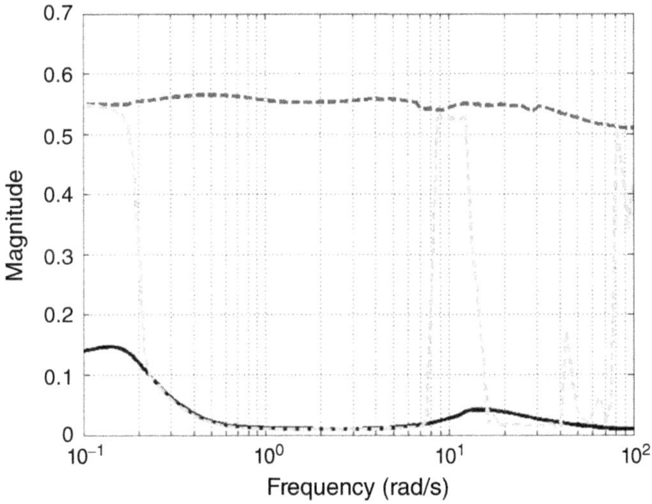

Figure 9.11 Nominal performance (solid) and robust performance (dashed).

9.5.3 Closed-Loop Nominal and Robust Performance

After designing K using the D–K iteration method, the maximum value of μ for RP is obtained: 0.57. To guarantee the RP with the obtained controller, the performance criterion (9.25) should be satisfied (infinity-norm to be smaller than the upper bound of μ). Figure 9.11 shows that the frequency response of NP and RP for the closed loop system is satisfied. It is noteworthy that $||\Delta||_{\infty} <1$, for every diagonal Δ.

$$\left\| \begin{bmatrix} W_e(I + F_U(G,\Delta)K^{-1}) \\ W_P K(I + F_U(G,\Delta)K^{-1}) \end{bmatrix} \right\|_{\infty} < 1 \tag{9.25}$$

Figure 9.12 shows the RP index for perturbed system sensitivity function. It is clear that the frequency response of dynamical perturbed MG system remains below 0.57.

9.5.4 Robust Stability

The upper and lower bounds of μ for the perturbed system are shown in Fig. 9.13. It is visible that the upper bound of μ is 0.922, so if the $||\Delta||_{\infty} < \frac{1}{0.922}$, the RS for the perturbed closed-loop system is achieved, and $||\Delta||_{\infty} < 1$ is satisfied. Figure 9.14 demonstrates the robust properties of system with the designed μ-controller. It is shown that the magnitude response of the perturbed sensitivity function over the given frequency range is below the magnitude of the inverse of weighting function W_e.

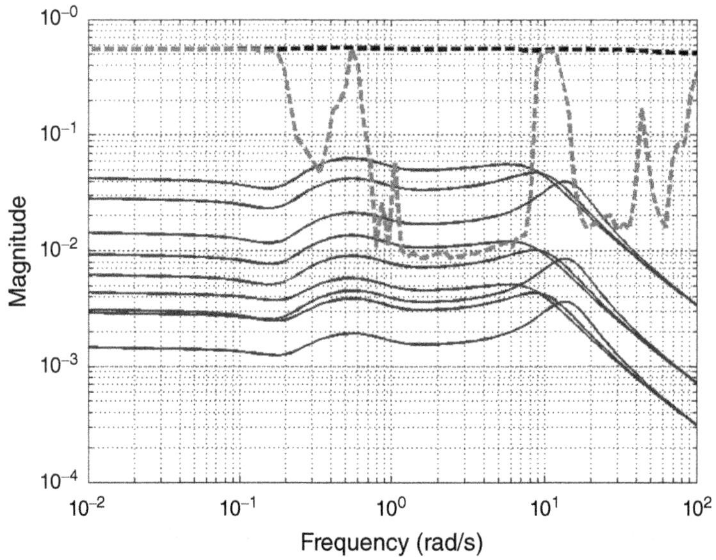

Figure 9.12 RP index of perturbed systems (solid) and μ upper and lower bounds of perturbed closed-loop (dashed).

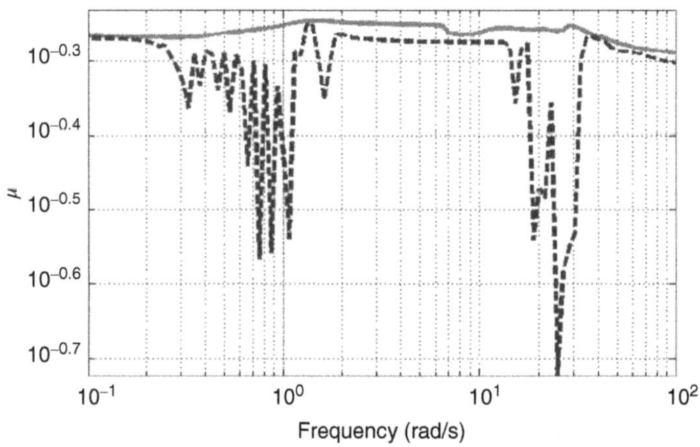

Figure 9.13 Robust stability of K, upper bound (solid) and lower bound (dashed).

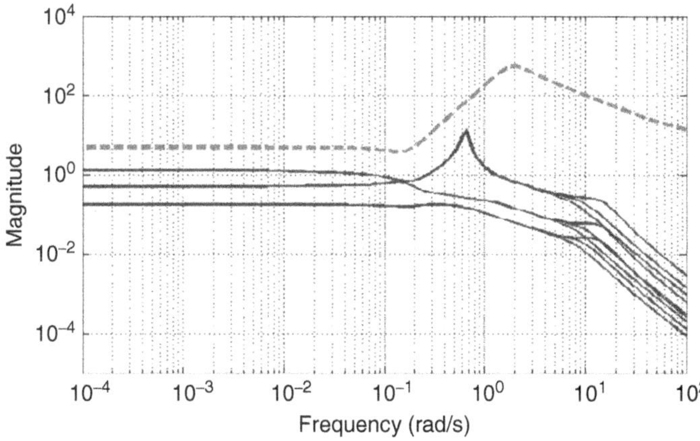

Figure 9.14 Sensitivity functions of perturbed systems with K (solid), W_e^{-1} (dashed).

9.6 Order Reduction and Application Results

9.6.1 Controller Order Reduction

High order of designed controllers is one of the main problems of H_∞ and μ-based robust control methods, especially for the high-order plants. Using these control methodologies, the obtained controller are higher than or at least equal to the order of the given plants. Different methods are introduced for reducing the order of robust controllers.

Generally, to reduce the order of stable systems/controllers, an absolute-error-based approximation method such as balanced truncation, singular perturbation approximation (balanced residualization), and Hankel-norm approximation can be used. Here, the Hankel-norm approximation approach [39] is used. This method can be easily handled in MATLAB environment using *sysbal* and *hankmr* commands. The *sysbal* command generates the balanced realization of a system (input argument). This command removes the unobservable and/or uncontrollable modes for un-minimal systems and returns the Hankel singular values. These values can be used to choose the order to be reduced, and then, the *hankmr* command reduces the controller order. However, one can use other relevant available methods.

The order of obtained controller, by H_∞ method, was 14, which is reduced to 6. Figure 9.15 shows the Bode diagram of full-order (original) and reduced-order for H_∞ controller. After three iterations in D–K method, the μ-synthesis controller order is 28, which is reduced to 7, using Hankel-norm approximation method. In Fig. 9.16, the Bode diagrams of full-order and reduced-order controllers are demonstrated. Figures 9.15 and 9.16 show that the resulting high-order controllers are reduced to low orders without performance degradation.

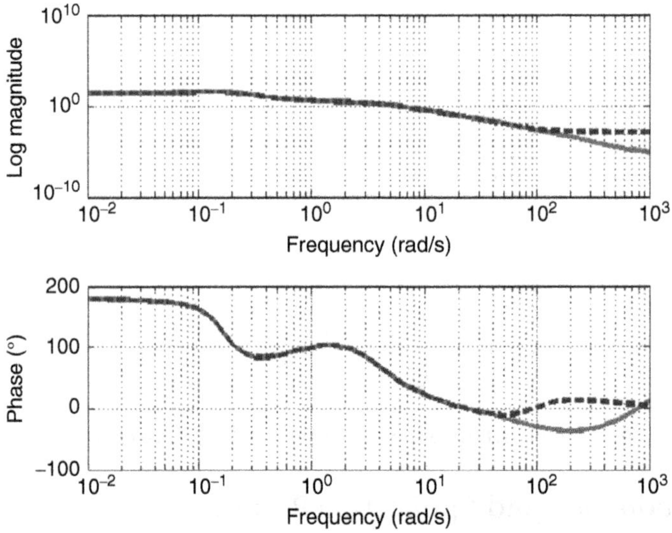

Figure 9.15 Comparison between original (solid) and reduced-order (dashed) H_∞ controller.

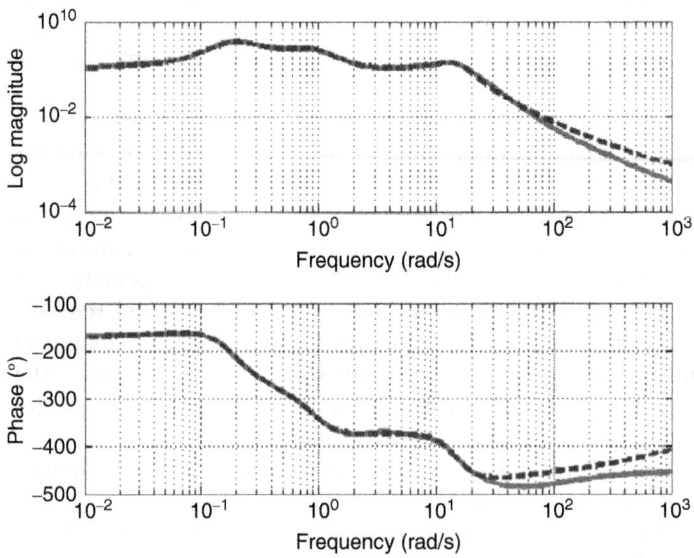

Figure 9.16 Comparison between original (solid) and reduced-order (dashed) μ-controller.

9.6.2 Application Results

In this section, the time-domain simulation of MG frequency, in the presence of ΔP_{Wind}, ΔP_{φ}, and ΔP_{Load} changes (as disturbances) and parameter perturbation, is presented. The H_∞ method and μ-synthesis (with structured and unstructured uncertainties) are compared. For investigating the MG frequency response, four sever test scenarios are performed.

Scenario 1: Wind Power Fluctuation (ΔP_{Wind})

A step change is considered in the wind power. Figure 9.17a shows the wind power step changes and Fig. 9.17b represents the MG output frequency response by comparing the proposed controllers.

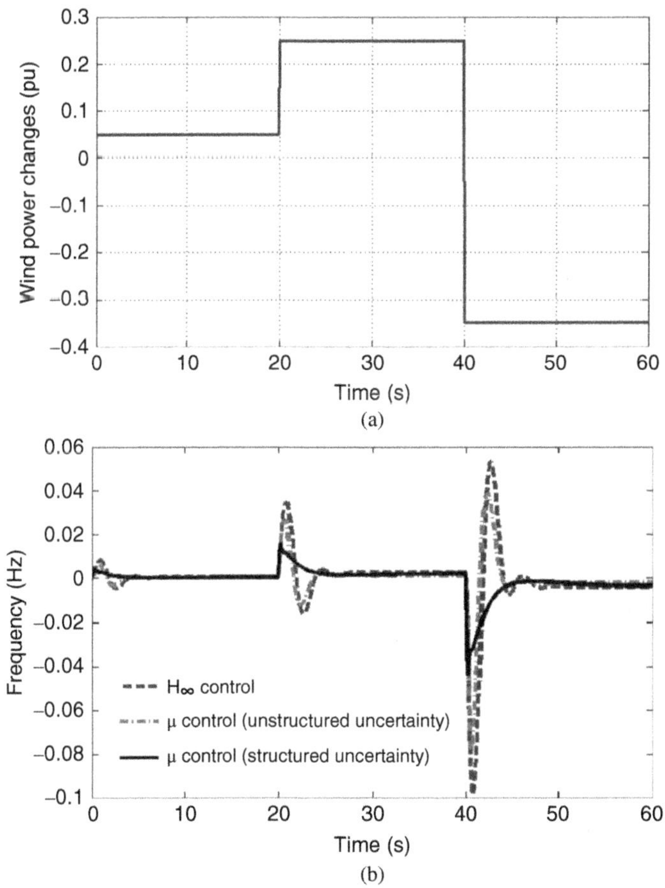

Figure 9.17 System response for step changes in wind power: (a) wind power change pattern and (b) MG output frequency.

Scenario 2: Load Fluctuation (ΔP_{Load})

A multiple-step load deviation is considered with changes at times 0, 20, and 40 s. Figure 9.18a shows the multiple-load-step change pattern. Figure 9.18b represents the MG frequency response by comparing between the performance of H_∞ controller and μ-controller (with structured and unstructured uncertainty).

Scenario 3: Solar Power Fluctuation (ΔP_φ)

As another serious test, a multiple-step change in the sun irradiation power is applied at times 1, 10, 2 0, 30, 40, and 5 0 s as shown in Fig. 9.19a. Figure 9.19b represents the MG frequency for the designed controllers.

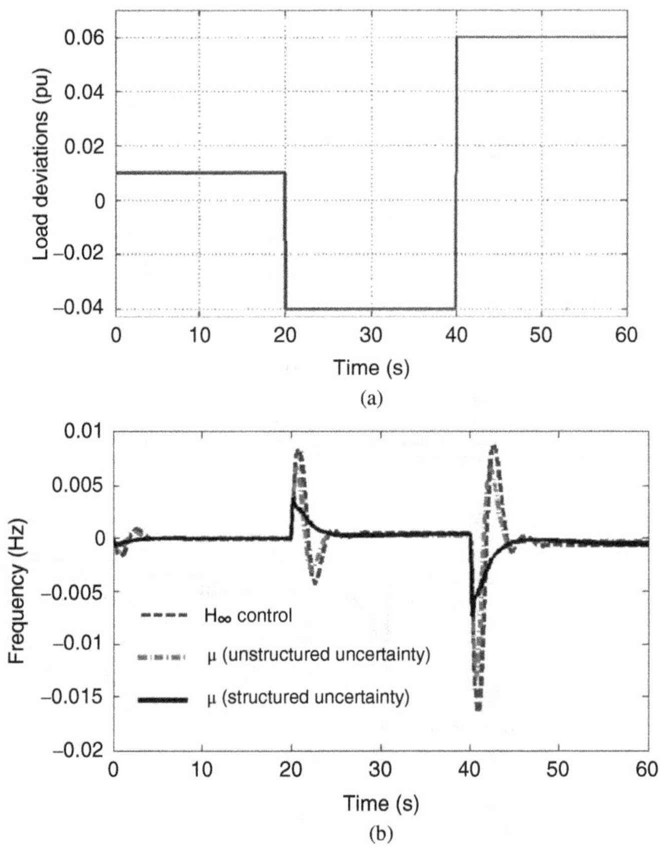

Figure 9.18 System response for step changes in load: (a) multiple load deviation and (b) MG output frequency.

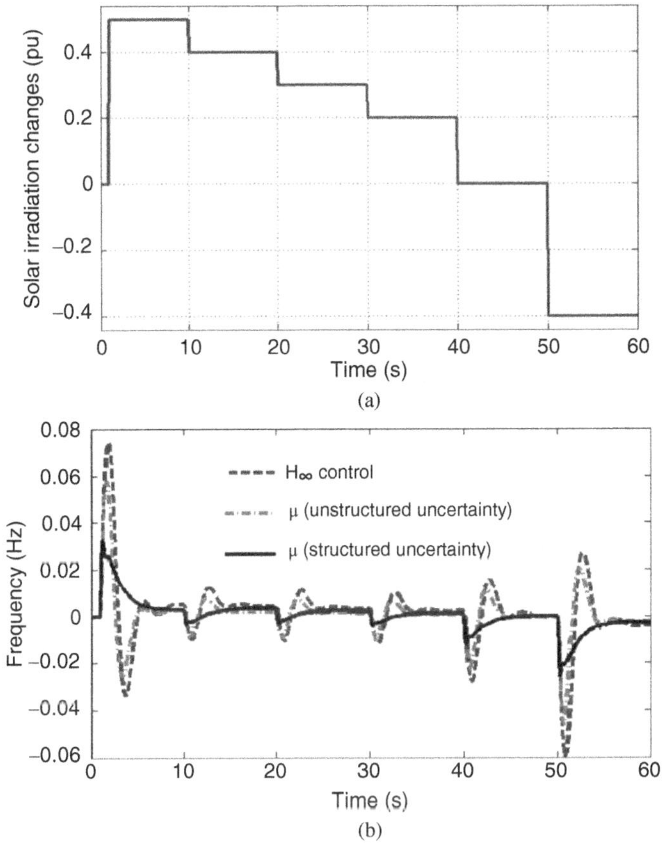

Figure 9.19 System response for step changes in solar power: (a) solar power change pattern and (b) MG output frequency.

Scenario 4: Simultaneous Changes in ΔP_{Wind}, ΔP_{φ}, ΔP_{Load}, and MG Parameters

In this scenario, firstly, simultaneous disturbances in wind power, sun irradiation power, and load are considered. Figure 9.20a shows the mixed changes in wind power, solar irradiation, and load. Figure 9.20b shows the MG frequency response behavior. Figure 9.21b shows the system response for a more smooth and nonstationary fluctuation pattern [40] in solar irradiation and wind powers in the presence of step load variation as depicted in Fig. 9.21b.

Then, the MG frequency response in the face of 50% decrease in H and D parameters (as uncertainties) is examined. The results are shown in Fig. 9.22.

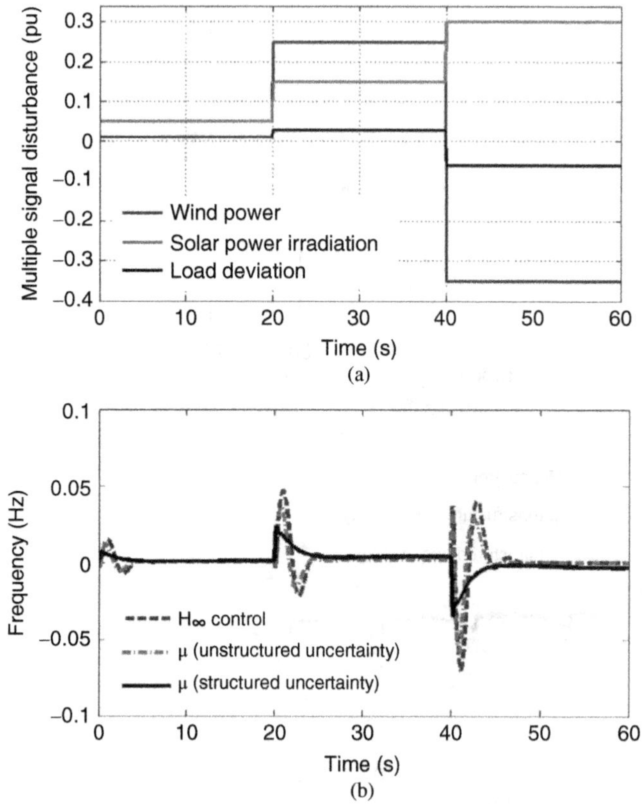

Figure 9.20 System response for step disturbances: (a) multiple disturbances in load, wind speed, and solar irradiation and (b) MG output frequency.

Simulation results show that μ-controller with structured uncertainty gives quite better performance compared to μ-controller with unstructured uncertainty as well as H_∞ controller, because the uncertainties in the MGs are of parametric/structured type; hence, using an unstructured uncertainty-based control technique provides nonexact conservative results.

9.6.3 Comparison with Well-Tuned Proportional-Integral (PI) Controllers

PI (or PID) controllers are the most popular and common type of controllers and highly used in power electric industry. Hence, in the most relevant literature, new advanced control synthesis techniques are compared with the PI controllers. Tuning of PI control parameters for general applications may not be so difficult, but finding a set of gains/parameters to ensure an optimal performance for a given control system is a complex task. Traditionally, the PI controllers are tuned either manually or using rule-based methods.

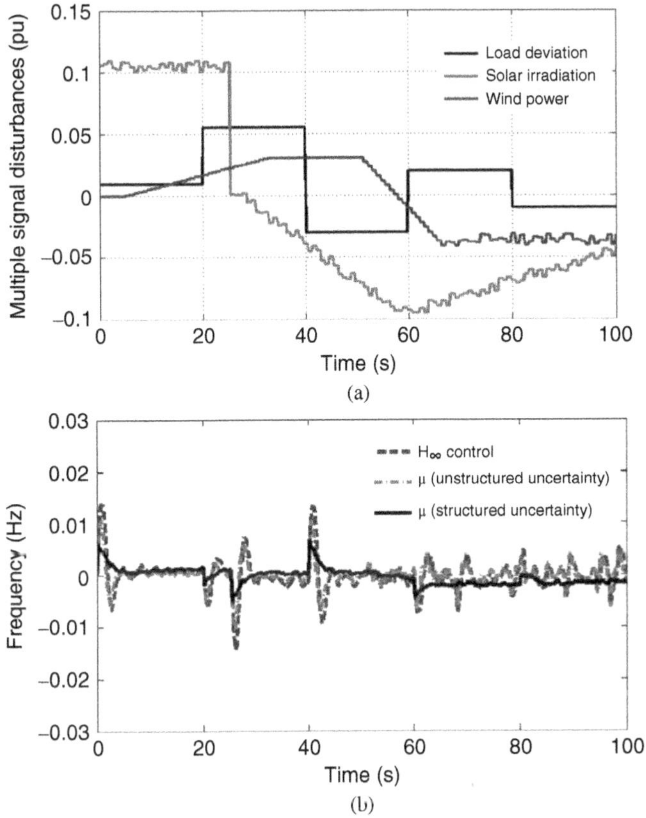

Figure 9.21 System response for nonstationary fluctuations: (a) multiple disturbances in load, wind speed, and solar irradiation and (b) MG output frequency.

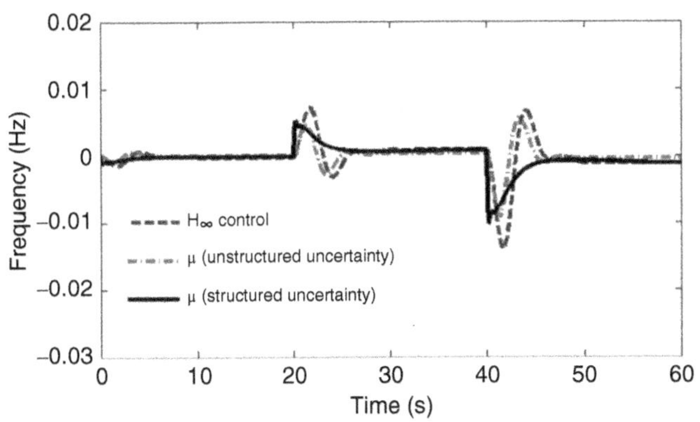

Figure 9.22 MG output frequency in the presence of 50% uncertainty in H and D parameters and disturbance signal in Fig. 9.20a.

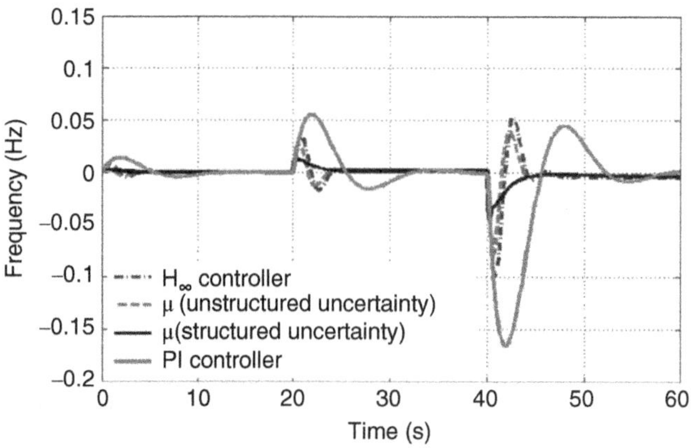

Figure 9.23 Frequency response comparison for the proposed robust control design and conventional PI control design.

Here, to investigate the performance of the proposed robust controllers, a conventional proportional–integral (PI) controller, in which its parameters are first determined by the Ziegler–Nichols method and then are tuned in the simulation environment in the presence of various test examples, is used. The result for the given wind power change pattern (Fig. 9.17a) is shown in Fig. 9.23.

One may say that the Ziegler–Nichols method is not known as a proper method for a well-tuned PI controller design especially for a complex system. To cover this gap and to show the effectiveness of the proposed robust approaches, the proposed H_∞ and μ-control syntheses are compared with two optimal PI controllers. These PI controllers are designed using the MATLAB-based tuning algorithm and internal-model-control-(IMC)-based tuning technique.

9.6.3.1 MATLAB-Based Tuning Algorithm

Manual tuning methods are iterative and time-consuming and may cause some damages to the system hardware in real-world applications. Rule-based methods also have some serious limitations; for example, they are usually applicable for a specific class of dynamic systems, such as stable and/or low-order plants.

For the sake of this challenge, MATLAB-based tuning algorithm (MATLAB PID tuner) provides a simple and friendly program for tuning the control parameters, automatically [41]. The mentioned algorithm follows an optimal solution to meet given design requirements, even for problems that the traditional rule-based methods are not straightly applicable. The automated PI/PID tuning workflow involves the following:

- Identifying plant model from input–output test data

- Modeling controllers in MATLAB using controller objects or in Simulink using controller blocks
- Automatically tuning the controller gains and then fine-tuning the design, interactively
- Tuning multiple controllers in batch mode
- Tuning single-input single-output PI/PID controllers as well as multiloop PI/PID controller architectures.

Closed-loop stability, adequate performance, and adequate robustness are known as typical tuning objectives for the mentioned algorithm in MATLAB software [41]. The closed-loop stability means that the closed-loop system output remains bounded for bounded input, and the adequate performance means that the closed-loop system tracks reference changes and suppresses disturbances as rapidly as possible. It is noteworthy that a larger loop bandwidth (the frequency of unity open-loop gain) causes a faster controller respond to changes in reference and/or disturbance signals. Finally, the adequate robustness means that the loop design has enough gain and phase margins to cover the modeling errors and variations in system parameters/dynamics.

The MATLAB-based tuning algorithm for tuning of PI/PID controllers meets the aforementioned objectives and can provide a desirable balance between the closed-loop system performance and robustness. For a given robustness, the tuning algorithm chooses a controller design that balances two performance measures that are reference tracking and disturbance rejection. One can direct the design to focus on one of these performance measures.

9.6.3.2 Internal Model Control-Based PI Tuning Method

The IMC design strategy can be used as a useful methodology for tuning PI/PID controllers [42]. The IMC approach has two important advantages: first, it explicitly takes into account the model uncertainty, and second, it allows the designer to make a trade-off between the control system performance and robustness to cover the perturbations and modeling errors. The IMC approach is well discussed in [42].

9.6.3.3 Comparison of Results

Comparison of the proposed robust controllers with the designed PI controllers using the MATLAB-based tuning approach and IMC-based tuning technique is illustrated in Figs 9.24–9.29. Response for the given wind power change and load deviation patterns in Figs 9.17a and 9.18a are shown in Figs 9.24 and 9.25, respectively. Figures 9.26–9.28 show the system frequency response to the given disturbances depicted in Figs 9.19a, 9.20a, and 9.21a, respectively. The comparison results in the presence of multiple disturbance signals (Fig. 9.20a) and 50% perturbation in system inertia and damping coefficients are shown in Fig. 9.28. Finally, the system response for simultaneous

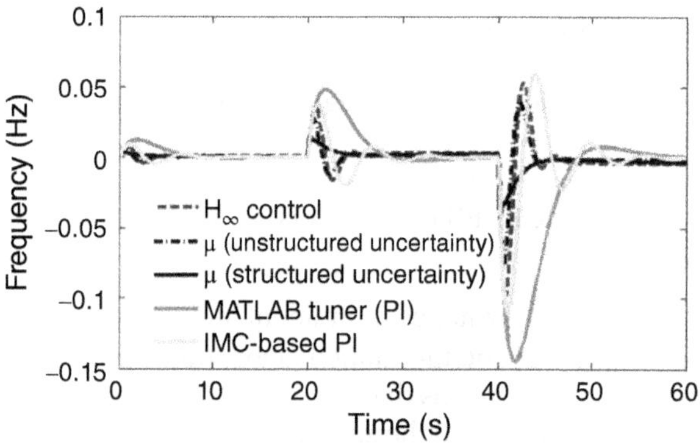

Figure 9.24 Frequency response for the shown wind power change pattern in Fig. 9.17a.

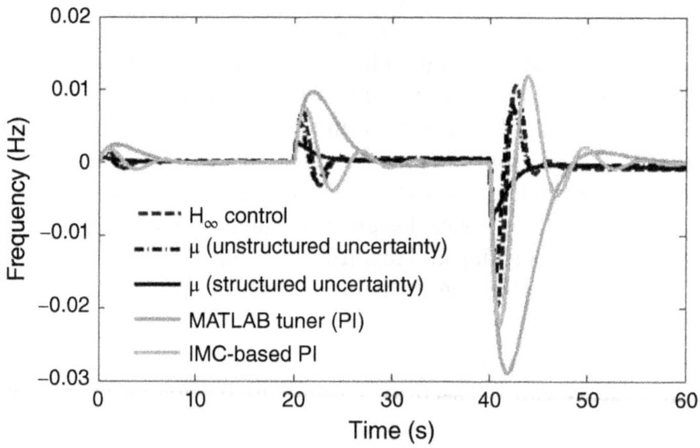

Figure 9.25 Frequency response for the shown multiple load change in Fig. 9.18a.

multiple disturbances in load, wind speed, and solar irradiation (Fig. 9.21a) is shown in Fig. 9.29.

As demonstrated in Fig. 9.22, for this test scenario, the IMC-based PI controller fails to stabilize the MG frequency. Whereas, in all scenarios, the proposed H_∞ and μ-approaches provide quite better performance.

9.6.3.4 Discussion

Functional complexity, diversity in generation/load, variable nature of RESs, and continuous change in structure (uncertainty) are known as some important characteristics of MGs. As shown in the present work, conventional controls may fail to meet the specified frequency control objective in the MGs. The aforementioned characteristics introduce robust control techniques as

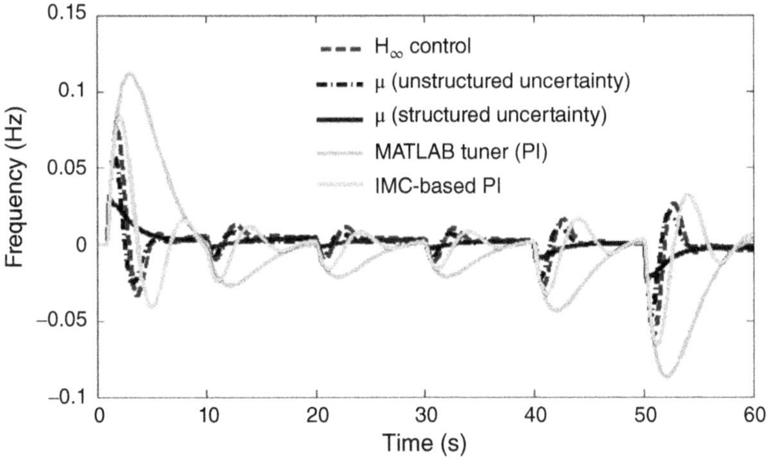

Figure 9.26 Frequency response for the shown solar power change pattern in Fig. 9.19a.

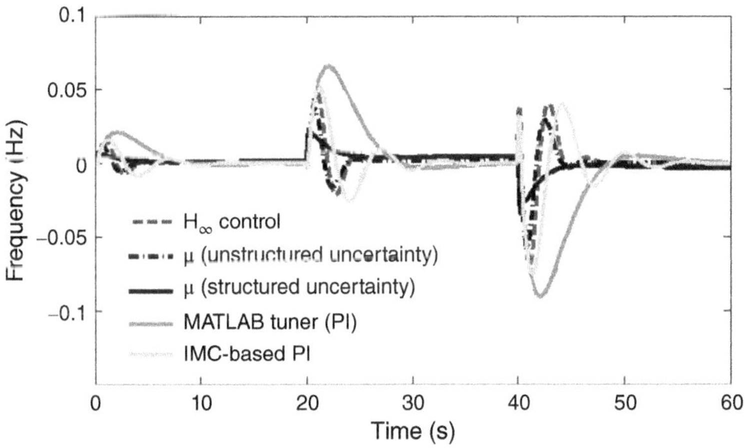

Figure 9.27 Frequency response for the multiple disturbances in load and solar wind power shown in Fig. 9.20a.

powerful and more suitable control tools for stability analysis and control synthesis problems in MGs and modern power grids.

The achievements of the present work are not only limited to applications of H_∞ and μ-synthesis control techniques for secondary frequency control design of MGs. The main outcomes can be summarized as follows:

1) Although the H_∞ and μ-synthesis approaches have already been applied to the power system, as discussed in [36], only few applications are related to

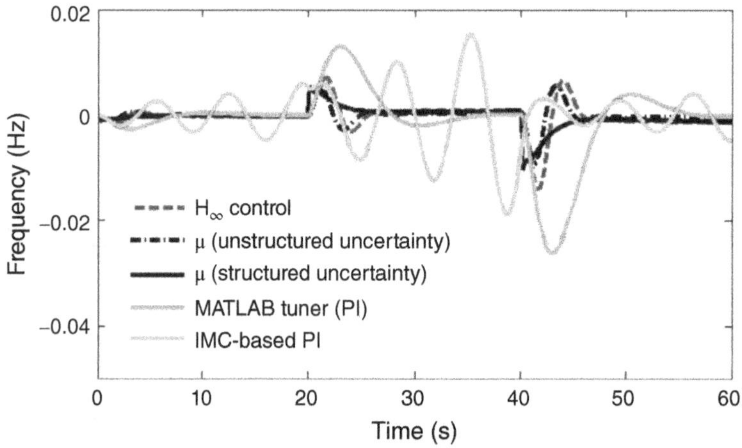

Figure 9.28 Frequency response in the presence of 50% uncertainty in *H* and *D* parameters and disturbance signal in Fig. 9.20a.

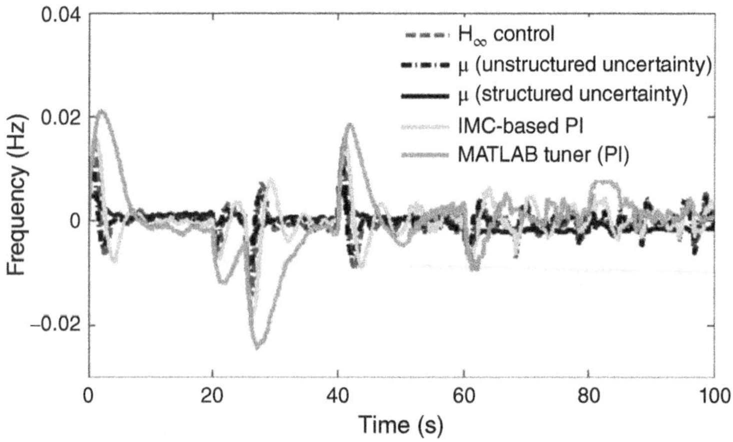

Figure 9.29 Frequency response for multiple disturbances in load, wind speed, and solar irradiation (Fig. 9.21a).

MGs. Dynamics, generation, and load characteristics in MGs may be different from those for conventional power grids. Hence, the application of a robust control theorem cannot be so easy and straightforward. On the other hand, in the previous published works, only one robust method is discussed, while in the present work, a comprehensive/precisely comparison between H_∞ and μ-synthesis methods is given and their advantages and disadvantages for the frequency control issue are clarified.

2) The type of most effective uncertainties in MGs, that is, structured and parametric uncertainties, is investigated. It is shown that the uncertainties in an MG can be modeled as structured/parametric type, and this investigation opens the way to use this work as a reference for further research works on robust MG control. In direction of this achievement, the impact of uncertainty modeling is illustrated, and the superiority of structured/parametric-based control theory such as μ-synthesis for MG control is emphasized.
3) In direction of the aforementioned achievement, the impact of uncertainty modeling is illustrated, and the superiority of structured/parametric-based control theory such as μ-synthesis for MG control is emphasized. It is shown that the structured/parametric-based robust control strategy dramatically increases the regulation performance and control accuracy.
4) Finally, simultaneously considering load change, various power fluctuations, and uncertainties in the MG frequency control problem in a single control framework/formulation via a systematic design approach can also be considered as a significant outcome of the present work.

9.7 Robust Multivariable Microgrid Control Design

In Section 4.7, a voltage-source-converter-(VSC)-based MG case study with two DGs in the islanded operation mode is analyzed as a multivariable dynamic system and the state-space model is obtained. This section presents robust multivariable control synthesis for the mentioned system using the H_∞ and H_2 control theorems.

9.7.1 Uncertainty Determination

Uncertainty formulation is the first step in the RS and performance analysis and control synthesis. In order to assess the RS and performance requirements for power grids, the parametric uncertainties are significant [36]. However, for the sake of H_∞ and H_2 control design, the unstructured uncertainty models are commonly used. Here, both types of uncertainty modeling for ±30% changes in the load resistance are considered.

9.7.1.1 Unstructured Uncertainty Modeling

The unstructured uncertainty can be modeled using the following four steps:

1) Parametric uncertainty determination by considering 30% perturbation in the load resistance as follows:

$$R_L = R_{Ln}(1 + r\delta_r) \tag{9.26}$$

where R_{Ln} is the nominal resistance load, $|\delta_r| \leq 1$ and $r = 0.3$.

2) Converting the parametric uncertainty to the unstructured uncertainty by considering an input multiplicative uncertainty presentation method:

$$\hat{G}(s) = G_n(s)(1 + \Delta_u(s))$$
$$\Delta_u(s) = G_n(s)^{-1}(\hat{G}(s) - G_n(s)) \tag{9.27}$$

where $\hat{G}(s)$ and $G_n(s)$ are the transfer function matrices of the uncertain and nominal systems, respectively, and $\Delta_u(s)$ is the uncertainty transfer function matrix block.

3) Curve fitting for finding the upper bound of all possible $\Delta_u(s)$ curves by a first- or second-order transfer function.
4) Generating the weighting function such that

$$\Delta_u(s) = W_u(s) \cdot \Delta(s), \quad \|\Delta(s)\|_\infty \leq 1 \tag{9.28}$$

$W_u(s)$ is a 4×4 weighting transfer function matrix, all entities of which are the low-order transfer functions that, for the problem at hand, can be obtained as follows:

$$W_u(s) = \begin{pmatrix} \frac{46s + 39450}{s + 852} & \frac{113s + 1.2 \times 10^5}{s + 2295} & \frac{43s + 27690}{s + 620} & \frac{77s + 57280}{s + 1148} \\ \frac{109s + 10^5}{s + 2102} & \frac{46s + 28770}{s + 612} & \frac{78s + 60290}{s + 1175} & \frac{43s + 32210}{s + 726} \\ \frac{46s + 22090}{s + 465} & \frac{80s + 49480}{s + 1035} & \frac{43s + 18380}{s + 413} & \frac{110s + 11040}{s + 2270} \\ \frac{78s + 49090}{s + 1017} & \frac{46s + 3992}{s + 85} & \frac{109s + 10^5}{s + 2049} & \frac{43s + 19050}{s + 429} \end{pmatrix} \tag{9.29}$$

9.7.1.2 Parametric Uncertainty Modeling

Figure 9.30 shows the block diagram of the system that shows only connections between the control inputs and the state variables (the control input–output variables are determined using the relative gain array (RGA) concept discussed in Chapter 4). As mentioned, the only uncertain parameter is load resistance. The load resistance block's inputs are and Δi_{d2} and its outputs are Y_1 and Y_2. Figure 9.31 shows the procedure of the parametric uncertainty modeling, where $\Delta_r = \delta_r I_2$ and the matrix M_{R_L} is obtained as follows:

$$M_{R_L} = \begin{bmatrix} R_{Ln} & 0 & rR_{Ln} & 0 \\ 0 & R_{Ln} & 0 & rR_{Ln} \\ 1 & 0 & 0 & 0 \\ 0 & 1 & 0 & 0 \end{bmatrix} \tag{9.30}$$

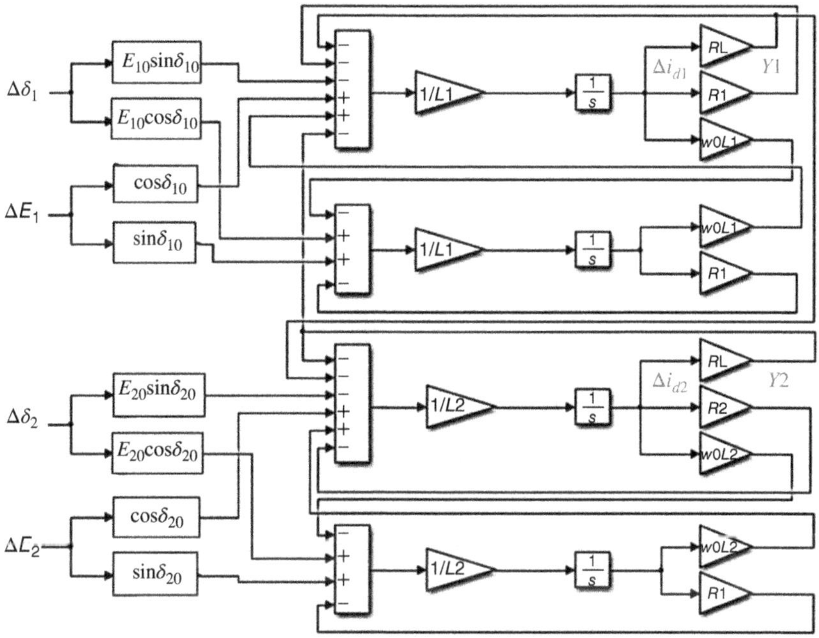

Figure 9.30 System block diagram.

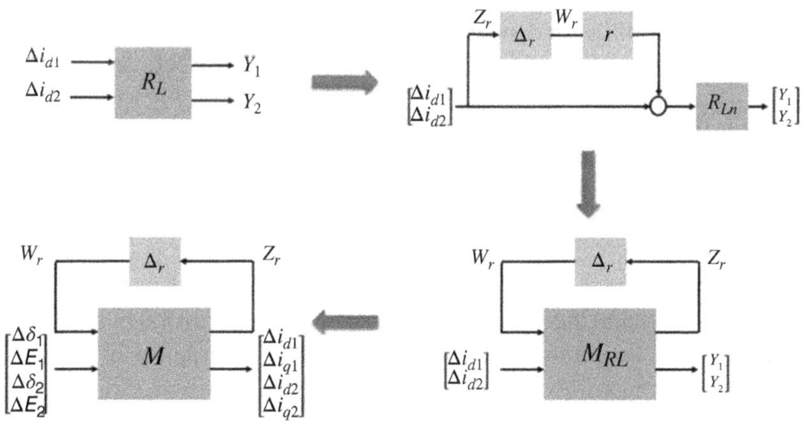

Figure 9.31 Parametric uncertainty block modeling.

The open-loop transfer function matrix M is realized with the following state-space equations:

$$\begin{aligned}\dot{x}(t) &= Ax(t) + B_u u(t) + B_w w(t)\\ y(t) &= x(t) = Cx(t) + D_u u(t) + D_w w(t)\\ z_r(t) &= C_r x(t)\end{aligned} \tag{9.31}$$

where

$$B_u = B, \quad C = \bar{I}_{4\times4}, \quad D_u = \bar{0}_{4\times4}, \quad D_{ur} = \bar{0}_{2\times4}, \quad D_{wr} = \bar{0}_{2\times2}$$

$$B_w = \begin{bmatrix} -\dfrac{rR_{Ln}}{L_1} & 0 & -\dfrac{rR_{Ln}}{L_2} & 0 \\ -\dfrac{rR_{Ln}}{L_1} & 0 & -\dfrac{rR_{Ln}}{L_2} & 0 \end{bmatrix}^T, \quad C_r = \begin{bmatrix} 1 & 0 & 0 & 0 \\ 0 & 0 & 1 & 0 \end{bmatrix}$$

9.7.2 Robust Stability and Performance

In this section, for the given case study, RS and nominal and RP requirements are represented. Robust H_∞ and H_2 controllers are designed, and the results are compared with a sequential PID control system [34].

9.7.2.1 Robust Stability Requirement

In order to use the standard H_∞ and H_2 control design procedures, the closed-loop structure, depicted in Fig. 9.32, is considered. Using the small gain theorem [39], a system is robustly stable if and only if $\|T_{wz}(s)\|_\infty \leq \gamma$ and $\|\Delta(s)\|_\infty \leq 1/\gamma$, for different dynamic and uncertainty test scenarios. Table 9.2 shows the γ and the maximum acceptable uncertainty disk radius for three scenarios. For instance, the maximum infinity-norm of $T_{wz}(s)$ for scenario 2 is 0.12. Hence, the maximum acceptable uncertainty disk radius to have a stable system is 8.33, that is, $\|\Delta\|_\infty$ must be less than 8.33. However, this uncertainty limitation is too conservative due to considering the unstructured uncertainty model. RP indices are also compared in Table 9.2 for the given scenarios.

9.7.2.2 Nominal and Robust Performance Requirement

For the problem at hand, having a desirable tracking, less control signal energy, and disturbance attenuation are considered as the performance objectives.

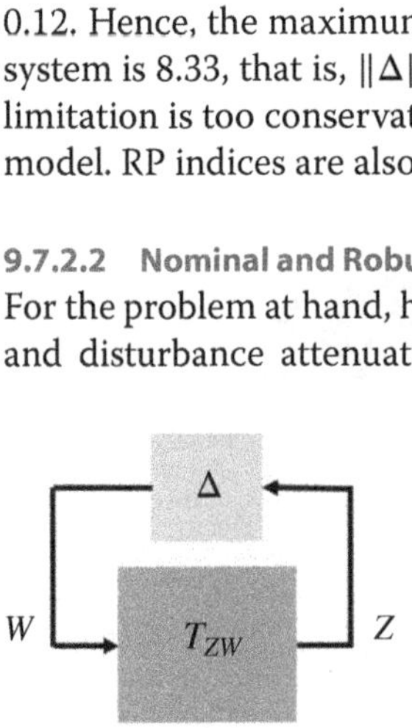

Figure 9.32 Closed-loop system for robust control analysis and synthesis.

Table 9.2 γ and maximum acceptable uncertainty disk radius for scenarios 1–3.

Scenarios	1	2	3
$(X/R)_1$, $(X/R)_2$	0.3, 0.4	1, 1.2	4, 5
γ	0.042	0.12	0.49
Uncertainty radius	23.81	8.33	2.05

Good tracking and disturbance attenuation behave in the same direction, but they should be compromised with the goal of less control signal energy. The next step in the performance assessment is weighting function determination. There is no any exact and straightforward law for finding weighting functions. One may start by an initial weight and continue the tuning process concerning the closed-loop performance in a simulation environment. Some guidelines for tuning of weighting functions are given in [36].

The closed-loop system with selected weighting functions is shown in Fig. 9.33. The W_1, W_2, and W_3 are the weighting functions for obtaining desirable tracking, less control signal energy, and disturbance attenuation, respectively.

In nominal open-loop performance analysis, the uncertainty and less control signal energy goals are not considered. Ignoring the second item is due to the absence of the controller. On the other hand, for the robust open-loop performance objectives, the uncertainty is considered, while the second item is ignored due to the absence of the controller. Then, the open-loop interconnection is found and the closed-loop transfer function matrices for NP $G_{cl\text{-}np}(s)$ and RP $G_{cl\text{-}rp}(s)$ are obtained as follows:

$$\begin{bmatrix} z_1 \\ z_2 \end{bmatrix} = G_{cl-np} \begin{bmatrix} r \\ d \end{bmatrix}, \quad \begin{bmatrix} z \\ z_1 \\ z_2 \end{bmatrix} = G_{cl-rp} \begin{bmatrix} w \\ r \\ d \end{bmatrix} \tag{9.32}$$

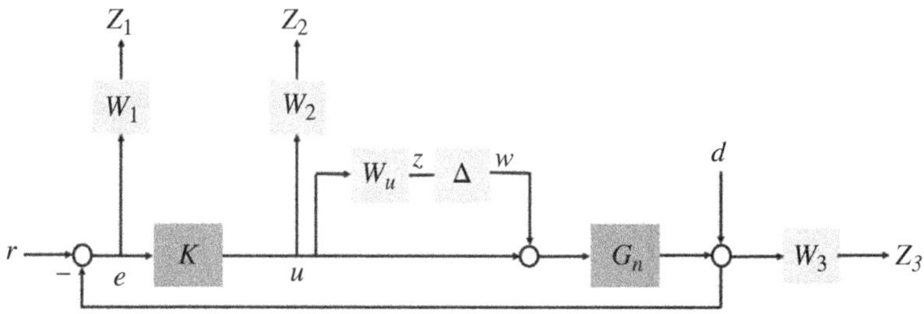

Figure 9.33 Closed-loop system diagram.

9.7.2.3 Robust H$_\infty$ Controller

The closed-loop structure shown in Fig. 9.33 is used for designing of robust H$_\infty$ controller. Here, the weighting functions are chosen as follows:

$$w_1(s) = \frac{2\times 10^5(s+120)}{s+15} \rightarrow W_1(s) = \text{diag}(w_1, w_1, w_1, w_1)$$

$$w_2(s) = \frac{s+2}{s+10} \rightarrow W_2(s) = \text{diag}(w_2, w_2, w_2, w_2)$$

$$w_3(s) = \frac{0.01(s+5)}{s+0.045} \rightarrow W_3(s) = \text{diag}(w_3, w_3, w_3, w_3)$$

In order to design an H$_\infty$ controller satisfying RS and RP requirements, the open-loop interconnection of the transfer function matrices G, W_1, W_2, W_3, and W_u is found according to Fig. 9.33, and then, the controller is designed using robust control tools in MATLAB software. Since the order of the resulting robust controller is high, the residualization and truncation order reduction methods are applied to reduce the order of the robust controller. Figure 9.34 shows the frequency responses of the original and reduced-order controllers using residualization and truncation methods. The orders of original and reduced-order controllers are 216 and 6, respectively. Generally, the residualization method in low frequencies and truncation method in high

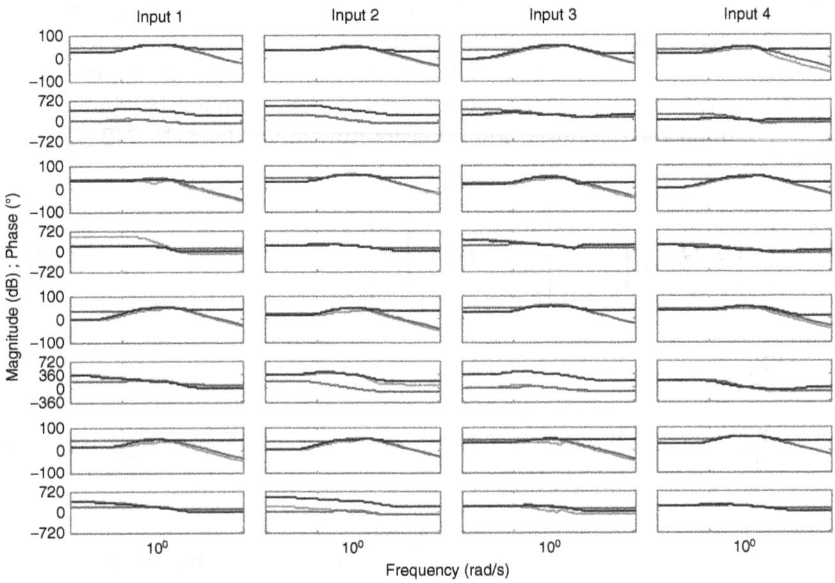

Figure 9.34 Frequency responses of the H$_\infty$ original controller and reduced-order controllers using residualization and truncation methods.

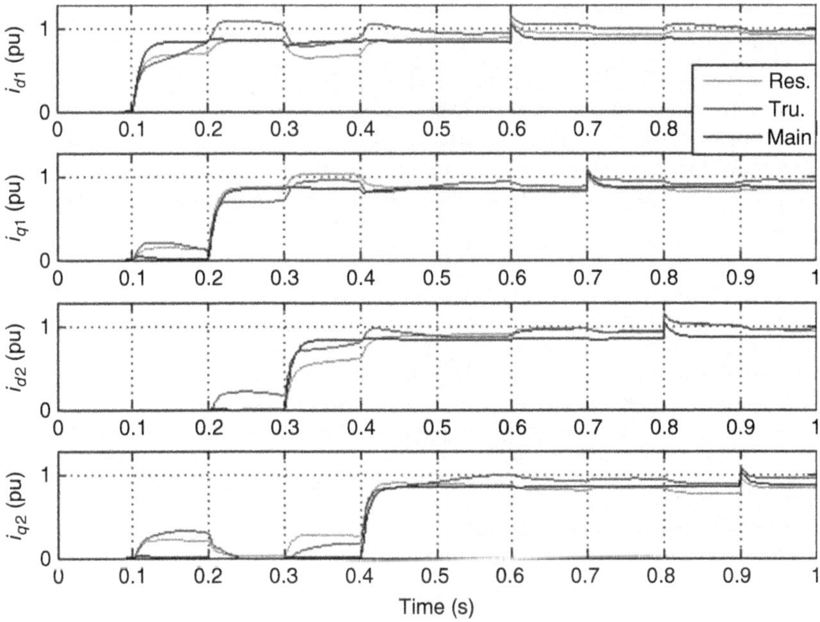

Figure 9.35 Closed-loop responses for the main H_∞ controller and reduced-order controllers using residualization and truncation methods.

frequencies have a response similar to that of the original system. Figure 9.35 shows the time responses of the closed-loop system with the original and reduced-order controllers following a sequence of events. The time of events is given in Table 9.3. According to Fig. 9.35, the residualization method has faster responses, but the steady-state error of the truncation method is less. In addition, the system tracks the reference input well using the original and residualized controllers, and the disturbance effects are less than 10% (0.02 pu) in all times. The interaction between i_d and i_q of the inverter is mutually obvious in the step responses. In fact, when i_d is set to the desired value, i_q has had a certain amount, and when output i_d and i_q of an inverter are set, other inverter has had a certain output. This behavior is due to the operation of MG in islanding mode.

Table 9.3 Time of the events.

Time (s)	**0.1**	**0.2**	**0.3**	**0.4**	**0.6**	**0.7**	**0.8**	**0.9**
Event	1 pu step in i_{d1}	1 pu step in i_{q1}	1 pu step in i_{d2}	1 pu step in i_{q2}	0.2 pu step in dist. 1	0.2 pu step in dist. 2	0.2 pu step in dist. 3	0.2 pu step in dist. 4

9.7.2.4 Robust H_2 Controller

All sections of the H_2 controller design are similar to those of the H_∞ controller design except the weighting functions. In H_2 controller design, strictly proper weighting transfer functions should be applied to have a strictly proper closed-loop transfer function, because the H_2 design method tries to minimize H_2-norm of the desired closed-loop transfer function, whereas H_2-norm of the proper transfer functions is infinity. Then, the weighting functions are obtained as follows:

$$w_1(s) = \frac{1.8 \times 10^6}{s+35} \rightarrow W_1(s) = \text{diag}(w_1, w_1, w_1, w_1)$$

$$w_2(s) = \frac{100}{s+30} \rightarrow W_2(s) = \text{diag}(w_2, w_2, w_2, w_2)$$

$$w_3(s) = \frac{0.05}{s+0.045} \rightarrow W_3(s) = \text{diag}(w_3, w_3, w_3, w_3)$$

Figure 9.36 shows the closed-loop response for the original and reduced-order controllers following the given step events in Table 9.3. The original controller is obtained with order of 368, and reduced-order controllers are of sixth order. As can be seen, the truncation method is not successful in this test scenario. It has an overshoot between 50% and 100% for different outputs, and its settling

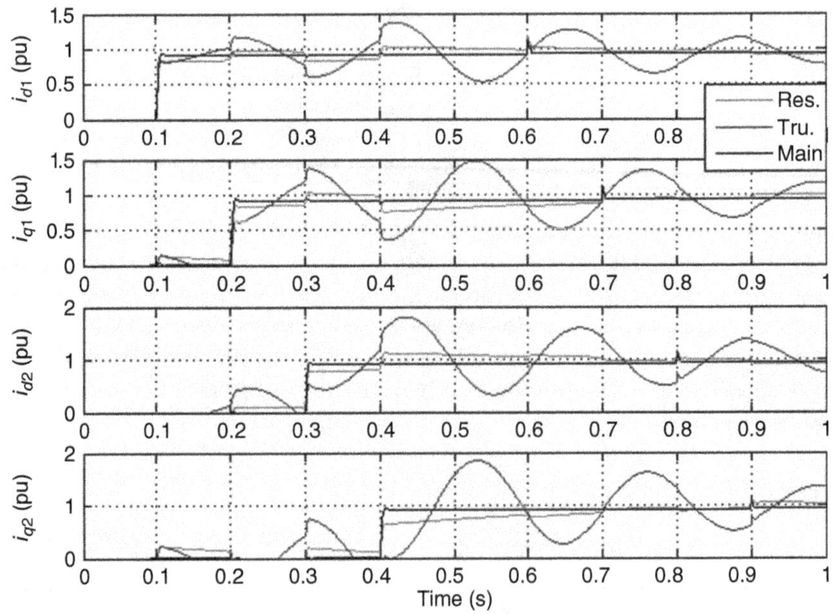

Figure 9.36 Closed-loop for the main H_2 controller and sixth reduced-order controllers using residualization and truncation methods.

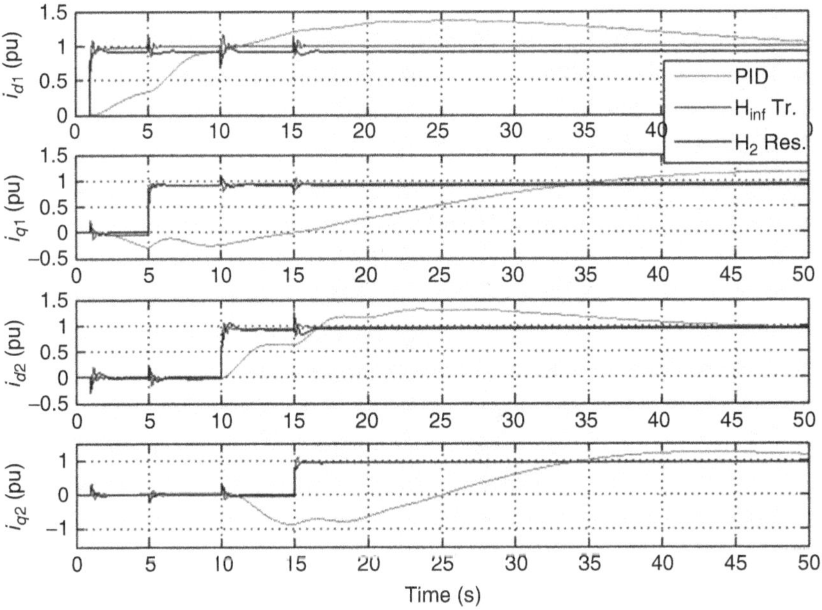

Figure 9.37 Closed-loop response comparison for the sequential tuned PID, robust truncated H_∞, and residualized H_2 controllers.

time is larger than 1 s. However, the residualization method provides faster responses with a minimum possible steady-state error.

The truncated controller cannot track the reference inputs for many test scenarios, while the residualized controller has performed a desirable performance. Figure 9.37 shows the comparison between sequential tuned PID, robust truncated H_∞, and residualized H_2 controllers. PID controller provides a large settling time. The performance of the robust truncated H_∞ and residualized H_2 controllers is much more better than that of the tuned PID controller. However, the robust truncated H_∞ and residualized H_2 controllers do not exhibit an impressive difference between each other. In both cases, the steady-state error and maximum overshoot and undershoot are less than 0.7%, 20%, and 25%, respectively.

9.8 Robust Tuning of VSG Parameters

The VSG mimics the behavior of conventional synchronous generators (SGs). A comprehensive survey on VSGs and the existing topologies is given in Chapter 7 and [7]. Almost all the VSG-based methods are common in virtually emulating the inertia and damping properties of conventional SG via the well-known swing equation, virtually. This section introduces the concept of *virtual inertia* for MGs via an EVSG. The EVSG uses the characteristics of *primary* and *secondary* controls, in addition to the inertia and damping properties. The H_∞ robust control is then used for the sake of optimal tuning of the EVSG parameters. The proposed approach is evaluated via an MG test system [35].

9.8.1 The Extended VSG Dynamics

The relation of frequency deviation and difference between generated active power and load is expressed by the well-known SG swing equation [6] as follows:

$$\Delta P_m(t) - \Delta P_l(t) = 2H\frac{d\Delta f(t)}{dt} + D\Delta f(t) \tag{9.33}$$

where Δf, ΔP_m, ΔP_l, H, and D are frequency deviation, mechanical power change, load change, inertia constant, and load damping coefficient, respectively. Block diagram of this equation in the Laplace domain is realized in Fig. 9.38a. When H and/or D decrease (increase), the rate of frequency deviation and frequency deviation in stable state will increase (decrease), respectively.

In the conventional SGs, frequency regulation is performed by combination of three blocks: (i) rotating mass/rotor (by providing inertia and damping properties), (ii) primary control loop via the droop characteristic (as governor natural response), and (iii) secondary control loop (via an integral function). Here, to improve the performance of the basic VSG, which is described in

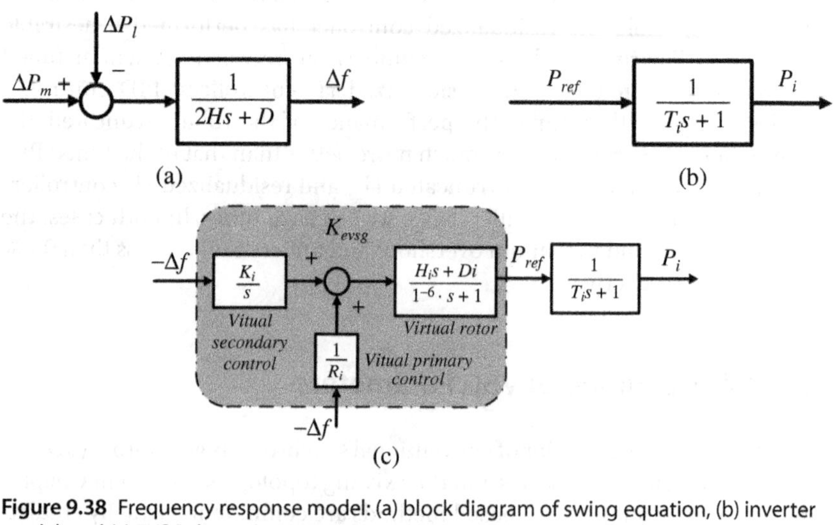

Figure 9.38 Frequency response model: (a) block diagram of swing equation, (b) inverter model, and (c) EVSG dynamics.

Chapter 7, in addition to emulating inertia and damping factor (9.33), the droop characteristic (in the primary regulation loop) and the integral function (in the secondary frequency regulation) are also emulated [43].

Generally, the inverters have two separate operation modes, acting as a current source or as a voltage source. Given the value of line/filter parameters, the output voltage and frequency, as well as real and reactive powers of the inverter, can be controlled using local feedback applied to the inverter. Therefore, inverters can follow their reference power with a small time constant; thus, they can be modeled as first-order transfer function, as shown in Fig. 9.38b.

According to the swing equation in (9.33), the power signal reference (P_{ref}) for an inverter emulating the basics SG characteristics can be obtained as

$$P_{ref} = -(H_i s + D_i)\Delta f \tag{9.34}$$

where H_i and D_i are virtual inertia constant and virtual damping factor, respectively. Negative sign in (9.34) shows negative feedback. Here, H_i and D_i are considered as *virtual rotor* parameters. Inverter can be considered as an SG with small time constant. Therefore, similar primary and secondary control loops can be virtually designed for inverters. Primary control, as initial governor response, and secondary control loop gains can be expressed as

$$\frac{1}{R_i} + \frac{K_i}{s} \tag{9.35}$$

where R_i and K_i are virtual droop characteristics and virtual integrator gain, respectively.

Combining the *virtual rotor* and the *virtual primary and secondary* controllers, the inverter behaves as a controlled SG that is expressed as an EVSG. The EVSG dynamics are shown in Fig. 9.38c. As shown in this figure, H_i, D_i, R_i, and K_i are *virtual parameters* of the EVSG unit (K_{evsg}).

9.8.2 Case Study and H$_\infty$ Control Synthesis

9.8.2.1 Case Study

Here, the MG described in [19] is considered as the case study. The MG test system consists of conventional DEG, PV systems, WTG, FC system, battery energy storage system (BESS), and flywheel energy storage system (FESS). DGs are connected to the MG by power electronic interfaces, that is, inverters. Inverters are used for synchronization with AC sources, that is, DEG and WTG, and to convert the DC output of sources such as PVs, FC, and energy storage devices. Here, the energy source for EVSG is provided by a PV system [35]. A simplified frequency response model for this case study is shown in Fig. 9.39, and the electrical parameters of the system can be found in [19].

The MG case study is controlled optimally and robustly in the presence of uncertainty and disturbances using the EVSG with optimal parameters. In this

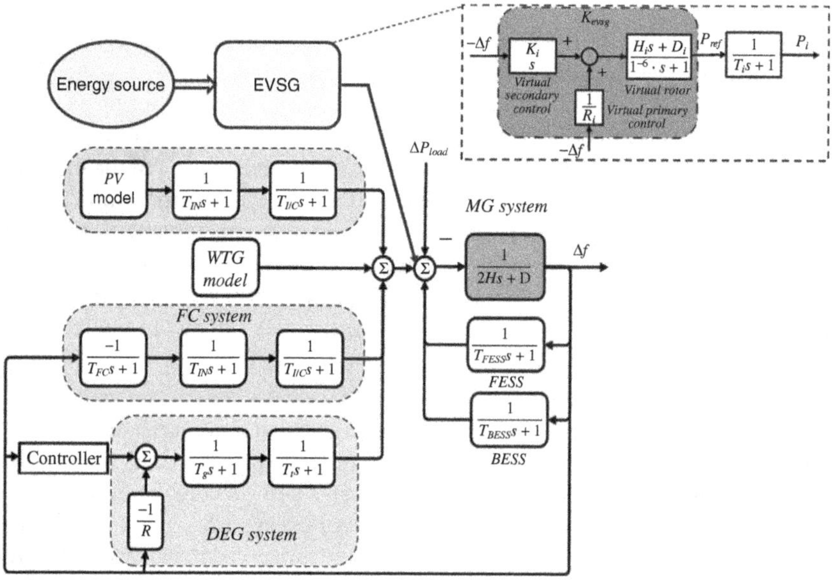

Figure 9.39 Frequency response model of the MG test system, including the EVSG.

section, an H_∞ robust controller is designed in order to calculate the RP index, for optimal tuning of the EVSG parameters in the next step.

9.8.2.2 H_∞ Control Synthesis

9.8.2.2.1 Uncertainty Modeling Uncertainty as difference between actual systems and its mathematical models can be classified as "structured uncertainty" and "unstructured uncertainty." The "parametric uncertainty" can be presented in the structured form using LFT. For the MG system in this study, H and D are considered as uncertain parameters (with $\pm 90\%$ changes). This uncertainty is extracted from the system in the form of structured uncertainty and arranged in a standard configuration of the upper LFT as shown in Fig. 9.40.

9.8.2.2.2 Performance Consideration The main goal of the frequency controller is to eliminate the frequency deviation and, accordingly, regulate the system frequency in the presence of disturbances. In this study, natural changes in wind speed (ΔP_W) and solar radiation (ΔP_S), as well as the load profile (ΔP_L), are the sources of disturbances. It should be noted that a trade-off between the minimum control action signal energy and disturbance rejection should be taken into account. Therefore, the whole system interconnections can be represented as shown in Fig. 9.40. To build the system and subsystems, the *sysic* command

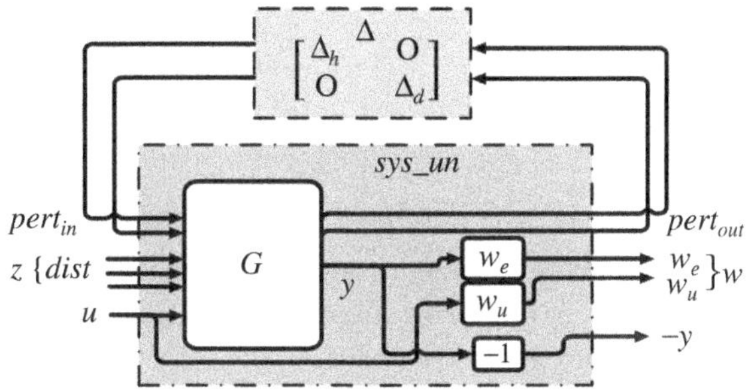

Figure 9.40 Standard structure of case study MG for H_∞ synthesis with uncertain block.

in MATLAB is used. The W_e and W_u are weighting functions that shaped and normalized the output and control signals. The effects of weighting functions in controller design are essential, but as mentioned, there is no analytical method to determine the weighting functions optimally [6]. In this study, the weighting functions are selected following a trial-and-error process as follows:

$$w_e = \frac{0.005s^3 + 0.05s^2 + 50s + 125}{s^3 + 100s^2 + 300s + 1}, \quad w_u = \frac{5s + 50}{s^2 + 2000s + 1.7e4}$$

9.8.2.2.3 *Optimal H_∞ Controller Design* The H_∞ control is an optimization control problem, which minimizes the H_∞-norm of transfer function (T_{wz}) from the disturbance signals (w) to the controlled output signals (y) in the closed-loop system. To solve this optimization problem, it is usually sufficient to find a stabilizing controller such that the H_∞-norm of T_{wz} satisfies

$$\|T_{zw}(s)\|_\infty = \|F_L(G, K)\|_\infty < 1 \tag{9.36}$$

For this purpose, the *hinfsyn* command in MATLAB can be used to design the H_∞ controller. Here, the order of designed controller is 22.

9.8.2.2.4 *Nominal Stability and Performance* The NS is satisfied because the closed-loop system T_{wz} is internally stable for the designed K_{hinf}. For evaluating the NP, the controller must satisfy the performance criterion for all frequencies as follows:

$$\|T_{zw}(s)\|_\infty = \sup_{w \in R} \overline{\delta}\,(T_{zw}(jw)) \leq 1 \tag{9.37}$$

9.8.2.2.5 *Closed-Loop Robust Stability and Performance* The RS and RP are satisfied for the closed-loop system if and only if the closed-loop system can be internally stable and meet the relevant performance criterion, respectively, for all possible plants in the presence of uncertainty. For the structured uncertainty, the RS and RP should be satisfied as described in the following.

1) *Robust stability:* Consider $M-\Delta$ configuration (Fig. 9.3), with the following M and Δ:

$$\Delta^* = \{\Delta(\cdot) \in RH_\infty, \Delta(s_0) \in \Delta^*, s_0 \in \mathbb{C}, Re(s_0) \geq 0\} \tag{9.38}$$

$$M = \begin{bmatrix} M_{11} & M_{12} \\ M_{21} & M_{22} \end{bmatrix} \tag{9.39}$$

where M is the system including the controller and Δ is uncertainty block that is expressed in general form of (9.11). As mentioned in Section 9.3, the closed-loop of system for all $\Delta \in \Delta^*$ and $\|\Delta\|_\infty \leq 1$ is internally stable if and only if the nominal system can be stable and

$$\sup_{w \in R} \mu_\Delta(M_{11}) \leq 1 \tag{9.40}$$

2) *Robust performance:* Consider a generalized Δ as follows:

$$\Delta_T = \left\{ \begin{bmatrix} \Delta_u & 0 \\ 0 & \Delta_p \end{bmatrix}, \quad \Delta_u \in \Delta^*, \ \Delta_p \in \mathbb{C} \right\} \tag{9.41}$$

where Δ_u and Δ_p (apocryphal uncertain block) represent uncertainty and performance requirements, respectively. The RP of the closed-loop system for all $\Delta \in \Delta_T$ and $\|\Delta\|_\infty \leq 1$ is guaranteed if and only if

$$\sup_{w \in R} \mu_\Delta(M) \leq 1 \tag{9.42}$$

According to this aforementioned theorem, the upper bound for RS and RP must be less than unity at all frequencies, as shown in Fig. 9.41. Therefore, the RS and RP (for ±90% change in H and D) are satisfied. The *mu* command in MATLAB is used to determine the mentioned upper and lower bounds. The frequency responses of the system following 0.1 pu load step change for the cases without any controller, with fine-tuned PI controller, under H_∞ robust controller (K_{hinf}), are shown in Fig. 9.42.

9.8.3 Robust Tuning of Extended VSG Parameters

Conventionally, the parameters of VSGs are defined using the parameters of conventional SGs or based on the trial and error method. A robust control method is used here for optimal tuning of these parameters. The basis of this method is that the frequency response of K_{evsg} matched with the frequency response of K_{hinf}. Notice that the structure of EVSG controller is simpler than a high-order robust controller, and this is a great advantage. Once both systems are minimum phase, by matching magnitude of the systems, phases will be matched automatically. Since the K_{hinf} and K_{evsg} are minimum phase, the magnitude of two controllers should be matched in the operating frequency range. Therefore, the purpose is minimization of frequency response magnitude difference of two controllers as follows:

$$\min_{w \in [w1, w2]} e = \| |K_{vsg}(jw)| - |K_{hinf2}(jw)| \|_2 \tag{9.43}$$

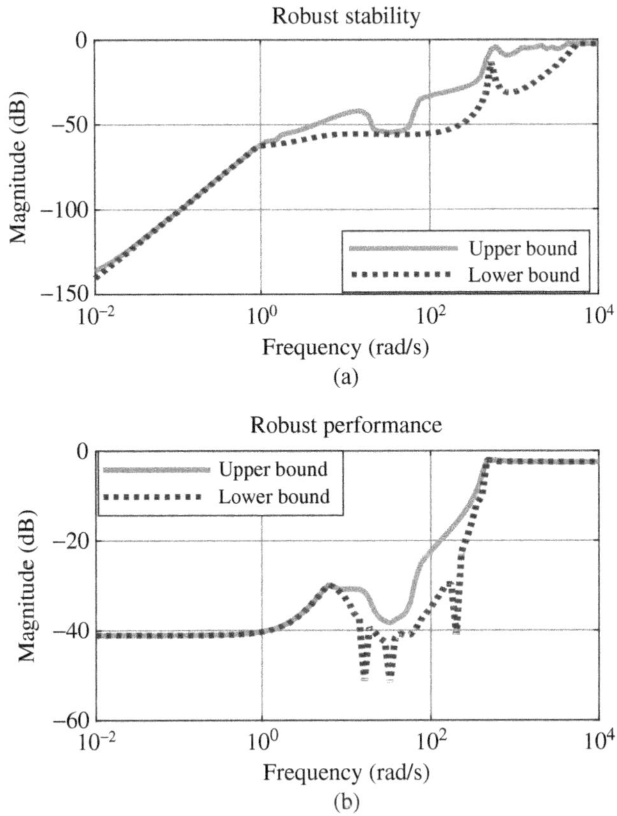

Figure 9.41 μ-Synthesis: (a) robust stability and (b) robust performance (upper bound of μ must be less than 1 at all frequencies).

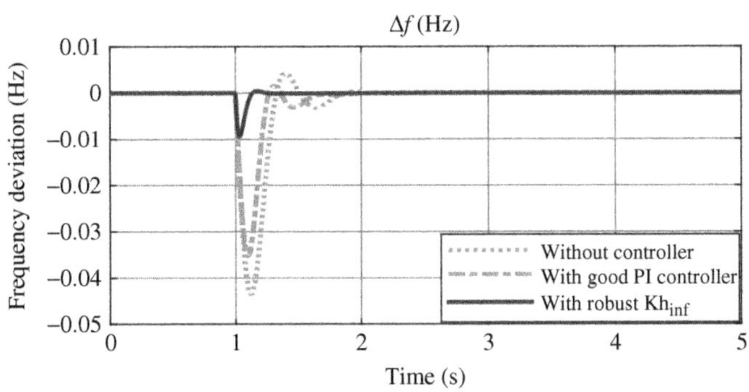

Figure 9.42 Frequency response of the test system following of the 0.1 pu step change in ΔP_L.

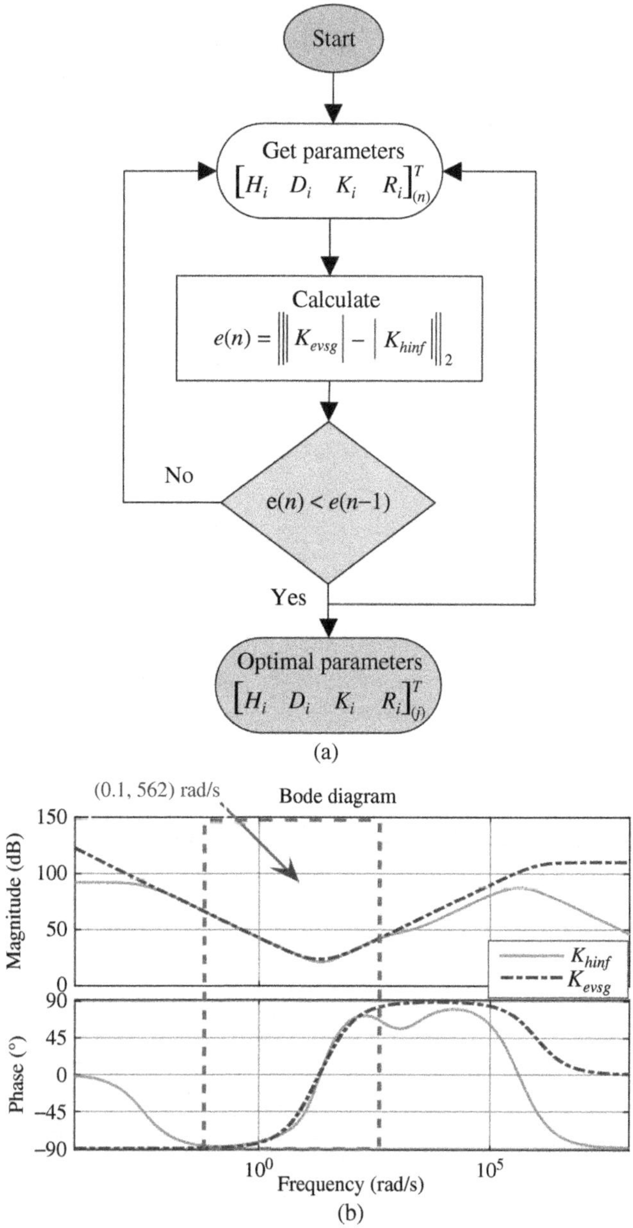

Figure 9.43 Robust tuning of EVSG: (a) proposed algorithm and (b) Bode diagram of K_{evsg} and K_{hinf}.

For this minimization, a simple method is proposed, whose algorithm is shown in Fig. 9.43a. Here, first, a large range is considered for parameters and then this range is limited to achieve a desirable accuracy. For $w \in (0.1, 562)\,\text{rad/s}$, and using the proposed method, the optimal virtual parameters are obtained as follows:

$$H_i = 0.9, \quad D_i = 10.4, \quad K_i = 12.9, \quad R_i = 2.8$$

Bode diagrams of K_{evsg} and K_{hinf} for the calculated optimal parameters are plotted in Fig. 9.43b. As depicted in this figure, the Bode diagrams of two controllers are well matched for the system operating frequency range.

9.8.4 Simulation Results

It is noteworthy that the design of K_{hinf} controller is only performed for the optimal tuning of EVSG system, that is, K_{evsg}. The presented MG case study

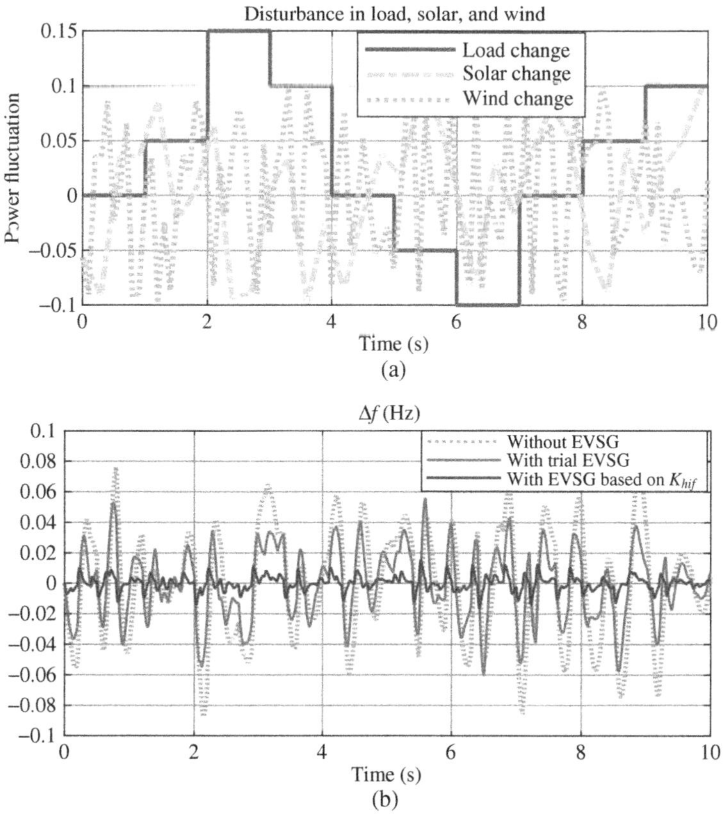

Figure 9.44 System response for simultaneous variation in load and renewable power generation: (a) disturbance pattern and (b) frequency deviation.

is used to verify the effectiveness of the proposed approach. The time-domain simulation results show the frequency response of the system in the presence of disturbances (ΔP_L, ΔP_W, and ΔP_S) and high perturbation of H and D parameters as system uncertainty.

As first scenario, the fluctuation in solar irradiation, wind speed, and load are considered. Simultaneous and random changes of ΔP_S, ΔP_W, and ΔP_L (shown in Fig. 9.44a) are applied to the system, and the results are shown in Fig. 9.44b. As second scenario, the performance of the proposed control method under system uncertainty is examined in this study. For this purpose, in addition to applying 0.1 pu step load change, the H and D parameters are decreased up to 75% and 90%. The results are shown in Fig. 9.45a,b, respectively. As can be seen, the results verify the effectiveness of the proposed method in the presence of severe disturbance and high parameter perturbations.

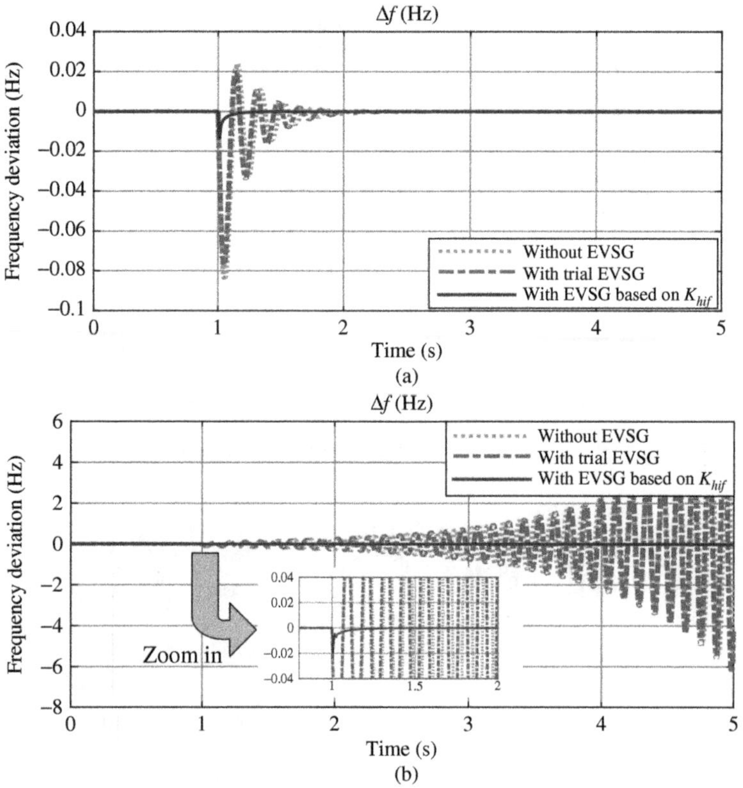

Figure 9.45 System response for simultaneous step load change and parameters perturbation: (a) 75% decrease in H and D and (b) 90% decrease in H and D.

9.9 Summary

In this chapter, robust H_∞ and μ-synthesis (via *D-K* iteration) methods were used for secondary frequency control design problem in an islanded MG. The H_∞ and μ-controllers were designed in a way to reduce the effects of ΔP_{Wind}, ΔP_{φ}, and ΔP_{Load} disturbances and dynamic perturbations. It was shown that in the case of using the structured uncertainty (μ-approach), the obtained controller demonstrates better performance. Time-domain simulation results showed that the proposed controllers can balance the power generation and load properly and regulate the MG frequency effectively. Due to the use of structured uncertainties, μ-controller has better performance than the H_∞ controller. The order of original robust controllers was reduced. In comparison with optimal PI controllers, the application of robust reduced-order controllers dramatically improved the MG frequency control performance.

In the next step, robust H_∞ and H_2 controllers designed for an islanded MG with two voltage-source-converter-based DGs. The performances of the reduced-order controllers were assessed, and it was shown that the reduced order H_∞ controller (using the truncation method) and the reduced-order H_2 controller (using the residualization method) have much better performance than the conventional PID controller. Finally, a new VSG scheme called extended VSG was proposed for the sake of frequency control in MGs. The H_∞ control concept is then used to tune the extended VSG parameters.

References

1 Al-Saedi, W., Lachowicz, S.W., Habibi, D., and Bass, O. (2012) Power quality enhancement in autonomous microgrid operation using particle swarm optimization. *International Journal of Electrical Power & Energy Systems*, **42** (1), 139–149.

2 Dekker, J., Nthontho, M., Chowdhury, S., and Chowdhury, S.P. (2012) Economic analysis of PV/diesel hybrid power systems in different climatic zones of South Africa. *International Journal of Electrical Power & Energy Systems*, **40** (1), 104–112.

3 Kamalapur, G.D. and Udaykumar, R.Y. (2011) Rural electrification in India and feasibility of photovoltaic solar home systems. *International Journal of Electrical Power & Energy Systems*, **33** (3), 594–599.

4 Alzola, J.A., Vechiu, I., Camblong, H. *et al.* (2009) Microgrids project, part 2: design of an electrification kit with high content of renewable energy sources in Senegal. *Renewable Energy*, **34** (10), 2151–2159.

5 Gu, W., Liu, W., Wu, Z. *et al.* (2013) Cooperative control to enhance the frequency stability of islanded microgrids with DFIG-SMES. *Energies*, **6** (8), 3951–3971.

6 Şerban, I. and Marinescu, C. (2011) Aggregate load-frequency control of a wind-hydro autonomous microgrid. *Renewable Energy*, **36** (12), 3345–3354.

7 Goya, T., Uchida, K., Kinjyo, Y. *et al.* (2011) Coordinated control of energy storage system and diesel generator in isolated power system. *International Journal of Emerging Electric Power Systems*, **12** (1), 1–22.

8 Bevrani, H., Watanabe, M., and Mitani, Y. (2014) *Power System Monitoring and Control*, Wiley-IEEE Press, New York, NY.

9 Adhikari, S. and Li, F. (2014) Coordinated Vf and PQ control of solar photovoltaic generators with MPPT and battery storage in microgrids. *IEEE Transactions on Smart Grid*, **5** (3), 1270–1281.

10 Guerrero, J.M., Loh, P.C., Lee, T.-L., and Chandorkar, M. (2013) Advanced control architectures for intelligent microgrids – Part II: Power quality, energy storage, and AC/DC microgrids. *IEEE Transactions on Industrial Electronics*, **60** (4), 1263–1270.

11 Guerrero, J.M., Vasquez, J.C., Matas, J. *et al.* (2011) Hierarchical control of droop-controlled AC and DC microgrids – a general approach toward standardization. *IEEE Transactions on Industrial Electronics*, **58** (1), 158–172.

12 Bevrani, H., Watanabe, M., and Mitani, Y. (2012) Microgrid controls, in *Standard Handbook for Electrical Engineers*, McGraw-Hill, New York, pp. 16.160–16.176.

13 Ngamroo, I. (2012) Application of electrolyzer to alleviate power fluctuation in a stand alone microgrid based on an optimal fuzzy PID control. *International Journal of Electrical Power & Energy Systems*, **43** (1), 969–976.

14 Olivares, D.E., Mehrizi-Sani, A., Etemadi, A.H. *et al.* (2014) Trends in microgrid control. *IEEE Transactions on Smart Grid*, **5** (4), 1905–1919.

15 Etemadi, A.H., Davison, E.J., and Iravani, R. (2012) A decentralized robust control strategy for multi-DER microgrids – Part I: Fundamental concepts. *IEEE Transactions on Power Delivery*, **27** (4), 1843–1853.

16 Mehrizi-Sani, A. and Iravani, R. (2010) Potential-function based control of a microgrid in islanded and grid-connected modes. *IEEE Transactions on Power Systems*, **25** (4), 1883–1891.

17 Shafiee, Q., Guerrero, J.M., and Vasquez, J.C. (2014) Distributed secondary control for islanded microgrid; a novel approach. *IEEE Transactions on Power Electronics*, **29** (2), 1018–1031.

18 Bidram, A., Davoudi, A., Lewis, F.L., and Guerrero, J.M. (2013) Distributed cooperative secondary control of microgrids using feedback linearization. *IEEE Transactions on Power Systems*, **28** (3), 3462–3470.

19 Bevrani, H., Habibi, F., Babahajyani, P. *et al.* (2012) Intelligent frequency control in an ac microgrid: online PSO-based fuzzy tuning approach. *IEEE Transactions on Smart Grid*, **3** (4), 1935–1944.

20 Vachirasricirikul, S. and Ngamroo, I. (2012) Robust controller design of microturbine and electrolyzer for frequency stabilization in a microgrid system with plug-in hybrid electric vehicles. *International Journal of Electrical Power & Energy Systems*, **43**, 804–811.

21 Vachirasricirikul, S. and Ngamroo, I. (2011) Robust controller design of heat pump and plug-in hybrid electric vehicle for frequency control in a smart microgrid based on specified-structure mixed H_2/H_∞ control technique. *Applied Energy*, **88**, 3860–3868.

22 Li, X., Song, Y.J., and Han, S.B. (2008) Frequency control in micro-grid power system combined with electrolyzer system and fuzzy PI controller. *Journal of Power Sources*, **180**, 468–475.

23 Packard, A. and Doyle, J. (1993) The complex structured singular value. *Automatica*, **29** (1), 71–109.

24 Singh, V.P., Mohanty, S.R., Kishor, N., and Ray, P.K. (2013) Robust H-infinity load frequency control in hybrid distributed generation system. *International Journal of Electrical Power & Energy Systems*, **46**, 294–305.

25 Vachirasricirikul, S. and Ngamroo, I. (2014) Robust LFC in a smart grid with wind power penetration by coordinated V2G control and frequency controller. *IEEE Transactions on Smart Grid*, **5** (1), 371–380.

26 Yang, S., Lei, Q., Peng, F.Z., and Qian, Z. (2011) A robust control scheme for grid-connected voltage-source inverters. *IEEE Transactions on Industrial Electronics*, **58** (1), 202–212.

27 Han, Y., Young, P.M., Jain, A., and Zimmerle, D. (2015) Robust control for microgrid frequency deviation reduction with attached storage system. *IEEE Transactions on Smart Grid*, **6** (2), 557–565.

28 Kahrobaeian, A. and Mohamed, Y.A.I. (2013) Direct single-loop μ-synthesis voltage control for suppression of multiple resonances in microgrids with power-factor correction capacitors. *IEEE Transactions on Smart Grid*, **4** (2), 1151–1161.

29 Li, P., Yin, Z., and Li, Y. (2014) *The Realization of Flexible Photovoltaic Power Grid-Connection μ-Synthesis Robust Control in Microgrid*. PES General Meeting Conference & Exposition, 2014 IEEE, pp. 1–5.

30 Hossain, M., Pota, H., Mahmud, M., and Aldeen, M. (2015) Robust control for power sharing in microgrids with low-inertia wind and PV generators. *IEEE Transactions on Sustainable Energy*, **6** (3), 1067–1077.

31 Babazadeh, M. and Karimi, H. (2013) A robust two-degree-of-freedom control strategy for an islanded microgrid. *IEEE Transactions on Power Delivery*, **28** (3), 1339–1347.

32 Lee, D.-J. and Wang, L. (2008) Small-signal stability analysis of an autonomous hybrid renewable energy power generation/energy storage system – Part I: Time-domain simulations. *IEEE Transactions on Energy Conversion*, **23** (1), 311–320.

33 Bevrani, H., Feizi, M.R., and Attaee, S. (2016) Robust frequency control in an islanded microgrid: H_∞ and μ-synthesis approaches. *IEEE Transactions on Smart Grid*, 7 (2), 706–717.

34 Naderi, M., Khayat, Y., Batmani, Y., and Bevrani, H. (2016) *Multivariable Control Based Modeling, Analysis and Robust Control Synthesis for an Islanded Microgrid*. 3rd International Conference on Power and Energy System Engineering (CPESE 2016), Kitakyushu, Japan, September.

35 Fathi, A., Shafiee, Q., and Bevrani, H. (2016) *Robust Frequency Control of Islanded Microgrids Using an Extended Virtual Synchronous Generator*. 1st International Conference on New Research Achievements in Electrical & Computer Eng., Tehran, May 12.

36 Bevrani, H. (2014) *Robust Power System Frequency Control*, 2nd edn, Springer.

37 Tang, X. *et al.* (2015) A novel frequency and voltage control method for islanded microgrid based on multienergy storage. *IEEE Transactions on Smart Grid*, 7 (1), 410–419.

38 Majumder, R. (2013) Some aspects of stability in microgrids. *IEEE Transactions on Power Systems*, **28** (3), 3243–3252.

39 Gu, D.-W., Petkov, P.H., and Konstantinov, M.M. (2005) *Robust Control Design with MATLAB®*, Springer-Verlag, London.

40 Pan, I. and Das, S. (2015) Kriging based surrogate modeling for fractional order control of microgrids. *IEEE Transactions on Smart Grid*, **6** (1), 36–44.

41 Available at: http://www.mathworks.com/help/control/getstart/pid-tuning-algorithm.html.

42 Morari, M. and Zafirion, E. (1989) *Robust Process Control*, Prentice-Hall, Englewood Cliffs, NJ.

43 Mohammadi, F. (2014) A new control strategy to produce virtual inertia. MSc thesis. University of Kurdistan, Kurdistan, Iran.

10

Intelligent Microgrid Operation and Control

Recently, intelligent system applications have received increasing attention in microgrid (MG) operation, planning, control, and management. Numerous research works indicate the applicability of intelligent approaches on the MGs. While many of these approaches are still under investigation, however, due to significant advances in metering, computing, and communication technologies, there already exist a number of practical implementations of intelligent systems in some real MGs across the world.

There exist some limitations for the application of conventional mathematically and analytically based methodologies for MG systems with many uncertainties and intermittency in both configuration and component, on the one hand, and the simplicity, smartness, and flexibility of artificial intelligent approaches, on the other, attract increasing interest in the applications of intelligent technologies such as knowledge-based expert systems, fuzzy systems, artificial neural networks (ANNs), genetic algorithms, search-based evolutionary algorithms, and other intelligent technologies. Recent relevant research works on MGs indicate that more emphasis has been placed on the combined usage of intelligent technologies for further improvement of the operation, control, and management.

This chapter is focused on some application examples of intelligent systems in MG operation and control. After an introduction, the most important intelligent control technologies for application in MGs are addressed. ANNs-based power and load forecasting examples in MGs are briefly described. Then, several examples of intelligence-based frequency and voltage control synthesis using fuzzy logic, particle swarm optimization (PSO), and multiagent systems (MASs) are given. The explained examples are supplemented by simulations and real-time experimental tests.

Microgrid Dynamics and Control, First Edition. Hassan Bevrani, Bruno Francois, and Toshifumi Ise.

10.1 Introduction

Over the years, many applications have been proposed in the literature to demonstrate the advantages of intelligent systems over conventional systems. A number of intelligent systems have been currently utilized in the power grids. Some application areas of intelligent technologies in Japanese power system utilities are reported in [1]. Recently, similar interests have also appeared in the area of MG planning, operation, control, and management.

The MG stability and control synthesis/analysis have been augmented with valuable research contributions in the past decade. Significant improvements have appeared in the area of control synthesis to cope with uncertainties, various load characteristics, changing structure, and integration of high rate of renewable energy sources (RESs) and energy storage systems (ESSs) [2]. Several analog and digital control schemes using nonlinear and linear optimal/robust, adaptive, and intelligent control techniques have been presented. The most recent advance in the MG control synthesis to tackle the difficulty of using complex/nonlinear MG models and/or insufficient knowledge about the system is the application of intelligent concepts such as ANNs, fuzzy logic, MASs, and evolutionary and heuristic optimization algorithms [3].

The intelligent technology offers many benefits in the area of complex and nonlinear control problems, particularly when the system is operating over an uncertain operating range [4]. Generally, for the sake of control synthesis, nonlinear systems such as MGs are approximated by reduced-order dynamic models, possibly linear, which represent the simplified dominant system characteristics [5–7]. However, these models are only valid within specific operating ranges, and a different model may be required in the case of changing operating mode and conditions. Therefore, more flexible and intelligent approaches are needed.

In recent years, following the advent of modern intelligent methods, such as ANNs, fuzzy logic, MASs, GAs, expert systems, simulated annealing, tabu search, PSO, ant colony optimization, and hybrid intelligent techniques, some new potentials and powerful solutions for MG control synthesis, operation, and energy management have arisen.

Numerous research works are proposed to apply the fuzzy logic to MG operation and control [8–17]. In [8], a fuzzy-based controller is applied to all distributed generators (DGs) in an MG, and the system performance is analyzed by numerical simulations. A distributed control framework for voltage regulation and reactive power sharing is discussed in [9]. The control system regulates the MG voltage and manages the reactive power sharing among the DGs using consensus protocols. A sparse communication network has been used to link the source controllers in order to exchange the information. The fuzzy logic is applied to fine-tuning of the consensus parameters.

In [10], a fuzzy control in addition to a proportional–integral (PI) controller is used to stabilize the rate of a diesel generation. An intelligent fuzzy-based secondary frequency control method is proposed in [11]. For this purpose, a supervisory fuzzy logic controller is developed in parallel with the existing conventional PI control loop. Thus, in case of system dynamic/parameter changes, the PI controller does not need to be retuned, and the supervisory fuzzy logic controller compensates the negative impact of the mentioned changes on the MG frequency. In [12], a fuzzy control is executing a maximum power point tracking control of a photovoltaic (PV) system in an islanded MG to improve the response against rapidly changing weather conditions. A fuzzy-logic-based intelligent generalized droop control for simultaneous frequency and voltage regulation in an AC MG has been developed in [13]. The addressed methodology fixes the dependency of general droop control (GDC) on the MG line parameters.

A fuzzy-logic-based operating current regulation strategy for improving the reliability of superconducting magnetic energy storage (SMES) systems in the MGs has been proposed in [14]. In [15], a methodology has been proposed to develop a fuzzy-logic-based supervisor for energy management and frequency control in a wind-generator-based MG. A robust energy management system (EMS) scheme for an MG based on model predictive control theory as a mathematical framework is presented in [16]. The robust EMS is formulated using a fuzzy prediction interval model as the prediction model. This model allows us to represent both nonlinear dynamic behavior and uncertainty in the available energy from nonconventional energy sources. In [17], the transient stability of MGs during fault and disconnection from the main grid is studied. A fuzzy-based PI controller is used to limit the current deviations of inverter-based DGs. The fuzzy unit adjusts the parameters of the PI controller.

The ANNs have also been widely used in the MG control, operation, and energy management [18, 19]. Reference [18] presents an ANN-based MG central control (ANN-based MGCC) to handle the MG voltage and frequency regulation tasks in an islanded MG. This objective is satisfied using online tuning of secondary controllers. An ANN-based intelligent technique for online tuning of PI parameters in an MG's frequency control loop is addressed in [19]. The impacts of DGs/RESs power fluctuation on the system frequency and physical constraints are considered, and the results are compared with the conventional PI-based frequency control designs.

Recently, the applications of MASs have attracted more attention [20–24]. A two-level architecture for distributed-energy-resource management for multiple MGs using MAS is proposed [20]. In fact, the MASs, which are developed by the Java Agent DEvelopment framework (JADE), are used to model the market scenario with energy buyers and energy sellers. Reference [21] presents a MAS using JADE for real-time operation of an MG, focusing on generation scheduling and demand-side management. A MAS control strategy to coordinate power sharing between heterogeneous energy storage devices distributed throughout a DC MG is proposed in [22]. A MAS-based control system consisting of three levels is proposed in [23]. The first level is based on local droop control, the second level compensates power balance between the supply and the demand optimally, and the third level is at the system level based on electricity market. The coordination among the agents is made to ensure power quality, voltage, and frequency of the MG by determining the set points that optimize the overall operation of the MG. In response to the drawback of traditional droop control methods for the inherent trade-off between power sharing and voltage and frequency regulation, an MAS-based control scheme for automatic generation control (AGC) and automatic voltage control is proposed in [24]. The control objectives are satisfied using the acquired distributed global information (such as the average voltage and the output active power of the MG) by the MAS.

Several papers have been also published on the application of other intelligent techniques, optimization, and evolutionary algorithms or a combination of the mentioned methods [25, 26]. An adaptive PSO-based droop control method for simultaneous MG frequency and voltage control is presented in [25]. The PSO is used to estimate the optimal virtual line parameters. The impacts of active and reactive power fluctuations on voltage and frequency of MGs are also analyzed. An intelligent dynamic EMS for MGs is proposed in [26], using an evolutionary adaptive dynamic programming and reinforcement learning framework. The ANNs are also used during the mentioned system design.

The same objective is followed in [27] using a neuro-fuzzy system to minimize voltage and frequency deviations in islanded MGs under severe changes in load. In islanded MGs, due to lack of the backup power, the imbalance between consumption and generation may lead to severe voltage/frequency fluctuations. The most important advantage of the proposed controller is independency from the MG structure and operating conditions. Reference [28] presents an adaptive/intelligent power control approach for MGs in the grid-connected operation mode. The proposed critic-based adaptive control system contains a neuro-fuzzy controller and a fuzzy critic agent. The fuzzy critic agent employs a reinforcement learning algorithm based on neuro-dynamic programming.

A combination of the *fuzzy logic* and PSO techniques is used in [7] for optimal tuning of the PI-based frequency controllers in the MG systems. In the proposed control strategy, the PI parameters are automatically tuned using fuzzy rules, according to the online measurements. In order to obtain an optimal performance, the PSO technique is applied online to determine the parameters of the membership functions. The result is compared with the *pure fuzzy PI* control method as well as classical PI control design using the *Ziegler–Nichols* technique.

A real-time implementation of GA for MG reconfiguration is proposed in [29], and the reconfiguration algorithm has been implemented in the real time using real-time test bed (MG). An intelligent algorithm for islanding and seamless reconnection of MG with the control of unified power quality conditioner (UPQC) has been presented in [30]. A programmable architecture for active and optimal distributed control of elements of the grid to achieve the desired behavior is given in [31]. It generates feedback laws that are adapted to the current status of the MG and responds to anomalous events in a resilient manner.

In this chapter, the application of intelligent techniques in the MG control synthesis and operation is emphasized, and basic control configurations with recent achievements are briefly discussed. The chapter is organized as follows: the most important intelligent technologies for application in the MG planning, control, and operation issues are introduced in Section 10.2. ANN-based power and load forecasting examples in the MGs are briefly described in Section 10.3. Then, several control synthesis examples using fuzzy logic, PSO, and MASs are explained in Section 10.4. The examples are supplemented by simulations and experimental results for various possible operation scenarios. Finally, the chapter is summarized in Section 10.5.

10.2 Intelligent Control Technologies

In this section, the most common control frameworks for applications of fuzzy logic, ANNs, GA, MASs, and PSO in the MGs control synthesis are reviewed. Several synthesis examples are briefly explained, and some important practical points in the area of intelligent MG control are emphasized.

10.2.1 Fuzzy Logic Control

Nowadays, fuzzy logic because of simplicity, robustness, model-free property, and reliability is used in almost all fields of science and technology, including for solving a wide range of control problems in power grids control and operation [4]. Unlike the traditional control theorems that are essentially based on the linearized mathematical models of the controlled systems, the fuzzy control methodology tries to establish the controller directly based on the measurements, long-term experiences, and the knowledge of domain experts/operators. Several studies have been reported for the fuzzy-logic-based control design in MGs [17]. There are many possible fuzzy logic control structures for the MG control purpose, some differing significantly from each other by the number and type of inputs and outputs, or less significantly by the number and type of input and output fuzzy sets and their membership functions, or by the type of control rules, inference engine, and the defuzzification method. In fact, it is up to the designer to decide which control structure would be optimal for a given MG control problem [4]. Similarly to conventional power grids, the applications of fuzzy logic in MG systems can be classified into three categories: (i) using fuzzy logic system as main controller [15], (ii) using fuzzy logic system as a primer for tuning the gains of the existing controller [9, 13, 17], and (iii) using fuzzy unit to support (and sometimes in parallel with) the conventional controller [11].

10.2.1.1 Fuzzy Logic System as Main Controller

A general scheme for fuzzy-logic-based MG control system is given in Fig. 10.1. As shown, the fuzzy controller has four blocks. Crisp input information (usually measured feedback signals) from the MG is converted into fuzzy values for each input fuzzy set with the fuzzification block. The universe of discourse of the input variables determines the required scaling/normalizing for correct per-unit operation. The inference mechanism determines how the fuzzy logic operations are performed and, together with the knowledge base, determines the outputs of each fuzzy *if–then* rule. Those are combined and converted to crispy values with the defuzzification block. The output crisp value can be calculated by the center of gravity or the weighted average; then, the scaled output as control signal is applied to the generating units.

Generally, a controller synthesis based on fuzzy logic for a dynamical system involves the following four main steps: (i) define the states and input/output control variables and their variation ranges; (ii) identify appropriate fuzzy sets and membership functions and create the degree of fuzzy membership function for each input/output variable and complete fuzzification; (iii) define a suitable inference engine and construct the fuzzy rule base, using the control rules that the system will operate under; and (iv) determine the defuzzification method and combine the rules and defuzzify the output [4].

In consistent with the MG control design, the first step of fuzzy controller design is to choose the correct input signals to the MG set point. The output

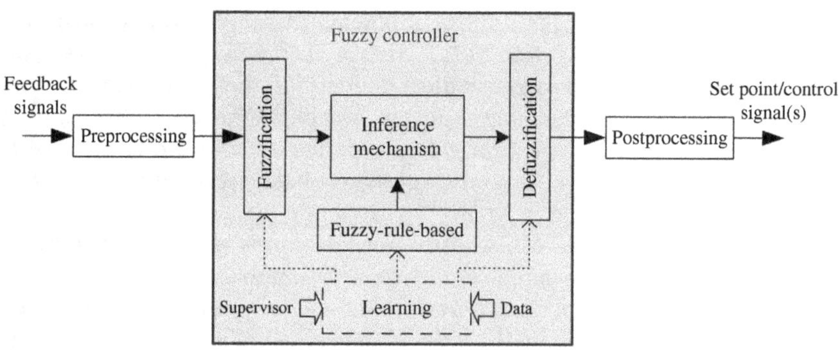

Figure 10.1 A general scheme for fuzzy-logic-based MG control.

frequency, voltage, current, power, and their derivative are commonly chosen MG signals as inputs of the fuzzy controller. These signals are then used as rule-antecedent (*if* part) in the formation of rule base, and the control output is used to represent the contents of the rule-consequent (*then* part) in the performance of rule base. *Normalization* or scale transformation, which maps the physical values of the current system state variables into a normalized universe of discourse, should be properly considered. This action is also needed to map the normalized value of control output variables into its physical domain (denormalization output). The normalization can be achieved by dividing each crisp input on the upper boundary value for the associated universe [4].

In real world, many phenomena and measures are not crisp and deterministic. *Fuzzification* plays an important role in dealing with uncertain information, which might be objective or subjective in nature. The fuzzification block in the fuzzy controller represents the process of converting crisp quantity into fuzzy. In fact, the fuzzifier converts the crisp input to a linguistic variable using the membership functions stored in the fuzzy knowledge base. Fuzziness in a fuzzy set is characterized by the *membership functions*. Using suitable membership functions, the ranges of input and output variables are assigned with linguistic variables. These variables transform the numerical values of the input of the fuzzy controller to fuzzy quantities. These linguistic variables specify the quality of the control. Triangular, trapezoid, and Gaussian are more common membership functions to use in fuzzy control systems. The fuzzy system is characterized by a set of linguistic statements in the form of *if–then* rules. Fuzzy conditional statements make the rules or the rule set of the fuzzy controller [4]. Finally, the *inference engine* uses the *if–then* rules to convert the fuzzy input to the fuzzy output. Two practical applications of the mentioned fuzzy control framework are used to provide voltage/frequency reference for MG voltage/frequency regulation in Section 10.4.

On the other hand, defuzzifier converts the fuzzy output of the inference engine to crisp using membership functions analogous to the ones used by the fuzzifier. For the *defuzzification* process, commonly the center of sums, mean-max, weighted average, and centroid methods are employed to defuzzify the fuzzy incremental control law. The parameters of fuzzy logic controller such as membership functions can be adjusted using an external tuning mechanism. The resulting controller is known as an adaptive, self-learning, or self-tuning fuzzy controller. An adaptive fuzzy controller has a distinct architecture consisting of two loops: an inner control loop, which is the basic feedback loop, and an outer loop, which adjusts the controller parameters [4]. This architecture is shown in Fig. 10.2.

The adaptive fuzzy controllers commonly use some other intelligent techniques such as neural networks in the tuning block, which have learning capability. This combination provides *neuro-fuzzy* controller. A neuro-fuzzy controller is a fuzzy controller that uses a *learning algorithm* inspired by *neural*

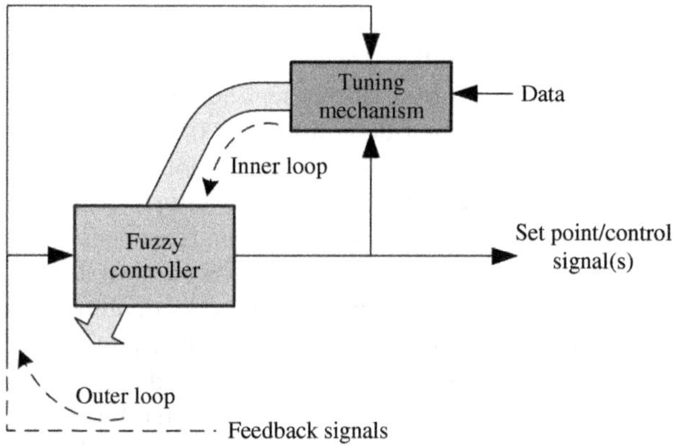

Figure 10.2 A general scheme for adaptive fuzzy logic control system.

network theory to determine its *parameters* by processing data samples. The combined intelligence-based MG control synthesis using ANN and fuzzy logic techniques is presented in several works to utilize the novel aspects of both designs to a single MG control design problem [25, 26].

10.2.1.1.1 Intelligent General Droop Control Example In Section 5.3.2, new formulations for GDCs in MGs are obtained as given in (5.35) and (5.36). Then, it was examined for the given MG example shown in Fig. 5.9. The main drawback of the addressed GDC control method is the strong dependency of the mentioned approach on the line parameters (R and X). To solve this problem, fuzzy logic system is used instead of the GDC in [32] to meet the same accuracy without any dependency on the MG structure. To increase the flexibility of the fuzzy system and adding the trainability, adaptive neuro-fuzzy inference system (ANFIS) is used instead of pure fuzzy logic. Therefore, Fig. 5.9 can be redrawn as shown in Fig. 10.3.

The validity of the neuro-fuzzy system has been investigated. The ANFIS model is created using the GDC, which is used in the MG shown in Fig. 10.3. For a simple three-bus MG example given in [32], the training data are obtained for the case of $K_R = 1$. For evaluating the performance of the proposed system, the collected data from simulations are divided into two sets of real data and training (test) data. These two sets are compared with each other. Figure 10.4a shows the ANFIS network outputs versus the real outputs. All two data sets are also drawn together in Fig. 10.4b. Comparison of the real and network data shows whether the training process is done accurately. As depicted in Fig. 10.5, the simulation results for a load change pattern, as the one shown in Fig. 5.10b, investigate that the developed control strategy is capable of regulating the MG frequency and voltage as accurately as the mentioned GDC [32].

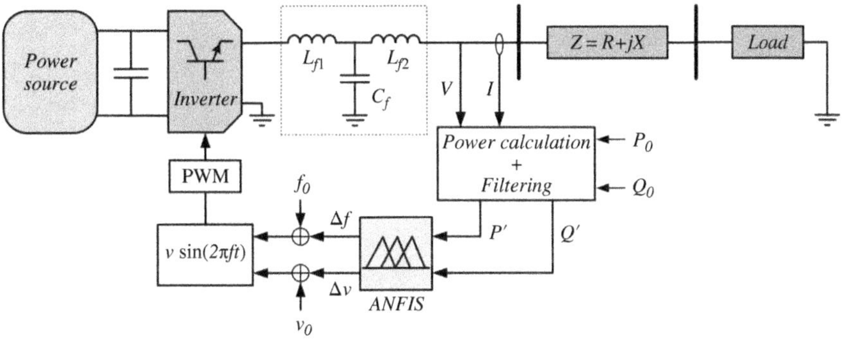

Figure 10.3 The GDC with neuro-fuzzy system.

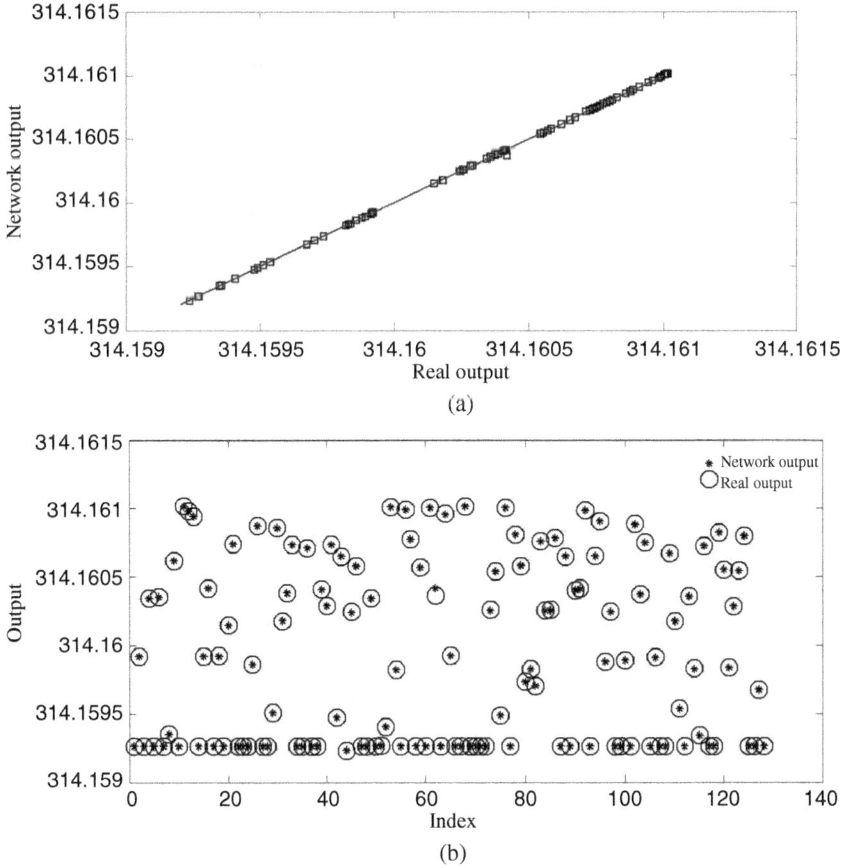

Figure 10.4 Validation of the ANFIS network: (a) trained network output versus real output and (b) both output together.

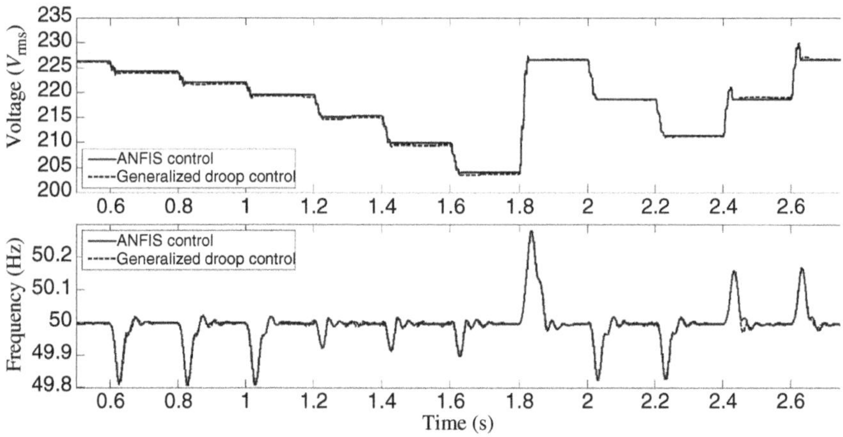

Figure 10.5 System voltage and frequency following a load disturbance.

10.2.1.2 Fuzzy Logic for Controller Tuning

The PI control structure has been widely used in the power grids as well as at MG control levels due to its structure simplicity and inexpensive cost. The success of the PI controller depends on the appropriate choice of its gains. Assume the feedback signal to feed the PI controller to be $e(t)$, then the control signal generated by a PI controller in the continuous-time domain is represented by

$$u(t) = k_P e(t) + k_I \int_0^t e(\tau)d\tau \tag{10.1}$$

where $u(t)$ is the control signal, and k_P and k_I are the proportional and integral coefficients, respectively. In practice, tuning the PI gains is usually realized by experienced human experts, which may not be capable of achieving an optimal performance for MG control applications due to nonlinearities, varying structure, and uncertainties. The fuzzy logic can be effectively used for tuning the PI controllers to improve the performance of the MG control systems in comparison with conventional PI tuning methods. In Section 10.4, it will be found that the MG control systems with fuzzy-logic-based PI controllers have better capabilities compared to the conventionally tuned PI controllers. The mentioned control scheme is conceptually shown for a PI controller in Fig. 10.6. Many research works use fuzzy logic to tune the gain parameters of existing PI controllers in the MG systems.

10.2.1.2.1 A Fuzzy-Based Tuning Control Example A fuzzy logic tuning system for supporting the MG secondary control (voltage, frequency, and reactive power sharing regulation) is shown in Fig. 10.7 [9]. The control layer regulates the voltage and frequency of the MG and simultaneously manages the reactive power sharing among the inverter-based DGs using consensus protocols

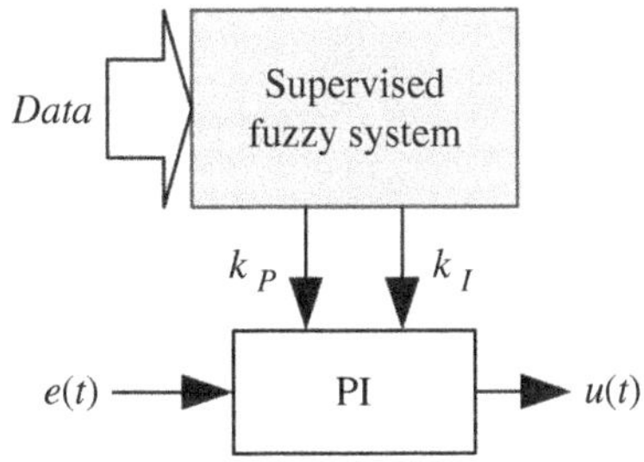

Figure 10.6 Fuzzy logic system for tuning of PI controller.

[33, 34]. Frequency regulation is performed by a dynamic consensus protocol to change the droop coefficient to reduce mismatching, and for the sake of voltage regulation, another consensus protocol is used to estimate the averaged voltage. Upon comparison of the local generation with neighbors, the reactive power sharing control shifts the droop coefficient to mitigate mismatching. As shown in Fig. 10.6, a cyber network is used to form a communication network to exchange data between the controllers of all DGs.

Parameters of consensus protocols may affect the system stability and performance; however, they are normally defined via trial and error method. Here, a fuzzy logic system is proposed for fine-tuning of the consensus parameters. In order to train the fuzzy logic system, it is necessary to know the ranges and effects of the consensus parameters on frequency/voltage stability and performance.

As illustrated in Fig. 10.7, a frequency correction term, ω_j, is defined to add the frequency reference to return the MG frequency to the nominal value. The frequency regulator at each node i receives the term $\delta\omega_j$ from all the neighbors, then, the frequency regulator produces the term $\delta\omega_i$ as in (10.2).

$$g_i \frac{d(\delta\omega_i)}{dt} = -\omega + \omega_0 - \sum_{j=1}^{N} a_{ij}(\delta\omega_i - \delta\omega_j) \tag{10.2}$$

where a_{ij} is the term for consensus parameters, g_i is a positive gain that has been used to control the secondary control speed, $\delta\omega_i$ is the frequency regulator control variable, and the couple (ω, ω_0) are the output frequency and its rated value. As seen in (10.2), the loading mismatch is fed to an integral controller to add the droop control to shift the droop coefficient by a different amount of $\delta\omega_i$. This term successfully boosts the frequency of all nodes, synchronizes the system frequency, and will guarantee active power sharing among DGs. As shown in Fig. 10.7, the proposed secondary controller at each source carries three modules. A voltage correction term adjusts the voltage set point for each node, that is, $E_{DGk} = E + \delta E^1_{DGk} + \delta E^2_{DGk}$, where E is the reference voltage of each node, and δE^1_{DGk} is the correction term produced by the voltage regulator. To find the term δE^1_{DGk}, the voltage regulator uses dynamic consensus to provide the averaged voltage amplitude of DGs in the system. The estimator at node i updates

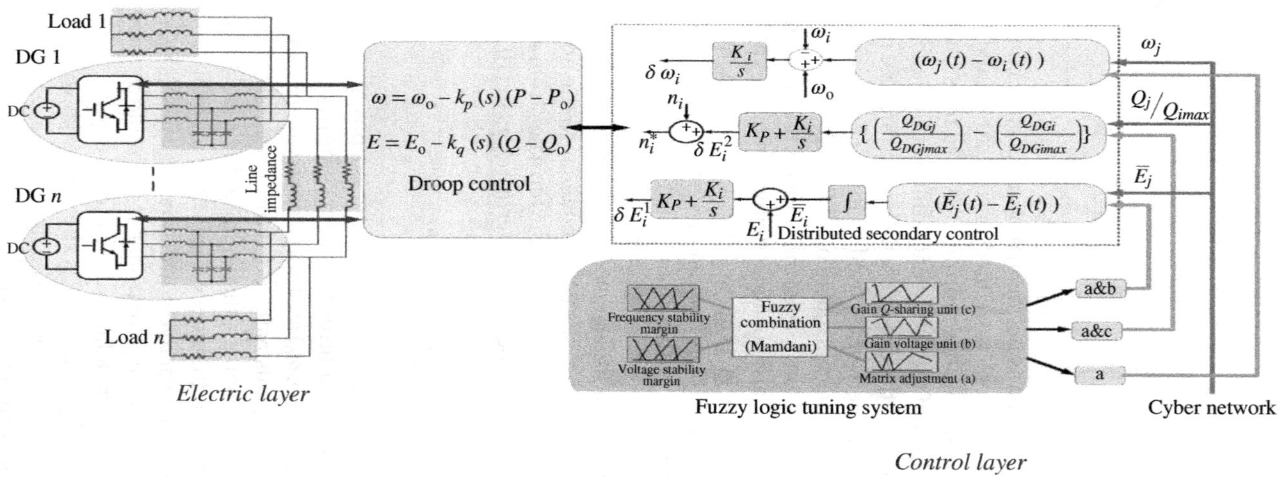

Figure 10.7 Fuzzy logic tuning system for supporting MG secondary control.

the averaged amplitude without directly measuring the other nodes as follows:

$$\overline{E}_i(t) = \int \left(\sum_{j=1}^{N} ba_{ij}\{\overline{E}_j(\tau) - \overline{E}_i(\tau)\} \right) d\tau + E_i(t) \tag{10.3}$$

where a and b are consensus parameters, E_i is the amplitude at node i, and $\overline{E}_i$ is the averaged voltage estimator. As shown in (10.3), the consensus protocol uses local voltages to estimate the voltage at each node i. Using a connected communication graph, the variation of $\overline{E}_i$ affects all other voltage estimators. The line impedance causes the voltage output of busses not equivalent with others. After producing $\overline{E}_i(t)$ by (10.3), the term δE^1_{DGi} is generated by the droop loops as follows:

$$\delta E^1_{DGi} = k_{pv}(E_{ref} - \overline{E}_i(t)) + k_{iv} \int (E_{ref} - \overline{E}_i(t)) d\tau \tag{10.4}$$

The other term, δE^2_{DGk}, is produced by the reactive power regulator that receives the term Q_{DGj}/Q_{DGjmax} from all neighbors, and again, it is added to the Q–V droop to shift up/down, which is presented by (10.5):

$$\delta E^2_{DGi} = k_{pQ} \left(\sum_{j=1}^{N} ca_{ij} \left(\frac{Q_{DGj}}{Q_{DGj\,max}} - \frac{Q_{DGi}}{Q_{DGi\,max}} \right) \right) + k_{iQ} \int \left(\sum_{j=1}^{N} ca_{ij} \left(\frac{Q_{DGj}}{Q_{DGj\,max}} - \frac{Q_{DGi}}{Q_{DGi\,max}} \right) \right) dt \tag{10.5}$$

Here, c is the coupling gain between the voltage and reactive power regulations, a_{ij} is the term for consensus protocol matrix, and k_{PQ} and k_{iQ} are proportional and integral controller. As seen in (10.5), the loading mismatch is fed to a *PI* controller to generate the term δE^2_{DGk}, then this term will be added to shift the droop characteristics up or down. This term manages the reactive power sharing between the agents.

To evaluate the proposed control methodology, it is applied to an MG case study with three inverter-based DGs [9]. The system response under frequent load changes is shown in Fig. 10.8. The desired values of a, b, and c are determined by the fuzzy logic system as $a = 218$, $b = 3.69$, and $c = 0.000399$. The terms "LS" and "NS" denote the low stability and normal stability, respectively, which are used to provide the fuzzy rules and membership functions [9].

10.2.1.3 Fuzzy Logic System as a Supplementary Controller

Concerning the limits, physical constraints, and uncertainty in an MG, it may be difficult to meet an optimal performance using a conventional PI controller. In response to this challenge in this category, the fuzzy logic system is used as a supplementing controller for producing a compensating signal to correct the control action signal provided by the conventional PI controller. The adding

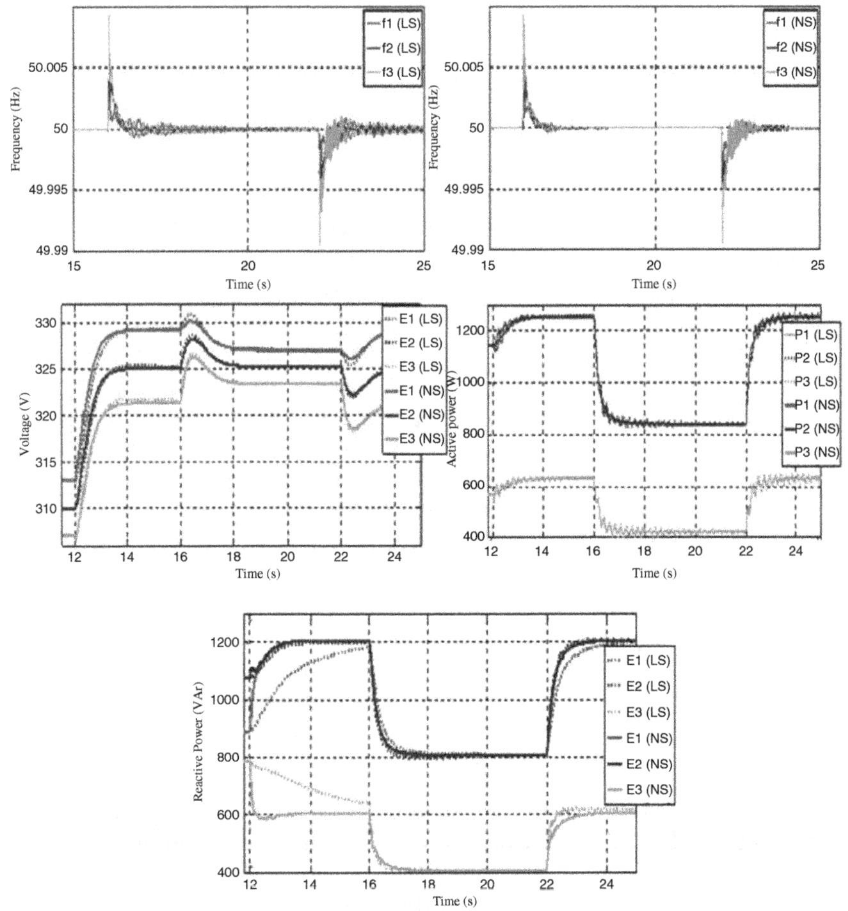

Figure 10.8 Smart tuning based on fuzzy logic compared with conventional methods.

signal slightly corrects the PI control output through the fuzzy rules, according to the on-line information. The performed complementary signal can perform an effective role in optimal tracking and compensating. The simplified control scheme is shown in Fig. 10.9.

In the proposed control scheme, the intelligent fuzzy system collects information about the MG response, that is, $e(t)$ and PI control output (u_{PI}), and then recommends an appropriate correction control signal Δu_{Fuzzy}. In fact, the fuzzy system performs as an automatic tuner. The main components of this system as an intelligent fine-tuner for the conventional PI controller include a feedback (through the $e(t)$ signal), extract knowledge about the performance of the existing PI controller (through the u_{PI} signal), and an embedded unit to suggest

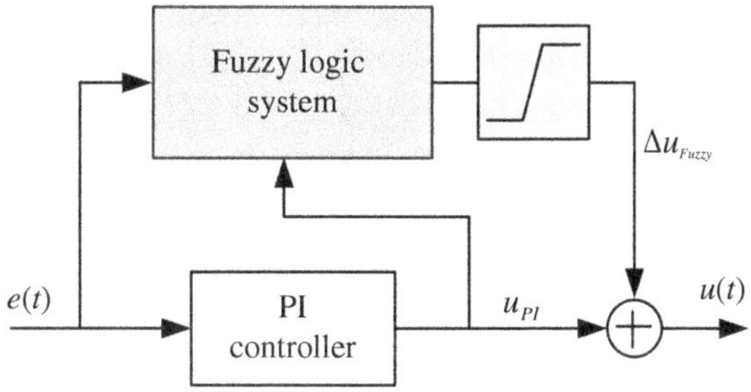

Figure 10.9 Fuzzy logic system for supporting the PI controller.

suitable correction (Δu_{Fuzzy}) to be made to the final control action signal (u) as follows:

$$u = u_{PI} + \Delta u_{Fuzzy} \tag{10.6}$$

and

$$u_{PI} = k_P e(t) + k_I \int e(t)dt, \quad \Delta u_{Fuzzy} = f(e, \Delta u_{PI}) \tag{10.7}$$

where k_P and k_I are the proportional and integral constant gains of the PI controller, respectively. The $f(\cdot)$ denotes the fuzzy logic function. Nonnecessity of the available PI controller can be considered as an advantage of the proposed control strategy in some practical applications [35].

10.2.1.3.1 An Example of Fuzzy Logic System as a Supplementary Controller A supervisory fuzzy system is developed for frequency control in an AC MG case study, the closed-loop frequency response model of which is shown in Fig. 10.10. System parameters and detailed design are given in [11]. The MG case study contains several DGs and storage devices including wind turbine generators (WTG), diesel engine generators (DEG), photovoltaic (PV) panels, fuel cells (FCs), battery energy storage (BES) and flywheel energy storage (FES) systems [11].

The fuzzy supervisory control system provides a coordination between the FC and the DEG units in the MG frequency control task. The FC has a high efficiency and low response time. Thus, the coordination between FC and DEG enhances the response time, decreases the DEG fuel consumption, and consequently decreases air pollution. As shown in Fig. 10.10, fuzzy logic controller produces two control signals for correcting the supplied control signal from PI controller to the DEG and FC units. The k is the participation factor of the FC. The result of the coordinated framework is compared with the results of conventional PI control shown in Fig. 10.11.

10.2.2 Artificial Neural Networks

ANNs are numerical model-free estimators, which can estimate how the output functionally depends on the input without the need for complex mathematical

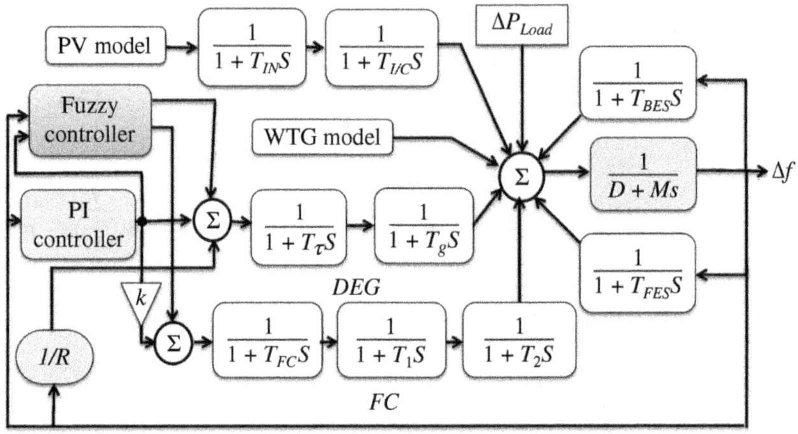

Figure 10.10 Closed-loop frequency response model.

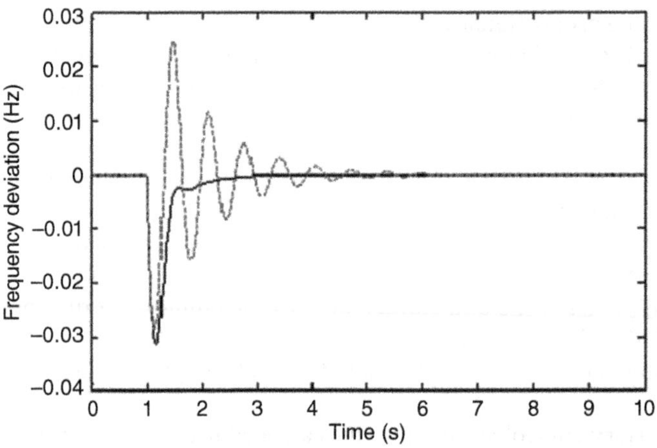

Figure 10.11 Frequency deviation; PI and fuzzy control (solid line), and only PI control (dashed line).

models. An ANN consists of a number of nonlinear computational processing elements (neurons), arranged in several layers including an input layer, an output layer, and one or more hidden layers in between. Every layer contains one or more neurons, and the output of each neuron is usually fed into all or most of the inputs of the neurons in the next layer. The input layer receives input signals, which are then transformed and propagated simultaneously through the network, layer by layer. A neuron accepts one or more input signals and produces one output, which is a nonlinear function of the weighted sum of inputs. The mapping from the input variables to the output variables can be fixed by setting all the weights associated with each neuron to some constants. In fact, the training of an ANN in a control structure is a procedure to adjust these values so that the ANN can map all the input control values to the corresponding output control values [4].

From the control configuration point of view, the most proposed ANN-based MG control designs can be divided into three general control structures that are conceptually shown in Fig. 10.12 [4]: (i) directly using ANN system as a controller to provide control command in the main feedback loop, (ii) using ANN for tuning the parameters of existing fixed-structure controller (e.g., PI), and finally, (iii) using ANN system as an additional controller in parallel with the existing conventional simple controller such as PI, to improve the closed loop performance.

The aforementioned three configurations are presented in Fig. 10.12a–c, respectively. *Backpropagation*, which is a gradient-descent learning algorithm, is one of the most popular supervised learning algorithms in all the mentioned configurations. It back-propagates the error signals from the output layer to all the hidden layers, such that their weights can be adjusted accordingly. Backpropagation is a generalization of the least mean square (LMS) procedure for feedforward, multilayered networks with hidden layers. It uses a

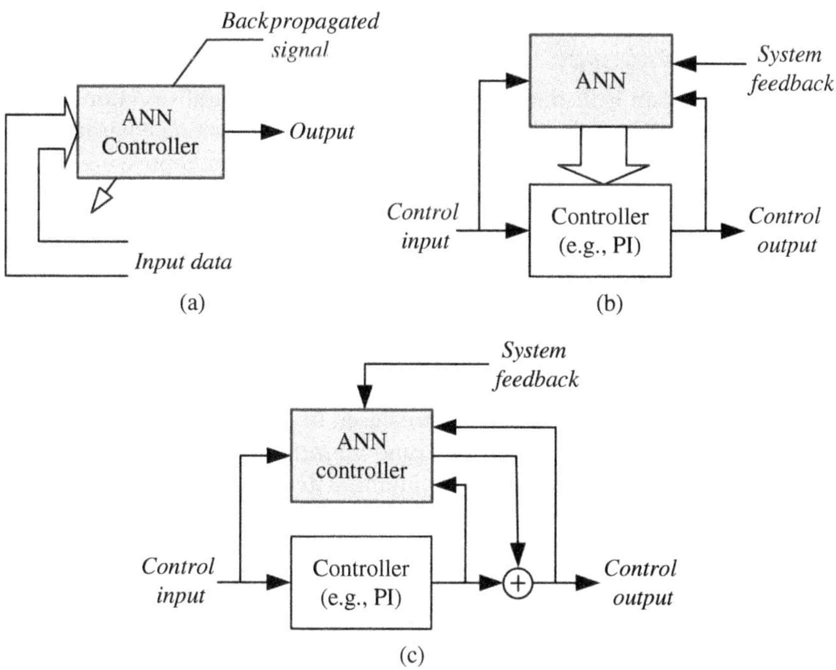

Figure 10.12 Common configurations for ANN-based control schemes.

gradient-descent technique, which changes the weights between neurons in its original and simplest form by an amount proportional to the partial derivative of the error function with respect to the given weight [4].

As shown in Fig. 10.12b, the ANN performs as an automatic tuner. The initial values for the parameters of the fixed-structure controller (e.g., k_P and k_I gains in PI) must first be defined. The trial and error and the widely used Ziegler–Nichols tuning rules are usually employed to set initial gain values according to the open-loop step response of the plant. The ANN collects information about the system response and recommends adjustments to be made to the controller gains. This is an iterative procedure until the fastest possible critical damping for the controlled system is achieved. The main components of the ANN tuner include a response recognition unit to monitor the controlled response and extract knowledge about the performance of the current controller gain setting and an embedded unit to suggest suitable changes to be made to the controller gains.

The application of ANN for PV power and load forecasting as well as for estimating the existing error (in forecasting results) is given in Section 10.3. Several ANN-based control synthesis examples for MG frequency and voltage control are given in [1, 19, 35].

10.2.3 Genetic Algorithm and Particle Swarm Optimization

10.2.3.1 Genetic Algorithm

The GA mechanism is inspired by the mechanism of natural selection where stronger individuals would likely be the winners in a competing environment. Normally in a GA, the parameters to be optimized are represented in a binary string. A simplified flowchart for the GA is shown in Fig. 10.13. The *cost function*, which determines the optimization problem, represents the main link between the problem at hand (system) and the GA unit and also provides the fundamental source to provide the mechanism for evaluating the algorithm steps. To start the optimization, the GA uses randomly produced initial solutions created by a random number generator. This method is preferred when a priori information about the problem is not available. There are basically three genetic operators used to produce a new generation. These operators are *selection, crossover,* and *mutation*. The GA employs these operators to converge at the global optimum. After randomly generating the initial population (as random solutions), the GA uses the genetic operators to achieve a new set of solutions at each iteration. In the selection operation, each solution of the current population is evaluated by its fitness normally represented by the value of some objective function, and individuals with higher fitness value are selected [4].

Different selection methods such as stochastic selection or ranking-based selection can be used. In the selection procedure, the individual chromosomes

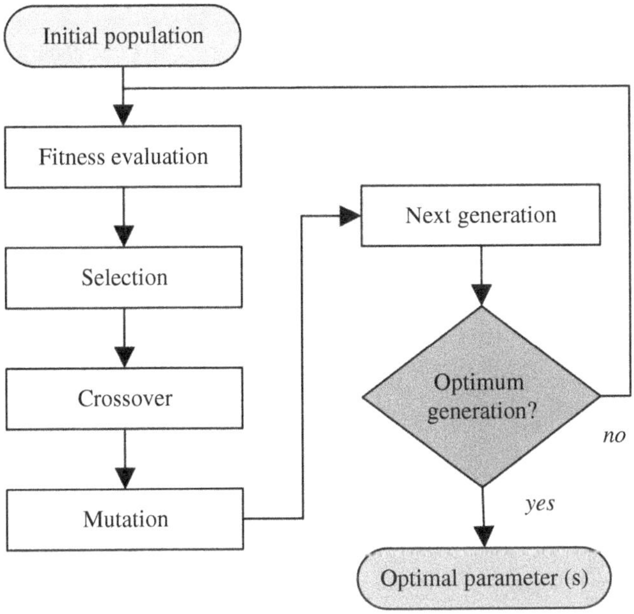

Figure 10.13 A simplified GA flowchart.

are selected from the population for the later recombination/crossover. The fitness values are normalized by dividing each one by the sum of all fitness values, named selection probability. The chromosomes with a higher selection probability have a higher chance to be selected for later breeding.

The crossover operator works on pairs of selected solutions with certain crossover rate. The crossover rate is defined as the probability of applying crossover to a pair of selected solutions (chromosomes). There are many ways to define the crossover operator. The most common way is called the *one-point crossover*. In this method, a point (e.g., for given two binary coded solutions of certain bit length) is determined randomly in two strings, and the corresponding bits are swapped to generate two new solutions. Mutation is a random alteration with small probability of the binary value of a string position and will prevent the GA from being trapped in a local minimum. The coefficients assigned to the crossover and mutation specify the number of children. Information generated by fitness evaluation unit about the quality of different solutions is used by the selection operation in the GA. The algorithm is repeated until a predefined number of generations have been produced [4].

Unlike the gradient-based optimization methods, the GAs operate simultaneously on an entire population of potential solutions (chromosomes or individuals) instead of producing successive iterates of a single element, and the computation of the gradient of the cost functional is not necessary. The GA

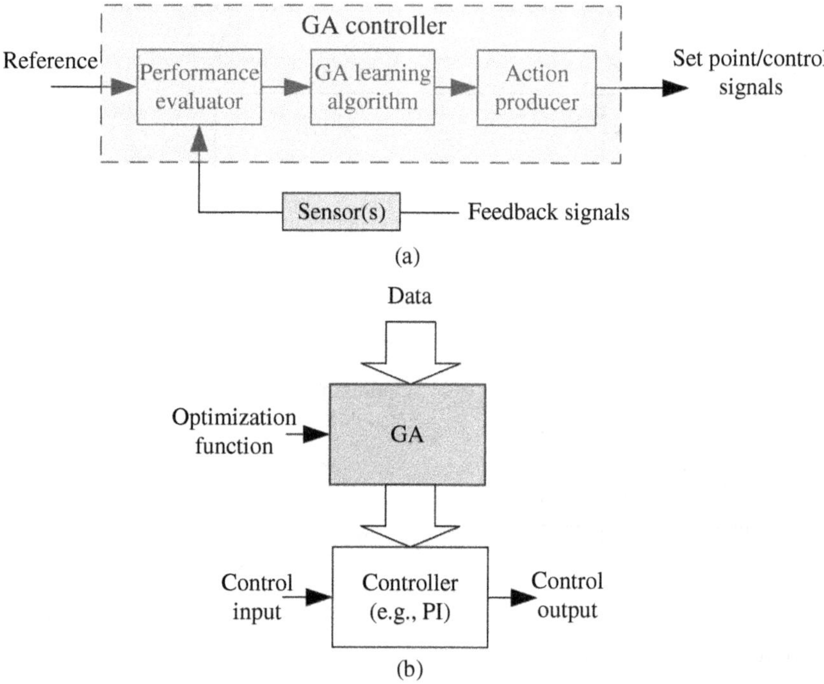

Figure 10.14 A GA-based control: (a) GA as main controller and (b) GA-based controller tuning scheme.

is one of the rapidly emerging optimization approaches in the field of control engineering and control system design [36]. Applications to optimal/adaptive tracking control, active noise control, multiobjective control, robust tuning of control systems via seeking the optimal performance indexes provided by robust control theorems and using in fuzzy logic and ANN-based control systems are some important applications of GA in control systems.

A simple GA-based control system is conceptually shown in Fig. 10.14a. The GA controller consists of three components: performance evaluator, learning algorithm, and control action producer. The performance evaluator rates a chromosome by assigning it a fitness value. The value indicates how good the chromosome is in controlling the dynamical plant to follow reference signal. The learning algorithm may use a set of rules in the form of "*condition then action*" for controlling the plant. The desirable action will be performed by control action producer when the condition is satisfied. The control structure shown in Fig. 10.14a is implemented for several control applications in different forms. The GAs can be effectively used to adjust the parameters of MG controllers (e.g., PI). The overall control framework for this purpose is shown in Fig. 10.14b.

10.2.3.2 Particle Swarm Optimization

The PSO is a population-based stochastic optimization technique. It belongs to the class of direct search methods that can be used to find a solution to an optimization problem in a search space. The PSO originally has been presented based on social behavior of bird flocking, fish schooling, and swarming theory [37]. In the PSO method, a swarm consists of a set of individuals, where each individual is specified by the *position* and *velocity* vectors ($x_i(t)$, $v_i(t)$) at each time or iteration. Each individual is named as a "*particle*," and the position of every particle represents a potential solution to the optimization problem under study. In an n-dimensional solution space, each particle is treated as a *n-dimensional* space vector and the position of the ith particle is presented by $v_i = (x_{i1}, x_{i2}, \dots, x_{in})$; then, it flies to a new position by velocity represented by $v_i = (v_{i1}, v_{i2}, \dots, v_{in})$. The best position for ith particle represented by $p_{best,i} = (p_{best,i1}, p_{best,i2}, \dots, p_{best,in})$ is determined according to the best value for the specified objective function [4].

Furthermore, the best position found by all particles in the population (global best position) can be represented as $g_{best} = (g_{best,1}, g_{best,2}, \dots, g_{best,n})$. In each step, the best particle position, global position, and the corresponding objective function values should be saved. For the next iteration, the position x_{ik} and velocity v_{ik} corresponding to the kth dimension of ith particle can be updated using the following equations:

$$v_{ik}(t+1) = w \cdot v_{ik} + c_1 \cdot rand_{1,ik}(p_{best,ik}(t) - x_{ik}(t)) + c_2 \cdot rand_{2,ik}(g_{best,k}(t) - x_{ik}(t)) \quad (10.8)$$

$$x_{ik}(t+1) = x_{ik}(t) + v_{ik}(t+1) \quad (10.9)$$

where $i = 1, 2, \dots, n$ is the index of particles, w is the inertia weight [37], $rand_{1,ik}$ and $rand_{2,ik}$ are random numbers in interval [0 1], c_1 and c_2 are learning factors, and t represents the iterations. Usually, a standard PSO algorithm contains the following steps [4]:

Step 1: All particles are initialized via a random solution. In this step, each particle position and associated velocity are set by randomly generated vectors. The dimension of position should be generated within a specified interval, and the dimension of velocity vector should also be generated from a bounded domain using uniform distributions.

Step 2: Compute the objective function for the particles.

Step 3: Compare the value of the objective function for the present position of each particle with the value of objective function corresponding to prespecified best position, and replace the prespecified best position by the present position, if it provides a better result.

Step 4: Compare the value of the objective function for the present best position with the value of the objective function corresponding to global best position, and replace the present best position by the global best position, if it provides a better result.

Step 5: Update the position and velocity of each particle according to (10.8) and (10.9).

Step 6: Stop algorithm if the stop criterion is satisfied. Otherwise, go to step 2.

In order to illustrate the application of PSO in the MGs, an example is described in Section 10.4.

10.2.4 Multiagent System

An MAS is a system comprising two or more intelligent agents to follow a specific task. Nowadays, the MAS technology is widely used in planning, monitoring, control, and automation systems. The MAS philosophy and its potential value to the power systems are discussed in [38, 39]. Several reports have been published on the application of MAS technology with different characteristics and intelligent cores for the MG control systems [20–24].

In MG control, the structural flexibility and having a degree of intelligence are highly important. In such systems, agents require real-time responses and must eliminate the possibility of massive communication among agents. In the synthesis of real-time MAS, the designer must at least denote the required number and type of agents in the system, the internal structure of each agent, and the communication protocol among the available agents. Each agent is implemented on a software platform that supports the general components of the agent. The software platform must provide a communication environment among the agents and supports a standard interaction protocol.

10.2.4.1 Multiagent System Concept

The MASs perform a subfield of (distributed) AI [4]. An MAS includes several agents and a mechanism for coordination of independent agents' actions. Various definitions for an agent are given in the computer science and AI literature [40]. An *agent* can be considered as an intelligent entity, which is operating in an *environment*, with a degree of autonomy, specific goal(s), and knowledge. An agent can alter the environment by taking some actions and can act autonomously in response to environmental changes. Autonomy means that the agent is able to fulfill its tasks without the direct intervention of a human, and the environment is everything (systems, hardware, and software) external to the agent. Of course, the agent is also a part of the environment and can alter the environment by taking some actions. In a single-agent system, if there are other agents, they are also considered as part of the environment.

Suitability for representing and control of interconnected/distributed systems, simplicity of mechanism, programming and implementation, the capability of parallel processing/computation, scalability (handling a numerous units), extensibility and flexibility (integrating of new parts and entities), maintainability (because of modularity due to using multiple-component agents), responsiveness (handling anomalies locally instead of propagating them to the whole system), robustness against failure, and reliability are some important reasons to use MASs in control system (specifically distributed control systems) designs.

Figure 10.15a illustrates a view of typical MAS, conceptually. Here, the agent is shown in black, which sends and receives messages and interacts (via sensors and actuators) with its environment, autonomously. The agents may also interact directly as indicated in the figure by the arrows between the agents. There may be numerous agents with different structures, local goals, actions, and domain knowledge, with or without the ability to communicate other agents, directly. In addition to autonomy, veracity, and rationality, the main characteristics that an agent may have, some or all of those, are social ability, responsiveness, proactiveness, adaptability, mobility, and learning. These characteristics are well defined in [41].

In an MAS, at least one agent is usually equipped with intelligent inference. Intelligent core can play a major role in an intelligent agent for reasoning about the dynamic environment. Various intelligent cores/inferences such as symbolic representation, *if–then* rules and fuzzy logic, artificial neural networks, reinforcement learning [42], and Bayesian networks [43] can be used in MASs.

MAS-based control systems represent control schemes that are inherently distributed and consist of multiple entities/agents. The control architecture for MASs can be broadly characterized as *deliberative control*, *reactive control*, and a combination of both. Deliberative control is based on planning, while reactive control is based on coupling between sensing and actuation. Complex control tasks can always be hierarchically decomposed into several simpler subtasks. This naturally leads to agent architectures consisting of multiple layers. Figure 10.15b shows a schematic diagram of a three-layer agent architecture. In real-time control applications, the agents should be capable of reasoning about the best possible action without losing too much time on sending or receiving data. In order to deal with the timing constraints caused by the real-time nature of the domain, it was therefore desirable that the agents could perform this high-level reasoning process. Adopting a somewhat hybrid approach thus seemed to be an appropriate choice.

Figure 10.15b shows the functional hierarchy of the agent architecture. As illustrated, the agent and environment form a closed-loop system. The bottom layer is the *interfacing layer*, which takes care of the interaction with the environment. This layer observes the details of environment as much as possible from the other layers. The middle layer is the signal *processing/modeling layer*,

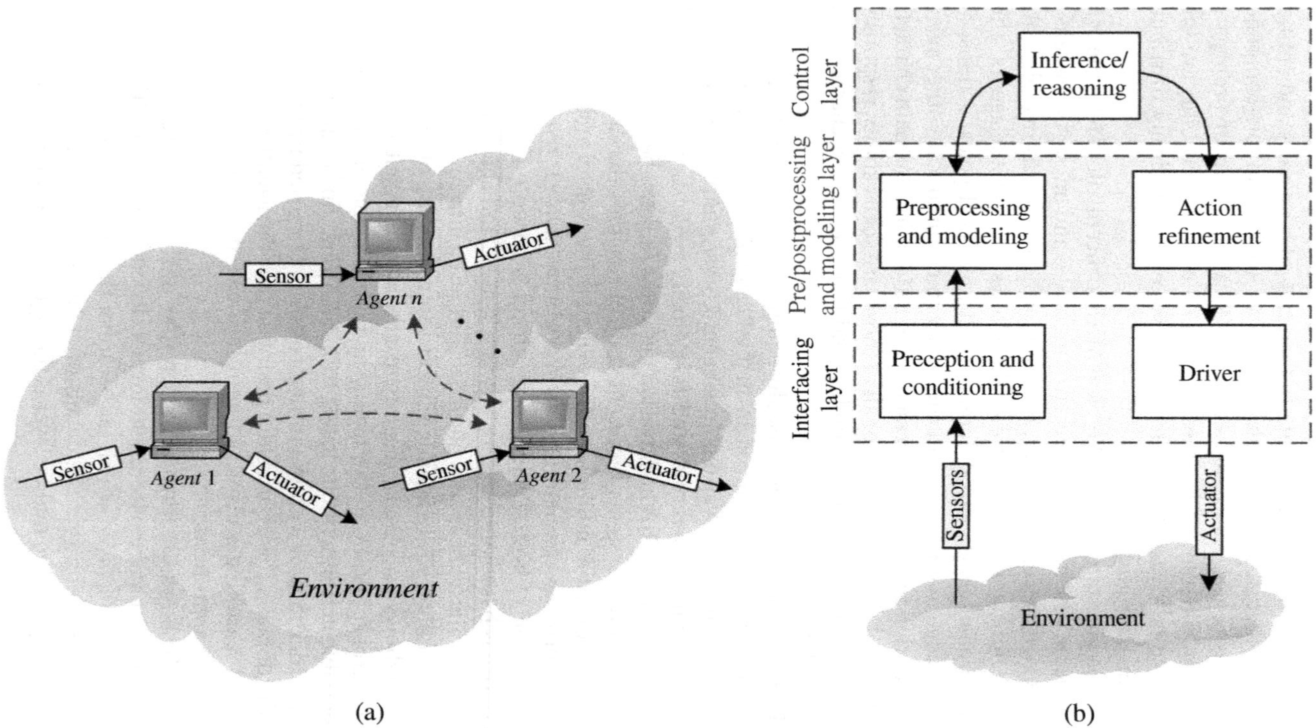

Figure 10.15 MAS: (a) a conceptual framework and (b) a typical intelligent agent architecture.

which simulates and provides a clear view of the world (environment) with a set of possible choices for the third layer. The highest layer in the architecture is the *control layer*, which contains the reasoning component of the system. In this layer, the best possible action is selected from the observation/modeling layer depending on the current environment state and the current strategy of the overall control system. The most recent environment state information is then used by the control layer to reason about the best possible action. The action selected by the control layer is subsequently worked out in the second layer, which determines the appropriate actuator command. This command is then executed by the actuator control module in the interfacing layer. A control synthesis example using MAS for MG frequency control is explained in Section 10.4.

10.2.4.2 Multiagent System Applications in Microgrid Power Management

As mentioned in Chapter 5, for designing decentralized power control system in an MG, intelligent MAS approach can be considered as an effective solution [44–47]. The MAS approach can solve specific operational problems in the MG. Since the DGs/RESs have different owners and several decisions should be locally made, a unique centralized control is difficult. Furthermore, because of a liberalized market, the decisions of the controller of each unit concerning the market should have a function of intelligence.

For decentralized power management, four kinds of agents are considered in [45]: production agent, consumption agent, power system agent, and MGCC agent. The MGCC agent has only coordinating tasks, and, more specifically, it announces the beginning and the end of a negotiation for a specific period and records final power exchanges between the agents in each period. A main feature of the MAS is that the software within each agent can embed local intelligence. Each agent uses its intelligence to determine future actions and independently influences its environment [48].

An intelligent MG requires a fairly advanced communication system with capabilities similar to the human speech; for example, the agent communication language (ACL) provides an environment for information and knowledge exchange. The need for a high-level communication environment can be shown by considering the communication needs of two agents within an MG. For instance, at a given time, one agent may have an instantaneous surplus of 1500 W and the other one may need 500 W. It is neither efficient nor required to provide the exact values since the situation can change within a short time. The agents exchange not only simple values and on–off signals but also knowledge, commands, and procedures to be followed through the ACL. For example, the agent that controls a load can participate in the local MG market by sending a request message to all DG/RES agents stating the amount of required energy. Furthermore, its object-oriented nature and data abstraction enable each agent to handle only the necessary or allowable information and knowledge.

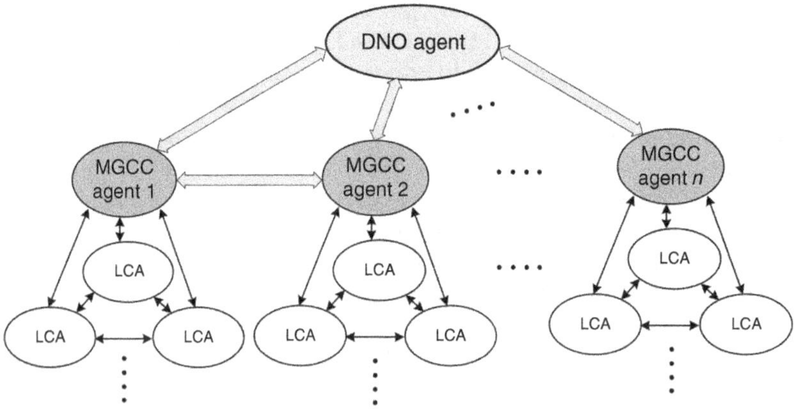

Figure 10.16 MAS architecture for a decentralized control of an MG in an interactive structure.

Figure 10.16 shows a decentralized MG control structure, which is organized with three agent types: distribution network operator (DNO) agent, MGCC agent, and local control agent (LCA). The higher level (DNO agent) corresponds to an MV power network (grid level), and its agent is responsible for communication between the MG and the DNO and the message exchange regarding the energy market. The medium level is the management level in which the agents of all MGCCs coordinate the DGs/loads controllers, market participation, and possible collaborations with the adjacent MGs.

In an MG, the MGCC is responsible for the communication among all the controllers inside of MG as well as the external communication with the other MGCCs and the DNO, to control all DGs concerning the economic and environmental aspects. The main challenge in this method is to develop communication functions such that a new functionality requires minimum changes in the agent-based software. To add a new functionality, all that should be required is to train the agents to deal with a new type of message or a new object. The method is also used to coordinate a large amount of production systems whose task cannot be easily clarified.

10.3 ANN-based Power and Load Forecasting in Microgrids

The increasing use of RES electricity, such as wind and PV power, contributes to the diversity of the energy supply portfolio and greatly reduces the risks of expanded usage of fossil fuel. However, energy productions from the RESs are characterized by uncertainty and intermittency, especially for PV power, which is the case in this section. Hence, new operational challenges appear for transmission grid, such as voltage ramping, regulation requirements, and impacts on system stability. Then, in order to maintain a secure grid operation and reliability, the amount of ancillary services and the required dynamics are increasing. More spinning reserve and backup units are required to perform the regulation needs.

Classically, in order to optimize the power operation and system cost, PV power generation forecasting is needed for system operation, scheduling, and dispatch of required hourly power reserve. It is also needed to perform the different regulation tasks [49, 50]. Some forecasting models have been obtained using simple physics-based methods, but the more common applied technique is the ANN [51, 52]. However, the predicted uncertainty associated to each particular forecast cannot be eliminated even with the best tool. In addition, the combination of power generation and consumption variability with uncertainty forecasting makes the levels reserve setting or scheduling more difficult for power system operators. Therefore, a correct stochastic model for power system management must take into account the uncertainties of generation and consumption. As in [53], a stochastic process is applied for considering wind and solar power generators and load uncertainty. Based on an uncertainty model, the decision-making approaches of power reserve can be computed by an uncertainty analysis of the forecasted error. Uncertainties of PV power generation and load demand are considered for calculating the MG operating power reserve.

In this section, a three-layer backpropagation ANN is applied for next 24 h PV power generation and load forecasting. The predicted results are used for predicting the errors of forecasting and uncertainty analysis. Compared with conventional forecasting schemes, the ANNs have some additional advantages, such adaptability to new conditions and a high input data error tolerance. Many correlated factors, such as solar irradiation, humidity, pressure, cloud cover percentage, temperature of ambient/panels, and wind speed, may affect the output of PV power generators, and different parameters have been used in the past works. In [51], irradiation and the sampling timestamp (date and time) are used for PV power forecasting. While in [54], humidity, average temperature, and global solar radiation are used for next 24 h solar radiation predict.

In the present work, an MG with 1 MW PV generator and three micro-gas turbines (MGTs) with 300 kW rated power (for each one) are considered. The MG is useful for residential consumers with a 1 MW peak load [20], as will be explained in Chapter 12. Here, first, a two backpropagation ANNs are used for PV power and load forecasting. Then, additional two ANNs have been applied to forecast errors of PV power and load prediction. For the sake of error analysis, an hourly probability density function (PDF) of all the predicted errors is used.

10.3.1 PV Power Prediction

In [55], a backpropagation ANN for next 24-h PV power generation prediction is developed based on last n hour PV power generation, solar irradiation, and air temperature on the forecast day. The ANN structure is shown in Fig. 10.17. In the performed study, the PV power forecasts are evaluated with data from a real 17 kW PV power plant.

Firstly, the ANN is trained with historical record data and then predicted weather data are used as input variables of the designed ANNs to get hourly PV power generation output prediction. The efficiency of the proposed method is validated by analyzing the maximum error and mean absolute percentage error between the predicted and measured values as well as standard deviations. The results are evaluated with data from a real 17 kW PV power plant. The input data for the first layer are last 24 hourly points of PV power and solar irradiation, as well as 24 points of hourly average temperature of the predicted day. A trial and error method has been used to determine the appropriate hidden layer number and the hidden neuron number in each hidden layer. A total of 24 hourly predicted points of PV power output are taken as output (output layer).

The root-mean-square error (RMSE) function is used to determine the number of hidden units and weights as follows:

$$RMSE = \sqrt{\frac{1}{n}\sum_{i=1}^{n}(y - y_{pred})^2} \tag{10.10}$$

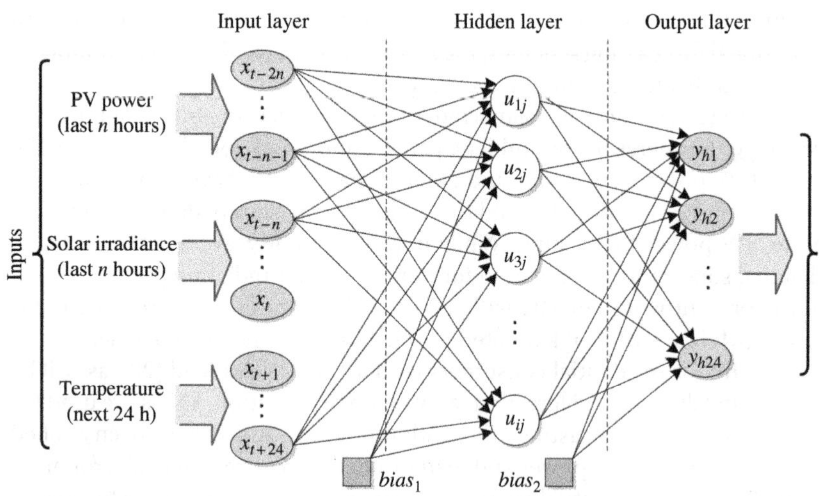

Figure 10.17 Structure of the ANN-based PV power forecasting system.

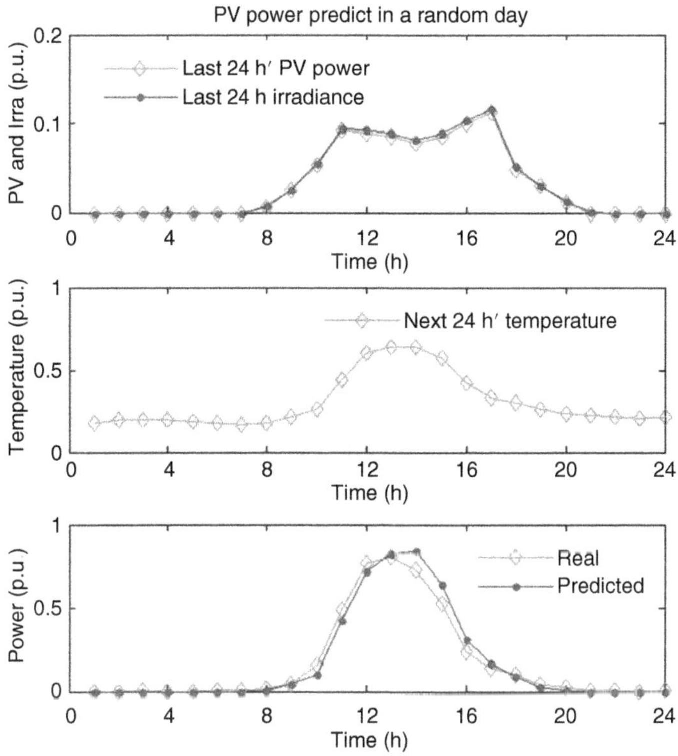

Figure 10.18 PV power prediction on the test sample on a random day.

Then, the mean absolute error (MAE) between the predicted value y_{pred} and the actual value y_{meas} has been calculated.

$$MAE = \frac{1}{n}\sum_{i=1}^{n}|y_{meas} - y_{pred}| \tag{10.11}$$

About 60%, 20%, and 20% of 2-year real data are used to create the training set, the validation set, and the test set, respectively. The predicted values and actual measurements on a random day (from the test sample) are given in Fig. 10.18. The obtained results of RMSE and MAE for next 24 h PV power predictions are given in Table 10.1.

10.3.2 Load Forecasting

Similarly to the PV power prediction, an ANN can also be used for the load forecasting. For example, consider a three-layer ANN, in which the input layer includes last n hours of load demand measurements and 24 h average

Table 10.1 Errors of the PV power forecasting with ANNs.

Data	RMSE (%)	MAE (%)
Training set	5.09	2.69
Validation set	5.58	3.13
Test set	5.95	3.12

temperature of the forecast day; an output layer gives next 24 h load demand prediction and a hidden layer.

Similarly to PV power forecast, the real load data are used, and for ANN training, validation, and test, 60%, 20%, and 20% of the existing data are applied, respectively. The RMSE has been minimized to calculate the hidden neuron number and weight. The minimum RMSE is found with 370 hidden layer units and with last 48 h load measurements.

The predicted values and actual measurements on the test samples on a random day are shown in Fig. 10.19. The inputs are historical load consumption

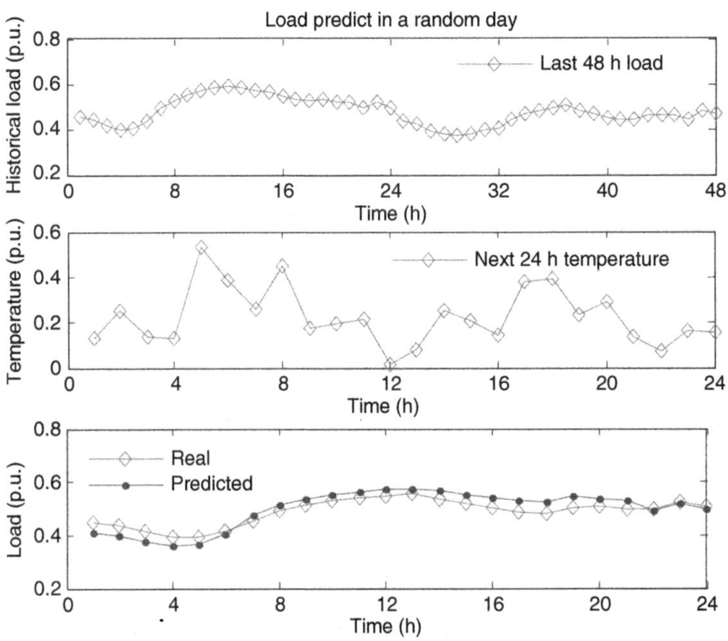

Figure 10.19 Load prediction on the test sample on a random day.

Table 10.2 Errors of the load demand forecasting with ANNs.

Data	RMSE (%)	MAE (%)
Training set	4.14	3.22
Validation set	4.75	3.77
Test set	4.90	3.84

of last 48 h and predicted temperatures for next 24 h. Outputs are next 24 h of load. The obtained results of RMSE and MAE for next 24 h load forecasting are listed in Table 10.2.

10.3.3 Forecasting Error

According to the prediction results from the PV power and load forecast, an ANN-based methodology can be used to predict the errors of forecasting as shown in Fig. 10.20. System uncertainty can be assessed through the analysis of estimated errors of PV power and load forecasting. An hourly error *pdf* has been obtained by collecting all the prediction errors.

10.3.3.1 PV Power and Load Forecasting Error Estimation

Each sample (i) of PV power and load for a given hour (h) is composed of two parts: forecasted value and predicted forecasting error.

$$Pv_h^n(i) = \widetilde{PV}_h(i) + \varepsilon_{Pv\_h}(i) \tag{10.12}$$

$$L_h(i) = \tilde{L}_h(i) + \varepsilon_{L\_h}(i) \tag{10.13}$$

where $\widetilde{PV}_h(i)$ is a PV power forecasted data, $\tilde{L}_h(i)$ is a load forecasted data, $\varepsilon_{PV\_h}(i)$ is the error of PV forecasting, and the $\varepsilon_{L\_h}(i)$ is the error of load forecasting.

In order to assess prediction uncertainties, two additional neural networks are applied to assess the errors of PV power and load forecasts ($\tilde{\varepsilon}_{PV\_h}(i)$ and $\tilde{\varepsilon}_{L\_h}(i)$, respectively). Outputs are 24 predicted forecasting errors corresponding to each hourly predicted PV power and load, while inputs are the last 24 h forecasting errors of PV power and loads. The forecasted results of error prediction on PV power and load forecasting are listed in Tables 10.3 and 10.4, respectively. As it can be seen, predicting the errors of each hourly of PV power and load forecasting has decreased the RMSE and MAE on the training, validation, and test sets. These obtained estimated errors have been used for uncertainty analysis.

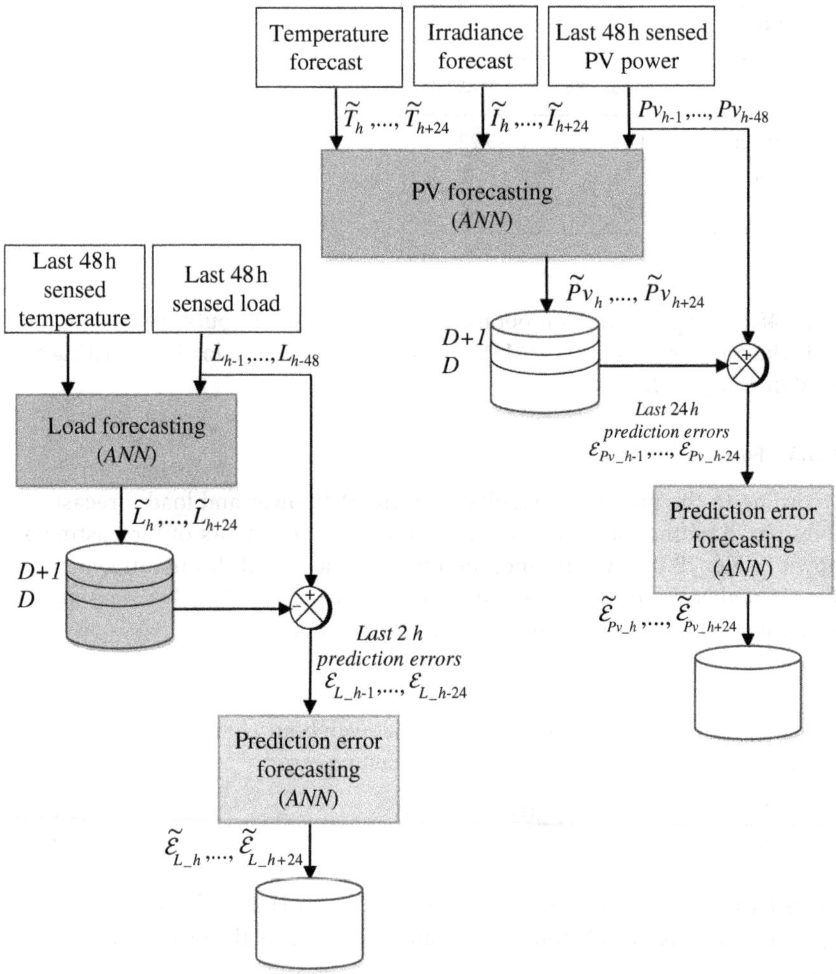

Figure 10.20 PV power production forecasting, load and error prediction with ANNs.

Table 10.3 Results of the error estimation for the PV power forecasting.

Data	RMSE (%)	MAE (%)
Training set	4.37	2.14
Validation set	5.28	2.93
Test set	5.01	2.72

Table 10.4 Results of the error estimation for the load forecasting.

Data	RMSE (%)	MAE (%)
Training set	4.00	3.05
Validation set	4.04	3.27
Test set	3.95	3.07

10.3.3.2 Errors Estimation Analysis

Figure 10.21 illustrates the error prediction results of PV power and load forecasting from the test samples at 2 p.m. in 160 days.

The distributions of predicted errors for PV power and load forecasting have been computed with a normal PDF for each of the 24 h. The PDF function was used with mean μ and standard deviation σ. Figure 10.22 shows the normal probability and pdf of predicted errors of PV power and load forecasting at 2 p.m. (14th hour).

For each hour, the mean average and the standard deviation of the predicted error are calculated. The forecasting uncertainty is determined as forecasted data with upper and lower bound margins of the predicted errors of forecasting. The inverse of the normal cumulative distribution function (CDF) is used to extract the upper and low bound margins with a desired probability. As an example, application for assessment of PV power uncertainty is shown in

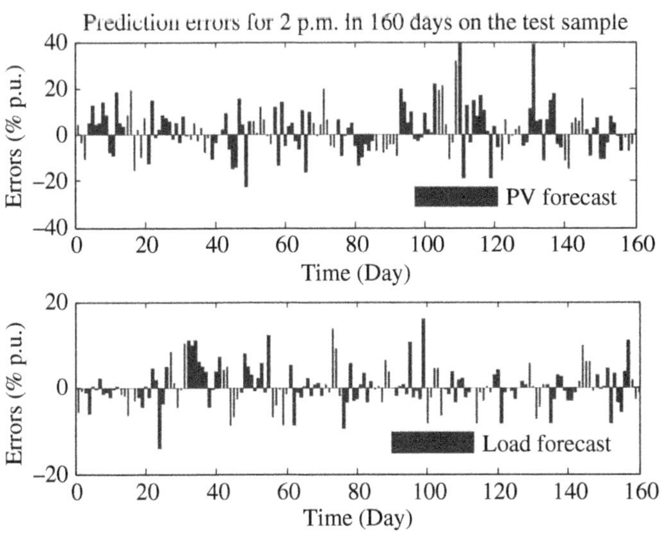

Figure 10.21 PV power and load prediction errors at 2 p.m.

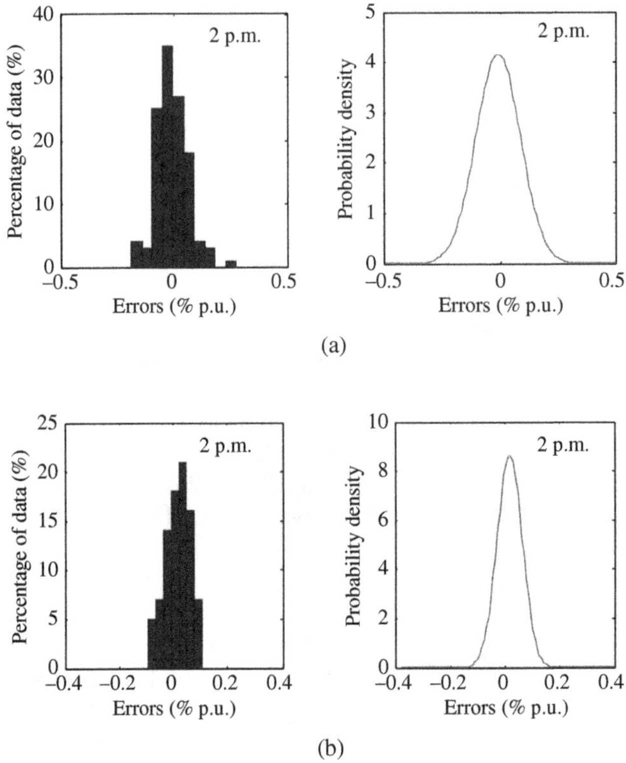

Figure 10.22 Normal probability and PDF of the (a) PV power errors for 2 p.m. ($\mu_1 = -0.0353$, $\sigma_1 = 0.1571$), and (b) load errors for 2 p.m. ($\mu_1 = -0.0015$, $\sigma_1 = 0.0113$).

Fig. 10.23, where $\widetilde{PV}_h^{(n)}$ and $\tilde{\varepsilon}_{PV\_h}^{(n)}$ are forecasted and predicted errors of PV power forecasting, respectively; the μ_{PV_h} and σ_{PV_h} are mean and standard deviation of forecasting errors.

Figure 10.24 presents the uncertainty of PV power and load forecasting with various probability indexes from 60% to 90% on a random day. As it can be seen, the uncertainty increases as the time horizon grows.

10.4 Intelligent Frequency and Voltage Control in Microgrids

In this section, four control design examples for voltage and frequency control in MGs using fuzzy logic, PSO, and MASs are given. In the first example,

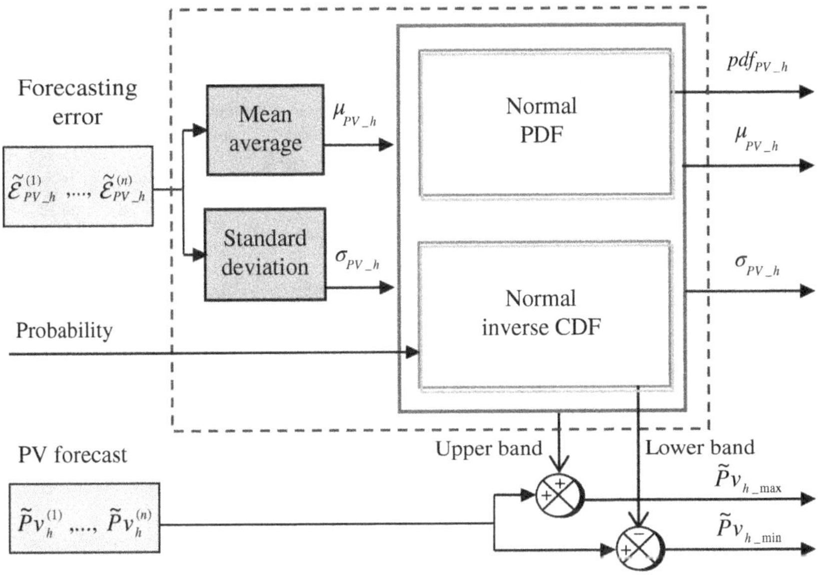

Figure 10.23 PV power calculation at hour h with a given probability.

a fuzzy logic supervisory (FLS) control in the MGCC is responsible for supporting the grid frequency regulation. The MG case study consists of an MGT and a combination of PV panels and supercapacitor (SC) as an active generator (AG). The second example addresses a fuzzy-based distribution voltage control in DC MG. The proposed control strategy combines the power of fuzzy logic and gain-scheduling controls, and the effectiveness is examined via laboratory experiments. The PSO algorithm is used for stability improvement of an MG with multiple Virtual Synchronous Generators (VSGs) in the third example, and finally, an MAS-based secondary frequency control is designed in the fourth example. The MAS-based control strategy provides a desirable coordination between the diesel unit and energy capacitor system (ECS) in the MG to track the probable load changes in the system.

10.4.1 Fuzzy-logic-based Supervisory Frequency Control

In this section, a FLS is developed to produce a reference power for frequency control, contributing an AG, which includes PV arrays, MGT, and SCs [56]. The developed supervisory strategy can be also applied when an MG supports the grid frequency control in the connected operation mode. Thus, it improves the grid integration of dispersed DGs/RESs through the MGs. The performance of the proposed FLS is evaluated in the simulation environment.

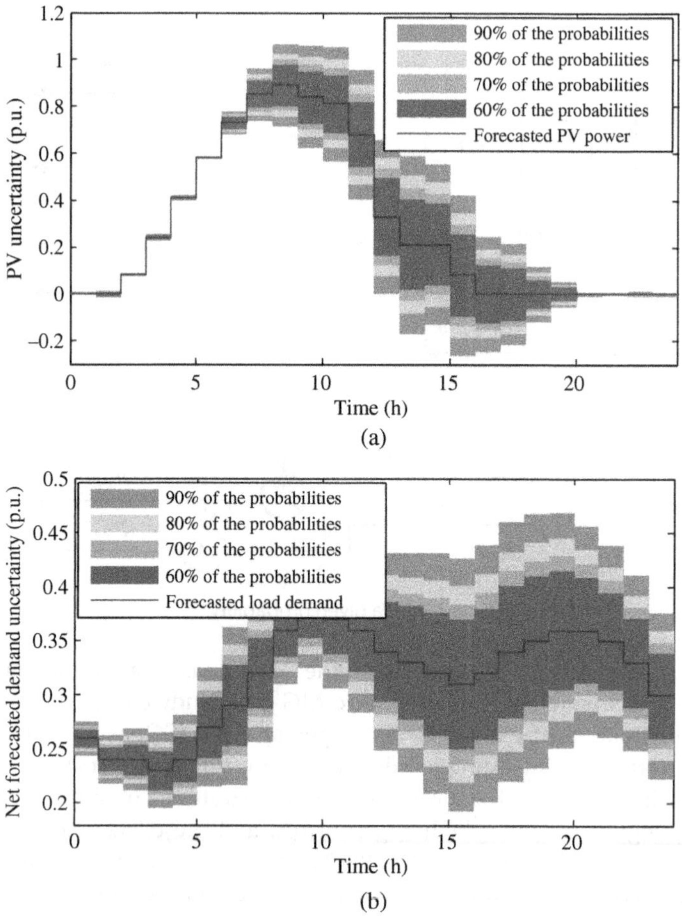

Figure 10.24 Forecasting with uncertainty (random day): (a) PV and (b) load.

10.4.1.1 Case Study

Figure 10.25 shows the MG case study. The MG includes PV generators (17.3 kW), an SC (as a short-term storage unit), and an MGT (28 kW) to be able to generate a reference power and to ensure a power reserve. The SC considered as a 16.5 F/480 V SC bank with a specific energy of 138.6 kWh/kg. These generating/storage units provide an AG, and the MG contains the mentioned AG and 10 kW local load. The MG is connected to a grid (distribution network) including a conventional 200 kVA generator and 130 kW load, as shown in Fig. 10.25.

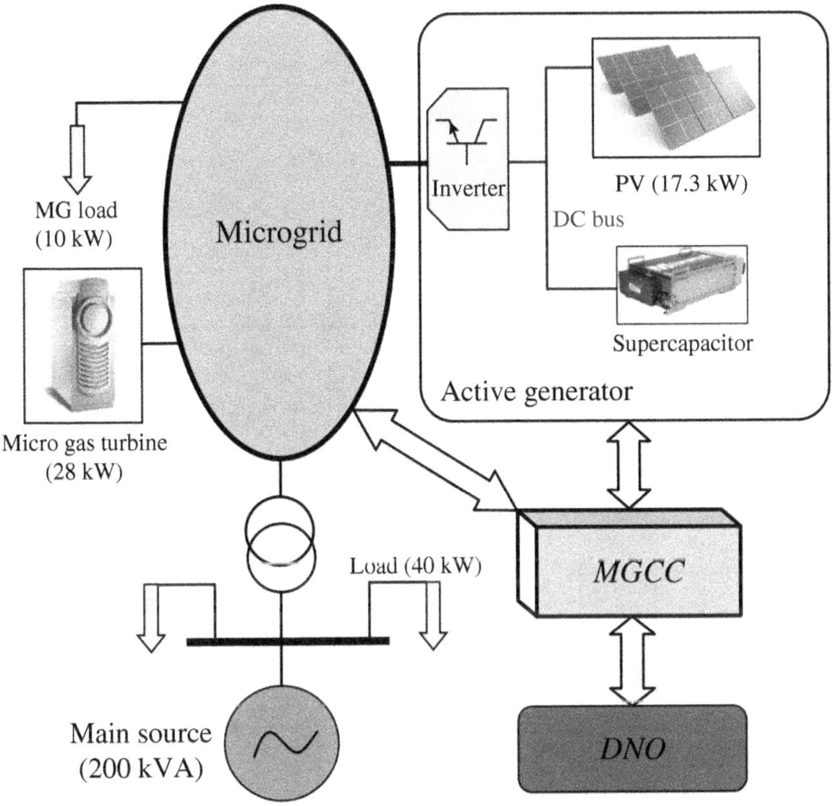

Figure 10.25 Case study.

A basic supervisory strategy for frequency control in order to achieve a smoothing power with the AG similar to the one shown in Fig. 10.26 has been developed in [57]. The input power is the fluctuating measured PV power ($P_{PV\text{-}mes}$). The reference power of the SCs enables to smooth the fast fluctuations of the power, thanks to a high-pass filter (HPF). Then, the smoothed power is compared to the reference power, and the MGT produces the required regulating power ($P_{MGT\text{-}ref}$). A feedback loop with a proportional gain is used to improve the MGT response. This idea has some drawbacks that are explained in [56].

10.4.1.2 Proposed Fuzzy-logic-based Control Strategy

The proposed FLS framework to follow the network reference power provided by DNO is shown in Fig. 10.27. As mentioned in Section 10.2, due to the system

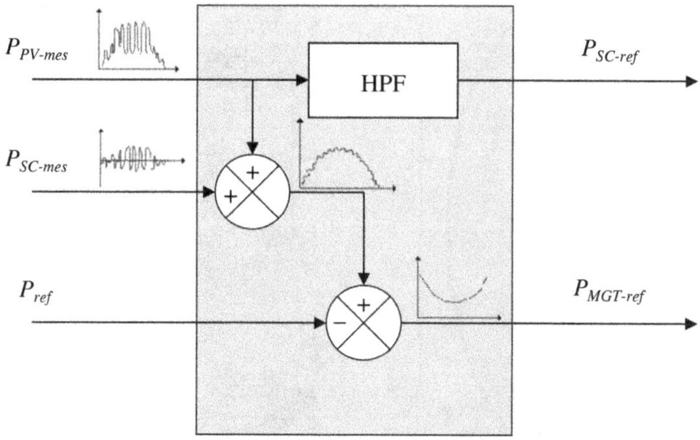

Figure 10.26 A basic idea for supervisory active power compensation.

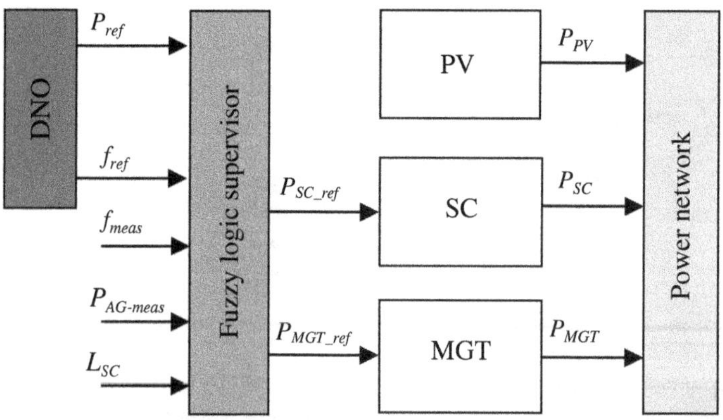

Figure 10.27 The proposed supervisory control framework.

uncertainty, as well as the random behavior of the PV irradiance and network load, the fuzzy logic could be considered as a desirable alternative for the mentioned supervisory control issue [58, 59].

Inputs of the supervisor are as follows: the network reference power P_{ref} and the reference frequency f_{ref}, the MG measured frequency f_{meas}, the AG power $P_{AG\text{-}meas}$, and the SC storage level L_{SC}. In the AG, the SCs have to filter the fast fluctuations of the generated power, and the MGT has to smooth the power and to avoid the SC saturation state level. For the primary frequency control, the SC is used to decrease the peak of frequency for few seconds due to its fast rise time. The MGT performs the primary frequency control automatically and the secondary frequency control following the received power reference

($P_{ref\text{-}S}$). The $P_{ref\text{-}S}$ is a signal called *secondary load frequency level,* lies between −1 and +1, which is sent to the generating units in order to change their power set-point values.

The structure of the FLS is shown in Fig. 10.28, which includes two parts: (i) a fuzzy logic unit to provide the reference power and to allow participation in the system frequency control and (ii) a droop characteristic, which performs the primary frequency control. The inputs of the fuzzy logic unit are the fast fluctuations of the power ΔP_{PV}, the SC level L_{SC}, the variation of frequency Δf, and the power ΔP with $\Delta P = P'_{ref} - P_{AG\text{-}meas}$. The reference power of the AG is obtained as follows:

$$P'_{ref} = P_{ref} + P_{ref\text{-}P} + P_{ref\text{-}S} \tag{10.14}$$

where $P_{ref\text{-}P}$ is the power reference for the primary frequency control, and $P_{ref\text{-}S}$ is used for the secondary frequency control. Two filters allow for

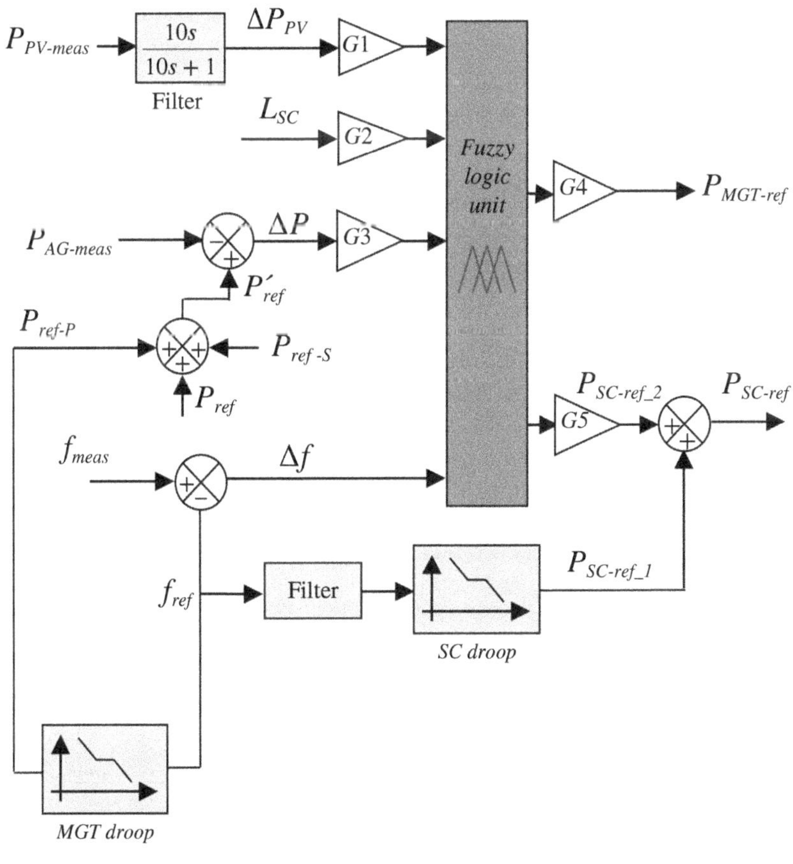

Figure 10.28 Block diagram of the proposed FLS.

maintaining only the fast fluctuations of power or frequency. The SC and MGT reference powers are the sum of the power references for the frequency control ($P_{sc\text{-}ref\_1}$, $P_{MGT\text{-}ref\_1}$) and the smoothing of power ($P_{sc\text{-}ref\_2}$, $P_{MGT\text{-}ref\_2}$), respectively.

As explained in Section 10.3, three parts characterize a fuzzy logic unit: (i) fuzzyfication part, which consists of converting the inputs with membership functions as shown in Fig. 10.29; (ii) inference part, which consists of defining the rules; and (iii) defuzzyfication part, which consists of converting the outputs with membership functions, as can be seen in Fig. 10.30.

10.4.1.3 Simulation Results

The simulation time took 3 h with a reference power for the AG changing each hour $P_{AG\text{-}ref} = \{24; 30; 26\}$ kW. Each hour, 40 kW load is disconnected for 400 s. The DNO sends the grid frequency reference 200 s after the load disconnection, which is equivalent to a reference power of −2 kW for 200 s. The results are compared to the basic supervision strategy, represented by the dotted light curves. Figure 10.31 shows the results of the total simulation, while Fig. 10.32 focuses on the first hour of simulation. Figure 10.31 shows the PV power (Fig. 10.31a), the MGT power (Fig. 10.31b), the SC power (Fig. 10.31c), the AG power (Fig. 10.31d), the SC level (Fig. 10.31e), and the grid frequency (Fig. 10.31f).

For the PV power, it is noteworthy that with the MGT and the SCs, the AG provides the reference power (Fig. 10.31d) in the MG and therefore the power requested by the DNO in the grid. Thus, the frequency regulation performance is improved in comparison with the basic strategy (Fig. 10.31f). For each reference power, the primary and secondary frequency controls are performed. The energy level of the SCs (in case of using FLS) is never in saturation state (except at the time of frequency control), thus the AG provides a more regular power reserve to contribute to the network frequency control. The small drop of power observed at $t = 6300$ s is due to the PV power being low; the MGT is in high saturation and cannot provide the power requested any more.

Figure 10.32 shows the grid frequency, the MGT power, the SC power and level, and the AG power. When the load is disconnected at $t = 800$ s, the SC and MGT power decrease to reduce the frequency deviation. The peak of frequency 50.97 Hz is decreased to 50.60 Hz as shown in Fig. 10.32d. In the second step, the inertia of the revolving machines allows it to regulate the frequency. At $t = 1000$ s, the DNO sends a reference power and the MGT power increases. In the same way, when a load is connected, the SC and MGT power increase (Fig. 10.32b and c). The peak of frequency is increased from 49.04 to 49.44 Hz. The power delivered by the AG source corresponds to the reference power when the frequency is equal to 50 Hz.

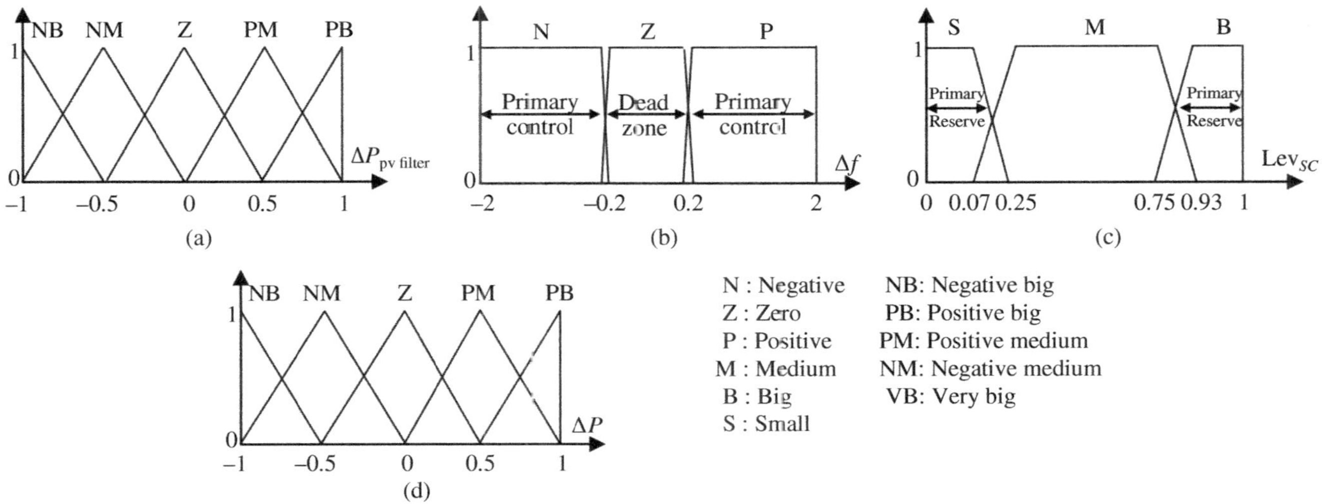

Figure 10.29 Fuzzyfication, membership functions of (a) ΔP_{PV}, (b) Δf, (c) L_{SC}, and (d) ΔP.

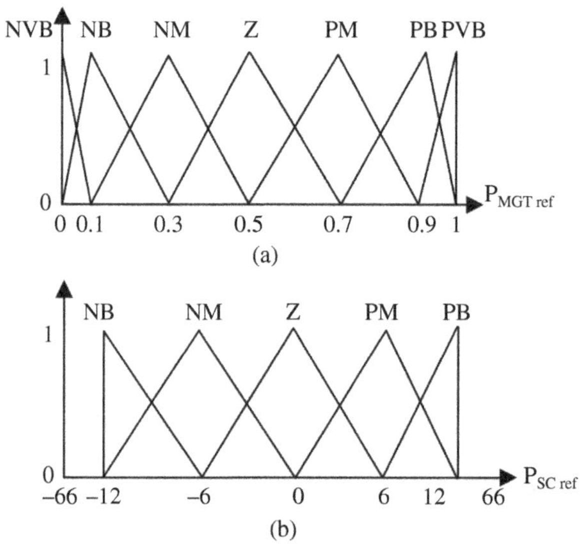

Figure 10.30 Defuzzyfication, membership functions of (a) MGT power and (b) SC power.

10.4.2 Fuzzy-based Distribution Voltage Control in DC Microgrids

In Chapter 6, a combined gain-scheduling control and droop control methodology is developed for distributed voltage control in a DC MG [60]. This section focuses on the application of fuzzy control to improve the gain-scheduling control and demonstrate the validity of the proposed control through experiments. The performed experimental system is the same as the one described in Chapter 6, and its configuration is redrawn in Fig. 10.33.

The electric double-layer capacitors (EDLCs) are used as the energy storage unit, and each DC/DC converter of EDLC controlled the DC distribution voltage when the system was operated in an intentional islanding mode. The experimental results show that the DC distribution voltages are within 340 V ± 5%, and the energy ratios of the storage units are approximately equal, implying that DC voltage regulation and stored energy balancing control can be realized simultaneously by applying the proposed control strategy.

10.4.2.1 Proposed Control Strategy

As a concrete system, a DC MG for a residential area is considered as shown in Fig. 6.1. There are around 50–100 houses in the system, and each house has a cogeneration system (CGS), such as microturbine combined heat and power (micro-CHP) unit, and gas engine. The CGSs are connected to DC distribution line (three-wire, ±170 V), and the electric power is shared among the houses.

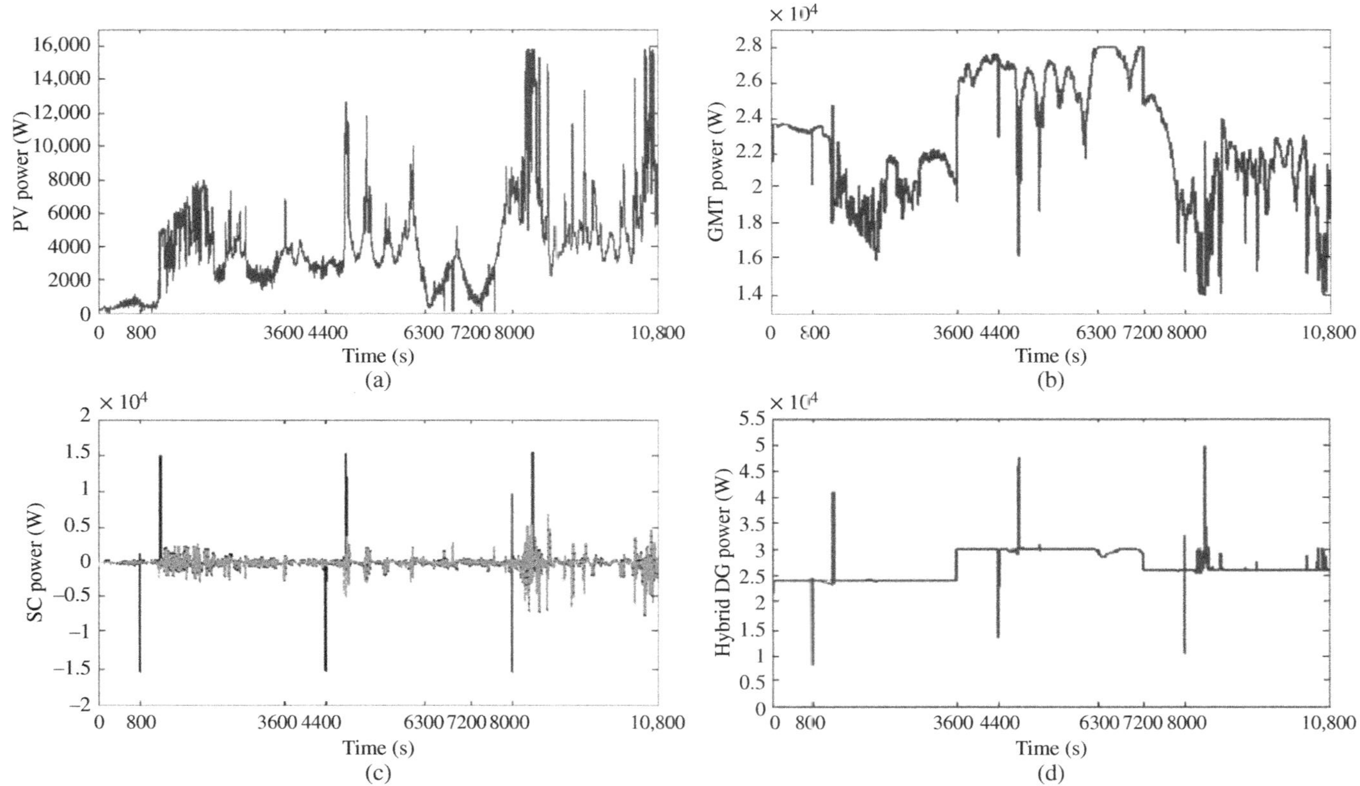

Figure 10.31 System response in comparison with basic strategy: (a) PV power, (b) MGT power, and (c) SC power and level, (d) AG power, (e) SC level, and (f) frequency.

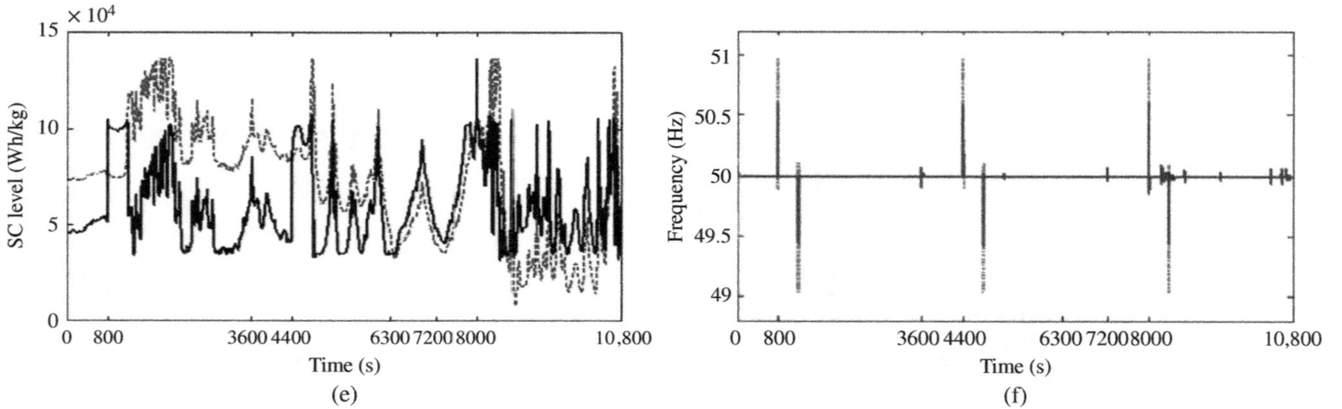

Figure 10.31 (*Continued*)

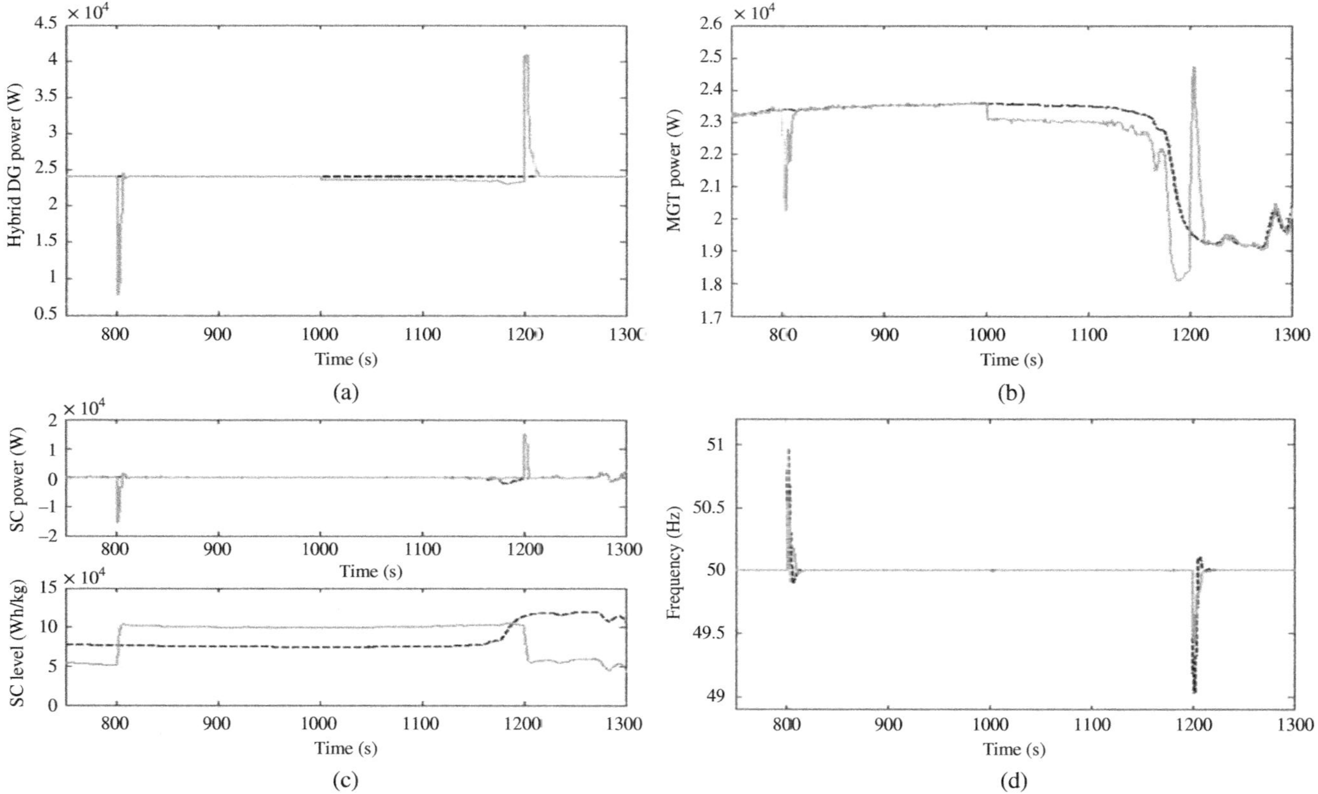

Figure 10.32 System response in comparison with basic strategy: (a) PV power, (b) MGT power, (c) SC power and level, and (d) AG power.

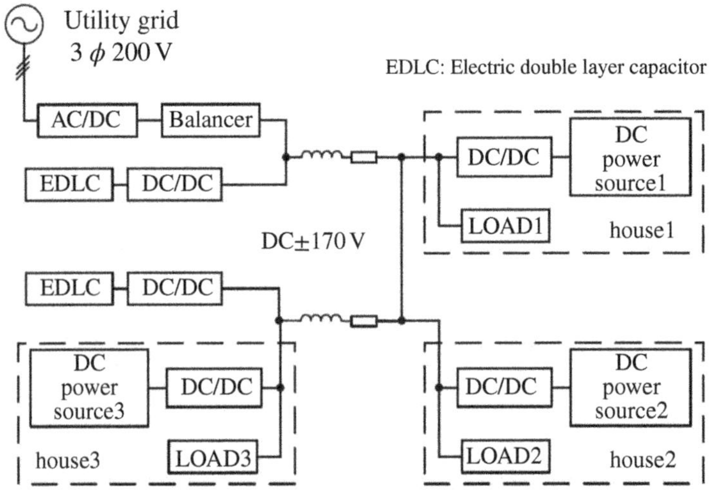

Figure 10.33 Configuration of the MG case study.

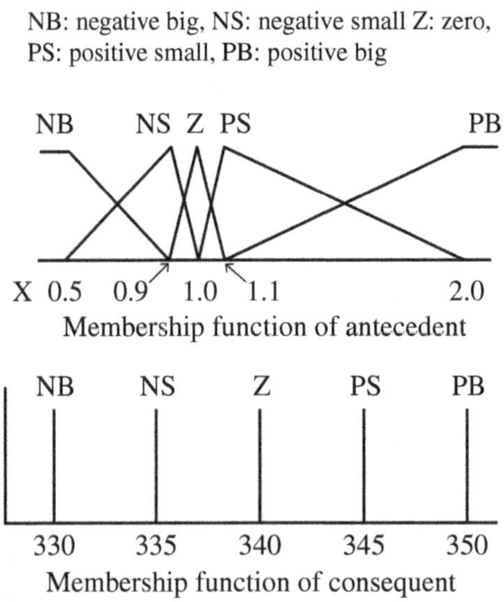

Figure 10.34 Control rule of fuzzy control.

A DC voltage control that incorporates a fuzzy control that changes the DC voltage reference to balance the stored energy is proposed. Figure 10.34 shows the membership functions of the developed fuzzy control unit. The input X (X_k) is the ratio of the stored energy and the average of all stored energies as shown in (6.5).

Then, a membership value is calculated from the membership function of PB, PS, Z, NS, and NB. Finally, the DC voltage reference is obtained from the resulting membership function. For example, if X_1 is 1.7, the values of PS and PB in the antecedent membership function are 0.4 and 0.6, respectively. Then, V^*_{DC} of EDLC1 is given by the consequent membership function as $V^*_{DC} = 345 \times 0.4 + 350 \times 0.6 = 348$ [V].

It is assumed that the DC distribution voltage has a tolerance of ±5% with reference to AC systems. In this case, the upper and lower references should be within 3% because the gain-scheduling control includes a tolerance of ±2%. Therefore, the upper and lower limits are selected as 350 and 330 V, respectively, in the membership function of the consequent. An isosceles triangle is usually used as a membership function. However, to realize a good voltage regulation, the peaks of PS and NS in the antecedent are shifted to 1.1 and 0.9, respectively. We first obtained this membership from computer numerical simulation and then applied it to experiments. Although the fuzzy rules are determined on the basis of experience, their effectiveness are demonstrated through simulations and experiments described in the following sections.

10.4.2.2 Simulation Results

To confirm the performance of the proposed method, simulations are conducted using PSCAD/EMTDC software. The circuit and the main parameters are given in Chapter 6 (Fig. 6.26 and Table 6.2). The configuration and the parameters are based on the experimental system shown in Fig. 10.33. The simulation parameters are set as described in Section 6.6.

Although the fuzzy control is chosen for the stored energy management in this work, the other control could be a candidate to obtain the voltage reference V^*_{DC}. For example, a simple droop control is illustrated in Fig. 6.25. The control diagram is simpler than the proposed fuzzy control, and the only gain K_v is needed as the control parameter. In other words, it has only one degree of freedom to adjust the control performance.

Figure 10.35a shows the simulation results when the proposed control is used. The stored energy ratio tended to be 1. The features of EDLC currents are different from those shown in Fig. 6.28, because EDLC2 has 1.3 times energy at the initial state. EDLC2 initially discharges its power, and then the energies of both EDLCs are balanced at around 80 s. The reason for the difference in the EDLC terminal voltages, V_{EDLC1} and V_{EDLC2}, despite an almost unity ratio of EDLC charge (W_2/W_1), is the difference in the maximum voltages of EDLC1 and EDLC2. Regarding the distribution voltages at the output converters of the EDLCs, the maximum and the minimum are 349.9 and 331.7 V, respectively. Although the range of the voltage is a little wider than the previous results, the range is within 340 V ± 3%, which is satisfied with the assumed specification (340 V ± 5%) described earlier.

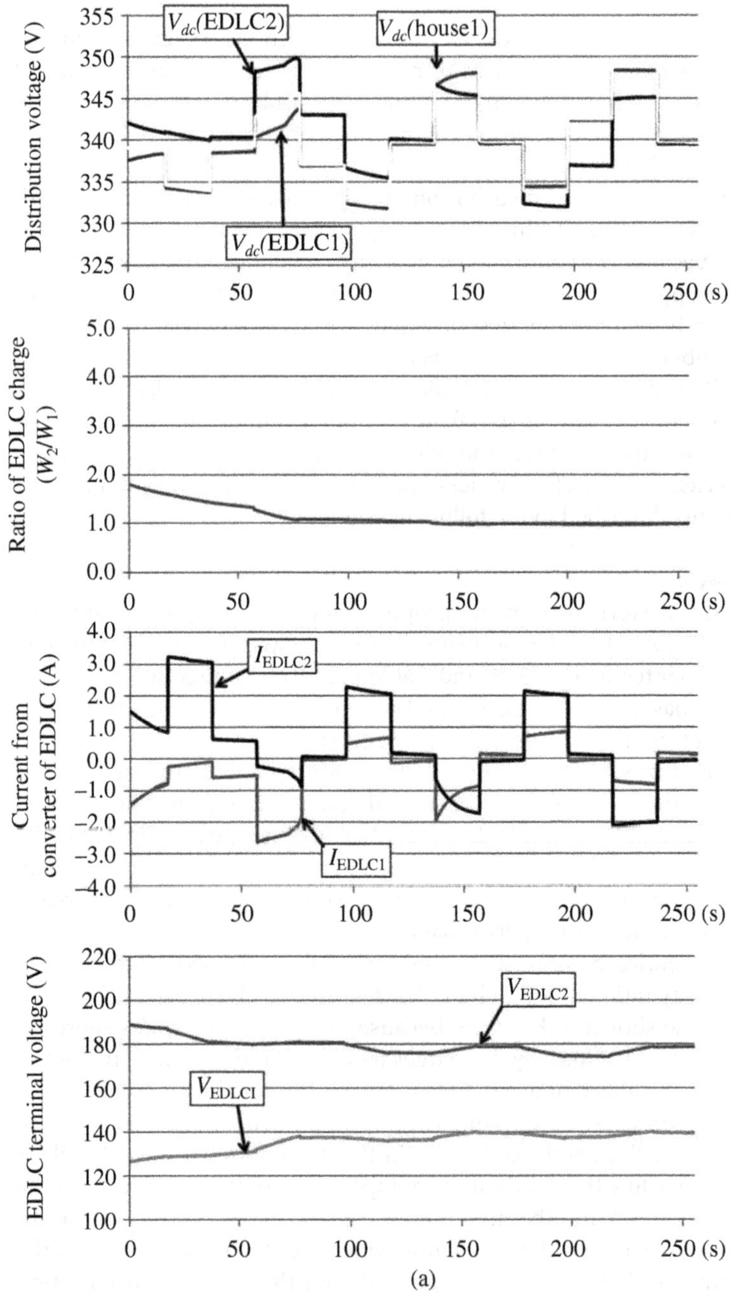

Figure 10.35 Simulation results (gain-scheduling control and fuzzy control) under different initial conditions: (a) $W_2/W_1 \approx 2$ and (b) $W_2/W_1 \approx 0.5$.

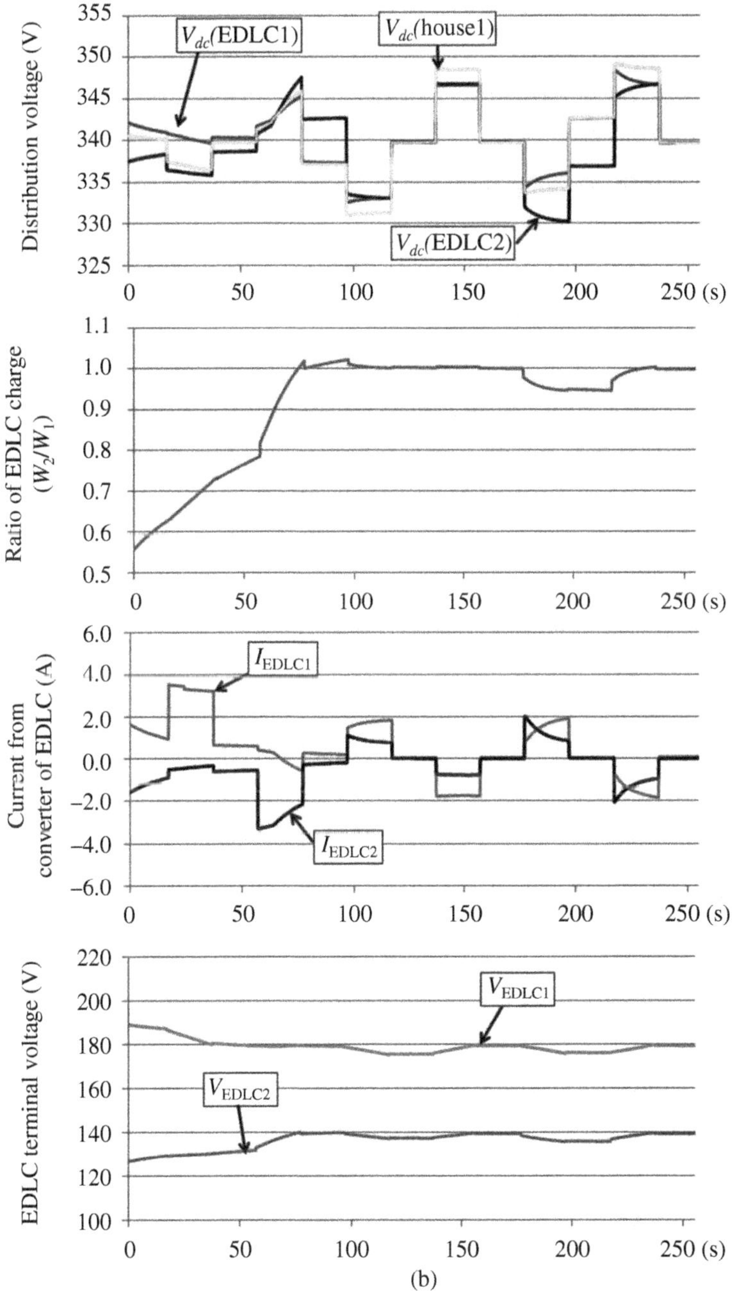

Figure 10.35 (*Continued*)

10.4.2.2.1 Comparison between Fuzzy Control and Droop Control In the performed simulation, the configuration, parameters, and events are the same as those in the previous simulations discussed in Chapter 6. The only difference is that all parameters of EDLC1 and EDLC2 are exchanged: EDLC1 (rated voltage, 216 V, and rated capacitance, 18.75 F) and EDLC2 (rated voltage, 160 V, and rated capacitance, 18 F). The minimum voltage of both EDLCs is also set to 100 V, and the initial voltages of EDLC1 and EDLC2 are 200 V and 125 V, respectively. Therefore, the initial stored energy ratio (W_2/W_1) is about 0.5. Figure 10.35b shows the simulation results when the fuzzy control was used. Although there is no overshoot as the results of the droop control, the results after 120 s are almost the same as those of the previous droop control. Figure 10.36 shows the relation between input X_k and the output V^*_{DC} in each control. The fuzzy control changes the slope in the range of X_k between 0.5 and 2.0. On the contrary, the droop control has one slope in the shorter range of X_k between 0.8 and 1.2, and the V^*_{DC} is fixed at the minimum (330 V) or the maximum (350 V) in the other area.

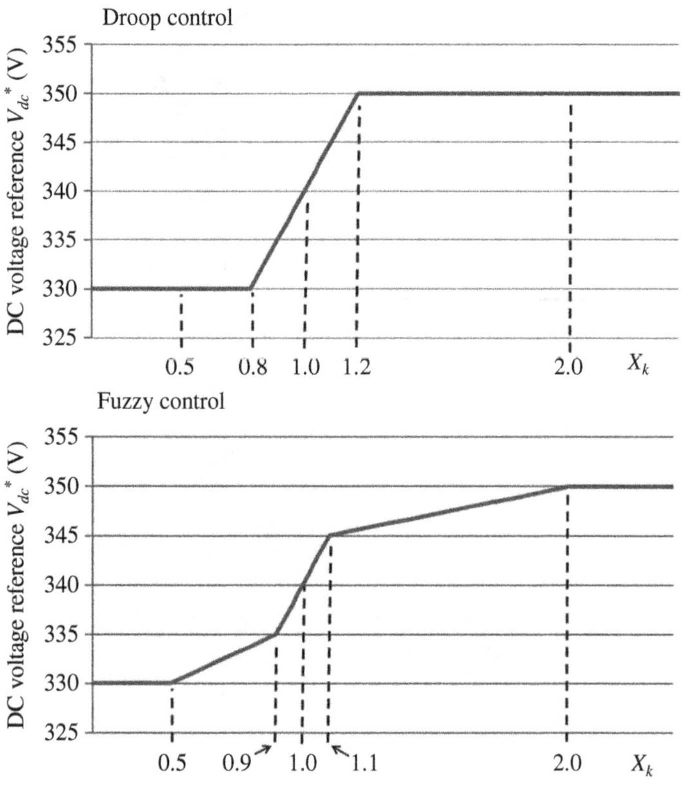

Figure 10.36 Relations between input X_k and voltage reference V^*_{DC}.

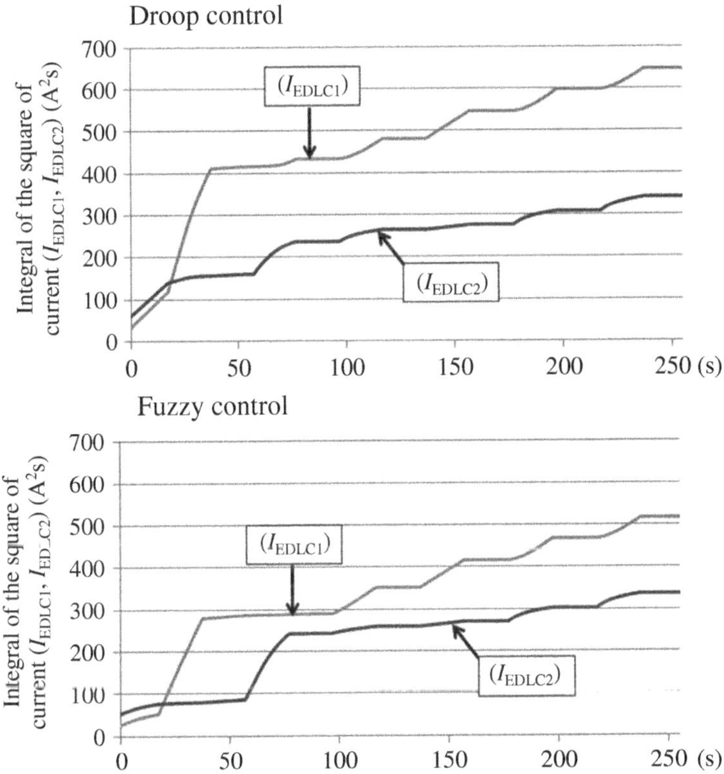

Figure 10.37 Integral of the square of the current (I_{EDLC1} and I_{EDLC2}).

The characteristics of the proposed control can be seen from the results shown in Fig. 10.35b. Since the initial value of X_k is set to 0.5, the power from EDLC1 is supplied to EDLC2 in order to balance the stored energy. The absolute values of I_{EDLC1} and I_{EDLC2} decrease during 0–20 s, which contributed to the reduction of losses due to the line resistances (R_{line1} and R_{line2}) and the inner resistances of both EDLCs. Figure 10.37 shows the integral of the square of I_{EDLC1} and I_{EDLC2} in each case. In case of the fuzzy control, the integral numbers of the square of I_{EDLC1} at 250 s are 20.1% lower than the results of the droop control, while the integral numbers of the square of I_{EDLC2} at 250 s are almost the same. It indicates that the loss in case of the fuzzy control is lower than that of the droop control. Therefore, this work proposes the fuzzy control as the energy balance control.

10.4.2.3 Experimental Results

Two experiments are conducted to confirm the performance of the proposed method practically. The practical circuit and the main parameters are shown in

Chapter 6 (Fig. 6.31 and Table 6.6). The system configuration and parameters are almost the same as the simulation and experimental circuits mentioned in Sections 6.7 and 6.8.

10.4.2.3.1 Case I The on/off events of the loads and DGs in case I are the same as those of the previous simulation shown in Fig. 6.27. The initial voltages of EDLC1 and EDLC2 are 125 and 200 V, respectively. Then, the initial stored energy ratio (W_2/W_1) is about 2. Figure 10.38a shows the experimental results when the proposed control is used. The energies of both EDLCs balanced at around 100 s. Both the experimental results are almost the same as the simulation results. Besides, the experimental results demonstrate the validity of the proposed methods. Regarding the performance during the change of loads/sources, there are no disturbances or instabilities due to the gain-scheduling control or the fuzzy control. In case of the fuzzy control, at the beginning, the signs of I_{EDLC1} and I_{EDLC2} are opposite to balance their energy; then, the envelopes of I_{EDLC1} and I_{EDLC2} are gradually close to each other during the change of loads/sources.

10.4.2.3.2 Case II In this experiment, Load 1 and the output of DG 3 are turned on from 0 to 1 kW at 5 s. Other loads and DGs are turned off through the experiment. The initial voltages of EDLC1 and EDLC2 are 130 and 180 V, respectively. Therefore, the initial stored energy ratio (W_2/W_1) is about 1.3.

Figure 10.38b shows the experimental results when the proposed control is used. The stored energy ratio (W_2/W_1) approaches 1, and the energy ratios of each EDLC become equal. The voltages of EDLC1 and EDLC2 decrease slightly because of the presence of switching losses of the converters and the line resistance. From this result, it can be confirmed that the stored energy ratio is converged by the proposed control. The response takes around 150 s in this case, which is satisfactory for the given DC MG example. If there is a case where the faster response is needed, the member ship function of antecedent has to be shrunk horizontally centering on $X = 1$.

10.4.3 Particle Swarm Optimization (PSO)-based Stability Enhancement in a Microgrid with Virtual Synchronous Generators

The direct method is one of main transient stability analysis tools for conventional power grids to evaluate the transient energy function after a large disturbance [61]. In this method, the transient energy or a relevant index is evaluated after a contingency or new dispatch configuration. If it is below the critical level, the stability is guaranteed. Evaluating the energy function requires off-line simulation of the system and is not suitable for online monitoring and control [62]. The *rotor angle deviation* (RAD) index can provide a real-time transient stability index and is useful for practical studies of the transient stability issues

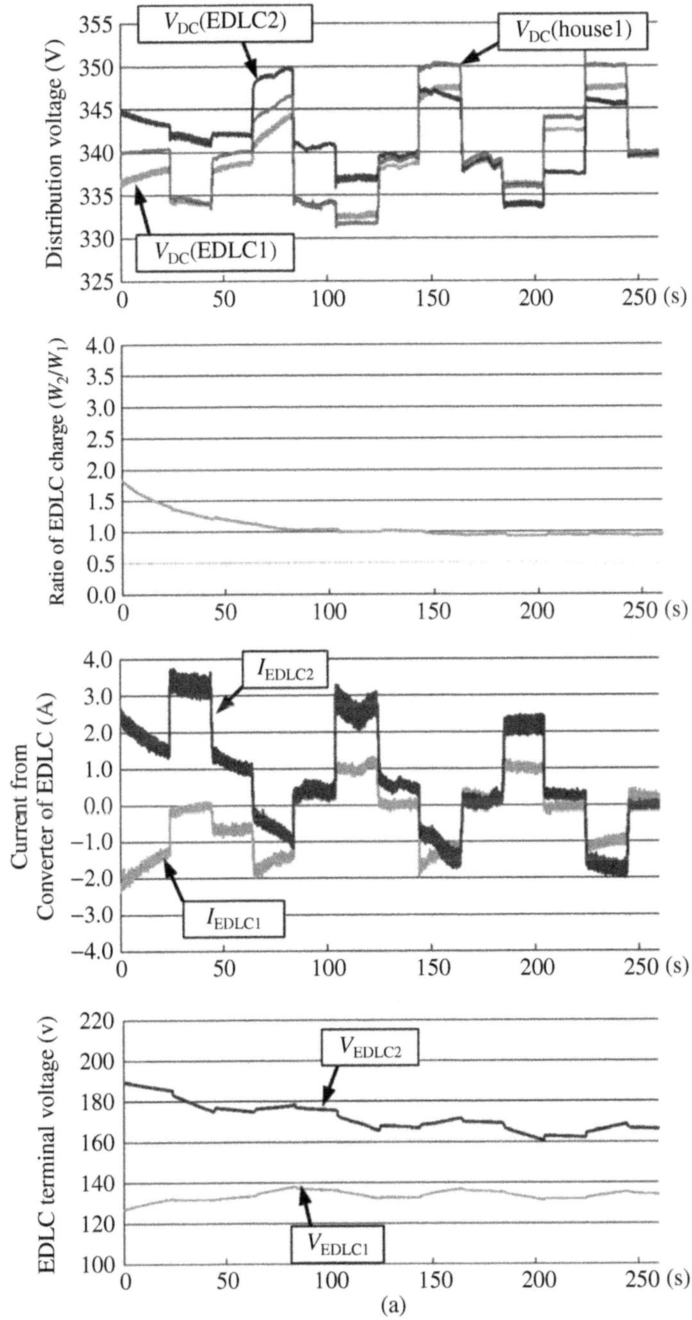

Figure 10.38 Experimental results (gain-scheduling control and fuzzy control) of (a) case I and (b) case II.

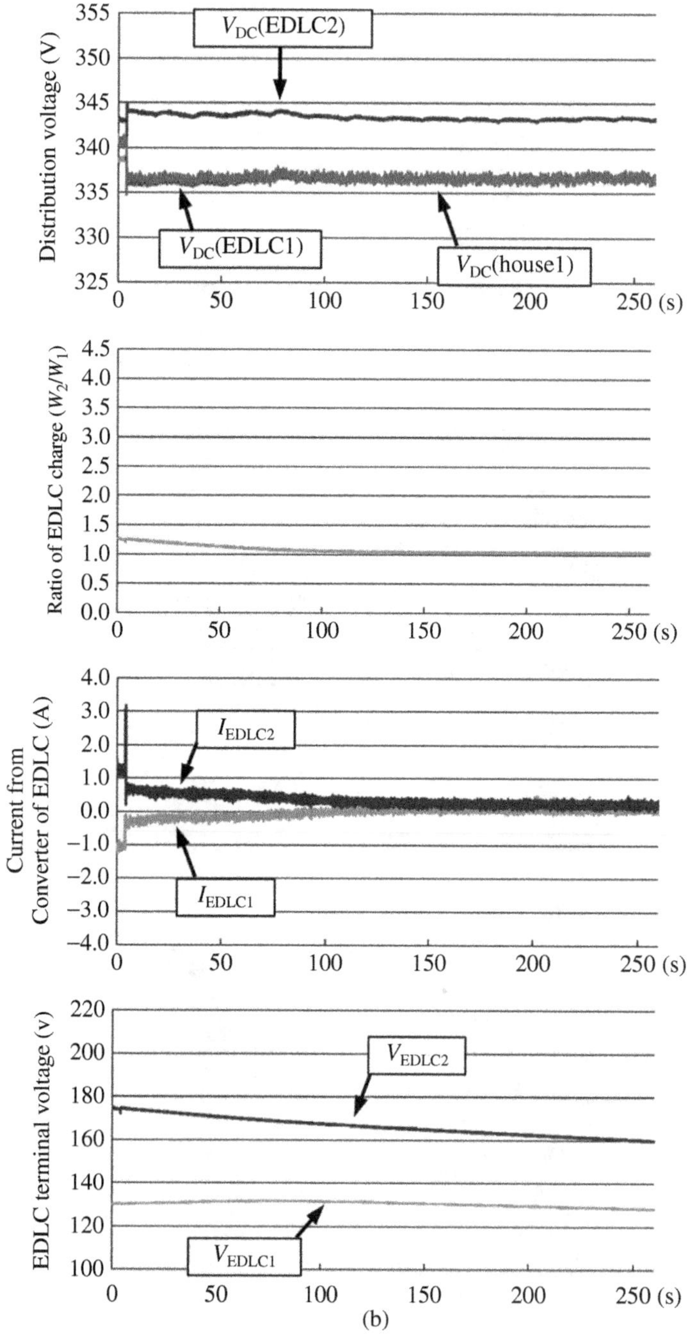

(b)

Figure 10.38 *(Continued)*

[63–66]. This criterion is consistent with industry practice and has been evaluated by the utility engineers [63].

However, in the case of VSG units, the rotor angle is not available to study the transient stability issue. To solve this problem, one may use the *voltage angle deviation* (VAD) of generators with respect to the angle of the center of inertia (COI) to generalize the RAD security constrain for the multi-VSG power grid applications.

In conventional power grids with synchronous generators (SGs), desired performance can be achieved by including the power system security criteria in optimal power dispatching [63–65]. Besides, power system stabilizer is embedded in the system to enhance the stability. Once the power system condition changes, the parameters of the controllers must be retuned, because the optimized parameters under a specific condition might not result in the desired response under another condition. The AI methods are implemented to readjust the parameters when system condition changes. Major state quantities of an SG such as mechanical angular frequency and voltage angle can be found in the VSG control issue. Moreover, the VSGs have an advantage with respect to the SGs that their model parameters can be tuned in a real-time manner to achieve a desirable performance.

In this section, to have a smooth transition after a change in operation or disturbance, the PSO is implemented to find appropriate parameters of VSG units. Then, one can apply a proper mechanism such as alternating inertia scheme (described in Chapter 8) for damping the oscillation and improve the stability of the system. In this section, first, an MG case study with multiple VSGs and the VAD constraint index (as a tool for transient stability assessment) is introduced. Then, the PSO is used to obtain the desired response of the given MG.

10.4.3.1 Microgrid with Multiple Virtual Synchronous Generators Units

A multi-VSG system shown in Fig. 10.39 is used as a case study in this section. The parameters of the system are represented in Table 10.5 with the base power and frequency of 250 kVA and 60 Hz, respectively. The VSGs and SGs are equipped with a governor similar to what is shown in Fig. 7.3b of Chapter 7. The automatic voltage regulator (AVR) shown in Fig. 10.40 is used to maintain the terminal voltage of the generators at the nominal value (2 kV).

The values of J and D determine the time constant of the VSG unit. Selecting these results is a challenging issue without a routine solution. Mimicking an SG, as explained in Chapter 7, the J can be given by

$$J = \frac{2HS_{base}}{\omega_0^2} \tag{10.15}$$

where H is the machine inertia constant, S_{base} is the base power of the machine, and ω_0 is the system frequency. The parameter H determines the period of time during which the machine is able to supply the nominal load based solely on the

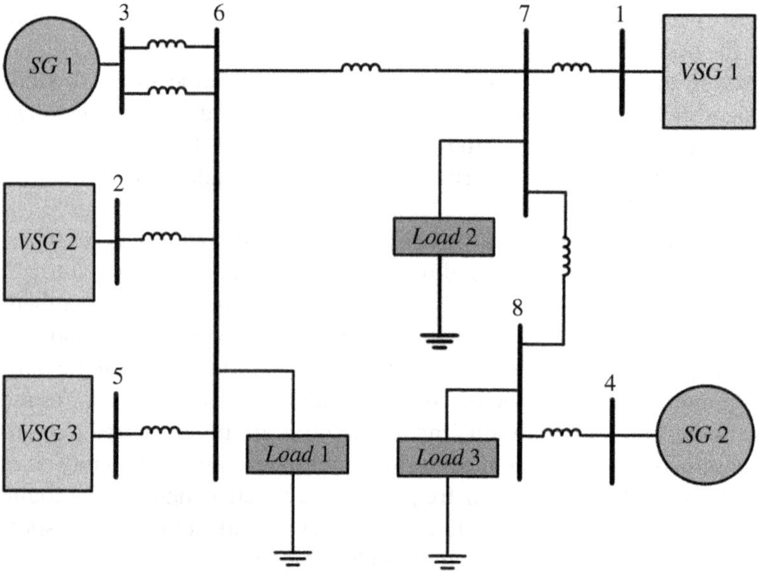

Figure 10.39 Multi-VSG system.

Table 10.5 The parameters of the multi-VSG system.

Line	X (pu)	Line	X (pu)
3-6	0.170	6-7	0.696
3-6	0.170	1-7	0.058
2-6	0.025	7-8	0.038
5-6	0.127	4-8	0.064
VSG filter inductance		0.353 pu	
H_{SG1} and H_{SG2}		4 s	

energy stored in the rotating mass. Higher H and larger time constant result in a slower response but a smaller frequency deviation after a change or disturbance. Although it depends on the machine size and power, for typical SGs, H varies between 2 and 10 s.

The VSG response at a specific output power and voltage is determined by parameters of its second-order differential equation built by the real part of its eigenvalues σ_i and the damping ratio ξ_i. These parameters are calculated for a single VSG connected to a grid in [67]. The following equations indicate that

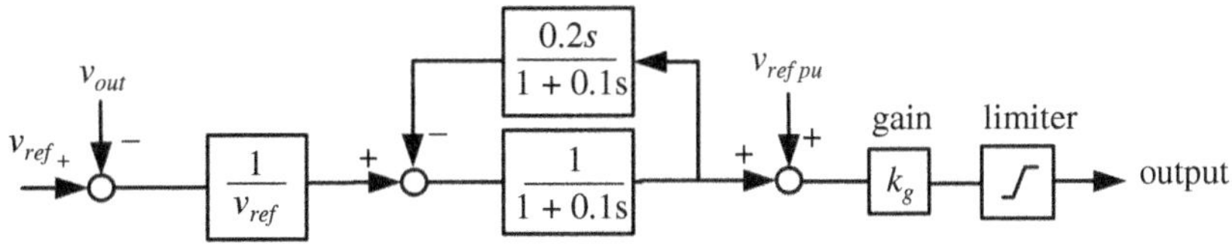

Figure 10.40 Automatic voltage regulator diagram.

the VSG response parameters are directly related to J and D:

$$\sigma_i = -\frac{D_i}{2J_i\omega_s}, \quad \omega_{ni} = \sqrt{\frac{P_{maxi}\cos(\theta_{0i})}{J_i\omega_s}}, \quad \xi_i = \frac{-\sigma_i}{\omega_{ni}} \tag{10.16}$$

where $P_{max\,i}$ is the maximum transferable power from ith VSG bus to the grid, θ_{0i} is the voltage angle of the ith VSG with respect to the grid, and ω_{ni} is the undamped natural frequency of the ith VSG. Under any working conditions, the desired system response parameters can be achieved by tuning J and D.

As a large disturbance scenario, a symmetrical fault with fault resistance of 1 Ω is established at Bus No. 8 of Fig. 10.39, and it lasted for 0.13 s. The system loading condition when fault occurred is presented in Table 10.6. The angular frequencies of generators and VADs are shown in Fig. 10.41. It can be observed that with such fault, the angular velocity of a group of generators increased and the ones of the other group are decreased until the two groups of generators go out of step. In this figure, the breakup of VADs of two groups of generators and instability after the fault is obvious.

Table 10.6 System condition when the fault occurs.

SG1	0.8 pu	200 kW
SG2	1 pu	250 kW
VSG1	1 pu	250 kW
VSG2	1 pu	250 kW
VSG3	1 pu	250 kW
Load 1 (pu)	R 0.494 pu	506.3 kW at rated voltage
	L 0.636 pu	310.6 kVA at rated voltage
Load 2 (pu)	R 0.706 pu	354.0 kW at rated voltage
	L 0.990 pu	205.7 kVA at rated voltage
Load 3 (pu)	R 0.393 pu	634.9 kW at rated voltage
	L 0.542 pu	373.2 kVA at rated voltage

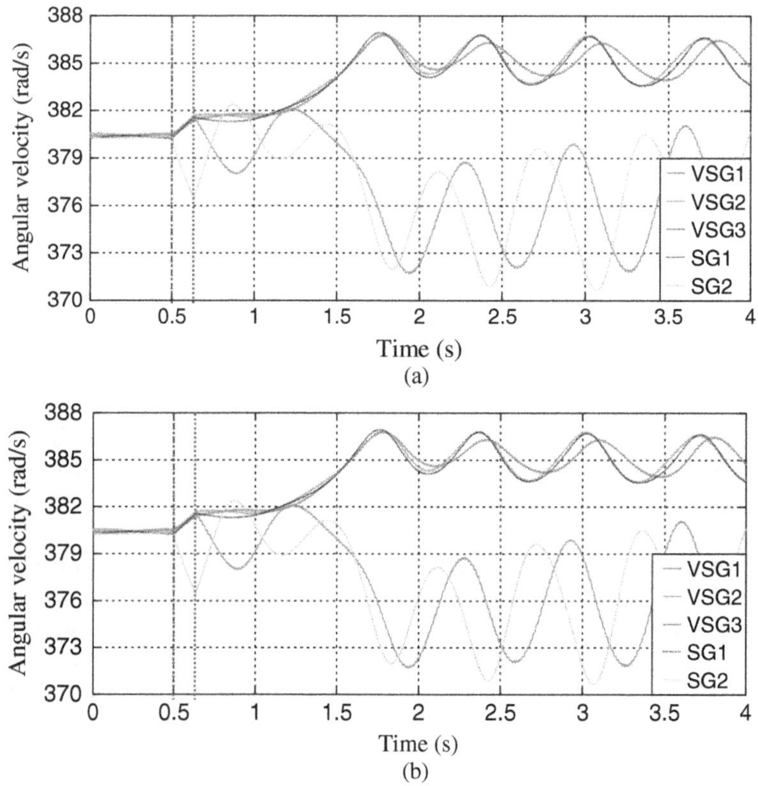

Figure 10.41 System generators response following a fault at bus 8: (a) angular frequency and (b) VAD.

10.4.3.2 Particle Swarm Optimization-based Virtual Synchronous Generators Optimization

The PSO algorithm used for multi-VSG system is shown as follows. In this algorithm, the behavior of each individual (VSG unit) is affected by the best local answer (best parameter values for VSG during one search) and the best global individual.

The best answer means the values of parameters result in the minimum value of a fitness function. The PSO algorithm can be implemented with two different objectives combined in its fitness function: first, the smooth transition after a large disturbance will be aimed, and second, the stability constraint will be included in the PSO algorithm fitness function to see whether the VAD constrained dispatch of the moment of inertia will improve the stability of the system. The embedded fitness function in the algorithm will aim for the objective. To have a smooth transition after a change or disturbance, the values of the parameters of each VSG unit that correspond to the desired values of quantities introduced in (10.16), should, be discovered by the algorithm. The penalty-type fitness function can be written as

$$\text{Min } fit_{1_i} = P_\xi \times |\xi_i - \xi_0| + P_\sigma \times |\sigma_i - \sigma_0||_{\sigma_i > \sigma_0} \quad (10.17)$$

where P_σ and P_ξ stand for the penalty factors, and σ_0 and ξ_0 are the desired real part of eigenvalues and the damping ratio, respectively (10.16). The proper values of constants in (10.17) are slightly different from the theoretical calculated values because of modeling approximations and simplifications. However, with a brief experience, the constants of (10.17) can be tuned finely.

1. For i = 1 to M (M = population size i.e. the number of VSG units)
   Initialize (J_i, D_i) for each VSG from interval
   Initialize $V_{(Ji,\,Di)} = 0$ (V = speed of each particle)
2. End For
3. For i = 1 to M
   $(J_i, D_i)_{BESTS} = (J_i, D_i)$ (Initialize the "memory" of each particle)
4. End For
5. Repeat
   For i = 1 to M
   $V_{(Ji,\,Di)} = w \times V[i] + C1 \times R1 \times ((J_i, D_i)_{BESTS} - (J_i, D_i))$
   $+ C2 \times R2 \times ((J, D)_{GBEST} - (J_i, D_i))$
   (Calculate the speed of each particle)
   (W = Inertia weight, $C1$ & $C2$ are positive constants)
   ($R1$ & $R2$ are random numbers in the range [0,1])
   $(J_i, D_i)_{new} = (J_i, D_i) + V_{(Ji,\,Di)}$
   If a particle gets outside the pre-defined limit
   then it is reintegrated to its boundaries
   Evaluate $(J_i, D_i)_{new}$ of each VSG by calculating the fitness function
   If $(J_i, D_i)_{new}$ is better than $(J_i, D_i)_{BESTS}$ then $(J_i, D_i)_{BESTS} = (J_i, D_i)_{new}$
   If $(J_i, D_i)_{new}$ is better than $(J, D)_{GBEST}$ then $(J, D)_{GBEST} = (J_i, D_i)_{new}$
   End For
6. Until stopping condition is reached

The other constraint is needed to maintain the VAD of all generators within the limits by finding the appropriate values of J of the VSG units. For this purpose, the second part of fitness function can be written as follows:

$$fit_{2i} = P_{VAD} \sum_{\substack{i=1 \\ |\Delta\theta_i| > \theta_{max}}}^{Ng} (|\Delta\theta_i| - \theta_{max}) \quad (10.18)$$

The overall fitness function that aims for the smooth transition after a change or disturbance and limiting the VAD of all generators is embedded in the PSO algorithm:

$$fit_i = fit_{1i} + fit_{2i} \tag{10.19}$$

By this fitness function, when the VAD of a generator exceeds the limit, the algorithm compromises between the smooth transition and maintaining the VAD of generators below the limit. The algorithm searches for the best values of J_i and D_i by evaluating the fitness function at each iteration until the maximum number of iterations is reached. Then, $(J_i, D_i)_{\text{BESTS}}$ are extracted as the optimum values for ith VSG parameters.

The algorithm is embedded into the case study system, and the same simulation scenario as before is applied. The optimum values of parameters are calculated by the algorithm via a low-pass filter with the cut-off frequency of 16 Hz to remove the chattering. Figure 10.42 depicts the values of J and D calculated by PSO during simulation. As it is observed, around $t = 1.5$ s, the values of J and D of VSG2 and VSG3 are reduced in order to maintain the VAD of SG2 within the specified limit, that is, 1.57 rad. Figure 10.43 shows the angular frequency and VAD index of generators. It is observed that the angular frequencies of generators converged at predisturbance system frequency value, and all generators remained in-step. Moreover, as shown in Fig. 10.43b, the VAD of SG2 is maintained at the limit via adjusting the inertia of VSGs until the transients are passed and the VAD of SG2 reaches the limit. In [67], the PSO performance is compared with the performance of alternating inertia-based approach, which is described in Chapter 8.

It is shown that for a smooth transition after the operation change or a disturbance, the PSO algorithm can provide a desirable response. It should

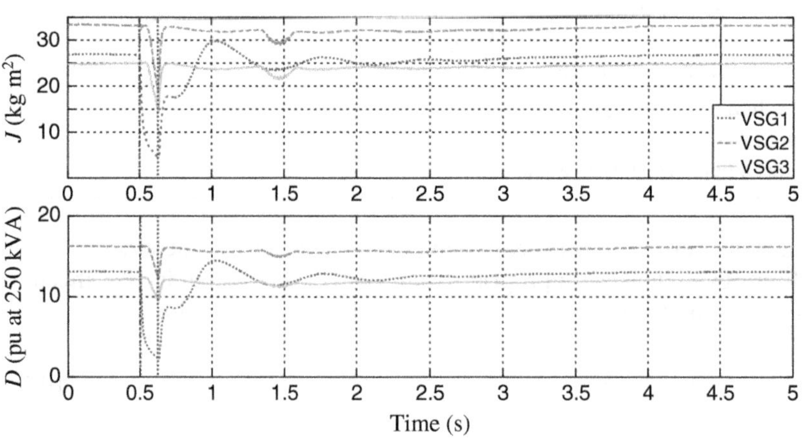

Figure 10.42 Calculated J and D by the PSO algorithm in the presence of fault.

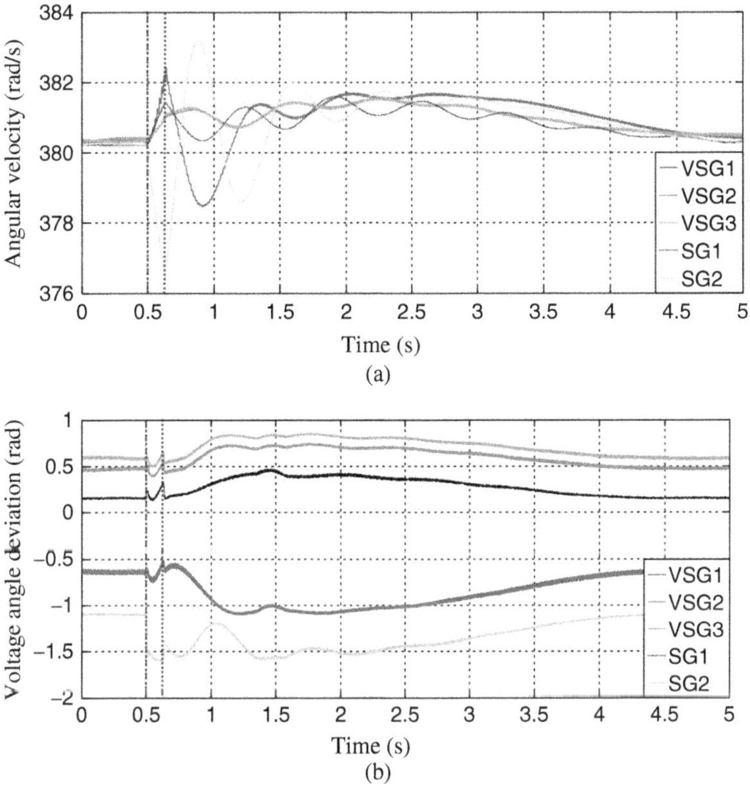

Figure 10.43 System generators response following a fault for the PSO-based scheme: (a) angular frequency and (b) VAD.

be mentioned that the PSO includes communication between VSG units and considered as a global control that the conditions of all system elements can be included in the control objectives. In addition, it is flexible, and various constraints can be implemented in it.

10.4.4 Multiagent-based Secondary Frequency Control

In this section, an MAS-based secondary frequency control scheme for an AC islanded MG with dispersed power sources such as PV units, wind generation units, diesel generation units, and an ECS for the energy storage is presented [4]. The addressed scheme has been proposed through the coordination of controllable power sources such as the diesel units and the ECS with small capacity. All the required information for the proposed frequency control is transferred between the diesel units and the ECS through computer networks. A basic configuration of the proposed frequency regulation scheme is shown in Fig. 10.44.

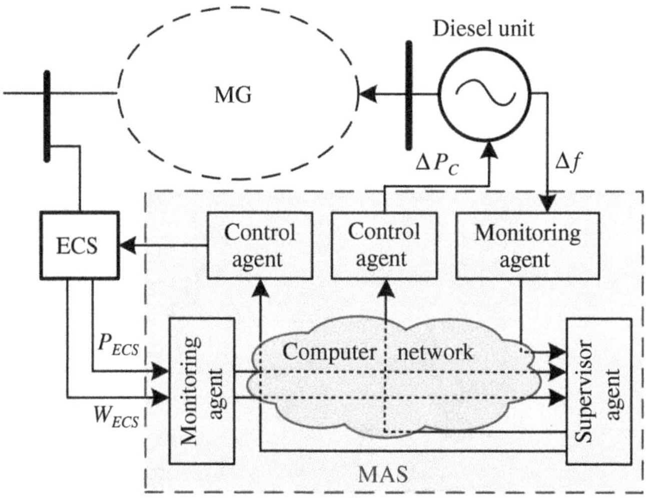

Figure 10.44 MAS-based MG frequency regulation scheme.

In this figure, W_{ECS} and P_{ECS} are the current stored energy and produced power by ECS unit, respectively. Experimental studies have been performed on the laboratory system to investigate the efficiency of the proposed MAS-based control scheme.

The required regulation performance is achieved through the charging/discharging operation on the ECS following the monitored frequency deviation Δf on the diesel unit. As shown in Fig. 10.44, three different types of agents are specified in the proposed MAS-based frequency system: *monitoring agents* for the distribution of required information through the computer network, the *control agents* for the charging/discharging operation on the ECS and also for the power regulation on the diesel units, and finally, a *supervisor agent* for the coordination between the ECS and the diesel units. These three agent types can communicate each other through a secure computer network to achieve a desirable frequency regulation performance. Here, agent controls operate as local controllers, while the monitoring agent and the supervisor agents operate in the MGCC as MG central control level.

As mentioned before, since the ECS units are able to provide a fast charging/discharging operation, the variations of power from the wind turbine units and also from the PV units can be absorbed through the charging or discharging operation of the ECS units. In addition, the variation of power consumption at the variable load can also be absorbed through the charging/discharging operation of the ECS units. A small-sized ECS is considered in this study; therefore, the continuous charging or discharging operation is not available on the ECS because of its restricted capacity. To overcome this situation, the regulation power on the diesel units is inevitable to maintain the stored energy of the

ECS in a proper range for the continuous control on the ECS. In the proposed control scheme, the ECS provides the main function of secondary frequency control, while the diesel units provide a supplementary function for this issue. Namely, a coordination between the ECS and the diesel units has been proposed for balancing the total power generation and the total power demand in the islanded MG. The proposed MAS consists of three types of agents: monitoring agents for the distribution of required information through the computer network, the control agents for the charging/discharging operation on the ECS and also for the power regulation on the diesel units, and finally, a supervisor agent for the coordination between the ECS and the diesel units. Experimental studies have been performed to demonstrate the efficiency of the proposed MAS-based secondary frequency control scheme.

Because of the specific feature of the ECS dynamics, the fast charging/discharging operation is possible to achieve in an ECS unit. Therefore, the variations of power generation from the wind turbine and PV units and also the variation of demand power on the variable loads can be efficiently absorbed through the charging/discharging operation of the ECS unit. A small-sized ECS is considered in this study. Therefore, an additional regulation power (from the diesel units) is required to maintain the stored energy level of the ECS in a proper range.

Figure 10.45a–c illustrates the feedback configurations of the diesel control, ECS control loop, and coordination loop for the diesel unit, respectively. The second and third control loops are implemented in the supervisor agent. As shown in these figures, W_r and W_{ECS} are the target and measured (current) stored energies. The P_{ECS} and P_m represent regulation powers provided by ECS and diesel units, respectively. In the proposed control scheme, the ECS provides the main function of frequency control and the diesel units provide a supplementary function to support the charging or discharging operation on the ECS unit. Namely, a coordinated control between the ECS and the diesel units has been performed to balance the power demand and the total power generation.

Here, T_{delay} and PS_{ECS} represent the communication time delay [68] and the control signal for output setting of ECS unit. Applying the control signal PS_{ECS} from the mentioned loop enables an appropriate charging/discharging operation to be performed on the ECS for the frequency regulation purpose.

The simplified single-line diagram of the studied laboratory system with the pictures of real devices in the laboratory is shown in Fig. 10.46. The laboratory system consists of a 5 kVA generator driven by a DC motor representing the diesel unit, a 70 Wh ECS with the maximum charging/discharging power of 3 kW, a variable load, and the transmission-line modules. The variation of power generation from the PV and the wind turbine units is represented by the variation of power consumption on the variable load. During the start-up process, the laboratory system is connected to the external commercial power

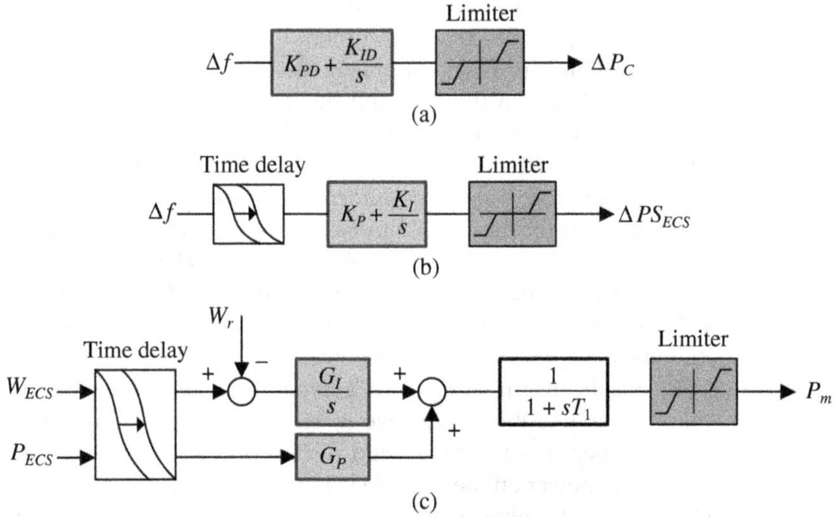

Figure 10.45 Feedback control system for (a) diesel unit, (b) ECS, and (c) coordination of diesel unit.

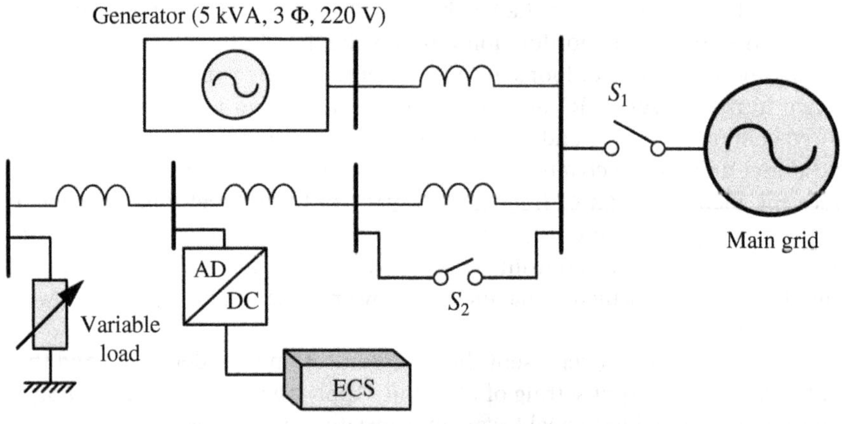

Figure 10.46 Single-line diagram of the laboratory MG system.

source, that is, the switch S_1 is closed. Following the start-up process, the switch S_1 is opened to change the MG operation from the grid-connected to the islanded mode.

The performance of the proposed control scheme is examined in various load disturbance scenarios. The tuning of control parameters is performed for the step load change scenario 1 shown in Fig. 10.47a. The tuned parameters for the ECS controller are $K_P = 4$ and $K_I = 10$. In this case, the step load change is

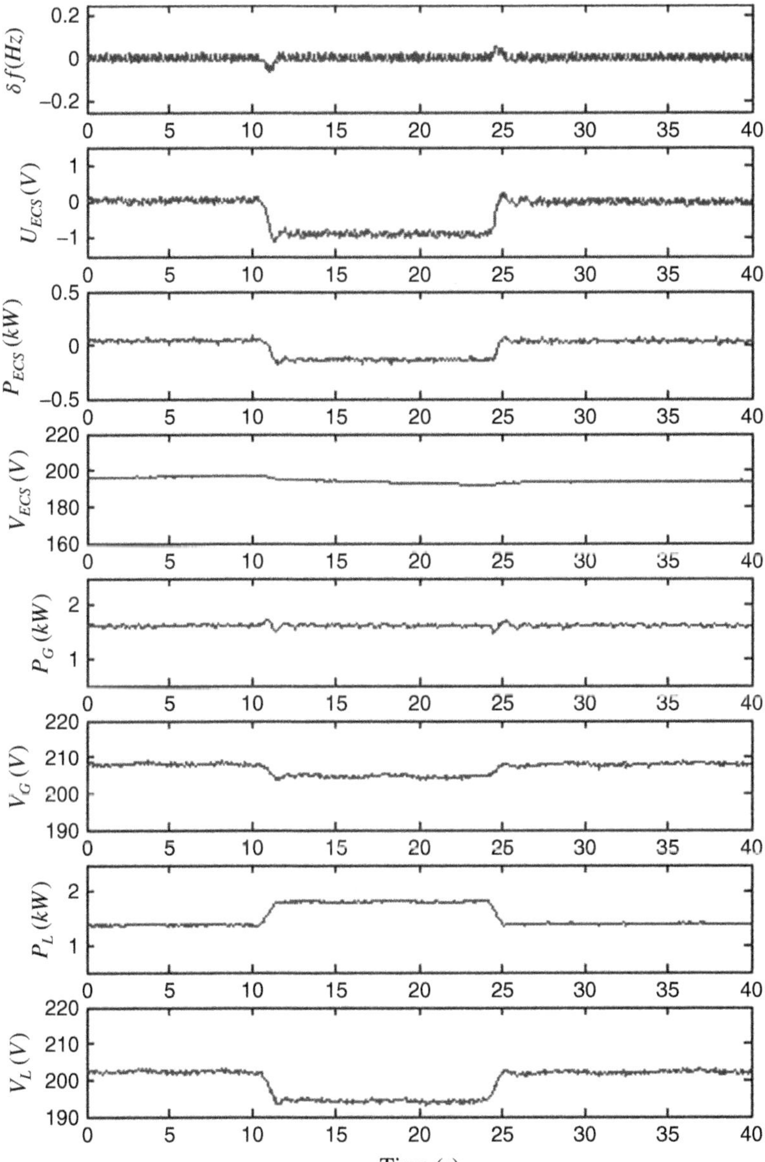

Figure 10.47 MG response for step load change: (a) scenario 1 and (b) scenario 2.

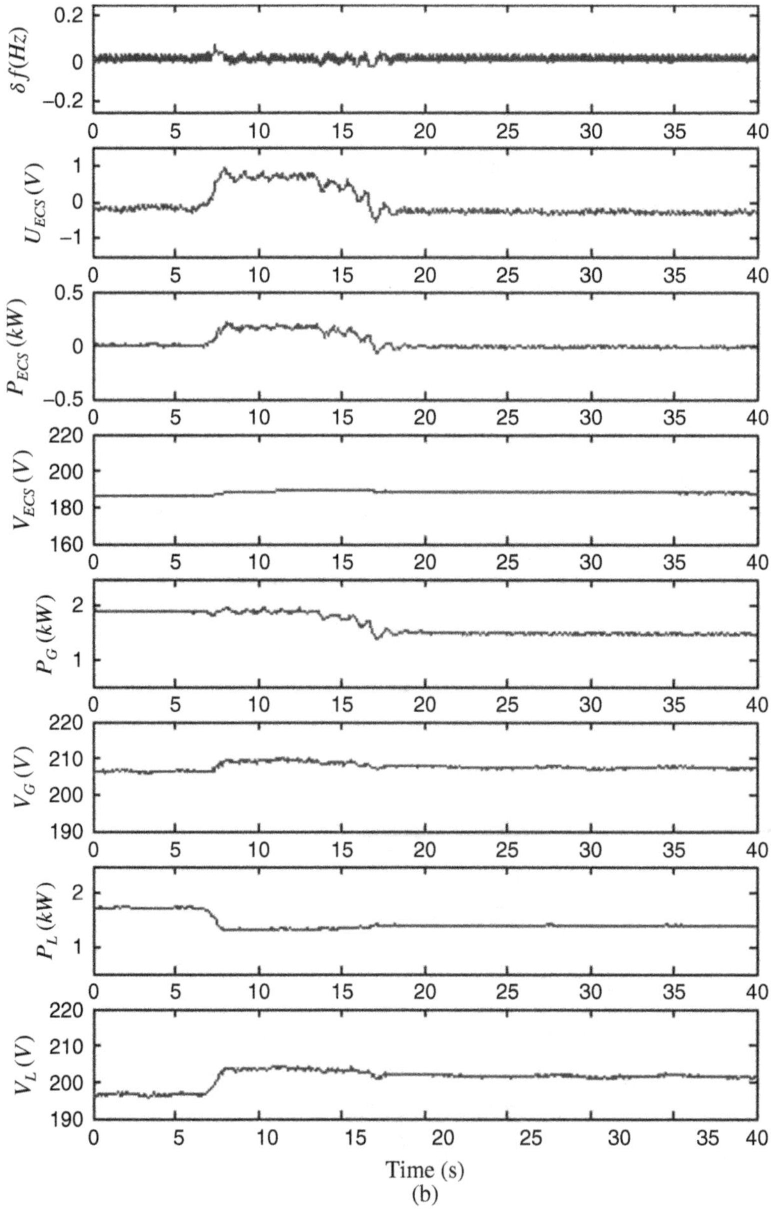

Figure 10.47 (*Continued*)

applied just for a specific time span. Therefore, the integration of the discharging power does not reach its critical level. Namely, the coordination from the diesel unit is not required in this case.

However, in the case of the step load change scenario 2, the step load change is not cleared; therefore, the coordination from the diesel unit is inevitable to maintain the stored energy of the ECS within a prespecified range. The diesel unit cannot follow the fast random load change because of its response speed; therefore, the coordination is not considered for the random load change. For the large disturbance caused by the line switching S_2, the coordination is not necessary because any load change is not applied to the laboratory system.

Typical experimental results are illustrated in Fig. 10.47. As shown in these figures, the frequency deviation Δf (Hz) monitored on the generator (representing the diesel unit), the control signal U_{ECS} (V) for the charging or discharging operation on the ECS, the charging/discharging power P_{ECS} (kW) on the ECS, the ECS terminal voltage V_{ECS} (V), the generator power P_G (kW), the generator terminal voltage V_G (V), the power consumption P_L (kW) on the variable load, and the terminal voltage V_L (V) of the variable load are illustrated from the top to the bottom.

The averaged frequency deviation and the maximum frequency deviation are summarized in Table 10.7 for both the conventional and the proposed MAS-based frequency control under different types of load changes and the large disturbance given by the line switching. As clearly indicated in Table 10.7, the MG frequency performance is highly improved by applying the proposed MAS-based control scheme. The estimated time delay is around 70 ms during the experiments. For the larger time delay, the compensation is inevitable to maintain a better control performance.

The stored energy on the ECS is easily monitored by the voltage V_{ECS} (V) measured at the ECS terminal. During the experiments, the operation range of the ECS terminal voltage V_{ECS} (V) is specified from 140 to 240 V. As shown in Fig. 10.47b, the ECS terminal voltage is maintained almost constant

Table 10.7 Averaged and maximum frequency deviation under different load change scenarios.

AGC scheme	Δf_{ave} (Hz)	Δf_{max} (Hz)
Scenario 1		
Conventional	0.2632	0.8019
Multiagent	0.0105	0.0676
Scenario 2		
Conventional	0.0623	0.4771
Multiagent	0.0085	0.0609

by the coordination from the generator representing the diesel unit. More experimental tests are discussed in [4, 69].

10.5 Summary

In this chapter, following an introduction and literature review, famous intelligent control technologies used in MGs are addressed. Then, a three-layer ANN is applied for next 24 h PV power and load forecasting. In the studied application, results have shown a desirable forecasting with an acceptable accuracy. Considering the uncertainty in the performed forecasting, a probabilistic model is then proposed for normal function density uncertainty distribution by integrating the uncertainties from both the PV power and load.

Then, several MG control design examples for frequency and voltage regulation using intelligent methodologies and evolutionary algorithms are presented. A fuzzy-logic-based distribution voltage control is developed for DC MGs. The proposed control combines a gain-scheduling technique with fuzzy logic control. The experimental results show that the DC voltage regulation and stored energy balancing control are realized, simultaneously. The PSO algorithm is used to find the optimum values of moments of inertia and damping factors of VSG units in a multi-VSG MG, following a large disturbance. The fitness function of the mentioned algorithm is modified to make a compromise between the smooth transition objective and maintaining the VAD within the stability limit. Finally, the MAS and fuzzy logic are used to design MG frequency synthesis. In both designs, an effective coordination is performed between the main generating units and the energy capacitor devices to reduce the MG frequency variation.

References

1 Bevrani, H. and Hiyama, T. (2011 Chapter 5) Intelligent power systems operation and control: Japan case study, in *Intelligent Automatic Generation Control*, CRC Press, New York, pp. 1–10.

2 Bevrani, H., Ghosh, A., and Ledwich, G. (2010) Renewable energy sources and frequency regulation: survey and new perspectives. *IET Renewable Power Generation*, **4** (5), 438–457.

3 Bevrani, H., Mitani, Y., and Watanabe, M. (2013) Microgrids controls, in *Standard Handbook for Electrical Engineers*, 16thSection 16.9 edn, McGraw-Hill Co., New York, pp. 159–176.

4 Bevrani, H. and Hiyama, T. (2011) *Intelligent Automatic Generation Control*, CRC Press, New York.

5 Lee, D.J. and Wang, L. (2008) Small-signal stability analysis of an autonomous hybrid renewable energy power generation/energy storage system, Part I: Time-domain simulations. *IEEE Transactions on Energy Conversion*, **23** (1), 311–320.
6 Bevrani, H., Feizi, M.R., and Ataee, S. (2016) Robust frequency control in an islanded microgrid: hinf and Mu synthesis approaches. *IEEE Transactions on Smart Grid*, 7 (2), 706–717.
7 Bevrani, H., Habibi, F., Babahajyani, P. *et al.* (2012) Intelligent frequency control in an AC microgrid: on-line PSO-based fuzzy tuning approach. *IEEE Transactions on Smart Grid*, **3** (4), 1935–1944.
8 Papadimitriou, C.N. and Vovos, N.A. (2010) *A Fuzzy Control Scheme for Integration of DGs into a Microgrid*. Proceedings of IEEE Mediterranean Electrotechnical Conference, April 2010, pp. 872–877.
9 Ahmadi, S., Shafiee, Q., Nazarpour, D., and Bevrani, H. (2016) *Fuzzy Logic Based Distributed Secondary Control for Islanded Microgrids*. Proceedings of 1st IEEE Conference on New Research Achievements in Electrical and Computer Engineering (CBCONF), Tehran, Iran, May 2016.
10 Elmitwally, A. and Rashed, M. (2011) Flexible operation strategy for an isolated PV-diesel microgrid without energy storage unit. *IEEE Transactions on Energy Conversion*, **26** (1), 235–244.
11 Feizi, M.R., Babahajiani, P., and Bevrani, H. (2015) Fuzzy-PI-based supervising frequency control design in a stand-alone AC microgrid. *Engineering Intelligent Systems*, **23** (1), 31–39.
12 Alajmi, B.N., Ahmed, K.H., Finney, S.J., and Williams, B.W. (2011) Fuzzy-logic-control approach of a modified hill-climbing method for maximum power point in microgrid standalone photovoltaic system. *IEEE Transactions on Power Electronics*, **26** (4), 1022–1030.
13 Ahmadi, S., Shokoohi, S., and Bevrani, H. (2015) A fuzzy logic-based droop control for simultaneous voltage and frequency regulation in an AC microgrid. *International Journal of Electrical Power & Energy Systems*, **64**, 148–155.
14 Gong, K., Shi, J., Liu, Y. *et al.* (2016) Application of SMES in the microgrid based on fuzzy control. *IEEE Transactions on Applied Superconductivity*, **26** (3).
15 Courtecuisse, V., Robyns, B., Petit, M., Francois, B., and Deuse, J. (2008) *Performance Comparison of Different Wind Generator Based Hybrid Systems*. 13th International Power Electronics and Motion Control Conference (EPE-PEMC), Poland, September 1–3, 2008, pp. 2184–2191.
16 Valencia, F., Collado, J., Sáez, D., and Marín, L.G. (2016) Robust energy management system for a microgrid based on a fuzzy prediction interval model. *IEEE Transactions on Smart Grid*, 7 (3), 1486–1494.
17 Shokoohi, S., Bevrani, H., Moshtagh, J., and Ahmadi, S. (2015) Transient stability enhancement in microgrids including inverter interfaced

distributed generations. *Journal of Iranian Association of Electrical and Electronics Engineers*, **12** (3), 1–7.

18 Shokoohi, S., Sabori, F., and Bevrani, H. (2014) *Secondary Voltage and Frequency Control in Islanded Microgrids: Online ANN Tuning Approach.* Smart Grid Conference-SGC, Tehran, Iran, December 2014.

19 Habibi, F., Shokoohi, S., and Bevrani, H. (2012) *Designing a Self-tuning Frequency Controller Using ANN for an Isolated Microgrid.* International Symposium on Smart Grid Operation and Control (ISSGOC 2012), Sanandaj, Iran, May 2012.

20 Kumar Nunna, H.S.V.S. and Doolla, S. (2013) Multiagent-based distributed-energy-resource management for intelligent microgrids. *IEEE Transactions on Industrial Electronics*, **60** (4), 1678–1687.

21 Logenthiran, T., Srinivasan, D., Khambadkone, A.M., and Aung, H.N. (2012) Multiagent system for real-time operation of a microgrid in real-time digital simulator. *IEEE Transactions on Smart Grid*, **3** (2), 925–933.

22 Morstyn, T., Hredzak, B., and Agelidis, V.G. (2016) Cooperative multi-agent control of heterogeneous storage devices distributed in a DC microgrid. *IEEE Transactions on Power Systems*, **31** (4), 2974–2986.

23 Logenthiran, T., Naayagi, R.T., Woo, W.L. *et al.* (2015) Intelligent control system for microgrids using multiagent system. *IEEE Journal of Emerging and Selected Topics in Power Electronics*, **3** (4), 1036–1045.

24 Li, Z., Zang, C., Zeng, P. *et al.* (2016) MAS based distributed automatic generation control for cyber-physical microgrid system. *IEEE/CAA Journal of Automatica Sinica*, **3** (1), 76–89.

25 Shokoohi, S. and Bevrani, H. (2012) *PSO Based Droop Control of Inverter Interfaced Distributed Generations.* Conference on Smart Electric Grids Technology (SEGT2012), Tehran, Iran, December 18–19, 2012, pp. 77–82.

26 Venayagamoorthy, G.K., Sharma, R.K., Gautam, P.K., and Ahmadi, A. (2016) Dynamic energy management system for a smart microgrid. *IEEE Transactions on Neural Networks and Learning Systems*, **27** (8), 1643–1656.

27 Shokoohi, S., Bevrani, H., and Naghshbandi, A.H. (2012) *Application of Neuro-fuzzy Controller on Voltage and Frequency Stability in Islanded Microgrids.* Conference on Smart Electric Grids Technology (SEGT2012), Tehran, Iran, December 18–19, 2012, pp. 62–67.

28 Khorramabadi, S.S. and Bakhshai, A. (2015) Intelligent control of grid-connected microgrids: an adaptive critic-based approach. *IEEE Journal of Emerging and Selected Topics in Power Electronics*, **3** (2), 493–504.

29 Shariatzadeh, F., Vellaithurai, C.B., Biswas, S.S. *et al.* (2014) Real-time implementation of intelligent reconfiguration algorithm for microgrid. *IEEE Transactions on Sustainable Energy*, **5** (2), 598–607.

30 Khadem, S.K., Basu, M., and Conlon, M.F. (2015) Intelligent islanding and seamless reconnection technique for microgrid with UPQC. *IEEE Journal of Emerging and Selected Topics in Power Electronics*, **3** (2), 483–492.

31 Kohn, W., Zabinsky, Z.B., and Nerode, A. (2015) A micro-grid distributed intelligent control and management system. *IEEE Transactions on Smart Grid*, **6** (6), 2964–2974.

32 Bevrani, H. and Shokoohi, S. (2013) An intelligent droop control for simultaneous voltage and frequency regulation in islanded Microgrids. *IEEE Transactions on Smart Grid*, **4** (3), 1505–1513.

33 Nasirian, V., Shafiee, Q., Guerrero, J.M. *et al.* (2016) Droop-free distributed control for AC microgrids. *IEEE Transactions on Power Electronics*, **31** (2), 1600–1617.

34 Simpson-Porco, J.W., Shafiee, Q., Dorfler, F. *et al.* (2015) Secondary frequency and voltage control of islanded microgrids via distributed averaging. *IEEE Transactions on Industrial Electronics*, **62** (11), 7025–7038.

35 Bevrani, H. (2014) *Robust Power System Frequency Control*, 2nd edn, Springer, Switzerland.

36 Man, K.F. and Tag, K.S. (1997) *Genetic Algorithms for Control and Signal Processing*. Proceedings of IEEE International Conference on Industrial Electronics, Control and Instrumentation-IECON, vol. 4, pp. 1541–1555.

37 Eberhart, R. and Kennedy, J. (1995) *A New Optimizer Using Particle Swarm Theory*. Proceedings of Sixth International Symposium, Micro Machine and Human Science, Nagoya, Japan, pp. 39–43.

38 McArthur, S.D.J., Davidson, E.M., Catterson, V.M. *et al.* (2007) Multi-agent systems for power engineering applications – Part I: Concepts, applications and technical challenges. *IEEE Transactions on Power Systems*, **22** (4), 1743–1752.

39 McArthur, S.D.J., Davidson, E.M., Catterson, V.M. *et al.* (2007) Multi-agent systems for power engineering applications – Part II: Technologies, standards and tools for multi-agent systems. *IEEE Transactions on Power Systems*, **22** (4), 1753–1759.

40 Wooldridge, M. and Weiss, G. (eds) (1999) *Intelligent Agents, in Multi-agent Systems*, MIT Press, Cambridge, MA, pp. 3–51.

41 Wooldridge, M. and Jennings, N. (1995) Intelligent agents: theory and practice. *Knowledge Engineering Review*, **10** (2), 115–152.

42 Daneshfar, F. and Bevrani, H. (2010) Load–frequency control: a GA-based multi-agent reinforcement learning. *IET Generation, Transmission & Distribution*, **4** (1), 13–26.

43 Bevrani, H., Daneshfar, F., and Hiyama, T. (2012) A new intelligent agent-based AGC design with real-time application. *IEEE Transactions on Systems, Man, and Cybernetics – Part C*, **42** (6), 994–1003.

44 Dimeas, A. and Hatziargyriou, N.D. (2005) Operation of a multi-agent system for microgrid control. *IEEE Transactions on Power Systems*, **20** (3), 1447–1455.

45 Bevrani, H., Daneshfar, F., Daneshmand, P. R., and Hiyama, T. (2010) *Reinforcement Learning Based Multiagent LFC Design Concerning the*

Integration of Wind Farms. IEEE International Conference on Control Applications, Yokohama, Japan, 2010.

46 Daneshfar, F., Bevrani, H., and Mansoori, F. (2012) Load-frequency control: a GA based Bayesian networks multi-agent system. *Iranian Journal of Electrical & Electronic Engineering*, **7** (2), 141–148.

47 Bevrani, H., Daneshfar, F., and Daneshmand, P.R. (2010) Intelligent power system emergency regulation concerning the integration of wind power units, in *Wind Power Systems: Applications of Computational Intelligence*, Springer Book Series on Green Energy and Technology (eds L.F. Wang, C. Singh, and A. Kusiak), Springer-Verlag, Heidelberg, pp. 407–437.

48 Dimeas, A. and Hatziargyriou, N.D. (2007) *Agent Based Control for Microgrids*. IEEE Power Engineering Society General Meeting, Tampa, USA, June 2007.

49 Mills, A. (2014) *Integrating Solar PV in Utility System Operations*. Argonne National Laboratory, LBNL-6525E, March 2014.

50 Romero-Cadaval, E., Francois, B., Malinowski, M., and Zhong, Q.C. (2015, ISSN 1932-4529, Institute of Electrical and Electronics Engineers, New York, USA) Grid-connected photovoltaic generation plants as alternative energy sources. *IEEE Industrial Electronic Magazine*, **9** (1), 18–32.

51 Cococcioni, M., D'Andrea, E., and Lazzerini, B. (2011) *24-Hour-Ahead Forecasting of Energy Production in Solar PV Systems*. 11th International Conference on Intelligent Systems Design and Applications (ISDA), 2011, pp. 1276–1281.

52 Hecht-Nielsen, R. (1989) *Theory of the Backpropagation Neural Network*. IJCNN, International Joint Conference on Neural Networks, 1989, pp. 593–605.

53 Sobu, A. and Wu, G. (2012) *Dynamic Optimal Schedule Management Method for Microgrid System Considering Forecast Errors of Renewable Power Generations*. 2012 IEEE International Conference on Power System Technology (POWERCON), 2012, pp. 1–6.

54 Yan, X., Abbes, D., and Francois, B. (2014) *Solar Radiation Forecasting Using Artificial Neural Network for Local Power Reserve*. 2014 International Conference on Electrical Sciences and Technologies (CISTEM).

55 Yan, X., Francois, B. and Abbes, D. (2015) *Operating Power Reserve Quantification Through PV Generation Uncertainty Analysis of a Microgrid*. PowerTech, 2015 IEEE Eindhoven, 2015, Eidhoven, pp. 1–6.

56 Herbreteau, J., Courtecuisse, V., Peng, L., Degobert, P., Robyns, B., and Francois, B. (2008) *Association of PV, Gas Micro Turbine and Short Term Storage System to Participate in Frequency Control*. International Conference on Renewable Energies and Power Quality: ICREPQ 08, Santander Espagne, March 12–14, 2008.

57 Degobert, Ph., S. Kreuawan, and X. Guillaud (2006) *Micro-grid Powered by Photovoltaic and Micro Turbine.* International Conference on Renewable Energy and Power Quality (ICREPQ'06), Palma de Mallorca, Spain, CDROM, April 2006.
58 Bevrani, H. *et al.* (2014) Intelligent LFC concerning high penetration of wind power: synthesis and real-time application. *IEEE Transactions on Sustainable Energy,* **5** (2), 655–662.
59 Bevrani, H. and Daneshmand, P.R. (2012) Fuzzy logic-based load-frequency control concerning high penetration of wind turbines. *IEEE Systems Journal,* **6** (1), 173–180.
60 Kakigano, H., Miura, Y., and Ise, T. (2013) Distribution voltage control for DC Microgrid using fuzzy control and gain-scheduling technique. *IEEE Transactions on Power Electronics,* **28** (5), 2246–2258.
61 Pai, M.A. (1989) *Energy Function Analysis for Power System Stability,* Kluwer, Norwell, MA.
62 Bevrani, H., Watanabe, M., and Mitani, Y. (2014) *Power System Monitoring and Control,* Wiley-IEEE Press, New York: USA.
63 Gan, D., Thomas, R.J., and Zimmerman, R.D. (2000) Stability-constrained optimal power flow. *IEEE Transactions on Power Systems,* **15** (2), 535–540.
64 Hakim, L., Kubokawa, J., Yuan, Y. *et al.* (2009) A study on the effect of generation shedding to total transfer capability by means of transient stability constrained optimal power flow. *IEEE Transactions on Power Systems,* **24** (1), 347–355.
65 Guangchao, G., Ajjarapu, V., and Quanyuan, J. (2014) A hybrid dynamic optimization approach for stability constrained optimal power flow. *IEEE Transactions on Power Systems,* **29** (5), 2138–2149.
66 Jiang, Q., Wang, Y., and Geng, G. (2014) A parallel reduced-space interior point method with orthogonal collocation for first-swing stability constrained emergency control. *IEEE Transactions on Power Systems,* **29** (1), 84–92.
67 Alipoor, J. (2015) Study on stabilization of power system with distributed generations using virtual synchronous generator. PhD thesis. Osaka University, January.
68 Bevrani, H. and Hiyama, T. (2009) On robust load-frequency regulation with time delays: design and real-time implementation. *IEEE Transactions on Energy Conversion,* **24** (1), 292–300.
69 Hiyama, T., Zuo, D., and Funabashi, T. (2002) *Automatic Generation Control of Stand Alone Power System with Energy Capacitor System.* Proceedings of the Fifth International Conference on Power System Management and Control (PSMC 2002), London, pp. 59–64.

11

Emergency Control and Load Shedding in Microgrids

Similarly to conventional power grids, the remaining system voltage and frequency in an acceptable range are necessary in a microgrid (MG). In the grid-connected operation mode, this issue is realized by the main grid, and because of that, distributed generations (DGs) often work based on active power and reactive power (PQ)-controlled inverters with specific amount of active and reactive powers. But in the islanded operation mode, the MG security strongly depends on the capability of existing controllers in the existing control levels. A nonsmooth transient state between two operation modes may also seriously affect the MG voltage and frequency. Stable performance of the MG in the transit state as well as in the islanded operation mode is dependent on the proposed protection plans and emergency control strategies applied on the MG loads and voltage source inverter (VSI)-based DGs. Following disconnection, control mode of one or several DGs may change from PQ control to VSI control. The local primary and secondary controllers may be unable to stabilize an islanded MG following large disturbances such as load change and generator trip, which changes the balance between generation and consumption. Thus, an emergency control such as load shedding is required under these conditions.

In the present chapter, following a general introduction on emergency control in MGs, the load shedding as the most important and well known emergency control strategy is described. Then, two load shedding algorithms are explained; and undervoltage-frequency load shedding is emphasized. Finally, some important factors on the voltage/frequency performance during the load shedding process are discussed.

11.1 Introduction

Renewable energy sources (RESs) such as wind turbines and solar cells are examples of clean energy options. For increasing penetration of RESs to reduce environmental pollution and to use less fossil fuel resources and some

Microgrid Dynamics and Control, First Edition. Hassan Bevrani, Bruno Francois, and Toshifumi Ise.

economical and marketing benefits, the concept of MG is introduced. An MG is a collection of DGs and loads that work together in different operation modes [1]. DGs use interfaces to connect to the MG. Depending on the source type, they are connected via synchronous generator, induction generator, or power electronic converter/inverter as interface [2]. Local loads of an MG include both critical and noncritical loads. Critical loads need continuous power with appropriate quality. This requirement is significant especially in the islanded operation mode when the main network is not available. In this state, the DGs of MG try to meet the need properly.

The MGs are located at low- or medium-voltage (MV) level and near to the consumers. Therewith, the generation capacity of resources in an MG is usually low. Furthermore, this vicinity brings with itself the advantages such as losses reduction and, as a result, increasing MG efficiency. In addition, the ability of operating in both grid-connected and islanded operation modes increases reliability. Despite these advantages, there are several problems such as the absence of adequate worldwide standards on power quality and frequency/voltage profiles, as well as the complexity of control and protection plans [3].

Low inertia and increasing high frequency harmonics as a result of the existence of inverter-based DGs differs for an MG and a conventional power system. In addition, a different structure of protection plans may be required because of different short-circuit levels and sometimes the meshed structure of MGs that leads to a multidirectional power flow. In these cases, MGs may need to use directional relays instead of overcurrent relays [4].

An acceptable performance and stable dynamic should be ensured in both operation modes. Normally, an MG works in the grid-connected operation mode and exchanges power with the main grid. In this operation mode, the DGs usually work based on a current control scheme. It is expected that DGs generate prespecified power in a different way from one system to another system [5, 6]. Under this condition, the main grid serves as backup, and any difference between generation and consumption, which causes variation in frequency and voltage out of specified ranges, is compensated by the main grid. Islanding by either planned or unplanned events separates MG from the main grid [7], via breaker at the point of common coupling (PCC).

In the islanded operation mode, control issues are more important due to the absence of the main grid as supplier. In this operation mode, the MG must supply the local loads or at least critical loads by itself. So, islanding is usually accompanied with the control mode change in one or several electronically interfaced DGs to set relevant parameters inside the MG [8]. DGs' control mode may be changed from current control mode to voltage control mode. Since some generating units have slow dynamic response for this mode change and contribution in transient state control [9] and also the generation–load imbalance following islanding is often considerable, an emergency control such as load shedding is also required.

Regardless of grid-connected and islanded operation modes for an MG, there is also a transient state. When the MGs are disconnected from the main grid, their status strongly depends on their conditions in the grid-connected operation mode. The amount of interchanged power is one of the important factors. In the grid-connected operation mode, the MGs usually absorb power from the main grid. As a result, they may be in face of power shortage in the islanded operation mode [10]. Therefore, DGs must change their production to meet the generation–consumption balance; otherwise, using storage devices, load shedding schemes and protection plans are necessary [9].

Similarly to conventional power systems, an MG can be in face of different events such as tripping generator or line, increasing load, and power quality issues. Therefore, various control loops in different control levels are needed to ensure a stable condition. Controllers maintain system integrity and restore the normal operation subjected to disturbances [11]. Small disturbances do not need critical actions, whereas large disturbances may need emergency control plans to prevent blackout events. This chapter is focused on the load shedding, which is one of the common emergency control plans in all power grids. Emergency control actions can be applied to both demand and generation sides. Load shedding plan is related to the demand side. It curtails the amount of load until available generation could supply the remind loads [12].

Due to some reasons such as low inertia, low electric distances, and RESs uncertainty; load shedding in the MGs is different than the load shedding in the conventional power systems [13]. Frequency and voltage measurements separately or together can be used for a load shedding plan [12–15]. The measurement thresholds determine the starting time of load shedding process [16]. At first, it is preferred to shed lower important loads or the loads that pay less cost for supplying during disturbance [13]. In addition, the buses with a higher rate of change of voltage (*dv/dt*) may be more suitable for load removal [15]. Economy, policy, and social issues may also influence the load shedding as described in [17, 18].

The rest of this chapter is organized as follows: following an introduction on load shedding in an MG as the most popular control strategy, two load shedding algorithms are explained in Sections 11.3 and 11.4, and undervoltage–frequency load shedding is emphasized in Section 11.5. Some important factors for the voltage/frequency performance during the load shedding process are discussed in Section 11.6, and, finally, the chapter is summarized in Section 11.7.

11.2 Load Shedding as a Well-known Emergency Control Strategy

The recent global blackouts have occurred due to different forms of system instability [19]. Special protective plans, underfrequency load shedding (UFLS) and undervoltage load shedding (UVLS) algorithms, are generally used independently to prevent system instability under emergency conditions [20, 21]. Since instability occurs in different forms and is not counteracted by either of the voltage and frequency instability phenomena, the effectiveness of using conventional UFLS and UVLS for a wide range of contingencies [22] is not guaranteed.

The MGs are small-scale power grids structured in distribution systems to support the efficiency, reliability, and expandability. On the other hand, using RESs in the MGs helps to reduce global warming and to speed up entering the power industry in deregulated environments [23].

The balance between generation and load and also the remaining voltage and frequency in a normal variation range are the most essential requirements in an electrical network, such as an MG. In the grid-connected operation mode, this issue is not challengeable because the main grid is met as a reference bus. Therefore, the main grid is responsible for compensating the imbalance between generation and consumption [8, 24], and the voltage and frequency reference values are also determined by the main grid.

The emergency control schemes and protection device dynamics are usually represented using incremented/decremented step behavior. Thus, for simplicity, the related blocks can be represented as a sum of incremental (decremental) step functions. For instance, as shown in Fig. 11.1, for fixed UFLS and UVLS schemes [25], the function of $\Delta P_{UFLS,\ UVLS}$ in the time domain could be considered as a sum of the incremental step functions of $\Delta P_j u(t-t_j)$. Therefore, for L load shedding steps,

$$\Delta P_{UFLS,UVLS}(t) = \sum_{j=0}^{L} \Delta P_j\ u(t-t_j) \tag{11.1}$$

where ΔP_j and t_j denote the incremental amount of load shed and time instant of the jth load shedding step, respectively. Using the Laplace transformation, it is possible to represent $\Delta P_{UFLS,\ UVLS}(s)$ in the following summation form:

$$\Delta P_{UFLS,UVLS}(s) = \sum_{l=0}^{N} \frac{\Delta P_l}{s}\ e^{-t_l s} \tag{11.2}$$

Here, the ΔP_l is the size of equivalent step load/power changes due to a generation/load event or a load shedding scheme at t_l.

A load shedding solution taking into account economic factors and MG transient stability is presented in [18]. After islanding, some disconnected

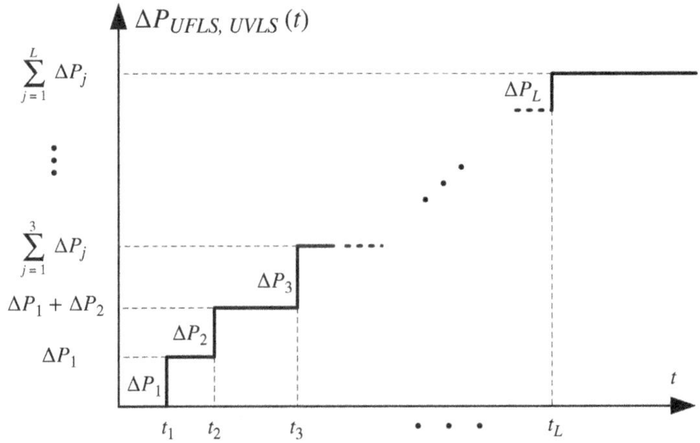

Figure 11.1 *L*-step load shedding scheme.

loads during load shedding might be reconnected again. In order to avoid large frequency/voltage deviations during the load reconnection, a specific set of steps for load reconnection should be defined. These steps can be changed with respect to the percentage of load shedding [22, 26]. In UFLS, frequency measurement is used to detect emergency conditions. In addition to frequency measurement, using the rate of change of frequency (ROCOF) increases the sensitivity and accuracy of the emergency control schemes in the presence of contingencies and improves detection quality of emergency conditions. Furthermore, it may also help to detect the source of disturbance and estimate the amount of power imbalance [27].

The type of shed load (active and reactive) is also important to address the MG stability during the load shedding process [28]. Considering both voltage and frequency measurements in a load shedding algorithm may be more effective to capture all postcontingency conditions. The necessity of considering both system frequency and voltage indices in emergency control is shown in [25, 29].

In this chapter, two load shedding scenarios are proposed to improve the stability of MGs under emergency conditions. Since the MGs are low-inertia power grids with the potential of high rate of voltage/frequency variation, detecting emergency conditions is very important. The proposed load shedding algorithms use voltage, frequency, and ROCOF indices for detecting emergency conditions.

Due to uncertainty in the nature of RESs, constraints in planning and electricity market, and frequent changing of operation state between two grid-connected and islanded operation modes, the number of contingencies in MGs may be remarkably higher than in the conventional bulk power grids.

Measuring appropriate parameters, recognizing emergency conditions such as islanding detection, and managing an appropriate emergency control procedure are vital [28]. In the connected operation mode, the main grid acts as reference bus for MG and accordingly compensates the imbalance power between generation and load. In the case of severe disturbances in the system, the MG turns into the islanded operation mode in order to avoid the probable collapse. In the new operation mode, the MG loads must be supplied locally; however, it might not be possible to supply all the MG loads because of economic, planning, and/or generation constraints; thus, load shedding must be applied [9, 10, 30].

The authors' investigations show that five important points should be considered in a load shedding algorithm: (i) the total load that needs to be shed, (ii) the number of shedding steps, (iii) the amount of load that must be removed at each step, (iv) the location of loads that should be shed, and (v) the time delays between the load shedding steps.

As already mentioned, the UFLS and UVLS are the most known load shedding methodologies in the power system emergency control literature [20, 22]. However, single-measurement-(voltage or frequency)-based load shedding may not meet satisfactory performance/reliability and may increase the possibility of overload shedding. In the UFLS, the impact of voltage variation on frequency deviation is usually ignored. There is opposite behavior for the UVLS. Since these two measurements are not independent particularly, a combination of both load shedding approaches based on simultaneous voltage and frequency measurements, that is, undervoltage–frequency load shedding (UVFLS), could be more effective [31, 32].

A new load shedding scheme that uses the rate of frequency drop and the voltage dip for estimating the amount of load to be shed among the load buses is proposed in [33]. A wide-area measurement-based load shedding via monitoring of both voltage and frequency is discussed in [32]. The aforementioned load shedding strategies are given for conventional power systems. It is shown that in an MG, due to the lower rate of inertia and faster dynamic, the frequency/voltage variation with normal load changes tends to be much higher than the main grid's frequency/voltage variation. Therefore, using low droop coefficient and small disturbances may lead to large variations in the MG frequency/voltage.

To run a load shedding algorithm, estimating the amount of unbalanced power could be very useful, because the required amount of curtailed load has to be managed according to the estimated power deficit [34]. In a conventional power grid, according to the swing equation for synchronous generators, we have [25]

$$\Delta P_m(t) - \Delta P_L(t) = \Delta P_D(t) = 2H\frac{d\Delta f(t)}{dt} + D\Delta f(t) \tag{11.3}$$

or, using Laplace transformation, we can obtain

$$\Delta f(s) = \frac{1}{2Hs + D}[\Delta P_m(s) - \Delta P_L(s)] \tag{11.4}$$

where Δf, ΔP_m, H, D, ΔP_D, and ΔP_L are frequency deviation, mechanical power change, inertia constant, damping coefficient, and load–generation imbalance proportional to the total load change, respectively. The magnitude of power deficit after disturbance is calculated as

$$\Delta P_D(t) = 2H\frac{d\Delta f(t)}{dt} \tag{11.5}$$

Assuming no speed governing at $t = 0^+s$ (i.e., $\Delta P_m = 0$), using (11.3), one can write

$$\Delta f(s) = \frac{-\Delta P_L(s)}{2Hs + D} \tag{11.6}$$

The given equation formulates the frequency deviation caused by load change. The power deficit is often classified in a specific load shedding table based on the load and its priority. In the mentioned table, the load shedding can be classified within different steps in terms of size and type [21].

As mentioned, unlike conventional power grids, in MGs, the DGs/RESs do not provide inertia or they have low inertia. Therefore, calculating the accurate amount of inertia is more difficult, so the conventional method for power deficit calculation may not be applicable and needs to be modified [35, 36].

11.3 Load Shedding Algorithm: Example 1

11.3.1 Proposed Algorithm

Since the MG loads contain active and reactive power parts, both voltage and frequency variables are affected in a load shedding process. Besides, the methods that are based on only the frequency or voltage measurement may be suitable for only frequency and voltage instability. But as discussed in [25], the instability phenomena do not always occur in pure forms.

The proposed load shedding example including the relevant process steps is illustrated in Fig. 11.2. The algorithm is based on measuring both frequency and voltage parameters. Following a disturbance, the frequency and voltage of system may change. Here, the amount of voltage reduction following disturbances is used to determine the number of load blocks to be shed. Hence, the voltage measurement is used first. A significant change in voltage triggers the mentioned algorithm to start. But to prevent unnecessary load shedding, the MG frequency and ROCOF are measured during a specified time window (K s). In the case of crossing the frequency and ROCOF thresholds, the load shedding is necessary. In this situation, the proposed algorithm will be initiated,

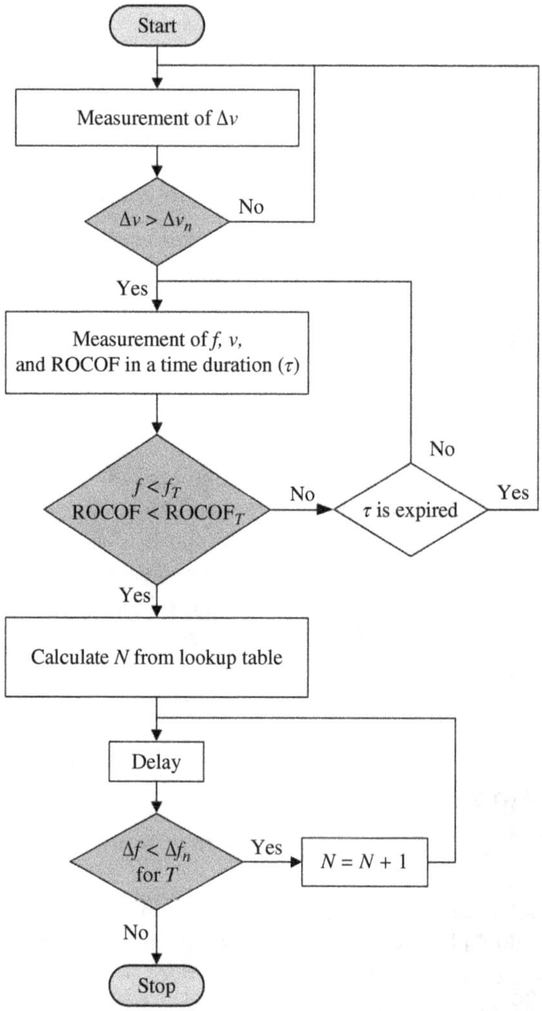

Figure 11.2 Flowchart of the proposed load shedding algorithm (Example 1).

and the threshold value and the time constant K are determined according to the occurred disturbance to shed the specified load blocks.

One can calculate the amount of required load to be shed in the load shedding steps based on the changing level in the frequency and/or voltage measurements. Here, the amount of voltage reduction is used to determine the severity of disturbance. If the occurred disturbance is large, the amount of voltage reduction will also be high. The number of load blocks for load shedding must be calculated, and for this purpose, a lookup table could be provided. Considering voltage drop records for different events, a lookup table can be performed by increasing loads until the system is stabilized by eliminating all needful load blocks for load shedding.

In the present example, when load shedding algorithm is initiated, N load blocks are shed based on the performed lookup table. To consider the necessity of more load shedding, after removal of the first group of load blocks, the load shedding algorithm needs a delay for breaker operation. To ensure that the frequency does not reach its nominal value, it is measured again after T time. The choice of T is dependent on the system dynamics. This process continues until the frequency returns to the allowed limits. The Δf_n and ΔV_n are the permitted amount of voltage and frequency deviations in a normal operation. The f_T and $ROCOF_T$ are the threshold values of frequency and ROCOF for starting the load shedding process.

11.3.2 Case Study

The single-line diagram of an MG system in MV level with its elements is shown in Fig. 11.3. The MG topology and its parameters are the same as those in the example in [5]. The MV system with a base voltage of 13.8 kV is connected to the 69 kV grid by a substation transformer and a breaker. The system includes four buses/feeders and three DGs. DG_2 and DG_3 are electronically interfaced to the MG, and DG_1 is a conventional synchronous generator. The available loads are

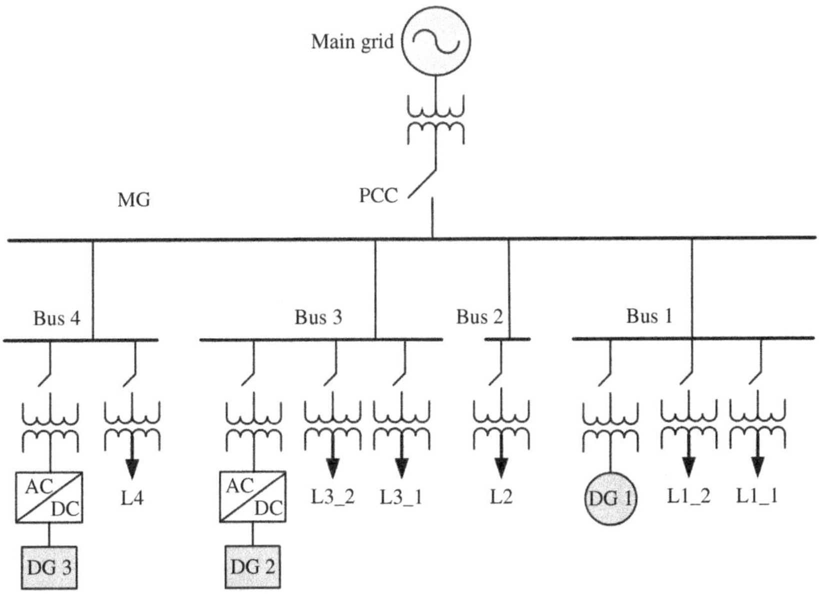

Figure 11.3 MG case study for load shedding (Example 1).

considered as RL type in both critical and noncritical forms. Only noncritical loads are shed under the necessary conditions.

In the grid-connected operation mode, the reference values for the MG voltage and frequency are provided by the main grid. Therefore, the DGs work in a specific control strategy such as PQ control strategy, which is described in Chapter 5. In the islanded operation mode, the measured voltage and rotor speed (frequency) of synchronous generator are compared with their reference values provided in the MG control system. These differences cause DGs to change their production to maintain the MG's voltage and frequency stability similarly to the grid-connected operation case.

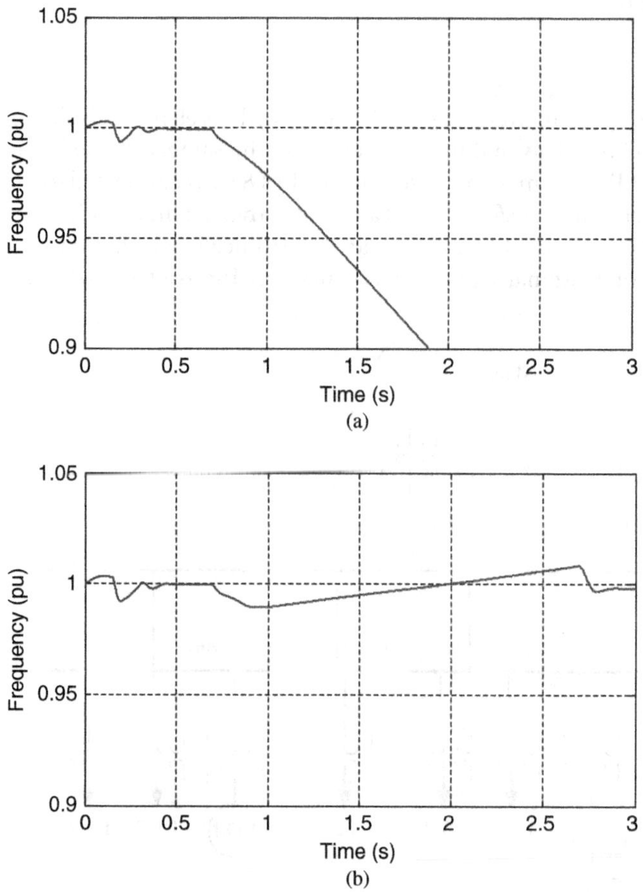

Figure 11.4 MG frequency response following loss of DG 2: (a) without load shedding and (b) with load shedding.

In the grid-connected operation mode, the DGs generate constant power, and the MG imports the necessary extra required power from the main grid. Following islanding, the MG's DGs increase the power to supply its loads. But due to the DGs' power limits, they may be unable to stabilize the system after an intentional islanding event. This issue is examined in a case study in [37].

11.3.3 Simulation Results

For the given case study, the proposed load shedding plan is applied following loss of DG 2 at $t = 0.7$ s. Figure 11.4 shows the MG frequency response with and without load shedding. It can be seen that the frequency leads to instability after $t = 0.7$ s. Simulation results for the voltage profiles [38] show that all bus voltages (except the bus connected to DG 2) are decreased and the system voltage finally collapses after a while [38].

Here, the load shedding lookup table, which is described in Section 11.3.1, is determined based on the voltage deviation range following a disturbance. The obtained results for the given case study are summarized in Table 11.1.

For the application of the proposed algorithm, the threshold of frequency f_T is considered at 0.99 pu. After tripping of DG_2 in the islanded operation mode, the voltage decreases to 0.774 pu. According to the lookup table, the number of load blocks for load shedding is one, so a load block is shed at $t = 0.886$ s. During T time delay, the frequency returns to the normal range. This means that it is not necessary to shed more load blocks. The effect of load shedding scheme is presented in Fig. 11.4.

The load shedding performance can also be affected by the location of shed loads. For instance, the priority of loads for shedding in the present case study is defined as $L_{1\text{-}2}$, L_2, $L_{1\text{-}1}$, $L_{3\text{-}1}$, $L_{3\text{-}2}$, L_4. If instead of load $L_{1\text{-}2}$ with the first priority, load L_4 with the last priority is shed, the frequency may not return to the normal range. In this situation, more than one load must be shed. This issue is shown

Table 11.1 Lookup table for load shedding.

Voltage deviation range (pu)	Number of load blocks for load shedding (N)
$v \geq 0.776$	0
$0.776 < v \leq 0.733$	1
$0.733 < v \leq 0.71$	2
$0.71 < v \leq 0.696$	3
$0.696 < v \leq 0.68$	4
$0.68 < v \leq 0.67$	5
$v \leq 0.67$	All load blocks

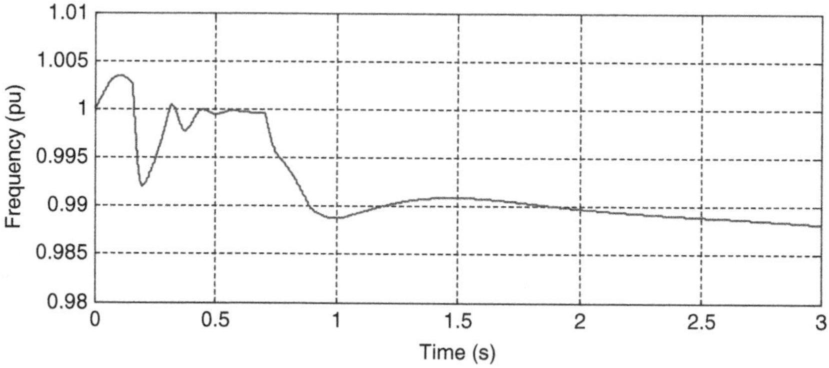

Figure 11.5 The effect of shedding L_4 on frequency of MG (pu).

in Fig. 11.5. By increasing the number of load blocks for shedding, the effect of location is minimized.

Under an emergency condition, the DGs often work at their maximum power. Hence, only the synchronous generator tries to support the MG stability concerning the injected inertia. Thus, the loads near to the conventional unit may have more impact on the load shedding. Unlike load shedding in the conventional power systems, the load far from the disturbance location (DG 2) is in first priority for shedding.

11.4 Load Shedding Algorithm: Example 2

11.4.1 Proposed Algorithm

The flowchart of the second proposed load shedding algorithm is shown in Fig. 11.6. The developed algorithm uses voltage and frequency measurement simultaneously. The number of load shedding steps is determined depending on the MG dynamics and the occurred disturbance.

The proposed algorithm can be applicable to a wide range of MG systems. The load shedding algorithm will be triggered if voltage or frequency measurement meets the defined corresponding threshold value. Threshold values for voltage and frequency are specified by v_T and f_T, respectively, and they are used to initiate the load shedding process after Δt delay. The short delay of Δt is used to ensure that the voltage/frequency drop does not occur following a measured noise.

Due to low inertia of MGs, following a disturbance, the system voltage and frequency indicators can fluctuate significantly, thus using the rate of change of frequency and voltage indices ($d\Delta f/dt$ and $d\Delta v/dt$) may not be suitable for making proper decisions during a load shedding process. Here, a new index

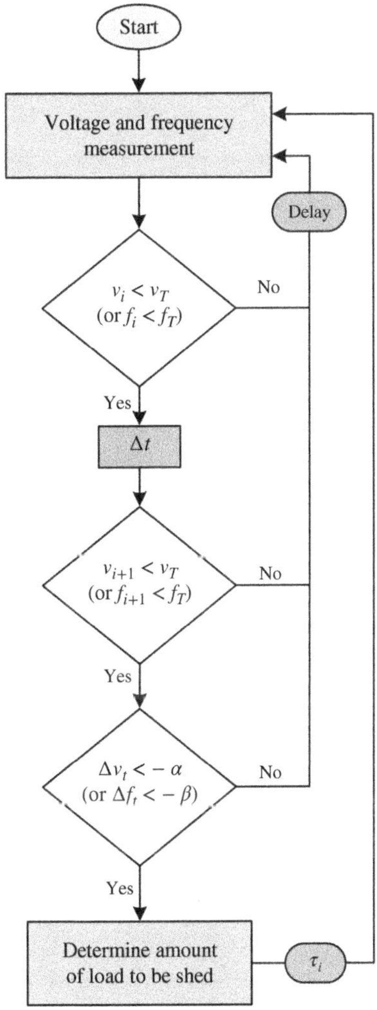

Figure 11.6 Flowchart of the proposed load shedding (Example 2).

average rate of drop in frequency and voltage variation is introduced, which is described in Fig. 11.7:

$$\Delta v_t = \frac{\Delta v}{\Delta t}, \quad \Delta f_t = \frac{\Delta f}{\Delta t} \tag{11.7}$$

The index of average rate of drop can be used for calculation of the time and amount of load shedding. Particularly, the average rate of drop concept is used to remove the impacts of measurement noises and determine the minimum required amount of load that should be removed at the right time. As shown in Fig. 11.6, the τ_i is the specified time delay between two steps of load shedding.

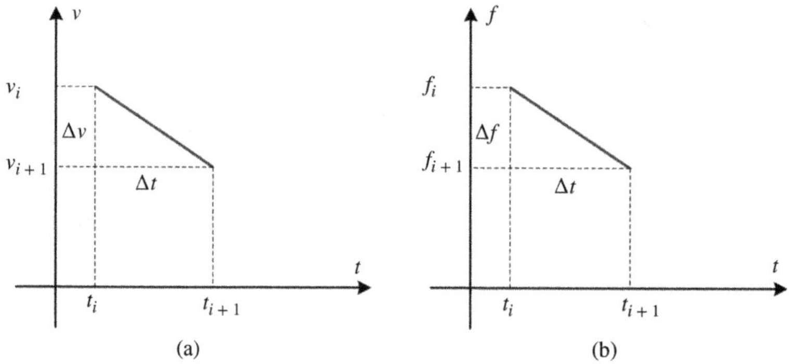

Figure 11.7 The concept of average rate of drop in (a) voltage and (b) frequency.

In the proposed load shedding algorithm, if the voltage or frequency value crosses the determined threshold (v_T or f_T), then the average rate of drop in voltage/frequency (Δv_t or Δf_t) is compared with the relevant threshold values ($-\alpha$ or $-\beta$). If one of the following inequalities is satisfied, the amount of load to be curtailed as well as the time will be calculated and the shedding process is executed.

$$\Delta v_t < -\alpha, \quad \Delta f_t < -\beta \tag{11.8}$$

11.4.2 Case Study

Figure 11.8 shows a single-line diagram of an MG case study for the evaluation of the proposed load shedding algorithm depicted in Fig. 11.6 [39]. The MG has the voltage base of 0.48 kV, which is connected to a main 200 kV grid at the PCC. For connecting/disconnecting the MG to/from the main grid, a circuit breaker (CB 1) is used. The main grid is modeled with a 20 kV three-phase grid voltage source with 12 MVA short-circuit power, which is linked to the MG through a 12.5 MVA power transformer. The MG uses four microsources connected to different buses. The microsources connected to feeders 1 and 2 are synchronous diesel generators (DG 1 and DG 2). The DGs are equipped with excitation and governor control system, and their nominal power is 150 and 100 kV, respectively.

Two sets of fixed-speed wind turbine with induction generator (WTG 1 and WTG 2) as asynchronous generating units are connected to buses 3 and 4. Their nominal power is 100 and 75 kV, respectively. Loads with different power factors have been placed at the feeders. Each load is connected to the relevant bus through a breaker in order to curtail the load under the necessary conditions.

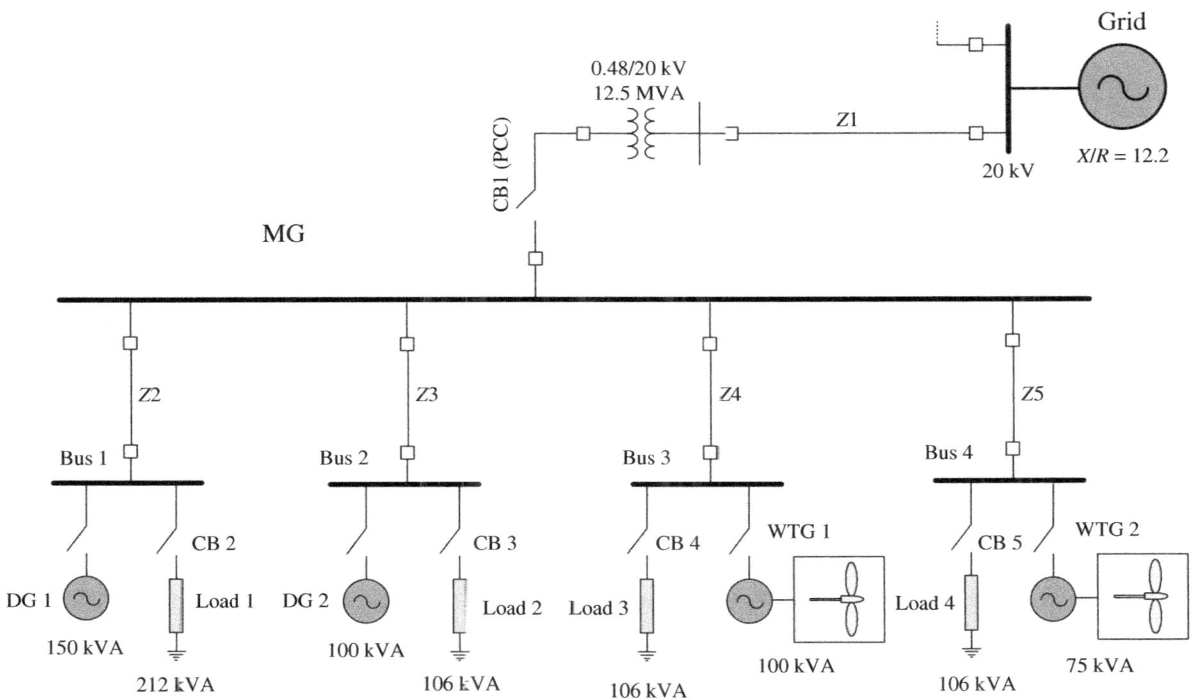

Figure 11.8 MG case study for load shedding (Example 2).

11.4.3 Simulation Results

As mentioned before, in the grid-connected operation mode, the main grid is responsible for providing the active and reactive power, and in the case of a load–generation imbalance in the MG, the main grid compensates it by injecting the power and hence supports the MG to maintain its voltage and frequency in normal operation. Transition state from the grid-connected to islanded mode due to the predetermined programs (such as planned maintenance and economic issues) or unplanned reasons (such as occurrence

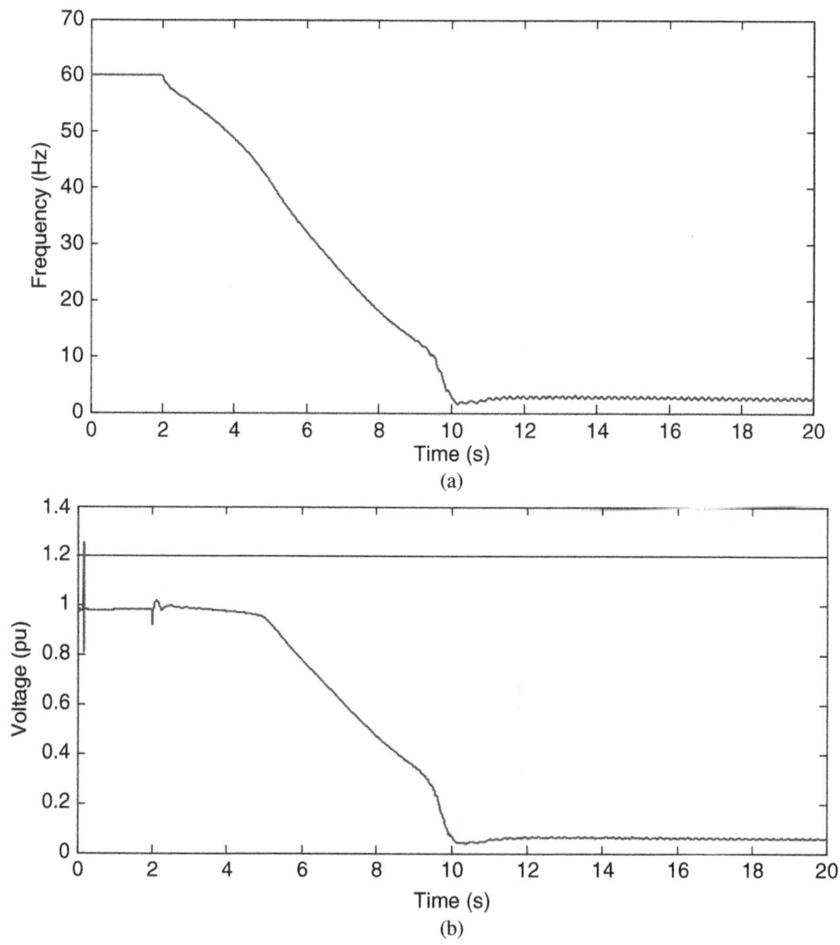

Figure 11.9 System response following islanding at 2 s: (a) frequency and (b) voltage measurements.

of errors and faults in the main grid) is a significant state that should be passed as smooth as possible. For the example at hand, it is assumed that due to an error in the main grid at $t = 2$ s, the CB 1 opens and the MG goes to the islanded operation mode. Figure 11.9 shows the system frequency and voltage measurements.

Following the event, the generating units react rapidly and increase their output power to prevent more drop in system voltage and frequency. However, since in the given case study, the required power for loads is greater than the produced power of the microsources, the imbalance between production and consumption leads to frequency and voltage instability. As shown in Fig. 11.9, drop in voltage occurs after a relatively long time delay rather than frequency drop, and this is due to the injection of reactive power by the wind turbines connected to the inductive generators (WTG 1 and WTG 2).

Using the proposed load shedding algorithm, the mentioned transition operation state has been passed safely. For the present case study, the voltage and frequency thresholds are considered as $v_T = 0.94$ pu and $f_T = 58.8$ Hz, respectively. The time and amount of load percentage to be shed in each step are given in Table 11.2. A total amount of 22% of the MG load must be curtailed in four steps. It can be shown that this is the minimum required load that should be disconnected for maintaining the system under a stable operation condition.

Figure 11.10 shows the frequency and voltage response following the application of developed load shedding algorithm. The lowest frequency experienced by the system is 56.5 Hz, and the frequency returns to the nominal value during $t = 12$ s.

Unlike frequency, voltage of the MG does not exhibit a large deviation, mostly because of injecting a considerable amount of reactive power by the WTGs (it is also noteworthy that the amplitude dimension is in pu). Since frequency drops earlier than voltage here, the load shedding process is started by crossing the frequency threshold.

Table 11.2 Load shedding table based on the proposed algorithm.

Load shedding steps	Load shedding time	Load shedding (%)
1	2.2127	7
2	2.4876	6
3	2.5973	5
4	2.7382	4

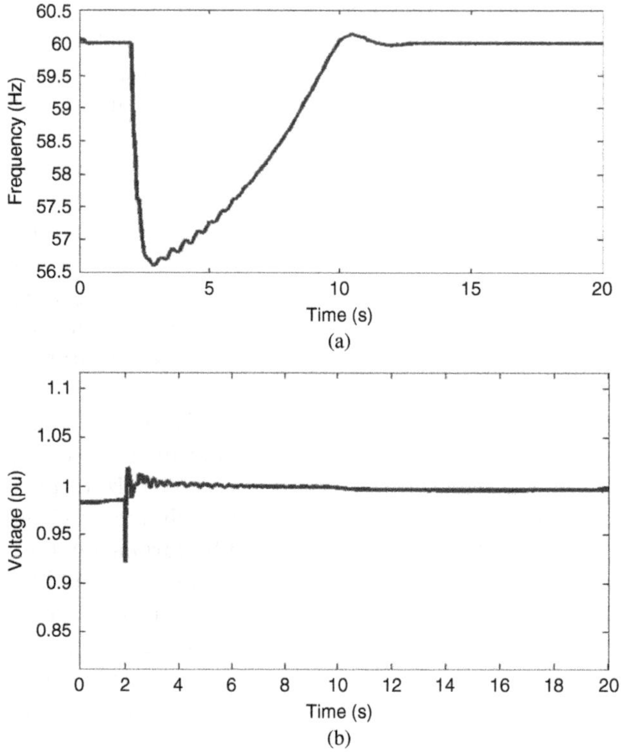

Figure 11.10 System response (following islanding at 2 s) using the proposed load shedding: (a) frequency and (b) voltage measurements.

11.5 Undervoltage–frequency Load Shedding

To evaluate the effectiveness of the undervoltage–frequency load shedding, which is explained in Section 11.4, and to discuss the important perspectives, a simple MG test system (Fig. 11.11) is considered. The MG test system includes four DGs and six load blocks. For the sake of first study, the system is considered without DG 4 and load 4. It is also assumed that DG 1 is a DFIG wind turbine and other DGs are diesel generator units. The MG parameters are given in [35] and some information are also available in [36].

Figure 11.12 depicts the system frequency and voltage magnitude in bus 6, when the MG is disconnected from the main grid at $t = 2$ s. As can be seen, without load shedding, the MG cannot support its local loads; as a result, the MG voltage and frequency may collapse. The proposed load shedding approach is then applied and verified to avoid this event; loads with lower priority are

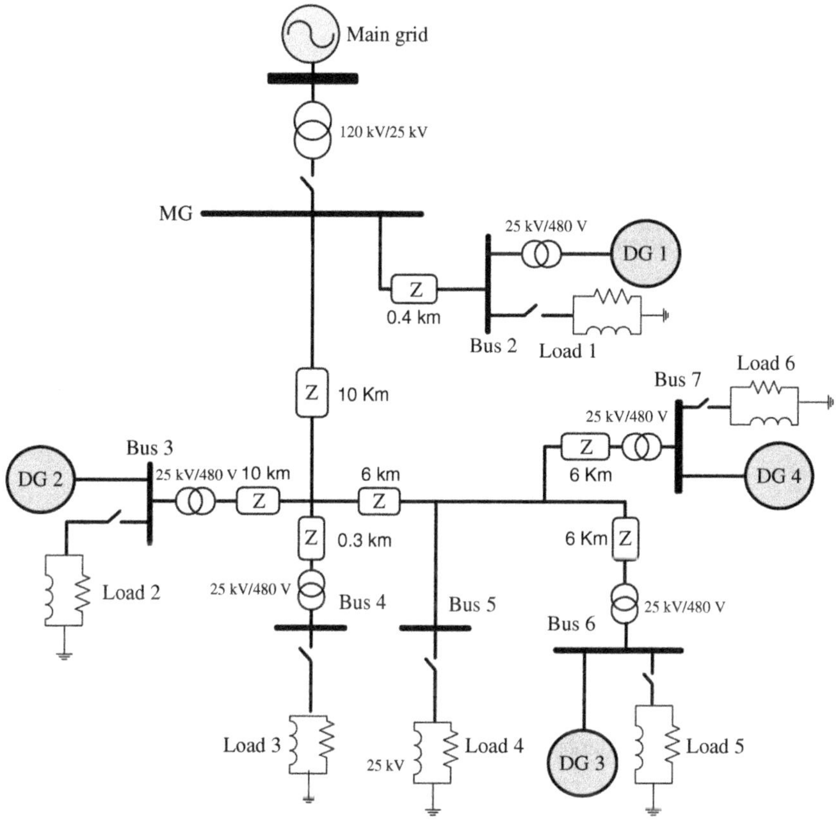

Figure 11.11 The case study for load shedding evaluation.

curtailed first. The load shedding process occurs in three steps, and the system frequency and voltage are returned to the normal range.

11.5.1 Δv–Δf Plane

In order to study the system voltage and frequency behavior following a disturbance, a Δv–Δf graphical analysis tool is introduced in [40]. This tool is useful for the postcontingency stability analysis. Following a contingency event, the power system operating point deviates from its stable precontingency state. If the contingency is not severe enough, the system may converge to another stable point. Otherwise, the operating point state may move to an unstable region. The states that are used in this trajectory are Δf and Δv, represented in the following complex statement [25]:

$$S = \widehat{\Delta f} + j\widehat{\Delta v} \tag{11.9}$$

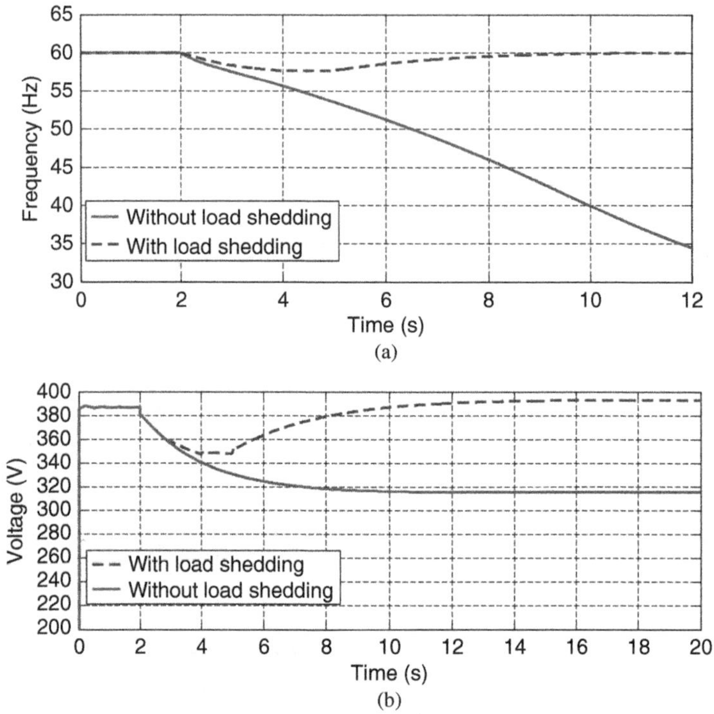

Figure 11.12 System response following islanding at 2 s: (a) frequency and (b) voltage at bus 6.

where

$$\widehat{\Delta f} = \frac{\Delta f}{f_0}, \quad \widehat{\Delta v} = \frac{\Delta v}{V_0} \tag{11.10}$$

The f_0 and V_0 are the normal values of frequency and voltage, respectively.

As mentioned before, for emergency control purposes, voltage and frequency are two suitable measurable variables that are used to illustrate the state of system following an event. Therefore, it seems that the Δv–Δf plane could be useful to see the state of MG system following a contingency. Figure 11.13 shows the frequency–voltage behavior in the given MG system following the mode change from the grid-connected to the islanded operation, with and without using the proposed load shedding. Unstable trajectory depicts the MG voltage–frequency response without load shedding, while the stabilized trajectory denotes the result of applied load shedding.

It is shown in [40] that in addition to analysis purpose, the mentioned Δv–Δf plane can be used in the process of an undervoltage–frequency load shedding synthesis based on the state variables given in (11.10). For this

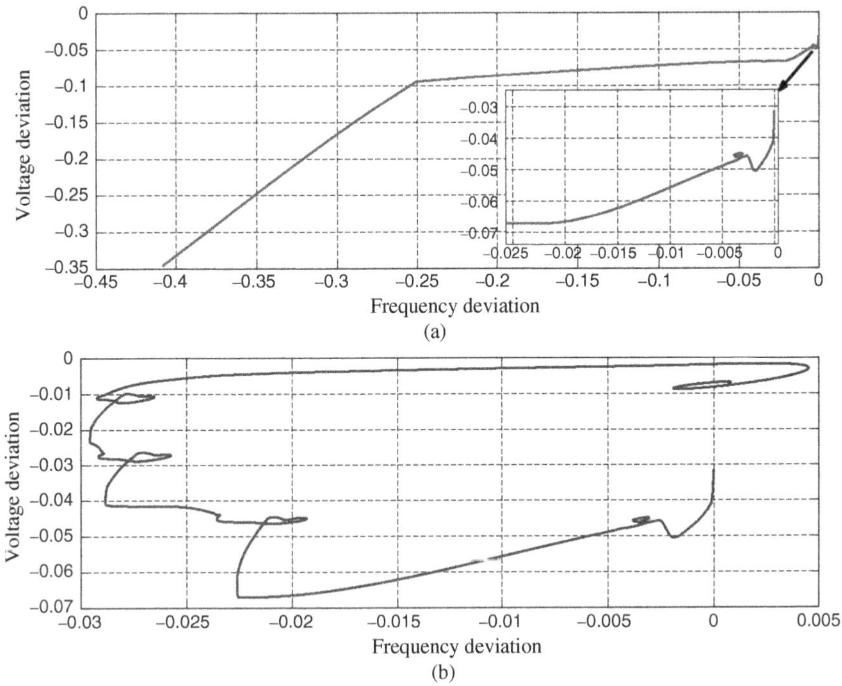

Figure 11.13 Δv–Δf trajectory postcontingency behavior: (a) without load shedding (unstable) and (b) using load shedding (stable).

purpose, two-dimensional threshold boundaries should be defined instead of one-dimensional voltage or frequency threshold plane that is common in conventional UFLS and UVLS algorithms. Since the threshold movement sizes in two directions are not equal, an elliptical shape for the existing boundaries could be considered [29].

11.5.2 Voltage and Frequency Performances

It is easy to show that the type of shed load affects the load shedding performance. Here, for evaluation of the impact of the removed load type on the response behavior, the high-reactive-power load blocks have been shed. The frequency response is shown in Fig. 11.14. It is assumed that the MG moves to the islanded operation mode at $t = 3$ s. Figure 11.14 shows quite better performance for frequency response with shedding high-reactive-power loads for the given case study. But, generally, the quality of this impact on the load shedding performance depends on the MG dynamics. Furthermore, this issue may make an inverse impact on the voltage performance [41] (similarly to that occurring in the present case study).

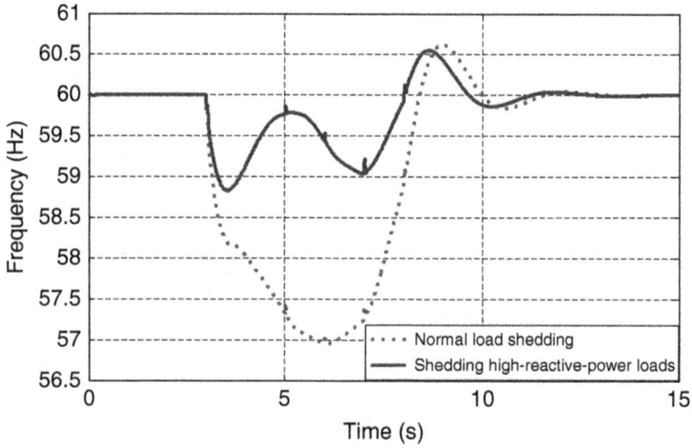

Figure 11.14 Impact of shedding high-reactive-power load.

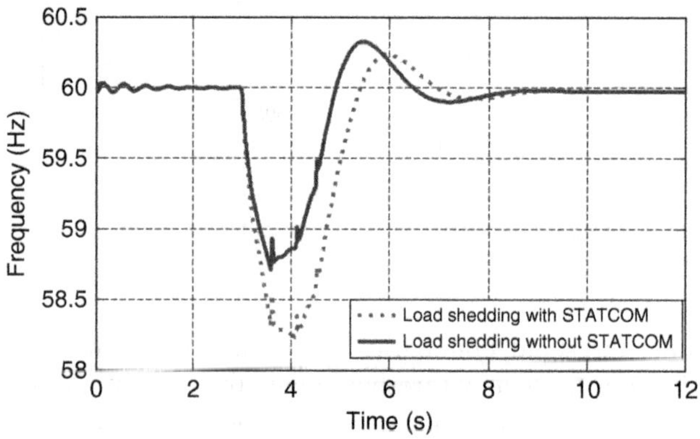

Figure 11.15 Load shedding with and without STATCOM.

Because a natural conflict exists between voltage and frequency/angle performances [42], any effort to improve the voltage profile in an AC MG may inversely affect the MG frequency. In order to show this fact, the impact of static compensator (STATCOM) on the load shedding result is evaluated. For this purpose, a STATCOM has been added to bus 5, and it is assumed that all four DGs work in the MG. It is clear that the STATCOM enhances the voltage performance in the load shedding process, while, as shown in Fig. 11.15, it degrades the system frequency to some extent.

Similar tests can be performed by increasing the active power in the MG. For the example at hand, the output active power of DG 2 has been increased and then the proposed load shedding algorithm is executed. The results are shown in Fig. 11.16. Figure 11.16a and b depicts the system frequency and Δv–Δf

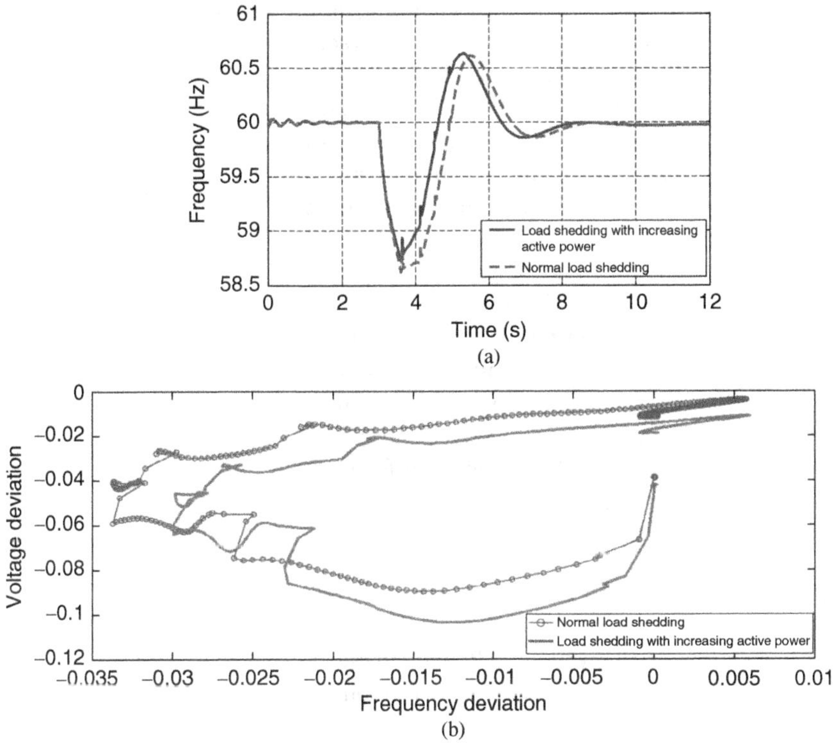

Figure 11.16 Load shedding following increasing active power: (a) MG frequency and (b) Δv–Δf plot.

trajectory. In comparison of the previous test, the loads in buses 4 and 5 have also been changed.

11.6 Summary

Similarly to conventional power grids, having acceptable voltage and frequency levels is necessary in an MG. In the grid-connected operation mode, this issue is realized by the main grid. But in islanded mode, the MG security strongly depends on the capability of existing MG controllers in primary, secondary, and emergency control levels. Stable performance of the MG in the transit and islanding states depends on the capability of the working control loops, particularly in the central and emergency control level.

Following a large disturbance, the MG's frequency and voltage may drop quickly if the remaining generation no longer matches the load demand. Significant loss of generation without adequate system response can produce extreme frequency/voltage drops. Depending on the size of frequency/voltage drop experienced, emergency control and protection schemes may be required to avoid the MG from a probable blackout. Load shedding is known as the most popular emergency control plan, under such conditions. Load shedding is an emergency control action designed to ensure MG stability, by curtailing local load blocks to match the remaining generation supply. The emergency load shedding would only be used if the frequency or voltage drops below the specified threshold.

In this chapter, following a general introduction on emergency control in MGs, load shedding, the most important and well-known emergency control strategy, was described. Two load shedding algorithms were explained, and undervoltage–frequency load shedding was emphasized. The potential benefits of the mentioned load shedding strategy compared to conventional UFLS and UVLS approaches were examined, and finally, some important factors for the voltage and frequency performance during the load shedding process were discussed.

References

1 Chowdhury, C., Chowdhury, S.P., and Crossley, P. (2009) *Microgrids and Active Distribution Networks*, The Institution of Engineering and Technology, London, UK, p. 24.

2 Bevrani, H. (2012) Microgrid controls, in *Standard Handbook for Electrical Engineers*, 16th edn (ed. H. Wayne Beaty), Section 16.9, McGraw-Hill, New York, pp. 160–176.

3 Willis, H. and Scott, W. (2000) *Distributed Power Generation: Planning and Evaluation*, CRC Press, New York, PA.

4 Bevrani, H., Gholami, M., and Hajimohammadi, N. (2013) Microgrid emergency control and protection: key issues and new perspective. *International Journal of Energy Optimization and Engineering*, **2** (1), 78–100.

5 Katiraei, F. and Iravani, M.R. (2006) Power management strategies for a microgrid with multiple distributed generation units. *IEEE Transactions on Power Delivery*, **21** (4).

6 Katiraei, F., Iravani, M.R., and Lehn, P.W. (2005) Micro-grid autonomous operation during and subsequent to islanding process. *IEEE Transactions on Power Delivery*, **20** (1), 248–257.

7 IEEE 1547 (2003) *IEEE Standard for Interconnecting Distributed Resources with Electric Power Systems*. IEEE 1547 Standard.

8 Piagi, P. and Lasseter, R.H. (2006) *Autonomous Control of Microgrids*. IEEE Power Engineering Society General Meeting, Montreal, Canada, 18–22 June.

9 Moreira, C.L. (2008) Identification and development of microgrids emergency control procedures. Ph.D. dissertation. Dept. Elect. Eng., Porto Univ., Porto, Portugal.

10 Pecas Lopes, J.A., Moreira, C.L., and Madureira, A.G. (2006) Defining control strategies for microgrids islanded operation. *IEEE Transactions on Power Systems*, **21**, 916–924.

11 Bevrani, H. (2014) *Robust Power System Frequency Control*, 2nd edn, Springer, Switzerland.

12 Tikdari, A. (2009) Load shedding in the presence of renewable energy sources. MSc dissertation. Kurdistan University.

13 Mahat, P., Chen, Z., and Bak-Jensen, B. (2010) Underfrequency load shedding for an islanded distribution system with distributed generators. *IEEE Transactions on Power Delivery*, **25**, 911–918.

14 Seethalekshmi, K., Singh, S., and Srivastava, S. (2011) A synchrophasor assisted frequency and voltage stability based load shedding scheme for self-healing of power system. *IEEE Transactions on Smart Grid*, **2**, 221–230.

15 Joshi, P. (2007) Load shedding algorithm using voltage and frequency data. MSc dissertation. Clemson University.

16 Bevrani, H. and Hiyama, T. (2011) *Intelligent Automatic Generation Control*, CRC Press (Taylor & Francis Group), New York, USA, April.

17 Kim, H., Kinoshita, T., and Lim, Y. (2011) Talmudic approach to load shedding of islanded microgrid operation based on multiagent system. *Journal of Electrical Engineering and Technology*, **6**, 284–292.

18 Du, P. and Nelson, J. (2009) *Two-step Solution to Optimal Load Shedding in a Microgrid*. IEEE Power System Conference and Exposition (PSCE 09), IEEE/PES, Vol. 9, pp. 1–9.

19 Kundur, P., Paserba, J., Ajjarapu, V. *et al.* (2004) Definition and classification of power system stability IEEE/CIGRE joint task force on stability terms and definitions. *IEEE Transactions on Power Systems*, **19** (3), 1387–1401.

20 Taylor, C.W. (1992) Concepts of undervoltage load shedding for voltage stability. *IEEE Transactions on Power Delivery*, **7**, 480–488.

21 Rudez, U. and Mihalic, R. (2011) Analysis of underfrequency load shedding using a frequency gradient. *IEEE Transactions on Power Delivery*, **26** (2), 565–575.

22 Concordia, C. and Fink, L.H. (1995) Load shedding on an isolated system. *IEEE Transactions on Power Systems*, **10**, 1467–1472.

23 Fathi, M. and Bevrani, H. (2013) Statistical cooperative power dispatching in interconnected microgrids. *IEEE Transactions on Sustainable Energy*, **4**, 586–593.

24 Lopes, J., Moreira, C., and Madureira, A. (2006) Defining control strategies for microgrids islanded operation. *IEEE Transactions on Power Systems*, **21**, 916–924.

25 Bevrani, H. (2014 Chapter 8, Section 4) *Robust Power System Frequency Control*, 2nd edn, vol. 85, Springer, New York, pp. 242–248.
26 Xu, Y. and Liu, W. (2011) Novel multiagent based load restoration algorithm for microgrids. *IEEE Transactions on Smart Grid*, **2**, 152–161.
27 Bevrani, H., Ledwich, G., and Ford, J.J. (2009) *On the Use of df/dt in Power System Emergency Control*. Proceedings of PSCE'09, IEEE/PES, March 15, pp. 1–6.
28 Gouveia, C., Moreira, J., Moreira, C., and Pecas Lopes, J. (2013) Coordinating storage and demand response for microgrid emergency operation. *IEEE Transactions on Smart Grid*, **4**, 1898–1908.
29 Bevrani, H. and Tikdari, A.G. (2010) *On the Necessity of Considering Both Voltage and Frequency in Effective Load Shedding Schemes*. Proceeding of IEEJ Technical Meeting, PSE-10-02, Fukui, Japan, January 21, pp. 7–11.
30 Kundur, P. (1994) *Power System Stability and Control*, McGraw-Hill, New York.
31 Bevrani, H. and Tikdari, A. (2009) *On the Necessity of Considering Both Voltage and Frequency in Effective Load Shedding Schemes*. Proceedings of IEE J Technical Meeting, PSE-10, January, pp. 7–11.
32 Ghaleh, A.P., Sanaye-Pasand, M., and Saffarian, A. (2011) Power system stability enhancement using a new combinational load-shedding algorithm. *IET Generation, Transmission & Distribution*, **5** (5), 551–560.
33 Abdelwahid, S., Babiker, A., Eltom, A., and Kobet, G. (2014) Hardware implementation of an automatic adaptive centralized underfrequency load shedding scheme. *IEEE Transactions on Power Delivery*, **29**, 2664–2673.
34 Terzija, V.V. (2006) Adaptive underfrequency load shedding based on the magnitude of the disturbance estimation. *IEEE Transactions on Power Systems*, **21**, 1260–1266.
35 Sarchami, O. (2015) An effective algorithm for load shedding in microgrids in emergency conditions. MSc thesis. University of Kurdistan, October.
36 Sarchami, O., Shafiee, Q., and Bevrani, H. (2016) *An Under Voltage-frequency Load Shedding Method for Emergency Condition of Microgrids*. 1st International Conference on New Research Achievements in Electrical and Computer Engineering, Tehran, May 12.
37 Hajimohammadi, N. and Bevrani, H. (2013) *Load Shedding in Microgrids*. 21th Iranian Conference on Electrical Engineering ICEE-2013, Mashad, Iran.
38 Hajimohammadi, N. and Bevrani, H. (2013). *On Load Shedding Design in Microgrids*. 18th Electric Power Distribution Conference, Kermanshah, April 30–May 1, Iran.
39 Azimi, M. (2014) Emergency control in microgrids. MSc thesis. University of Kurdistan, March.

40 Bevrani, H., Watanabe, M., and Mitani, Y. (2014, Chapter 08) *Power System Monitoring and Control*, John Wiley & Sons, Hoboken, New Jersey, pp. 158–185.

41 Golpira, H., Bevrani, H., and Naghshbandi, A.H. (2012) An approach for coordinated automatic voltage regulator power system stabiliser design in large-scale interconnected power systems considering wind power penetration. *IET Generation, Transmission & Distribution*, **6** (1), 39–49.

42 Bevrani, H., Hiyama, T., and Mitani, Y. (2008) Power system dynamic stability and voltage regulation enhancement using an optimal gain vector. *Control Engineering Practice*, **16** (9), 1109–1119.

12

Microgrid Planning and Energy Management

Planning and energy management are critical issues in distribution power network and microgrids (MGs). Several functions such as power prediction of renewable energy sources (RESs), load forecasting, energy storage reserve, peak shaving, reduction of CO_2 emissions, and maximization of the use of renewable energy options should be considered in the MGs planning and energy management.

Determining structure and topology of the grid as well as topology, type, and optimal size of microsources to satisfy the technical constraints, economic, environmental, and reliability indices are important issues in the MG planning problem. Central control in islanded operation mode and cooperation of central control and global control in grid-connected operation mode are the main responsible control levels in the energy management and power control.

This chapter addresses some significant aspects in MG planning and management and relevant topics. After an introduction, an MG planning example for a real MG case study is briefly explained. The PV power and demand forecasting are given in detail. Energy management concerning the technical, environmental, and economic issues, supplemented by real-time experimental studies, is described. The day-ahead optimal operation and power reserve (PR) dispatching are emphasized, and a robust energy consumption scheduling (ECS) methodology for interconnected MGs in the presence of demand uncertainty is presented.

12.1 Introduction

Recent oil crisis and, more importantly, the risks of global warming caused by the massive release of greenhouse gas emissions impose new obligations in energy saving and reduction of home electrical consumption. The evolution in the building industry tends to be the construction of healthy buildings, intelligent, high environmental quality with positive energy. The goal of positive

Microgrid Dynamics and Control, First Edition. Hassan Bevrani, Bruno Francois, and Toshifumi Ise.

energy buildings is to design long-term buildings that do not consume more than they produce, with a view to increase the energy autonomy of individual homes. Currently, the annual consumption in the residential sector in France is about 190 kWh/year/m^2 and the objective for new homes is 80 kWh/year/m^2. The research project at the European level aims to reach 40 kWh/year/m^2 in 2025 and positive energy in 2050 [1].

Another challenging factor is the liberalization of the electricity market and the opening to new producers. Home sources (meeting the needs of the building sector) concern only the modular small power generation (less than 1 kW to several tens of kW) and are connected to the grid at low voltage (LV). Most interesting technologies for the residential/urban sector in many countries are photovoltaic (PV) generators and cogeneration, that are also the concern of this chapter. The PV systems have now a relative maturity even if new improvements of energy performances are awaited and that their current deployments will induce a significant reduction of costs. The increase in PV generators connection and decentralized electricity producers into power grids causes strong structural changes within the electrical energy supply.

The increasing number of decentralized PV producers generates many local troubles in the grid. Their system-compliant integration into the supply process must be performed in order to enable their extensive expansion. For example, in a residential/urban network with all homes equipped with PV panels, the total produced power may be higher than the consumed power and so induce overvoltage. In this case, a part of PV power has to be switched off by the grid operator. Hence, the arrival of 100% green energy production induces new requirements to manage the network automation processes in an efficient manner, technically and economically.

Because of the intermittent and fluctuated power generation of the renewable energy-based distributed generators (DGs), these sources alone cannot actually provide services to the electric network [2]. However, as described in Chapter 2, the integration of energy storage units in a renewable energy-based power generation system makes these generators more active for the ancillary services supply. An alternative to grid reinforcement could be a restructuring of the power system architecture and increasing the share of DGs and RESs that generate electricity at the local scales. For instance, micro gas turbines (MGTs)-based combined heat production (CHP) and PV generators can play an essential role in domestic small-scale electricity generation.

For several decades, utilities try to partially adjust loads to market constraints (economic) and network (technical) through a demand side management. A demand side management refers to a set of scenarios whereby the consumer can reduce energy consumption during peak usage or other critical energy use periods. It is based on different payment/incentive schedules that can be classified into two families:

1) Demand response is based on price signals that are used to motivate customers to change their load profile. In this scheme, customers manage voluntarily their loads by curtailing/shedding some electrical loads at peak times. The price is higher during peak hours for customers to move their loads during hours when electricity is cheaper. Demand response can be implemented by a dynamic pricing and then customers pay for the used electricity according to a specific pricing structure (hourly pricing, time of use, critical peak pricing, etc.). Demand response can be also implemented by a (day-ahead) notification program and customers are paid an incentive or pay a penalty based on a calculated load reduction. Higher flexibility of demand response technically facilitates new possibility for the grid ancillary services and regulation tasks [3, 4].
2) Another option is the direct control of some loads. In this case, customers offer the ability to switch "off" and "on" loads under certain circumstances. In this context, electrical vehicles and their batteries will play an important role in the future as a possible flexible load and a possible storage of the local PV electrical production. Figure 12.1 shows a home application with PV panels and storage units and a demand side management capability via some controllable loads.

During the day, this home application may be a power producer or a consumer and is also known as a "prosumer." Recently, energy-boxes (E-boxes) were developed in order to follow energy consumption. They have been upgraded to increase consumer satisfaction with various options for automatic control of some loads. This customer-enabled management

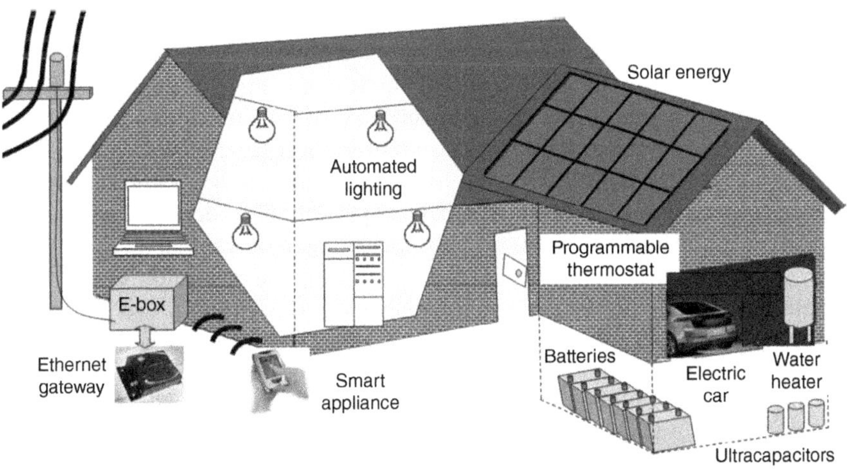

Figure 12.1 Prosumer with load demand response and electrical production capabilities.

provides opportunities for consumption adaptation to time pricing and new grid services for higher quality power supply. The concept of placing a smart driver at the grid coupling point allows the integration of all technical components (processor card, relays, communication devices, sensors, etc.) required implementing strategies for energy management [5, 6]. The home has a dedicated connection for the injection of energy to the distribution network. The E-box integrates at least three functions: a load manager, an advanced meter, and a local energy management.

The load manager enables customers to automatically preprogram appliances to turn on when prices are lower or to create energy consumption habits, such as uninterrupted supply of critical loads and time programmable use. Moreover, it can reduce a part of the home power demand when the grid is under stress by disconnecting offered controllable loads [7]. An advanced meter feeds the local energy management system (EMS) as well as the load manager. A further strategy is to transform the actual PV generators into controllable active generators (AG) in order to offer new flexibilities for energy management of electrical networks. Thanks to embedded storage technologies and a dedicated local control algorithm, an AG is able to deliver prescribed power references and power system services and can be dispatched to the distribution system.

PV panels associated with hybrid storage systems perform a PV-based AG. All components in an AG are connected to a common inner DC bus, which is interfaced to the electrical network through a three-phase inverter. The storage units are used to create a long-term energy reserve and to supply power with fast dynamics. The inner instantaneous power balancing and power dispatching among the PV source and storage units according to the storage level capacity and to the specific requirements/limitations of each source are performed by a local controller, whose functions are detailed in Chapter 5. Real and reactive power references are received from the microgrid central control (MGCC) as described in Chapter 2. The state of charge (SOC) of batteries and the sensed powers at the connection point are also sent to this control center.

In this chapter, a DC coupling of the PV system and storage systems is considered to form a home active PV generator as presented in Chapter 2 and shown in Fig. 12.2. The interesting aspect of this hybrid generator is that it can deliver a prescribed power level (P_{AG_ref}) such as a conventional generator (e.g., a diesel engine or MGT). The local energy management (EM) thus allows the use of PV energy according to the grid operator requirement also when the sun is not shining. In this case, batteries are tapped to provide the required power. To highlight the difference with conventional PV panels, this concept of AG is used in this system. Figure 12.3 shows a residential/urban MG with several prosumers that use PV-based AGs, controllable loads, and MGTs.

In the mentioned MG, the MGCC measures the MG state variables and dispatches orders to microsources through the communication bus. Local controllers (of the MGT and the PV-based AG) receive power set points from

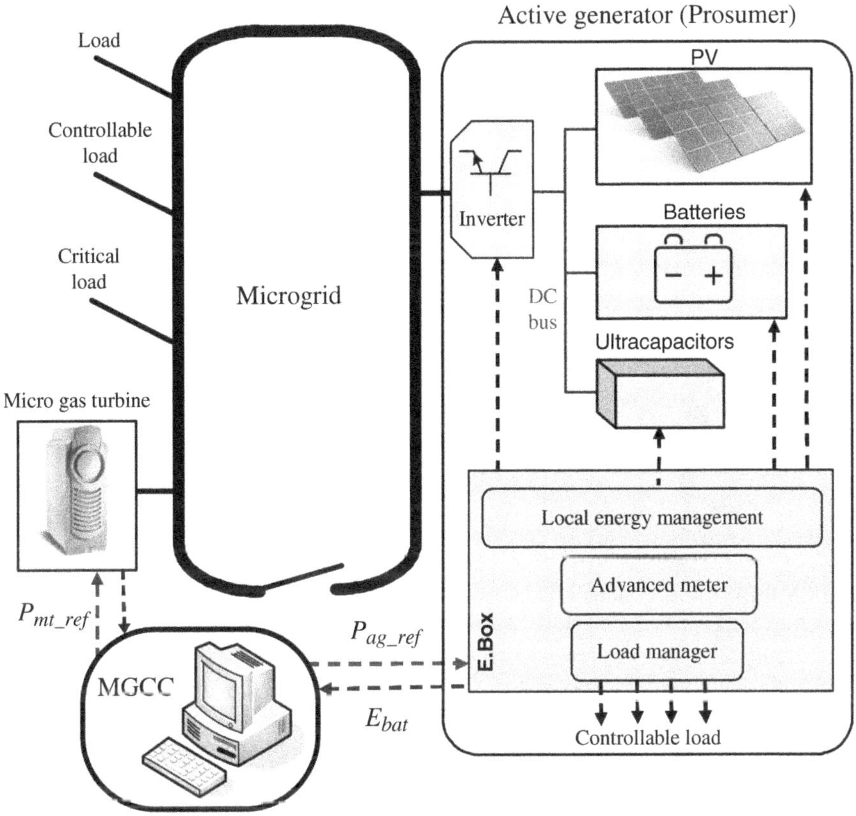

Figure 12.2 An MG with a PV-based AG.

the MGCC. At the same time, they send various information, for example, the sensed power production at the coupling point.

Several functions in the MG planning and EMS have to be considered as power prediction of the renewable energy, load forecasting, energy storage reserve, peak shaving, maximized use of RESs, reduction of CO_2 emissions, and new power planning. In the EM point of view, the global objective can be considered to use in priority PV-based AGs to power the MG, while the MGTs are operating as backup generators. In the given MG case study in Section 12.3, to facilitate the presentation of theoretical developments, a single prosumer and an MGT are considered in the MG. A deterministic approach is studied for the design of the MG EMS, by having information on the load demand and PV production.

The rest of this chapter is organized as follows: An MG planning example for a real MG case study is briefly explained in Section 12.2. An introduction

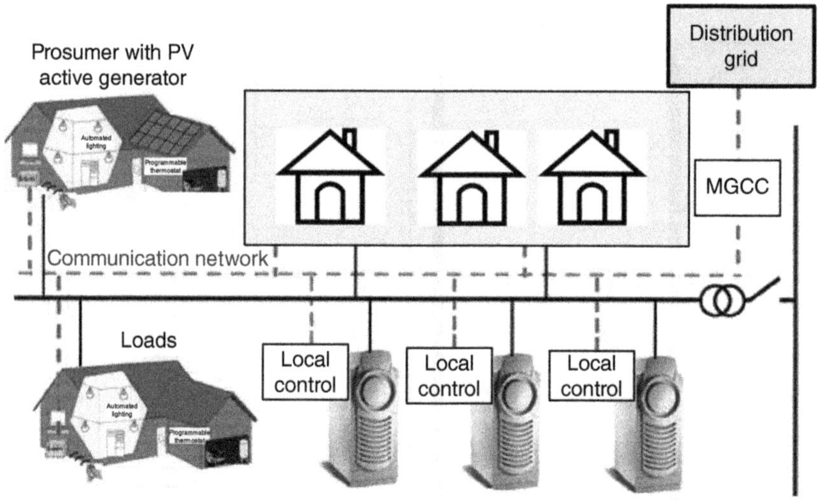

Figure 12.3 A typical residential/urban MG.

on PV power and demand forecasting as well as the relevant data processing is given in Section 12.3. Section 12.4 describes energy management in the MGs in the technical point of view, which is supplemented by a real-time experimental study. This work is updated in Section 12.5 considering the environmental and economic aspects. Section 12.6 presents the day-ahead optimal operation and PR dispatching in the MGs. A robust ECS methodology for interconnected MGs in the presence of demand uncertainty is given in Section 12.7, and, finally, the chapter is summarized in Section 12.8.

12.2 Microgrid Planning: An Example

Nowadays, one of the most important environmental issues is global warming and constantly increasing air pollution by fossil fuels [8]. The RESs can be an appropriate solution for this problem according to the researchers and policy makers' outcomes [9]. On the other hand, sensitive loads are increased due to industrialization, social welfare improvement, and technology development. Computer numerical control (CNC) machines have been known as very common and also sensitive loads in the lathing and milling industries in which computers play an integral part of these controllers.

DGs/RESs are introduced as appropriate resources to meet the loads especially sensitive loads locally. These resources are integrated with local loads in the form of MGs [10], which can operate in both grid-connected and islanded operation modes. The industrial estates are of interest because of their enormous greenhouse gas emissions and existence of the sensitive loads such as induction furnaces and CNC machines. In fact, the sensitive loads importance is another reason to plan and construct a new MG in the industrial estates.

One of the important issues in MG planning is determining the optimal size of their resources to meet the load so that the objective function is minimized/maximized and technical constraints are met. Economic, environmental, and reliability indices are used as objective functions in the planning formulation of the MGs. Different software and optimization techniques are used for optimal sizing of MGs in the literature. HOMER (hybrid optimization model for electric renewable) is one of the powerful software, which are used by many researchers for optimal planning of MGs [11]. The main objective of the present example is the optimal planning of MGs for industrial estate, and the idea is enhancing power quality, reliability, and security of CNC workshops as sensitive loads by implementing an appropriate MG.

Here, the MG planning objective includes assurance of continued and secure operation of CNC machines as sensitive loads, minimizing total costs of MG construction and operation. The HOMER software is used to specify the technical feasibility of MG planning and to select the best plan economically and environmentally.

12.2.1 Description of Input Parameters

In this study, a set of eight CNC workshops is selected as the load to test the feasibility of the designed MG. This set is located in a real industrial estate. Every workshop has certain and individual CNC machines and other equipment. However, all of them are similar in the viewpoint of consuming energy type. Nevertheless, their consuming energy amounts are different. An average load is considered for workshops instead of their different amounts to facilitate the load estimation. In this study, based on the nearby workshops visit, certain equipment and consumptions are considered for all workshops consisting of CNC machines, manual lathing/milling machines, lighting, and miscellaneous consumptions such as air-conditioning, computers, and refrigerators. On average, every workshop has three CNC machines, two manual lathe/milling machines, ten 46 W duplex lamps, and 2 kW equivalent consumption of miscellaneous equipment. Since the majority of the CNC machines, whether milling or lathing, are three-axis or can expand to it, they are considered three-axis on average. According to the detailed forecasted consumptions, the total consumption of a CNC workshop is calculated in Table 12.1. Table 12.2 shows the typical data of daily load forecasting of the workshops set in July 2016. The scaled annual average energy demand of the CNC workshops set as simulated by HOMER software is 3294 kWh/day, which is equivalent to 1.2 GWh/year. A sample of the daily load profile for weekdays and weekends is shown in Fig. 12.4.

The data of the solar radiation and ambient temperature was obtained from National Aeronautics and Space Administrative (NASA) [12]. It is a 22-year

Table 12.1 CNC workshop consumption in a normal working hour.

Electrical load type	Consumption calculation	Number	Load (kW)
CNC machine	$(2+2+2)\times 0.5+(11+3+0.5+1)\times 1+0.2\times 0.16+1\times 0.2+1\times 0.3=$ 19 kW	3	57
Manual lathing/milling machine	$(5+1)\times 1+0.5\times 0.3=6.15$ kW	2	12.3
Lighting	$2\times 0.046=0.092$ kW	10	0.92
Miscellaneous equipment	2 kW	–	2
Total CNC workshop load			72.22

Table 12.2 Daily load forecast data of the workshops.

Time zone (h)	0–8	8–12	12–13	13–17	17–21	21–24
Percent of daily high load (weekends) (%)	10 (3)	90 (27)	60 (18)	100 (30)	50 (15)	20 (6)
Weekday load (kW)	42.6	383.2	255.5	425.8	213	85.2
Weekend load (kW)	12.7	115	76.6	127.7	64	25.6

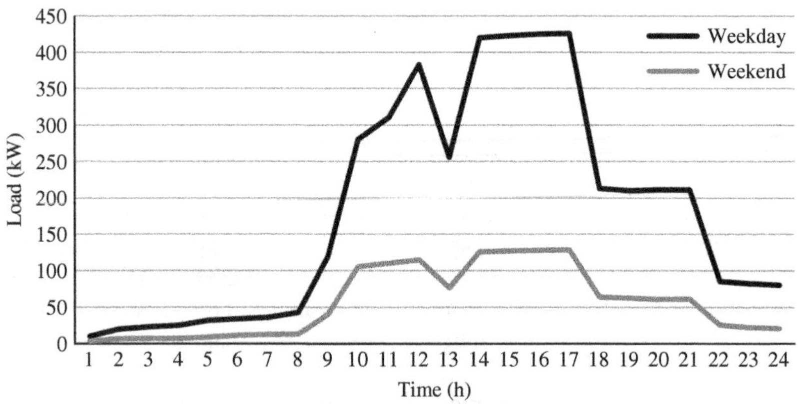

Figure 12.4 A sample of the daily load profile for the studied CNC workshops set in weekday and weekend.

monthly average solar radiation and ambient temperature data. Also, the wind speed data was obtained from the NASA database for the given location [12]. It is a 22-year monthly average wind speed data measured at 10 m above the surface of the earth. Figure 12.5 shows the monthly average of solar radiation according to the clearness index, as well as the wind speed.

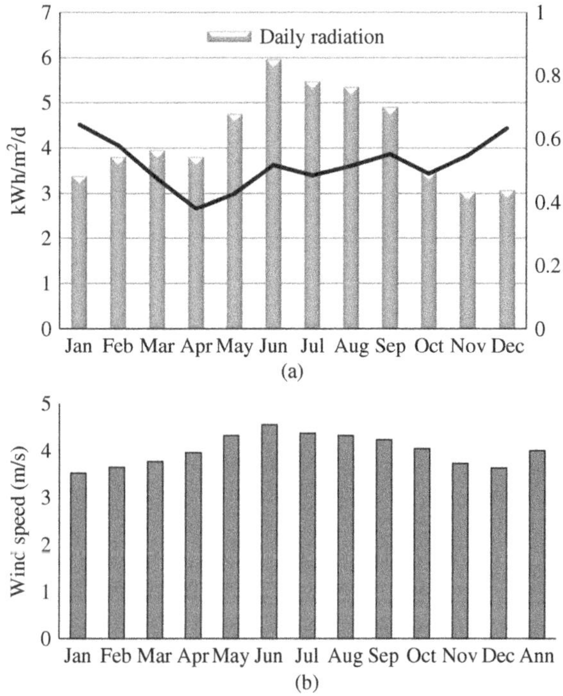

Figure 12.5 The monthly averaged data: (a) solar radiation and the clearness index and (b) wind speed at 10 m above the surface of the earth.

The updated diesel price in the mentioned area at the time of planning was US$0.18/L. The project lifetime is considered to be 25 years with an annual discount rate of 4%. The system fixed capital cost for stores managing, labor wages, different civil constructions, required licenses, administration and government approvals, and other miscellaneous costs [13] is estimated to be $30,000 for the whole project and the system fixed operation and maintenance cost is estimated at $500/year.

12.2.2 System Description and Specification

The performed MG is made up of eight main components including DG, a fuel cell (FC), electrolyzer, hydrogen tank, PV modules, wind turbines (WTs), batteries, and power converters as shown in Fig. 12.6a. The MG can operate in the grid-connected or islanded operation modes. Figure 12.6b shows the configuration of the MG in the HOMER software environment. The descriptions of the selected components are represented in Table 12.3.

Three rates off-peak, normal, and peak are defined for the region load. Since the understudy MG is in an industrial region, peak hours are 9–12 and 14–17,

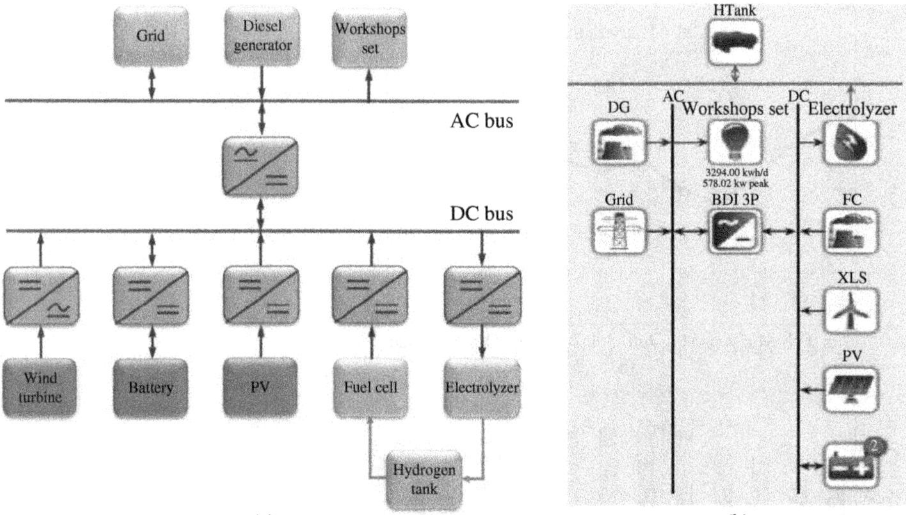

(a) (b)

Figure 12.6 MG case study: (a) block diagram and (b) the overall configuration in HOMER environment.

normal hours are 8–19 without peak hours, and other hours are considered off-peak hours. Buying/selling tariffs from/to the grid for the MG are given in Table 12.4.

12.2.3 Numerical Results and Discussion

Some simulations are executed for the MG shown in Fig. 12.6, and the HOMER search space is given in Table 12.5. The simulations are done with a project lifetime of 25 years. Nine reasonable and preferable plans that provide desirable optimization results are shown in Table 12.6. The HOMER simulation results include a number of feasible but not reasonable results such as existence of the hydrogen tank without the electrolyzer and FC or existence of the hydrogen tank and electrolyzer without the FC. These results have been eliminated to access reasonable plans. The plans have been sorted from 1 to 9 based on the MG total net present cost (NPC) in Table 12.6. One of the first deductions is the less NPC for plans without battery than equivalent plans with battery.

Figure 12.7 shows per unit values of the NPC, the levelized cost of energy (COE), DG pollution, and renewable fraction that are calculated by dividing real values by the maximum value of each index. This diagram shows the variation of different indices with respect to each other for the selected plans. The NPC and COE indices have approximately the same increase because of their similar behavior in the economic assessment. The DG pollution (equivalent to DG production) index shows plans 3 and 4 have the maximum DG production

Table 12.3 Description and economic and technical specification for the components of the proposed MG.

Description	Specification	Description	Specification
PV modules		*Fuel cell*	
Model	PV-MF100EC4	Type	PEMFC
Rated power	250 kW	Rated power	200 kW
Capital cost	\$7300/kWh	Capital cost	\$3000/kW
Replacement cost	\$2974/kWh	Replacement cost	\$2500/kW
Operating and maintenance cost	\$10/kWh	Operating and maintenance cost	\$0.02/kWh
Lifetime	25 years	Lifetime	50,000 h
Temperature coefficient	−0.5%/°C	Efficiency	50%
Derating factor	80%	*Electrolyzer*	
Wind turbine		Rated power	200 kW
Type	BWC Excel-S	Capital cost	\$1900/kW
Rated power	10 kW DC	Replacement cost	\$1400/kW
Capital cost	\$30,730	Operating and maintenance cost	\$20/year
Replacement cost	\$22,900	Lifetime	15 years
Operating and maintenance cost	\$458/year	Efficiency	90%
Lifetime	15 years	*Hydrogen tank*	
Storage battery		Rated power	300 kg
Type	Surrette 6CS25P	Capital cost	\$1200/kg
Capital cost	\$1229/single cell	Replacement cost	\$100/kg
Replacement cost	\$1229/single cell	Operating and maintenance cost	\$15/year
Operating and maintenance cost	\$10/year	Lifetime	25 years
Diesel generator		*Inverter*	
Type	Perkins	Type	Leonics GTP519S
Rated power	250 kVA	Rated power	900 kW
Capital cost	\$182.5/kW	Capital cost	\$300/kW
Replacement cost	\$175/kW	Replacement cost	\$300/kW
Operating and maintenance cost	\$0.03/h	Operating and maintenance cost	\$10/year
Lifetime	15,000 h	Lifetime	10 years
Minimum load ratio	30%	Efficiency	90%

Table 12.4 Grid buying and selling tariffs.

State	Buying energy cost ($/kWh)	Selling energy cost ($/kWh)
Peak	0.0155	1.634
Normal	0.0047	1.634
Off-peak	0.0011	1.634

Table 12.5 Values of all optimization variables.

Converter (kW)	Grid purchase (kW)	H Tank (kg)	Electrolyzer (kW)	Battery 1 (n)	Battery 2 (n)	FC (kW)	DG (kW)	PV (kW)	WG (n)
0	0	0	0	0	0	0	0	0	0
300	1000	100	100	40	40	100	50	60	7
600		200	200	80	80	200	100	120	14
900				120	120		150	180	21
							200	250	

Table 12.6 The best optimization results of the MG planning.

Plan	PV (kW)	WT (n)	FC (kW)	DG (kW)	Battery (n)	Grid (kW)	Electrolyzer (kW)	H Tank (kg)	Converter (kW)	NPC ($)
1	60	21	–	50	–	1000	–	–	300	1,679,098
2	60	21	–	50	80	1000	–	–	300	1,871,337
3	250	–	–	100	–	1000	–	–	300	2,126,354
4	250	–	–	100	80	1000	–	–	300	2,353,911
5	250	7	–	–	–	1000	–	–	300	2,568,618
6	180	21	–	–	80	1000	–	–	300	2,779,870
7	120	21	100	50	–	1000	200	100	300	3,272,760
8	120	21	100	50	80	1000	200	–	300	3,316,866
9	120	21	200	50	80	1000	200	100	300	3,672,796

and plans 5 and 6 do not have any DG capacity. Another interesting result, which this index shows, is the less DG pollution of plan 2 than plan 1. The renewable fraction index has uniform rate except in plans 5 and 6. In these two plans, RESs compensate DG absence.

Consequently, plans 1 and 2 have the minimum NPC with the difference of \$0.2 M and plans 5 and 6 have the maximum renewable fraction, zero air pollution, and greater NPC. Plans 1 and 5 have no battery that is a good reason

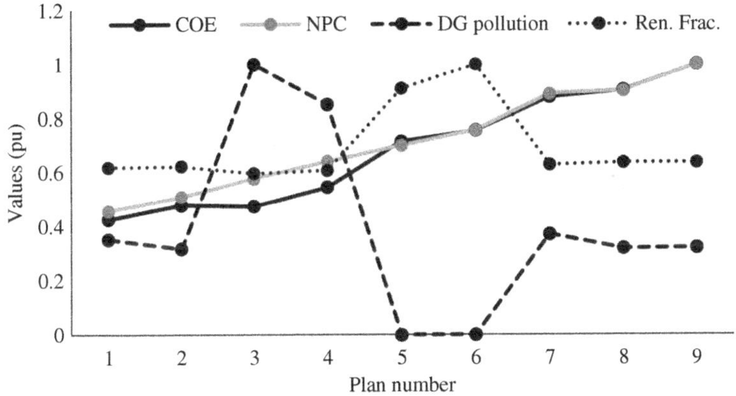

Figure 12.7 Per unit values of NPC, COE, DG pollution, and renewable fraction for the selected nine plans.

for their deletion. Finally, it can be said that plan 2 is the best plan with almost minimum NPC in the presence of battery and DG for ensuring sensitive loads electrification.

12.3 Forecasting Techniques

12.3.1 PV Power Prediction

The naturally poor predictability of the solar energy level is a weakness in using the PVs in an electric system. The PV panels provide electrical power only in the daytime with a power peak around midday. Meanwhile, huge production variations may occur. Another disadvantage happens in winter, when the maximum power consumption peak in the grid meets with the moment of sunset. However, according to the historic data, the obtained power prediction can be adapted more and more accurately to the real situation by means of predictive models and intelligent methods.

The produced power output of PV panels as a function of time can be written as the sum of two functions: a "typical" curve that corresponds to an irradiation profile with a clear sky and a "stochastic" curve that is a function of the cloud coverage. The least squares method (LSM) can be applied to calculate the desirable possible prediction of a local PV power production based on irradiation predictions of four meteorological stations. The success of the prediction is strongly dependent on the weather type.

For the presented residential application, the day-ahead and real-time PV power forecasting are used. The day-ahead PV power forecasting is used for the 24-h planning of the long-term energy management and is based on the

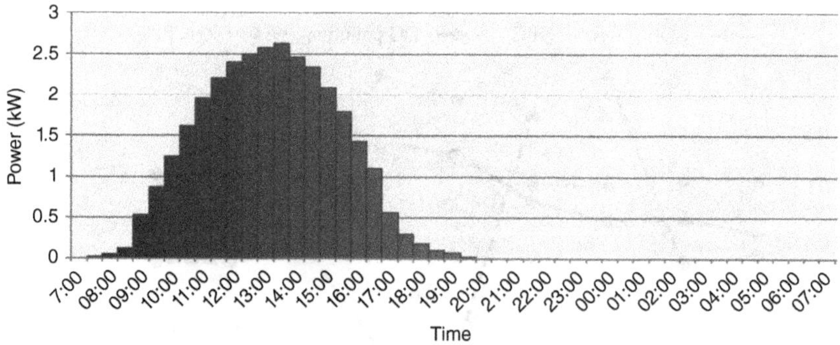

Figure 12.8 24-h-ahead PV power forecasting.

day-ahead weather forecasting and the historic data. The forecasting gives a 24- h PV power prediction with a constant value for each half hour as shown in Fig. 12.8.

The real-time PV power forecasting is used for the maximum power point tracking (MPPT) and for the real-time set-point correction in the inner layer of the short-term EM. The real-time PV power forecasting is based on the real-time weather condition detections (solar irradiation, panel temperature, cloud identification, etc.). Using artificial neural networks (ANNs) is a well-known forecasting methodology for PV power, which has been widely used in the literature, and an example is also given in Chapter 10.

12.3.2 Load Forecasting

Load forecasting is important for energy purchasing and generation, load switching, contract evaluation, and infrastructure development. In this chapter, the load forecasting is used for the generation power planning and the EM inside the MG. Load forecasting can be performed during different timescales according to the needs [14]. A forecasting longer than a year is used for the sizing of power systems. The purchasing and generation of electricity as well as the EM require a load forecasting from an hour to a week or from a week to a year. The characteristics of these forecasts are different.

Most forecasting methods use statistical techniques or artificial intelligence algorithms such as regression, ANNs, fuzzy logic, and expert systems. Two methods the so-called end-use and econometric approaches are widely used for medium-/long-term forecasting. A variety of methods, various regression models, time series, neural networks, statistical learning algorithms, fuzzy logic, and expert systems have been developed for short-term forecasting [15, 16]. An ANN-based example for load forecasting is given in Chapter 10.

The development and improvements of mathematical tools will lead to develop more accurate load forecasting techniques.

For short-term load forecasting, several factors should be considered, such as time factors, weather conditions, and power demand behaviors. The time factors include the season of year, the day of week, and the hour of day. There are important differences in load between weekdays and weekends. The load in different weekdays can also behave differently. For example, Mondays and Fridays, being close to weekends, may have structurally different loads than Tuesday through Thursday. This is particularly true during the summer time. Holidays are more difficult to forecast than nonholidays because of their relative infrequent occurrence [14].

Weather conditions are also important in the load forecasting issue. Forecasted weather parameters must be used for the short-term load forecasts. Various weather variables could be considered for load forecasting. Temperature and humidity are the most commonly used load predictors. Most electric utilities serve different types of customers such as residential, commercial, and industrial. The electric usage pattern is different for customers belonging to different classes. Therefore, most utilities distinguish load behavior on a class-by-class basis.

During 1 year, differences are attributable to heating, which explains the difference in power peak levels between winter and other seasons in the year, and additional lighting consumption. In winter, the dark causes a peak in a time later in summer and mid-season. Similarly, the lighting need in the morning in winter locates the maximum consumption at a specific time, while it tends to shift to another time when the sun rises earlier.

Therefore, a set of information can be used to estimate load profiles and can be used to make decisions. Based on historic electrical power production requirements for the residential homes in a region, the behavior of the loads in an MG can be forecasted and estimated. Several important factors influence a load in an electrical network: weather situation (temperature, cloudiness, etc.), economic activity (huge modifications of load forecasting in the holidays), the legal working time, and so on. In this chapter, the load forecasting method is based on the meteorological data and the historical consumption data.

A 24-h-ahead load forecasting profile ($\tilde{P}_{Load\_24\,\mathrm{h}}$) is given in Fig. 12.9 with data each half hour. Here, the pattern of the studied home power consumption in a day has been characterized by five timing domains: (i) the morning peak (Zone A) related to the resumption of activities in homes, rail transport, industry, and tertiary sector; (ii) the lunch peak (Zone B); (iii) the sag in the afternoon (Zone C); (iv) the evening peak (Zone D) due to end activities of the day in office, the recovery of the residential consumption, and the activity transportation peak; and (v) the night sag (Zone E). At that time, consumption is restricted

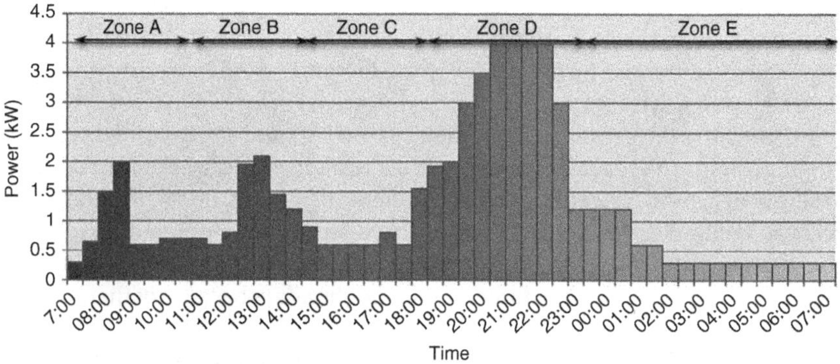

Figure 12.9 24-h-ahead load forecasting ($\tilde{P}_{Load\_24\,h}$).

to continuous industrial processes, public lighting, electronic stand by devices, and the winter heating.

12.3.3 Energy Estimation

In order to plan the power generation and implement the EM algorithms, an energy analysis is necessary. The maximum available energy generated by the PV panels in daytime, the consumed electric energy during daytime, and the stored energy in the batteries are all estimated values and are used to decide how much complementary energy generated by the backup generator (e.g., MGT) is needed. Therefore, the energy estimation from the forecasting data (PV power prediction and load forecasting) is essential.

For the energy estimation, the initial time point is the starting time in day (t_0) and the day duration (Δt). The start of sunrise is usually chosen as the initial time (t_0). However, for cloudy or rainy days, the initial time may be changed because the solar irradiation and the temperature influence the PV power production. Both parameters depend on the season and weather conditions. So in the morning, the initial time is set as soon as the PV panels start to generate the power and the time $t_0 + \Delta t$ is fixed at the start of sunset.

12.3.3.1 Estimation of the Available PV Power

The long-term PV power prediction is based on a half hour forecasting (Fig. 12.8). The scheduled energy of the PV production during each 1/2 h interval can be calculated using the data from the PV power prediction ($\tilde{P}_{PV\_24\,h}$) as follows:

$$\tilde{E}_{PV\_1/2\,h} = \int_{t_0+nTe}^{t_0+(n+1)Te} \tilde{P}_{PV\_24\,h}(t)dt = Te \cdot \tilde{P}_{PV\_24\,h}(t_0 + n \cdot Te),$$
$$\text{for } t_0 + nTe \le t \le t_0 + (n+1)Te \qquad (12.1)$$

where $Te = 30$ min and $n \in \{0, 1, \dots, 47\}$. The total PV energy for 1 day is the sum of all the previous calculated half-hour PV energies:

$$\tilde{E}_{PV} = \sum_{n=0}^{n=23} \tilde{E}_{PV\_1/2\,\mathrm{h}}(t_0 + nT_e) \tag{12.2}$$

In order to plan the production program, the energy, which is demanded by the load, can be also estimated during the same interval:

$$\tilde{E}_{Load\_1/2\,\mathrm{h}} = \int_{t_0+nTe}^{t_0+(n+1)Te} \tilde{P}_{Load\_24\,\mathrm{h}}(t)dt = Te \cdot \tilde{P}_{Load\_24\,\mathrm{h}}(t_0 + n \cdot Te) \tag{12.3}$$

The energy demanded by the load during daytime is also necessary to estimate the energy difference between the PV power and the demanded power for the MG system shown in Fig. 12.2. If the PV power is not enough, the MTG has to generate more power to assure the power supply in the MG. If the PV power is more than the load during the daytime, the surplus PV power can be stored in the energy storage units (batteries). The energy difference ($\tilde{E}_{diff\_1/2\,\mathrm{h}}$) for each half hour can be estimated from the PV power prediction and the load forecasting:

$$\tilde{E}_{diff\_1/2\,\mathrm{h}}(t) = \tilde{E}_{PV\_1/2\,\mathrm{h}}(t) - \tilde{E}_{Load\_1/2\,\mathrm{h}}(t) \tag{12.4}$$

Thanks to the energy storage systems (ESSs), this energy difference ($\tilde{E}_{diff\_1/2\,\mathrm{h}}$) can be absorbed or compensated. The size of the storage device determines the capacity of the power absorption and the power compensation. Because of the rated power limit, the exchanged energy with the batteries during the half hour is limited. Considering a constant rated value (P_{bat_max}), the maximum battery energy in the half hour can be calculated as follows:

$$\tilde{E}_{bat\_1/2\,\mathrm{h}\_max} = \int_0^{Te} P_{bat\_max}\ dt = P_{bat\_max} \cdot T_e \tag{12.5}$$

12.4 Energy Management

12.4.1 Daily Power Management and Setting of Power References

As the first case study, consider the MG shown in Fig. 12.2. In this MG, a PV-based AG and an MGT are available. Because of the renewable energy benefits (less gas emission and low operating cost), the PV-based AG and MGT are considered prior and backup sources, respectively. The objective of the daily power planning is to calculate the power reference setting for each power source and controllable load. Here, the controllable loads are not considered and the system management functions that directly related to the technical information are emphasized [1].

12.4.1.1 Constraints

12.4.1.1.1 Battery Power Constraints The first constraint concerns the optimal charging/discharging of the batteries to increase their lifetime. Generally, the battery cycle life is defined as the number of complete charge–discharge cycles that a battery can perform before its nominal capacity falls below 80% of its initial rated capacity. Lifetimes of 500–1200 cycles are typical. The actual aging process results in a gradual reduction in capacity over the time, and some algorithms for charging and discharging may be required for improving the battery life.

The simplest and most obvious way of getting the maximum life out of a battery is to ensure that it always works well within its designed operating limits. There are, however, some further actions that can be taken to increase the battery life. Most battery failures are due to an inappropriate charging. Using a flexible charging algorithm and a protection system to prevent the overcharging may not extend the battery life but can prevent it from being cut short.

In PV applications, the battery charging source is the PV panel, which is an intermittent current source. So the intermittency of the PV power cannot ensure a smooth process of charging due to the fluctuation in current. To ensure the battery cycle life, it is not discharged in the daytime. Although the PV power is not enough for the power supply, only the MGT is asked for the power compensation. At night, the batteries are discharged. This charging and discharging algorithm not only assures a high charging/discharging efficiency but also increases the battery cycle life [17]. Classically, the depth of discharge is maintained between 10% and 90% during a normal operation to increase the battery lifetime for a Li-ion battery and 50% to 90% for a lead acid battery. Here, the storage battery capacity is 106 Ah and the rated battery power (P_{bat_max}) is also considered.

12.4.1.1.2 MGT Constraint The rated power of the MGT (P_{MGT_max}) is 33 kW. Starting of an MGT needs much more fuel than a continuous operation with a constant generated power and it causes a lot of air pollution. Moreover, starting may take several minutes if the turbine is cold [18]. In order to minimize energy losses and the gas emission at each start time, the MGT should continuously work. Therefore, in the case of low power demand, the MGT is forced to work with a low power level (P_{MGT_min}) corresponding to the minimum energy:

$$\tilde{E}_{MGT\_1/2\,\mathrm{h}\_min} = \int_0^{Te} P_{MGT\_min} dt = P_{MGT\_min} \cdot Te \tag{12.6}$$

12.4.1.2 Determinist Algorithm

According to daily predictions of the available power and energy from the PV ($\tilde{P}_{PV\_24\,\mathrm{h}}$, $\tilde{E}_{PV\_1/2\,\mathrm{h}}$) and the required power and energy of the loads ($\tilde{P}_{Load\_24\,\mathrm{h}}$, $\tilde{E}_{Load\_1/2\,\mathrm{h}}$), a power production planning for the prosumer

(P_{AG_ref0}) and for the MGT (P_{MGT_ref0}) must be determined. The MGCC refreshes the power references each 30 min. As no power is available from PV panels at night, power references are calculated separately for the nighttime and for the daytime.

12.4.1.2.1 *Daytime* In the day ($t_0 < t < t_0 + \Delta t$), and for each 1/2 h period, two cases are considered:

Case 1: If the sum of available PV energy and the minimum MGT energy is less than the demanded load energy ($\tilde{E}_{PV\_1/2\,h} + \tilde{E}_{MGT\_1/2\,h\_min} < \tilde{E}_{Load\_1/2\,h}$), the PV panels can work with an MPPT algorithm and all PV power is injected to the grid. The MGT has to generate the required additional power:

$$P_{AG\_ref0} = \tilde{P}_{PV\_24\,h},\ P_{MGT\_ref0} = \tilde{P}_{Load\_24\,h} - P_{AG\_ref0} \tag{12.7}$$

where the $P_{AG\_ref0}, P_{MGT\_ref0}$ are the initial power references (24-h power planning) for the AG and MGT, respectively.

Case 2: If the sum of available PV and the minimum MGT energy is more than the demanded load energy, the priority is given to the renewable energy for the electrical production so that the MGT works with a minimum power and the AG power is limited to the required additional power:

$$P_{AG\_ref0} = \tilde{P}_{Load\_24\,h} - P_{MGT\_min},\ P_{MGT\_ref0} = P_{MGT\_min} \tag{12.8}$$

where the P_{MGT_min} is the minimum power production of the MGT. The excess PV energy will be managed by the local controller.

12.4.1.2.2 *Nighttime* The energy management during the nighttime ($t_0 + \Delta t < t < t_0 + 24\,h$) depends on the available energy from the batteries. This energy ($\tilde{E}_{bat}$) can be estimated or communicated by the E-box to the MGCC. At night, two cases can be distinguished (cases 2 and 4). For both cases, batteries have to be discharged in order to be ready for charging the next day at t_0. According to the stored energy ($\tilde{E}_{bat}(t_0 + nTe)$) and the rated energy ($\tilde{E}_{bat\_1/2\,h\_max}$), the available energy of batteries during the next 1/2 h is obtained as follows:

$$\tilde{E}_{bat\_1/2\,h\_rest}(t) = \min[\tilde{E}_{bat\_1/2\,h\_max},\ \tilde{E}_{bat}(t_0 + nTe)] \tag{12.9}$$

where the $\tilde{E}_{bat\_1/2\,h\_rest}$ is the available battery energy for the next half hour, the $\tilde{E}_{bat\_1/2\,h\_max}$ is the maximum battery exchanged energy for a half hour, and the $\tilde{E}_{bat}(t_0 + nTe)$ is the stored energy in the batteries.

Case 3: If the sum of available stored battery energy and the minimum MGT energy is more than the demanded energy ($\tilde{E}_{bat\_1/2\,h\_rest} + \tilde{E}_{MGT\_1/2\_min} > \tilde{E}_{Load\_1/2\,h}$), priority is given to the AG for the electrical production since it has enough previously stored energy from PV panels. The MGT will work with the minimum power as follows:

$$P_{AG\_ref0} = \tilde{P}_{Load\_24\,h} - P_{MGT\_min},\quad P_{MGT\_ref0} = P_{MGT\_min} \tag{12.10}$$

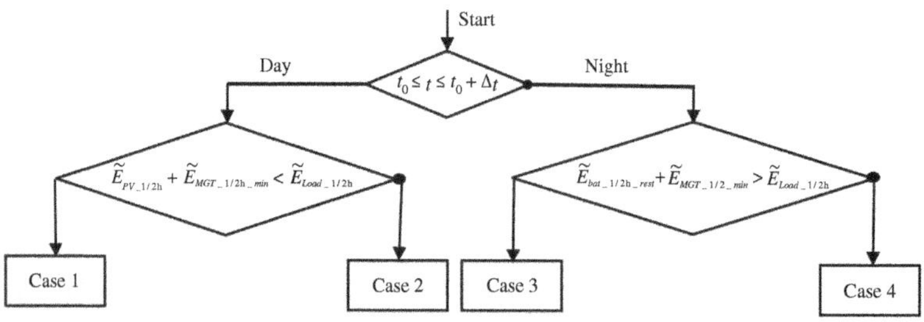

Figure 12.10 Determination of operating cases.

Case 4: If the sum of stored battery energy and the minimum MGT energy is less than the demanded energy, the power reference of the AG is calculated in order to discharge the batteries and the MGT must generate the required additional power:

$$P_{AG\_ref0} = \frac{\tilde{E}_{bat\_1/2\,h\_rest}}{Te}, \quad P_{MGT\_ref0} = \tilde{P}_{Load\_24\,h} - P_{AG\_ref0} \tag{12.11}$$

The case where the battery energy is insufficient is then included. The determination of operating cases is summarized in Fig. 12.10.

12.4.1.3 Practical Applications

In order to illustrate the theoretical results, the 24-h-ahead power management is calculated with the practical load forecasting, as well as the estimated PV power (Fig. 12.9) for $t_0 = 7\text{h}00$ and $\Delta t = 12\text{h}30$. Estimations of the required energy during the day for the loads ($\tilde{E}_{Load\_day}$) and the available energy from the PV panels ($\tilde{E}_{PV\_day}$) show that enough much renewable energy is available (Fig. 12.11a)

As the MGT must produce the minimum energy during the daytime, the surplus energy is estimated as

$$\Delta\tilde{E} = \tilde{E}_{PV\_day} - \tilde{E}_{Load\_day} + \tilde{E}_{MGT\_day\_min} < E_{bat\_max} \tag{12.12}$$

$$\tilde{E}_{MGT\_day\_min} = \int_{t_0}^{t_0+24\,h} P_{MGT\_min}\, dt \tag{12.13}$$

This energy can be stored in batteries. The required energy from the MGT for the night operation is then deduced (Fig. 12.11b) as follows:

$$\tilde{E}_{MGT\_night} = \tilde{E}_{Load\_night} - \tilde{E}_{bat\_night} \tag{12.14}$$

where $\tilde{E}_{bat\_night} = \Delta\tilde{E}$.

As a communication network exists, the estimated value of the battery energy can be replaced by a sensed value, which is sent by the E-box at the start of night ($t_0 + \Delta t$). Calculated power references from the determinist algorithm are shown in Fig. 12.12.

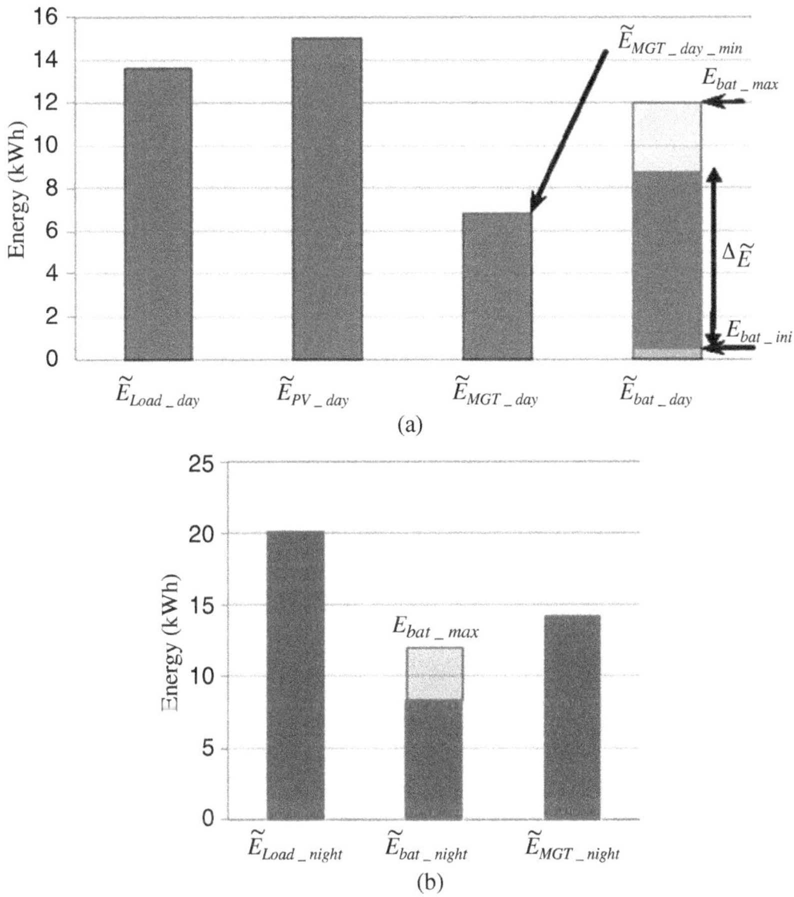

Figure 12.11 Energy analysis for (a) the daytime and (b) the nighttime.

Between 7h00 and 7h30, the PV power is not enough (Fig. 12.8) and the power reference for the MGT is equal to the load demand (Fig. 12.9). Between 9h00 and 17h00, the MGT power reference is controlled with a minimum power.

12.4.2 Medium-term Energy Management

12.4.2.1 Reduction of the Uncertainty

Power references from the long-term EM are calculated by 24-h-ahead planning from the load and PV production forecasting. Sometimes, the real situations (weather conditions, power demand by loads) are different from the forecasted conditions. In practice, considering the current forecasted loads ($\tilde{P}_{Load\_1/2\,h}$) and forecasted PV production ($\tilde{P}_{PV\_1/2\,h}$) for the next half an hour,

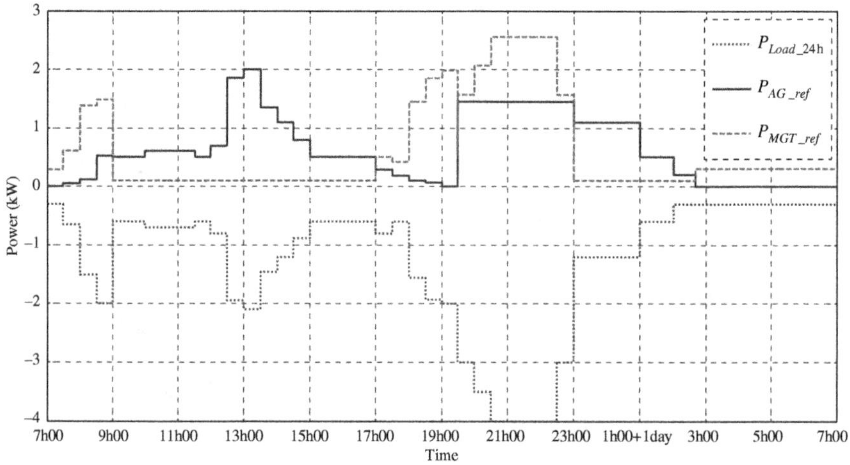

Figure 12.12 Power references from the power planning in the central energy management.

the following deviation from the 24-h-ahead forecasted data can be obtained:

$$\Delta P_{PV\_1/2\,h} = \tilde{P}_{PV\_1/2\,h} - \tilde{P}_{PV\_24\,h},\ \ \Delta P_{Load\_1/2\,h} = \tilde{P}_{Load\_1/2\,h} - \tilde{P}_{Load\_24\,h} \tag{12.15}$$

The medium-term EM takes into account these different conditions by modifying locally power references according to the real situations. This operating is quite similar to the secondary control in large power systems [19]. Here, the refreshment is done each half hour in the MGCC as follows:

$$P_{AG\_ref1} = P_{AG\_ref0} + \Delta P_{AG1\_1/2\,h} \tag{12.16}$$

where the P_{AG_ref1} is the power reference after the medium-term EM, the P_{AG_ref0} is the power reference after the daily power planning, and the $\Delta P_{AG1\_1/2\,h}$ is the power modification of the medium-term EM.

The modified reference of the medium-term EM depends on the correction of the PV prediction and load forecasting as follows:

$$\Delta P_{AG1\_1/2\,h} = \Delta P_{PV\_1/2\,h} + \Delta P_{Load\_1/2\,h} \tag{12.17}$$

In a similar way, the power reference for the MGT is updated as

$$P_{MGT\_ref1} = P_{MGT\_ref0} + \Delta P_{MGT\_1/2\,h} \tag{12.18}$$

where the P_{MGT_ref1} is the power reference after the medium-term EM, the P_{MGT_ref0} is the power reference after the daily power planning, and the $\Delta P_{MGT\_1/2\,h}$ is the power modification of the medium-term EM. After the correction of the power references, the MGCC sends these corrected power references to the local controller for each generator (PV-based AG and MGT) through the communication bus.

12.4.2.2 Energy Management of Batteries

In the AG, the critical sources that are impacted by the medium-term management are the batteries. Deep discharging, undercharging, and overcharging may damage batteries and shorten their lifetime. So in order to optimize the use of batteries, only one charging/discharging cycle in 24 h has been set. Moreover, during the daytime, the battery charging power reference (P_{bat_ref0}) is set as a constant during each half hour according to the AG power reference (P_{AG_ref0}), the local short-term PV power prediction ($\tilde{P}_{PV\_1/2\,h}$), and the estimated state of charge of batteries ($\tilde{SOC}(t)$) during this half hour. Moreover, the batteries are not set in the discharging mode during the daytime. The algorithm of the daytime charging is started as soon as the sensed PV power is higher than the load power demand.

$$\tilde{P}_{PV\_1/2\,h} - P_{AG\_ref1} > 0 \tag{12.19}$$

If the predicted PV power ($\tilde{P}_{PV\_1/2\,h}$) is lower than the load reference during a half hour (P_{AG_ref1}), the batteries are not charged and are in standby mode as shown in Fig. 12.13. Otherwise, they are charged in the same half hour, while the battery energy constraint is considered. If the batteries are fully charged, they will remain in standby mode.

If the surplus PV power (the remaining PV power after satisfying the load) is less than the rated battery power, the batteries can be charged with their maximum power; otherwise, the batteries are in standby. At night, the batteries are discharged. The discharging power reference is calculated each half hour according to the SOC. In the same way, the energy capacity limit and the discharging power limit should also be considered as depicted in Fig. 12.14.

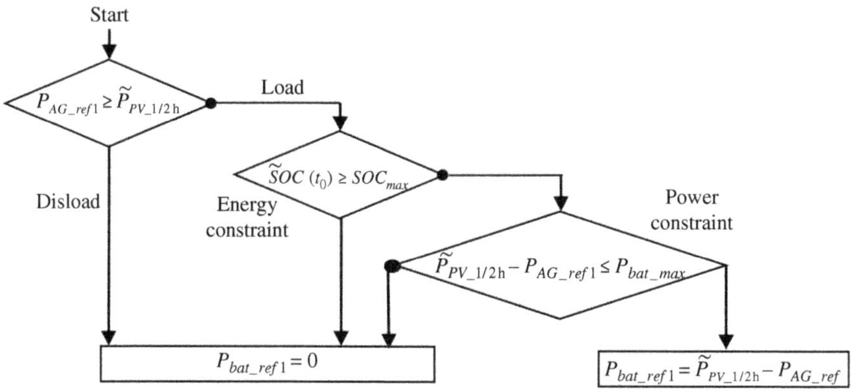

Figure 12.13 Flow diagram for the battery charging algorithm during the daytime.

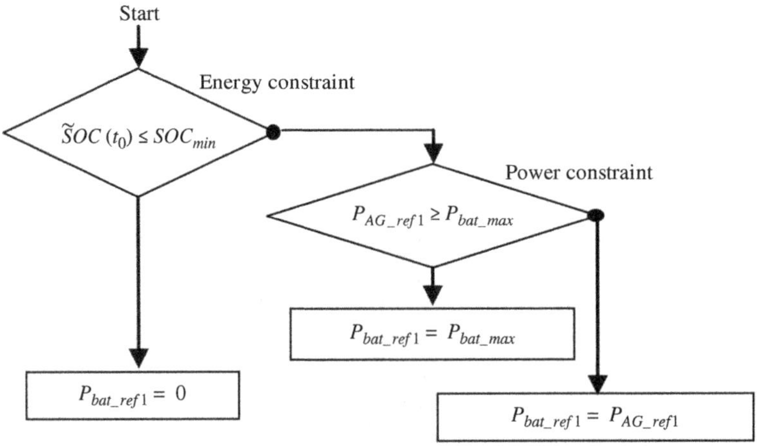

Figure 12.14 Flow diagram for the battery discharging algorithm at night.

12.4.3 Short-term Power Management

12.4.3.1 Primary Frequency Regulation

The MGCC sends desired power references (P_{AG_ref1} and P_{MGT_ref1}) each half hour. These quantities are the refreshed and planned exchanged powers for this duration. As real-time variations exist in the load and also in the produced PV power, a primary frequency control must be used to adjust the power production of generators in order to achieve the real-time power balancing [20]. The two generators inside the MG can share this control function. The AG should use a primary frequency control in order to ensure the short-term power balancing as discussed in Chapters 3 and 5. When the frequency deviation exceeds a predefined threshold value, the controller is activated to increase/decrease the power for restoring the power balance. A droop constant gives the additional power that is supplied as a function of the frequency deviation.

12.4.3.2 Power Balancing Strategies for the AG

The AG has to provide the real-time power reference $p_{AG\_ref}(t)$, which is the sum of the secondary power reference (P_{AG_ref1}) and the real-time primary power reference $\Delta P_{AG\_ref}(t)$. For the problem at hand, the battery power is assumed to be constant during a half hour. The real-time power balancing must be implemented by a power buffer with fast dynamic capabilities. As described in Chapter 1, supercapacitors (SCs) are fast dynamic storage systems with high power exchange capabilities. So, they are suitable for the optimal charging of the battery as well as for supplying peak power to the grid if necessary, but their energy density is low.

Since the batteries are charged/discharged with a constant power reference, the power sharing level of the control system must be adapted as shown in

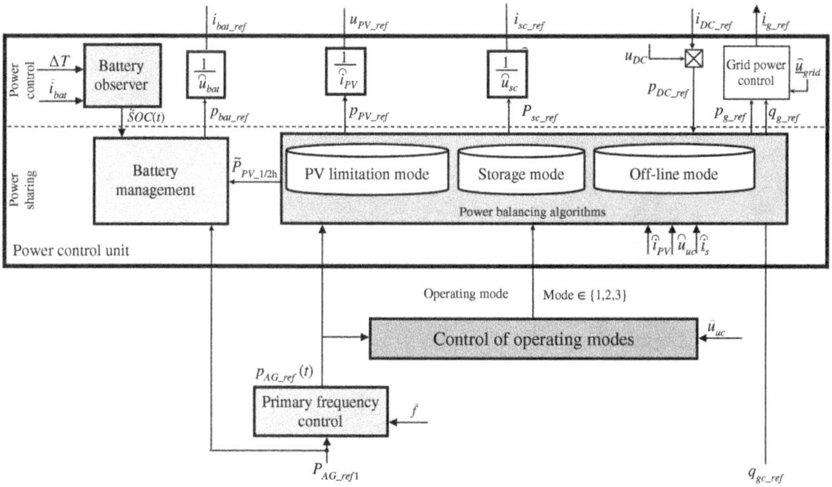

Figure 12.15 Modified control system for the application.

Fig. 12.15. One algorithm is still executed and calculates the power references for the SCs, PV panels, and the inverter according to a selected operating mode and the measured quantities. Then, the power references are transformed to current or voltage references via a feedback system. The SOC is also calculated and then in case of necessity the loading in the last 1/2 h to be stopped.

12.4.3.2.1 Control of Operating Modes Here, the power reference of the batteries is different since a constant battery power reference replaces the low-pass filter used in Chapter 2. In this application, the SOC of batteries is managed by the MGCC in order to maximize the PV power (Fig. 12.15). The PV limitation mode of the PV-based AG does normally not happen if batteries are already well sized. However, this operating mode is kept even if it is not required for this application.

12.4.3.2.2 Energy Management of the SCs Now the batteries are dispatched by the MGCC and so the controllability of the PV-based AG relies on the state of charge of SCs. As batteries are now unavailable for the load/disload of SCs, energy from PV panels will be used to load them, while the injection to the grid will be used to disload them.

12.4.4 Experimental Tests

12.4.4.1 Experimental Platform

Control algorithms of the MGCC have been tested and validated through real-time simulations under conditions similar to that the residential/urban MG may face in reality. These tests thus validating the preanalysis are

Figure 12.16 MG platform at L2EP.

performed through off-line simulations. Real-time analysis consists of using real parts of the electrical component under operating conditions very close to reality. Hardware in the loop (HIL) is useful for real time simulation of the MG operating conditions. Consequently, this type of analysis provides more realistic results from the off-line simulation regarding the operation of the understudy devices.

In order to verify the aforementioned algorithms for the residential/urban MG application, the proposed EMS has been implemented on an experimental platform with the real-time simulator (RTS). An MG platform has been set up with a central control host as shown in Fig. 12.16. The objective of the experimental work is to validate the MGCC by the HIL test and to show the coordinated operation of different sources (PV-based AG and MGT) in the MG. The HIL is organized as in the structure given in Fig. 5.16. Test devices are divided into two classes: the MG devices (the DG units, loads, etc.) and the control devices (the local controllers, MGCC, etc.).

Three types of devices are used: (i) the real PV-based AG, which is fed by a DC current source for emulating identical PV power productions in order to compare various control algorithms in the same illumination conditions; (ii) virtual devices are simulated under the Real-Time Simulator (RTS). The RTS includes the MGT, 4 kW noncritical loads, and 6 kW passive loads – passive loads represent the critical loads and must remain connected to the MG during the test; and (iii) a power amplifier, which is used as an interface to create the point of coupling of the AG; it is interfaced to the RTS via analog-to-digital conversion (ADC) cards and digital-to-analog conversion (DAC) cards.

The control devices are composed of four parts: (i) a DSP card is used to implement the local controllers of the PV-based AG, (ii) the MGCC is implemented in a PC with a SCADA software. It communicates with the local controllers, (iii) the local controllers of the MGT unit and the load controllers are integrated into the RTS, and (iv) the communication bus exchanges the signals between the MGCC and local controllers.

An Ethernet with a Modbus protocol for the communication is used and enables the connection of various devices, for example, a system that measures currents and voltages and the communication of data to a computer. Modbus is suited to connect a supervisory computer with a remote terminal unit (RTU) in SCADA systems. To make comparisons, several different tests are carried out with different sizes of PV panel installation and different sizes of loads.

12.4.4.2 Analysis of the Self-consumption of One House

The first case is studied with PV installation of 3 kW peak power and 3 kW rated local loads for one house. 106 Ah batteries are used with 800 W maximum charging power and a 500 W maximum discharging power. The 24-h-ahead PV power prediction and the load forecasting for the test are given in Figs 12.17 and 12.18 with $t_0 = 7h00$ and $t_0 + \Delta t = 18h00$.

As depicted in Fig. 12.19, during the daytime the available energy from PV panels is larger than the required one to feed the loads. The stored energy in batteries corresponds then to the obtained energy from the MGT, which remains switched on with a delivered minimum power. This stored energy is retrieved for the night operation. Power references (accordingly the amount of generated power) are calculated by the MGCC and are sent to the local controllers as shown in Fig. 12.20. It is clear that between 12h30 and 13h30, PV panels are able to supply the loads, but the MGT must not be switched off and must operate with a minimum power. The same situation appears in the morning between 8h00 and 9h00. The obtained calculated power references show an appropriate use of the PV-based AG during these durations. It is also shown that the MGT power reference is reduced to the minimum value.

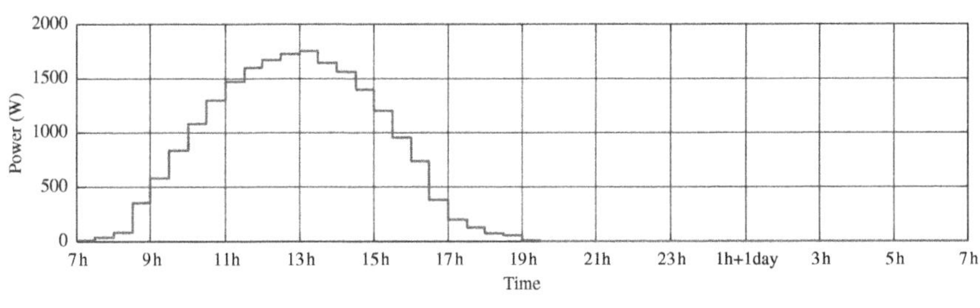

Figure 12.17 24-h-ahead PV power forecasting for the self-consumption of one house.

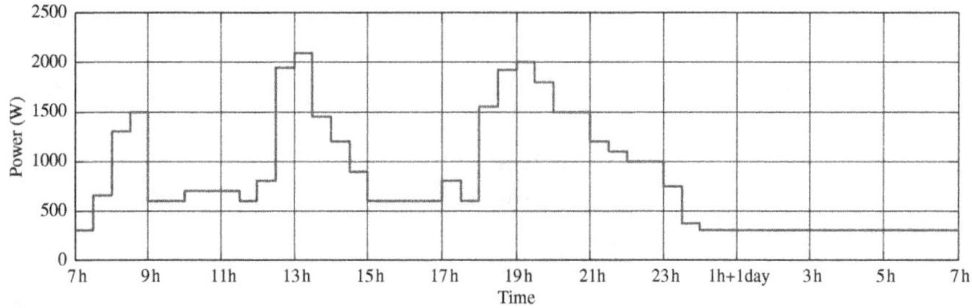

Figure 12.18 24-h-ahead load forecasting for the self-consumption of one house.

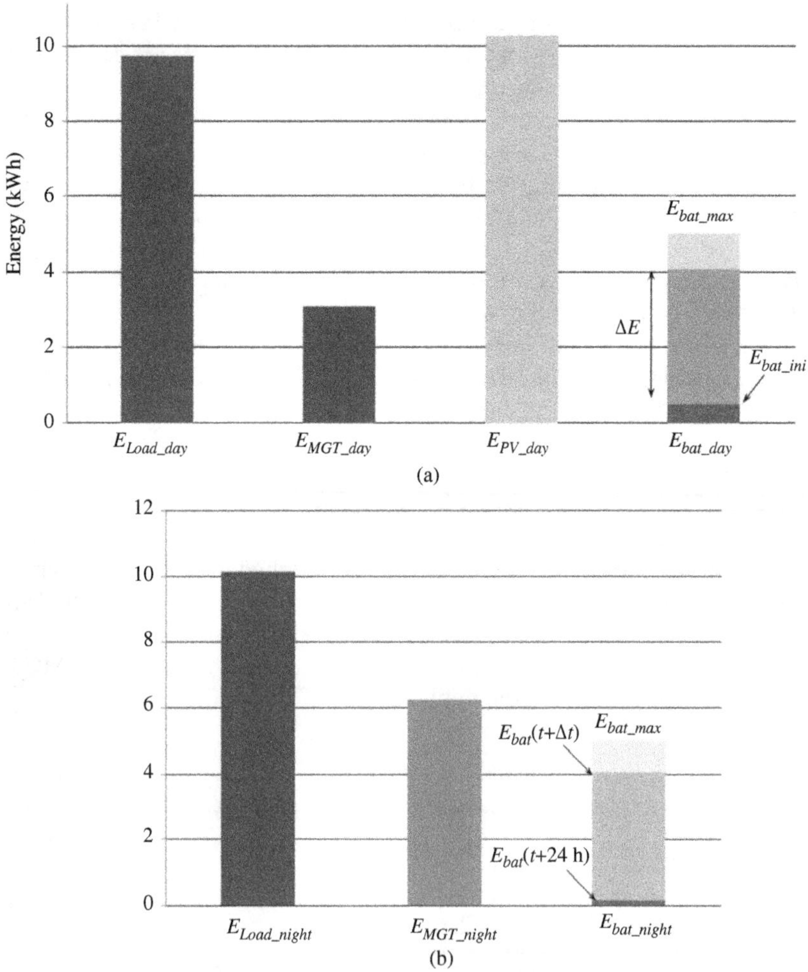

Figure 12.19 Energy analysis the self-consumption of one house: (a) daytime and (b) nighttime.

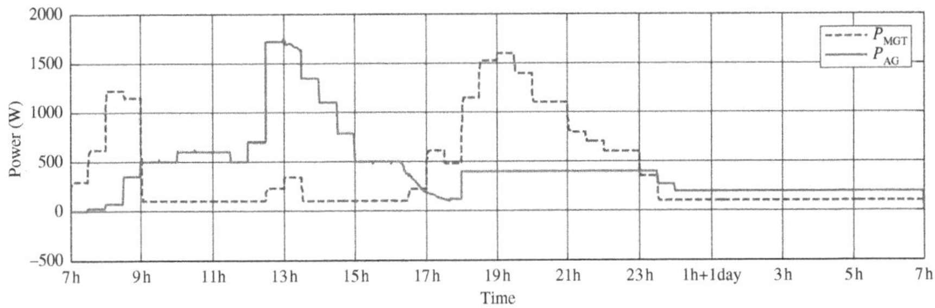

Figure 12.20 Generated powers in the MG for the self-consumption of one house.

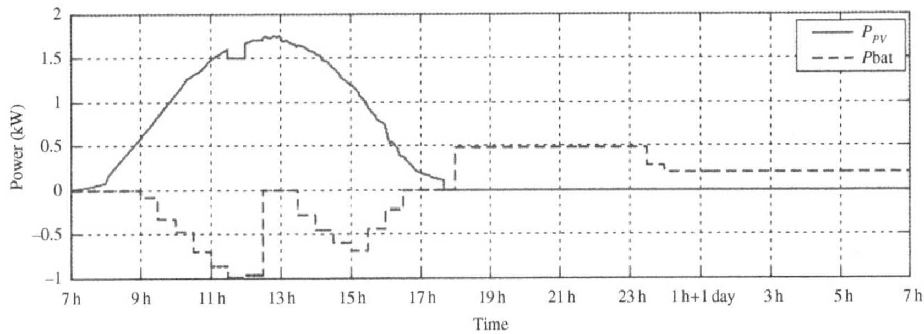

Figure 12.21 Sensed powers inside the PV-based AG for the self-consumption of one house.

At 18h00 (start of the nighttime), batteries are loaded and the power reference is then calculated to disload them with a maximum rated current until 23h30. As mentioned, the rated current of batteries is less than that from PV panels. Then, the load is decreased, the power reference for batteries is modified until the end of the nighttime.

The sensed PV and battery power inside the AG are shown in Fig. 12.21. A part of the PV power is lost at 11h30 because the PV power is reduced. At the same time, the battery current is at the rated value in a loading mode. For this test, increasing batteries may avoid this loss. During the nighttime, the batteries deliver a nearly constant power with a rated value between 18h00 and 23h30 (the rated value in disloading mode is not the same as in the loading mode) since the load demand is high. Then, battery power is decreased and the turbine power is on the minimum value (after 24h00). Figure 12.22 shows the evolution of the battery SOC for the self-consumption of one house.

From Fig. 12.22, one charging and discharging cycle of the batteries during 24 h are effectively set. The power planning is determined for a period of half an hour. Then deviations occur in this time base. Nevertheless, even if the obtained average value of powers is desirable, the instantaneous values of power can have considerable differences with the planned references.

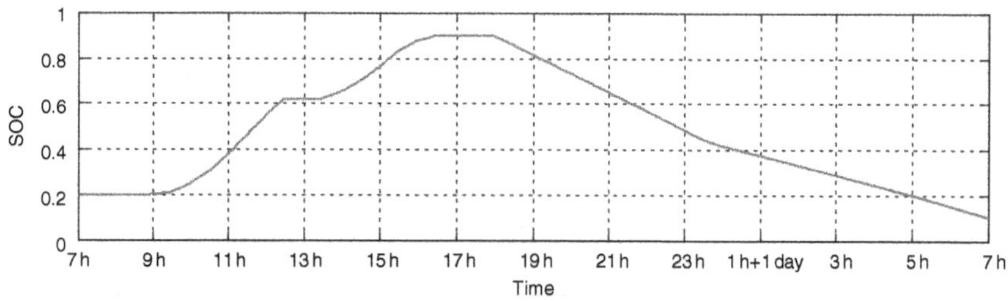

Figure 12.22 Time evolution of batteries SOC for the self-consumption of one house.

A zoom is shown in Fig. 12.23a during a constant generated PV power. When an instantaneous load increase occurs, the battery cannot provide the full power compensation because of its slow dynamic characteristic. Under this condition, the SCs are used for the fast power compensation.

In Fig. 12.23b, the grid load remains constant. When the instantaneously generated PV power increases, the battery is charged with a constant power, but it cannot absorb all the surplus PV power. In this case, the SCs are used to achieve the real-time power balancing. So for both cases, the use of SCs enables the production in real time of a constant power during half hour. For the sake of simplification, the primary frequency control is disabled for these tests. According to experimental results, the consumed electrical energies of each generator in the MGT can be deduced. So for case 1, the PV power provides 53% of the total power supply and the MGT power covers the remaining 47% power.

12.4.4.3 Increasing the Penetration Ratio

12.4.4.3.1 Single PV-based Producer The previous 3 kW prosumer (PV-based AG) is now located in a residential MG with three other consumers. A 12 kW rated local loads for the four houses is considered. The PV power prediction and the load forecasting for the test are given in Fig. 12.24. According to the forecasting information, the PV energy is much lower than the demanded power of the total loads. According to this day-ahead analysis, batteries will not be used.

Reference powers from the MGCC show a full contribution of the PV generation (without power limitation) but less than the MGGT as shown in Fig. 12.25. For this case, the storage devices are not locally used because the required load power is always more than the PV power during the daytime. This is confirming that for a small penetration of PV production, all the production can be sent to the grid and so storage is unnecessary.

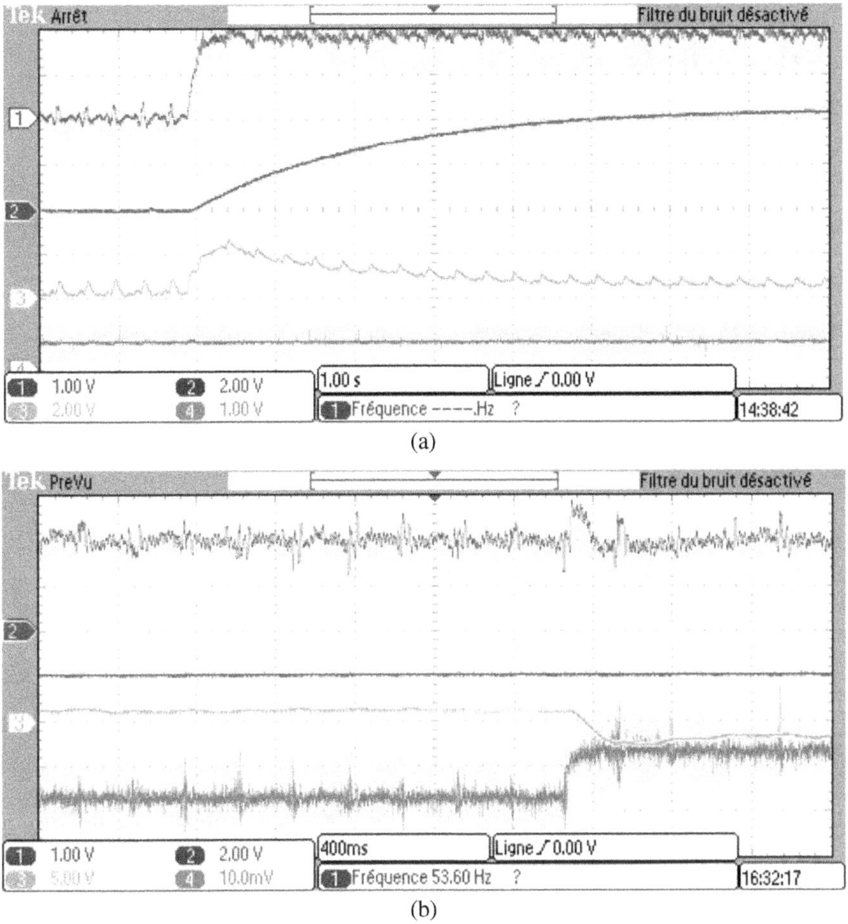

Figure 12.23 The SCs dynamic response: (a) power compensation and (b) power absorption. p_{AG_mes} (Ch 1): 100 W/div; p_{bat_mes} (Ch 2): 100 W/div; p_{uc_mes} (Ch 3): 150 W/div; p_{PV_mes} (Ch 4) : 150 W/div.

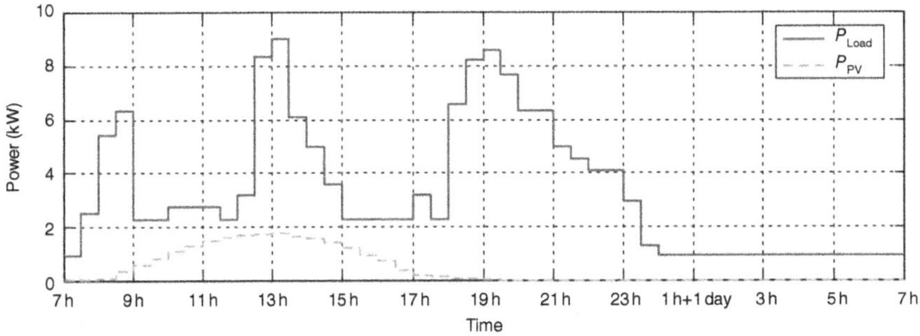

Figure 12.24 PV power and load forecasting for 24 hour.

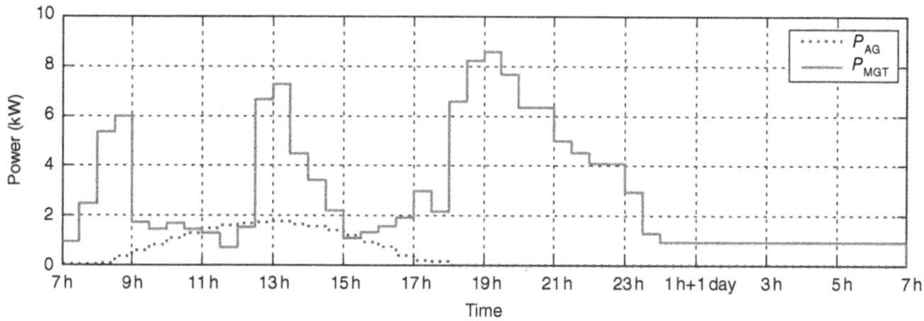

Figure 12.25 MGCC-based power references setting for the case of one producer.

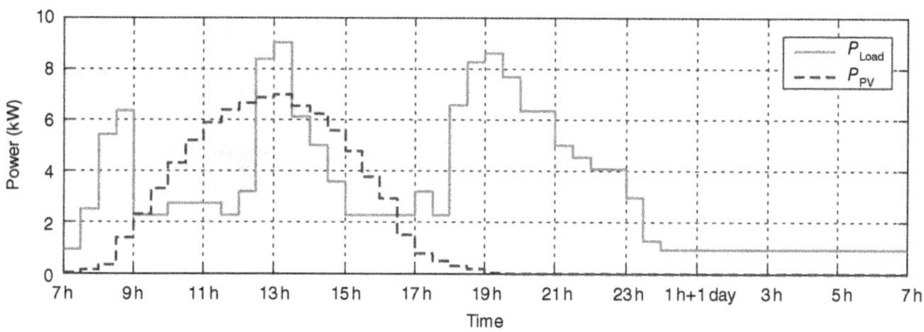

Figure 12.26 24-h-ahead PV power and load forecasting.

According to the experimental results, the produced electrical energies of each generator can be deduced. The PV power provides 17% of the total power supply and the MGT power covers the remaining power 83%.

12.4.4.3.2 Full-scaled PV-based Producers In this studied case, four houses have each one 3 kW peak power PV panel. The considered load profile is the same as the previous one. For this test, there is no storage installation in local. As houses are closed, PV productions are nearly equal and the total PV power is shown in Fig. 12.26; the load forecasting remains the same as before.

In order to make easier comparisons, the four power references for prosumers have been summed in Fig. 12.27a. Sensed PV powers are also summed in Fig. 12.27b. Between 9h00 and 12h30 and between 13h30 and 16h30, the PV power is limited and the MGT is derived with its minimum value since the loads are not enough. Hence, a part of the available PV power is lost. Moreover, the turbine has to provide alone the power peak at 19h00. The PV power provides 36% of the total power supply and the MGT power covers the remaining power 64%.

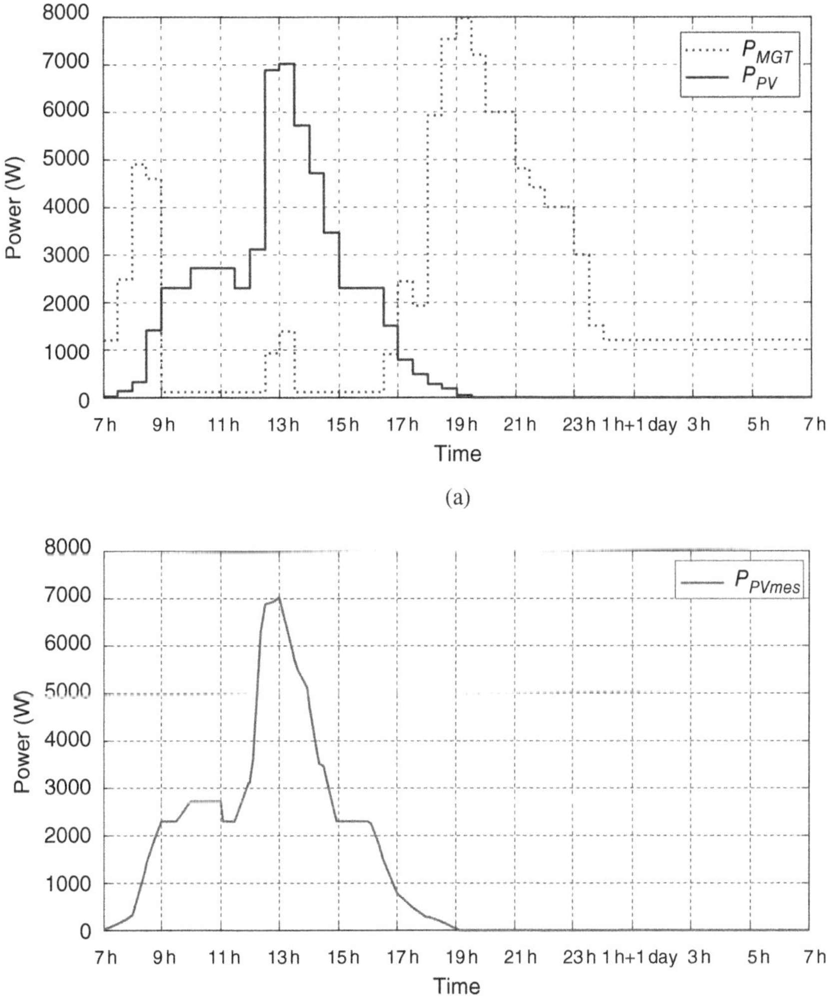

(b)

Figure 12.27 Full-scaled PV-based production: (a) MGCC power reference and (b) total sensed PV power.

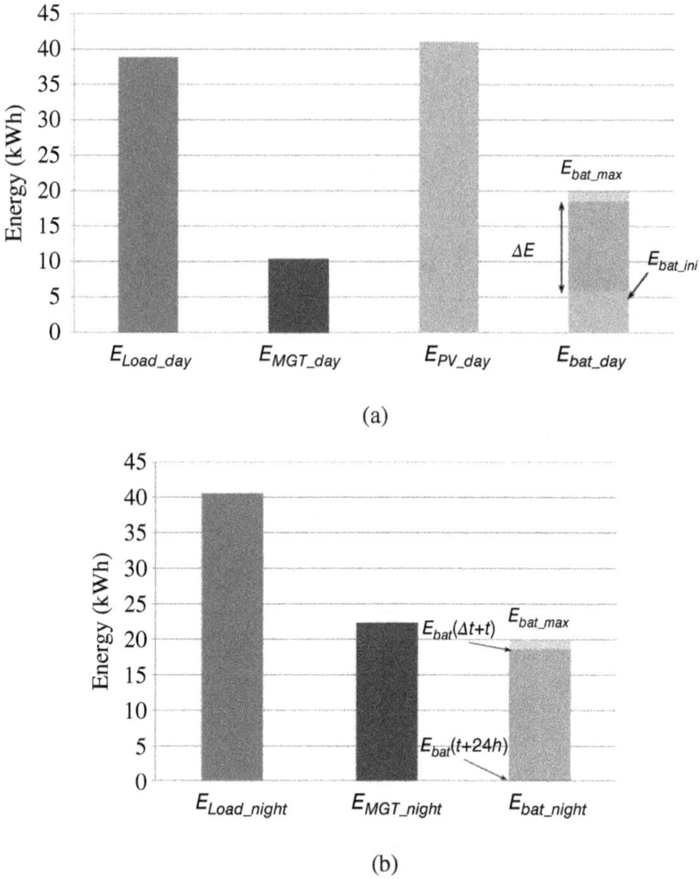

Figure 12.28 Energy analysis for full-scaled PV-based producer: (a) daytime and (b) nighttime.

12.4.4.3.3 Full-scaled PV-based Prosumers with Energy Storages Four sets of 106 Ah batteries are now added with a total 3200 W maximum charging power and a total 2000 W maximum discharging power. The same 24-h-ahead PV power prediction and load forecasting are used. The energy analysis (Fig. 12.28) shows that the total available PV energy is larger than the requested one from loads during the daytime. The available energy for battery storage is nearly equal to the energy of the MGT during the day. This energy is used for the night operation and so reduces the energy from the MGT during the night operation. During the day, the PV power is same as the previous one. The

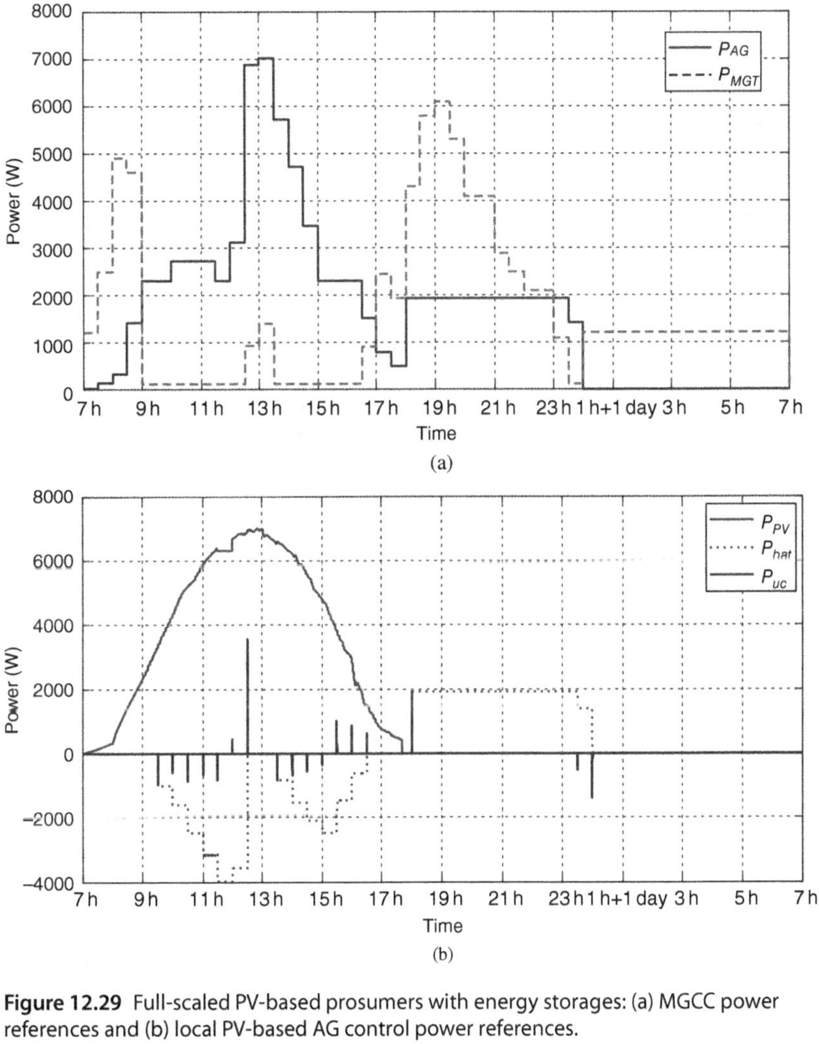

Figure 12.29 Full-scaled PV-based prosumers with energy storages: (a) MGCC power references and (b) local PV-based AG control power references.

experimental results are shown in Fig. 12.29. At the start of the nighttime, power references for prosumers are set to disload batteries, and the turbine peak power is little bit decreased. Thanks to the batteries, more PV power is used and the penetration of the PV power generation is increased to 41%. (Therefore, energy consumption percentages for full-scaled PV-based producer are 41% by E_{MT} and 59% by E_{AG}.)

12.5 Emission Reduction and Economical Optimization

In this section, to take into account environmental and economic issues by the MGCC, two stages are presented [21]. The first proposed stage is a day-ahead operational planning for minimizing CO_2 emissions and fuel consumption of an urban MG. This algorithm solves the unit commitment problem (UCP) by means of a dynamic programming (DP) with predictions of the available energy from PV generators, the power demand from the loads and the state of charge (SOC) of batteries inside the PV-based AGs. Based on these data, this stage gives power references for MGTs while maximizing the use of PV generators and reducing equivalent CO_2 emissions and the economic operating cost.

The second proposed stage is implemented during the day and consists of reducing variations due to the power uncertainty (from the PV production and load demand). This stage enables to retrieve day-ahead calculated optimal economic costs and/or CO_2 emissions. An adjustment algorithm corrects, each 30 min, references coming from the day-ahead operational planning, if changes in forecasted values occur. This adjustment algorithm is based on a sequential quadratic programming method. Both energy management functions can be implemented thanks to a communication network. The consideration of a cluster of generators and consumers reduces the complexity of algorithms.

As described in the previous section, the MGCC must assign power references and also other appropriate control signals to the DG units, conventional production units, and controllable loads. It was emphasized in Chapter 5 that the goals of the MGCC algorithm for EM are to ensure uninterruptable power supply to the loads, to maximize renewable generators in the energy mix, and to minimize the economic costs and the CO_2 equivalent emissions of MGTs by setting their power references such that they produce the minimum pollution and have minimum startups and shutdowns.

The MG management is analyzed through various functions that can be classified in a timescale. The long-term EM elaborates a 24-h-ahead operational planning including (i) the RESs/DGs production forecast taking into account the time dependency of the prime source, environmental impacts, and cost of generation [22]; (ii) the management of nonsensitive loads that may be disconnected/shed according to the supervision requirement; (iii) the provision of an appropriate level of PR capacity according to the electricity market and the load demand forecast; and (iv) the maintenance intervals.

The medium-term EM operates during the daytime and includes the adjustment of forecasts for the power available from DGs and power demanded by the loads, and the adjustment of the long-term operational planning, based on deviations in the above-mentioned forecasts from those predicted 24-h-ahead. The short-term power balancing is performed in the local controllers and includes the primary voltage and frequency control [23]. The long-term operation schedule and the EM can be mathematically expressed as an UCP. Due to the complexity of the problem, the required computation time may vary according to optimization tools used, but the DP approach remains a suitable compromise choice [24].

12.5.1 Micro-Gas Turbine (MGT) Fuel Consumption and Emissions

Fuel consumption of MGT can be assessed by using their partial load efficiency characteristics [25]. The energetic efficiency η_i (between thermal energy (F_{MGT_i} in $kWh_{thermal}$) supplied to the MGT combustion chamber and electric energy output (E_{MGT_i} in $kWh_{electric}$)) and the efficiency characteristic α_i (which is a nonlinear function depending on the partial load ratio) are defined as follows:

$$\eta_i = \frac{E_{MGT\_i}}{F_{MGT\_i}}, \quad \alpha_i(t) = \frac{P_{MGT\_i}(t)}{P_{MGT\_i\_max}} \tag{12.20}$$

where i is the unit number (here such as Fig. 12.3, three MGTs are considered). The P_{MGT_i} (kW) is the generated MGT electric power and $P_{MGT\_i\_max}$ (kW) is the rated MGT electric power. Based on (12.20), the consumed fuel thermal energy of MGTs can be obtained, using the following equation:

$$F_{MGT\_i}(t) = \frac{\alpha_i(t)\, P_{MGT\_i\_max}}{\eta_i}\tau \tag{12.21}$$

By considering a 0.04 €/$kWh_{thermal}$ cost for the consumed gas, the operational cost of each generator (i) is obtained from (12.21) as a function of the generated electric power $P_{MGT\_i(t)}$:

$$C_i = f(P_{MGT\_i}(t)) \tag{12.22}$$

The CO_2 equivalent emissions are calculated by applying to the pollutant gases different weights corresponding to their global warming potential. For the assessment of emissions, the masses of three exhaust gases, NO_x, CO, and CO_2, are evaluated in g/kWh as a mathematical function of the generated useful power [26]:

$$m_x(t) = \mu_x\, E_{MGT\_i}(t) = \mu_x\, \alpha_i(t)\, P_{MGT\_i\_max}\, \tau \tag{12.23}$$

where the μ_x (mg/$kWh_{electric}$) is the emission factor (also called specific emissions) for the pollutant gas x to produce the generic useful electrical energy output E_{MGT_i} and m_x is the mass of the emitted pollutant gas x. The characteristic of CO_2 equivalent emissions of each MGT is expressed as a nonlinear function of its power output through a polynomial interpolation [21]. In order to apply a multiobjective optimization procedure for a trade-off between pollutant emissions and consumed fuel price, a price per ton of CO_2 equivalent emissions is considered.

12.5.2 Day-ahead Optimal Operational Planning

12.5.2.1 Unit Commitment Problem Formulation

The 24-h-ahead operational planning is discretized in 48 time steps (t) of 30 min and power references are considered constant during each step. For all time steps, the operational planning consists of selecting MGTs to be used, determining the instant they should be committed, and calculating the optimal power references of each generator. Power references of the three studied MGTs are gathered in a vector:

$$X(T) = [P_{MGT\_1}(T),\ \ P_{MGT\_2}(T),\ \ P_{MGT\_3}(T)] \tag{12.24}$$

where δ_i is the state of each MGT during each time step (1 if the plant is running or 0 if the plant is shut down). Boolean MGT states are gathered in a vector as follows:

$$u(t) = [\delta_1(t),\ \ \delta_2(t),\ \ \delta_3(t)] \tag{12.25}$$

The general objective of unit commitment is to minimize the total operating cost of an electrical system, while satisfying all system constraints.

As the power industry goes to new restructured forms, the UCP must be adapted and applied to small DG clusters including RES. So here the UCP is used to formulate and solve our objective functions for cost minimization and CO_2 equivalent emission reduction as described in Fig. 12.30.

12.5.2.2 Objective Functions and Nonlinear Constraints

The CO_2 equivalent emissions of each generator and the cost of the consumed gas are expressed as a nonlinear function of its power output and, respectively, called $CO_{2\_i}(P_{MGT\_i}(t))$ and $C_i(P_{MGT\_i}(t))$. Penalties for startup and shutdown of the units are also considered. If one unit is shut down in the current time step and will run in the next time step, a startup penalty is applied. If one plant will not be committed at $t+1$ and it is running at t, a shutdown penalty is applied. The penalties avoid switching on and off the units because it increases emissions and shortens the lifetime of the units. In this study, startup penalty is considered equal to the consumed fuel cost during 5 min operation at full load. Shutdown penalty is considered equal to 2.5 min operation at full load. The startup and shutdown penalties are expressed by cost and CO_2 functions: $C_{pe\_c\_i}(\delta_i(t+1), \delta_i(t))$ and $C_{pe\_CO_2 i}(\delta_i(t+1), \delta_i(t))$.

The two objective functions to be minimized are defined as

$$J_C(t) = \sum_{t=1}^{48}\sum_{i=1}^{3} \delta_i(t)\cdot C_i(P_{MGT\_i}(t)) + C_{pe\_c\_i}(\delta_i(t+1), \delta_i(t)) \tag{12.26}$$

$$J_{CO_2}(t) = \sum_{t=1}^{48}\sum_{i=1}^{3} \delta_i(t)\cdot CO_{2\_i}(P_{MGT\_i}(t)) + C_{pe\_CO_2\_i}(\delta_i(t+1), \delta_i(t)) \tag{12.27}$$

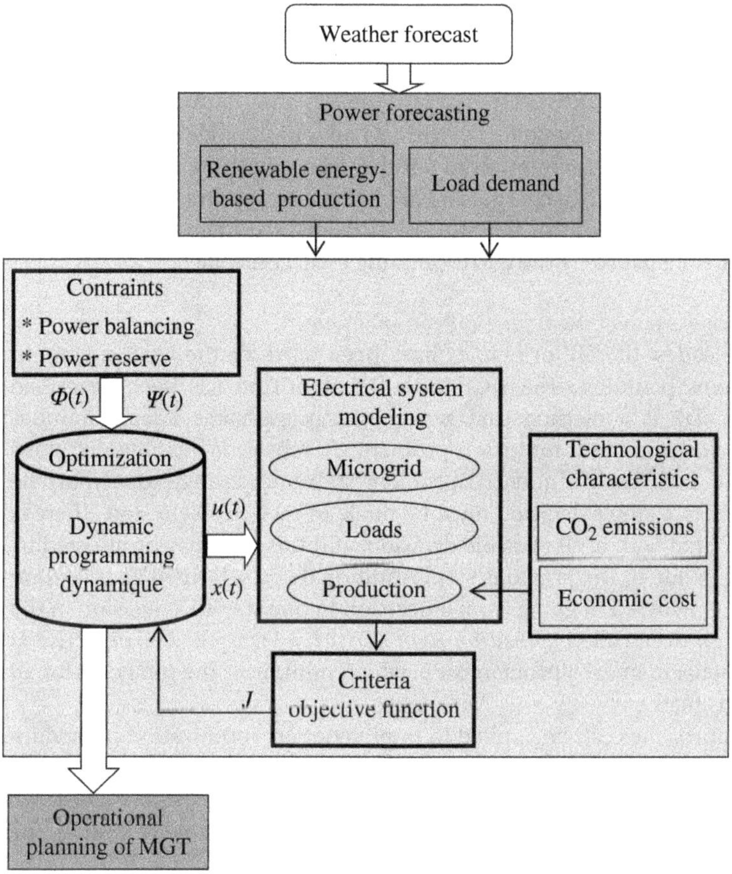

Figure 12.30 Scheme of the day-ahead optimal operational planning.

$P_{MGT\_i}(t)$ is the generated power. Maintenance and management costs are not considered in the present study.

Day-ahead planning strategies use renewable-energy-based production and load forecasts. With N AGs and M MGTs, the power balancing between the loads (P_L) and the generators (P_{AG_n} and P_{MGT_i}) in each time step must be performed with a maximum use of the "clean" PV energy. This is expressed as an equality constraint:

$$\psi(t) = P_L(t) - \sum_{n=1}^{N} P_{AG\_n}(t) - \sum_{i=1}^{3} \delta_i(t) \cdot P_{MGT\_i}(t) = 0 \tag{12.28}$$

The MGT loading level has to be higher than 50% of the MGT's rated power for improving efficiency and reducing CO_2 equivalent emissions. Moreover, the power in reserve must be equal to or larger than 10% of the generator rated power. The corresponding inequality constraint ($\Phi(t)$ in Fig. 12.30) is expressed as

$$50\%P_{MGT\_max\_i} \leq P_{MGT\_i} \leq 90\%P_{MGT\_max\_i} \tag{12.29}$$

A last group of constraints refers to the MG operation mode. The constraints differ from one mode of operation to another (day/night, PV power available or not, AG's battery state of charge) and are taken into account in Section 5.2.5. The MGCC does not control the SOC of embedded storage units in various AGs since it is managed by the corresponding local controller.

12.5.2.3 Application of the Dynamic Programming

In recent decades, the DP principles have been used for the UCP solving in power systems planning. The UCP with DP algorithm has been described in [23]. The DP is a method that systematically evaluates a large number of possible decisions in a multistep problem. A subset of possible decisions is associated with each sequential problem step and a single one must be selected, that is, a single decision must be made in each problem step. There is a cost associated with each possible decision, and this cost may be affected by the decision made in the previous step. Additional costs, termed as *transition costs*, may be incurred in going from a decision in one step to a decision in the following step over, called *transition path* ($Tr(u(t-1), u(t))$). The objective is to make a decision in each problem step, which minimizes the total cost for all the decisions made.

Several approaches can be applied to implement an optimization procedure to solve the UCP by minimizing equations (12.28) and (12.29). The optimal solution of Bellman's recursive equation for all time steps is used to construct the optimal solution of the overall problem. For one state of MGT's ($u(t)$), the operational cost is expressed as

- the cost of electricity production during the time step $[(t-1), t]$

$$\sum_{i=1}^{3} \delta_i(t) \cdot C_i(P_{MGT\_i}(t)) \tag{12.30}$$

- and the cost during the previous time step taking into account the transition cost due to the start or stop of generators

$$Tr(u(t-1), u(t)) = F(t-1, u(t-1)) + \sum_{i=1}^{3} C_{pe\_i\_c}(\delta_i(t-1), \delta_i(t)) \tag{12.31}$$

At step t, the UCP formulation for the studied system can be expressed in the form of the following recursive DP equation:

$$F(t, u(t)) = \sum_{i=1}^{3} \delta_i(t) \cdot C_i(P_{MGT\_i}(t)) + Tr(u(t-1), u(t)) \tag{12.32}$$

The optimal solution of the overall problem is obtained by selecting the optimal values of control variables in the previous equation for all time steps recursively from $t = 48$ to $t = 1$ as depicted in Fig. 12.31 [27]. In the same way, a similar formulation can be found to evaluate CO_2 emissions:

$$F(t, u(t)) = \sum_{i=1}^{3} \delta_i(t) \cdot CO_{2-i}(P_{MGT\_i}(t)) + Tr(u(t-1), u(t)) \tag{12.33}$$

where

$$Tr(u(t-1), u(t)) = F(t-1, u(t-1)) + \sum_{i=1}^{3} C_{pe\_CO_2\_i}(\delta_i(t-1), \delta_i(t)) \tag{12.34}$$

12.5.2.4 Maximization of Renewable Penetration and Online Adjustment

In order to give a priority to the AG, their power references are first calculated. Then, power references of MGTs will satisfy the remaining power balancing if necessary. According to the available PV power and stored energy in batteries, three cases have been considered (Fig. 12.32). During the daytime, if the predicted PV power ($\tilde{P}_{PV\_24\,h\_n}$) is enough to feed the predicted load demand ($\tilde{P}_{Load\_24\,h}$), then power references of N AGs are set to limit the total generated power (all AGs are closed and have the same PV power sizing) as follows:

$$P_{ref\_AG\_24\,h\_n}(t) = \frac{\tilde{P}_{Load\_24\,h}(t)}{N} \tag{12.35}$$

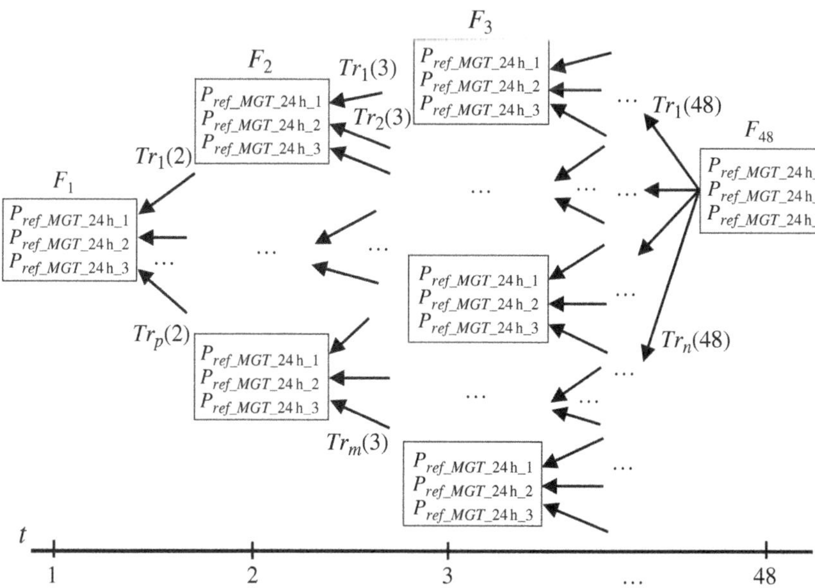

Figure 12.31 Principle of optimal path by DP.

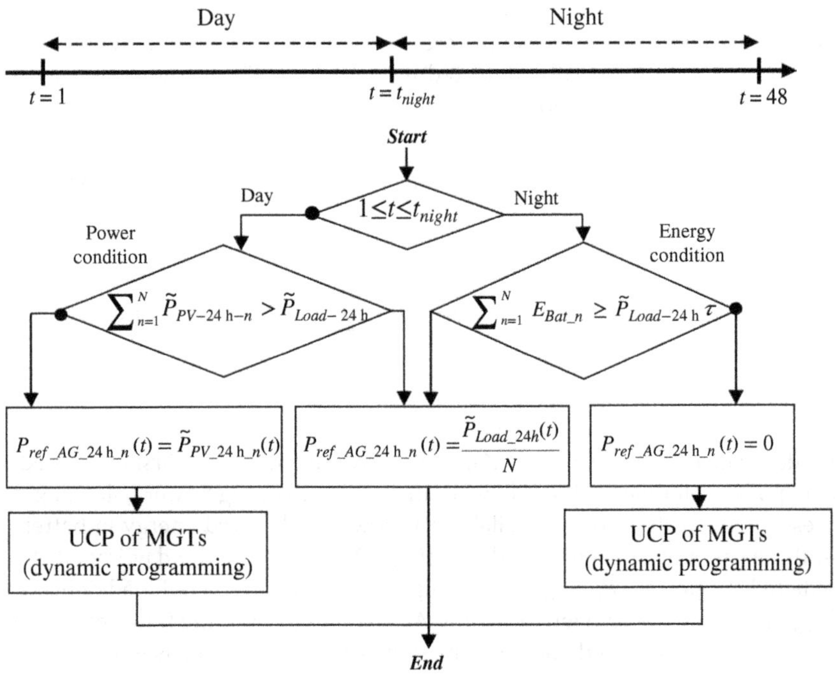

Figure 12.32 Power reference calculation and dispatching.

The energy surplus is automatically saved in batteries by the local control system of AGs and can be estimated as

$$\sum_{n=1}^{N} E_{bat\_n}(t+1) = \sum_{n=1}^{N} E_{bat\_n}(t) + \left[\tilde{P}_{Load\_24\,h}(t) - \sum_{n=1}^{N} P_{ref\_AG\_24\,h\_n}(t)\right] \times 1800 \tag{12.36}$$

Otherwise, power references of AGs are set to the predicted values as follows:

$$P_{ref\_AG\_24\,h\_n}(t) = \tilde{P}_{PV\_24\,h\_n}(t) \tag{12.37}$$

And power references of MGT must balance the remaining power according to

$$\sum_{i=1}^{3} P_{ref\_MGT\_24\,h\_i}(t) = \tilde{P}_{Load\_24\,h}(t) - \sum_{n=1}^{N} P_{ref\_AG\_24\,h\_n}(t) \tag{12.38}$$

Then, the DP algorithm considers all possible states of MGTs and, for each one, power references are calculated by a quadratic optimization algorithm (satisfying the (12.38)). At night, if the stored energy in the batteries is enough to feed the predicted load demand, the MGTs are switched off and power references of AGs are set to limit the total generated power:

$$P_{ref\_AG\_24\,\mathrm{h}\_n}(t) = \frac{\tilde{P}_{Load\_24\,\mathrm{h}}(t)}{N} \tag{12.39}$$

Calculation of battery SOC can be refreshed as follows:

$$\sum_{n=1}^{N} E_{bat\_n}(t+1) = \sum_{n=1}^{N} E_{bat\_n}(t) - \tau \tilde{P}_{Load\_24\,\mathrm{h}}(t) \tag{12.40}$$

Otherwise, power references of MGTs must balance the remaining power:

$$\sum_{i=1}^{3} P_{ref\_MGT\_24\,\mathrm{h}\_i}(t) = \tilde{P}_{Load\_24\,\mathrm{h}}(t) \tag{12.41}$$

Once again, the DP algorithm considers all possible states of MGTs and their power references are calculated by a quadratic optimization algorithm. The AG is capable of maintaining a prescribed power reference ($P_{ref\_AG\_24\,\mathrm{h}\_n}$), received from the MGCC, in the limits imposed by the actual SOC of the batteries. According to the daily predictions of the available power from the PV ($\tilde{P}_{PV\_24\,\mathrm{h}}$) and the required power of the loads ($\tilde{P}_{Load\_24\,\mathrm{h}}$), a day-ahead power production planning for the AG and for MGTs is determined by the MGCC. Sometimes, the real situation (weather conditions, power demand by loads) could differ from the forecasted values. As a 10% PR is scheduled in power references of MGT (12.21), instantaneous deviations due to the power unbalancing can be handled by the primary controllers of local controllers.

Current forecasting techniques allow us to have an updated forecast every 30 min. In practice, the current forecasted loads ($\tilde{P}_{Load\_t+1}$) and forecasted PV production ($\tilde{P}_{PV\_t+1}$) for the next 30 min (time step $t+1$) are considered. The deviation from the day-ahead forecasted data ($\tilde{P}_{PV\_24\,\mathrm{h}(t+1)}$ and $\tilde{P}_{Load\_24\,\mathrm{h}(t+1)}$) is expressed as

$$\Delta P_{PV\_t+1} = \tilde{P}_{PV\_24\,\mathrm{h}\_t+1} - \tilde{P}_{PV\_t+1}, \quad \Delta P_{Load\_t+1} = \tilde{P}_{Load\_24\,\mathrm{h}\_t+1} - \tilde{P}_{Load\_t+1} \tag{12.42}$$

During the daytime, the online adjustment of the MGCC consists of modifying AG power references each 30 min so that primary PRs of MGTs are not used and so that their produced power are as close as possible to the optimal operational point as calculated 1 day ahead. Deviations are taken into account by modifying power references of the generators according to the new situation:

$$P_{AG\_ref\_t+1} = \tilde{P}_{AG\_ref\_24\,\mathrm{h}\_t+1} + \Delta P_{PV\_t+1} \tag{12.43}$$

where $\tilde{P}_{AG\_ref\_24\,\mathrm{h}\_t+1}$ is the AG power reference, calculated 1 day ahead by the MGCC. A correction in power references for the MGTs in the system may be also necessary. If the power deviation from forecasted values is larger than 10%, the algorithm has to switch on a turbine that was not planned to be run at $t + 1$ in order to compensate the lack of PV power.

12.5.3 Experimental Results

The implementation of the proposed MGCC via a SCADA system needs to add centralized remote control and TCP/IP network. So, a test in real scenarios is required to evaluate the performance of the developed solution and to detect necessary adaptations before it will be implemented in the distribution system. For security reasons, this test cannot be done directly in real situations onto a power system under operation. Moreover, precise energy scenarios have to be played in order to validate the correct interactivity between the MGCC and RTUs. So a test bed has been developed based on the RTS, which is connected to the laboratory grid through a 30 kW power amplifier as shown in Fig. 12.33.

The RTS is also connected to the SCADA, which executes the MGCC, via a Modbus communication network. The three MGTs and the loads are simulated in the RTS and other equipment are real: the communication network, one AG, and the SCADA system. A real prototype of the AG is also connected to the grid and monitored by the SCADA. The feedback loop of local control systems is closed via current/voltage measurements at the residential loads, but also RTU, local controllers, and models of MGTs are embedded in the RTS and executed. This test bed enables to reproduce identical demand scenarios under various sun conditions in order to validate the day-ahead operational planning and the adjustment algorithm and to test various optimization goals.

The studied urban MG includes residential loads, two 30 kW MGTs, one 60 kW MGT, and twelve homes with 3.6-kW-PV-based AGs with embedded storage as shown in Fig. 12.3. All power generators and electrical loads are locally connected. So line losses and voltage drops can be ignored. As 12 generators must be considered, which are very close in the urban network (so they receive the same solar irradiation), the measured output power of the real single AG is multiplied by 12. The MGCC measures the MG state variables and dispatches power references (refreshed by the adjustment algorithm every half hour) to microsources through a communication network. The local controller receives these power set points. At the same time, they send various data, for example, the sensed power production at the coupling point.

First, a basic dispatching strategy without PV generators and optimization has been experimented with the load demand forecast that is given in Fig. 12.34. It consists of setting MGT power references proportional to the generator rated power while the added power references correspond to the remaining power required by loads. Then, PV-based AGs are considered and their calculated global power references from the strategy for the maximization of the renewable penetration are shown in Fig. 12.35.

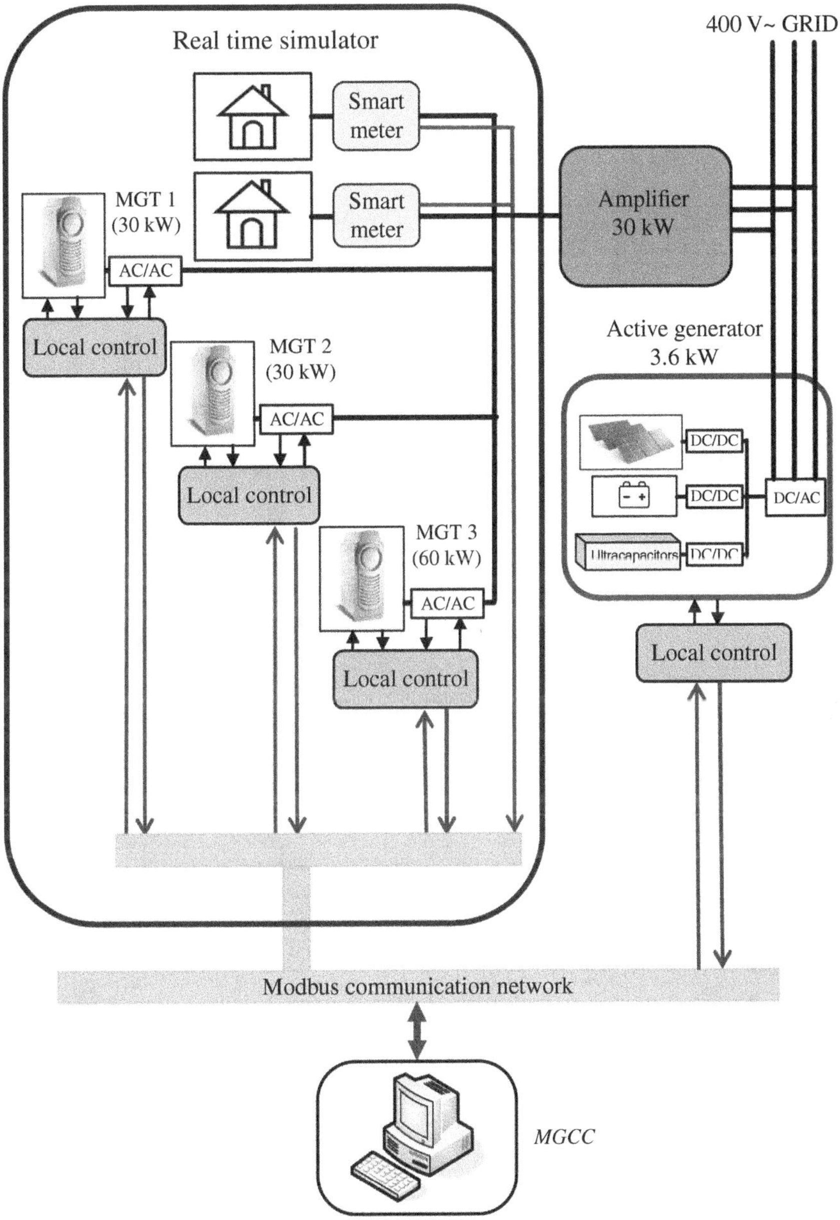

Figure 12.33 Schematic diagram of the experimental setup.

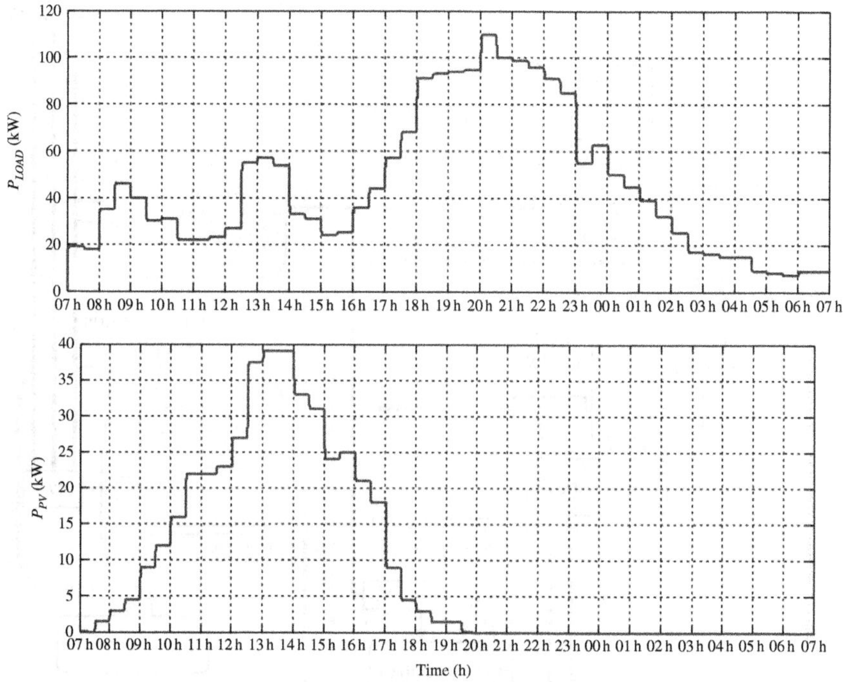

Figure 12.34 Day-ahead load forecast (kW) and PV power forecast in MPPT (kW).

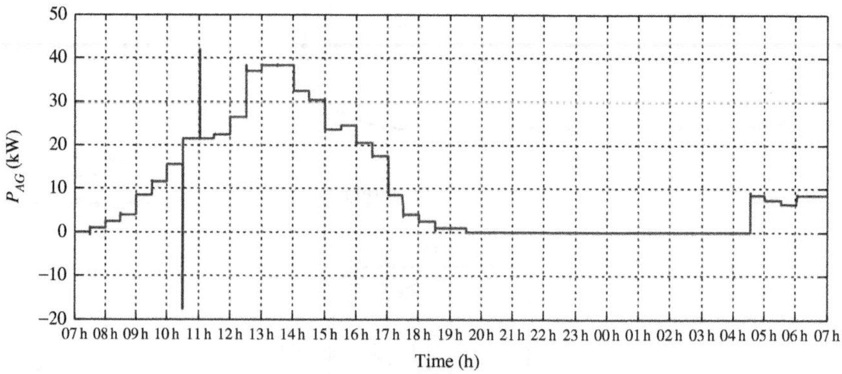

Figure 12.35 Global power reference of PV-based AGs.

Table 12.7 Day-ahead operational planning results.

	CO_2 cost	Fuel cost	Total cost
Without optimization	€50	€153	€203
Mono-objective (CO_2)	€35 (−30%)	€147 (−4.2%)	€181 (−10.7%)
Mono-objective (fuel)	€36 (−28.7%)	€146 (−4.6%)	€183 (−10.1%)
Multiobjective	€35 (−29.3%)	€147 (−4.1%)	€182 (−10.3%)

The DP-based optimization procedure has been performed with three different objective functions: minimization of the CO_2 equivalent emissions from the MGTs, the consumed fuel, and a trade-off between these two functions. Results in Table 12.7 show that each objective function causes about 10% reduction in the total operational cost (the cost of CO_2 equivalent emissions and the cost of consumed fuel), compared to the same system without optimization of the operational planning.

There is not a great difference in the overall cost between the different objective functions because, in fact, the CO_2 equivalent emissions and the consumed fuel are not really independent functions: the fuel consumption is used in the calculation of the CO_2 emissions, which is then used in the calculation of CO_2 equivalent emissions. Thus, minimizing either the CO_2 equivalent emissions or the consumed fuel is enough for approximating very well the optimal system operation.

A statistical analysis of sensed powers from MGTs has been done without optimization and with an optimized operational planning under the same scenario (Fig. 12.36). Results prove that the optimization algorithm selects power references such that MGTs operate more frequently in the domain between 0.7 and 0.8 p.u. Operation in this region causes less CO_2 equivalent emissions, as described in [21].

12.6 Day-ahead Optimal Operation and Power Reserve Dispatching

In this section, the objective function of UCP with DP is applied to minimize the CO_2 emissions and cost according to the given operational planning scheme in Fig. 12.30. To solve the objective function, different power and reserve dispatching strategies are performed under several nonlinear constraints [28].

The general objective is to minimize the total operating cost while satisfying all system constraints [29, 30]. In the presence of N AGs and M MGTs in an

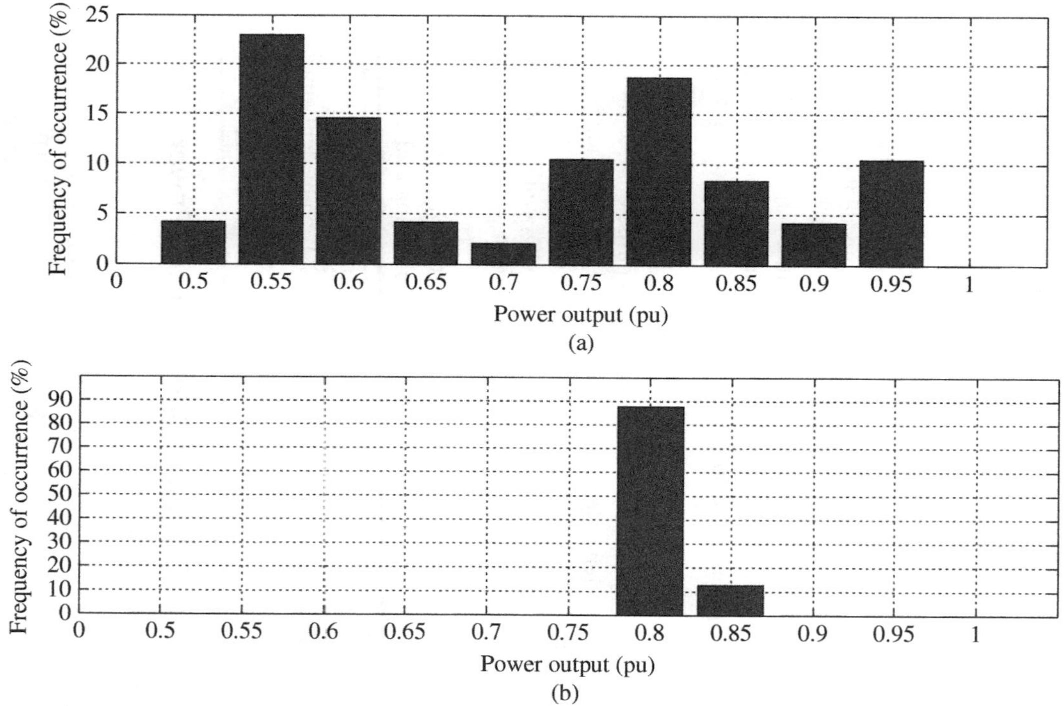

Figure 12.36 Occurrence of MGT 2 power set points: (a) without optimization and (b) using the CO_2 equivalent emissions as objective function.

MG, it is subjected to an equality constraint for AGs (P_{AG}), MGT generation (P_M), PR (P_R), and loads (P_L) balance. Restriction limits of the output related to each unit are as follows:

1) Quantified PR ($P_R(t)$) with a security level of x% of loss of load probability (LOLP)
2) Power balancing among loads and power generators as

$$\psi(t) = P_L(t) + P_R(t) - \sum_{n=1}^{N} P_{AG\_n}(t) - \sum_{i=1}^{M} (\delta_i(t) \cdot P_{M\_i}(t)) = 0 \quad (12.44)$$

3) MGT corresponding inequality constraint as

$$50\% P_{M\_max\_i}(t) \leq P_{M\_i}(t) \leq 100\% P_{M\_max\_i}(t) \quad (12.45)$$

4) The battery constraints are (i) the SOC of batteries has to be kept in a safety range $SOC_{max} > SOC > SOC_{min}$; (ii) the charge and discharge power of batteries cannot exceed the power limit $|P_{bat}| < P_{\_bat\_max}$; and (iii) if batteries are fully charged, the PV panels must not work in maximum power point.
5) Maximizing the PV energy usage, such that in the daytime, the extra PV energy will be stored in batteries as long as the batteries are not full, and then it will be discharged during the nighttime.

As a case study, consider again the MG shown in Fig. 12.2. In this MG, two power sources are considered: PV-based AG and MGT. The PV-based AG is preserved as a prior source because of its low operating cost and no gas emission features, while the MGT is set as a backup source for the additional energy needs. In the first strategy, the PR is only provided by the MGT and at least one of them is set in operation mode. In the second strategy, PR is distributed to the MGT and AG. Therefore, in some periods all the MGTs will be shut down and the load and reserve power will be provided only by the PV-based AG.

12.6.1 Scenario 1: Power Reserve Provided by MGTs

In this situation, all PRs are provided by MGT, so at least one of them works continuously. While the PV power is used to cover the load demand, the remaining PV energy will be stored into the batteries for nighttime. The deficit power is provided by the MGTs. Since no power is available from PV panels at night, power references are calculated separately for the daytime and for nighttimes.

12.6.1.1 Daytime

a) If the predicted PV power is more than the predicted load demand during the time step t ($\tilde{P}_{PV}(t) \geq \tilde{P}_L(t)$), then similar to (12.35), the PV power reference is set to a limited power:

$$P_{ref\_AG\_n}(t) = \frac{\tilde{P}_L(t)}{N} \quad (12.46)$$

where N is the total number of available PV-based AGs. The PR needs to be covered by the MGTs such that (i) if a PR is less than the minimum MGT power output ($P_{ref\_M} < P_{MGT\_min}$), then the MGT must operate at the minimum operating set point and in this situation the reserve will be more than the required power; and (ii) if the power reference of MGTs is more than the minimum MGT power output ($P_{ref\_M} > P_{MGT\_min}$), then the MGT must cover this reserve power: $P_{ref\_M} = P_R$. The energy surplus, if it can be taken by batteries, is automatically stored by local control of the AG according to (12.47), where the parameter τ is the duration of the available constant power:

$$\sum_{n=1}^{N} E_{bat\_n}(t+1) = \sum_{n=1}^{N} E_{bat\_n}(t) + \tau \cdot [P_{PV}(t) - P_L(t)] \tag{12.47}$$

b) If the predicted PV power is not enough for the predicted load during the time step t ($\tilde{P}_{PV}(t) \geq \tilde{P}_L(t)$), the power references of AGs are set to the predicted values as (12.48) and the MGT must balance the remaining load demand and PR according to (12.49).

$$P_{ref\_AG\_n}(t) = \tilde{P}_{PV\_n}(t) \tag{12.48}$$

$$\sum_{i=1}^{M} P_{ref\_M\_n}(t) = \tilde{P}_L(t) - \tilde{P}_{PV\_n}(t) + P_R(t) \tag{12.49}$$

12.6.1.2 Nighttime (Discharge the Battery)

To cover the load demand, the AG is considered as the prior source and the SOC of battery is calculated first. This energy storage can be estimated by the E-box to the MGCC. In this situation, two cases can be distinguished:

a) If the energy storages are enough to feed the predicted load, then MGTs are only used to provide the reserve power reference ($P_{ref\_M} = P_R$). At each time step, equation (12.47) can be used as battery discharge reference with $P_{PV}(t)$ equals to 0.
b) Otherwise, the MGT must balance the remaining load demand as follows:

$$\sum_{i=1}^{M} P_{ref\_MGT\_n}(t) = \tilde{P}_L(t) + P_R(t) \tag{12.50}$$

12.6.2 Scenario 2: Power Reserve Provided by Micro Gas Turbines and PV-based Active Generator

In this case, the reserve and operational power are distributed into both MGT and AG. As the AG is set as the prior energy source, if the AG energy (PV power during the daytime and battery at night) is more than the load and PR, then all the MGTs can be shut down.

12.6.2.1 Daytime

a) If the predicted PV power exceeds the predicted load demand added with the necessary PR during the time step t, $\tilde{P}_{PV}(t) \geq \tilde{P}_L(t) + P_R(t)$, then the power reference is set as limited by (12.51). The energy surplus, only if it can be assimilated, is saved automatically in batteries by local controller as given by (12.52).

$$P_{ref\_AG\_n}(t) = \frac{\tilde{P}_L(t) + P_R(t)}{N} \tag{12.51}$$

$$\sum_{n=1}^{N} E_{bat\_n}(t+1) = \sum_{n=1}^{N} E_{bat\_n}(t) + \frac{1}{2}(P_{PV}(t) - P_L(t)) \tag{12.52}$$

b) If the predicted PV power is not enough for the predicted load and the PR, $\tilde{P}_{PV}(t) < \tilde{P}_L(t) + P_R(t)$, the power reference of AG is set to the predicted values (12.53). Then, the MGTs must balance the remaining load demand as described in (12.54).

$$P_{ref\_AG\_n}(t) = \tilde{P}_{PV\_n}(t) \tag{12.53}$$

$$\sum_{i=1}^{M} P_{ref\_M\_n}(t) = \tilde{P}_L(t) - \sum_{n=1}^{N} P_{ref\_AG\_n}(t) \tag{12.54}$$

12.6.2.2 Nighttime

a) If the energy storage is enough to feed the predicted load and PR, then MGTs are turned off. The SOC of batteries must be refreshed and checked again at each time step:

$$\sum_{n=1}^{N} E_{bat\_n}(t+1) = \sum_{n=1}^{N} E_{bat\_n}(t) - \tau \cdot (\tilde{P}_L(t) + P_R(t)) \tag{12.55}$$

b) Otherwise, the MGTs operate to balance the remaining load demand, as in (12.51).

12.6.2.2.1 Formulation of the Unit Commitment Problem The 24-h-ahead operational planning is discretized with T periods. Power references are considered constant during each time step. This issue is handled by the short-term power balancing functions in the local control system. In the performed study, power references of the studied MGTs are represented as a vector $x(t)$, and the states of each MGT during each time step is shown a vector $u(t)$:

$$x(t) = [P_{M\_1}(t), P_{M\_2}(t), \ldots, P_{M\_M}(t)] \tag{12.56}$$

$$u(t) = [\delta_1(t), \delta_2(t), \ldots, \delta_M(t)] \tag{12.57}$$

12.6.2.2.2 Optimization Strategies For the proposed MG EM, the optimization process tackles to find an output vector $u(t)$ providing the generation set points of MGTs to guarantee the minimum fuel cost or CO_2 equivalent emissions, while satisfying the load balance within the settled time interval. The fuel costs ($C_{\_M\_i}(t)$) are expressed as a nonlinear function of its power output. The startup and shutdown penalties increase the emissions and shorten the lifetime of the MGT units. They can be expressed with a CO_2 function as $C_{P\_C\_i}(\delta_i(t+1), \delta_i(t))$. In this situation, consumed fuel cost during 5 and 2.5 min at full load are declared as startup and shutdown penalties, respectively. Therefore, the total fuel cost for each MGT can be expressed as follows:

$$J_{C\_i}(t) = \delta_i \cdot C_{M\_i}(t) + C_{P\_C\_i}(\delta_i(t+1), \delta_i(t)) \tag{12.58}$$

Then, the objective function for the whole system is to minimize the total fuel cost after 24 h operation. In this study, if the capital cost and maintenance cost of PV and MGT generators are taken into consideration, the proposed MG EM objective function can be defined as

$$J_{Cost} = \sum_{t=1}^{T}\sum_{i=1}^{M}(\delta_i \cdot C_{M\_i}(t) + C_{P\_C\_i}(\delta_i(t+1), \delta_i(t))) \tag{12.59}$$

The equivalent CO_2 emissions for each MGT at time step t can be expressed as (12.60). Then, the global proposed objective function can be defined as only for CO_2 equivalent emissions (12.61).

$$J_{CO_2\_i}(t) = (\delta_i \cdot CO_{2,M\_i}(t) + C_{P\_CO_2\_i}(\delta_i(t+1), \delta_i(t))) \tag{12.60}$$

$$J_{CO_2} = \sum_{t=1}^{T}\sum_{i=1}^{M}(\delta_i \cdot CO_{2,M\_i}(t) + C_{P\_CO_2\_i}(\delta_i(t+1), \delta_i(t))) \tag{12.61}$$

12.6.2.2.3 Dynamic Programming for the UCP For a system with T time steps, if at time step T, the minimum running cost is $F(T, x(T), u(T))$, then the total cost of the remaining step $(T-1)$ should be carried by the remaining time steps also at minimum cost [13]. Mathematically, the objective function $J(T)$ can be written as (12.62), where the $F(T, x(T), u(T))$ is represented by (12.63).

$$J(T) = F(T, x(T), u(T)) + \sum_{t=1}^{T-1} F(t, x(t), u(t)) \tag{12.62}$$

$$F(t, x(t), u(t)) = Tr(F(t), x(t), u(t)) + F(t+1, x(t+1), u(t+1)) \tag{12.63}$$

The objective is to find the vector of the generator states $u(t)$ and the power references of MGTs $x(t)$, in (12.56) and (12.57), which minimize the total system cost. The optimal solution of the overall problem is obtained by selecting the optimal $u(t)$ for all time steps recursively from $t = T$ to $t = 1$ as illustrated in Fig. 12.37.

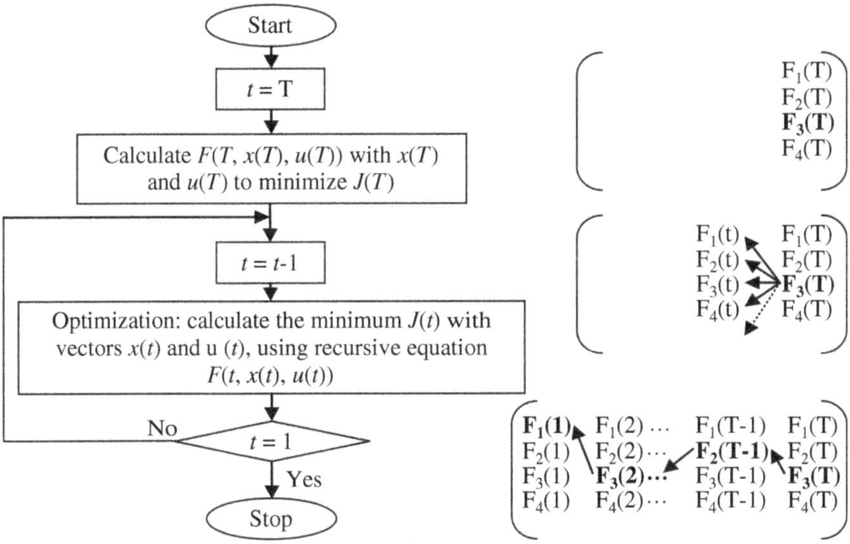

Figure 12.37 DP algorithm.

For each MGT state $u(t)$, the operational cost (economic criteria) is composed of two parts: (i) the total fuel cost during the time step t is $\sum_{i=1}^{M} \delta_i \cdot Cost_{MGT\_i}(t)$ and (ii) the cost during the previous time step taking into account the transition cost due to the startup or shutdown of generators is as follows:

$$Tr(u(t-1), u(t)) = F(t-1, u(t-1)) + \sum_{i=1}^{M} C_{Pe\_C\_i}(\delta_i(t+1), \delta_i(t)) \tag{12.64}$$

At step t, the UCP formulation for the studied system can be expressed in the form of a recursive **DP** equation as (12.65). A similar formulation is found to evaluate CO_2 emissions by (12.66) and (12.67):

$$F(t, x(t), u(t)) = \sum_{i=1}^{M} \delta_i \cdot C_{M\_i}(t) + Tr(u(t-1), u(t)) \tag{12.65}$$

$$Tr(u(t-1), u(t)) = F(t-1, u(t-1)) + \sum_{i=1}^{M} C_{Pe\_CO_2\_i}(\delta_i(t+1), \delta_i(t)) \tag{12.66}$$

$$F(t, x(t), u(t)) = \sum_{i=1}^{M} \delta_i \cdot CO_{2,M\_i}(t) + Tr(u(t-1), u(t)) \tag{12.67}$$

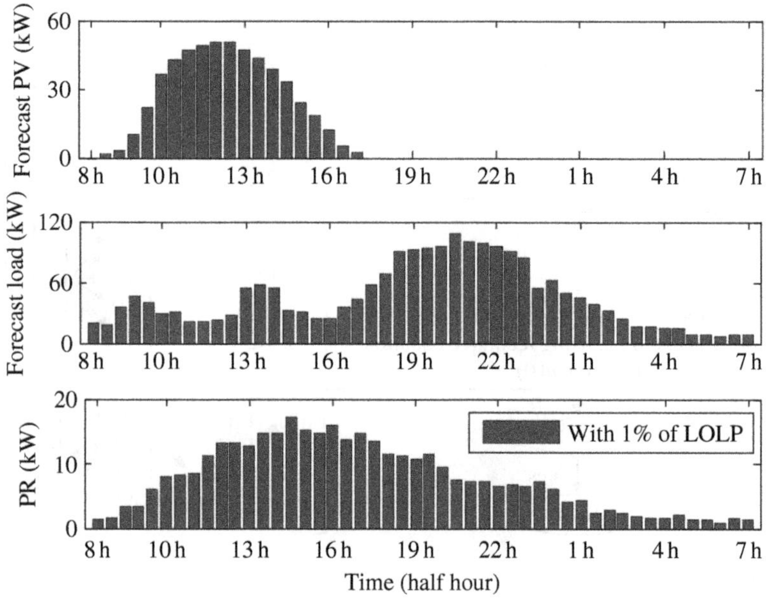

Figure 12.38 Day-ahead PV power forecast, load forecast, and power reserve with 1% of LOLP.

12.6.3 Optimal Reserve Power Dispatching Application for Unit Commitment Problem

In the present case study, a system with 110 kW of rated load, 55 kW of rated PV power, three MGTs with rated power of 30, 30, 60 kW as well as a PR, with 1% of LOLP (coming from the uncertainty assessment of load and PV forecasting) are considered, as shown in Fig. 12.38. The 24-h-ahead optimal operational planning is discretized with 48 periods ($T = 48$) for each 30 min ($\tau = 0.5$).

12.6.3.1 Dynamic Programing Application Scenarios

The DP-based optimization procedure has been performed with two objective functions, CO_2 equivalent emission minimization and total fuel cost minimization, under three different scenarios: (i) without PV generators: all the load demand and PR are covered by the MGT, (ii) PR provided by the MGT (method 1), and (iii) PR provided by both PV-based AG and MGT (method 2).

Results in Table 12.8 illustrate that compared with scenario 1, the total cost and pollution in scenarios 2 and 3 are decreased about 18% and 15%, respectively (thanks to the PV power usage). And among each scenario, using the optimization, the system cost and pollution are lower. In scenario 2, there is no PR dispatched in PV-power-based AG, but a bigger battery is needed in comparison of scenario 3, which has 38.9% of PR provided by the AG.

Table 12.8 Day-ahead operational planning results.

Scenario	Optimize strategy	Cost (€)	Pollution (kg)	PR on AG (%)	$E_{battery\_max}$ (kWh)
1	None	219	1392	0	0
	Environmental	212	1196	0	0
	Economic	210	1263	0	0
2	None	183	1156	0	80.2
	Environmental	181	1067	0	80.2
	Economic	178	1120	0	80.2
3	None	182	1098	38.9	54.1
	Environmental	179	991	38.9	54.1
	Economic	177	1061	38.9	54.1

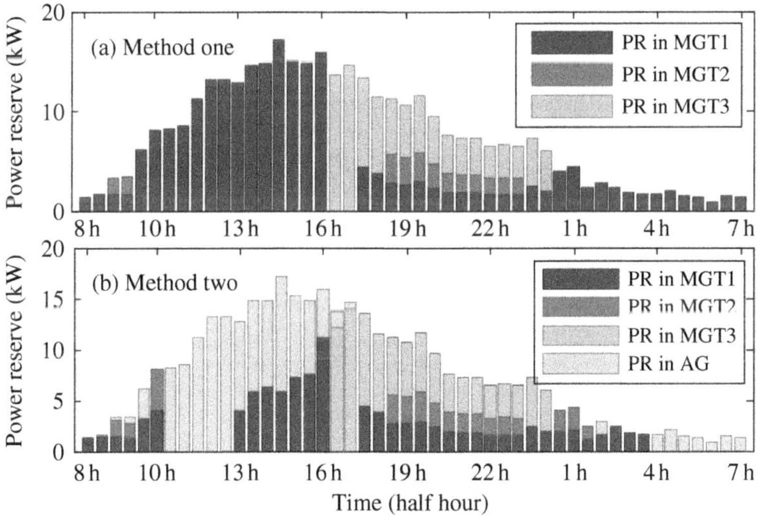

Figure 12.39 Power reserve dispatching in scenario 3.

12.6.3.2 Comparison of Power Reserve Dispatching

Figure 12.39 demonstrates the PR dispatching among different power generators in scenarios with two methods. To simplify the explanation, the results are chosen for the economic-based strategy. As it can be seen in second graph of Fig. 12.39, from 11 to 13 h and from 4 to 7 h (next day), the PR is provided by only AG, as well as in some other time steps it is covered by both MGT and AG. Instead, in first graph of Fig. 12.39, all the PR comes from MGTs only.

Figure 12.40 gives an example of reference power of AG, battery charge/discharge, and SOC. As it can be seen, PV-power-based AG provides the power both during the daytime and night. In the daytime, the energy comes from PV panels, while at night it comes from the discharge of batteries. The maximum charging and discharging power are 18 and 15 kW, respectively. The maximum storage is 54.1 kWh, which is much smaller than 80.2 kWh in

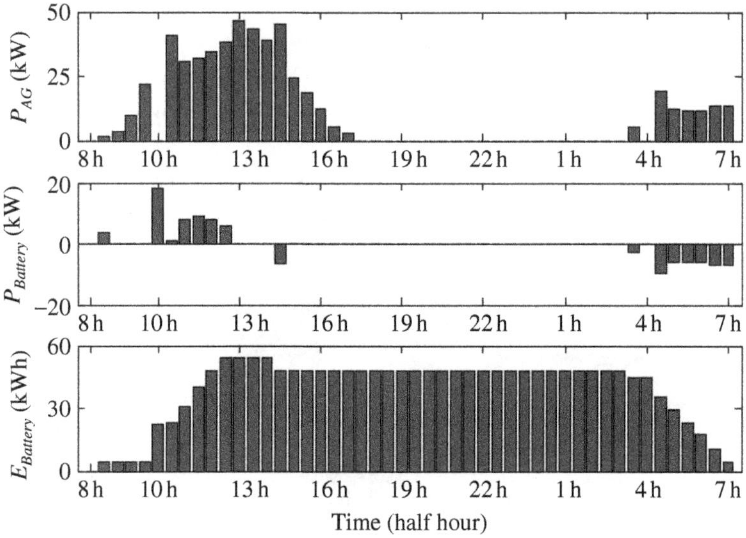

Figure 12.40 Reference power of AG, battery power, and energy.

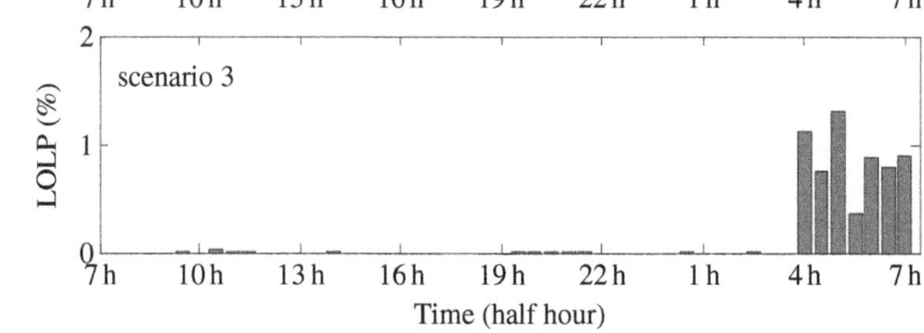

Figure 12.41 LOLP for each hour with scenarios 2 and 3.

scenario 2 (in Table 12.8). The reason is that in scenario 2 much more PV energy needs to be stored into the batteries during the daytime.

12.6.3.3 Security Level Analysis

Figure 12.38 indicates that the PR is calculated with 1% of LOLP. However, after operational and reserve dispatching, the MGTs usually cannot reach their maximum output. So with the remaining power plus original reserve power, a novel LOLP can be calculated. The novel results are presented in Fig. 12.41. It is clear that scenario 2 has a great improvement on security level as LOLP, at each time step, decreases from 1% to almost below 0.1%. While in scenario 3, some improvement appears except in the steps when all of the power is coming from batteries. It represents the risk that the electrical system needs to be faced with in this situation.

12.7 Robust Energy Consumption Scheduling in Interconnected Microgrids

As mentioned before, each MG can operate in the islanded and grid-connected operation modes. The performance measure in the islanded operation mode is the reliability of stand-alone operation. However, in the grid-connected mode, the MG operates while connected to the main grid and other MGs. This is especially characterized by the fact that each MG can sell a portion of its generated power to neighborhood grids and at the same time is able to purchase a portion of its demand from those grids under supervision of the global control center or distribution network operator (DNO). As a result of power sharing in this mode, load demand supplement is guaranteed in all time by the interconnected grid [31].

Increasing the amount of demand for electrical energy along with growing environmental concerns motivate the idea of establishing new power systems with flexible and intelligent programs of generation as well as demand side management. These programs run by utility companies aim at providing consumers with a reliable and cost–efficient energy and, at the same time, making efficient use of generation and transmission infrastructure. While many of these programs are still under investigation, there already exist a number of practical applications in many countries across the world [5, 32]. In this direction, ECS and power dispatching can be considered as important global control issues in distribution networks with interconnected MGs.

Demand side can be managed by either reducing or shifting the consumption of energy. While the former can be efficient to some extent, the latter proposes shifting of high-load household consumptions to off-peak hours in order to reduce *peak-to-average ratio* (PAR). The high PAR might lead to degradation of power quality, voltage problems, and even potential damages to utility and consumer equipment. With the advancement of smart metering technologies and increasing interest in power distribution networks with two-way communications capability, load management has been appeared in the form of ECS. In the ECS, the power consumption time of connected units is optimally scheduled so that some interesting measures such as generation cost and PAR can be optimized efficiently.

Consequently, customers will be encouraged to shift their heavy loads to off-peak hours. These issues motivate the design of ECS with the aim of minimizing power generation cost and PAR. The proposed ECS schemes in the literature mainly perform network-wide load management with the assumption of the knowledge of the whole network demand *a priori* or at least with known statistical characteristics. In other words, a network operator should be aware of the whole network demand in some way. Due to the diversity of power customers ranging from household to industrial domains with uncertain demands, however, this case is not *mostly* valid. Alternatively, an operator who is aware of the demand in a local area (LA), not other neighbor areas (NAs), might be interested in ECS within this area. The fact that aggregate power generation cost depends on the network-wide demand necessities considering the impact of uncertain demands in the design of the ECS.

To investigate the mentioned difficulty, consider a distribution network organized by a distribution company (Disco) as shown in Fig. 12.42. This distribution network together with connected N MGs perform an LA with known demand on average and other NAs with uncertain demands. The network operator performs ECS of demand in the LA considering NAs demand as a random variable. This ECS is formulated with two stochastic optimization problems, one with the objective of the network-wide power generation cost minimization and the other with the objective of PAR minimization. While these two objectives are correlated in some extent, optimizing one does not necessarily imply the optimality of the other.

These objectives are compared using optimal, adaptive, and uniform scheduling schemes in terms of generation cost and PAR. In the optimal one, the optimal solutions of two underlying problems are achieved with the assumption of the knowledge of NAs demand in advance. Without this assumption for practical purposes, an adaptive scheme with *online* stochastic iterations to capture the randomness of uncertain demands over the time horizon continually is proposed. Finally, in uniform scheduling the demand of MGs in LA is uniformly distributed over the time horizon regardless of NAs demand.

The pricing of electricity can be used as a mechanism to encourage customers to follow a specified load scheduling. Various pricing schemes have been proposed by economists and regulatory agencies such as flat pricing, critical-peak pricing, time-of-use pricing, and real-time pricing. Among them, real-time pricing is motivated to be used in the next-generation power systems

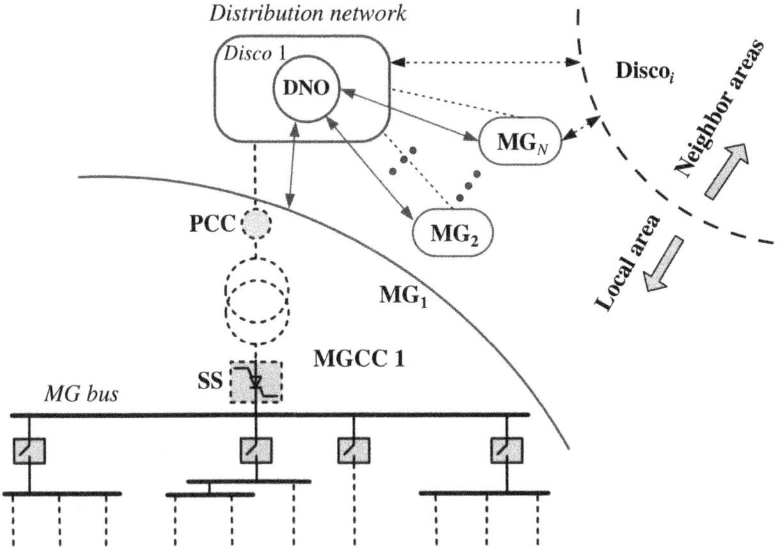

Figure 12.42 A distribution network with connected MGs.

concerning its environmental and economical gains. Accordingly, here to establish a real-time pricing scheme, an energy scheduling approach based on real-time generation cost is proposed.

The network shown in Fig. 12.42, owned by a Disco, is connected to a set $\Lambda \equiv \{n : n = 1, \dots, N\}$ of MGs operating in LA and other MGs working as NAs, which may belong to a different Disco. It is assumed that all MGs operate in the grid-connected mode. The DNO performs ECS for MGs in the LA during a time horizon $\tau \equiv \{t : t = 1, \dots, T\}$. The aim of ECS is to optimally manage and to shift the LA demand to reduce power generation cost and PAR within the power network. The demand of each MG$_n$ in LA during this interval is assumed to be a known value E_n on average. However, the demand by MGs in NAs is assumed to be an unknown value, denoted by γ.

Let P_n^t be the power provided to MG$_n$ in LA during time slot t. The objective of ECS is to determine a power set $P \equiv \{p_n^t\}_{n\in N}^{t\in\tau}$ to optimize a target performance measure, and at the same time to provide each MG$_n$ with a determined demand E_n on average. Since the generation cost in the distribution network depends on both LA and NAs demands, the uncertainty of γ should be taken into account in the determination of P. Moreover, these powers have strict minimum and maximum power levels. This imposes a constraint that each p_n^t must be within p_n^{min} and p_n^{max}, minimum and maximum power levels, respectively. In the mentioned network, the objective of DNO by implementing ECS could be either to minimize power generation cost or to minimize PAR. These objectives are discussed in subsequent sections.

12.7.1 Cost Minimization Formulation

Let $p^t \equiv \sum_{n=1}^{N} p_n^t + \gamma(t)$ be the total amount of power generated at time t to be delivered to LA and NAs. The power generation cost at this time can be denoted as a differentiable and convex function of p^t, denoted by $C(p^t)$. Accordingly, minimizing the average generation cost during the time horizon is formulated as

$$\min_{P} \frac{1}{T} \sum_{t=1}^{T} C\left(\sum_{n=1}^{N} p_n^t + \gamma(t)\right) \tag{12.68}$$

$$\text{s.t. } \frac{1}{T} \sum_{t=1}^{T} p_n^t \geq E_n \qquad \forall n \in \Lambda \tag{12.68a}$$

$$p_n^{min} \leq p_n^t \leq p_n^{max} \qquad \forall n \in \Lambda, \ \forall t \in \tau \tag{12.68b}$$

Constraint (12.68a) satisfies the required average demands by MGs in LA. Constraint (12.68b) restricts the power levels within some upper and lower bounds. This problem is convex and can be solved using convex optimization techniques such as interior point method (IPM) [33]. This requires the knowledge of $\gamma(t)$ for all in the beginning of the time horizon τ. However, this assumption is not valid in practice as the DNO is not aware of NAs demand *a priori* (i.e., in the beginning of the horizon).

Alternatively, we consider as an uncertain parameter in the form of a random variable varying over time, but without any assumption on its probability density function (PDF). With this assumption, (12.68) can be rewritten as

$$\min_{P} \overline{E}_{\gamma} \left[C\left(\sum_{n=1}^{N} p_n + \gamma\right)\right] \tag{12.69}$$

$$\text{s.t. } \overline{E}_{\gamma}[p_n] \geq E_n \quad \forall n \in \Lambda \tag{12.69a}$$

$$p_n^{min} \leq p_n^t \leq p_n^{max} \quad \forall \text{n} \in \Lambda \tag{12.69b}$$

where $\overline{E}_{\gamma}$ denotes the expectation with respect to γ. The aforementioned problem is also convex. However, we are interested in solving this problem progressively over time, when γ is realized at each time instant t. The challenge in the solution of problem (12.69) is due to the expectations that couple the scheduling over time. The solution would be straightforward if one decouples the demand constraints over time. This motivates the incorporation of (12.69a) into the objective function and forms a Lagrangian function as

$$L(P,Z) = \overline{E}_{\gamma} \left[C\left(\sum_{n=0}^{N} p_n + \gamma\right)\right] - \sum_{n=1}^{N} \lambda_n(\overline{E}_{\gamma}[p_n] - E_n) \tag{12.70}$$

and the corresponding dual function as

$$D(Z) = \inf_{P}\{L(P,Z) : \ p_n^{min} \leq p_n^t \leq p_n^{max} \ \forall n \in \Lambda\} \tag{12.71}$$

where $Z = \{\lambda_n \geq 0\}_{n\in\Lambda}$ is the set of Lagrange multipliers and "inf" represents infimum operation. The dual function provides a lower bound on the optimal solution of (12.69). The best lower bound is surely achieved by the corresponding dual problem as

$$\min_{Z\geq 0} D(Z) \tag{12.72}$$

Prior to solving this problem in the dual domain, we first need to evaluate $D(Z)$. $L(P, Z)$ can be rewritten as

$$L(P,Z) = \overline{E}_\gamma \left[C\left(\sum_{n=1}^{N} p_n + \gamma \right) - \sum_{n=1}^{N} \lambda_n p_n \right] + \sum_{n=1}^{N} \lambda_n E_n \tag{12.73}$$

Therefore, to evaluate $D(Z)$ in (12.71), we solve

$$\min_{P} \; \overline{E}_\gamma \left[C\left(\sum_{n=1}^{N} p_n + \gamma \right) - \sum_{n=1}^{N} \lambda_n p_n \right] \tag{12.74}$$

$$\text{s.t. } p_n^{min} \leq p_n^t \leq p_n^{max} \quad \forall n \in \Lambda \tag{12.74a}$$

For each value of γ, this problem is convex and can be solved using IPM [33] to obtain the optimal values $\{p_n^*(\gamma)\}_{n\in\Lambda}$. Having obtained $\{p_n^*(\gamma)\}_{n\in\Lambda}$, the dual problem in (12.72) can be solved using the subgradient method [34]. Beginning with an initial $\lambda_n(0)$, given $\lambda_n(t)$ at time t, the optimal values $\{p_n^{t*}(\gamma)\}_{n\in\Lambda}$ can be obtained from (12.74). Based on our experience, we usually choose $\lambda_n(0)$'s values such that the solution of the problem with relaxed constraints using these initial multipliers lies within the feasible region of decision variables, rather than the margin of the feasible region. We then update the Lagrange multiplier as

$$\lambda_n(t+1) = \lambda_n(t) + \alpha(E_n - \overline{E}_\gamma[p_n^{t*}(\gamma)])^+ \tag{12.75}$$

where $E_n - \overline{E}_\gamma[p_n^{t*}(\gamma)]$ is the subgradient of $D(Z)$ with respect to λ_n, α is a step size, and $(x)^+ \equiv \max(x, 0)$. The gradient iteration (12.75) is efficient to find the optimal scheduling. A key knowledge we need in (12.75) is the PDF of γ, only with which the expected value $\overline{E}_\gamma$ can be evaluated. The assumption of a known PDF of γ may be reasonable for theoretic studies. However, the importance of practical energy scheduling schemes motivates the optimal strategy by *learning* NAs demand on the fly. Interestingly, a stochastic gradient iteration can be developed to solve (12.72) without the PDF of γ *a priori*. To this end, we consider dropping $\overline{E}_\gamma$ from (12.75) to devise online iterations for *adaptive* decisions, based on per slot realization $\gamma(t)$, as

$$\hat{\lambda}_n(t+1) = \hat{\lambda}_n(t) + \alpha(E_n - p_n^{t*}(\gamma(t)))^+ \tag{12.76}$$

where hats are to emphasize that these iterations are stochastic estimates of those in (12.75). Provided that the random NAs demand process is

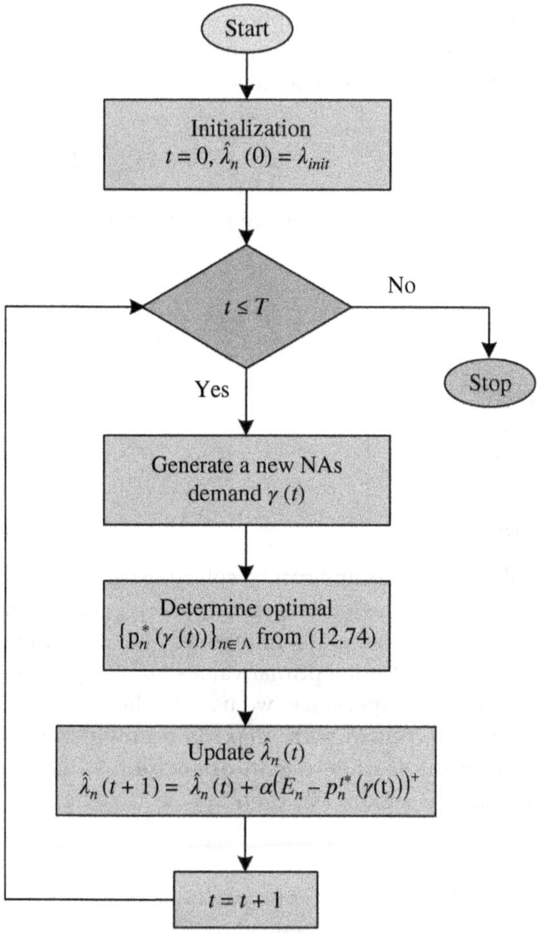

Figure 12.43 RCA-ECS algorithm.

stationary and ergodic, the stochastic gradient iteration (12.76) and the ensemble gradient iterations (12.75) consist of a pair of primary and averaged systems [35]. The convergence of such a stochastic gradient iteration can be established statistically, provided that α is small enough [36, 37]. The above described solution can be summarized as a robust cost-aware energy consumption scheduling (RCA-ECS) algorithm as shown in Fig. 12.43.

12.7.2 Peak-to-Average Ratio Minimization Formulation

In order to minimize PAR of the total instantaneous power delivered to LA and NAs during the time period τ, a minmax formulation is proposed with the objective of minimizing the peak of this power. The objective is expressed as

$$\min_{P} \min_{t\in\tau} \sum_{n=1}^{N} p_n^t + \gamma(t) \tag{12.77}$$

Due to the unavailability of $\gamma(t)$ *a priori*, this objective, along with the aforementioned constraints, can be translated into the problem

$$\min_{P} \ s \tag{12.78}$$

$$\text{s.t.} \ \ s \geq \sum_{n=1}^{N} p_n + \gamma \tag{12.78a}$$

$$\overline{E}_\gamma[p_n] \geq E_n \quad \forall n \in \Lambda \tag{12.78b}$$

$$p_n^{min} \leq p_n^t \leq p_n^{max} \quad \forall n \in \Lambda \tag{12.78c}$$

where s is an auxiliary variable. The solution is mostly similar to that of the cost minimization. The Lagrangian can be formed as

$$L(P, \mu) = s - \sum_{n=1}^{N} \mu_n(\overline{E}_\gamma[p_n] - E_n) \tag{12.79}$$

and dual function as

$$D(\mu) = \inf_{P} \left\{ \text{L(P, } \mu) : \ \ s \geq \sum_{n=1}^{N} p_n + \gamma, \ p_n^{min} \leq p_n^t \leq p_n^{max} \quad \forall n \in \Lambda \right\} \tag{12.80}$$

where $\mu = \{\mu_n \geq 0\}_{n\in\Lambda}$ is the set of Lagrange multipliers. To evaluate $D(\mu)$, the following minimization problem should be solved:

$$\min_{P} \ s - \sum_{n=1}^{N} \mu_n \overline{E}_\gamma[p_n] \tag{12.81}$$

$$\text{s.t.} \ \ s \geq \sum_{n=1}^{N} p_n + \gamma, \ \ p_n^{min} \leq p_n^t \leq p_n^{max} \quad \forall n \in \Lambda \tag{12.81a}$$

Similar to the mentioned online learning iteration (12.76), the stochastic estimation of each μ_n (i.e., $\hat{\mu}_n$) can be learned over time using

$$\hat{\mu}_n(t+1) = \hat{\mu}_n(t) + \alpha(E_n - p_n^{t*}(\gamma(t)))^+ \tag{12.82}$$

By these iterations, problem (12.81) at each iteration t can be rewritten as

$$\min_{P} \ s - \sum_{n=1}^{N} \hat{\mu}_n(t) p_n^t \tag{12.83}$$

$$\text{s.t.} \ \ s \geq \sum_{n=1}^{N} p_n^t + \gamma(t), \ \ p_n^{min} \leq p_n^t \leq p_n^{max} \quad \forall t \in \tau, \quad \forall n \in \Lambda \tag{12.83a}$$

The first constraint term in (12.83a) can be expressed as $s = s' + \sum_{n=1}^{N} p_n^t + \gamma(t)$, where $s' \geq 0$ is an auxiliary variable. Substituting s with s' in (12.83), we obtain

$$\min_{P} s' - \sum_{n=1}^{N} (1 - \hat{\mu}_n(t)) p_n^t \tag{12.84}$$

$$\text{s.t. } s' \geq 0,\ p_n^{min} \leq p_n^t \leq p_n^{max} \quad \forall t \in \tau,\ \forall n \in \Lambda \tag{12.84a}$$

The solution for this problem is trivially achieved when $s' = 0$ and each term $(1 - \hat{\mu}_n(t))p_n^t$ is minimized for $p_n^{min} \leq p_n^t \leq p_n^{max}$. The latter is absolutely dependent on the sign of $(1 - \hat{\mu}_n(t))$. Doing so, it can be concluded that $p_n^t = p_n^{min}$ if $\hat{\mu}_n(t) < 1$, $p_n^t = p_n^{max}$ if $\hat{\mu}_n(t) > 1$, and $p_n^t = (p_n^{min} + p_n^{max})/2$ as a midpoint between these two extremes if $\hat{\mu}_n(t) = 1$. This solution results in a robust peak-to-average ratio-aware energy consumption scheduling (RPAR-ECS) algorithm, which is summarized in Fig. 12.44.

12.7.3 Simulation Results

For investigation of the proposed methods, a distribution network is considered in connection with an LA consisting of $N = 10$ MGs and an NA. The ECS located in the DNO schedules energy consumption of MGs in LA during a time horizon of length 6 h. The scheduling is updated every 1 min (i.e., 360). Average demands of MGs in LA are $E = [1/T, 2/T, \ldots, N/T]$, where n/T is the average demand of MG_n in kilowatt-hours per unit of time (i.e., E_n in constraints (12.68a) and (12.78b)). Minimum and maximum power levels are $p^{min} = [0,1, \ldots, 9]$ and $p^{max} = [5,6, \ldots, 14]$, respectively. The power generation cost is considered to be quadratic function (i.e., $C(\cdot) = (\cdot)^2$ in (12.68)), even though the proposed solutions are valid for any convex cost function. The performance of the proposed algorithms is evaluated in the following sections.

The impact of the proposed RCA-ECS algorithm on the time domain curvature of the total grid demand in the presence of an *unknown* NAs demand is evaluated and the result is compared with an optimal technique. A typical realization of $T = 360$ samples of NAs demand with a standard deviation of $\sigma = 20$ kWh and the corresponding optimal total demand are shown in Fig. 12.45a. As observed, the optimal solution schedules LA demand such that the system-wide total demand becomes smooth suitable for cost minimization. In fact, scheduling the LA demand provides a diversity for the ECS to mitigate the stochastic nature of NAs demand. Total demand using RCA-ECS scheme and the corresponding Lagrange multipliers [38] are also shown in Fig. 12.45b. Intuitively, after some initial time slots, the behavior of the total demand curve approximately converges to that of the optimal solution in Fig. 12.45a.

Performance measures of the RCA-ECS and the optimal ECS schemes, such as generation cost per kWh and PAR, versus the randomness of the NAs demand could be interesting in the following. As another scheduling

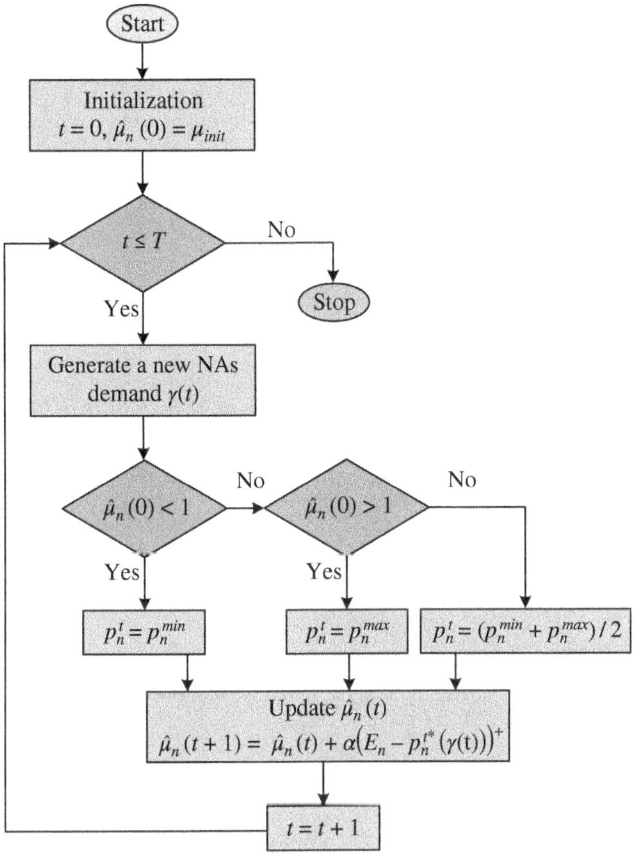

Figure 12.44 RPAR-ECS algorithm.

scheme, the results of uniform ECS scheme are also included. In this scheme, the demand of each MG in LA is *uniformly* distributed over the whole time horizon, independent of the NAs demand. This can also be considered as a deterministic solution. Cost and PAR performances versus the standard deviation are shown in Fig. 12.45c and d, respectively. For each instance, similar to the time-domain performance, first a data set with $T = 360$ samples is generated. This set is used to obtain the optimal ECS solution once in the beginning of the time horizon as well as to provide the RCA-ECS scheme with instantaneous realized NAs demand. As shown in the first part of Fig. 12.45a, there is a typical generated data set with $\sigma = 20$ kWh. As a common observation in Fig. 12.45c and d, performance measures are getting worse as σ increases. In the case of cost measure, this is due to the fact that the considered squared cost function results in higher cost per kWh for high

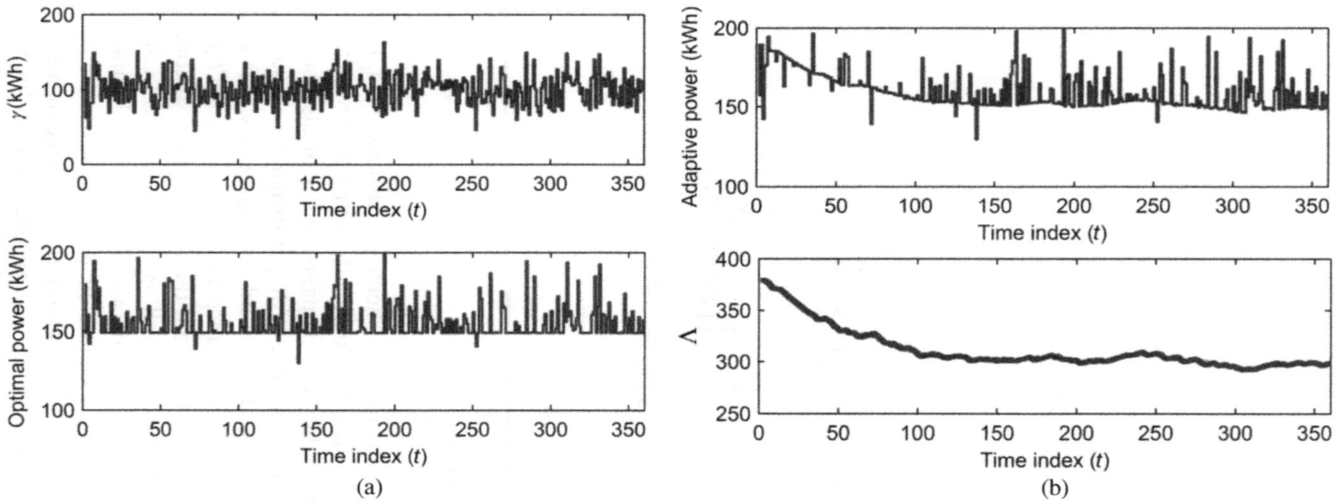

Figure 12.45 Simulation results: (a) generated NAs demand and the system-wide optimal demand, (b) adaptive system-wide demand and Lagrange multipliers, (c) generation cost per kWh in cost formulation, (d) PAR in cost formulation, (e) generation cost per kWh in the PAR formulation, and (f) PAR in PAR formulation.

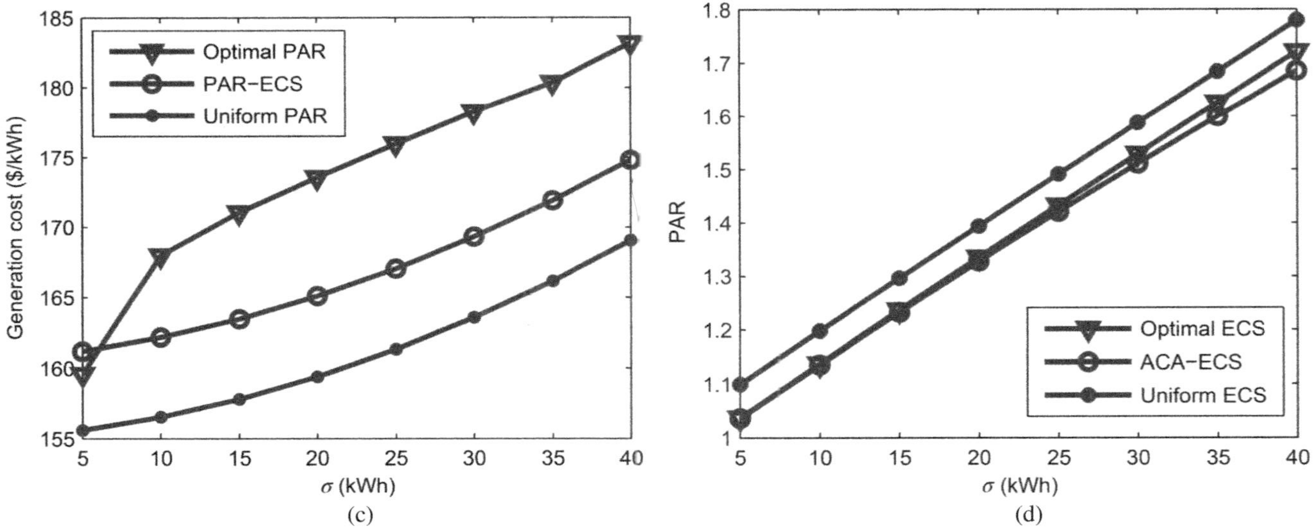

Figure 12.45 *(Continued)*

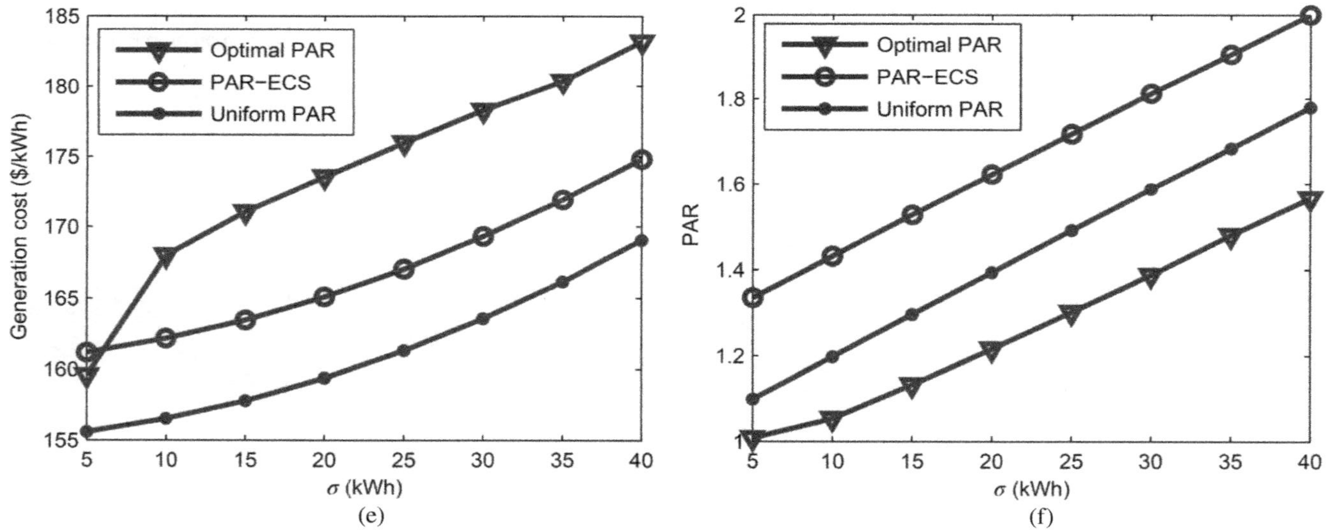

Figure 12.45 (*Continued*)

demand values in comparison with low demand values. The results in PAR are based on the fact that the averages of both LA and NAs demands are made constant when σ increases. Considering PAR as a fractional term of the peak demand over the average demand, it is reasonable to conclude that PAR increases as σ increases. Moreover, in Fig. 12.45c, with increase in σ, the performance gap between the compared ECS schemes and the optimal one increases. In the case of RCA-ECS, this is due to the fact that the stochastic estimator (given in [3]) would be far from optimality with the increase in the randomness of the uncertain part of power generation. In the case of uniform ECS, the degradation effect of high randomness would be more severe since this scheme does not take care of NAs demand in the scheduling decisions.

Furthermore, in Fig. 12.45c and d, the generation cost and PAR performances of the RCA-ECS scheme outperform those of uniform ECS. This is reasonably expected as RCA-ECS takes advantage of the diversity in NAs demand to smooth the total demand and therefore achieves a better performance. In the comparison between RCA-ECS and the optimal solution, it is observed that the optimal solution achieves lower cost. This is due to the fact that this solution fully takes into account the knowledge of NAs demand at the beginning of the time horizon for the scheduling of LA demand. However, RCA-ECS makes a scheduling decision adaptively per a time unit, when the demand of NAs is available in that unit. Remarkably, the PAR of RCA-ECS is comparable to that of the optimal solution. This implies that the optimality of generation cost does not necessarily imply the optimality of PAR too. This observation motivates the performance evaluation of PAR formulation in the following.

In order to evaluate the efficiency of PAR formulation, the generation cost per kWh and PAR performances of this formulation are illustrated in Fig. 12.45e and f, respectively. Similar to the cost formulation [38], the results of optimal solution in PAR formulation (optimal PAR) and uniform scheduling (uniform PAR) scheme are also included. Since the scheduling of the uniform strategy is independent of the objective function, the achieved results are the same in both cost and PAR formulations. We take advantage of this equality and take uniform strategy curves as references for comparison between these formulations.

Comparing Fig. 12.45c and e, it is observed that uniform scheduling was worst in the former, whereas it is best in the latter. Considering the results of uniform scheduling as reference in both figures, we conclude that cost minimization formulation is more cost-efficient in comparison with PAR formulation. In terms of PAR, the optimal solution in the PAR formulation achieves the lowest PAR. This is reasonably expected as this solution takes NAs demand into account *a priori*. In comparison with the uniform strategy, the PAR of RPAR-ECS scheme is high. More importantly, this implies that PAR performance of the proposed adaptive approach in cost minimization formulation even outperforms its equivalent one in the PAR minimization formulation.

This observation along with the lower generation cost in cost formulation demonstrates that our proposed adaptive approach achieves more *efficient* results with this formulation compared with PAR one. Also the proposed robust approach is a trade-off between the optimal (full NAs demand) and uniform (no NAs demand) schemes in terms of generation cost and PAR minimization.

In [31], the proposed methodology is compared with the game theory. In continuation of the present work in this section, the load demand management of interconnected MGs is formulated as a power dispatch optimization problem in [36]. Real-time pricing is employed as a motivation for interactions between the MGs. The objective is to minimize the network operational cost and at the same time to satisfy the stochastic demands within the MGs in average. With the solution of this problem, a cooperative power dispatching algorithm between MGs is proposed under the assumption of a communication infrastructure within the grid. The core parameter in this algorithm is a defined dynamic *purchase price* per unit of power at each MG. Considering their demands and supplies, the MGs progressively update and broadcast their prices throughout the grid. Every MG adaptively regulates its transactions with the rest of the grid by taking into account its realized demand as well as already announced prices from the other MGs. This strategy results in a semidistributed and reliable load management within the grid in a cost–efficient manner. The required communication network for the mentioned power management strategy is emphasized in [39].

12.8 Summary

The main objective of this chapter is to describe the planning and an innovative EMS for the MGs. An MG planning example for a real MG case study is briefly explained. Then, a coordinated management of energy resources is proposed through a communication network. The energy management is organized in three parts: the long-term energy management, the medium-term energy management, and the short-term power management. The proposed EMSs and the performed scenarios are tested on a real-time simulator.

A day-ahead operational planning has been proposed to calculate power references of DGs in order to optimize CO_2 emissions and the operational economic cost. A DP approach is used for solving the UCP in the studied MG. For the sake of real-time simulation, the performed MG central control has been implemented. Then, a day-ahead optimal operational planning has been applied to calculate power references with two different dispatching strategies of operational and reserve power. A DP approach has been used for solving the UCP in order to minimize equivalent CO_2 emissions and the total fuel cost. Finally, a robust ECS methodology for interconnected MGs in the presence of demand uncertainty is presented.

References

1 Lu, D. (2010) Design and control of a PV active generator with integrated energy storages: application to the aggregation of producers and consumers in an urban micro smart grid. PhD thesis. Ecole Centrale de Lille, France.

2 Bevrani, H., Ghosh, A., and Ledwich, G. (2010) Renewable energy sources and frequency regulation: survey and new perspectives. *IET Renewable Power Generation*, **4** (5), 438–457.

3 Babahajiani, P., Shafiee, Q., and Bevrani, H., Intelligent demand response contribution in frequency control of multi-area power systems. *IEEE Transactions on Smart Grid*, in press, DOI: 10.1109/TSG.2016.2582804.

4 Babahajiani, P., Shafiee, Q., and Bevrani, H. (2016) *Intelligent Coordination of Demand Response and Secondary Frequency Control in Multi-area Power Systems.* 1st IEEE Conference on New Research Achievements in Electrical and Computer Engineering (CBCONF), Tehran, Iran, May.

5 Bevrani, H., Watanabe, M., and Mitani, Y. (2014) *Power System Monitoring and Control*, IEEE-Wiley Press, New York.

6 Liu, Q., Bevrani, H., and Mitani, Y. An enhanced WAMS-based power system oscillation analysis approach, in *Dynamic Vulnerability Assessment and Intelligent Control for Sustainable Power Systems* (eds J.R. Torres and F.G. Longatt), Chapter 7, IEEE-Wiley, USA, to be submitted.

7 Lu, D., Zhou, T., Fakham, H., and François, B. (2010) Application of Petri nets for the energy management of a photovoltaic based power station including storage units. *Renewable Energy*, **35** (6), 1117–1124.

8 Leung, D.Y.C., Caramanna, G., and Maroto-Valer, M.M. (2014) An overview of current status of carbon dioxide capture and storage technologies. *Renewable and Sustainable Energy Reviews*, **39**, 426–443.

9 Ngan, M.S. and Tan, C.W. (2012) Assessment of economic viability for PV/wind/diesel hybrid energy system in southern Peninsular Malaysia. *Renewable and Sustainable Energy Reviews*, **16** (1), 634–647.

10 Sorensen, B. (2011) *Renewable Energy: Physics, Engineering, Environmental Impacts, Economics & Planning*, Elsevier.

11 Bahramara, S., Parsa Moghaddam, M., and Haghifam, M.R. (2016) Optimal planning of hybrid renewable energy systems using HOMER: a review. *Renewable and Sustainable Energy Reviews*, **62**, 609–620.

12 *Surface Meteorology and Solar Energy*, http://eosweb.larc.nasa.gov/sse/ (accessed August 01, 2010).

13 Sen, R. and Bhattacharyya, S.C. (2014) Off-grid electricity generation with renewable energy technologies in India: an application of HOMER. *Renewable Energy*, **62**, 388–398.

14 Feinberg, E.A. and Genethliou, D. (2005) Chapter 12: Applied mathematics for power systems: load forecasting, in *Applied Mathematics for Restructured Electric Power Systems*, Springer, pp. 269–285.

15 Mcmenamin, J.S. and Monforte, F.A. (1998) Short term energy forecasting with neural networks. *Energy Journal*, **19** (4), 43–61.

16 Ahmadi, S., Bevrani, H., and Jannaty, H. (2012) *AFfuzzy Inference Model for Short-term Load Forecasting*. 2nd Iranian Conference on Renewable Energy and Distributed Generation (ICREDG), Tehran, Iran, pp. 39–44.

17 Lu, D., Zhou, T., Fakham, H., and Francois, B. (2008) *Design of a Power Management System for a PV Station Including Various Storage Technologies*. 13th International Power Electronics and Motion Control Conf. (EPE-PEMC 2008), Poznan, September 1–3, 2008, CD-ROM.

18 Pilavachi, P.A. (2002) Mini- and micro-gas turbines for combined heat and power. *Applied Thermal Engineering*, **22** (18), 2003–2014.

19 Bevrani, H. and Hiyama, T. (2011) *Intelligent Automatic Generation Control*. New York, USA: CRC Press (Taylor & Francis Group).

20 Bevrani, H. (2014) *Robust Power System Frequency Control*, 2nd edn, Springer, Switzerland.

21 Kanchev, H., Colas, F., Lazarov, V., and Francois, B. (2014) Emission reduction and economical optimization of an urban microgrid operation including dispatched PV-based active generators. *IEEE Transactions on Sustainable Energy*, **5** (4), 1397–1405.

22 Ahn, S., Nam, S., Choi, J., and Moon, S. (2013) Power scheduling of distributed generators for economic and stable operation of a microgrid. *IEEE Transactions on Smart Grid*, **4** (1), 398–405.

23 Li, P., Degobert, P., Robyns, B., and Francois, B. (2009) Participation in the frequency regulation control of a resilient microgrid for a distribution network. *International Journal of Integrated Energy Systems*, **1** (1), 61–67.

24 Longenthiran, T. and Srinivasan, D. (2010) *Formulation of Unit Commitment (UC) Problems and Analysis of Available Methodologies Used for Solving the Problems*. IEEE International Conference on Sustainable Energy Technologies (ICSET), December 6–9, Kandy, Sri Lanka.

25 Boicea, A., Chicco. G., and Mancarella, P. (2009) *Optimal Operation of a Microturbine Cluster with Partial-load Efficiency and Emission Characterisation*. IEEE Powertech Conference, Bucharest, Romania.

26 Canova, A., Chicco, G., and Mancarella, P. (2007) *Assessment of the Emissions due to Cogeneration Microturbines under Different Operation Modes*. POWERENG 2007, April 12–14, Setubal, Portugal

27 Singhal, P.K. and Sharma, R.N. (2011) *Dynamic Programming Approach for Solving Power Generating Unit Commitment Problem*. 2nd International Conference on Computer and Communication Technology (ICCT'11), September 15–17, Allahabad, India, pp. 298–303.

28 Yan, X., Francois, B., Abbes, D., and Bevrani, H. (2016) *Day-ahead Optimal and Reserve Power Dispatching in PV Based Urban Microgrid*. 18th European Conference on Power Electronics and Applications-EPE'16 ECCE, September 5–9, Karlsruhe, Germany.

29 Ataee, S., Feizi, M.R., and Bevrani, H. (2016) *Optimal Operation of Renewable Energy-Based Grid-Connected Microgrid.* 2nd National Conference of Technology, Energy and Data on Electrical & Computer Engineering, May, Kermanshah, Iran.

30 Mohammadi, S., Bevrani, H., Moshtagh, J., and Bahramara, S. (2015) *Techno-economical Evaluation of Stand-Alone Hybrid Renewable Energy Systems for Urban Area in Sanandaj (Iran).* National Conference of Technology, Energy, and Data on Electrical and Computer Engineering, May, Kermanshah, Iran.

31 Fathi, M. and Bevrani, H. (2013) Adaptive energy consumption scheduling for connected microgrids under demand uncertainty. *IEEE Transactions on Power Delivery,* **28** (3), 1576–1583.

32 Liu, Q., Mitani, Y., and Bevrani, H. (2016) *An Enhanced WAMS-based Power System Oscillation Analysis Approach.* The International Conference on Electrical Eng. (ICEE-2016), July, Okinawa, Japan.

33 Boyd, S. and Vandenberghe, L. (2004) *Convex Optimization,* Cambridge University Press, Cambridge, UK.

34 Berteskas, D. (1999) *Nonlinear Programming,* Athena Scientific, Boston.

35 Solo, V. and Kong, X. (1995) *Adaptive Signal Processing Algorithms: Stability and Performance,* Prentice Hall, Englewood. Cliffs, NJ.

36 Fathi, M. and Bevrani, H. (2013) Statistical cooperative power dispatching in interconnected microgrids. *IEEE Transactions on Sustainable Energy,* **4** (3), 586–593.

37 Fathi, M. and Bevrani, H. (2012) *Adaptive Price-based Power Flow in next Generation Electric Power Systems.* International Symposium on Smart Grid Operation and Control (ISSGOC), Sanandaj, Iran.

38 Weers, D.D. and Shamsedin, M.A. (1987) Testing a new direct load control power line communication system. *IEEE Transactions on Power Delivery,* **2** (3), 657–660.

39 Fathi, M. and Bevrani, H. (2015) *Wireless Networking of Smart Meters in Next Generation Power Systems.* 2nd International Scientific Conf-University of Human Development (UHD Comp15), April, Sulaimaniyah, Iraq.

A

Appendix

Table A.1 Parameters for the case of single-inverter operation.

Parameters	Values	Parameters	Values
$E_{dg} = E_0$ (kV)	6.6	$k^*_{p\_dg}$ (pu)	20
S_{base_dg} (MVA)	1	$X^*_{f\_dg}$ (pu)	0.1298
$P^*_{0\_dg}$ (pu)	1	$X^*_{line\_dg}$ (pu)	0.0076
ω_0 (rad/s)	376.99	$\cos\delta_{dg}$	≈1
M^*_0 (s)	8	V_{bus} (kV)	6.57
D^*_0 (pu)	17	ΔP_{load} (MW)	0.0095

Table A.2 Parameters of SG.

Parameters	Values	Parameters	Values
V_{base_sg} (kV)	6.6	$X^*_q = X'^*_q$ (pu)	0.770
S_{base_sg} (MVA)	1	X''^*_q (pu)	0.375
$P^*_{0\_sg}$ (pu)	1	T''_d (s)	0.0348
ω_0 (rad/s)	376.99	T''_q (s)	0.0346
M^*_0 (s)	8	T_{d_sg} (s)	0.1
$k^*_{p\_sg}$ (pu)	20	I_{sg} (A)	43.1
$X^*_{line\_sg}$ (pu)	0.0076	φ_{sg} (rad)	0.00655
X^*_d (pu)	1.90	V_{bus} (kV)	6.57
X'^*_d (pu)	0.314	ΔP_{load} (MW)	0.0194
X''^*_d (pu)	0.280		

Table A.3 Experimental parameters for the case of single-inverter operation.

Parameters	Values	Parameters	Values
E_{dg} (V)	207	J_0 (kg m^2)	0.2815
S_{base_dg} (kVA)	5	D_0^* (pu)	17
P_{0_dg} (pu)	1	δ_{dg}	0
ω_0 (rad/s)	376.99	V_{bus} (V)	447.05
$k_{p\_dg}^*$ (pu)	20	ΔP_{load} (kW)	0.34

Table A.4 Experimental parameters for the case of parallel operation with an SG.

Parameters	Values	Parameters	Values
V_{base_sg} (V)	230	X_q (pu)	1.31
S_{base_sg} (kVA)	10	X_q' (pu)	0.55
$P_{0\_sg} = P_{0\_dg}$ (pu)	0	X_q'' (pu)	0.27
ω_0 (rad/s)	376.99	X_q'' (s)	0.01
J_{sg} (kg m^2)	0.563	T_q'' (s)	0.02
$k_{p\_sg}^*$ (pu)	20	I_{sg} (A)	4.77
X_d (pu)	1.35	φ_{sg} (rad)	0.76
X_d' (pu)	0.48	V_{bus} (V)	454.8
X_d'' (pu)	0.27	ΔP_{load} (kW)	1.29

Other DG parameters are the same as those listed in Table 3.3.

Table A.5 Simulation parameters of VSG.

Parameters	Values
Base power P_{base} (MW)	1.0
Base voltage V (kV)	6.6
Per-unit inertia constant M (s)	8
Moment of inertia J (kg m^2)	56.3
Damping factor D	0.045
Interconnecting reactance X_L (%)	26
Interconnecting resistance R (%)	7.6
PI control gain k_p	0.05
PI control time constant k_i (s)	0.001

Index

Microgrid Dynamics and Control, First Edition. Hassan Bevrani, Bruno Francois, and Toshifumi Ise.

e

f

g